2nd Conference of Computational Methods in Offshore Technology and 1st Conference of Oil and Gas Technology (COTech & OGTech 2019)

IOP Conference Series: Materials Science and Engineering
Volume 700

Stavanger, Norway
27 - 29 November 2019

ISBN: 978-1-7138-1030-8
ISSN: 1757-8981

Printed from e-media with permission by:

Curran Associates, Inc.
57 Morehouse Lane
Red Hook, NY 12571

Some format issues inherent in the e-media version may also appear in this print version.

Institute of Physics
Dirac House, Temple Back
Bristol BS1 6BE UK

Phone: 44 1 17 929 7481
Fax: 44 1 17 920 0979

techtracking@iop.org

Additional copies of this publication are available from:

Curran Associates, Inc.
57 Morehouse Lane
Red Hook, NY 12571 USA
Phone: 845-758-0400
Fax: 845-758-2633
Email: curran@proceedings.com
Web: www.proceedings.com

TABLE OF CONTENTS

WIND ENGINEERING AND RENEWABLE ENERGY

ADVANCED COMPUTATIONAL METHODS AND APPLICATIONS IN MARINE AND OFFSHORE TECHNOLOGY

COMPUTATIONAL MECHANICS AND DESIGN OPTIMIZATION

STRUCTURAL INTEGRITY MANAGEMENT AND LIFE EXTENSION OF STRUCTURES

OIL AND GAS FIELD DEVELOPMENT IN THE COLD CLIMATE REGION

ENERGY RESOURCES DEVELOPMENT IN THE ARCTIC

TECHNOLOGIES IN THE UPSTREAM SECTOR IN OIL AND GAS INDUSTRY WITH EMPHASIS ON THE HIGH NORTH

INDUSTRIAL ENGINEERING IN INDUSTRY 4.0

Second Conference of Computational Methods in Offshore Technology and First Conference of Oil and Gas Technology (COTech & OGTech 2019)

November 27 – 29, 2019, University of Stavanger, Norway

Preface

This conference is organized as a joint event of the COTech (Computational Methods in Offshore Technology) and OGTech (Oil and Gas Technology) conferences. The COTech conference started as part of the research and dissemination activities of the Program Area for research "COTech - Computational methods in Offshore Technology" at Faculty of Science and Technology, University of Stavanger (UiS). This Program Area for Research was founded in 2015 with seven professor, four associate professors, two adjunct professors and five research (PhD) students from the Department of Mechanical and Structural Engineering and Materials Science (IMBM), whose expertise and competence lies primarily within use of computational methods such as finite element methods, boundary and volume element methods, computational fluid dynamics and the like in marine and subsea technology, marine operations, design and analysis of mechanical systems, integrity and reliability of offshore structures and mechanical systems, renewable energy and wind engineering. In the offshore-related engineering area in particular, numerical computation approach is nowadays not only serving as a means to cultivate and realize innovative ideas, but also it is becoming the primary choice to solve complex engineering problems for the harsh and unfriendly environment in the Arctic.

The OGTech conference is organized as part of a collaborative project called UTFORSK between a team of researchers from University of Stavanger and Gubkin Russian University of Oil and Gas. The overall aim of the project is to make the team stronger and more sustainable. Among others, the project focuses on building a bridge of collaboration in research and education between the two countries, Norway and Russia, that share the Arctic region and to strengthen the research aspects of the Offshore Technology field in Arctic environment. By facilitating mobility of researchers and staff in both directions, the project aims to provide a common and successful learning environment for young researchers (Masters and PhDs) to make sure that students have skills and knowledge required in order to face the challenges that the Offshore industry meets in the North - such as environmental aspects and Offshore Technology within subsea/ marine structures in cold climate.

In general, the conference is intended to provide a platform for academics and professionals working within both the Offshore and Oil and Gas technologies to come together, present their recent works in the area, exchange ideas, and establish professional networks. It will serve as a forum for a multidisciplinary research and brings together

Norwegian and invited foreign researchers to enable them exchange their research experience and disseminate their results within the involved fields. The conference is organized under 8 thematic areas, which will also serve as conference sessions.

1. *Wind Engineering and Renewable Energy*
2. *Advanced Computational Methods and Applications in Marine Technology*
3. *Computational Mechanics and Design Optimization*
4. *Structural Integrity Management and Life Extension of Structures*
5. *Oil & Gas Field Development in the Cold Climate Region*
6. *Energy Resources Development in the Arctic*
7. *Technologies in the Upstream Sector in Oil- and Gas Industry with Emphasis on the High North*
8. *Industrial Engineering in Industry 4.0*

Among the submitted manuscripts, 66 full papers passed the review process. The review work was conducted according to the review policy of IOP conference series where each paper was reviewed by at least two reviewers involving both national and international reviewers. Eight known experts in the area were invited to present keynote papers from different countries.

The conference committee and editors of this proceeding would like to thank all the reviewers and authors of the papers in this proceeding for their valuable contributions to COTech & OGTech 2019 conference. The financial support from Department of Mechanical and Structural Engineering and Materials Science at University of Stavanger as well as the project funding from Diku (Norwegian Egency for International Cooperation and Quality Enhancement in Higher Education) through the UTFORSK project are highly appreciated.

Conference Organizing Committee

Prof. Hirpa G. Lemu
Prof. Emeritus Ivar Langen
Prof. Emeritus Bjørn H. Hjertager
Prof. Emeritus Ove Tobias Gudmestad
Prof. Anatoly Zolotukhin
Prof. Dimitrios Pavlou
Prof. Muk Chen Ong,
Assoc. Prof. Ove Mikkelsen
Mrs. Vlada Streletskaya

Topic Area Coordinators

Prof. Jasna B. Jakobsen
Prof. Muk Chen Ong
Prof. Dimitrios Pavlou
Adjunct Prof. Gerhard Ersdal
Prof. Emeritus Ove T. Gudmestad
Prof. Anatoly Zolotukhin
Prof. Tor Henning Hemmingsen
Assoc. Prof. Idris El-Thalji

IOP Conf. Series: Materials Science and Engineering **700** (2019) 011001 doi:10.1088/1757-899X/700/1/011001

Invited Keynote Speakers

Prof. Jørgen Amdahl, Norwegian University of Science and Technology, NTNU, Norway
Chief Scientist Arne Dugstad, Institute for Energy Technology (IFE), Norway
Prof. dr. sc. Hrvoje Jasak, University of Zagreb, Croatia
Prof. Dimitris A Saravanos, University of Patras, Greece
Prof. Svein Sævik, Norwegian University of Science and Technology, NTNU, Norway.
Prof. Jakob Mann, DTU Wind Energy, Denmark
Prof. Anatoly Zolotukhin, Gubkin Russian State University (NRU) of Oil and Gas, Russia
Professor Hojjat Adeli, The Ohio State University, USA

Technical Committee Members and Reviewers

1. Ass. Prof. Mustafa Serkan Abdüsselam, Giresun University, Turkey
2. Prof. Henning Agerskov, Technical University of Denmark (DTU), Denmark
3. Assoc. Professor Paul Akangah, North Carolina A&T State University, USA
4. Dr. José António Correia, University of Porto, Portugal
5. Prof. Sergey Arsenyev-Obraztsov, Gubkin University, Russia
6. Prof. Ivar Austvoll, University of Stavanger, Norway
7. Assoc. Prof. Vladimir P. Balitskiy, Gubkin Oil & Gas University, Russia
8. Prof. Abbas Barabady, The Arctic University of Norway (UiT), Norway
9. Prof. Kostas Belibassakis, National Technical Univ. of Athens, Greece
10. Prof. André De Boer, University of Twente, The Netherlands
11. Prof. Robert Brown, Memorial University of Newfoundland, Canada
12. Prof. Hermes Carvalho, Federal University of Minas Gerais, Brazil
13. Ass. Prof. José Miguel Castro, University of Porto, Portugal
14. Dr. Etienne Cheynet, University of Stavanger, Norway
15. Dr. Nianzhong Chen, University of Newcastle, UK
16. Dr. José António Correia, University of Porto, Portugal
17. Prof. Sreekanta Das, University of Windsor, Canada
18. Mr. Stanislav Duplinsky, Subsea 7, Norway
19. Assoc. Prof. David Fuhrman, Technical University of Denmark (DTU), Denmark
20. Prof. Alexander Ermakov, Gubkin University, Russia
21. Adjunct Prof. Gerhard Ersdal, University of Stavanger, Norway
22. Prof. Tiago Fazeres-Ferradosa, University of Porto
23. Dr. Stephen Fox, VTT Technical Research Centre of Finland, Finland
24. Prof. Koh Chan Ghee, National Univ. of Singapore (NUS), Singapore
25. Assoc. Prof. Knut Erik Giljarhus, University of Stavanger, Norway
26. Adj. Prof. Torfinn Havn, University of Stavanger, Norway
27. Prof. Emeritus Bjørn Hjertager, University of Stavanger, Norway
28. Prof. Lars Erik Holmedal, Norwegian Univ. of Sci. and Technology (NTNU), Norway
29. Adj. Assoc. Prof. David Häger, University of Stavanger, Norway
30. Prof. Jasna Jakobsen, University of Stavanger, Norway
31. Prof. Hrvoje Jasak, University of Zagreb, Croatia
32. Assoc. Prof. Tore M. Jonassen, Oslo University College, Norway
33. Adjunct Prof. Daniel Karunakaran, University of Stavanger, Norway

IOP Conf. Series: Materials Science and Engineering **700** (2019) 011001 doi:10.1088/1757-899X/700/1/011001

34. Prof. Uday Kumar, University of Luleå, Sweden
35. Prof. Emeritus Ivar Langen, University of Stavanger, Norway
36. Dr. Gunnar Lian, Equinor, Norway
37. Prof. Jayantha Prasanna Liyanage, University of Stavanger, Norway
38. Dr. Inge Lotsberg, DNV GL Oslo, Norway
39. Assoc. Prof. Lin Li, University of Stavanger, Norway
40. Dr. Pavel Liferov, Statoil AS, Norway
41. Dr. Yuzhu Li, Technical University of Denmark, Denmark
42. Dr. Shengnan Liu, Subsea 7, Norway
43. Assoc. Prof. STILIANOS MARKOLEFAS, University of Athens, Greece
44. Prof. Tore Markeset, University of Stavanger, Norway
45. Assoc. Prof. Ove Mikkelsen, University of Stavanger, Norway
46. Dr. Per Olav Moslett, DNV, Norway
47. Prof. Mohamad Mustafa, The Arctic University of Norway (UiT), Norway
48. Prof. Håkan Nilsson, Chalmers University of Technology, Sweden
49. Prof. Ove Njå, University of Stavanger, Norway
50. Assoc. Prof. Pantelis Nikolakopoulos, University of Patras, Greece
51. Assoc. Prof. Charlotte Obrai, University of Stavanger, Norway
52. Dr. Piotr Omenzetter, University of Aberdeen, UK
53. Dr. Theodosios Papathanasiou, Brunel University, London, UK
54. Dr. Luca A. Piciaccia, VEAS, Norway
55. Prof. Demosthenes Polyzos, University of Patras, Greece
56. Prof. Jose Victor Ramos, University of Porto
57. Prof. Paulo Rosa-Santos, University of Porto, Portugal
58. Prof. Raviliy Safieva, Gubkin Oil & Gas University, Moscow, Russia
59. Assoc. Prof. Hilde Sandhåland, Western Norway University of Applied Sciences, Norway
60. Prof. John V Sharp, Brunel University London, UK
61. Prof. Jonas T. Snærbjørnsson, Rykjavik University, Iceland
62. Assoc. Prof. Sudath Siriwardane, University of Stavanger, Norway
63. Assoc. Prof. Yanyan Sha, University of Stavanger, Norway
64. Assoc. Prof. Hodjat Shiri, Memorial Univ. of Newfoundland, Canada
65. Dr. Srinivas Sriramula, University of Aberdeen, UK
66. Prof. Francisco Taveira-Pinto, University of Porto, Portugal
67. Prof. Michalis Vrahopoulos, University of Athens, Greece
68. Dr. Jungao Wang, University of Stavanger, Norway
69. Prof. Yihan Xing, University of Stavanger, Norway
70. Prof. Qian Xudong, National University of Singapore (NUS), Singapore
71. Dr. Guang Yin, University of Stavanger, Norway
72. Prof. Stefanos Zaoutsos, University of Thessaly, Greece
73. Prof. Anatoly B. Zolotukhin, Gubkin University, Russia
74. Assoc. Prof. Sigmund Kyrre Ås, Norwegian University of Science and Technology (NTNU), Norway

Peer review statement

All papers published in this volume of *IOP Conference Series: Materials Science and Engineering* have been peer reviewed through processes administered by the proceedings Editors. Reviews were conducted by expert referees to the professional and scientific standards expected of a proceedings journal published by IOP Publishing.

The performance of structured and unstructured grids on wind simulations around a high-rise building

T-O Hågbo[1*], K E T Giljarhus[1], S Qu[2] and B H Hjertager[1]

[1] University of Stavanger, Norway
[2] Harbin Engineering University, China

Corresponding author: trond-ola.hagbo@uis.no

Abstract. In this study, computational fluid dynamics has been applied to analyze the performance of both structured and unstructured meshes when used in steady-state simulations of wind in an urban area. This case specifically examines the wind environment around a single high-rise building with the geometry of 1: 1: 2 (length, width, height) within the atmospheric boundary layer.The Reynolds-averaged Navier-Stokes simulations were conducted using OpenFOAM, with grids generated with *blockMesh*, *snappyHexMesh*, and *Pointwise*. A set of coarse meshes and a set of fine meshes were used for comparison.Minor differences in the flow pattern were observed depending on the mesh type used, including the reattachment lengths, the drag coefficients, the velocity profiles, and the positioning of the vortex leeward of the high-rise building. However, the main flow features were the same, indicating that for simulations where overall flow features are of importance there is high flexibility in choice of meshing technique.

1. Introduction

In the recent decade, Computational fluid dynamics (CFD) has been increasingly applied to solve wind engineering problems, such as pedestrian wind discomfort and wind loads on buildings and bridges. In validation exercises of a single building, a structured mesh is often used. However, performing simulations on complex geometries, such as urban areas, requires complex meshes. Due to the indispensably high irregularity of these meshes, they can be time-consuming, complicated, and sometimes impossible, to create using a structured type of mesh. Alternatively, an unstructured mesh can be made. Their notable features include flexibility and adaptation capability[1].

As the usage of unstructured meshes is necessary for some applications, it is essential to know their performance in flow simulations and their associated accuracy. For some applications, a relatively coarse and unstructured mesh is sufficient and the most effective usage of computational power and time. There are cases where only the main flow structure is of interest, and when there is high uncertainty associated with other parameters of the simulation. For example, when simulating wind through an actual urban area to study pedestrian wind comfort.

In this study, CFD has been applied to analyze the performance of both structured and unstructured meshes when used in steady-state simulations of wind in an urban area. This case specifically examines the wind environment around a single high-rise building with the geometry

IOP Conf. Series: Materials Science and Engineering **700** (2019) 012001 doi:10.1088/1757-899X/700/1/012001

of 1: 1: 2 (length, width, height) within the atmospheric boundary layer (ABL). For the sake of detailed examinations, the flow past a simple shape, such as a high-rise building, seems to be more practical than in an actual urban environment.

Wind tunnel experiments on the high-rise building have been conducted by Meng and Hibi[2], with numerous other sources examining wind simulation studies of identical geometry. Examples include: Yoshie et al.[3], Tominaga et al.[4], Shao et al.[5], Gousseau et al.[6], Dadioti and Rees[7], and Liu and Niu[8]. The papers listed mostly focus on comparing various turbulence models and or transient simulations.

While unstructured meshes are necessary for some applications, the majority of existing research on the high-rise building benchmark case utilizing structured meshes. Therefore, this study will evaluate the performance of various mesh types, both structured and unstructed, in urban wind simulations to improve the understanding of the mesh type dependency.

The paper will evaluate mesh performance by first defining the computational domain. Second, the differences in grids used are illustrated by presenting coarse meshes. Then, the numerical setup is described in detail. In the last part of the methodology section, all the simulation cases are listed subsequently. The results are then presented and discussed, starting with the differences in reattachment lengths and drag coefficient, then moving towards other general flow features such as the flow structure and velocity profiles.

2. Methodology

This section describes the computational methodology. As the accuracy and reliability of CFD simulations can easily be compromised, best practice guidelines (BPG) have been developed for properly conducting CFD analysis of the wind environment in urban areas. These guidelines are based on cross-comparison between CFD simulations, wind tunnel experiments, as well as fields measurements on several cases. The BPG provided by Franke et al.[9] and Tominaga et al.[10] were carefully followed to set up the simulations.

2.1. Computational domain

The computational domain size is identical for all the simulations conducted. The dimensions are $21.5b \times 13.75b \times 11.25b$ in the x (length), y (width) and z (height) directions correspondingly and an overview is illustrated in Figure 1. b is the width and length of the building which height is $2b$. In the experiments, b was 0.08 m[2], and the same scale was used in the simulations.

Both the width and the height of the domain replicate the geometry of the wind tunnel setup, as recommended by the BPG. Also, the resulting blockage ratio is approximately 1.3 %, which is well within the suggested maximum value of 3 % recommended by the BPG, as proposed by Beatke et al.[11].

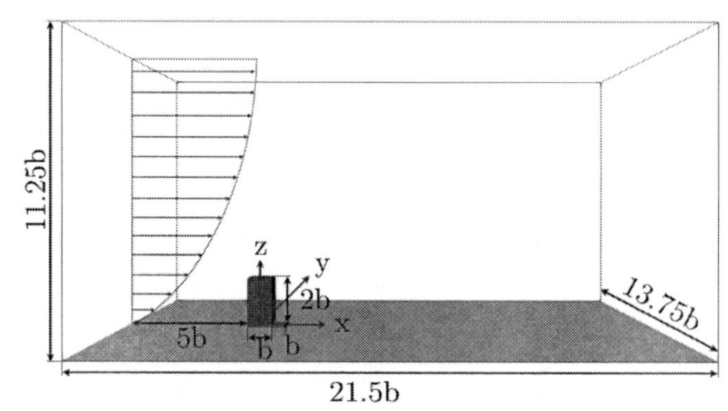

Figure 1 Overview of the computational domain.

IOP Conf. Series: Materials Science and Engineering **700** (2019) 012001 doi:10.1088/1757-899X/700/1/012001

The length upstream of the building is $5b$, which is consistent with previous numerical studies such as [3, 4, 5, 7, 8]. Except for Tominaga et al.[4], all the other studies listed above were performed using a length downstream of the building of $21b$. Tominaga et al. used $21.5b$, which is used in this study.

2.2. Computational grids

The main parameter reviwed in this study is the mesh type and its performance when used in wind simulations in urban areas. The grids used in this study were generated with either a built-in mesh generator in OpenFOAM, *blockMesh* or *snappyHexMesh*, or the commercial mesh generation software *Pointwise*[12]. The different meshing utilities are described below.

blockMesh is suitable for highly simple geometries, such as the case at hand. It works by dividing the computational domain into multiple hexahedral blocks. These blocks can be stretched and bent to ensure proper cell density near walls and in specific areas where high flow gradients are emerging. All the meshes generated with *blockMesh* are structured.

snappyHexMesh is a hex-dominant, unstructured mesh generator suitable for more complex geometries. It required an already existing base mesh, generated in this case by *blockMesh*, and an object file holding information of the triangulated surface geometries. *snappyHexMesh* snaps the mesh to the surface of the object file and iteratively refines the mesh close to the surfaces of choice.

Pointwise is an advanced and highly flexible commercial mesh generation tool with the ability to create a variety of mesh types, both structured and unstructured, as well as hybrid types. For this study, *Pointwise* was used to generate both structured meshes entirely consisting of hexagons, and unstructured meshes constructed of prisms.

Figure 2 and Figure 3 illustrate the main differences in the mesh types used. Only examples of the coarse grids are illustrated.

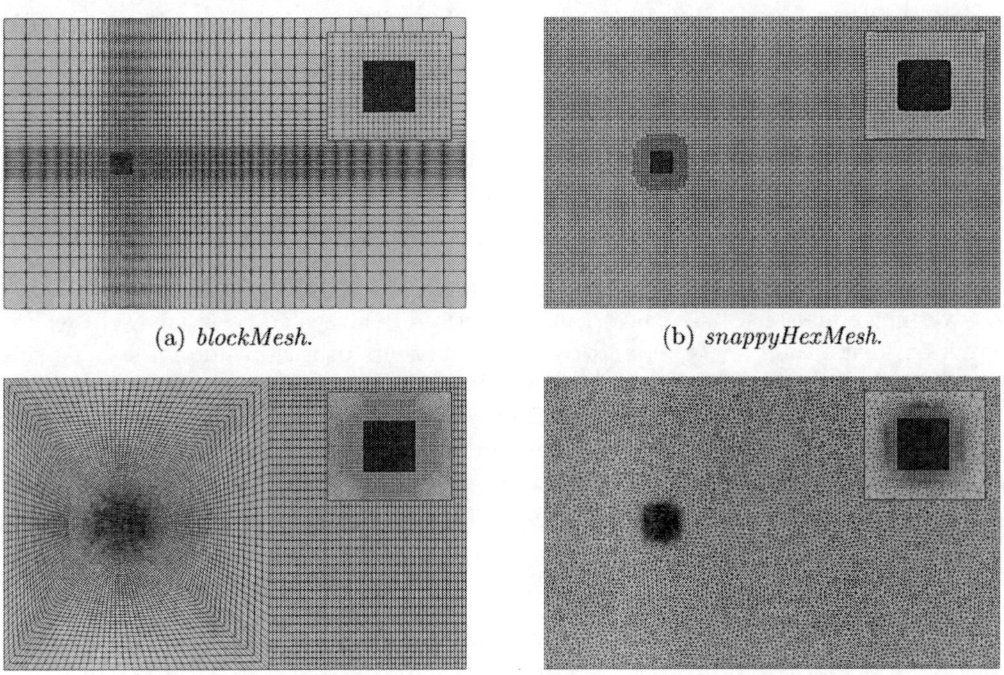

(a) *blockMesh.* (b) *snappyHexMesh.*

(c) *Pointwise, structured.* (d) *Pointwise, unstructured.*

Figure 2. Top view of the grid types showing the grid on the ground and the roof of the building. Inset shows a close-up near the building.

IOP Conf. Series: Materials Science and Engineering **700** (2019) 012001 doi:10.1088/1757-899X/700/1/012001

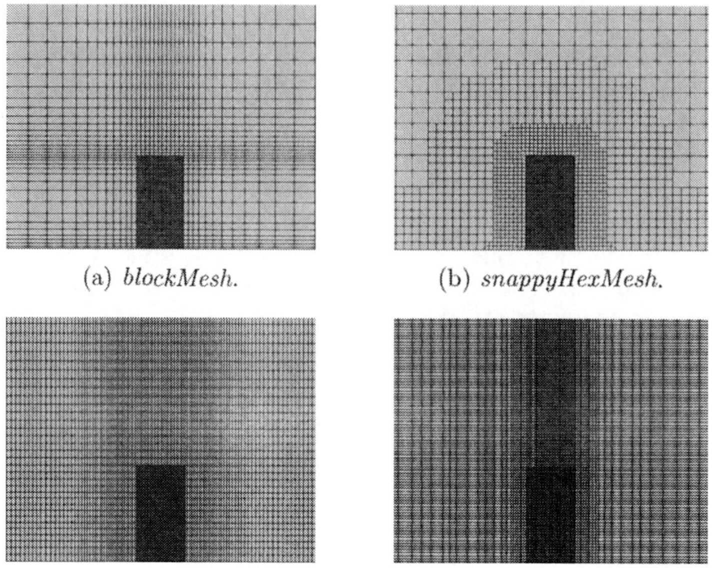

(a) *blockMesh.* (b) *snappyHexMesh.*

(c) *Pointwise, structured.* (d) *Pointwise, unstructured.*

Figure 3. Front view of the grid types at $x = 0$.

The configurations of the fine grid versions are similar, but with more cells and smaller cell widths for the cells in contact with the building or the ground. For the meshes generated with *blockMesh*, an even higher stretching has been applied. For the meshes generated with *snappyHexMesh*, proper cell width is obtained by adding more refinement levels near the building and ground. The computational cases with different meshes are summarized in Table 1. For some of the mesh types have a few different versions within the type been used in the simulations for comparison.

2.3. Numerical setup and other parameters

The turbulent wind flow pattern around the high-rise building was obtained by solving the incompressible, three-dimensional steady Reynolds-averaged Navier-Stokes (RANS) equations with the finite volume method. The Reynolds number based on H (building height) and U_H (inflow velocity at $z = H$) was 2.4×10^4. The pressure-velocity coupling was handled by the SIMPLE algorithm, and second-order discretization schemes were used for the convection terms of the governing equations. The convergence criteria for the scaled residuals were set to drop by four orders of magnitude for U, k, and ϵ and three orders of magnitude for p. The simulations were conducted using the OpenFOAM toolbox, which is an open-source CFD software package.

The vertical distributions of the quantities at the inflow boundary were set on the basis of the wind tunnel experiment[2] to assure similar conditions. As the vertical profile of the streamwise velocity approximately obeyed a power law, the velocity at the inlet was set to be $U = U_H(z/z_H)^{0.27}$, where U_H is the wind speed at reference height z_H. z_H was set to be the height of the building, $2b = 0.16\,\mathrm{m}$, with an U_H of $4.5\,\mathrm{m\,s^{-1}}$ following the measurements. The profiles of both the turbulent kinetic energy, k, and the dissipation rate, ϵ, were interpolated from the experimental data.

The outlet of the domain was set to zero static pressure with zero gradient for the remaining variables. For the lateral and top boundaries, the slip condition was used for all the variables. No-slip conditions were applied to the building and the ground. The roughness length z_0 on the ground was set to 1.8×10^{-4}, in accordance with Yoshie et al.[3].

IOP Conf. Series: Materials Science and Engineering **700** (2019) 012001 doi:10.1088/1757-899X/700/1/012001

Two near-wall treatment methods were used. For the simulation cases using the fine meshes ("the fine simulations"), no wall functions were applied. Wall functions were implemented for the cases using a coarse mesh ("the coarse simulations"). A logarithmic law for smooth walls was used on the building, and a logarithmic law, including a roughness parameter was applied for the ground.

Turbulent closure was provided by three different turbulence models for incompressible flows. For the "coarse simulation", both the standard k-ϵ model (SKE)[13] and the realizable k-ϵ model (RKE)[14] were used. The RKE was chosen as a reference since it is recommended by Blocken et al.[15] in the application of pedestrian wind environment simulations. In the "fine simulations", the cells close to the walls are presumably mostly within the viscous sublayer, and a low-Reynolds number model is more appropriate. The Lien and Leschziner low-Reynolds number k-ϵ model (LLKE)[16] was chosen.

2.4. Simulation cases

The setup for all the simulations was identical except for three changes:

- Grid type – coarse/fine, structured/unstructured, and generated with different meshing utility.
- Wall functions - the coarse simulations used wall functions as opposed to the fine simulations that did not.
- Turbulence models - the coarse simulations used either the SKE or the RKE while the fine simulations used the LLKE.

In Table 1, the "Case name" column the "C" stands for coarse mesh while the "F" indicates that a fine m esh h as b een u sed. T he fi rst tw o le tters in th e "M esh" co lumn re presents the meshing utility used for creating the final m esh, w ith " BM" s tanding f or *b lockMesh*, " SH" for *snappyHexMesh* and "PW" for *Pointwise.*

Table 1. List of simulation cases.

Case name	Mesh	Cells	Turbulence model	y^+ (average) building	ground
C1	BMC1	104 k	SKE	152	99.3
C2	— " —	— " —	RKE	152	98.5
C3	BMC2	— " —	SKE	29.5	38.0
C4	— " —	— " —	RKE	29.1	37.5
C5	SHC1	191 k	SKE	30.0	117
C6	— " —	— " —	RKE	29.7	117
C7	SHC2	519 k	SKE	29.9	58.7
C8	— " —	— " —	RKE	29.6	58.5
C9	PWC1	674 k	SKE	117	61.5
C10	— " —	— " —	RKE	114	61.4
C11	PWC2	3.66 M	SKE	13.7	50.1
C12	— " —	— " —	RKE	13.8	56.9
F1	BMF1	20.6 M	LLKE	7.43	2.83
F2	SHF1	16.7 M	— " —	2.93	4.10
F3	SHF2	27.2 M	— " —	2.14	2.62
F4	PWF1	3.59 M	— " —	3.65	0.97
F5	PWF2	8.68 M	— " —	2.22	1.20

The mesh BMC1 has no stretching of the cells whatsoever, but BMC2 has, which is the mesh presented in Figure 2a and Figure 3a. SHC2/SHF2 has one more refinement level to the ground compared to SHC1/SHF1, which is presented in Figure 2b and Figure 3b. PWC1/PWF1 are structured meshes, while PWC2/PWF2 are unstructured. PWC2 is here categorized as coarse, even though the number of cells is high. It was decided to include C11 and C12, although the average y^+ values at the building are far lower than recommended for the turbulent models used.

3. Results and Conclusion

3.1. Reattachment lengths and drag coefficients

Table 2 gives the reattachment lengths and the drag coefficients for all the simulated cases, a few results from the literature, and the experimental data. X_R is the distance from the roof edge facing the wind where flow separation occurs to the point where the flow reattaches to the roof. X_F is the length leeward of the building where the flow reattaches to the ground.

Flow reattachment on the roof was obtained for four cases: C11, C12, F3, and F5. For the coarse simulations, it was achieved only when a mesh consisting of far more cells than the other coarse meshes was used. The two cases that obtained flow reattachment on the roof for the fine meshes were also the two cases with the lowest average y^+ values at the building. In general, a large number of grid cells are needed to fully resolve the roof vortex, which is not achieved for the coarse meshes. X_F was 9 to 12 % longer for all the coarse simulations when using the turbulence model RKE instead of SKE. When using SKE X_F was in the range $2.74b$ to $2.87b$ when excluding C11. The range for RKE was $2.99b$ to $3.12b$, excluding C12. For LLKE and the fine simulations, the range was considerably larger, $2.23b$ to $3.42b$.

The drag coefficients, C_D, for the coarse simulations are somewhat lower than for the fine simulations, 0.809 to 1.076 compared to 1.071 to 1.347. When excluding the results from F5, the variance within the fine simulations is much less compared to the coarse simulations. This is expected since these meshes are closer to mesh-independent results.

Table 2. Reattachmenth lengths and drag coefficients.

Case	X_F/b	X_R/b	C_D	Case	X_F/b	X_R/b	C_D
C1	2.84	-	0.994	F1	3.42	-	1.071
C2	3.12	-	0.855	F2	2.65	-	1.081
C3	2.74	-	1.030	F3	2.53	0.16	1.113
C4	2.99	-	0.948	F4	2.83	-	1.118
C5	2.80	-	0.880	F5	2.23	0.21	1.347
C6	3.11	-	0.811				
C7	2.80	-	0.881	C SKE[4]	2.7	-	N/A
C8	3.14	-	0.809	C SKE[5]	2.4	-	N/A
C9	2.87	-	1.076	C SKE[7]	2.68	0.50	N/A
C10	3.21	-	1.002	C RKE[7]	5.37	0.50	N/A
C11	2.61	0.42	1.061	F SKE[8]	3.50	0.58	N/A
C12	2.85	0.41	0.972	Exp.[2]	1.42	0.52	N/A

IOP Conf. Series: Materials Science and Engineering **700** (2019) 012001 doi:10.1088/1757-899X/700/1/012001

3.2. Flow pattern and velocity distribution

Figure 4 shows the flow pattern in a vertical slice through the centre of the building, viewed from the side. All simulations show similar flow features, with a stagnation point in front of the building at approximately 3/4 of the building height, and a large recirculation zone behind the building.

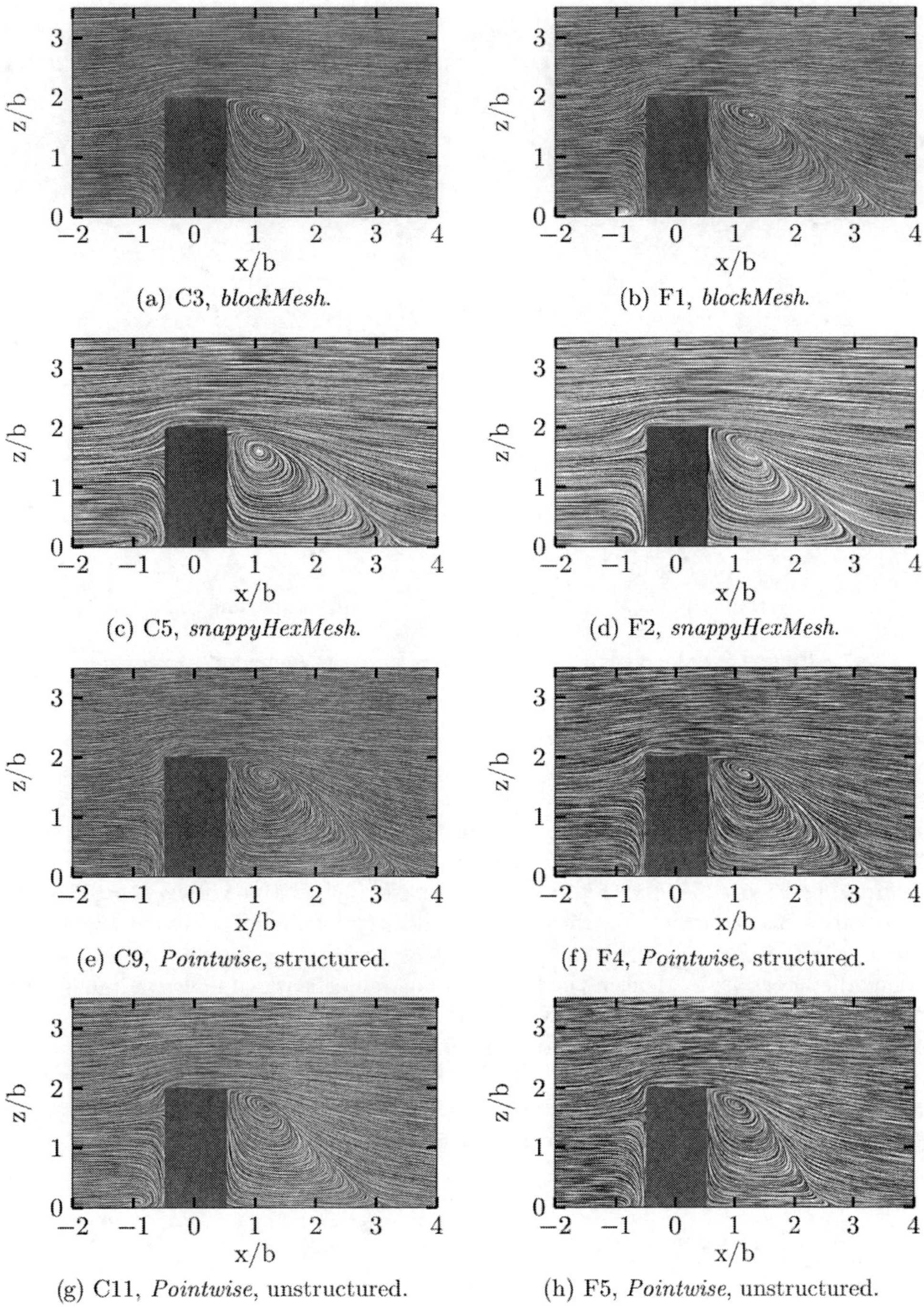

Figure 4. Flow patterns using line integral convolution, side view at $y/b = 0$.

Figure 5 shows the horizontal distributions of the mean wind speed near the ground surface at $z/b = 0.125$. Only contour plots of two simulations were included, in addition to numerical and experimental results from the literature for reference. The other simulations gave similar results. Notice that the shape and placement of the contour lines are comparable between these simulations and the references. However, the wind velocity close to the ground is somewhat lower for all the simulations compared to the experiments and the C SKE case[5].

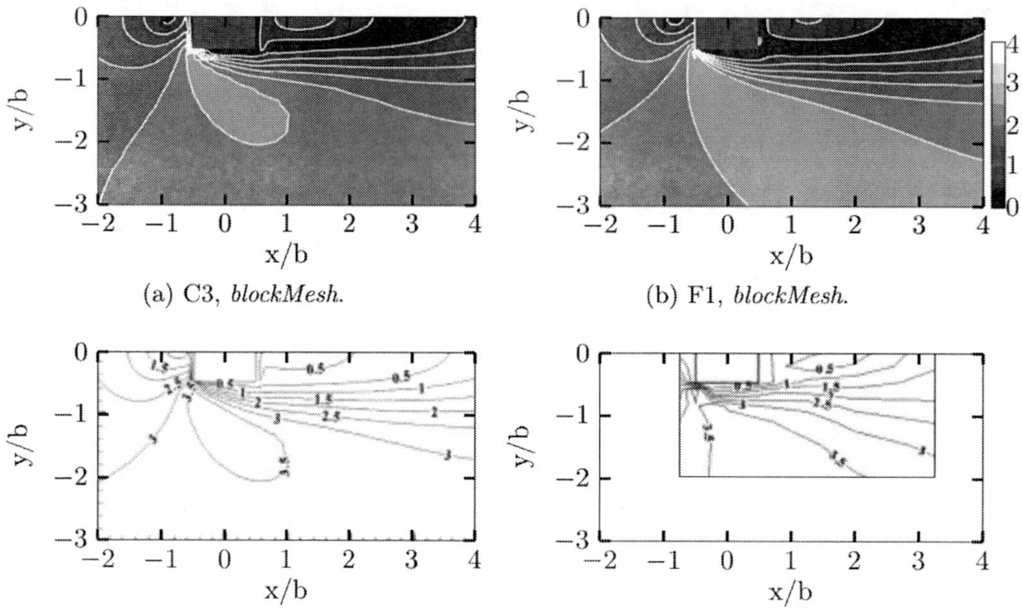

(a) C3, *blockMesh*.

(b) F1, *blockMesh*.

(c) C SKE[5].

(d) Experimental data[2].

Figure 5. Velocity contour lines in m s^{-1}, top view at $z/b = 0.125$.

Figure 6 is a collection of vertical profiles of wind velocity in the x-direction placed on top of the measurement regions (indicated with dotted black lines). The crosses represent the experimental values, while the colored lines represent data sampled from the simulations. The simulation cases included are the same as in Figure 4. Lines and crosses on the left side of its sampling line represent negative values, hence that the wind is moving in the negative x-direction.

By comparing Figure 6a and Figure 6b, it is clear that the velocity profiles in the fine simulations vary less in the region leeward of the building roof compared to the coarse simulations. It indicates that the coarse grids have too few cells to have reached a resolution-independent result. The differences are evident in the region in particular, as high flow gradients are emerging, which need more cells to resolve the flow behavior.

When comparing the profiles to the experimental values, one cannot conclude that one mesh type is superior to the others. However, C5, F2, and F5 appear to produce velocity profiles the closest to the experimental values. Both C5 and F2 have unstructured meshes generated with *snappyHexMesh*, and F5 uses an unstructured mesh made with *Pointwise*.

IOP Conf. Series: Materials Science and Engineering **700** (2019) 012001 doi:10.1088/1757-899X/700/1/012001

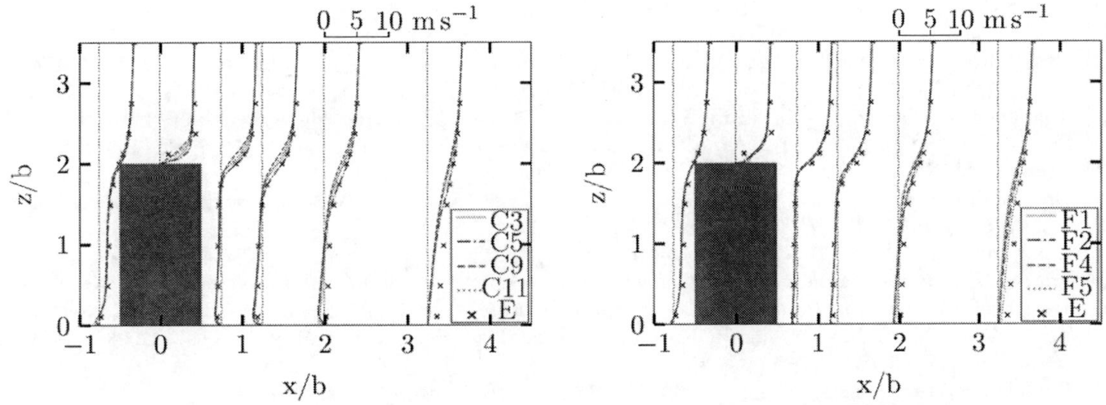

(a) Coarse simulations. (b) Fine simulations.

Figure 6. Vertical velocity profile comparison, side view at $y/b = 0$.

3.3. Concluding remarks

This study compared the use of different meshing techniques for simulating wind flow around a rectangular, high-rise building.

All flow simulations gave similar flow patterns, indicating that for simulations where overall flow features are of importance there is high flexibility in choice of meshing technique. The difference between the use of wall functions and wall-resolved simulations were also minor. Hence, it appears that a coarse and unstructured type of mesh can be sufficient in the applications where only the general flow features are of interest. This is the most flexible gridding technique for complex urban areas.

The reattachment lengths behind the building were considerably longer than in the experiments for all the simulations. This is in agreement with other RANS simulations in the literature. In general, the reattachment length is a highly sensitive parameter due to the transient nature of the flow. More advanced simulations using, for example, large eddy simulations, have been able to accurately predict the reattachment length, but even for such simulations there is high variability in the reported results [4, 6, 8].

Acknowledgments

This research is part of the Future Energy Hub Project funded by The Norwegian Research Council (project no.: 280458), The University of Stavanger, and local industry partners. We thank the anonymous reviewers for insightful and much-appreciated feedback.

References

[1] Kim S E and Boysan F 1999 *J Wind Eng Ind Aerodyn* **81** 145–158
[2] Meng Y and Hibi K 1998 *J Wind Eng* **1998** 55–64
[3] Yoshie R, Mochida A, Tominaga Y, Kataoka H, Harimoto K, Nozu T and Shirasawa T 2007 *J Wind Eng Ind Aerodyn* **95** 1551–1578
[4] Tominaga Y, Mochida A, Murakami S and Sawaki S 2008 *J. Wind Eng Ind Aerodyn.* **96** 389–411
[5] Shao J, Liu J and Zhao J 2012 *J Built Environ* **57** 145–155
[6] Gousseau P, Blocken B and Van Heijst G 2013 *J Comp Fluids* **79** 120–133
[7] Dadioti R and Rees S J 2015 Validation of open source CFD applied to building external flows *Proc. BS2015, 14th Conf. Int. Build. Perform. Simul. Assoc.*
[8] Liu J and Niu J 2016 *J Built Environ* **96** 91–106
[9] Frank J, Hellsten A, Schlünzen H and Carissimo B 2007 *Quality Assurance and Improvement of Meteorological Models. University of Hamburg, Meteorological Institute, Center of Marine and Atmospheric Sciences*
[10] Tominaga Y, Mochida A, Yoshie R, Kataoka H, Nozu T, Yoshikawa M and Shirasawa T 2008 *J Wind Eng Ind Aerodyn* **96** 1749–1761
[11] Baetke F, Werner H and Wengle H 1990 *J Wind Eng Ind Aerodyn* **35** 129–147
[12] Pointwise The choice for CFD meshing `https://www.pointwise.com`
[13] El Tahry S 1983 *J Energy* **7** 345–353
[14] Shih T H, Liou W W, Shabbir A, Yang Z and Zhu J 1995 *J Comp Fluids* **24** 227–238
[15] Blocken B, Roels S and Carmeliet J 2004 *J Wind Eng Ind Aerodyn* **92** 849–873
[16] Lien F S and Leschziner M 1993 *J Fluids Eng* **115** 717–725

Mooring systems analysis of floating wind turbines in Italian seas

P Re[1], G Passoni[1], O T Gudmestad[2*]

[1] Politecnico di Milano, Italy
[2] University of Stavanger, Norway

* Corresponding author: Ove.t.gudmestad@uis.no

Abstract. In recent years, interest in exploiting wind power in deep waters has grown, with the realization of floating structures that could accommodate a turbine and produce electricity. One of the many challenges that these projects face is the position keeping of the system in a defined part of the sea: this is carried out with mooring systems, designed to withstand extreme marine conditions. The first aim of this work is to analyze these systems, to find a general solution for a floating wind farm in the Italian seas. Of importance is to suggest a design that could be fabricated in an Italian dry dock and towed to site. A numerical model of the floating structure and the environment has been realized using the software, OrcaFlex, and simulating the 50- and 100-year return period conditions in the Adriatic Sea and in the Strait of Sicily, with different mooring combinations of present-day technologies. The model's implementation requires environmental and structural parameters: the former have been obtained with a statistical analysis of actual recordings of wind speed and wave height and with the official Italian cartography. Structural details have been derived by looking at existing floaters, finding the most suitable one for the Italian environment. After simulations, the results are interpreted with a comparison of tension in the lines, vertical forces at the anchors, wire length and mooring footprint, finding out that depth has the greatest impact on these factors. With respect to this, the proposed numerical model can provide a simple indication for the mooring of floating wind turbines in two different areas of the seas surrounding the Italian peninsula: for the Adriatic Sea (shallow waters), suction anchors with chains should be preferred, while, for the Strait of Sicily (deeper waters), traditional anchors may be considered, with lines composed of chains or a combination of chains with synthetic fibers.

1. Introduction

The principal sources of energy in most parts of the world are non-renewable resources like oil, coal, natural gas and nuclear power. However, these resources present disadvantages and critical issues: burning fossil fuels produces greenhouse gases and other pollutants harmful to the environment, and these sources are subjected to limitations and price volatility. Nor is nuclear power free from serious problems: accidental radioactive leaks, nuclear proliferation and safe storage of wastes are only a few of the challenges to be considered.

On the other hand, renewable sources like wind have great potential and advantages: they are virtually inexhaustible, they produce power without the pollutant emissions, and they are freely available in every part of the world. This is in accord with the global goal of reducing greenhouse gas emissions, following international agreements, like the Paris one [1] or the European Union Renewable Energy directive [2]. However, renewable sources present disadvantages for power production, such as intermittency and predictability.

In recent years, installations of wind farms have grown, not only on the mainland but also offshore, with a series of advantages:

- the turbine's size is not limited by the existing infrastructure in the manufacturing process, when it can be assembled near the shore;
- large sea areas are available, meaning that the installations do not cause land consumption or soil degradation;
- the wind generally blows more strongly and steadily over the sea surface, meaning higher and more constant power production;
- the visual impact and noise pollution can be eliminated by the turbines' installation at a sufficient distance from the coastline.

However, the previous factors are balanced by some issues:

- the turbines must not only withstand the wind load, they must also resist environmental challenges like waves and currents, resulting in an increased design complexity, which also affects the cost;
- the turbines have to be marinized, and the support structure must be designed, installed and, in the future, decommissioned;
- offshore wind farms are harder to access than those on land, increasing the maintenance challenges and costs.

Before 2017, all commercial offshore wind farms were placed within 50 meters of sea depth, connected to the seabed with structures like monopiles and jackets. These solutions are not economically profitable in deeper waters, where floating structures take the lead. In recent years, the research has focused on developing the concept of floating wind turbine farms, unlocking the potential of the wind in open seas and exploiting the oil and gas industry's experience in terms of mooring and floating structures' design.

The technical realization and profitability of floating structures has been demonstrated by the oil and gas industry with long-term floating structures. However, the development of floating wind turbine designs, capable of conquering a part of the energy market, requires extensive analysis. In the offshore environment, in addition to wind load, new aspects must be considered: hydrodynamic loads are the most evident, but debris, ice and marine growth build-up must also be covered. The analysis has to account for the coupled motion of the floater and the turbine and the mooring systems.

2017 was a turning point: the world's first floating offshore farm began electricity production near the Scottish shores in October. The project, known as "Hywind Scotland", was developed and constructed by Equinor in Norway and towed off the Aberdeenshire coast during the summer. It consists of five spar-type turbines floating in the deep waters of the North Sea, moored to the seafloor and connected to the Scottish electricity grid through a submarine cable.

The Italian environment lacks a future or potential project of the Hywind-kind [3], [4], even though the potential, that the seas surrounding the Italian peninsula offer, is great. This, coupled with the reduced visual impact (which is and always has been the main opposition to wind turbine installation in Italy), could lead to the creation of a new and renewable power source, with economic and occupational advantages.

A numerical model was developed using the commercial software, OrcaFlex, a fully 3D, nonlinear, time-domain, finite element program, capable of dealing with a wide range of offshore applications, like floating wind turbines and their moorings.

In the process of achieving the main target, several objectives were set:

- identifying the mooring difference between two locations with different sea depths, wind speeds and wave heights;

- finding the environmental design parameters concerning wind speed and wave height, with the application of extreme value statistic theory to a dataset of real recordings in the selected areas;
- identifying the most suitable floating wind turbine concept, with respect to characteristics of the Italian seas;
- exploring the possibility of the realization in Italy of the chosen concept from the inception to the construction phase.

The paper is divided into four main chapters, after the introduction, followed by the conclusions. Chapter 2 presents the process for obtaining the environmental design parameters, for the selected locations of the study in the Mediterranean Sea: the Adriatic Sea near Ancona and the part of the Strait of Sicily near the southern shores of the island. First, descriptions are given of the sources of the data for sea depth, wind speed and wave height. Thereafter, the results of the statistical analysis, in which the theory of the extreme values is applied, are given.

At the beginning of Chapter 3, a general overview of Italian power production is provided, along with the potential for the realization of floating farms. After that, the characteristics of the most prominent types of wind farm are summarized. Finally, the choice of the structure to be modelled is made, considering compatibility with the Italian seas and potential construction in an Italian shipyard.

In Chapter 4, the theory behind the numerical model is explained: mooring line types and their characteristics. The real mooring hardware to be modelled is also presented: wire, chains, synthetic ropes, anchors and suction anchors.

In Chapter 5, the results and the previously presented theory come together in the actual model and its results. Static and dynamic simulations follow; tension in the wires and forces at the anchors are examined, and results are produced. Finally, looking at the previous steps, different locations and simulations are compared and discussed, with the purpose of finding the most suitable mooring solutions. The final discussion in the conclusions (Chapter 6) leads to considerations with respect to the original objectives, in relation to the actual results. Suggestions for further work, based on the issues and limitations that were highlighted during the investigations, are also described.

2. Environmental design parameters

A reliable data source for wind speed and wave height has been found in the Italian agency, ISPRA (Istituto Superiore per la Protezione e la Ricerca Ambientale), which runs an extensive monitoring network across the Italian peninsula and the seas around it. This agency operates several monitoring networks, two of which have been used in this work: The Rete Mareografica Nazionale (National Sea Network, also known as RMN), for the wind speed, and the Rete Ondametrica Nazionale (National Wave Network, also known as RON), for the wave height.

We have looked at wind data (1998 to mid-2017) near to an intermediate depth location in the Adriatic, in the harbor of Ancona (LAT: 43° 37' 29.16" LONG: 13° 30' 23.46") [5] and data for the period 1999 to mid-2017 close to a deep water location in the Mediterranean, at Porto Empedocle, at the southern shores of Sicily (LAT: 37° 17' 8.72" LONG: 13° 31' 36.64") [6]. For months with a ratio of data measured over data expected being less than 75%, the data are disregarded [x thesis]; however, we consider the data available to be representative for extreme statistics analysis. Two wave buoys are available for retrieving wave data at the locations studied: from 1999 to 2014 near Ancona, in the Adriatic Sea (LAT: 43° 49' 55.2" LONG: 13° 42' 36") [5] and from 1998 to 2014 near Mazara del Vallo, in the Mediterranean Sea, south of Sicily (LAT: 37° 31' 4.8" LONG: 12° 31' 58.799") [7].

In this work, the Gumbel distribution (concerning the maximum) was used to obtain the values of wind speed and significant wave height, considering return periods of 50 and 100 years. To estimate the distributions' parameters, the maximum goodness of fit (MGE) method was used, by including significant waves with heights larger than 3.0m. The waves corresponding to the design heights have a peak period with the same magnitude as those recorded by the stations for the biggest waves.

The resulting design wind conditions at 10 m height (10 min averaging) are given in Table 1, and the resulting wave conditions are given in Table 2. These values are used in the mooring analysis for the

IOP Conf. Series: Materials Science and Engineering **700** (2019) 012002 doi:10.1088/1757-899X/700/1/012002

floating wind turbines. Note again that the main goal of the present work is to analyze and to find a solution for a mooring system connected to floating wind turbine structures in the Italian seas.

Table 1. Design wind speed (at 10 m height above SWL) at the selected measurement stations.

Return period	Design wind speed (m/s), Ancona	Design wind speed (m/s) Porto Empedocle
50 years	20.90	22.41
100 years	22.64	23.83

Table 2. Design significant wave conditions (Jonswap spectra) at the selected measurement stations.

Return period	Design wave height (m)		Design peak period (s)	
	Ancona	Mazara del Vallo	Ancona	Mazara del Vallo
50 years	5.62	6.21	9.5	11
100 years	6.01	6.66		

We have investigated the potential of offshore wind power in a simplified 3D numerical model of the floater. The mooring and the environmental design conditions has been developed and the combined loads of wind and waves in the open sea simulated. The results from the simulations are evaluated, considering several factors: wire length, mooring footprint, tension in the lines, forces at the anchors and their interactions.

3. Wind turbine foundation design, construction and installation

We have investigated the potential of offshore wind in Italian waters and have reviewed the capacity of Italian and Mediterranean yards to construct the different floating wind turbine concepts. The WindFloat design is the most flexible to construct and install, as it can be fabricated in a dry dock, towed to site and installed without the use of large crane vessels. Mooring is by a traditional mooring system with the use of chains, wires or polyester lines. The concept is composed of a semi-submersible floater with plates at the base of each column; see Figure 1.

Figure 1. WindFloat hull and turbine [8, 9].

Dry docks play a great part in the construction of the semi-submersibles, and the possibility of the realization of a possible floating wind turbine project in Italy has been explored by collecting and comparing the dimensions of Italian dry docks. To improve the number of suitable construction sites and to create competition between shipyards, similar information has been searched and collected in the Mediterranean area [10]. The larger Italian and other Mediterranean docks (with a width of more than 50m) are listed in Tables 3 and 4, respectively.

For the realization of a semi-submersible structure, a selection of EU shipyards is available, and Italian construction is possible: the larger availability of dry docks means increasing the competition between companies. For example, considering the Palermo dry dock and the ideal 5 MW WindFloat [8], [9] at the same time, four semi-submersible structures can be built in the same shipyard, with conventional cranes and techniques. For towing to locations from Italian construction sites, see Fig. 2.

Table 3. Dry docks in Italy with a width of more than 50 m.

Location	Name	Length (m)	Width (m)	Depth (m)
Palermo	Graving dock 4	370	68	11.35
Livorno	Accosto 77	350	56	9.7
Trieste	Graving dock 4	295	56	11

Table 4. Dry docks in Mediterranean Area with a width of 50 m or more.

Location	Country	Name	Length (m)	Width (m)	Depth (m)
Marseille	France	Forme 10	465	85	9.2
Skaramagas	Greece	Dock 1	421	75	5.5
Valletta	Malta	Graving dock 6	362	62	9.3

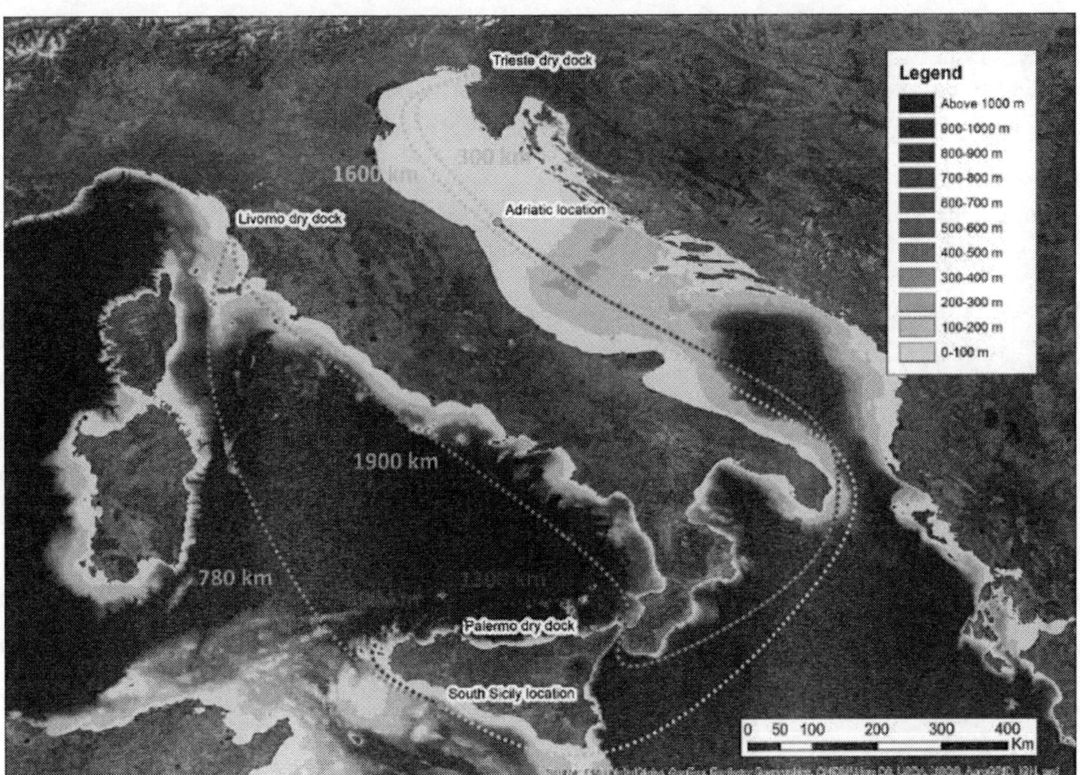

Figure 2. Italian ports for WindFloat design construction, and towing distances to the study locations.

4. Mooring

Over the years, steel chains, steel wires and synthetic ropes have been used to connect floating objects to the seabed: the lines compose a catenary shape, that acts on its tension (increasing or decreasing it as the catenary lifts off or lies on the seabed) to stabilize the platform subjected to sea forces. The catenary mooring system has been deployed with success in shallow waters. With the need to exploit resources in deep waters, the suspended weight of the lines is becoming a limiting factor; steel chains are currently less used in deep water applications, now being replaced by synthetic fiber ropes [11], which reduce the mooring line length, the overall tension and the total costs. The many advantages that this type of mooring has over traditional catenary systems should be realized: the lines are considerably lighter, more flexible and can absorb dynamic motions through extension without causing excessive tension. Another significant advantage of taut rope mooring is a reduction in the line length and in the space occupied on the seabed. Anchoring of the floating wind turbines is achieved by drag anchors or suction anchors.

The mooring spread could be symmetrical or asymmetrical: even if the former is the simplest in terms of design, it might not necessarily be the best in terms of performance. Some considerations must be made [8]:

- directionality of the weather: if the weather actions come from a dominant direction, it could be useful to use an asymmetrical mooring system to balance those forces;
- directionality of the currents: the same consideration of the directionality of the weather can be made for currents; an asymmetrical mooring system can balance the forces applied by strong currents;
- subsea spatial layout: equipment positioned on the seabed could restrict the available positions for the moorings;
- other space restriction: it could be useful to couple mooring lines to gain further space.

5. Mooring modelling

The environmental conditions at the two sites have been defined and then simulated in terms of water depth, wave representation with JONSWAP spectrum and mean wind speed values while distinguishing the conditions for return periods of 50 and 100 years [10]. Then, a description of the implemented floater is given, which includes the chosen WindFloat design with all its characteristics, such as columns, connections and water entrapment plates. The water depth chosen is 100m for the Adriatic Seas and 400m for the Sicilian location.

The floater's response to waves is provided through the Response Amplitude Operators (RAOs), and the wind response is implemented in the software, using the areas subjected to the wind forces and specifying the application point on the structure. Since the RAO's generation is a long and elaborate process, and in view of the fact that the present study is focused on the mooring, an existing RAO has been used. This set of equations has been taken from an example available on the Orcina website, regarding a structure like WindFloat one: a generic semi-submersible platform used in the oil industry [13], with RAO specified for 0°, 45° and 90° directions. The structural similarities along the wind response modelling are taken to give reliable answers regarding the mooring systems and their loads. It should be noted that, for a real design, the specific RAOs must be developed and possibly tested in a wave tank. While the wind load analysis in our model is based on the simplified mean wind load, the wave response analysis does account for all the hydrodynamic effects. The overall response analysis is therefore considered adequate for the preliminary investigation of the mooring line loads. See Figure 3 for OrcaFlex Model of the floater and Figure 4 for a suction anchor mooring system.

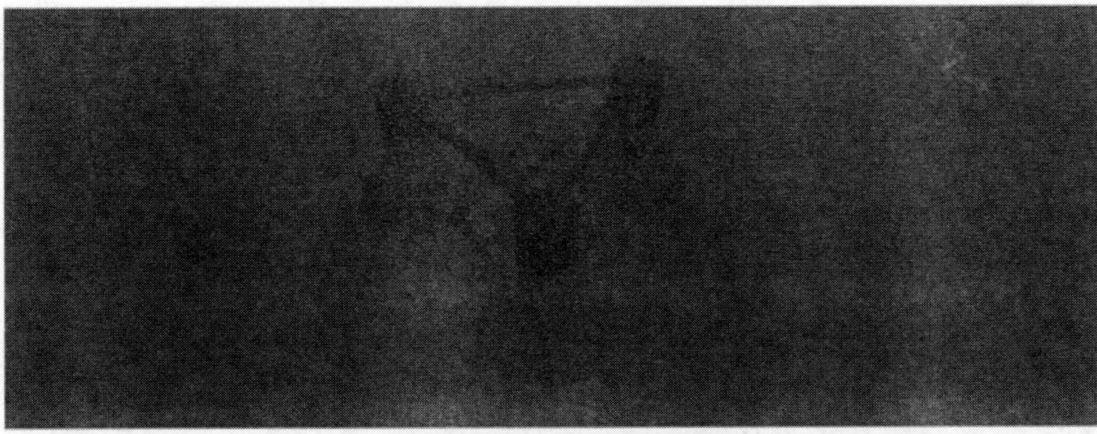

Figure 3 3D OrcaFlex model

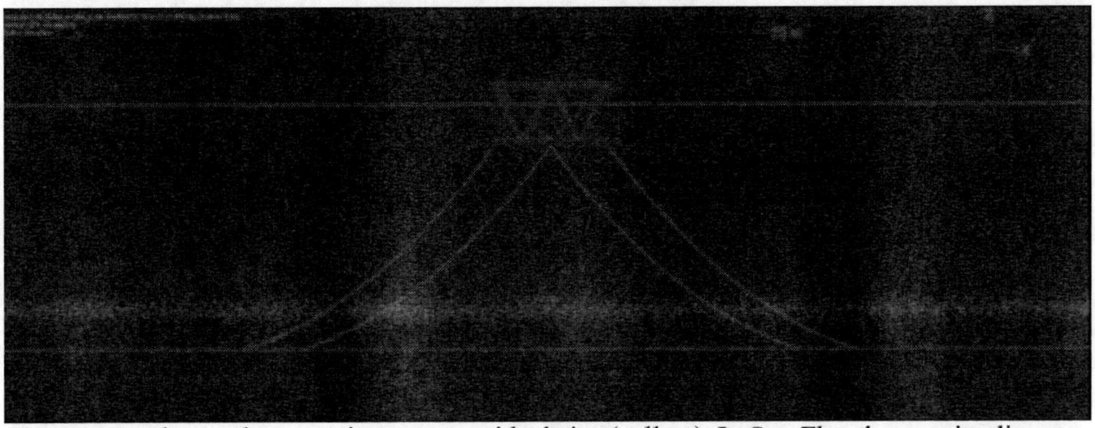

Figure 4 Suction anchor mooring system with chains (yellow). In OrcaFlex the mooring lines are modelled as finite line elements.

Wind loads are proportional to the areas subjected to the flow; in the OrcaFlex model we specify the surge area, the sway area and the yaw area moment. The surge area comprehends the whole area covered by the rotor's blades and the yaw area moment, see Table 5. Here, Asurge is the surge area, Asway is the sway area, Y is the yaw area moment and L is the length between perpendicular of the floater. The wind loads are applied at the position of the nacelle, to influence realistically the floater response in such situations. Note that the wind load is based on extrapolating the wind speed at 10 m to the appropriate height. A comprehensive discussion regarding statistical analysis of wind velocities is given in [10], Chapters 2.4 and 5.5. In the finding of the environmental design parameters concerning wind speed and wave height, the best results were obtained with a Gumbel distribution, having parameters estimated with maximum goodness of fit procedure. Since the difference concerning wind speed between the values found for 50 years and 100 years of return period is very little in both locations, the analysis of the result focused on 100 years return period scenarios. The wind speed at 10 m height above SWL used was 23.8 m/s.

Table 5. Wind load calculation.

Parameter	Values
A_{surge} (m^2)	12561
A_{sway} (m^2)	394
Yaw area moment (m^3)	15681
Length between perpendiculars (m)	40

IOP Conf. Series: Materials Science and Engineering **700** (2019) 012002 doi:10.1088/1757-899X/700/1/012002

Traditional anchors and suction anchors have been represented in the model. Different solutions concerning the mooring lines have been explored: chains, chains with synthetic fibers and synthetic fibers only. The first two cases led to a catenary design for the wires and, in the last one, the possibility of a taut mooring has been tested. Dimensions and the material's parameters are implemented, considering the environmental differences between locations. Table 6 and give parameters for mooring lines.

Table 6. Mooring chain parameters [14, 15].

Parameter	Value	Value
Mooring type	Chain	Synthetic rope
Bar diameter (m)	0.05	0.08
Link type	Stud-link	8-strand polyester
Grade/ Standard	R4	ISO 10556
Weight in air (kN/m)	0.537	0.050
Displacement (m)	0.071	0.037
Weight in water (kN/m)	0.466	0.013
Breaking load (kN/m)	2740	1091

The mooring lines follow a catenary shape, and two simulations for each return period have been performed with two different wire compositions: the first using a chain and the second using chains, in proximity of the anchor and the floater, and synthetic ropes in the middle; see Table 7.

Table 7. Mooring wire lengths for chain-only simulation (top) and for chain and synthetic fiber combination.

Length	Value (m)
Adriatic Sea, Chain length	500
South Sicily, Chain length	2000
Adriatic Sea	
Top chain length	70
Synthetic rope length	360
Bottom chain length	70
South Sicily	
Top chain length	200
Synthetic rope length	1600
Bottom chain length	200

Static and dynamic mooring analysis are carried out [10]. The first provides the equilibrium position of the system and the starting point for the dynamic simulation, in which the wave loads are applied. In the static results, the vertical component of the forces at the anchor is analyzed to obtain a starting parameter for a possible design for the anchor itself. Concerning the case with a chain-only simulation, an attempt to reduce the vertical forces has been made, progressively increasing the horizontal spread of the mooring lines from one time to three times the water depth. The same line characteristic has been kept: the touchdown point has always been maintained at the suction anchor. Results show a clear pattern: increasing the mooring spread and decreasing the angle between seabed and wire at the proximity of the suction anchor, the vertical forces decrease. In this process, the chain becomes more and more close to a traditional anchor system. The decrease in vertical forces permits the use of a less performant (and so less expensive) suction anchor, at the cost of using more material and forming a larger footprint on the seabed. This process is confirmed in both locations and can be seen in Figures 5 and 6, along with the angle between the wire and the seabed.

IOP Conf. Series: Materials Science and Engineering **700** (2019) 012002 doi:10.1088/1757-899X/700/1/012002

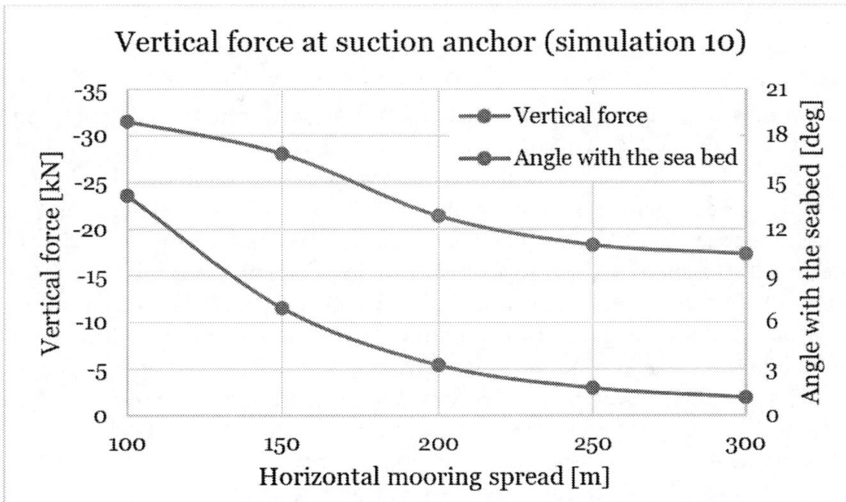

Figure 5. Vertical force at suction anchor (Adriatic Sea, suction anchor mooring system with chains, 100 years return period).

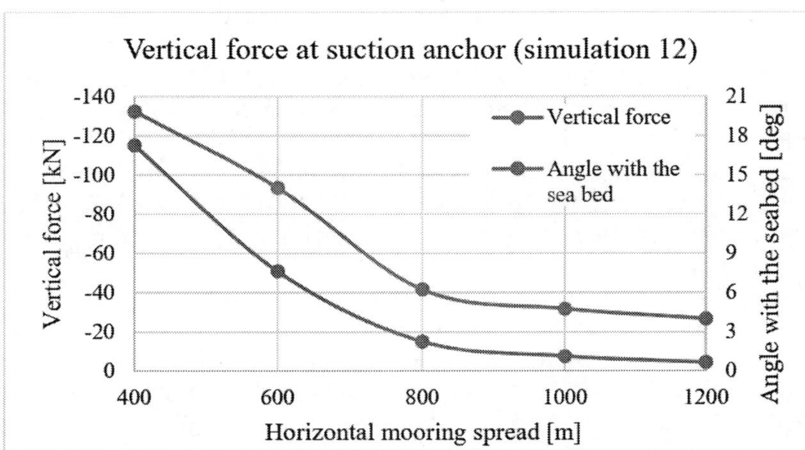

Figure 6. Vertical force at suction anchor (south Sicily, suction anchor mooring system with chains, 100-year return period).

In the dynamic simulations [10], the forces in both the wires/ropes and at the anchors, are analyzed for the two values of the return period of the environmental loading. In the dynamic analysis, OrcaFlex carries out a time simulation of the response of the system to waves and wind, based on user-defined inputs. The dynamic simulation uses the static analysis as its initial configuration, and time then evolves forward from there. The analysis of the result has focused on the tension in the wires and on the forces at the anchors: simulation has been performed for both return periods and for every wire and anchor, then the obtained design data have been compared.

These results have been obtained through the analysis of the time history of the simulation that produces cycles of loading and unloading on the wires and the anchor: the peak values have been implemented in a statistic concerning the extremes, using the OrcaFlex option, "extreme values statistics" [13]. In this procedure, the Generalized Pareto distribution gave results and their consequent confidence limits, for both tensions and anchor forces, simulating two storms, lasting for three and six hours.

The resulting values for tension present common patterns for both the Adriatic Sea and South Sicily: the most stressed part in the wire is in the proximity of the floater, the limit loads are never exceeded. The difference between a simulation with a 50-year return period and one with a 100-year return is negligible, and the variation of the storm duration affects the results poorly.

The values concerning the loads at the anchors, present differences between traditional and suction anchors: in the first case, they are almost not affected by the simulation, and the values are very close to the static analysis ones; in the second case, they present peaks of loading and unloading. There is also a difference for simulated suction anchors with chains and taut: the vertical load from the dynamic analysis in the first case differs from the static values more than for the second case [10]. As for the tension values, the difference between results obtained with a storm lasting three hours and one lasting six hours, and with environmental loads with 50- and 100-year return periods, is negligible. The values used as our design values are the return parameters (tension or vertical load), found by using the maximum likelihood fitting procedure. The values, along with confidence intervals (CI) with a confidence level set to 95%, mean that the reported return parameter is defined as the parameter whose expected number of exceedances in the specified storm duration is one [13]. In the present study, the values of CI come from a peak over threshold method analysis, in which the threshold used is the one suggested by the software, with no further considerations. An exceedance probability of one per storm should be acceptable, as the actual design of the mooring lines takes into account load factors to be multiplied with the load and material factors, reducing the allowable capacity of the mooring line tension.

In the found design values, some considerations are to be made considering the loads at the anchor: a traditional anchor must, in every case, avoid vertical loads, and a suction anchor must be designed to withstand a certain dynamic vertical load. This capacity will depend on soil conditions and anchor design, aspects not considered in this study. These considerations, along with the uncertainty of the statistical analysis, must be considered during the detailed design of the mooring system. For the traditional anchors, the dynamic simulations [10] showed loads close to the static ones: the difference between the mean load values and the peaks is around 10 N. In addition, most of the load is distributed, like in the static conditions, in the horizontal directions with no sign of vertical forces that could lead to an uplift of the anchors.

A discussion of the results is then given, comparing mooring length, footprint, tension and vertical forces at the anchors, along with a proposal for mooring solutions for the locations of the study. As the sea depth increases, the mooring footprint is heavier (Figure 7); the area used for suction anchors in the South Sicily location is 15 times larger than that used in the Adriatic Sea. A remarkable difference is present between traditional anchors and suction anchors: the former requires a much larger area than the latter, in every case.

The small suction anchor footprint is increased when we require a reduction in the vertical forces at the anchor: keeping the touchdown point of the mooring wire at the anchor, then a rise of the mooring spread and a contemporary decrease of the angle between the line and the seabed occurs, causing the vertical force to drop. So, at the cost of a larger footprint and a longer wire line, which uses more material, the decrease in vertical forces allows the use of a less performant and cheaper anchor. The decision of whether or not to apply this solution of vertical forces' reduction must be analyzed in terms of costs and environmental impact: the enlarged footprint of the suction anchor is near to that of a traditional anchor, and the increase in wire length could not be compensated by a cheaper anchor. Ultimately, the advantages of a suction anchor could disappear, and this could make the use of a traditional anchor more convenient in terms of cost-efficiency.

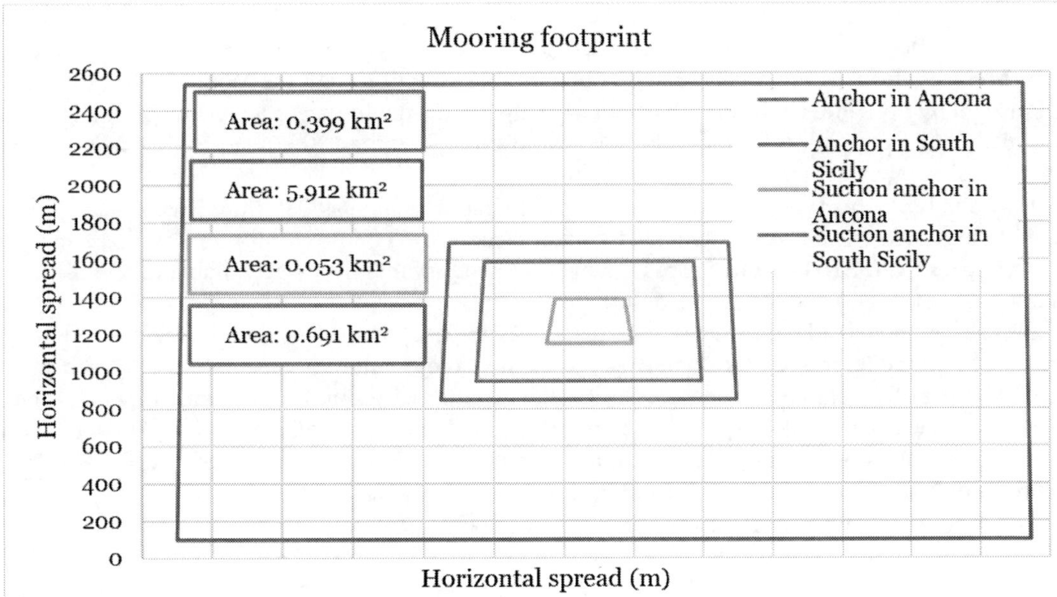

Figure 7. Mooring footprint in the Adriatic Sea and south Sicily.

6. Conclusions

This paper deals essentially with finding the most suitable mooring solutions for the realization of a floating wind turbine farm in the Italian seas, focusing on two different locations. This was obtained through the implementation of a 3D numerical model in the OrcaFlex software, which comprehends a common geometry of the floater, environmental loads and site characteristics in the two chosen study areas.

For the last two aspects, it emerged that a different return period for wind speed and wave height made little difference in the results. The parameter that most affected the outcomes, differentiating the two areas, was sea depth: when the depth increased, the lines' length, the mooring footprint, the mooring lines' tension and the vertical forces at the anchors, for the suction ones, were higher.

A similar consideration can be drawn from the simulations for the mooring solutions: in the same depth conditions, when a traditional anchor was used, the lines' length and the mooring footprint were higher, while wires' tension was lower than a suction anchor. A suction anchor must also cope with vertical forces, not present in the traditional ones.

A general indication from the simulations was also obtained for suction anchors: vertical forces at the anchors could be reduced with the increase of the horizontal spread of the moorings and the consequent reduction of the angle between seabed and lines.

Considering the advantages and disadvantages of the previous aspects and the technical challenges of the installation and the design, finally, the proposed numerical model was successful in providing a simple indication for the mooring of floating wind turbines in two different areas of the seas surrounding the Italian peninsula: for the Adriatic Sea (shallow waters), suction anchors with chains and, for the Strait of Sicily (deeper waters), traditional anchors with chains or chains with synthetic fibers.

Generally, it can be derived that, for shallow waters, the suction anchors should be the preferred solution: the smaller occupied area and the shorter lines compensate for the design difficulties of a similar device and the higher tensions of the wires. On the other hand, for deeper waters, the enlarged footprint and longer wires are the price to pay for the simplicity and the lower tension of traditional anchors.

In the findings of the environmental design parameters concerning wind speed and wave height, the best results were obtained with the Gumbel distribution, having parameters estimated with maximum goodness of fit procedure. An important issue emerged in the available data from the wave-monitoring networks: the recordings are often discontinuous, due to operational problems for the "Rete Ondametrica Nazionale", which has since been dismantled.

The modelled structure is the one that appears to be the most suitable in the Italian environment, with the current grade of development: a semi-submersible structure with the WindFloat design, developed in Portugal by Principle Power. It is particularly suitable for the Italian seas: its low draft allows transport and installation in shallow waters, like the Adriatic ones, without the need for specific vessels, apart from simple tugboats, or good weather conditions.

Italy can build this type of floating wind turbine in ports, such as Palermo, Livorno and Trieste, with suitable drydocks for the construction of similar structures. The possibility of commissioning in other EU countries in the Mediterranean area has also been explored: ports are available in Marseille (France), Skaramagas (Greece) and Valletta (Malta).

Numerical models like the one used in this work can be a powerful tool in the assessment of a specific aspect of a possible project, like moorings. Their accuracy is mainly based on the hypothesis made at the beginning of the elaboration process and the data gathered about the environment, or, in other words, creating a numerical model means recreating a simplified version of the reality, but, for the moment, some aspects are still missing. These features can be a starting point for future works with the goal of improving the results, and they are listed below:

- two main environmental variables have not been analyzed: currents and seafloor composition; once located in a more definite area, these aspects should be implemented in future works. Currents will be part of the external loading on the structure, and seafloor composition will influence the anchors' performance, especially influencing the design of the suction ones;
- other approaches could be followed in the statistical analysis of environmental factors: the use of all records instead of the monthly maxima. If the data coverage is good enough, considering all the recordings could improve the data at one's disposal. This may require additional monitoring networks;
- a more detailed modelling of the floater with a wind turbine and a dynamic response analysis should be performed in a real project. If the natural frequency of the floater is close to the peak wave frequency, the resulting resonance could endanger the structural integrity. Further analysis of blades and tower are also needed, in order to ensure limited response due to turbulence as well as to the mean wind speed.
- the RAO used in this paper has not been computed specifically for a semi-submersible floating wind turbine: future works should face this issue with a specific numerical analysis and testing in tanks with scaled models of the floater-turbine system;
- this work does not address system economics: manufacturing, installation or decommissioning considerations or optimization of the floating wind turbine mooring system. These aspects could be a crucial factor in the realization of a floating wind farm project.

Acknowledgements
The first author acknowledges the University of Stavanger for the opportunity to work in Stavanger during the preparation of his Master thesis, autumn 2017.

References

[1] UNFCCC; The Paris agreement. 4 November 2016. http://unfccc.int/paris_agreement/items/9485.php
[2] European Commission; CO2 time series 1990-2013 per region/country. 2014. http://edgar.jrc.ec.europa.eu/overview.php?v=CO2ts1990-2013&sort=des9
[3] Skaare B 2017 *Development of the Hywind concept.* Proceedings of ASME 2017 36th International Conference on Ocean, Offshore and Arctic Engineering, Trondheim, Paper No: OMAE2017-62710
[4] Skaare B Nielsen F G, Hanson T D, Yttervik R, Havmøller O and Rekdal A 2015 Analysis of measurements and simulations from the Hywind Demo floating wind turbine. *Wind Energy,*

18(6) 1105-22.

[5] RON Ancona. n.d.; http://dati.isprambiente.it/id/buoy/ron/ancona-09/html

[6] RMN Porto Empedocle. n.d.; http://www.mareografico.it/?session=0S658683622AQB89T658771&syslng=ita&sysmen=-1&sysind=-1&syssub=-1&sysfnt=0&code=STAZ&idst=1S&mode=INFO

[7] RON Mazara del Vallo. n.d.; http://dati.isprambiente.it/id/buoy/ron/mazara-del-vallo-09/html

[8] Roddier D, Aubault A, Peiffer A and Weinstein J 2011 *A generic 5 MW windfloat for numerical tool validation & comparison against a generic spar.* ASME 2011 30th Int. Conf. on Ocean, Offshore and Arctic Engineering. Rotterdam, 2011.

[9] Roddier D Cermelli C Aubault A and Peiffer A 2017 *Summary and Conclusions of the Full Life-Cycle of the WindFloat FOWT Prototype Project.* Proceedings of ASME 2017 36th International Conference on Ocean, Offshore and Arctic Engineering, Trondheim, Paper No: OMAE2017-62561.

[10] Re P 2018 *Mooring systems analysis for floating wind turbines in the Italian seas*, MSc thesis School of Civil, Environmental and Land Management Engineering, Politecnico di Milano, Italy, 2018.

[11] Weller S D, Johanning L, Davies P and Banfield S J 2015 Synthetic mooring ropes for marine renewable energy applications, *Renewable Energy,* **83**, 1268-78.

[12] Chakrabarti S *Handbook of Offshore Engineering (2-volume set).* Elsevier Science, 2015.

[13] Orcina. OrcaFlex. n.d.; https://www.orcina.com/SoftwareProducts/OrcaFlex/

[14] DNV GL; *Offshore mooring chain.* https://rules.dnvgl.com/docs/pdf/dnvgl/OS/2015-07/DNVGL-OS-E302.pdf

[15] DNV G L 2016 *Synthetic fibre ropes for towing, mooring and anchoring.* March 2016. https://rules.dnvgl.com/docs/pdf/DNVGL/CP/2016-03/DNVGL-CP-0100.pdf

IOP Conf. Series: Materials Science and Engineering **700** (2019) 012003 doi:10.1088/1757-899X/700/1/012003

CFD simulations of a suspension bridge deck for different deck shapes with railings and vortex mitigating devices

I Kusano[1*], J B Jakobsen[1] and J T Snæbjörnsson[1, 2]

[1] University of Stavanger, Stavanger, Norway
[2] Reykjavik University, Reykjavík, Iceland

[*] Corresponding author: ibuki.kusano@uis.no

Abstract. Aerodynamic response of a single-box suspension bridge girder is investigated using CFD simulations. The importance of including railings and vortex mitigating devices such as guide vanes and a spoiler in the simulations is explored since they alter the flow field around the deck greatly. The paper compares results from a bare deck section and a section with mitigating devices. A clear vortex shedding observed for the bare deck section is suppressed by the use of mitigating devices and the aerodynamic force coefficients are very different between these cases. The effect of varying deck width is also studied. Flutter derivatives are defined based on quasi-steady formulation and flutter velocity is computed for each section. The section with the largest width to depth ratio has the best performance against flutter. This study is an initial phase of multi-fidelity optimization of bridge deck shape considering aerodynamic constraints.

1. Introduction

With the advance of technologies, the achievable span of suspension bridges becomes longer and longer. The Great Belt Bridge and Akashi Bridge, which both opened in 1998, have main span of more than 1.6 km. Other suspension bridges, such as Xihoumen Bridge in China, which opened in 2009 and the recent construction of Osman Gazi Bridge in Turkey follow this trend. As the structures become longer and more flexible, they are more vulnerable to wind-induced phenomena such as flutter, buffeting and vortex shedding.

For the design of such structures, it is essential to consider their aerodynamic response, since it may lead to collapse. The shape of a bridge deck is one of the key factors influencing the structural response under wind load. Among different deck shapes, the stream-lined single box girder has become popular after its successful application to the Severn Bridge in UK in 1966. It possesses good performance against vortex shedding and flutter instabilities. Since the box section provides larger torsional inertia than open sections, the deck can be designed shorter in height, which results in more slender and aesthetic appearances. The box section is also more economic in construction as well as corrosion maintenance [1].

The characterization of aerodynamic response of a bridge deck has been historically performed in wind tunnel tests. However, according to Mannini at al. [2], Sarkic et al [3] among others, the improvement in Computational Fluid Dynamics (CFD) techniques in the last decade has made this method more increasingly reliable to study aerodynamic behaviour of bridge deck cross sections. For instance, Ge and Xian [4] used this technique to obtain flutter response of a bridge numerically.

This work focuses on the use of 2D unsteady Reynolds Averaged Navier-Stokes (URANS) CFD simulations to obtain aerodynamic coefficients for different deck shapes. Although this method presents some limitations such as perfect correlation of the flow along the span-wise direction, it can accurately provide the mean pressure distribution around the geometry and consequently force coefficients as demonstrated by researchers such as Vairo [5], Brusiani [6] and Nieto [7].

The aerodynamic coefficients are essential to characterize bridge deck section since they can be used to carry out flutter and buffeting analyses based on quasi-steady theory. This formulation is appropriate for high reduced wind velocities and streamlined deck cross sections according to Wu and Kareem [8] and Tubino [9].

In this work, the aerodynamic chatacteristics of different deck sections are studied. First of all, two deck sections, one with a bare deck and a second one with vortex mitigating devices are studied and compared. Then four sections of different deck width and a constant height are investigated. Flutter derivatives based on the evaluated force coefficients are compared as well as the flutter speed for each section. Although many researchers have worked with CFD simulation of a bare deck [10,11,12], there are are limited number of numerical investigations of deck sections with vortex mitigating devices [13,14,15]. Some previous studies have addressed the influence of aerodynamic appendages such as flaps on flutter velocity [16] or porosity of railings on critical flutter velocity [17].

This study is an initial phase of multi-fidelity optimization of a suspension bridge deck shape under flutter constraint. Julsundet Bridge in Norway was used as an application example.

2. Formulation

2.1 Force coefficients

The aerodynamic force coefficients and Strouhal number of a bridge deck section under wind load are expressed as:

$$C_L = \frac{L}{0.5 \cdot \rho U^2 B}; \quad C_D = \frac{D}{0.5 \cdot \rho U^2 H}; \quad C_M = \frac{M}{0.5 \cdot \rho U^2 B^2}; \quad S_t = \frac{fH}{U} \tag{1}$$

where C_L, C_D and C_M are lift, drag and moment coefficients, L, D and M are the time-averaged lift and drag forces and moment per unit of length, ρ is the flow density, U is the undisturbed wind velocity, f is the vortex shedding frequency and B and H are the width and the height of the deck cross-section. The force and sign convention used for the study is shown in Figure 1.

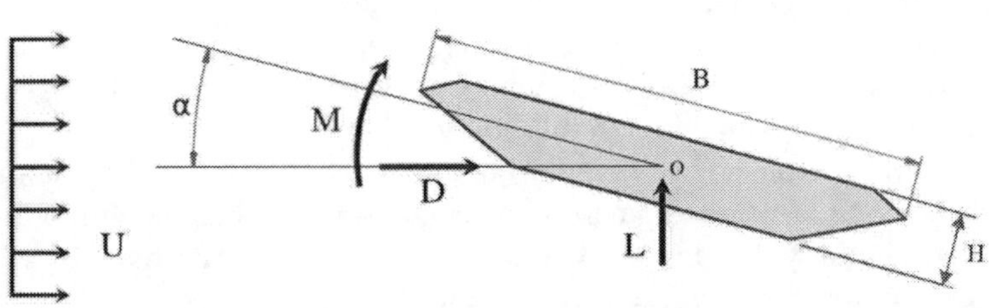

Figure 1. Force and sign convention of a bridge deck under constant velocity.

2.2 Computation of flutter speed based on quasi-steady theory

Fully-computational approach of flutter velocity computation permits the substitution of wind tunnel tests by equivalent numerical simulations [18]. This fully computational method is briefly described next.

Scanlan and Tomko [19] established three aeroelastic force components based on frequency dependent function called flutter derivatives as follows.

$$\mathbf{f}_a = \begin{Bmatrix} D_a \\ L_a \\ M_a \end{Bmatrix} = \frac{1}{2}\rho UKB \cdot \begin{pmatrix} P_1^* & -P_5^* & -BP_2^* \\ -H_5^* & H_1^* & BH_2^* \\ -BA_5^* & BA_1^* & B^2A_2^* \end{pmatrix} \begin{Bmatrix} \dot{v} \\ \dot{w} \\ \dot{\varphi}_x \end{Bmatrix} + \frac{1}{2}\rho U^2 K^2 \cdot \begin{pmatrix} P_4^* & -P_6^* & -BP_3^* \\ -H_6^* & H_4^* & BH_3^* \\ -BA_6^* & BA_4^* & B^2A_3^* \end{pmatrix} \begin{Bmatrix} v \\ w \\ \varphi_x \end{Bmatrix} \quad (2)$$

where B is the deck width, U is the acting wind speed, $K = B\omega/U$ is the reduced frequency with ω as the response frequency, whereas A_i^*, H_i^* and P_i^* ($i=1,\ldots,6$) are the flutter derivatives.

The quasi-steady model relates the displacements and rotations of the deck and velocities with the relative wind velocity components. The flutter derivatives may be expressed in terms of force coefficients by comparing the quasi-steady and Scanlan's model as:

$$H_1^* = -\frac{C'_{L,0}+C_{D,0}}{K}; \quad H_2^* = \frac{C'_{L,0}+C_{D,0}}{K}\cdot\mu_H; \quad H_3^* = -\frac{C'_{L,0}}{K^2};$$
$$A_1^* = \frac{C'_{M,0}}{K}; \quad A_2^* = \frac{C'_{M,0}}{K}\cdot\mu_A; \quad A_3^* = -\frac{C'_{M,0}}{K^2} \quad (3)$$

where $C_{L,0}$, $C_{D,0}$, $C_{M,0}$, are lift, drag, and moment coefficients at 0 degree of angle of attack, $C'_{L,0}$, $C'_{D,0}$, $C'_{M,0}$, are their derivatives while μ_H and μ_A can be estimated according to Larose and Livesey [20] as:

$$\mu_H \approx \frac{A_1^*}{H_1^*} \text{ and } \mu_A \approx \frac{A_3^*}{H_3^*}$$

The dynamic equilibrium of a deck under aeroelastic forces, $\mathbf{f_a}$ can be written in a matrix form as:

$$\mathbf{M\ddot{u}} + (\mathbf{C}+\mathbf{C_a})\mathbf{\dot{u}} + (\mathbf{K}+\mathbf{K_a})\mathbf{u} = \mathbf{0} \quad (4)$$

where **M**, **C**, and **K** are mass, damping, and stiffness matrices, $\mathbf{K_a}$ and $\mathbf{C_a}$ are aeroelastic stiffness and damping matrices. By applying modal analysis to Eq. (4), the solution is a combination of the most relevant m modes. This leads to the following eigenvalue problem:

$$(\mathbf{A} - \mu\mathbf{I})\mathbf{w}_\mu e^{\mu t} = 0 \quad (5)$$

where the matrix **A** is a function of the matrices, $\mathbf{K_a}$ and $\mathbf{C_a}$, and $\mathbf{w}_\mu$ is an eigenvector. In order to solve this problem, we need mode shapes and natural frequencies of the bridge under study, which can be obtained by a finite element model. The solution to Eq. (5) is expressed as $\mu_j = \alpha_j + i\beta_j$ where α is related to structural damping and β is the damped frequency ($j = 1,\ldots m$ where m is number of frequencies). Flutter occurs when α becomes null with increasing wind speed.

3. CFD analysis of the bridge deck

CFD simulations were performed to obtain force coefficients of the bridge deck for varying shapes. 2D URANS approach was adapted to perform the analyses. The open source CFD code of OpenFoam v.2.4 [21] was used to carry out CFD analyses in this study.

The k-ω Shear-Stress Transport (SST) turbulence model was selected among many available turbulence models because of its good performance and wide acceptance within the bridge engineering field. This model was initially developed by Menter [22] and later improved by Menter and Esch [23]. It takes advantage of robust and accurate formulation of Wilcox k-ω model [24] at the near wall region as well as a good performance of the k-ε turbulence model [25] in the far field. See for example, Sarkic et al. [3], De Miranda et al. [26] and Montoya [27].

In this present work, the bridge planned for crossing of the Julsundet Fjord in Norway is taken as a case study. It is a suspension bridge with a main span of 1.6 km and the deck section is 32 m wide and 4 m high aerodynamic single box girder. The baseline deck geometry in this study follows that of the 1:50 scale model of the planned bridge shown in Figure 2. The further description of the bridge can be found in a report from the Norwegian Public Road Administration [28].

The flow domain and the boundary conditions are illustrated in Figure 3. The flow domain size is $30B$ x $41B$ where B is the deck width. The simulation was performed at Re = 3.5e+5. Uniform velocity of 8 m/s at the inlet boundary was imposed while atmospheric pressure was imposed at the outlet.

Turbulent intensity of 5% was considered for the incoming flow with 0.1 B turbulent length scale. This length scale is also adopted by Ribeiro [29] and Nieto [7]. Both upper and lower walls have slip wall condition while the wall around the deck was non-slip. Linear interpolation of values from cell centres to face centres are carried out. The gradient terms are discretized by the Gauss scheme with linear interpolation. For divergence, the Gauss scheme with linear upwind and the limited linear interpolation schemes were applied. The 2D unstructured mesh was generated using software PointWise [30].

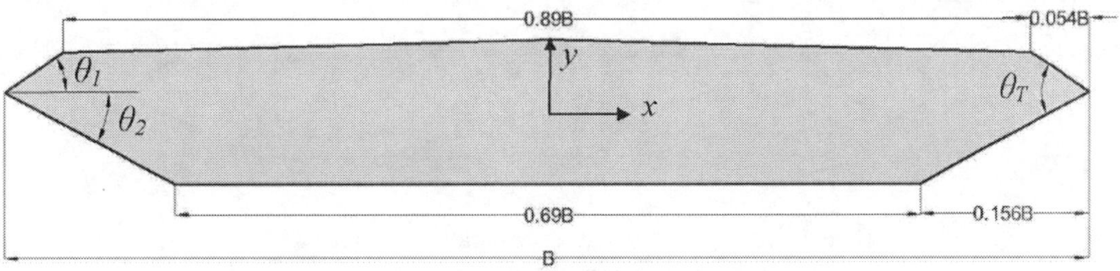

Figure 2. Base geometry of the bare deck section.

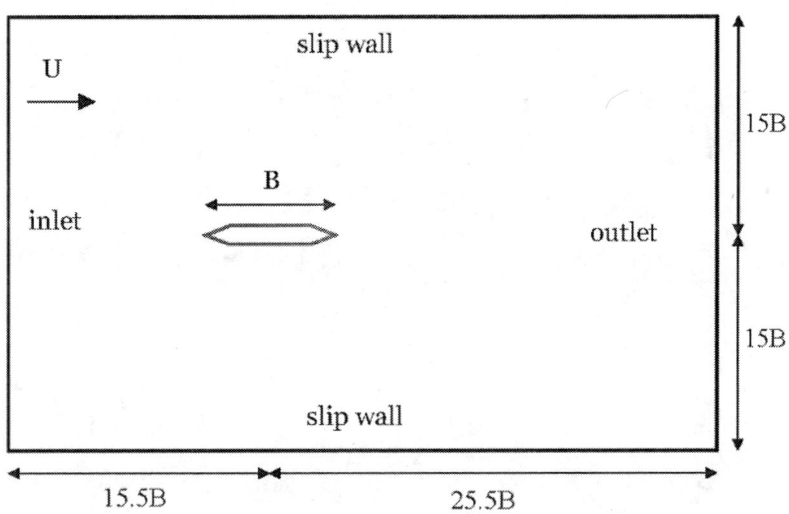

Figure 3. Flow domain size and boundary conditions.

A boundary layer mesh was applied around the deck wall in a way that gave a maximum y^+ value less than 4 where $y^+ = u_\tau \cdot h_1 / \nu = \sqrt{\tau_w / \rho} \cdot h_1 / \nu$, u_τ is wall friction velocity, h_1, the distance from the wall to the first cell centre, τ_w, wall shear stress, ν, kinematic viscosity. The first element height was set to h_1/B=1.14e-4 while Courant number was set to 2 for the simulation time-step. The time integration was verified by Courant number 1 and 0.5 prior to the study, and no significant difference in results were found. A mesh convergence study was conducted using a bare deck section with three different mesh densities in order to verify the discretization. Three types of meshes were considered: a coarse mesh with 112,564 cells, a medium mesh with 196,126 cells and a fine mesh with 321,435 cells. The coarse mesh underestimates the lift coefficient while the medium and fine mesh gave similar results. The medium mesh scheme was therefore selected for the simulations.

3.1 Case study 1
In order to assess the importance of mitigating devices and railings on the force coefficients, the bare deck model with original deck shape in Figure 2 was compared with the model with railings and the mitigating devices shown in Figure 4. The railings geometry is a simplified representation of that adopted in wind tunnel tests [28]. The model has various railings, two guide vanes as well as a spoiler at the bottom center part of the deck. The model scale is 1:50 with respect to the actual bridge dimension of B=32 m. Figure 5 shows the details of mesh for the section with aerodynamic devices. The number of cells for this model is approximately 240,000.

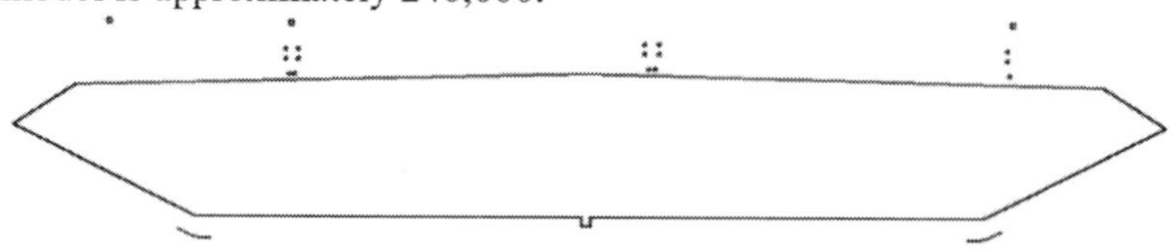

Figure 4. Deck geometry with mitigating devices and railings.

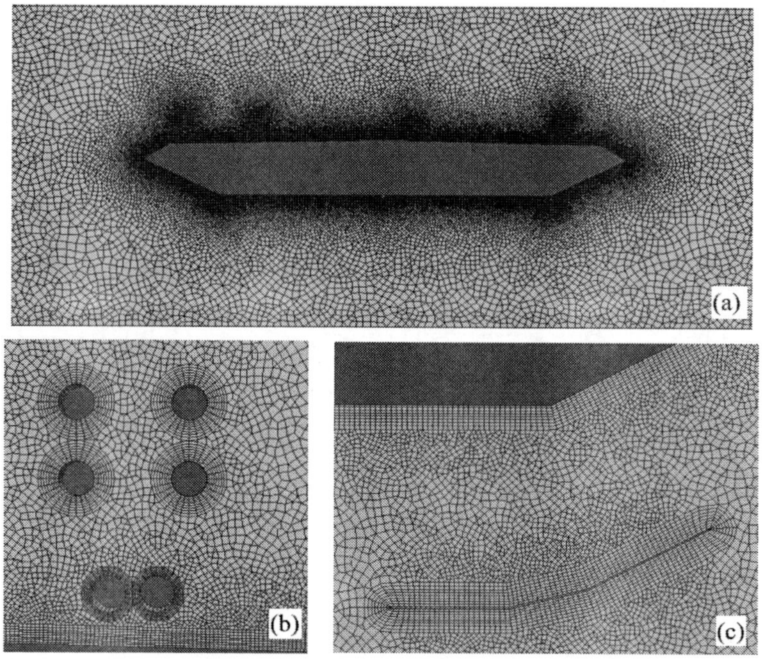

Figure 5. Meshing details of the model (a) with railings (b) and guide vanes (c).

The evaluated force coefficients are summarized in Table 1, along with standard deviation, Strouhal number and the gradient with respect to the angle of attack. Figure 6 shows the force coefficient history of both sections. The bare deck presents periodic oscillations with a single peak frequency while the amplitude of oscillation for the section with mitigating devices are one order of magnitude smaller.

Table 1. Comparison of parameters for the bare deck and the mitigating devices models.

Geometry	C_L	C_D	C_M	σ_{CL}	σ_{CD}	σ_{CM}	C_L'	C_D'	C_M'	St
Bare deck	-0.00043	0.391	0.0234	0.016	0.0035	0.0034	5.98	0.20	1.45	0.25
Deck & devices	-0.148	0.650	0.0021	0.002	0.0126	0.0005	4.64	0.43	1.29	-

IOP Conf. Series: Materials Science and Engineering **700** (2019) 012003 doi:10.1088/1757-899X/700/1/012003

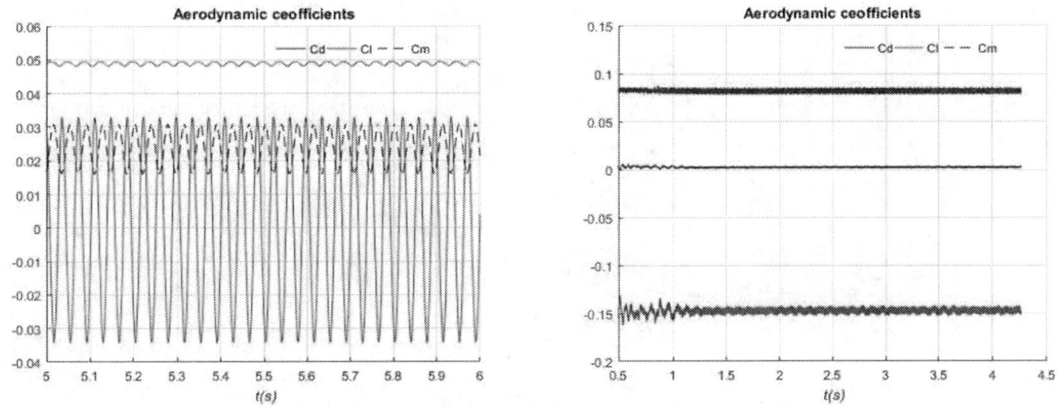

Figure 6. Force coefficients history of bare deck (left) and the section with mitigating devices (right).

Figure 7 shows flow fields for the two cases. The flow field for the bare deck section shows that there is a flow separation at the first corner after the leading edge both at the top and bottom of the deck, but the flow reattaches again close to the edge. From the Figure 7 b1), it is seen that the suction has similar value at the top as well as the bottom plates resulting in overall small lift coefficient. From Figure 7 c1), a vortex shedding phenomenon is clearly observed for this section, which is verified by the standard deviation of the force coefficients (Table 1).

On the other hand, the flow for the section with mitigating devices is rather complex due to the elements added to the deck. From the pressure plot, it can be seen that the pressures on the bottom part are largely negative while the top part experiences rather positive because the flow is slowed down by the railings. This results in a large negative lift coefficient. The vortex shedding behaviour seen for the bare deck section is suppressed for the deck with mitigating devices as could be expected.

The drag coefficient for the section with devices is 66% higher than for the bare deck section because of the various railings and mitigating devices while the moment coefficients are much smaller. The most notable difference is the lift coefficients; -0.00043 for the bare deck and -0.148 for the section with aerodynamic devices. The gradient of the lift coefficient with respect to the angle of attack is 22% smaller for the section with devices, which is favourable for the aerodynamic response of the bridge deck.

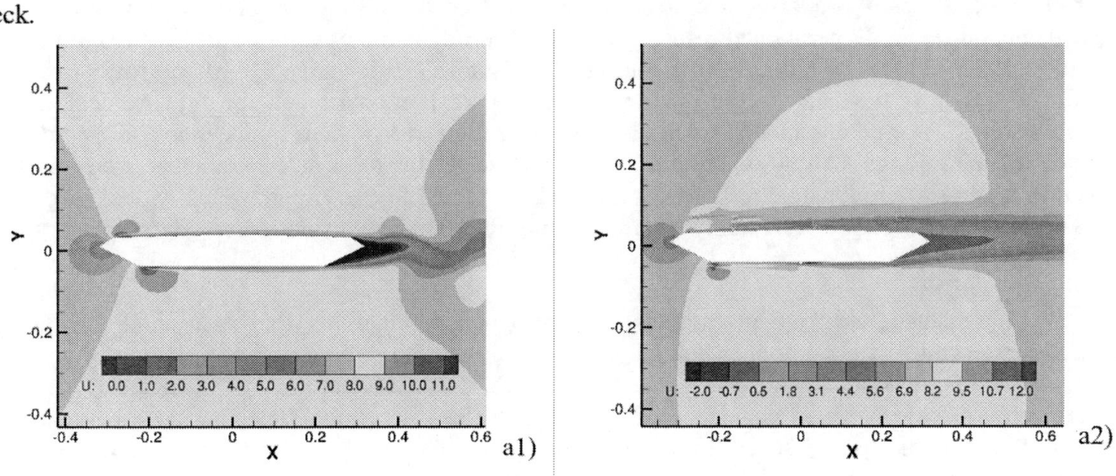

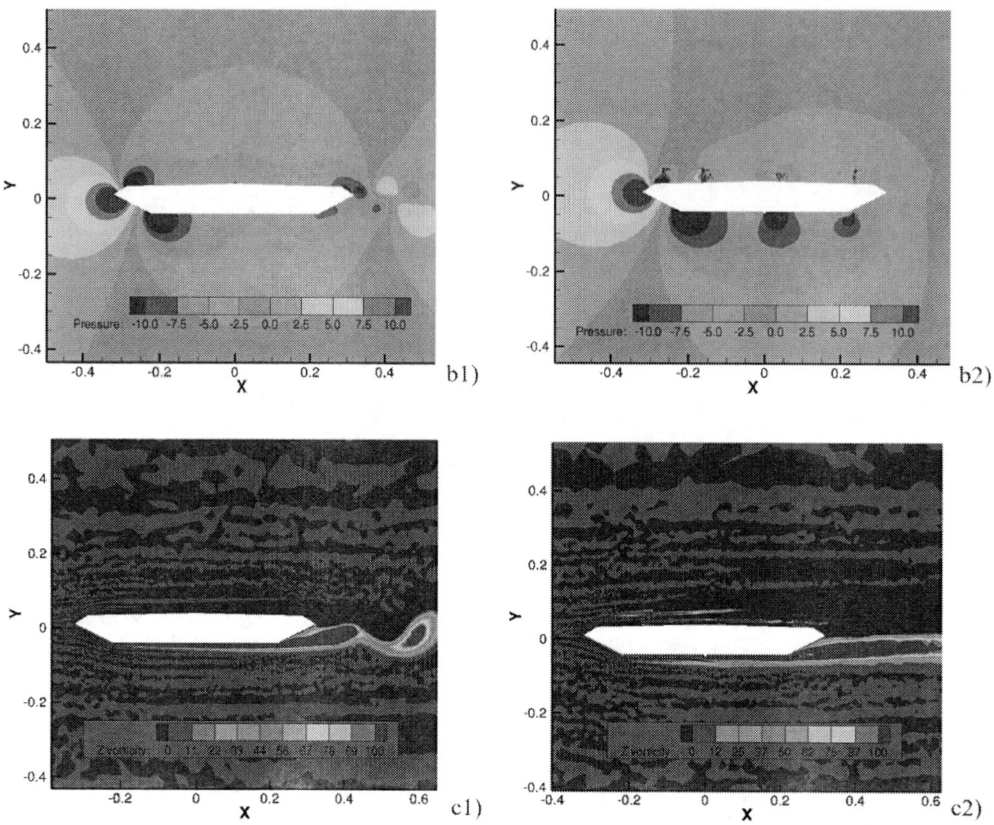

Figure 7. Comparison of instantaneous flow fields of the bare deck (left) and the section with mitigating devices (right): a) velocity *Ux* (m/s), b) pressure (range -10 to +10 Pa) c) z vorticity.

The corresponding force coefficients obtained in the wind tunnel tests are C_D=0.967, C_L=-0.309, C_M=-0.017, C_L'=3.624, C_M'=1.182 [28]. The difference between the simulation and the experimental results is considered primarily due to a higher density of railings on the section model in the experimental set up. Experimental results for another aerodynamic box girder [28] with depth to width ratio of 1:6.2 and similar railing configurations as Julsundet Bridge gave C_D =0.360 for the bare deck and C_D =0.845 for the same section with guard rails. Our drag coefficient for the bare deck model is comparable to these results and the simulation for the section with devices can be improved by additional details of the railings. On the other hand, the objective of this work is a parametric study of a deck section, in which aerodynamic response of the deck is studied for varying the deck width. This concept is later used as a part of multi-fidelity optimization of deck shape considering aerodynamic constraints.

3.2 Case study 2
In this Case study, we have investigated the influence of changing the deck width on the force coefficients while maintaining the deck height. The section with railings and mitigating devices is used for this parametric study. As we change the deck width, we are also changing the angles of the leading and trailing edges. We have considered four cases besides the original design, whose geometries are illustrated in Figure 8 and summarized in Table 2. Each geometry is expressed in terms of a non-dimensional width, $b/0.5B_0$, where b is the horizontal distance from the leading edge to the upper left corner of each deck geometry (Figure 8) and B_0 is the initial deck width.

IOP Conf. Series: Materials Science and Engineering **700** (2019) 012003 doi:10.1088/1757-899X/700/1/012003

Table 2. Geometrical properties of the simulations (See Figure 2 for the description of θ_1, θ_2, θ_T).

s. num	$b/0.5B_0$	θ_1	θ_2	θ_T
0	0.121	33	27	60
4	0.076	46	30	70
7	0.107	36	28	64
14	0.168	25	24	49
20	0.231	19	21	40

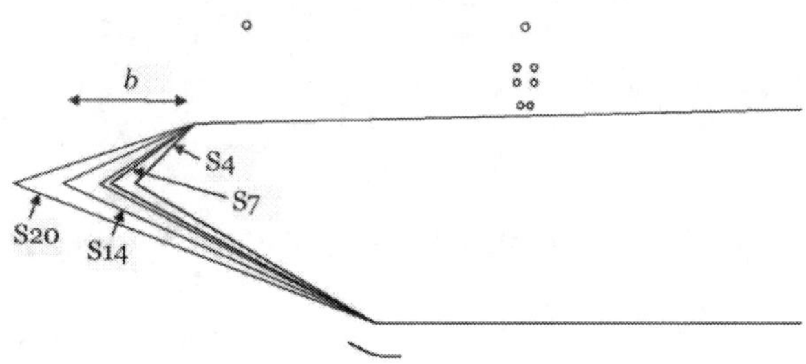

Figure 8. The deck shapes considered in case study 2 (the blue line shows the original design, s_0).

The force coefficients obtained for different deck shapes are summarized in Table 3 and plotted with respect to the nondimensional width in Figure 9. The lift and moment coefficients refer to the deck width considered in each case. Examples of the evaluated flow fields are shown in Figure 10 for the two extreme section shapes, s_4 and s_{20}.

For the most bluff section of *s4*, a large flow separation is observed because of the large angle of the leading edge, while for the most aerodynamic section of s_{20}, the flow is more attached to the deck. Large negative pressure is observed on the top left corner of the section *s4*, which makes the lift coefficient smaller compared to that of section s_{20}.

As the deck width increases and the leading-edge angle decreases, the lift coefficient and the moment coefficient decrease. The drag coefficient is also slightly reduced with decreasing leading-edge angle. The reduction in the gradient of the lift coefficient is most notable with increasing deck width.

Table 3. Force coefficient results of different geometries

geometry	$b/0.5B_0$	C_L	C_D	C_M	C_L'	C_M'
s4	0.076	-0.120	0.650	0.0090	5.67	1.36
s7	0.107	-0.138	0.643	0.0043	4.55	1.23
s14	0.168	-0.164	0.650	-0.0003	3.98	1.22
s20	0.231	-0.181	0.638	-0.0023	3.23	1.17

4. Computation of flutter velocity based on quasi-steady theory

From the force coefficients obtained in the CFD simulations for different deck sections, flutter derivatives were computed based on quasi-steady formulation. Four of the most important flutter derivatives, A_2, A_3, H_2, H_3 are plotted for the sections, *s4*, *s0* and *s20* in Figure 11. Other sections are omitted for clarity of representation. For A_2 and A_3 flutter derivative, the gradient C_M' is one of the key parameters and for A_2, the ratio of the gradient C_M'/C_L' is important. On the other hand, for H_2 and H_3, the gradient, C_L' is the dominating parameter.

IOP Conf. Series: Materials Science and Engineering **700** (2019) 012003 doi:10.1088/1757-899X/700/1/012003

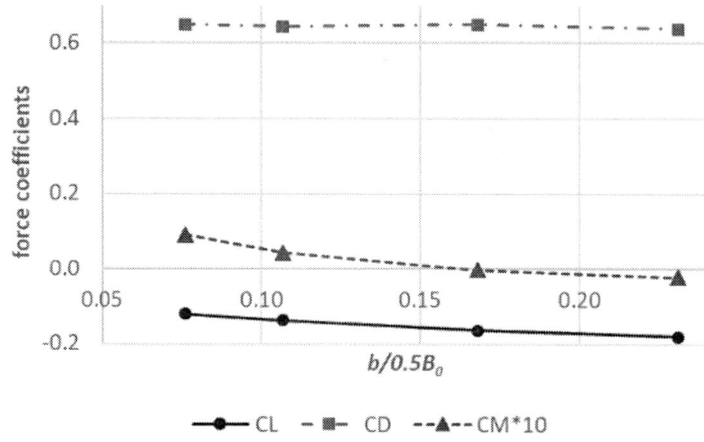

Figure 9. Force coefficients vs nondimensional width $b/0.5B_0$ (C_M is multiplied by 10).

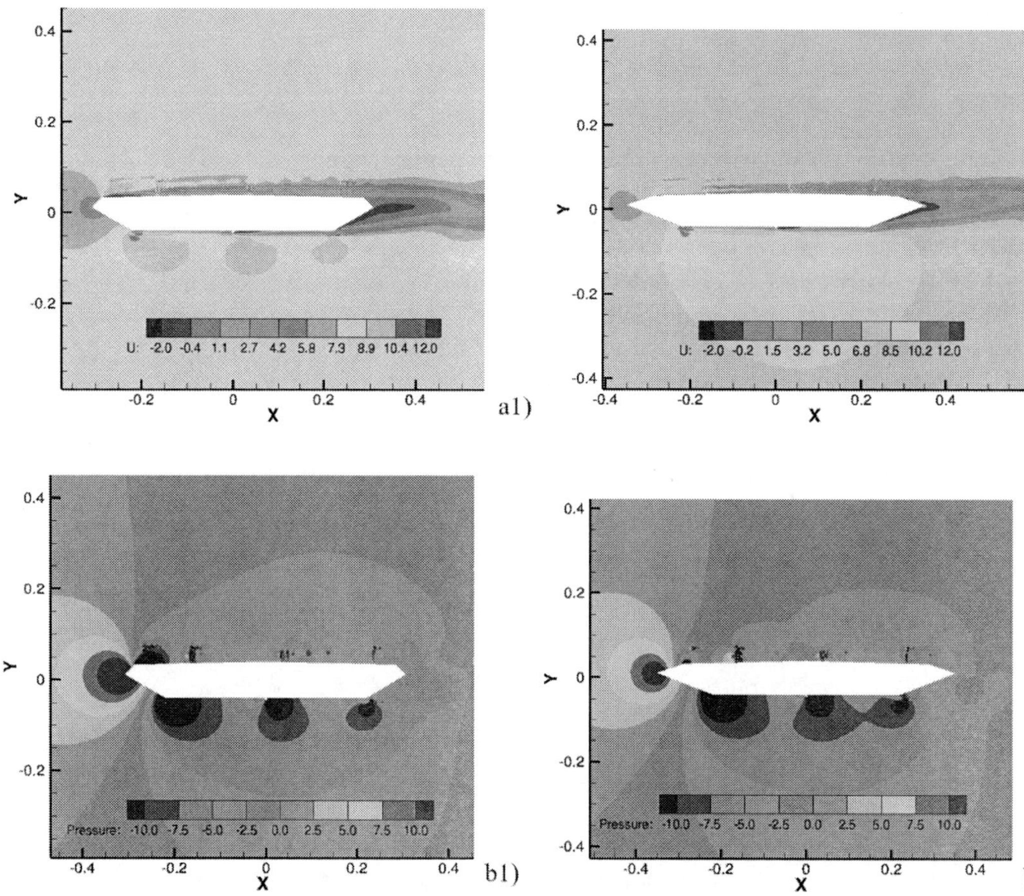

a1) a2)

b1) b2)

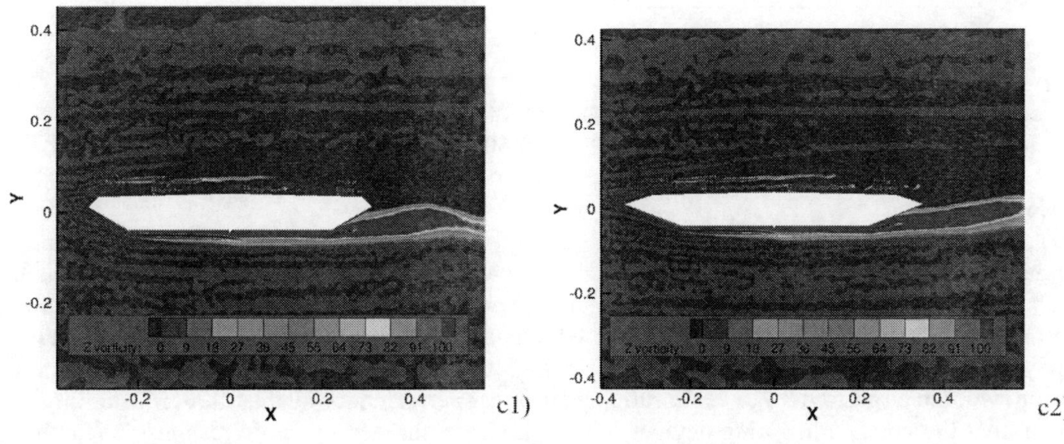

Figure 10. Instantaneous flow fields of *s4* (left) and *s20* (right): a) velocity *Ux* (m/s), b) pressure (range -10 to +10 Pa) c) z vorticity

Flutter velocity of different sections was computed using the first and the second vertical symmetric and the first torsional modes, which are 0.121 Hz, 0.166 Hz and 0.255 Hz respectively [28]. The structural damping ratio was assumed to be 0.3% [31]. The resulting flutter velocity and the corresponding reduced frequency for each section is summarized in Table 4. The flutter velocity of the original design section *s0* is 68.95 m/s, which is similar to the experimental results of 71 m/s [28]. As the deck width increases and the deck shape becomes more aerodynamic, flutter velocity increases. Although the vortex mitigating devices such as guide vanes and spoiler are primarily used to suppress vortex shedding, they also contribute favorably to increase flutter velocity by altering the flow field around the deck.

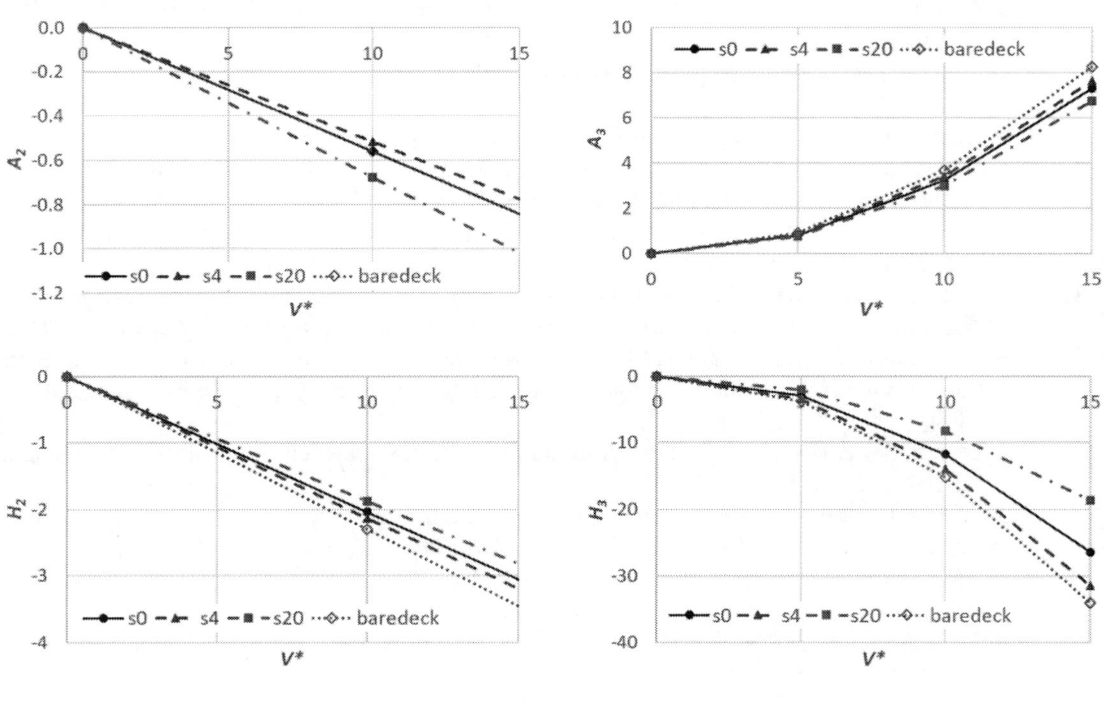

Figure 11. Flutter derivatives, A_2, A_3, H_2 and H_3 for different bridge sections.

Table 4. Flutter velocity V_f and the corresponding reduced frequency K for different deck sections.

geometry	*s4*	*s7*	*s0*	*s14*	*s20*	bare deck
$b/0.5B_0$	0.076	0.107	0.121	0.168	0.231	0.121
V_f (m/s)	63.44	67.71	68.95	75.39	79.82	61.21
K	0.68	0.62	0.60	0.52	0.46	0.71

5. Conclusions

CFD analysis of a single-box deck section was carried out to study the importance of railings and vortex mitigating devices on the bridge response under wind loading. Different deck sections were studied such as a bare deck section and sections with mitigating devices and varying deck width. Aerodynamic force coefficients obtained by CFD simulations were then used to compute flutter derivatives based on quasi-steady formulation. Flutter velocity of different sections was calculated for different sections. The section with the largest width to depth ratio has the best performance against flutter instability, while it is seen that the vortex mitigating devices tend to improve the aerodynamic response of the bridge by altering the flow field around the deck. This CFD study will be used as a part of multi-fidelity optimization of a bridge deck shape considering aerodynamic constraint.

Acknowledgement

This research is financed by the European Union's Horizon 2020 research and innovation program under Marie Sklodowska-Curie individual fellowship, grant agreement 794646-MFOptBF-H2020-MSCA-IF-2017. The authors would like to thank the Norwegian Public Road Administration and Svend Ole Hansen ApS for the information on the bridge deck geometry and the results of the wind tunnels tests with a section model of Julsundet Bridge. The authors also would like to thank Dr. Guang Yin at the University of Stavanger for his valuable advices on CFD simulations.

References

[1] Gimsing NJ and Georgakis CT 2011 *Cable suppered bridges: concept and design*. Wiley Inc.
[2] Mannini C, Soda A, Vob R, and Schewe G 2010 Unsteady RANS simulatios of flow around a bridge section. *J. Wind Eng. Ind. Aerod.* **98** 742-53.
[3] Sarkic A, Fisch R, Hoffer R, and Bletzinger K 2012 Bridge flutter derivatives based on computed, validated pressure fields. *J. Wind Eng. Ind. Aerod.* **104-106** 141-51.
[4] Ge YJ and Xiang HF 2008 Computational models and methods for aerodynamic flutter of long-span bridges. *J. Wind Eng. Ind. Aerod.* **96** 1912-1924.
[5] Vairo G 2003 A numerical model for wind loads simulation on long-span bridges. *Simulat. Model. Pract.Theor.* **11** 5-6, 315-51.
[6] Brusiani F, de Miranda S, Patruno L Ubertini F and Vaona P 2013. On the evaluation of bridge deck flutter derivatives using RANS turbuelnce models. *J. Wind Eng. Ind. Aerod.* **119** 39-47.
[7] Nieto F, Owen JS, Hargreaves DM and Hernandez S 2015 Bridge deck flutter derivatives: efficient numerical evaluation exploiting their interdependence, *J. Wind Eng. Ind. Aerod.* **136** 138-50.
[8] Wu T and Kareem A 2013 Bridge aerodynamics and aeroelasticity: a comparios of modelling schemes. *J. Fluid Struct.* **43** 347-70.
[9] Tubino F 2005 Relationships among aerodynamic admittance functions, flutter derivatives and static coefficients for long-span bridges. *J. Wind Eng. Ind. Aerod.* **93** 929-50
[10] Jeong W, Liu S, Jakobsen JB and Ong MC 2019 Unsteady RANS simulations of flow around a twin-box bridge girder cross section. *Energies*. **12**(14) 2670.
[11] Naimul H, Katsuchi H, Yamada H and Nishio M 2015 Flow field analysis of a pentagonal-shaped brige deck by unsteady RANS. *Eng. Appl. Comput. Fluid Mech.* **10**(1) 1-16
[12] Alvarez AJ, Nieto F and Hernandez S 2019 CFD based study on the effect of aerodynamic appendages in the aerodynamic response of a twin-box deck. 2nd National conference on wind

engineerng, Bucharest, Romania.
[13] Sarwar MW and Ishihara T 2010 Nuerical study on suppresion of vortex-induced vibrations of box girder bridge section by aerodynamic coutermeasures. *J. Wind Eng. Ind. Aerod.* **98** 701-11.
[14] Watanabe S and Fumoto K 2008 Aerodynamic study of slotted box girder using computational fluid dynamics. *J. Wind Eng. Ind. Aerod.* **96** 1885-94.
[15] Cobo del Arco D and Aparicio AC 1999 Improving suspension bridge wind stability with aerodynamic appendages. J. Struct. Eng. **125** 1367-75.
[16] Wilde K, Omenzetter P. and Fujino Y 2001 Suppression of bridge flutter by active deck-flaps control system. *J. Eng. Mech.* **127** 1 80-89
[17] Wang Q, Liao H, Li M and Ma C 2011 Influence of aerodynamic configuration of a streamline box girder on bridge flutter and vortex-induced vibration. *J. Modern Transp.* **19** 261-67.
[18] Larsen A and Walther JH 1998 Discrete vortex simulation of flow around five generic bridge deck section. *J. Wind Eng. Ind. Aerod.* **77–78** 591–602.
[19] Scanlan RH and Tomko JJ 1971 Airfoil and bridge deck flutter derivatives. *J. Eng. Mech. Div.* **97**(6), 1717–37.
[20] Larose, G.L., Livesey, F.M., 1997. Performance of streamlined bridge decks in relation to the aerodynamics of a flat plate. *J. Wind Eng. Ind. Aerod.* **69–71** 851– 60.
[21] OpenFoam Foundation Ltd, 2015. OpenFoam user guide.
[22] Menter F 1994 Two-equation eddy viscousity turbulence models for engineering applications. *AIAA J.* **32(8)** 1598-605.
[23] Menter F and Esch T 2001 Elements of industrial heat transfer prediction. Sixteenth Brazilian congress of mechanical engineering
[24] Wilcox DC 1988 Reassessment of the scale-determining equation for advanced turbulence models. *AIAA J.* **26**(11) 1299-310
[25] Wilcox DC 2006 *Turbulence modelling for CFD*. Third ed. DCW industries Inc, La Canada, CA, USA.
[26] De Miranda M and Bartoli G 2001 Aerodynaic optimization of decks of cable-stayed bridges. Cable-supported bridges-challenging technical limits, proceddings of IASBE conference.
[27] C. Montoya M, Nieto F, Hernandez S, Kusano I, Alvarez AJ and Jurado JA 2018 CFD-based aeroelastic characterization of streamlined bridge deck cross sections subject to shape modifications using surrogate models. *J. Wind Eng. Ind. Aerod.* **177** 405-28.
[28] Norwegian Public Road Administration, 2015. Forprosjekt Hengebru over Julsundet. technical report.
[29] Ribeiro AFP 2011 Unsteady RANS modelling of flow past a rectangular 5:1 cylinder: investigation of edge sharpness effects. 13th Int'l conference on wind engineering
[30] PointWise Inc. 2018. version 18.1 R2
[31] Caetano E 2015 Innovative Bridge Design Handbook. Alessio Pipinato.

IOP Conf. Series: Materials Science and Engineering **700** (2019) 012004 doi:10.1088/1757-899X/700/1/012004

Testing of a new transport and installation method for offshore wind turbines

J Haugvaldstad[1] and O T Gudmestad[2, *]

[1] Green Entrans AS, I-park I5, Prof. Olav Hansensvei 7A, 4021 Stavanger, Norway
[2] University of Stavanger, Norway

[*] Corresponding author: Ove.t.gudmestad@uis.no

Abstract. Transport and installation of offshore wind turbines require considerable efforts and can be time consuming, in particular when jackups are used for lifting operations. This is mainly due to the specific weather window requirements for towing and locating the jackups. A goal is ultimately to finalize all turbine assembly work onshore and at quayside and to transport the completed turbine to the site for easy installation. We suggest that this can be achieved by a transport and installation method we denote "the MINT method" whereby the wind turbine foundation is located on a barge while the tower is resting on a support system, a "MINT" system that is located at the stern of a mid-size standard offshore support vessel. The MINT is specifically designed for the transport and installation operations allowing relative motions and rotations between the tower and the transport vessel. In order to qualify the "MINT installation method", a numerical study of the feasibility of the transport and installation was carried out by Sintef Ocean using their SIMO software package [1]. The method was found feasible in favourable weather conditions. To further document feasibility, it was decided to carry out a *proof of concept test in a wave tank* to identify concerns with respect to the suggested method for transport and installation. This paper reports the findings of this test. The test is to be considered to represent an initial activity in the concept qualification process, see [2], and is not a full wave tank test carried out to measure stresses and strains for member sizing. The *objective of the paper* is therefore, to show how an initial wave tank test can be useful for concept qualification.

1. Introduction

For bottom fixed offshore wind turbine concepts in shallow water, state of the art is to use jack-up type vessels for offshore operations. The jack-up technology is, however, vulnerable to waves and the spud cans could leave large footprints in softer soils. The effects of these limitations may be high downtime during installation and heavy maintenance operations. It should, in particular, be noted that the need to change location of the service vessel very often during wind farm installation and maintenance work highlights the requirement for suitable weather windows.

For bottom fixed offshore wind turbine concepts in deeper waters, state of the art is to use crane vessels for offshore installation and maintenance operations; however, the use of these vessels might not be attractive in view of the costs of employing sufficiently sized vessels that can operate under a variety of weather conditions. It should be noted that smaller barge type crane vessels might have movement characteristics (resonance between waves and the heave, pitch and roll periods of these vessels) that will make such operations difficult or impossible during long periods of the year, in particular in North Sea wave climate conditions. It would be greatly advantageous if offshore service vessels developed for the offshore oil and gas industry could be used for offshore operations. A recent study of the details of 87 wind farms installed between 2000 and 2017 is presented in [3]. For discussions of the State of Art technology, see [4 -8]. More details of actual installation of offshore wind turbines is disused in [9] and [10]. Reference is furthermore given to [11].

The associated costs of using cranes are large. Considerable savings can be achieved in case completed structures/ facilities can be manufactured in a dry dock or near to a quayside and transported to the offshore site with use of transport vessels/ barges. Thereafter, installation should be achieved by ballasting only, for the vertical transfer onto the bottom.

Green Entrans AS has developed a method (herein called "MINT") for transport and installation (Figure 1) of one-mono-tower offshore structures, termed MC-7, see also Figure 2. The offshore support structure MC-7 is consisting of a mono-tower which penetrates the surface being supported by a tripod foundation. Figure 1 (upper part) shows the structure in the transport mode. This geometry is suggested to represent a minimum support structure for the topside functions.

The transport and installation method is considered to be particularly interesting for water depths in the range from 50 to 120m (Figure 3); outside the range of using mono-pile offshore structures and outside the range of the smaller installation jackups used in the wind industry for monopile foundations.

The main objective of the wave tank testing conduced at MaREI Centre, Lir - National Ocean Test Facility in Cork, was to document that the newly developed transport and installation method for the MC-7 offshore structures is practically feasible under specified wave conditions. The test was, therefore, part of concept qualification process as required by [2].

The following two complete MC-7 structural solutions were tested in scale model 1:50:

- "MC-7 Wind" complete for 70 meters water depth with an 8 MW / 380-ton wind turbine corresponding to a total full-scale weight of 3 425 t.
- "MC-7 Petro" complete offshore oil and gas wellhead platform for 120 meters water depth with a top-side corresponding to a full-scale weight of 600 tons and a total weight of 4750 t.

This paper reports on the test of the solution intended for wind energy generation, "MC-7 Wind" see Figure 4. The wind turbine substructure is higher than a support structure for oil and gas facilities as a wind turbine mono-tower must be extended to the height of the nacelle. The challenges to transport and install a concept for wind energy generation are, therefore, larger than the challenges to transport and install an oil and gas wellhead platform. Of largest concern is the stability associated with the high centre of gravity of the nacelle and the rotor during the installation phase, see Figure 5. For reference to the test report, see [12].

Figure 1. The MC-7 structure for water depths 50 -120 meters transported and installed as a complete unit by the MINT method.

Figure 2. MC-7 Wind turbine for 50 to 120 m water depth.

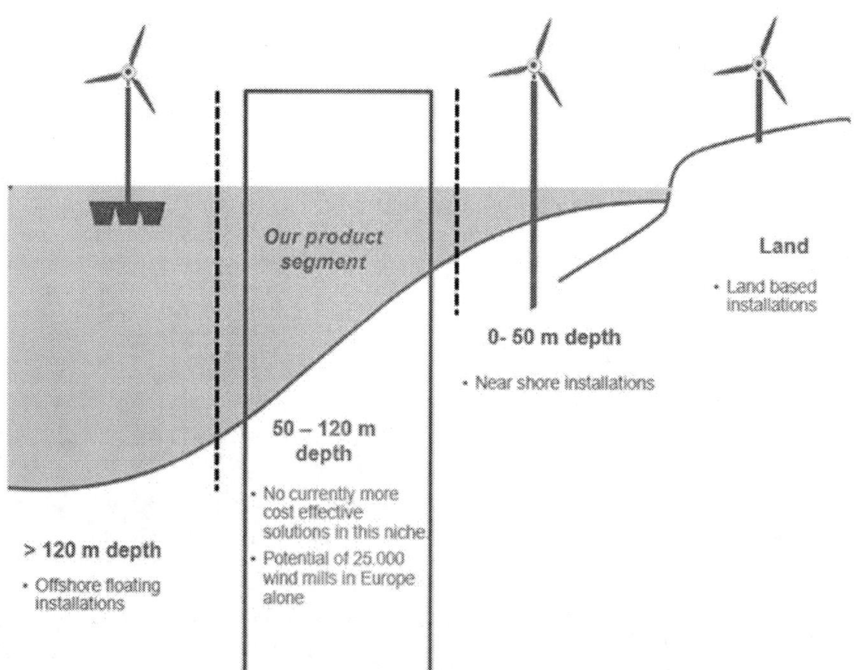

Figure 3. MC-7 Product segment.

Figure 4. Arrangement of test facilities for the testing of the MINT transport and installation method for wind turbines.

Figure 5. MC-7 Installation when the structure is being upended.

2. The MC-7 Wind unit and the MINT transport and installation concept

The basic idea behind the MINT concept for transport and installation of one-column offshore units is shown in Figure 2. The basic idea to simplify the transportation and installation is to use a mid-size support vessel and a barge for the transport. The two units have different Response Amplitude Operators (RAO's) and will move differently in wave conditions at resonance where the vessel has a larger resonance period than the barge in most degrees of freedom. Of particular concern would be large pitch response, creating a considerable bending moment in the tower.

Sintef Ocean prepared a verification report [1] of the feasibility of transporting and installing the concept using the SIMO software. A limited sea-state for transport of a complete MC-7 structure in an Hs of 2.5m was recommended. Sintef Ocean's verification report, however, concluded that an installation by direct transfer of the structures from the barge was not rotationally stable and therefore not feasible, as the wind turbine blades could come in contact with the sea. The launching procedure from the barge was modified according to Sintef's recommendations, and the MC-7 center of buoyancy and center of gravity as well as the separation procedure were adjusted. The new solution was implemented in the testing at Cork.

The objective to carry out the test was to further qualify the concept in accordance with the requirements of DNVGL [2]. It was, considered very important to document the feasibility of transport and installation by wave tank testing as a test would provide the required knowledge about the following two maritime operational aspects:

- How does a marine tow behave when a structural load is supported on 2 hulls located at different distances?
- How is the movement pattern when the unit is floating, but held over the water at one end of a floating hull?
- The test should also, by turning the structure from horizontal to vertical position, reveal a safe handling procedure to ensure the stability of the installation operation.

3. The tank test set up and test execution

The model tested was fabricated by Green Entrans AS at a 1:50 scale using Froude similitude scaling factors. The main parts are listed below:

- An installation vessel
- An installation barge
- A wind turbine named "MC-7 Wind" with a mass of 27.4 kg
- An installation system named MINT (Mono column Installation Technology) was located at the stern of the installation vessel to hold and control the MC-7 Wind column during transport and installation.

The test structure represented a full-scale wind turbine for the following conditions:

- Turbine size: 8 MW
- Turbine weights: 380 tons
- Hub height: 109 meter over sea
- Water depth: 70 meters
- Substructure weight including Tower: 2300 tons

The model assembly was composed of the installation vessel supporting the MINT system, the installation barge and the MC-7 Wind. It was placed in the basin with the installation vessel moored to the basin instrument bridge using 5 horizontal bungee cords located close to the water surface, 1 on the bow and 2 on each side. The barge was moored to the stern with a horizontal line attached to the end beach of the basin at water level. The vessel and the barge were only linked by the MC-7 structure being tested (as shown in Figure 6). The MC-7 Wind was equipped with water ballast chambers which could be fully filled or emptied depending on the towing or installation sequence.

Figure 6. The test set up in the wave tank. (Note the heel of the barge, indicating roll motion).

IOP Conf. Series: Materials Science and Engineering **700** (2019) 012004 doi:10.1088/1757-899X/700/1/012004

A total of 245 wave tank tests as required by the authors of this paper were carried out in regular scaled waves (1:50) with heights from 2.5 to 7.4 m and with wave periods from 6 to 12 seconds. Wave directions were 0 to 30 degrees relative to the test object. The tests in regular waves corresponded to swell wave conditions. Additional tests in irregular waves, which correspond to real sea states, were thereafter conducted to simulate swell wave conditions.

Note that transport and installation of the "MC-7 Petro" solution for use as wellhead platform in the oil and gas industry was also tested during the testing period. The results are not included in this paper, however, the stability during installation is easier obtainable for the oil and gas solution due to the lower centre of gravity for the topsides facilities of an offshore oil and gas platform compared with the high centre of gravity for the nacelle and the rotor of wind turbine facilities. The following "MC-7 Wind" model configurations were tested:

- Transport test: in regular and irregular waves, the vessel was moored to an instrument bridge spanning the wave tank, the barge was moored to the beach at water level and the "MC-7 Wind" was placed on both vessel and barge with its transport ballast characteristics. There was no bridge movement.
- Separation test: the test was started as for the transport test. Additional ballast was placed in the bottom of the foundation of the MC-7 model and once the required ballast was reached, the instrument bridge was moved at a constant speed of 0.2m/s to implement separation between the MC-7 and the barge. After separation, the MC-7 was actively controlled to bring it to vertical position and landing it onto the tank floor. The active control included ballasting by adding water in the model whilst controlling the MINT and winch systems on the installation vessel.

4. The test results

4.1 Transportation phase

Vessel and barge motion displacement data and load data (measured by load cells) were collected during all tests. The pitch RAOs were found to be the most critical for both regular and irregular waves. Results for 0-degree orientation are shown in Figure 7 representing the installation vessel pitch motion, in Figure 8 representing the "MC-7 Wind" pitch motion and in Figure 9 representing pitch motion of the "MC-7 Wind" in relation with the installation vessel. The relative motion is the most critical to avoid contact between the"MC-7 Wind" column and the stern of the vessel or the "MC-7" Wind nacelle and the vessel deck. For further presentation of the results of the testing, see [12].

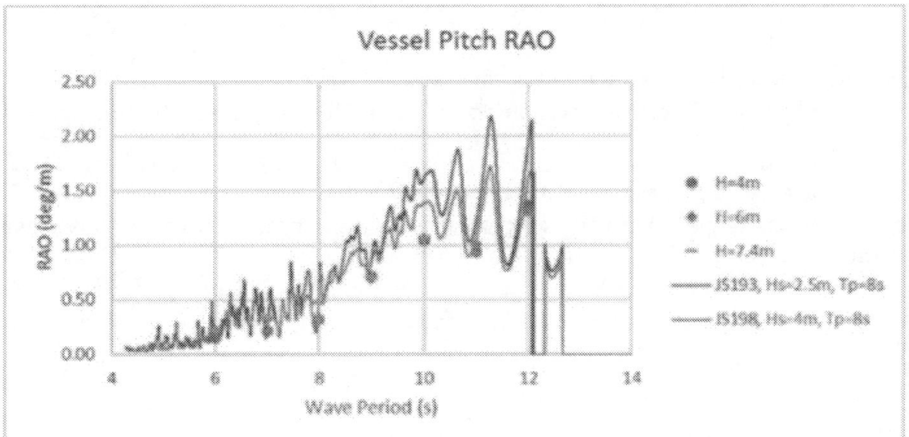

Figure 7. Pitch RAO of the installation vessel with the MC-7 Wind structure (JS = Jonswap spectrum)

Continuous measurement.

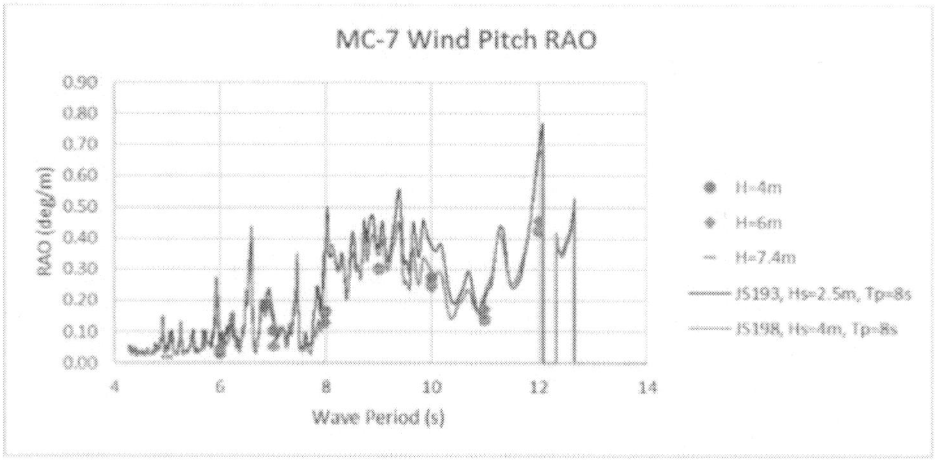

Figure 8. Pitch RAO of MC-7 Wind (JS = Jonswap spectrum). Continuous measurements.

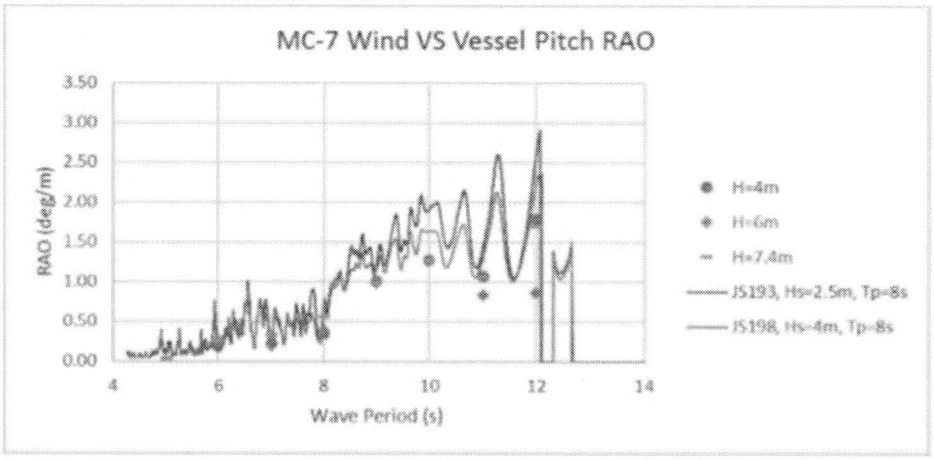

Figure 9. Pitch RAO of MC-7 Wind relative to the installation vessel with the Wind structure (JS = Jonswap spectrum). Continuous measurements.

4.1.1 Regular wave test results, see also [12]: Transportation phase, waves from front (0 degrees):

Barge

- The barge showed resonance in pitch for 6 second waves, RAO – 0,5
- The barge has large RAO values in pitch and heave for waves with 9 -10 s periods, RAO – 1.0

Vessel

- The vessel starts to pitch in 7 – 8 s waves, RAO – 0,3
- The vessel has large RAO values (RAO – 1.0) in pitch and heave for waves with 10 – 11 s periods.
  Comment: A vessel with an X-bow vessels should in principle do better with respect to pitch as the X-bow is designed to minimize vessel pitch. This will be investigated during later analysis.

Combined

- The combined unit's motion at the MINT location has large RAO values (RAO – 1.0) in pitch in regular waves above 8 seconds

- The amplitude of the motions at MINT location are reasonable for regular waves of up to 6 m height
- The behavior of the platform at the MINT location is sensitive to swell waves of 8 seconds and more, this is becoming critical for waves of height 6 m and more

Waves coming in at 15- and 30-degrees angle: The unit is starting to roll at the same periods as above. In case of heavy swells, waves from an angle should be avoided.

4.1.2 Irregular waves test results: Transportation phase, waves from front (0 degrees)

- The motion in irregular waves are smaller than in regular waves with similar wave heights, as is expected.
- The wave conditions for transport should through the test be demonstrated for Hs of 3 m with safe survival at Hs of 4m, provided the waves are not too long.
- The limiting wave conditions are based on spectra with peak periods of 7 to 8 seconds. For wave conditions with longer periods the accepted significant wave heights are reduced

The tests showed that the transport restrictions can be increased from the level recommended by SINTEF Ocean [1] to 6 meters (Hs 3 m), however, the wave periods must be assessed against vessel / barge sizes for the MC-7 structural solution in order to avoid resonance.

4.2 Test of installation phase

The MC-7 model stability during separation was thereafter tested when the side mooring bungee links were removed so the installation vessel was only moored on its bow. General comments on the separation motion stability are as follows:

- The pitch angle is -90 degrees when the MC-7 is vertical (up ended). This value may vary by a few degrees because the 0-degree value corresponds to the MC-7 structure in its transport position on both the installation vessel and barge and the column may not be held exactly horizontal.
- For all testing, a large roll, up to 40 degrees, is observed just after separation. This could cause some damage to the structure or the MINT. It is caused by the side bungees used in the mooring system. However, the large roll did not happen when the mooring bungees were removed.

As the large roll motion at separation was removed when testing without the side mooring bungee cords, the stability of the structure is acceptable in the wave conditions tested. However, the MINT system does not control the roll motion of the MC-7 structure and the commercial scale operation should ensure no external element, such as ballast pipes, has the potential to generate roll motions.

The vertical loads on vessel deck are shown in Figure 10. It should be noted that the vertical loads on the vessel deck after the separation from the barge are smaller than the loads during the transport. Following the separation from the barge, the wind turbine is floating and resting on the MINT (Figures 1 and 5) until it is installed on the bottom. The results from this test are now being used to improve on the transport and installation procedures. The large break-out loads in the winch and the barge during lift off from the barge were due to lack of procedures for this activity and the results are very useful for designing safe procedures.

The tests were carried out successfully and after some model modifications including trial and errors to adjust the ballasting procedures, all operation modes gave satisfactory results. Two main parameters should be taken into account for full scale operations:

- The maximum allowed pitch angle between the MC-7 structure and the installation vessel and barge should not be exceeded.
- Special attention should be given to reduce any mechanical loading capable of inducing roll motion on the MC-7 structure during the separation ~~installation~~ as the MINT installation system does not control this motion. The structure up-right itself when the upper leg structures are engaged in the water.

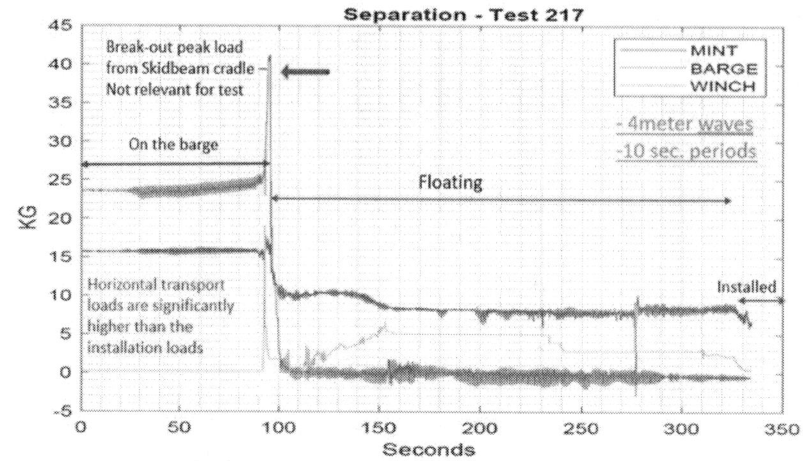

Figure 10. Vertical loads on vessel deck during the installation phase.

5. Conclusions

In general, the wave tank tests confirmed the initial assessments made by Sintef Ocean [1] for transportation in Hs 2,5 to 3 meter: With adjustment of the ballasting procedure during the upending phase, the transport and installation method is feasible.

The transport function was tested with both regular and irregular waves with wave heights from 2,5 to 7,4 meter. Heading variations from 0 to 30 degrees were investigated. 3 hours Jonswap spectra wave tests were also run with following set wave heights and periods: Hs = 2,5 m/ Tp = 8s and Hs = 4,0 m / Tp = 8s. Strain gauges and loadcells installed on the mono-tower and the barge did not reveal any large stresses during the transport and installation conditions except during the lift off situation of the structure from the barge where procedures must be developed to limit the loads.

5.1 Transport; recommended maximum sea-state criteria:

- Hs 3 meter
- Wave periods 8-9 seconds

Note: The transport survives Hs 4meter in 30 deg. heading.
The designed Transport solution is concluded robust and ready for full scale detail engineering.

5.2 Installation; Recommended maximum criteria:

- Hs 2,5 meter
- Wave periods 8-9 seconds

Note: The installation survives Hs 3 meter in 30 deg. heading
Note: The recommended transport and installation criteria are based on specific vessel and barge sizes /designs. The criteria may change when using other vessels / barges.

Data availability

On request to Green Entrans AS, test data will be made available, provided the results of the data analysis be shared with the data owner.

Author contributions

The test was managed by J Haugvaldstad, Green Entrans AS in collaboration with O. T. Gudmestad – University of Stavanger and the University College of Cork-Ireland. The basin tests were executed in the MaREI – LiR test center in Cork – Ireland under the leadership of Senior Marine Engineer F. Thiebaut.

IOP Conf. Series: Materials Science and Engineering **700** (2019) 012004 doi:10.1088/1757-899X/700/1/012004

Disclaimer
The results from the wave tank tests are presented in this paper. However, any misunderstandings regarding the presentation of the results reported in this paper are not the responsibility of the authors. Any use of the technology reported herein should be clarified with the technology owner, Green Entrans AS.

Acknowledgements
The test was sponsored by VRI Rogaland, the Norwegian Research Council's Regionalt Forskningsfond Vest. Contributions to the test was also granted from Equinor, Stavanger and Ocean Installer, Stavanger. The test was professionally carried out in the wave basin of University College of Cork, Ireland.

References
[1]. SINTEF Ocean AS (Solaas and Sandvik), Report 017 F-025, "SIMO simulations of towing and launch phases. MINT installation tool", June 2017.
[2]. DNVGL RPA203 2017 Technology qualification, Recommended Practice. Høvik, Oslo, Norway
[3]. Lacal-Arántegui, L, Yusta J M and Domínguez-Navarro J A 2018 Offshore wind installation: analysing the evidence behind improvements in installation time, *Renewable Sustain. Energy Rev.* **92**, 133-145.
[4]. SBM, GustoMSC. 2018 Equipment for Offshore Wind Turbine installation
[1] https://www.kivi.nl/uploads/media/5644b984995fc/Offshore%20Windturbine%20Installatie% B%20hoe%20hoog%20kun%20je%20gaan%20-%20ir.%20R.%20Noordermeer.pdf
[5]. Ahlström C, Andersson A and Blomgren N 2014 Optimus Pråm - Semi-submersible wind farm installation vessel for Blekinge Offshore https://publications.lib.chalmers.se/publication/208702.
[6]. Gintautas T, Sørensen D J and Vatne, R S 2016 Towards a risk-based decision support for offshore wind turbine installation and operation & maintenance, *Energy Procedia* **94**, 207 - 217, Presented at 13th Deep Sea Offshore Wind R&D Conference, EERA DeepWind' 2016, 20-22 Jan., Trondheim, Norway.
[7]. Seyr H and Muskulus M 2016 Safety indicators for the marine operations in the installation and operating phase of an Offshore wind farm, *Energy Procedia* **94**, 72 – 81. Presented at 13th Deep Sea Offshore Wind R&D Conference, EERA DeepWind' 2016, 20-22 Jan., Trondheim, Norway.
[8]. Paterson J, D'Amico F, Thies P R, Kurt R E and Harrison G 2018 Offshore wind installation vessels: A comparative assessment for UK offshore rounds 1 and 2, *Ocean Eng.* **148**, 637-47.
[9]. Gudmestad OT, Grønli J and Straume H 2010 *Marine operations for installation, intervention and decommissioning of Offshore Windmills*, Presented at The Royal Institution of Naval Architects conference Marine & Offshore Renewable Energy, 21 -23 April 2010, London, UK, Proceedings ISBN 978-1-905040-70-4, 175-184.
[10]. Bennett Offshore. Jack up units, a technical primer for the offshore industry professional, 2005. http://www.bennettoffshore.com/primer.html
[11]. Lloyds Register. Guidance Notes for Wind Turbine Installation Vessels, 2014. London. https://www.lr.org/en/guidance-notes-for-wind-turbine-installation-vessels/
[12]. Green Entrans AS, "Test Report, MINT Transport and Installation Tests, Technology Qualification". RFV report 04, Stavanger, December 2018.

IOP Conf. Series: Materials Science and Engineering **700** (2019) 012005 doi:10.1088/1757-899X/700/1/012005

The influence of an unstable turbulent wind spectrum on the loads and motions on floating Offshore Wind Turbines

J M Knight* and C Obhrai

Department of Mechanical and Structural Engineering and Material Science, University of Stavanger, Norway

*Corresponding author: jamknight333@gmail.com

Abstract. Floating offshore wind turbines (FOWT) are exposed to harsh environmental conditions and often experience unstable atmospheric conditions. Including the effects of atmospheric stability should improve the accuracy of fatigue load calculations, and subsequently, the design of the wind turbine. The current standards recommend two turbulence spectral models that are valid for neutral atmospheric conditions only. The objective of this study is to investigate the influence of Højstrup's 1981 Unstable Spectra Model on the loads and motions on a spar and a semi-submersible FOWT. This study focuses on the effect of unstable atmospheric conditions in the free stream wind and does not include the effect of wakes. The most significant differences observed in this study were for the tower top torsion loads, where very unstable conditions gave 47% larger loads than neutral conditions for OC3-Hywind turbine and 30.4% larger for OC4-DeepCwind turbine. Since very unstable conditions corresponded to the highest turbulence intensities and larger turbulent fluctuations, they also resulted in higher fatigue loads. The blade root flap-wise loads were also observed to be higher under unstable conditions compared to neutral conditions, but the differences were smaller with 7.5% for OC3-Hywind and 23% for OC4-DeepCwind.

1. Introduction

Offshore renewable energy has gained significant momentum throughout the world recently, which strengthens the need for further research of wind fields and load simulation techniques. To develop better designs and maintenance strategies, it is beneficial to look at the influence of atmospheric stability and turbulent wind modelling. Presently, most research studies relating to turbulent wind spectra are often focused on neutral wind conditions. The International Electrotechnical Commission (IEC) standards recommend two turbulence models: the IEC Kaimal Spectra & Exponential Coherence Model and the Mann Spectral Tensor Model [1], neither of which consider the effect of atmospheric stability. The objective of this study is to investigate how turbulent wind fields and the generated wind coherence influence the motions and loads on a floating offshore wind turbine (FOWT), specifically using Højstrup's 1981 Unstable Spectra Model and two FOWT designs, the OC3-Hywind spar-buoy and the OC4-DeepCwind semisubmersible. The Højstrup model takes into account the effect of boundary layer height and buoyancy generated turbulence on the velocity spectra under unstable conditions, which is of particular interest to floating offshore structures, such as wind turbines.

The effect of atmospheric stability on the loads of bottom fixed wind turbines has been studied previously [2], [3] and [4]. In the study by [3], atmospheric stability and its relationship to the fatigue loads of a fixed offshore wind turbine were investigated using measurements from the offshore wind farm, Alpha Ventus. The study specifically looked at the impact of atmospheric stability on the blade root bending and the tower base moments, which were found to be the highest under unstable conditions

IOP Conf. Series: Materials Science and Engineering **700** (2019) 012005 doi:10.1088/1757-899X/700/1/012005

in wind speeds between cut-in speed and approximately 14 m/s. They also noted that fatigue loads on the tower base were primarily influenced by atmospheric turbulence, unlike the fatigue loads on the blade root bending, which were influenced by a combination of wind shear for stable wind conditions and turbulence for neutral and unstable wind conditions. Sathe and Bierbooms [5] calculated fatigue loads at the blade root using non-neutral wind profiles and steady winds. They concluded that fatigue loads increase when using non-neutral wind profiles in comparison with those obtained under neutral conditions, with the highest loads observed for stable conditions. In the study by [4], the wind turbine fatigue loads were studied as a function of atmospheric stability without a classification system, but instead, atmospheric conditions were described by a continuous joint probability distribution of wind speed and stability. This study used the Kaimal neutral spectral model but varied the mean wind profiles and turbulence intensities based on measurements from the offshore meteorological mast Ijmuiden in the North Dutch Sea.

Finally, the study by [2] used data from a met mast at Høvsøre to derive parameters for the Mann Spectral Tensor Model [1] for different atmospheric stabilities. This resulted in higher simulated turbulence intensities and higher fatigue loads for neutral conditions. This is thought to be due to the fact that the Mann Spectral Tensor Model [1] is not able to simulate buoyancy generated turbulence, which is present in unstable atmospheric conditions. The Mann Spectral Tensor Model [1] was modified by [6] & [7] to account for non-neutral stability using an additional parameter z/L. The model parameters were obtained by fitting the model to measured one-dimensional velocity spectra, cospectra between horizontal and vertical velocity components, and temperature. However, the model had deficiencies in reproducing spectra and cospectra in unstable regimes at lower wavenumbers for larger lapse rates [7].

In addition to atmospheric stability, wind coherence is an important parameter to consider in turbulent wind modeling for offshore wind turbines. The effect of atmospheric stability on vertical coherence has been studied previously [8]. Using measurements from FINO 1, [8] derived a two-parameter decay function for modeling the coherence of the vertical velocity component in non-neutral conditions. In this study, the simpler Davenport Coherence Model was used with fixed decay coefficients, valid for neutral conditions only. Including the effect of atmospheric stability on vertical and horizontal coherence is a topic for further work. Until now, most studies have focused on the effect of atmospheric stability on the loads and motions of bottom fixed wind turbines and do not include wind spectra suitable for non-neutral conditions. A previous study by the authors [9] involved a short parametric study using the Højstrup Unstable Spectra Model and then used this to simulate the loads and motions of a spar-type FOWT. The present study extends the analysis of this previous work to investigate the effect of unstable conditions on the loads and motions of a spar type and a semi-submersible type floating wind turbine in free wind conditions.

2. Theoretical background

2.1. Atmospheric stability

Atmospheric stability can be characterized as stable, neutral, or unstable based on the "tendency of air particles to move vertically" relative to the temperature of their surroundings [10]. In stable conditions, the air particles are cooler than the surrounding air, causing them to sink or remain where they are. This stratification leads to less mixing and typically results in lower turbulence intensities. In unstable conditions, the air particles are warmer than the surrounding air, which causes them to rise and leads to more vertical mixing. Hence, this is called "buoyancy generated" turbulence. Due to the enhanced vertical mixing, one would expect larger turbulent fluctuations and higher fatigue loads under unstable conditions, which are typically more dominant at offshore sites [2]. However, stability conditions are largely dependent on the site location [11]. Overall, atmospheric stability tends to reduce turbulence in stable conditions and increase turbulence in unstable conditions.

In some cases, atmospheric stability is classified using the Monin-Obukhov length (L), or the length scale of energy-containing eddies. The Monin-Obukhov length is an important part of classifying thermal stratification in the surface layer, and is typically associated with the height above ground (z). The ratio between the height above ground and the Monin-Obukhov length is recognized as an important stability parameter which reflects the impact of varying height and stability conditions [12].

2.2. *Turbulent wind spectra*

The IEC standards currently recommend two turbulence models: the IEC Kaimal Spectra & Exponential Coherence Model and the Mann Spectral Tensor Model [1]. Since these two models were originally developed for neutral conditions, this paper will focus on the Højstrup 1981 Unstable Spectra Model and use the original Kaimal Spectra model as a comparison for neutral conditions. The Højstrup spectra model is based on a modification of the Kaimal spectral model such that when L tends to infinity (i.e. neutral conditions) the Højstrup model reduces to the Kaimal model. In this study, we compare the Kaimal Spectral model to the Højstrup model to study the influence of unstable conditions on the loads and motions of FOWT's.

2.2.1. Kaimal Spectra Model. The most commonly used equations for the Kaimal spectra in engineering applications are based on the neutral stability Kansas measurements [13] and are adjusted to account for the 4/3 ratio that is expected in the inertial subrange [12]. These equations are as follows [14]:

$$\frac{nS_u}{u_*^2} = \frac{105f}{(1+33f)^{5/3}} \tag{1}$$

$$\frac{nS_v}{u_*^2} = \frac{17f}{(1+9.5f)^{5/3}} \tag{2}$$

$$\frac{nS_w}{u_*^2} = \frac{2f}{1+5.3f^{5/3}} \tag{3}$$

where n is frequency in Hertz, S_u, S_v, S_w are the velocity spectra in the along wind, cross wind, and vertical direction respectively, u_* is friction velocity, and $f = nz/\bar{u}$, a nondimensional reduced frequency.

2.2.2. Højstrup 1981 Unstable Spectra Model. The goal of Højstrup's 1981 Unstable Spectra Model was to develop a simple model that could resemble velocity spectra in unstable conditions downwind of a change in surface roughness and heat flux [15]. Since atmospheric turbulence consists of both a buoyancy generated component and a mechanically generated component, Højstrup found it important to create a velocity spectrum that involved both aspects and could be modelled as the sum of two semi-empirical spectra. This can be seen in Equation (4) [15]:

$$S(n) = S_L(n) + S_M(n) \tag{4}$$

with $S_L(n)$ corresponding to the low frequency part of the spectra and $S_M(n)$ corresponding to the Kaimal Spectra Model. The backbone of the model is the addition of a buoyancy-produced part (low frequencies) and a shear-produced part (high frequencies), thus creating the following equations [15]:

$$\frac{nS_u}{u_*^2} = \frac{0.5f_i}{1+2.2f_i^{5/3}}\left(\frac{z_i}{-L}\right)^{2/3} + \frac{105f}{(1+33f)^{5/3}} \tag{5}$$

$$\frac{nS_v}{u_*^2} = \frac{0.32f_i}{1+1.1f_i^{5/3}}\left(\frac{z_i}{-L}\right)^{2/3} + \frac{17f}{(1+9.5f)^{5/3}} \tag{6}$$

$$\frac{nS_w}{u_*^2} = \frac{32f}{(1+17f)^{5/3}}(\frac{z}{-L})^{2/3} + \frac{2f}{1+5.3f^{5/3}}$$
(7)

For neutral conditions, when L= ∞, Equations (5), (6), and (7) reduce to the Kaimal spectrum. The key variables of the model are the three scaling lengths: height (z), inversion height (z_i), and Monin-Obukhov length (L). The reduced frequency parameters are defined as, $f = nz/\bar{u}$ and $f_i = nz_i/\bar{u}$, which provide knowledge of the spectra's variation with stability [16].

2.3. *Davenport Exponential Coherence Model*

A suitable model for vertical coherence, when the separation between points are small in comparison to the length scale of turbulence, is the Davenport Exponential Coherence Model. Equation (8) includes both vertical and horizontal cross-flow separations [17]:

$$\gamma_i(d_y, d_z, n) \approx \exp(-\frac{n}{\bar{u}} * \sqrt{(C_y^i\, d_y)^2 + (C_z^i\, d_z)^2})$$
(8)

where $i = \{u, v, w\}$, d_y and d_z are the horizontal and vertical separation distances between points, n refers to the frequency in Hz, $\bar{u}$ is the horizontal mean velocity, and C_y^i and C_z^i are decay coefficients in the y- and z-direction, respectively.

3. Wind turbine characteristics

The simulations conducted in this study included the spar-buoy type FOWT, a ballast stabilized concept, from Phase IV of the IEA Annex XXIII Offshore Code Comparison Collaboration (OC3) project [18], as well as the semi-submersible design from Phase II of the Offshore Code Comparison Collaboration Continuation (OC4) project [19], an extension of the OC3 project. Both phases make use of the National Renewable Energy Laboratory's (NREL) offshore 5 MW standard wind turbine, with alterations to the foundation type depending on the design concept. Basic parameters of NREL's 5 MW turbine are displayed in Table 1 and further details of the OC3 and OC4 FOWTs can be found in [18] and [19], respectively. Table 2 displays the eigen frequencies for the first 10 modes for the OC3-Hywind [18] and OC4-DeepCwind [19] FOWTs, which are relevant for the platform motion analysis.

Table 1. Specifications for NREL 5 MW Offshore Wind Turbine.

Parameter	NREL 5 MW Wind Turbine
Power Production Rating	5 MW
Number of Blades	3
Rotor Orientation	Upwind
Rotor Diameter	126 m
Hub Height	90 m
Cut in, Rated, Cut out Wind Speed	3 m/s, 11.4 m/s, 25 m/s
Cut in, Rated Rotor Speed	6.9 rpm, 12.1 rpm

Table 2. Eigen frequencies for first 10 modes of the OC3-Hywind & OC4-DeepCwind FOWTs.

Mode	Platform Motion	Spar (Hz)	Semisubmersible (Hz)
1	Surge	0.008	0.01
2	Sway	0.008	0.01
3	Heave	0.032	0.058
4	Roll	0.034	0.04
5	Pitch	0.034	0.04

6	Yaw	0.11	0.012
7	1st Tower fore-aft	0.43	0.43
8	1st Tower side-side	0.44	0.43
9	Drivetrain torsion	0.63	0.63
10	Blade flap-wise	0.64	0.64

3.1. *Wind Simulation Model*

To generate the velocity spectra of the Højstrup model, a MATLAB script was used to determine the spectra for the u, v, and w components based on Equations (5), (6), and (7). The spectral representation approach was then used to simulate turbulent wind fields using the method from [20]. A function called WindSimFast [21], available on MathWorks File Exchange, was used to simulate the random wind fields from the Højstrup spectra model. To allow for the simulations to closely resemble the stochastic nature of wind and waves and to minimize uncertainty, six random seeds were defined within the script to generate the velocity histories. The necessary input parameters are displayed in Tables 3, 4, and 5. Within this study, atmospheric stability was classified using Monin-Obukhov length, as presented in Table 5, which coincides with the atmospheric stability classes as suggested by [22]. The logarithmic wind speed profile corrected for unstable conditions was chosen to represent the mean wind speed profile for the simulations as given in the DNV standards [23]:

$$U(z) = U_{ref}\frac{\ln(z/z_o) - \psi_m(z/L)}{\ln(z_{ref}/z_o) - \psi_m(z_{ref}/L)}$$
$$\psi_m(z/L) = 2\ln[1+x] + \ln[1+x^2] - 2\tan^{-1}(x) \qquad (9)$$
$$x = \left(1 - 19.3(z/L)\right)^{\frac{1}{4}}$$

where z_{ref} is a reference height, U_{ref} is the mean wind velocity at z_{ref}, z_o is sea surface roughness length, and ψ_m is the stability function (shown specifically for unstable conditions). In this study z_o was fixed and did not vary with wind speed this could be done in future work

Table 3. Input parameters.

Sampling Frequency	Duration	XYZ grid	u_{*0}	z_i	z_o	κ
9.1 Hz	1hr	32768 x 32 x 32	0.4 m/s	1000 m	0.00014	0.4

Table 4. Decay coefficients used for Davenport Coherence Model

Coefficient	c_y^u	c_y^v	c_y^w	c_z^u	c_z^v	c_z^w
Value	7	7	6.5	10	10	7

Table 5. Atmospheric stability classes.

Atmospheric Stability Class	Monin-Obukhov Length (m)
Very Unstable	-90
Unstable	-180
Neutral	∞

IOP Conf. Series: Materials Science and Engineering **700** (2019) 012005 doi:10.1088/1757-899X/700/1/012005

Figure 1 shows the target spectra for the Højstrup spectra model for varying values of L, at hub height, where L = -90 m represents very unstable conditions, L = -180 m represents unstable conditions, and L = ∞ is the Kaimal neutral wind spectra. Given that the high frequency part of the Højstrup model is identical to the Kaimal model, it makes sense that the resulting spectra would converge at high frequencies, as can be seen in Figure 1.

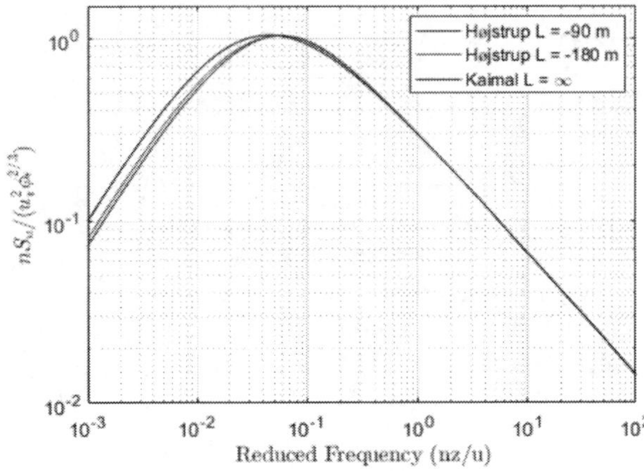

Figure 1: The target spectra for the Højstrup spectra model with varying L at hub height (90 m).

3.2. SIMO-RIFLEX AeroDyn

A FOWT with the characteristics in Table 1 was implemented into SIMA, a simulation and analysis tool developed by SINTEF Ocean [24]. Within SIMA, there is a SIMO-RIFLEX coupling tool that allows for the simulation of multi-body hydrodynamics. For this study, the OC3 spar-buoy FOWT [18] stored within SIMO-RIFLEX was used and adjusted to model wind fields using the Højstrup spectra model. The OC4 semisubmersible type FOWT [19], supported by the NREL 5 MW baseline wind turbine [25], was provided by SINTEF. The turbine tower and blades were modelled with nonlinear beam elements, and the mooring lines were modelled with nonlinear bar elements, which allowed for the rotation of each element [26].

The environmental loads implemented in this study include both wind and wave loads. The waves were defined using irregular airy waves based on the Joint North Sea Wave Project (JONSWAP) wave spectrum [27], with a peak parameter (γ) = 3.3, significant wave height (H_s) = 6 m, and a peak period (T_p) = 12 s. The peak period corresponds to a peak wave frequency (f_p) of 0.083 Hz.

The synthetic/generated turbulent wind fields, determined by the wind simulation model described previously, were stored in binary format and then used to create the wind loads applied within the coupled SIMO-RIFLEX simulations. Below rated (8 m/s), rated (11.4 m/s), and above rated (15 m/s) wind speed scenarios at hub height were defined. The logarithmic wind speed profile corrected for unstable conditions was chosen to represent the mean wind speed profile for the simulations as given in the DNV standards [23]. An example of the mean wind profile across the rotor at rated wind speed in very unstable conditions is shown in Figure 2.

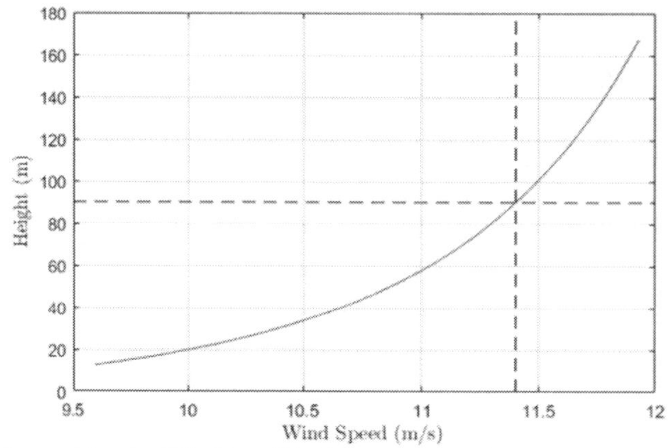

Figure 2: Corrected Logarithmic Wind Profile for very unstable conditions (L = -90 m) for rated wind speed (11.4 m/s) at the 90 m hub height.

Each scenario was simulated for a 1-hour time series, with a 0.02 s time step. The properties of the wind input also included an air density (ρ) = 1.225 kg/m^3, tip loss correction defined through the Prandtl tip loss method, drag force on the turbine tower using the Potential flow tower shadow method, and transient aerodynamics developed through Beddoes-Leishmann dynamic stall method [28].

4. Results and discussion

The results of this study are presented as damage equivalent loads and motion responses using the Højstrup 1981 Unstable Spectra Model under varying atmospheric stability, for both the OC3-Hywind and OC4-DeepCwind FOWT.

4.1. Damage equivalent loads

For a FOWT, it is particularly important to investigate the fatigue damages caused by repetitive loading. Based on the material of the component, the stress level, and the number of load cycles, the corresponding deterioration will vary and has the potential to continue until failure. To quantify the level of deterioration, it is common to consider the accumulated damage for each cycle based on Palmgren-Miner linear damage rule (Miner's rule) [29], which assumes that the damage accumulated from each load range can be added linearly [2]. In order to determine the load ranges (σ_i) and the subsequent number of cycles (n_i), the rain-flow counting method was used in conjunction with Miner's rule and applied to calculate fatigue damage for each load range using Equation (10) [30]:

$$D = \sum_i \frac{n_i}{N_i} = \sum_i \frac{n_i}{C} \sigma_i^b \tag{10}$$

where n_i is the subsequent number of cycles, N_i is the total number of cycles until failure at the load range, σ_i, $C = N_i \sigma_i^b$, and b is the characteristic parameter of the S-N curve or the Wöhler exponent. For this analysis, b was set as 3 for the tower, tower top, and mooring lines, since they are all made from steel, and 12 for the blades, since they are made from fiberglass [25].

Using the relationship in Equation (10), an expression for the equivalent alternating stress, σ_{eq}, can be derived, corresponding to the same fatigue damage if the loading were applied for the duration of N_{eq} (equivalent number of load cycles), which was assumed as 10^7 for this study, or approximately 20 years of operation [2]. To quantify the DEL from the load time series, the relation between the occurring stress and its number of cycles is required. This relation is commonly known as S-N curve and is usually obtained by experiments for different materials. However, since it is difficult to define the S-N curve for a particular component, fatigue damage can instead be quantified using the concept of damage equivalent load (DEL), which can be found with Equation (11) [30]:

IOP Conf. Series: Materials Science and Engineering **700** (2019) 012005 doi:10.1088/1757-899X/700/1/012005

$$\sigma_{eq} = (\sum_i \frac{n_i \sigma_i^b}{N_{eq}})^{1/b} \tag{11}$$

where N_{eq} is the equivalent number of load cycles.

4.1.1. OC3-Hywind spar-buoy FOWT. The normalized DELs (average from 6 simulations) for the blade root flap-wise moment and tower top torsion are displayed in Figure 3.

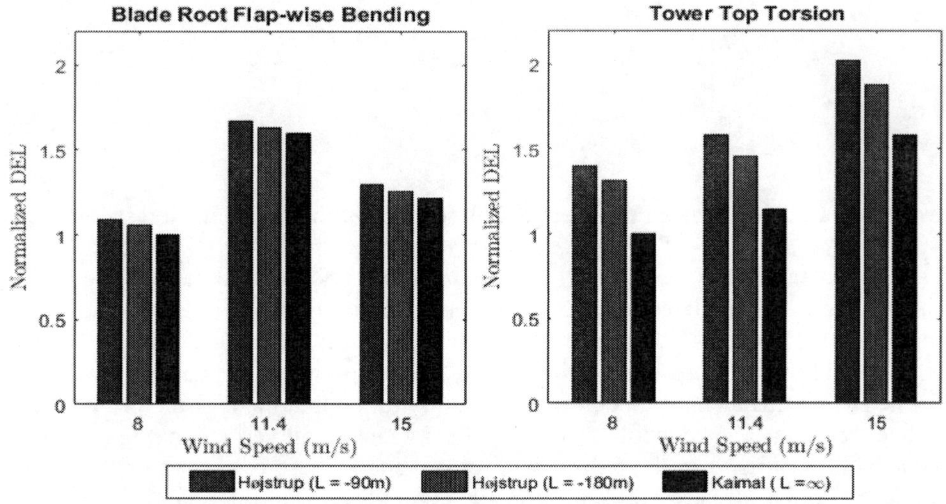

Figure 3. Normalized DELs for blade root flap-wise bending (left) and tower top torsion moment (right), normalized by the Kaimal model at 8 m/s, with the spar-buoy foundation.

Theoretically, unstable atmospheric stability conditions correlate to higher turbulence intensity and larger turbulent fluctuations than in neutral and stable conditions. Figure 3 shows that the highest DELs corresponded to very unstable conditions, followed by unstable and neutral conditions, although the differences appears to be minor for the blade root flap-wise bending moment. The maximum difference for the blade root flap-wise bending moment when considering the same wind speed was approximately 7.5%. In a study by [28], it was also found that the blade root flap-wise loads were not significantly influenced by atmospheric stability, with only a 6.5% difference. However, this previous study showed that neutral conditions gave the largest DEL, since the simulations were based on the fitted Mann model. In the study by [2], results showed that atmospheric stability had little influence on the blade root flap-wise bending moment, with only a 3% difference in dynamic loads between non-neutral and neutral wind conditions. Similar to [28], the simulations in [2] incorporated the fitted Mann model, and neutral conditions resulted in the largest DEL.

A similar trend was found for tower top torsion, with very unstable conditions giving the largest DELs. However, in this case, atmospheric stability, and its relationship to turbulence intensity, seems to have a significant influence on the resulting DELs. For the tower top torsion, the maximum difference between very unstable conditions and neutral conditions, considering the same wind speed, was approximately 47%. Figure 4 shows the simulated turbulence intensity at hub height for each of the wind speed scenarios defined in this study. The Højstrup spectra model resulted in the highest turbulence intensities for very unstable conditions, as expected.

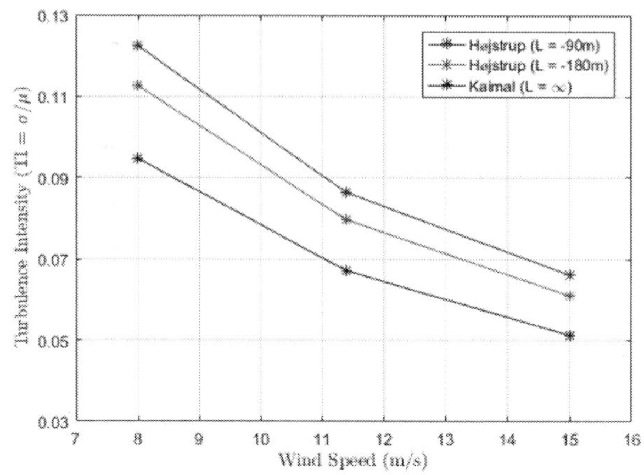

Figure 4: Simulated turbulence intensities at hub height (90 m) for each wind speed scenario.

In previous studies where the Mann Spectral Tensor Model was fitted to measurements, the highest turbulence intensities were simulated for neutral atmospheric conditions [2], [28] and [31]. As a result, these studies found that neutral atmospheric conditions resulted in the highest fatigue loads, followed by unstable and stable conditions. This is most likely because unstable conditions were not fully represented when using this method. In contrast, this study found that the highest turbulence intensities corresponded to very unstable conditions due to the inclusion of vertical mixing associated with buoyancy generated turbulence in the Højstrup spectra model. These results show the importance of including the low frequency energy present in unstable conditions on the resulting tower top torsion fatigue loads.

4.1.2. OC4-DeepCwind semisubmersible FOWT. The average normalized DELs for the blade root flap-wise bending moment and tower top torsion, normalized with the Kaimal model below rated scenario, are displayed in Figure 5.

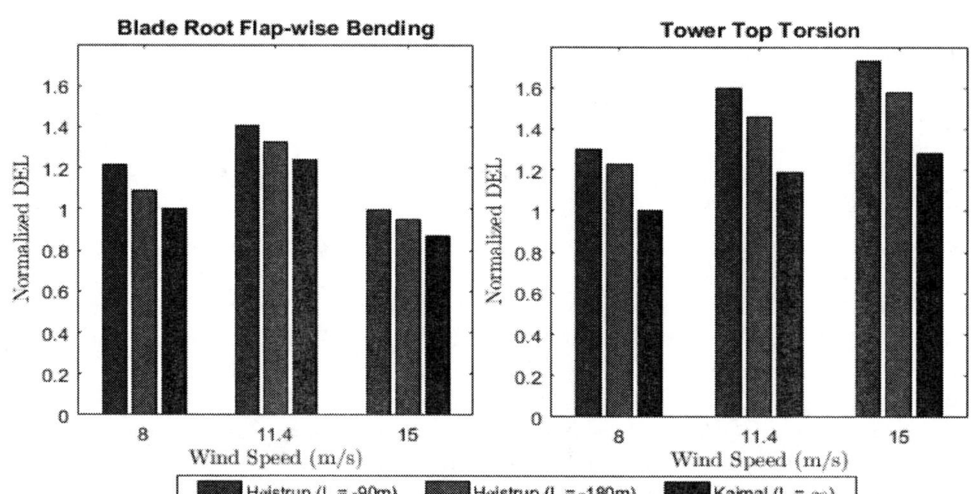

Figure 5. DELs for blade root flap-wise bending (left) and tower top torsion (right), normalized by the Kaimal model at 8 m/s, with the semisubmersible foundation.

The Højstrup wind spectra model under very unstable conditions resulted in the largest fatigue DELs for each wind speed for the semisubmersible foundation, similar to the results for the spar-buoy FOWT shown previously. The maximum difference between very unstable conditions and neutral conditions, considering the same wind speed, was approximately 23% for the blade root flap-wise moment and 30.4% for tower top torsion.

4.2. *Platform Motions*

In order to fully analyze the response of a FOWT, it is important to evaluate the motions in six degrees of freedom that the structure will experience due to both wind and wave loadings. In this study, the pitch and yaw rotations were highlighted.

4.2.1. OC3-Hywind spar-buoy FOWT. The spectral density for platform pitch and yaw rotation are displayed in Figure 6. The other platform motions were found to be relatively unaffected.

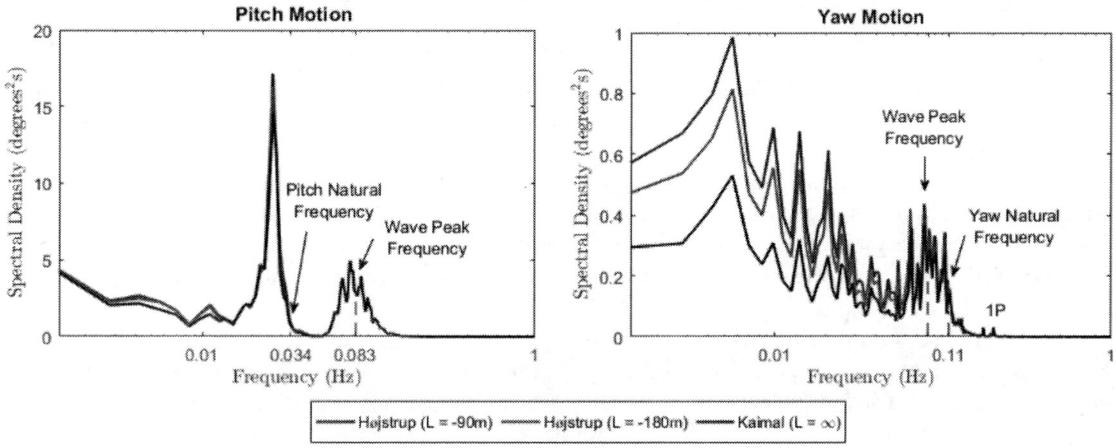

Figure 6. Spectral density of pitch rotation (left) and yaw rotation (right) for varying stability at rated wind speed (11.4 m/s) with the spar-buoy foundation.

Continuous and excessive platform motion may lead to fatigue damage if the frequency of the platform coincides with the frequency of the turbulent wind input. By looking at Figure 6, the platform pitch rotation appears to have been excited by the wave peak frequency, as well as frequencies just below the pitch natural frequency (0.034 Hz). Often, the platform pitch motion is related to negative damping associated with the blade pitch control system, which stems from the reduction in rotor thrust as the wind speed increases above rated [32]. However, modifications were made to the original control system for the NREL 5-MW turbine for the OC3-Hywind and OC4-DeepCwind projects by [19] in an effort to avoid the negative damping issue. These modifications included a reduction of gains in the blade-pitch-to-feather control system and a change in the generator torque control strategy when operating above rated power. This modified controller has been used in all simulations in this study, which seemingly reduced the impact of negative damping.

The differences between the turbulence models were noticeable in the results for the yaw rotation. Significant energy can be seen at low frequencies, around the wave peak frequency, at the OC3-Hywind yaw natural frequency (~0.11 Hz), and a very small response at the 1P natural frequency (0.2 Hz) (Figure 6). Since the spar-buoy foundation is characterized by high mooring stiffness and a low moment of inertia in yaw about the center of mass, it makes sense to see a "quasi-static" yaw response at low frequencies [26]. Additionally, results seen in [28] revealed that unstable conditions gave the largest yaw rotation for rated wind speed, followed by neutral and stable conditions. In this study, it is clear that very unstable conditions resulted in the largest excitation response in yaw, followed by unstable and neutral conditions.

4.2.2. OC4-DeepCwind semisubmersible FOWT. The spectral density for platform pitch and yaw rotation using the semisubmersible foundation are displayed in Figure 7.

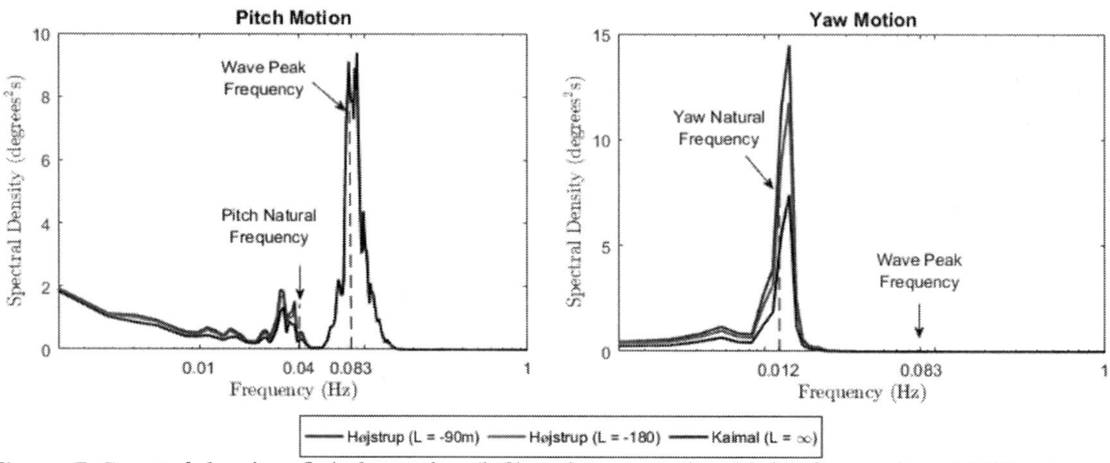

Figure 7. Spectral density of pitch rotation (left) and yaw rotation (right) for varying stability at rated wind speed (11.4 m/s) with the semisubmersible foundation.

The spectral energy plot of the platform pitch motion showed peaks in energy at the pitch motion natural frequency (0.04 Hz) and the wave peak frequency (0.083 Hz). Whereas, the platform yaw motion spectral energy plot showed significant energy at the yaw motion natural frequency (0.012 Hz), but very little energy at the wave peak frequency. Since a semisubmersible foundation is typically characterized by a soft mooring tension and a large moment of inertia in yaw [26], it seems likely that the yaw response would primarily occur at the yaw natural frequency. For each platform motion, the Højstrup model under very unstable conditions maintained the largest spectral energy. Since each degree of freedom for the OC4-DeepCwind FOWT has a fairly low natural frequency, there was no significant excitation response at frequencies higher than the wave peak frequency.

5. Conclusion

In this study, the influence of unstable turbulent wind fields on the loads and motions of floating offshore wind turbines was studied using the Højstrup 1981 Unstable Spectra Model and the Davenport Exponential Coherence function. The most significant difference between the turbulent wind cases was observed for the tower top torsion moment DEL's. Very unstable conditions resulted in fatigue loads for the tower top torsion that were 47% larger than neutral conditions for OC3-Hywind and 30.4% larger for OC4-DeepCwind. As expected, very unstable conditions corresponded to the highest turbulence intensities and the largest turbulent fluctuations due to the inclusion of buoyancy generated turbulence. In previous studies, the fitted Mann model was unable to fully simulate unstable conditions, which may explain why the results from both [2] and [28] found that the highest DEL's corresponded to neutral conditions when using this approach.

The purpose of using the Højstrup spectra model is to include the low-frequency energy that is present in unstable atmospheric conditions, and determine its influence on the resulting loads and platform motions. However, the Højstrup model was developed using onshore wind measurements and has not currently been validated for offshore conditions. A composite spectral model developed by [8], which uses local similarity theory and a combination of a pointed and a blunt model to describe the wind turbulence spectra for different atmospheric stabilities, could be used in further studies. The resulting spectra incorporated measurements collected from the FINO 1 platform and would most likely result in a better depiction of unstable conditions for an offshore environment. However, care needs to be taken when applying wind spectra at different offshore sites to ensure that they are correctly normalized and site specific turbulence intensities are used. Jonkman and Veers [11] concluded that specific turbulence intensity levels are needed to accurately represent offshore sites, which can be dominated by different atmospheric stability conditions. Therefore, implementation of the Pointed-blunt model would be best suited for a location with similar site conditions to those found at the FINO 1 platform.

Acknowledgments
The authors would like to express their sincere gratitude to Rieska Putri and Dr. Etienne Cheynet for their key insights, collaboration, and abundant knowledge that was needed in completing the objectives of this paper.

References
[1] International Electrotechnical Commission 2005 IEC 61400-1 Wind Turbines- Part 1: Design Requirements. Geneva: International Electrotechnical Commission, IEC.
[2] Sathe A, Mann J, Barlas T, Bierbooms W A A M and van Bussel G J W 2013 Influence of atmospheric stability on wind turbine loads. *Wind Energy*, **16** 1013-32.
[3] Kretschmer M, Shwede F, Guzmán R F, Lott S, and Cheng P W 2018 Influence of atmospheric stability on the load spectra on wind turbines at alpha ventus. *J. Phys.* Conf. Ser. **1037** 05200.
[4] Holtslag M C, Bierboom WA A M and van Bussel GJW 2016 Wind turbine fatigue loads as a function of atmospheric conditions offshore. *Wind Energy*, **19** 1917-32.
[5] Sathe A and Bierbooms W 2007 The influence of different wind profiles due to varying atmospheric stability on the fatigue life of wind turbines. *Journal of Physics: Conference Series*, **75** 12056–62.
[6] Chougule A 2017 Modelling atmospheric turbulence via rapid distortion theory: Spectral tensor of velocity and buoyancy. *J. Atmos. Sci.* **74** 949-74.
[7] Chougule A, Mann J, Kelly M and Larsen G 2018 Simplification and validation of a spectral-tensor model for turbulence including atmospheric. *Boundary Layer Meteorol.* **167** 371-97.
[8] Cheynet E, Jakobsen J and Reuder J 2018 Velocity spectra and coherence estimates in the marine atmospheric boundary layer. *Boundary-Layer Meteorol.* **169**(3) 429 - 60.
[9] Putri, R M, Obhrai, C, & Knight, J M (2019, October). Offshore Wind Turbine Loads and Motions in Unstable Atmospheric Conditions. In Journal of Physics: Conference Series (Vol. **1356**, No. 1, p. 012016). IOP Publishing.
[10] Atmospheric Stability: Stability of Dry Air. [online]. PowerPoint PPT Presentation. Available at: https://www.slideserve.com/Jimmy/atmosphericstability [Accessed January 2019]
[11] Jonkman J and Veers P 2019 DNV GL joint industry project on validation of turbulence models. *Cooperative Research and Development Final Report*, CRD-17-673. NREL/TP-5000-73167.
[12] Kaimal J C and Finnigan J J 1994 *Atmospheric boundary layer flows: Their structure and measurement.* Oxford University Press, Inc.
[13] Haugen, D A, Kaimal, J C and Bradley, E F 1971 An experimental study of Reynolds stress and heat flux in the atmospheric surface layer. *Quart. J. R. Met. Soc.*, **97** 168-80.
[14] Kaimal J C, Wyngaard J C, Izumi Y and Coté O R 1972 Spectral characteristics of surface-layer turbulence. *Quart. J. R. Met. Soc.*, **98** 563-89.
[15] Højstrup, J 1981 A simple model for the adjustment of velocity spectra in unstable conditions downstream of an abrupt change in roughness and heat flux. *Boundary-Layer Meteorol.* **21** 341-56.
[16] Olesen H R, Larsen S E and Højstrup J 1984 Modelling velocity spectra in the lower part of the planetary boundary layer. *Boundary-Layer Meteorol.* **29** 285-312.
[17] Davenport A G 1961 The Spectrum of Horizontal Gustiness near the Ground in High Winds. *Quart. J. R. Met. Soc.*, **87** 194-211.
[18] Jonkman J, Larsen T et al. 2010 *Offshore Code Comparison Collaboration within IEA Wind Task 23: Phase IV Results Regarding Floating Wind Turbine Modeling.* Conference Paper. NREL/CP-500-47534.

[19] Robertson A, Jonkman J, Masciola M, Song H, Goupee A, Coulling A and Luan C 2014 *Definition of the Semisubmersible Floating System for Phase II of OC4*. National Renewable Energy Laboratory. Technical Report. NREL/TP-5000-60601.

[20] Shinozuka M and Deodatis G 1991 Simulation of stochastic processes by spectral representation. *Appl. Mech. Review*, **44**(4) 191-204.

[21] Cheynet E 2018 Wind field simulation (the fast version). MathWorks file exchange. Available at: https://se.mathworks.com/matlabcentral/fileexchange/68632-wind-field-simulation-the-fast-version [Accessed September 2018]

[22] Gryning S, Batchvarova E, Brümmer B, Jørgensen H and Larsen S 2007 On the extension of the Wind Profile over Homogeneous Terrain beyond the Surface Boundary Layer. *Boundary Layer Meteorol.* **124** 251 – 68.

[23] Det Norske Veritas AS 2010 DNV-RP-C205: Environmental conditions and environmental loads. Oslo: Veritas Offshore Technology and Service AS, DNV.

[24] Karimirad, M. MARINTEK Numerical Tools for Coupled Aero-Hydro-Servo-Elastic Simulations of Offshore Wind Turbines. SINTEF. MARINTEK. [Accessed 2019].

[25] Jonkman J, Butterfield S, Musial W and Scott G 2009 Definition of a 5-MW Reference Wind Turbine for Offshore System Development. NREL/TP-500-38060.

[26] Bachynski E and Eliassen L 2019 The effects of coherent structures on the global response of floating offshore wind turbines. *Wind Energy*, **22** 219-38.

[27] Hasselman K, Barnett T P, et al, 1973 Measurements of wind-wave growth and swell decay during the Joint North Sea Wave Project (JONSWAP). Deutsche Hydro. Zeitschr. Riehe, A8

[28] Putri R 2016 A study of the coherences of turbulent wind on a floating Offshore Wind Turbine. Master Thesis, University of Stavanger,.

[29] Driscoll F, Jonkman J, Robertson A, Sirnivas S, Skaare B and Nielsen F 2016 Validation of a FAST model of the Statoil-Hywind Demo Floating Wind Turbine. *Energy Procedia*, **94** 3-19.

[30] Lalanne C 2009 *Fatigue Damage. Mechanical Vibration and Shock Analysis*, Second Edition, **4**.

[31] Pena A, Gryning S E and Mann J 2010 On the length scale of the wind profile. *Quart. J. R. Met. Soc.*, **136** (653) 2119–31.

[32] Lackner M 2013 An investigation of variable power collective pitch control for load mitigation of floating offshore wind turbines. *Wind Energy*, **16** 519-28.

Assessment of wave runup and wave rundown based on observed long-term wave conditions

D Myrhaug* and T Sunde

Dept. of Marine Technology, Norwegian University of Science and Technology, Trondheim, Norway

* Corresponding author: dag.myrhaug@ntnu.no

Abstract. The article presents an analytical method that can be applied to provide first estimates of wave runup (*RU*) and wave rundown (*RD*) on shorelines and coastal structures based on observed long-term wave statistics and is supplementary to [12] who based their results on long-term wind statistics. Some recently published wave *RU* and wave *RD* formulae are used, together with joint statistics of significant wave height and spectral peak period from the Northern North Sea. Results are exemplified for the average statistical properties of wave *RU* and wave *RD* in terms of the expected values and the standard deviations, as well as values estimated from 1-, 10- and 100-years return period contour lines.

1. Introduction

When waves approach a coastline, they usually break and run up on the coast, be it a structure or a beach. For low laying land or areas particularly exposed and vulnerable to waves this can represent critical situations leading to e.g. flooding, coastal erosion, and damage to coastal infrastructure such as breakwaters, seawalls, artificial reefs and sand barriers. Some recent works addressing these issues are those of [1], [2], [3] and [4]. The research attention has increased in the recent years driven by the focus on climate change and consequently possible sea level rise and more extreme weather. For shorelines and coastal structures, it is essential to reliably assess the maximum wave runup (*RU*) and the maximum wave rundown (*RD*) for safe design and cost-efficient coastal protections.

The wave *RU* height and the wave *RD* height are defined as the vertical difference between the highest and the lowest point, respectively, and the still water level. Two components contribute to the wave *RU*; the wave set-up and the swash. Here the wave set-up is the mean surface elevation level with reference to the mean surface elevation in deep water caused by the radiation stress (see [5]); the swash oscillates from the wave set-up corresponding to the interception between the water and the shoreline (or structure); see [1] for further details. Common design formulae for wave *RU* and wave *RD* use a 2% exceedance value of the *RU* maxima at the toe of the shoreline (or structure), $R_{2\%}$, as well as a 2% exceedance value of the *RD* maxima, $R_{d2\%}$. Most of these design formulae are given in terms of the surf (Iribarren) number, which is defined in terms of the significant wave height H_s in deep water, the spectral peak period T_p in deep water, and the slope of the shoreline (or the structure). Two other commonly used wave *RU* formulae are those of [6] and [7], where the first formula recently has been applied by [8]. However, these general formulae can not be treated by the present analytical approach.

Results from some recent studies will be advocated here and described briefly in Section 2; from [9], [1], [2], [3] and [4]. The latter reference gives a review and summary of wave *RU* formulae. Some of these recent wave *RU* formulae were applied by [10], [11] and [12] using long-term variation of wave and wind conditions. Here [10] and [11] based their results on long-term wave statistics, applying the wave *RU* formulae by [1] and the wave *RU* and wave *RD* formulae by [2], respectively. Further, [11] also demonstrated how 100-years return period values of wave *RU* and wave *RD* can be calculated based on a joint distribution of significant wave height and spectral surf parameter. The present article is supplementary to [12] who based their results on long-term wind statistics, applying eight wave *RU* formulae and one wave *RD* formula. The main purpose here is to demonstrate how similar and some additional results of wave *RU* and wave *RD* can be derived by using long-term wave statistics.

In this article it is demonstrated how long-term wave statistics in deep water can be applied to provide estimates of the wave *RU* and wave *RD* on shorelines or coastal structures. Results are obtained by adopting published formulae; eight wave *RU* and two *RD* formulae, together with a joint distribution of H_s and T_p from the Northern North Sea. A procedure of estimating the wave *RU* and wave *RD* from 1-, 10- and 100-years return period contour lines and the corresponding values of H_s, T_p and the spectral surf parameter are also provided. The present analytical method should provide a useful tool that can be used for initial estimation of wave *RU* and wave *RD* in for instance early feasibility studies and risk analysis, or for estimations in the field.

2. Background

Extreme *RU* and *RD* events have been extensively investigated over the last decades in small- and large-scale laboratory studies as well as in field campaigns. The results are given in terms of empirical formulae representing estimates of the 2% exceedance values of *RU* and *RD* maxima. The formulae used in this article are adopted from some recently published works and are summarized in the following.

Two *RU* and one *RD* formulae were presented by [2]; the following two formulae for maximum wave *RU*

$$R_{2\%} = 1.165 H_s \xi_p^{0.77} \tag{1}$$

$$R_{2\%} = (0.39 + 0.795 \xi_p) H_s \tag{2}$$

and for maximum wave *RD*

$$R_{d2\%} = (0.21 - 0.44 \xi_p) H_s \tag{3}$$

Here ξ_p is the spectral surf parameter defined as

$$\xi_p = m\left(\frac{H_s}{\frac{g}{2\pi} T_p^2}\right)^{-\frac{1}{2}} \tag{4}$$

where $m = \tan\alpha$ is the slope with an angle α with the horizontal, and g is the acceleration due to gravity. The wave *RU* formulae are based on the formulations by [13] and [14], with new empirical coefficients fitted to the equations; Equations (1) and (2) are modified versions of the original formulae by [13] and [14], respectively. Equations (1) to (3) were based on large-scale laboratory experiments on a prototype-scale sand barrier with a median grain size diameter d_{50} = 0.42 mm, and are valid for m in the range 0.088 – 0.154 and ξ_p in the range 1 – 2.9. Furthermore, [2] suggested that the linear model in ξ_p (Eq. (2)) is the most easily applicable to *RU* data for the range of beach slopes they considered (see [2] for further details).

The following wave *RD* formulae based on data from a large-scale laboratory experiment was presented by [15]

$$R_{d2\%} = -0.1\xi_p^{2.21} H_s \tag{5}$$

The experiment was carried out using both the uniform slopes 1/6 and 1/12 and composite slopes with steepness 1/3 for the lower slope and 1/6 for the upper slope. For the composite slopes an appropriate average slope was calculated. The slope surface was covered by an asphalt concrete layer. This formula is based on data in the spectral wave steepness range $H_s/((g/2\pi)/T_p^2)$ of 0.001-0.031, and $0.5 < \xi_p < 2.5$.

The following wave *RU* model for $\xi_p < 0.6$, based on data from small-scale laboratory experiments for the three slopes 1/50, 1/30 and 1/20, and a sand seabed with d_{50} = 0.7 mm was proposed by [1]

$$R_{2\%} = 4\,m^{0.3} H_s\,\xi_p \tag{6}$$

Assessment of wave *RU* predictions on beaches on the South-East Australian coast considering 11 existing empirical formulae was performed by [4]. The data used in this comparison represent sand bottom with d_{50} in the range 0.25 – 0.5 mm, *m* in the range 0.02 – 0.16, ξ_p in the range 0.32 – 1.65; the wave conditions in the assessment covered a range with average values of approximately $H_s = 1.5$ m and $T_p = 8.9$ s, which are slightly below the typical mean conditions of the region of $H_s = 1.6$ m and $T_p = 9.5$ s. However, the proposed models in Equations (9) and (10) are also considered to be valid for higher wave conditions because the formulae they are based on, cover a range up to $H_s = 5$ m and $T_p = 15$ s. Among the 11 formulae [4] considered (in addition to Equations (9) and (10)) were those by [16] and [9]. The wave *RU* formula by [16] is given as

$$R_{2\%} = (0.2 + 0.83\xi_p) H_s \tag{7}$$

and based on data from measurements at Duck, North Carolina, USA.

The wave *RU* formula by [9] is given as

$$R_{2\%} = (0.58\,m + 0.53\xi_p) H_s + 0.45 \tag{8}$$

and based on field measurements on the south coast of Algarve, Portugal representing the European Atlantic coast. The data they used were based on manually selected wave *RU* maxima representing *m* in the range 0.04 - 0.15, ξ_p in the range 0.3 – 2.9, sand seabed with d_{50} = 0.50 mm.

The formulae by [4] are

$$M1\ R_{2\%} = 0.99\,H_s\,\xi_p \tag{9}$$

$$M2\ R_{2\%} = (0.16 + 0.92\xi_p) H_s \tag{10}$$

and are based on the best fit to the 11 formulae they considered, where the one in Equation (9) is forced through the origin (i.e. with the origin in $\xi_p = 0$ for $R_{2\%}/H_s$ versus ξ_p).

Moreover, [3] proposed a parameterization of wave *RU* from field measurements on gravel beaches and numerical calculation given as

$$R_{2\%} = C\,m^{0.5} T_p H_s\ ;\ C = 0.33 \tag{11}$$

The field data they used were obtained during a 2-year period including storm conditions with H_s in the range 1 – 8 m, *m* in the range 0.05 – 0.20, ξ_p in the range 0.2 -1.9, d_{50} in the range 2 -50 mm (see [3] for more details).

As a summary, Equations (1) to (3) and (5) to (10) can be represented as

$$R_2 = a\,H_s + b\,H_s\,\xi_p^c + d \tag{12}$$

Table 1. Wave *RU* and *RD* formulae according to Equation (12) for models 1 to 9 and Equation (11) for model 10.

Model nr	Abbreviation	Ref.	a	b	c	d	C
		[2]					
1, RU	Bl 1		0	1.165	0.77	0	-
2, RU	Bl 2		0.39	0.795	1	0	-
3, RD	Bl d		0.21	-0.44	1	0	-
4, RD	Sc	[15]	0	-0.1	2.21	0	-
5, RU	Pe	[1]	0	4m0.3	1	0	-
6, RU	Ho	[16]	0.2	0.83	1	0	-
7, RU	Vo	[9]	0.58m	0.53	1	0.45	-
		[4]					
8, RU	At1		0	0.99	1	0	-
9, RU	At2		0.16	0.92	1	0	-
10, RU	Po	[3]	-	-	-	-	0.33

where R_2 represents $R_{2\%}$ for wave *RU* and $R_{d2\%}$ for wave *RD*, and the coefficients *a, b, c, d* are given in Table 1.

As referred to in Section 1, the wave *RU* formulae by [6] and [7] are frequently used for assessing wave *RU*, but can not be treated analytically. In particular, the formulations by [7] cover a wide range of conditions on different slopes on shallow and very shallow foreshores with different surface roughnesses, although some of the formulae, e.g. their Equations (5.1), (5.2), (5.5) and (5.6) have the forms of Equation (12). Alternative methods are, for instance; to apply Monte Carlo simulations, or simple methods as those of [17] or [18].

Thus, according to Equations (11) and (12), R_2 is given in terms of the deep water sea state parameters H_s and T_p. Different parametric models for the joint probability density function (*pdf*) of H_s and T_p are available in the literature; see [19] for a review. The joint *pdf* used to exemplify the results in this article is taken from [20], given as

$$p(H_s,T_p) = p(T_p \mid H_s)p(H_s) \tag{13}$$

Here $p(H_s)$ is the marginal *pdf* of H_s, given by the following combined lognormal and Weibull distributions

$$p(H_s) = \begin{cases} \dfrac{1}{\sqrt{2\pi}\kappa H_s} \exp\left[-\dfrac{(ln\, H_s - \theta)^2}{2\kappa^2} \right], H_s \le 3.25\,\text{m} \\ \beta \dfrac{H_s^{\beta-1}}{\zeta^\beta} \exp\left[-(\dfrac{H_s}{\zeta})^\beta \right], H_s > 3.25\,\text{m} \end{cases} \tag{14}$$

where $\theta = 0.801$, $\kappa^2 = 0.371$ are the mean value and the variance, respectively, of $ln\, H_s$ and $\zeta = 2.713$, $\beta = 1.531$ are the Weibull parameters.

$p(T_p \mid H_s)$ is the conditional *pdf* of T_p given H_s, given by the lognormal *pdf*

$$p(T_p \mid H_s) = \frac{1}{\sqrt{2\pi}\sigma T_p} \exp\left[-\frac{(ln\, T_p - \mu)^2}{2\sigma^2} \right] \tag{15}$$

where μ and σ^2 are the mean value and the variance, respectively, of $ln\, T_p$ given as

$$\mu = a_1 + a_2 H_s^{a_3}\ ,\ (a_1, a_2, a_3) = (1.780, 0.288, 0.474) \tag{16}$$

$$\sigma^2 = b_1 + b_2 e^{b_3 H_s}, (b_1, b_2, b_3) = (0.001, 0.097, -0.255) \tag{17}$$

This joint *pdf* of H_s and T_p was obtained as a best fit to 29 years of wave data from the Northern North Sea representing wind waves, swell waves and combinations; see [20] for more details. The unit of H_s in Equations (16) and (17) is in meter.

It should be noted that the wave *RU* and wave *RD* formulae used here are based on data from other locations than the wave data. Therefore, in a proper analysis the wave data should be from the same location as the wave *RU* and wave *RD* models are based on. However, here the present wave data are used since the main purpose is to demonstrate the application of the method.

3. Statistical properties of *RU* and *RD*

The statistical properties of R_2, from which the statistical properties of $R_{2\%}$ and $R_{d2\%}$ follow, are derived from the joint *pdf* of H_s and T_p, giving the joint *pdf* of R_2 and H_s. This is achieved from Equations (11) and (12) by a change of variables from (H_s, T_p) to (H_s, R_2), yielding

$$p(H_s, R_2) = p(R_2 \mid H_s)\, p(H_s) \tag{18}$$

From Equations (11), (4), (12) and (13) it is noticed that this change of variables only affects $p(T_p \mid H_s)$.

First, consider Equations (4) and (12) from which

$$T_p = \left[\frac{R_2 - aH_s - d}{bm^c \left(\frac{g}{2\pi} \right)^{\frac{c}{2}} H_s^{1-\frac{c}{2}}} \right]^{\frac{1}{c}} \tag{19}$$

This gives the Jacobian

$$\left| \frac{\partial T_p}{\partial R_2} \right| = \frac{\frac{1}{c}(R_2 - aH_s - d)^{\frac{1}{c}-1}}{\left[bm^c (\frac{g}{2\pi})^{\frac{c}{2}} H_s^{1-\frac{c}{2}} \right]^{\frac{1}{c}}} \tag{20}$$

Thus, by using this Jacobian and that $p(R_2 \mid H_s) = p(T_p \mid H_s) \mid \partial T_p / \partial R_2 \mid$, the following lognormal *pdf* of R_2 given H_s is obtained

$$p(R \mid H_s) = \frac{1}{\sqrt{2\pi}\, \sigma_R R} \exp\left[-\frac{(lnR - \mu_R)^2}{2\sigma_R^2} \right] \tag{21}$$

where

$$R = R_2 - aH_s - d \tag{22}$$

and μ_R and σ_R^2 are the conditional mean value and conditional variance, respectively, of $ln\, R$, given as

$$\mu_R = c\mu + ln[bm^c (\frac{g}{2\pi})^{\frac{c}{2}} H_s^{1-\frac{c}{2}}] \tag{23}$$

$$\sigma_R^2 = (c\ \sigma)^2 \tag{24}$$

where μ and σ^2 are given in Equations (16) and (17), respectively.

The conditional cumulative distribution function (*cdf*) of R given H_s is given by the standard Gaussian *cdf* Φ as ([21])

$$P(R \mid H_s) = \Phi\left[\frac{ln\,R - \mu_R}{\sigma_R}\right] \tag{25}$$

$$\Phi(v) = \frac{1}{\sqrt{2\pi}} \int_{-\infty}^{v} e^{-t^2/2} dt \tag{26}$$

Then, the expected value and the standard deviation of R given H_s are given in Equations (27) and (28), respectively, as ([21])

$$E[R \mid H_s] = \exp(\mu_R + \frac{1}{2}\sigma_R^2) \tag{27}$$

$$\sigma[R \mid H_s] = \left[(e^{\sigma_R^2} - 1)\exp(2\mu_R + \sigma_R^2)\right]^{1/2} \tag{28}$$

Thus, from Equation (22) it follows that

$$E[R_2 \mid H_s] = E[R \mid H_s] + aH_s + d \tag{29}$$

$$\sigma[R_2 \mid H_s] = \sigma[R \mid H_s] \tag{30}$$

Similarly, by taking Equation (11) as $R_2 = C\,m^{0.5}\,H_s\,T_p$, a change of variables from (H_s, T_p) to (H_s, R_2) only affects $p(T_p \mid H_s)$. Then, by using the Jacobian $|\partial T_p / \partial R_2| = 1/(C\,m^{0.5}\,H_s)$, this yields a lognormal *pdf* of R_2 given H_s in the form

$$p(R_2 \mid H_s) = \frac{1}{\sqrt{2\pi}\sigma_{R_2} R_2} \exp\left[-\frac{(ln\,R_2 - \mu_{R_2})^2}{2\sigma_{R_2}^2}\right] \tag{31}$$

where the conditional mean value and the conditional variance, respectively, of $ln\,R_2$, are given as

$$\mu_{R_2} = \mu + ln(C\ m^{0.5}\ H_s) \tag{32}$$

$$\sigma_{R_2}^2 = \sigma^2 \tag{33}$$

where μ and σ^2 are given in Equations (16) and (17), respectively.

4. Example of results

4.1. Conditional statistical values of RU and RD within a sea state

Here the average statistical properties of *RU* and *RD* expressed in terms of the expected values and the standard deviations are provided. Examples of results are given for

- Significant wave height in deep water, $H_s = 3\,\text{m}$
- Slope, $m = 1/10$

These parameters are within the validity range of the models (except for model 5 which is valid for $m = 1/50,\ 1/30,\ 1/20$), and is used here to serve the purpose of demonstrating the application of the method.

Table 2. Conditional expected values of *RU* for models 1, 2, 5 to 10 and *RD* for models 3, 4, for $H_s = 3\,\mathrm{m}$ and slope $m = 1/10$ (column 3); the corresponding expected value (EV) ± 1 standard deviation (SD) (column 4); see also Table 1.

Model nr	Abbreviation	EV (m)	EV ± 1 SD (m)
1, RU	Bl 1, Eq. (1)	2.7	2.2, 3.1
2, RU	Bl 2, Eq. (2)	2.9	2.5, 3.2
3, RD	Bl d, Eq. (3)	-0.31	-0.51, -0.10
4, RD	Sc, Eq. (5)	-0.15	-0.23, -0.07
5, RU	Pe, Eq. (6)	4.3	3.3, 5.2
6, RU	Ho, Eq. (7)	2.4	2.0, 2.8
7, RU	Vo, Eq. (8)	1.8	1.5, 2.0
8, RU	At 1, Eq. (9)	2.1	1.7, 2.6
9, RU	At 2, Eq. (10)	2.4	2.0, 2.9
10, RU	Po, Eq. (11)	3.1	2.4, 3.8

For these conditions the expected value (EV) and the expected value plus and minus one standard deviation (SD), i.e. $\mathrm{EV} \pm 1\ \mathrm{SD}$, of *RU* and *RD* can be calculated. The results are given in Table 2 showing that EV for wave *RU* ranges from 1.8m to 4.3m, with the lowest value based on model 7 and the highest value based on model 5; the coefficient of variation (SD/EV) varies from about 0.1 to about 0.2 with the lowest value for model 2 and the highest for model 5. Furthermore, EV for wave *RD* are -0.31m and -0.15m for models 3 and 4, respectively, while SD/EV is about 0.05 and 0.5, respectively. Based on the results for $\mathrm{EV} \pm 1\ \mathrm{SD}$ it appears that there is overlap between the values obtained by the two wave *RD* models 3 and 4, suggesting that the results are consistent. For the values obtained by the eight wave *RU* models there is overlap between models 1, 2, 6, 8, 9, 10; between models 7 and 8; between models 5 and 10. It should also be noted that the $\mathrm{EV} + 1\ \mathrm{SD}$ value of model 7 coincides with the $\mathrm{EV} - 1\ \mathrm{SD}$ values of models 6 and 9. Overall, this suggests that the results from models 1, 2, 6, 8, 9, 10 are consistent. It should be recalled that these results are not general, but valid for $H_s = 3\,\mathrm{m}$ and the slope $m = 1/10$ (which is outside the validity range of model 5 and thus the result for this model is less reliable). Moreover, these results are associated with the present $p(H_s, T_p)$, and thus other joint *pdfs* from the same locations as the wave *RU* and wave *RD* models are based might give different results.

4.2. Estimation of RU and RD from n-years return period contour lines of $p(H_s, T_p)$

N-years return period values of *RU* and *RD* can be estimated from the n-years return period contour lines of $p(H_s, R_2)$ in Equation (19). However, an alternative is to estimate these values of *RU* and *RD* from the n-years return period contour lines of $p(H_s, T_p)$ in Equations (13) to (17).

IOP Conf. Series: Materials Science and Engineering **700** (2019) 012006 doi:10.1088/1757-899X/700/1/012006

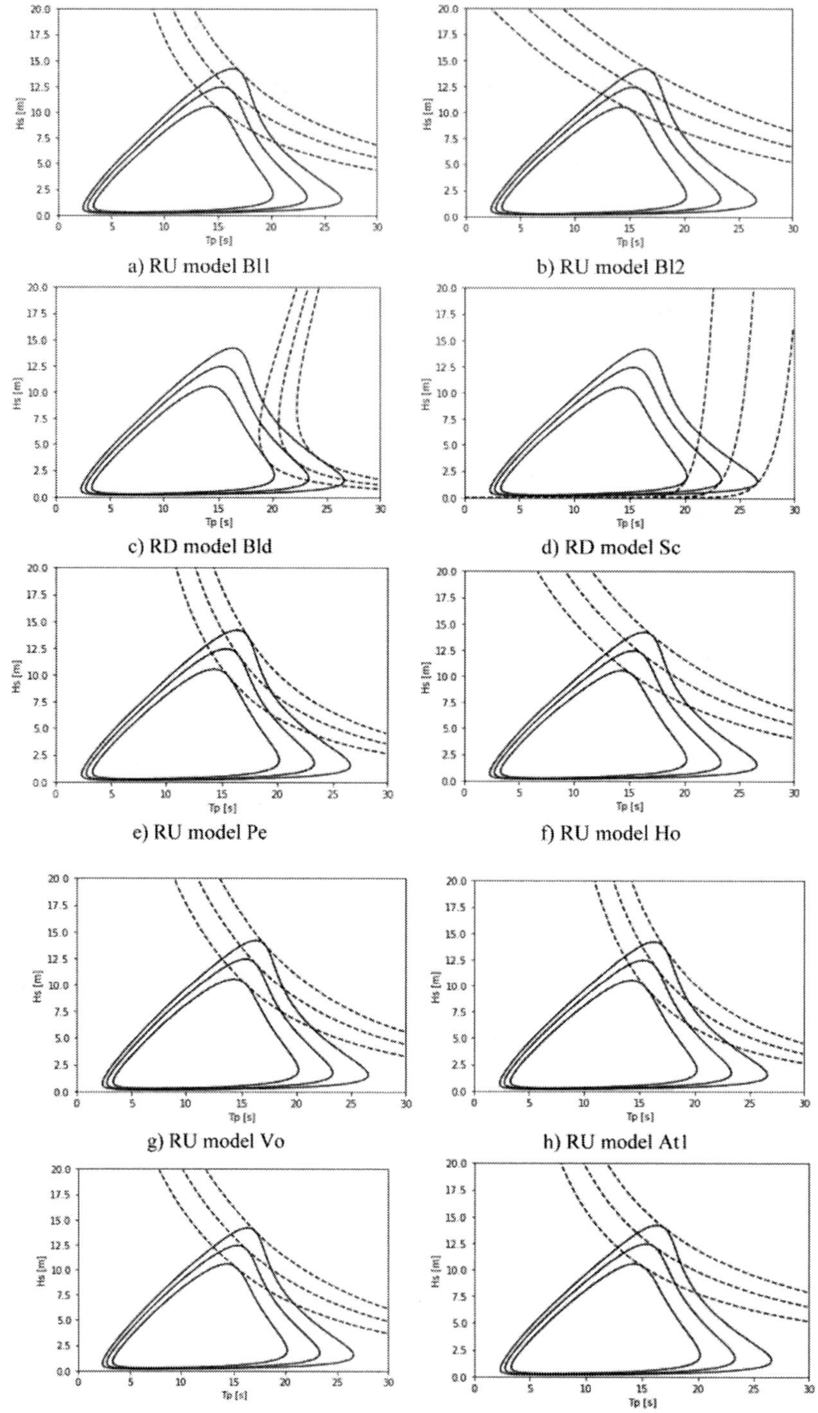

a) RU model Bl1
b) RU model Bl2
c) RD model Bld
d) RD model Sc
e) RU model Pe
f) RU model Ho
g) RU model Vo
h) RU model At1
i) RU model At2
j) RU model Po

Figure 1. 1-year, 10-years and 100-years contour lines of H_s (m) (vertical axes) and T_p (s) (horizontal axes) from the inner to outer contours, respectively, with the tangent lines corresponding to: (a) *RU* model 1; (b) *RU* model 2; (c) *RD* model 3; (d) *RD* model 4; (e) *RU* model 5; (f) *RU* model 6; (g) *RU* model 7; (h) *RU* model 8; (i) *RU* model 9; (j) *RU* model 10. See also Tables 1 and 3.

Figure 1 depicts the 1-year, 10-years and 100-years return period contour lines by assuming a sea state duration of 3 hours, represented by the inner to the outer contours, respectively, obtained using the IFORM method described in [22]. This method is based on transformation of the joint *cdf* of H_s and T_p to the standard Gaussian *cdf*, from which the return period contour lines are determined as circles.

First, consider Equation (11) which by solving for H_s gives

$$H_s = \frac{R_2}{C\, m^{0.5}} T_p^{-1} \tag{34}$$

For a given value of R_2, this is a curve in the (H_s, T_p) plane. The values of R_2 which implies that these curves will have tangent points with the 1-year, the 10-years and the 100-years contours can be determined by iteration, and thus the corresponding tangent points. The results are shown graphically in Figure 1j by the three curves giving tangent points to the 1-, 10-, 100-years contours for R_2 = 16.2 m, 20.5 m and 24.7 m respectively, corresponding to the 1-, 10-, 100-years return period values of R_2 based on model 10 and the slope $m = 1/10$. However, how realistic these large values of R_2 are, is questionable; more discussion is given subsequently. It is observed that these values of R_2 are governed by H_s in the sense that the tangent points are located close to the maximum values of H_s along the respective return period contours. The corresponding coordinates of the tangent points for H_s, T_p are given in Table 3, together with the corresponding values of ξ_p according to Equation (4).

Second, consider Equation (12), which for a given value of R_2 represents a curve in the (H_s, T_p) plane. Also in this case the values of R_2 corresponding to the tangent points to the 1-, 10-, 100-years contours are found by iteration. The results are shown graphically in Figures 1a, b, e – i for *RU* and in Figures 1c, d for *RD* by the three curves giving tangent points to the 1-, 10-, 100-years contours. The values of R_2 and the corresponding values of H_s, T_p and ξ_p for each of the *RU* and *RD* models are given in Table 3 for the slope $m = 1/10$. It is observed that these values of R_2 corresponding to *RU* are governed by H_s. However, the values of R_2 corresponding *RD* (Figures 1c, d) are related to tangent points given by combined lower H_s and higher T_p values.

Table 3. Extreme values of *RU* (R_2) for models 1, 2, 5 to 10 and *RD* (R_2) for models 3, 4 corresponding to the 1- 10, 100-years contours of (H_s, T_p) and corresponding values of H_s, T_p, ξ_p (for the slope $m = 1/10$), from left to right for each variable, respectively; see also Table 1.

Model nr	R_2 (m)	H_s (m)	T_p (s)	ξ_p
1, RU	8.0, 9.3, 10.5	10.2, 12.0, 13.9	15.2, 16.3, 17.0	0.60, 0.59, 0.57
2, RU	8.8, 10.3, 11.7	10.3, 12.3, 14.1	15.0, 15.9, 16.8	0.58, 0.56, 0.56
3, RD	-1.3, -1.5, -1.8	3.6, 3.4, 3.6	19.4, 22.1, 24.4	1.3, 1.5, 1.6
4, RD	-0.72, -1.0, -1.4	1.7, 1.6, 1.3	20.1, 23.3, 26.5	2.0, 2.3, 2.9
5, RU	12.2, 14.2, 16.0	9.8, 11.8, 13.7	15.6, 16.5, 17.2	0.62, 0.60, 0.58
6, RU	7.1, 8.3, 9.4	10.2, 12.2, 13.9	15.2, 16.1, 17.0	0.60, 0.58, 0.57
7, RU	4.3, 4.9, 5.5	10.0, 12.0, 13.7	15.4, 16.3, 17.2	0.61, 0.59, 0.58
8, RU	6.0, 7.0, 7.9	9.8, 11.8, 13.7	15.6, 16.5, 17.2	0.62, 0.60, 0.58
9, RU	7.2, 8.4, 9.5	10.2, 12.0, 13.9	15.2, 16.3, 17.0	0.60, 0.59, 0.57
10, RU	16.2, 20.5, 24.7	10.3, 12.2, 13.9	15.0, 16.1, 17.0	0.58, 0.58, 0.57

It should be noted that some of the estimated *RU* and *RD* values are obtained by exceeding the surf parameter validity bounds stated in Section 2, but is used here with the purpose of demonstrating the application of the method. This is the case for models 1 and 2 for all predictions, since the models are valid for ξ_p in the range 1 to 2.9; for model 4 for the 100-years return period value since $\xi_p = 2.9 > 2.5$; for model 5 for the 1-year return period value since $\xi_p = 0.62 > 0.6$. Furthermore, it is also noticed that the *RU* values from models 5 and 10 are large compared to the other model predictions. As referred to earlier, model 5 is valid for smaller slopes than $m = 1/10$ used here, and thus the results for this model are less reliable. If a smaller slope (i.e. in the range 1/50-1/20) is applied for the same wave conditions, it is expected to predict lower *RU* values (see Equation (6)). The model 10 results should be reasonable since Equation (11) is used within its range of validity. However, a firm statement on the validity of the return period results is not possible without comparison with measurements.

A similar procedure can be applied for other joint distributions of H_s, T_p as well as for other $n-$ years return period contour lines to obtain the wave *RU* and wave *RD* values and the corresponding values of H_s, T_p and ξ_p. However, as mentioned in Section 2, the wave data should be from the same location as the wave *RU* and wave *RD* models are based on.

4.3. Discussion

This section provides some comments on this approach versus a procedure which applies more complete computational demanding methods without using empirical formulae. For assessing the maximum wave *RU* and wave *RD* heights, common practice would be to start with available data on joint statistics of H_s and T_p (or other characteristic wave periods); preferably within directional sectors at a nearby location offshore (in deep water). The next step would be to apply a wave simulation model that includes effects of dissipation due to bottom friction and wave breaking, and thereby to obtain the joint statistics of H_s and T_p at the coastal (shallow water) site; then finally to use this as input for computing the maximum wave *RU* and wave *RD* heights. In general this practice would also include sites exposed to sea states with combined wind waves and swell waves from different directions. In this article an alternative is presented providing an analytical method which can be used to estimate maximum wave *RU* and wave *RD* heights from given deep water values of H_s and T_p, exemplified by including results using a joint distribution of H_s and T_p representing wind waves, swell waves and combinations.

The transition from deep water to the coastal site is assumed to be smooth, i.e. by neglecting wave energy dissipation effects over changing bed conditions with varying intermediate and shallow water depths. As a result, several effects affecting the estimated wave *RU* and wave *RD* heights are neglected, e.g.: that the wave field is inhomogeneous; from where the waves are coming and the location of the assessment site; return flows from dissipation effects which in turn will affect the local wave conditions. However, the approach enables analytical estimates of the maximum wave *RU* and wave *RD* heights, which are appropriate for making quick estimates. Then, these estimates can be used to compare with more complete computationally demanding methods. Under field conditions such an easily accessible and simple tool might also be useful as there is usually limited time and access to computational resources. Although the presented results are based on specific wave *RU* and wave *RD* formulae and a joint (H_s, T_p) distribution from another location than the wave *RU* and wave *RD* models are based on, the method can also be applied for other wave *RU* and wave *RD* formulae that can be treated analytically, joint distributions of sea state wave parameters, or for a given deep water wave spectrum including directional spreading effects. However, in such cases numerical calculations are most probably required. It is important, however, to assess the accuracy of this simple approach versus common practice, which is only possible to quantify by comparing with such methods over a wide parameter range, also including a sensitivity analysis of the results regarding the assumptions considered, but this is beyond the scope of this article.

5. Conclusions

An analytical method that can be applied to provide estimates of wave *RU* and wave *RD* on shorelines and coastal structures is presented. Results are achieved by adopting eight wave *RU* formulae and two wave *RD* formulae together with observed long-term wave statistics from deep water in the Northern North Sea. Expected values and variances of the 2% exceedance values of the wave *RU* maxima and the wave *RD* maxima are calculated. The main purpose is to demonstrate how long-term wave statistics can be used to obtain first estimates of wave *RU* and wave *RD*.

Examples of results for typical realistic conditions are also provided by presenting the conditional expected value (EV) plus and minus one standard deviation (1 SD) of the 2% exceedance values of the wave *RU* maxima and the wave *RD* maxima within a sea state. By comparing the results from the eight wave *RU* models and the corresponding $\mathrm{EV} \pm 1\ \mathrm{SD}$, it is overlap between models 1, 2, 6, 8, 9, 10; between models 7 and 8; between models 5 and 10; as well as between the two *RD* models 3 and 4. Overall, this suggests that the results from the *RU* models 1, 2, 6, 8, 9, 10 as well as from the *RD* models 3, 4 are consistent. However, a firm statement of this is not possible without comparison with measurements.

A procedure of estimating the 2% exceedance values of the wave *RU* maxima and the wave *RD* maxima as well as the corresponding values of significant wave height, spectral peak period and spectral surf parameter from 1-, 10-, and 100-years return period contour lines is also demonstrated. The estimated return period values seem to be reasonable, except for the estimates based on the wave *RU* model 5 which is valid for smaller slopes than that used here. However, a firm statement of this also requires comparison with measurements.

Overall, the estimates of wave *RU* and wave *RD* are associated with a large amount of uncertainty which is partly because the wave data are from another location than the wave *RU* and wave *RD* models are based on, and partly because the models are based on data from different locations. However, the present analytical method is convenient due to its simplicity, but it should only be used for early estimates due to the uncertainties related to the results. The method should provide an appropriate and useful tool for initial estimates of wave *RU* and wave *RD* in for instance early feasibility studies and risk analysis related to the assessment of climate change effects in coastal vulnerability studies, or for estimation in the field.

References

[1] de la Pena J M, Sanches-Gonzales J F and Diaz-Sanches R 2014 *J. of Waterways, Port, Coastal and Ocean Eng.* **140(4)** 04014015-1-04014015-11.
[2] Blenkinsopp C E et al. 2016 *Coastal Eng.* **113** 88-103.
[3] Poate T G, McCall R T and Masselink G 2016 *Coastal Eng.* **117** 176-190.
[4] Atkinson A L et al. 2017 *Coastal Eng.* **119** 15-31.
[5] Dean R G and Dalrymple R A 1984 *Prentice-Hall*, Upper Saddle River, NJ, USA.
[6] Stockdon H F et al. 2006 *Coastal Eng.* **53** 573-588.
[7] EurOtop 2018 Van der Meer J W et al., www.overtopping-manual.com.
[8] Serafin K A et al. 2019 *Coastal Eng.* **150** 1-17.
[9] Vousdoukas M I, Wziatek D and Almeida L P 2012 *Ocean Dynamics* **62** 123-137.
[10] Myrhaug D 2015 *J. of Ocean Eng. and Marine Energy* **1** 193-197.
[11] Myrhaug D and Leira B J 2017 *Coastal Eng.* **120** 75-77.
[12] Myrhaug D and Sunde T 2018 *Proc. of the Institution of Civil Engineers – Maritime Eng.* **171(1)** 40-46.
[13] Mase H 1989 *J. of Waterways, Port, Coastal and Ocean Eng.* **115(5)** 649-661.
[14] Hedges T S and Mase H 2004 *J. of Waterways, Port, Coastal and Ocean Eng.* **130(3)** 109-113.

[15] Schüttrumpf H, Bergman H and Dette H H 1994 *Proc. 24th Conf. on Coastal Eng.*, Kobe, Japan, pp. 553-564.
[16] Holman R A 1986 *Coastal Eng.* **9(6)** 527-544.
[17] Divoky D and McDougal W G 2006 *Proc. 30th Conf. on Coastal Eng.*, San Diego, USA, pp. 5291-5301.
[18] Bosom E and Jimenez J A 2011 *Nat. Hazards Earth Syst. Sci.* **11** 475-484.
[19] Bitner-Gregersen E M 2015 *Applied Ocean Res.* **51** 279-292.
[20] Moan T, Gao Z and Ayala-Uraga E 2005 *Marine Structures* **18(3)** 359-382.
[21] Bury K V 1975 *John Wiley & Sons*, New York, NY, USA.
[22] DNV 2014 Recommended Practice DNV- C205 Environmental Conditions and Environmental Loads. DNV GL AS.

IOP Conf. Series: Materials Science and Engineering **700** (2019) 012007 doi:10.1088/1757-899X/700/1/012007

Investigation of transitional turbulence models for CFD simulation of the drag crisis for flow over a sphere

S M Nakhostin and K E T Giljarhus*

Department of Mechanical and Structural Engineering and Materials Science, University of Stavanger, Stavanger, Norway

* Corresponding author: knut.e.giljarhus@uis.no

Abstract. The drag crisis is an interesting physical phenomenon for flow over bluff bodies, where the drag coefficient suddenly drops as the Reynolds number increases. This behavior is caused by the flow transitioning from laminar to turbulent flow in the boundary layer. The turbulent flow remains attached to the surface longer than the laminar flow, thereby reducing the size of the wake behind the body. The phenomenon has been extensively studied experimentally for simple geometries such as flow over a sphere or a cylinder. Numerical simulations, however, have mainly been performed in either the subcritical or the supercritical region. In this study, we investigate the ability of steady-state RANS CFD models to predict the drag crisis for flow around a sphere. Experiments using oil-film visualization are performed to determine the type of transition for flow over a sphere. Simulations are performed using OpenFOAM, with the k-ω shear stress transport (SST) turbulence model. For transition modelling, the Langtry-Menter modification to the SST model is considered. The simulations show improved prediction of the drag coefficient in the subcritical regime, but the implementation of the model is found to be unstable in the critical and supercritical range.

1. Introduction

The boundary conditions applied to solid surfaces in industrial CFD (computational fluid dynamics) simulations are typically assuming a fully turbulent boundary layer. This is often sufficient since relevant conditions for engineering-type structures typically are such that the boundary layer is turbulent for most of the structure. However, there are several cases in e.g. aerospace, wind turbines and turbomachinery where the transition from laminar to turbulent flow in the boundary layer can be relevant. Failing to account for the transition can lead to erroneous prediction of engineering quantities such as heat transfer coefficients and skin friction coefficients.

For flow in a boundary layer, the phenomena leading to transition are typically divided into three different modes; bypass transition, natural transition and separation-induced transition. Bypass transition is caused by turbulence from the free stream diffuses into the boundary layer, generating disturbances that eventually lead to transition. Natural transition is only possible for very smooth walls and extremely low disturbances in the mean flow. Here, initial disturbances develop into streamwise, unstable waves called Tollmien-Schlitling waves. The growth and breakdown of these waves eventually lead to a fully turbulent flow. Separation-induced transition typically occur in adverse pressure gradient flows where the laminar boundary layers separates, causing turbulence to be generated in the shear layer and subsequent turbulent reattachment.

Transition is a complex physical phenomenon and transition modelling is still an active area of research. No model has yet reached the level of maturity necessary for routine engineering simulations. The most popular turbulence models used in engineering are RANS (Reynolds-Averaged Navier-Stokes) models, which simulate the average flow field and account for turbulence by adding an additional turbulent viscosity to the Navier-Stokes equations. In the pioneering work of [1], the so-called local correlation-based transition model (LCTM) concept was introduced for transition modelling. These models attempt to solve additional transport equations to give a local criterion for transition. Hence, instead of modelling the physics of transition, they rely on empirical correlations to model the behavior. These models are suitable for implementation into a RANS framework and is therefore promising for use in general-purpose CFD simulations. Recently, a review of transition models based on LCTM showed good agreement for a wide range of test cases [2].

Most simulations with transition models have been performed on streamlined structures such as turbine blades, airfoils and wing turbines. Bluff bodies, characterized by large adverse pressure gradients and significant regions of separated flows, have been comparatively less studies in the context of transition modelling. Transition effects can have a crucial effect on such structures. A well-known phenomenon is the drag crisis, where the drag coefficient suddenly drops as the Reynolds number increases. This phenomenon can in some cases be exploited to create more efficient designs. The classical example is the dimples applied to a golf ball, which makes the ball fly longer due to initiating the drag crisis at a lower Reynolds number [3].

This study is an initial study of the capability of transition models to capture the drag crisis for flow over a sphere. The flow over a sphere is a fundamental problem in fluid dynamics, and has been studied extensively in the literature. A thorough review was recently performed in [4, 5]. Experimentally, the full range of Reynolds numbers have been studied. Most recently, the existence of a laminar separation bubble was found in the supercritical region [6]. When it comes to numerical methods, the lower range of Reynolds numbers before transition has been studied by several authors with good results. A few studies have been made in the super-critical regime [7-9]. Notably, only a single study has been performed in the drag crisis range [10]. This was done using a high-order Lattice-Boltzmann method, without applying any sub-grid turbulence model or wall model. Even with large unstructured grids with more than 100 million grid cells, they do not resolve the boundary layer completely, but their results indicate that the drag crisis can be reproduced by numerical simulations. Since such direct simulation methods are unsuitable for most engineering-type simulations, this study investigates the ability of steady-state RANS methods with transition modelling to reproduce the drag crisis phenomenon. The CFD simulations are performed with the open source CFD software OpenFOAM. Additionally, experiments using oil-film visualization are performed to see if the results from [6] could be reproduced and to obtain further insight into the transition behavior.

Section 2 introduces the numerical model and the transition turbulence model. Next, Section 3 describes the computational setup with details on the computational mesh and numerical settings. Sections 4 provides information on the experimental setup. Section 5 presents and discusses the results, while in Section 6 conclusions are drawn and recommendations for further work proposed.

2. Numerical model

The numerical model is based on the steady-state Reynolds-Averaged Navier-Stokes equations,

$$\frac{\partial u_i}{\partial x_i} = 0 \tag{1}$$

$$u_j \frac{\partial u_i}{\partial x_j} = -\frac{1}{\rho}\frac{\partial P}{\partial x_i} + \nu \frac{\partial^2 u_i}{\partial x_j^2} - \frac{\partial \overline{u_i' u_j'}}{\partial x_j} \tag{2}$$

Here, u_i represents the averaged velocity components, P is the dynamic pressure, ρ is the density and ν the kinematic viscosity. The influence of turbulence is accounted for by the Reynolds stress tensor, $\overline{u_i'u_j'}$. Here, u_i' denotes the fluctuating part of the velocity. The Reynolds stress component is modelled using the Boussinesq approximation, where it is expressed in terms of a turbulent viscosity, ν_t, and the mean flow gradients,

$$-\overline{u_i'u_j'} = \nu_T\left(\frac{\partial u_i}{\partial x_j} + \frac{\partial u_j}{\partial x_i}\right) - \frac{2}{3}k\delta_{ij} \qquad (3)$$

Here, k is the turbulent kinetic energy and δ_{ij} is the Kronecker delta function. The k-ω SST turbulence model is used to find the turbulent viscosity [11].

The Langtry-Menter model is used to model transition. This model is an addition to the k-ω SST model. The model does not attempt to model the actual physics of the transition process, but instead couples the general turbulence model with correlation-based transition modeling. Since the model contains a large number of correlations, only a brief overview of the model is given here, more details can be found in [1]. The model is also known as the γ-$Re_{\theta t}$ –SST model, because it makes use of two additional variables; the intermittency, γ, and the transition momentum thickness Reynolds number, $Re_{\theta t}$, in addition to the k and ω equations. The intermittency value is used to trigger the transition locally. It is coupled with the production of turbulent kinetic energy and used to turn on production downstream of the transition. The transition momentum thickness Reynolds number is used to capture the effect of the freestream behavior on the boundary layer. The two extra equations for $Re_{\theta t}$ and γ that are solved in this model are

$$\frac{\partial(\rho\gamma u_j)}{\partial x_j} = P_\gamma - E_\gamma + \frac{\partial}{\partial x_j}\left[\left(\mu + \frac{\mu_t}{\sigma_f}\right)\frac{\partial\gamma}{\partial x_j}\right] \qquad (4)$$

$$\frac{\partial(\rho u_j Re_{\theta t})}{\partial x_j} = P_{\theta t}(Re_{\theta t0} - Re_{\theta t}) + \frac{\partial}{\partial x_j}\left[\sigma_{\theta t}(\mu + \mu_t)\frac{\partial Re_{\theta t}}{\partial x_j}\right] \qquad (5)$$

The production term in the intermittency equation is designed to be switched on whenever the local strain rate exceeds the transition criteria. It contains multiple correlations that are tuned based on experiments performed of flow over a flat plate. The sink term ensures that the intermittency is close to zero in the laminar boundary layer and can also be used to model relaminarization. The equation for $Re_{\theta t}$ is used to transfer information from the free-stream to the boundary layer to give a local criterion for transition. The production term in this equation contains a correlation for the transition onset based on experiments.

The only modification to the SST model is that the production and dissipation terms in the equation for turbulent kinetic energy now include the intermittency to control the production/dissipation as follows,

$$\overline{P_k} = \gamma P_k$$

$$\overline{D_k} = \min(\max(\gamma, 0.1), 1.0)\, D_k$$

3. Computational setup

The above numerical model is solved using the OpenFOAM toolbox, version 6.0. The computational domain has size 20D x 10D x 10D, where D is the sphere diameter. The sphere is placed 5D from the inlet. At the inlet, constant inlet values as used, with a turbulence intensity of 0.5 %, and zero gradient for pressure. At the outlet, constant pressure is set with zero gradient for the remaining variables. Slip conditions are used at the sides of the domain.

The computational mesh is generated using the snappyHexMesh mesh generator. Three levels of refinement are used in the wake region and around the sphere, and layers are added near the surface to obtain a sufficiently fine resolution in the boundary layer. Around 20-30 cells are used in the layer region, with a stretching factor of 1.2. The y+ value was less than 1 for all cases. Figure 1 shows the overall mesh as well as a close-up of the mesh near the sphere surface.

The discretization scheme used for the convective term is the limited linear scheme, which is a central-upwind scheme using the Sweby limiter. For the remaining Laplacian, gradient and surface normal gradient terms the linear schemes are used, which apply the Gauss divergence theorem with explicit non-orthogonal correction. The SIMPLE algorithm is used for pressure-velocity coupling and the iteration procedure is stopped when the drag coefficient has converged to four significant digits.

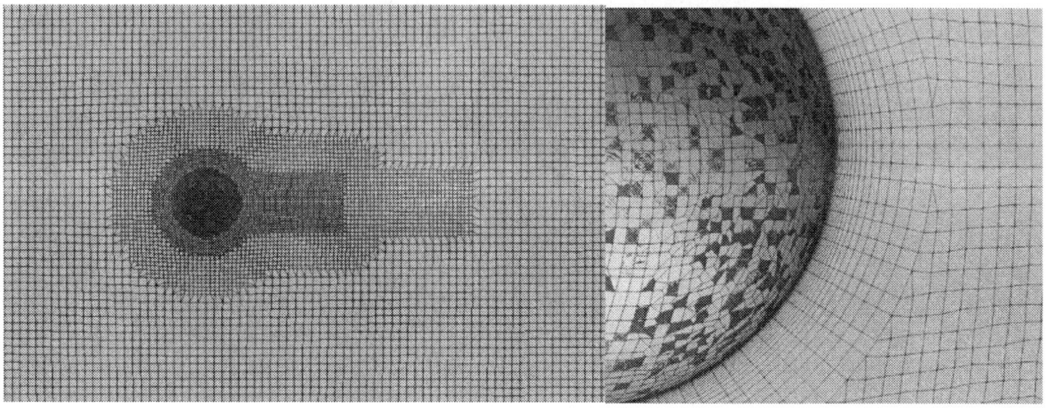

Figure 1. Illustration of computational mesh (left) and close-up of layers near the surface (right).

3.1. Mesh convergence study

To determine the sufficient size of the computational mesh, a mesh convergence study was performed for the case with Reynolds number 5×10^5. For the convergence study, the distance to the first cell is kept constant to maintain $y^+ = 1$ and the overall grid resolution elsewhere is increased. Figure 2 shows the drag coefficient for four grid sizes. For the two last grids, the coefficient converges to the value of 0.166. Hence, the grid size of approximately 500 000 cells is found to be sufficient for Reynolds number 5×10^5. For the sake of simplicity, this resolution is kept for all cases and only the distance to the first cell and number of cells in the near-wall layer is modified for the remaining cases.

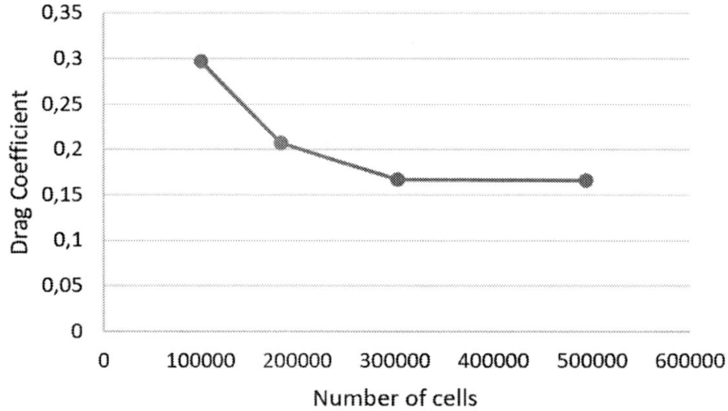

Figure 2. Mesh convergence study.

4. Experimental setup

The experiments were performed at the fluid mechanics lab at the Norwegian University of Science and Technology in Trondheim, Norway. A sphere with diameter 170 mm was placed in a closed-circuit wind tunnel with test section dimensions 1 m by 0.5 m. The wind tunnel is capable of speeds up to 35 m/s at low turbulence intensities of approximately 0.2 %. .The sphere was placed on a rod with 19 mm diameter, and with 130 mm distance from sphere to the floor of the wind tunnel. The setup is shown in Figure 3. The rod is connected to a force balance to measure the forces on the geometry. Note that since the rod is relatively thick compared to the sphere, and placed under the sphere, the rod will significantly influence the drag and the force results will not be directly comparable to other results in the literature. To investigate the transition behaviour, the sphere was covered in a mixture of motor oil and canola oil mixed with white dry pigment.

Figure 3. Sphere placed in wind tunnel.

5. Results

The experimental results are summarized in Figure 4. The drop in drag coefficient can be clearly seen at a Reynolds number of approximately 2.8×10^5. The oil film visualization for a Reynolds number of 3.2×10^5 illustrates the transition and separation behaviour. At the first line, there is laminar separation from the sphere, causing a laminar separation bubble. At the second line, there is turbulent reattachment before the flow finally fully separates from the sphere at the third line. This is in agreement with the results from [6]. In the development of the Langry-Menter transition model [1], the model is compared against several flat-plate test cases, both with zero pressure gradient and non-zero pressure gradients. Only one of these cases, T3C4, is designed to test separation-induced transition. In [1], they report good agreement for the skin friction for this case, but in [12], it is shown that for the Reynolds shear stress and turbulence kinetic energy, the results are rather unsatisfactory.

Figure 5 compares the simulated drag coefficients with the experiments of [13] and the empirical correlation from [14]. In the subcritical region, there is good agreement between the experiments and the simulations. However, in the transition region, the fully turbulent model deviate significantly from the experiments. This is expected, as for the fully turbulent model, the turbulent boundary layer will lead to later separation and thereby smaller wake and pressure drag. In the supercritical region, however, the results are in reasonable agreement with the experiments again. The transition model gives better agreement to the experiment in the transition region, but the deviation is still significant. Unfortunately, the simulations for higher Reynolds numbers with the transition model showed numerical instabilities so they are not included here.

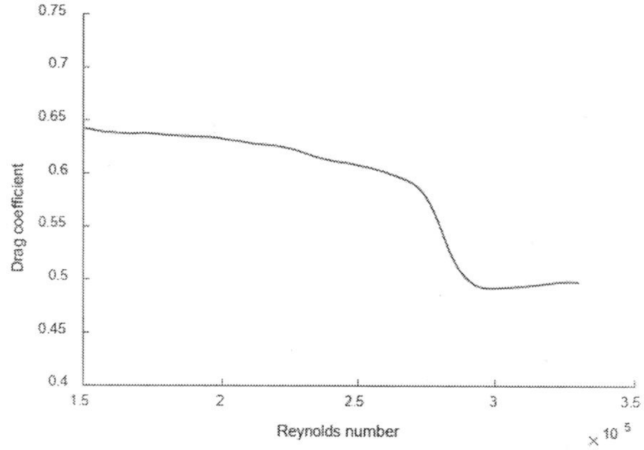

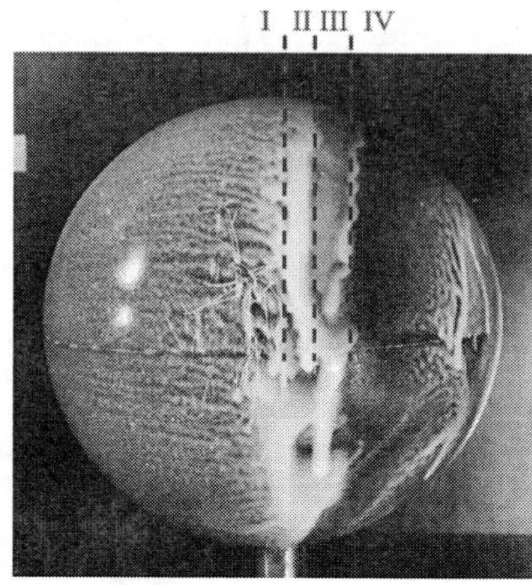

Figure 4. Experimental results. Left: Drag coefficients. Right: Oil film visualization of flow over sphere after drag crisis. I: Laminar boundary layer. II: Laminar separation bubble. III: Turbulent reattachment. IV: Turbulent separation.

Figure 6 show contours of velocity for the two turbulent models at $\mathrm{Re} = 1 \times 10^5$. The fully turbulent model has a smaller wake area than the transition model which explains the difference in drag values. This effect is further illustrated in Figure 7, which shows the pressure coefficient along the sphere surface. The simulation values are compared against the experiments of [13]. The experiment at $\mathrm{Re} = 1.62 \times 10^5$ are in the subcritical range while the experiment at $\mathrm{Re} = 3.18 \times 10^5$ are in the supercritical range. The fully turbulent simulation agree well with the pressure in the supercritical range. The transitional model is slightly closer to the subcritical pressure, but there are still significant differences. This corresponds to the discrepancies observed for the drag coefficient values.

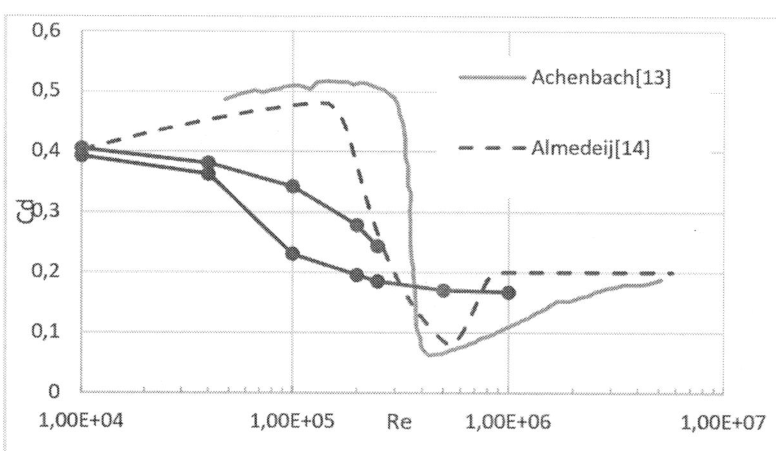

Figure 5. Comparison of experimental results of Achenbach [13], empirical relationship of Almedeij [14] and numerical results of this work with and without transition model.

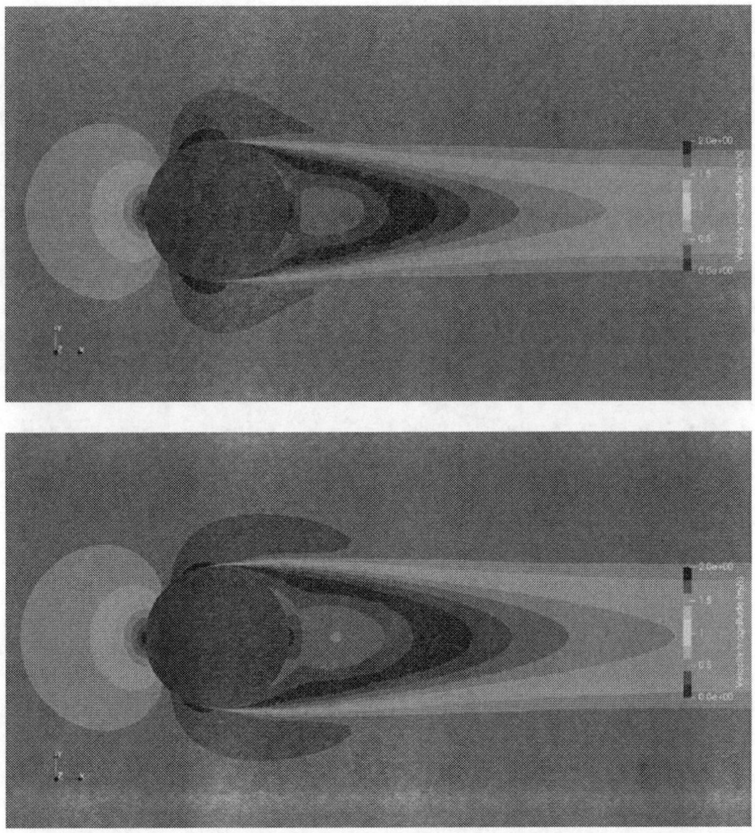

Figure 6. Comparison of velocity magnitude for Re $= 1 \times 10^5$ for fully turbulent model (top) and transition model (bottom).

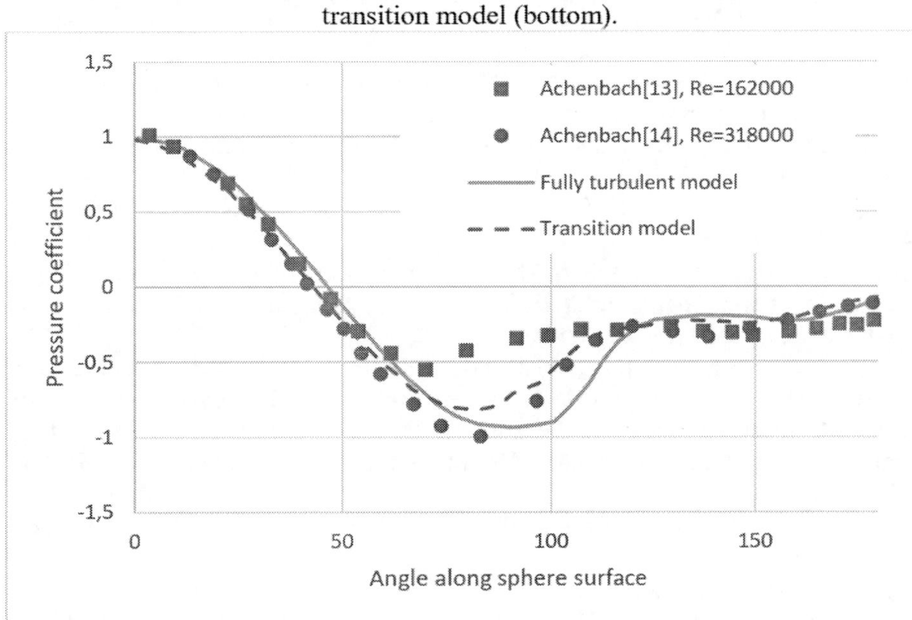

Figure 7. Pressure coefficient along sphere surface. Comparison of fully turbulent and transition model with experimental data from Achenbach[13].

6. Conclusions

In this study, we investigated the ability of CFD models to predict the drag crisis for flow around a sphere. Simulations were performed using the k-ω SST model with the Langtry-Menter transition model. Experiments using oil-film visualization were performed to determine the type of transition for flow over a sphere. The experiment showed the existence of a laminar separation bubble, with subsequent turbulent reattachment and separation. This is in agreement with recent experiments performed in [6]. The simulations showed improved prediction of the drag coefficient in the subcritical regime, but still with significant errors compared to the experiments. Due to numerical instabilities at higher Reynolds numbers, the model could not be applied to the critical Reynolds number range.
Future work will perform more experiments in the subcritical range, as well as investigate the numerical instabilities in more detail. Additionally, newer transition models such as the fully algebraic model introduced in [15] will be implemented in OpenFOAM and applied to this problem.

Acknowledgments
We thank Lars Morten Bardal and Ola Elfmark at SIAT – Centre for Sports Facilities and Technology for their kind assistance with the wind tunnel experiments.

References

[1] Menter F R, Langtry R and Völker S 2006 Transition modelling for general purpose CFD codes, *Flow turbul. combust.* **77**(1-4) 277-303
[2] Kaynak U, Bas O, Cakmakcioglu S C and Tuncer I H 2019 Transition Modeling for Low to High Speed Boundary Layer Flows with CFD Applications *In Boundary Layer Flows-Theory, Applications and Numerical Methods.* DOI: 10.5772/intechopen.83520
[3] Bearman P W and Harvey J K 1976 Golf ball aerodynamics *Aeronaut. Quart.* **27**(2) 112-122
[4] Tiwari S S, Pal E, Bale S, Minocha N, Patwardhan A W, Nandakumar K and Joshi J B 2019 Flow past a single stationary sphere, 1. Experimental and numerical techniques, *Powder Technol.*
[5] Tiwari S S, Pal E, Bale S, Minocha N, Patwardhan A W, Nandakumar K and Joshi J B 2019 Flow past a single stationary sphere, 2. Regime mapping and effect of external disturbances, *Powder Technol.,* In Press
[6] Deshpande R, Kanti V, Desai A and Mittal S 2017 Intermittency of laminar separation bubble on a sphere during drag crisis, *J. Fluid Mech.* **812** 815-840
[7] Constantinescu G and Squires K 2004 Numerical investigations of flow over a sphere in the subcritical and supercritical regimes, *Phys. Fluids* **16** 1449–66
[8] Jindal S, Long L, Plassmann P and Uzol N 2004 Large eddy simulations around a sphere using unstructured grids, *In 34th AIAA Fluid Dyn Conf Exhib*
[9] Jones D A and Clarke D B 2008 Simulation of flow past a sphere using the fluent code ADA494935
[10] Geier M, Pasquali A and Schönherr M 2017 Parametrization of the cumulant lattice Boltzmann method for fourth order accurate diffusion Part II: Application to flow around a sphere at drag crisis, *J. Comput. Phys.* **348** 889-98
[11] Menter F R, Kuntz M and Langtry R 2003 Ten years of industrial experience with the SST turbulence model *In Proc. fourth int. symp. Turbu., heat and mass transfer* 625–632
[12] Suluksna K, Dechaumphai P and Juntasaro E 2009 Correlations for modeling transitional boundary layers under influences of freestream turbulence and pressure gradient, *Int. J. Heat Fluid Flow* **30**(1) 66-75
[13] Achenbach E 1972 Experiments on the flow past spheres at very high Reynolds numbers, *J. Fluid Mech.* **54**(3) 565-75
[14] Almedeij J 2008 Drag coefficient of flow around a sphere: Matching asymptotically the wide trend, *Powder Technol.* **186**(3) 218-23
[15] Cakmakcioglu SC, Bas O, Kaynak U 2018 A correlation-based algebraic transition model, *Proc. Ints. Mech. Eng. C-J. Mech. Eng. Sci.* **232**(21) 3915-29

Effects of an inserted circular cylinder on a steady lid-driven rectangular cavity flow

J Zhu, L E Holmedal and D Myrhaug

Dept. of Marine Technology, Norwegian University of Science and Technology, Trondheim, Norway

E-mail: jianxun.zhu@ntnu.no

Abstract. This paper provides numerical investigations of flow in a steady lid-driven cavity of depth to width ratio 1/3 containing a circular cylinder. The inner cylinder is treated using a direct-forcing immersed boundary method, and a project method is applied to solve the incompressible Navier-Stokes equations. Three different Reynolds numbers and positions of the cylinder are considered; for a lower Reynolds number, flow structures are weakly affected by the cylinder near the left wall while two clockwise vortices attached to the cylinder are formed as the cylinder moves rightwards; for a moderate Reynolds number, an anticlockwise bottom vortex is formed for the cylinder near the left wall, and it disappears as the cylinder is shifted rightwards; for the largest Reynolds number, two anti-clockwise vortices attached the cylinder are formed for the cylinder close to the left wall, and as the cylinder moves gradually closer to the left wall a bottom vortex is formed and disappears together with the clockwise vortex to the right side of the cylinder.

1. Introduction

Vortex structures, e.g. elongated primary vortices and corner vortices, within a steady lid-driven cavity are controlled by two parameters; *i*) Reynolds number based on the lid motion velocity and the depth of the cavity; *ii*) the aspect ratio of cavity. These flow structures will change due to an inserted square or circular cylinder within the cavity. Understanding the influence of the inserted cylinder on the flow structures is of fundamental importance in engineering applications like heat exchangers and electronic coolers.

Oztop et al.[1] and Khanafer and Aithal[2] numerically investigated mixed convection and heat transfer in a steady lid-driven square cavity containing a circular cylinder by using a finite volume method and a finite element formulation, respectively. They found that for the forced convection flow an increase in the cylinder's size leads to the primary vortex breaking up into two vortices, and various locations of the cylinder result in the movement of the vortex cores. Moreover, a large gap between isotherms which affects the heat transfer was present due to strong flow circulation (i.e. the vortex formed in the cavity).

The lid-driven cavity flow with multiple embedded obstacles were also investigated and applied to test the accuracy of some numerical methods, e.g. the immersed boundary method by Su et al.[3] and the immersed interface method by Ito et al.[4] due to its low computational cost. Moreover, in the absence of the lid motion, many investigations were conducted for the natural convection (dominated by the temperature gradient) in a cavity containing obstacles with different locations and geometries (see e.g. [5, 6, 7]).

A thorough investigation of the vortex structures for steady lid-driven cavity flow containing a cylinder has, to the limit of author's knowledge, not been conducted in detail for purely forced convection flow. The present paper provides a detailed investigation of vortex structures in a two-dimensional steady lid-driven cavity of depth to width ratio 1/3 with an embedded circular cylinder with a dimensionless radius of 0.2 (i.e. the ratio between the cylinder radius and the cavity depth). Three different Reynolds numbers (100, 500 and 1000) based on the depth of the cavity and three different locations of the cylinder have been considered. A direct forcing immersed boundary method combined with the finite difference method proposed by e.g. Fadlun et al. [8] and improved by Peller et al.[9] is applied due to its easy meshing and efficiency of solving the Navier-Stokes equations in Cartesian grids.

2. Problem formation and numerical method

The flow with a constant density ρ and kinematic viscosity ν is governed by the two-dimensional Navier-Stokes equations described as follows:

$$\frac{\partial u_i}{\partial x_i} = 0 \tag{1}$$

$$\frac{\partial u_i}{\partial t} + \frac{\partial u_i u_j}{\partial x_j} = -\frac{\partial p}{\partial x_i} + \frac{1}{Re}\frac{\partial^2 u_i}{\partial x_j \partial x_j} \tag{2}$$

where the Einstein notation using repeated indices is applied. Here $u_i = (u,v)$ and $x_i = (x, y)$ for $i = 1$ and 2, are the velocity and Cartesian coordinates, respectively, whilst t, p and $Re = UH/\nu$ denote the dimensionless time, dimensionless pressure and Reynolds number, respectively, where H is the depth of cavity and U is the lid motion velocity. The time scale, pressure scale and length scale is H/U, $0.5\rho U^2$ and H, respectively.

Equations (1) and (2) are solved using a semi-implicit second order scheme in conjunction with a projection method, where Adams-Bashforth and Crank-Nicolson are applied to the convective terms and diffusive terms, respectively. A second-order central difference scheme is applied on a staggered mesh arrangement.

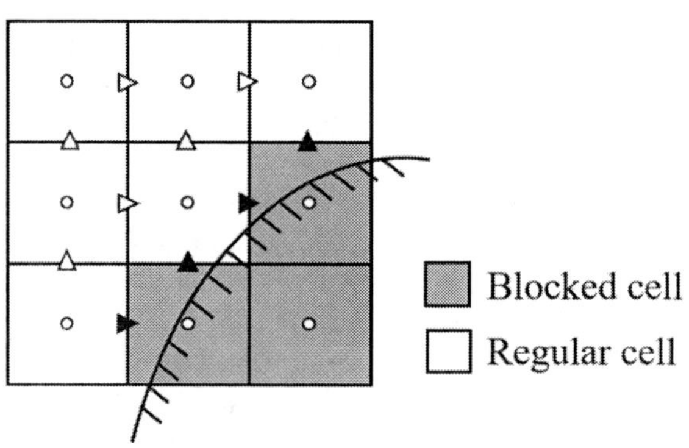

Figure 1. Definition of the blocked cell (grey square), the regular cell (white square), the active (white arrows) and inactive (black arrows) velocity points.

The immersed boundary method is based on a direct forcing approach combined with finite difference method. The relationship between the grid and the immersed boundary is shown in figure 1. A cell with the pressure point inside the solid is defined as a blocked cell while the rest cells inside the fluid are denoted by the regular cells. The velocity components at the connection face between the blocked and regular cells are set as inactive velocity points (▶ and

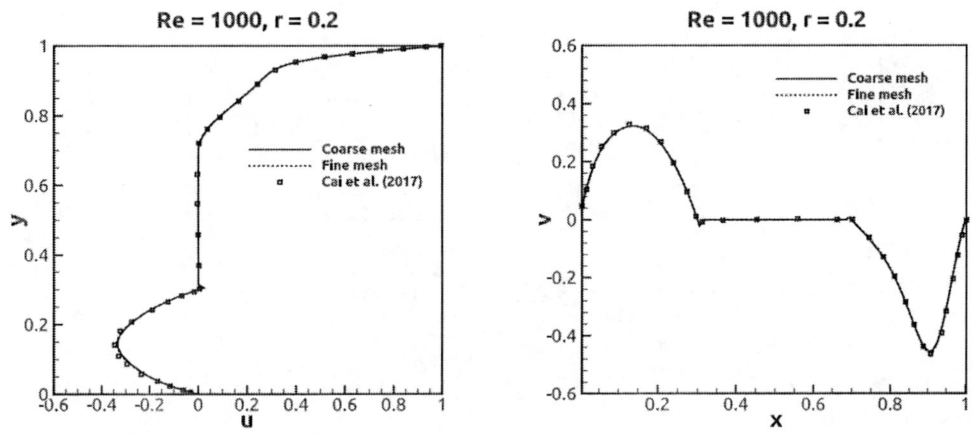

Figure 2. Comparison of the velocity profiles of the steady lid-driven cavity flow containing a centered cylinder of $r = 0.2$ for $Re = 1000$ obtained by the present method and by Cai et al.[11]: left image, distribution of velocity component u along $x = 0.5$; right image, distribution of velocity component v along $y = 0.5$.

▲), and those connecting two regular cells are considered as active velocity points (▷ and △). The inactive velocity points is updated by interpolation of its adjacent active velocity points. A linear interpolation applied in directions is given as follows

$$u_i = \frac{x_i - x_\Gamma}{x_{i+1} - x_\Gamma} u_{i+1} + \frac{x_{i+1} - x_i}{x_{i+1} - x_\Gamma} u_\Gamma, \quad i = 1, 2 \tag{3}$$

where u_Γ and x_Γ denote the velocity components and the position of the solid surface, respectively.

A weighting factor λ is introduced in Eq.(2) for the inactive velocity components which are interpolated from two directions[9, 10].

$$u_i = \lambda_x u_i^x + \lambda_y u_i^y \tag{4}$$

where the superscripts x and y denote the interpolation in x and y-directions, respectively. Moreover, a Neumann condition is applied for the pressure correction at the inactive velocity points.

3. Results and discussion

3.1. Verification against previous numerical results

A numerical simulation for flow in a steady lid-driven square cavity containing a centered circular cylinder has been conducted for $Re = 1000$. The velocity of the lid is given by $u = 1$, and no-slip conditions are imposed on the side and bottom walls as well as the cylinder. The dimensionless cylinder radius r equals to 0.2 ($= r'/H$). Two grid resolutions of 100×100 and 200×200 uniform cells are considered for the grid convergence test. Figure 2 shows the velocity profiles for u (left) and v (right) along $x = 0.5$ and $y = 0.5$, respectively. Previous numerical results obtained by Cai et al.[11] using an immersed boundary method which takes the body force as a Lagrange multiplier were included for comparison. A good agreement is obtained using both the coarse and fine meshes. It appears that the coarse mesh is sufficient to obtain the grid independent results.

IOP Conf. Series: Materials Science and Engineering **700** (2019) 012008 doi:10.1088/1757-899X/700/1/012008

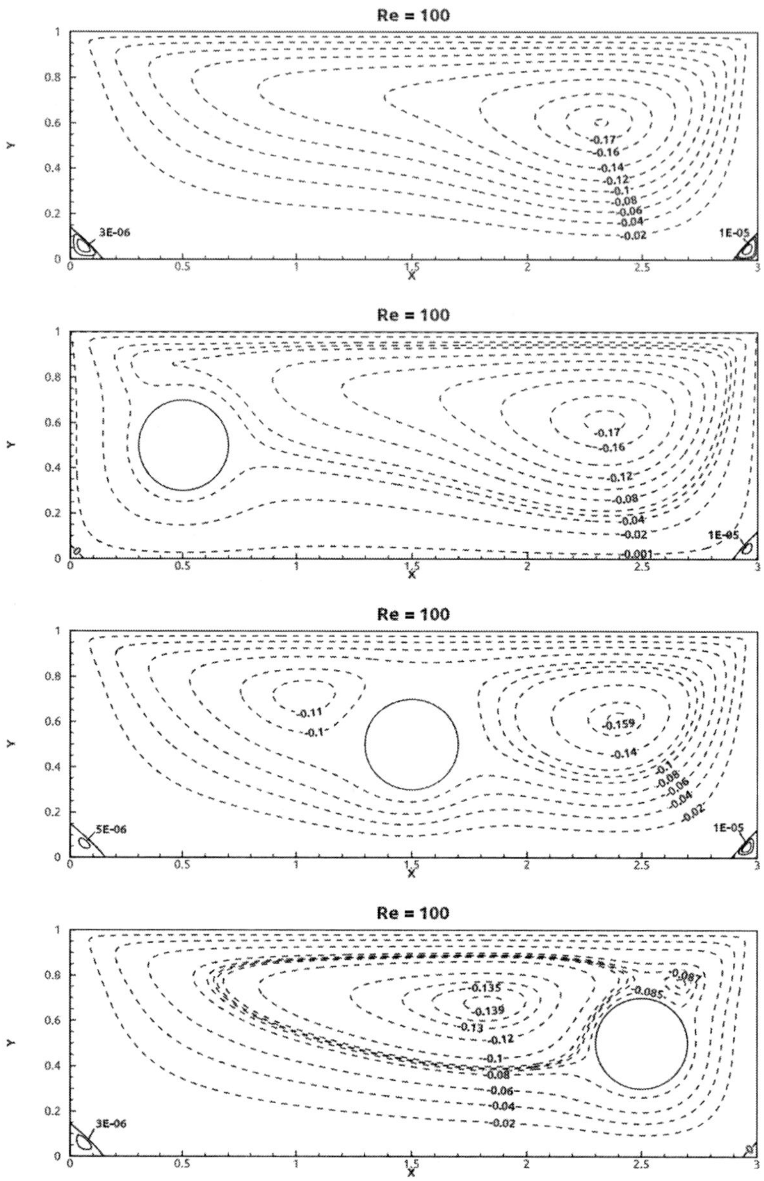

Figure 3. Streamfunctions within a steady lid-driven cavity of AR = 1/3 with and without a circular cylinder for Re = 100. The cylinder is positioned at (x, y) = (0.5, 0.5), (1.5, 0.5) and (2.5, 0.5).

3.2. Flow in a steady lid-driven rectangular cavity containing a circular cylinder

Predictions for flow in a steady lid-driven rectangular cavity containing a circular cylinder are presented for $Re = 100$, 500 and 1000 in this section. The depth to width ratios (AR) of the cavity is 1/3, and three different locations of the cylinder with $r = 0.2$ are considered to investigate their effects on flow structures. Moreover, the steady lid-driven cavity flows at respective Re are predicted for comparison purpose.

Figure 3 shows the streamfunctions within a steady lid-driven cavity of $AR = 1/3$ containing

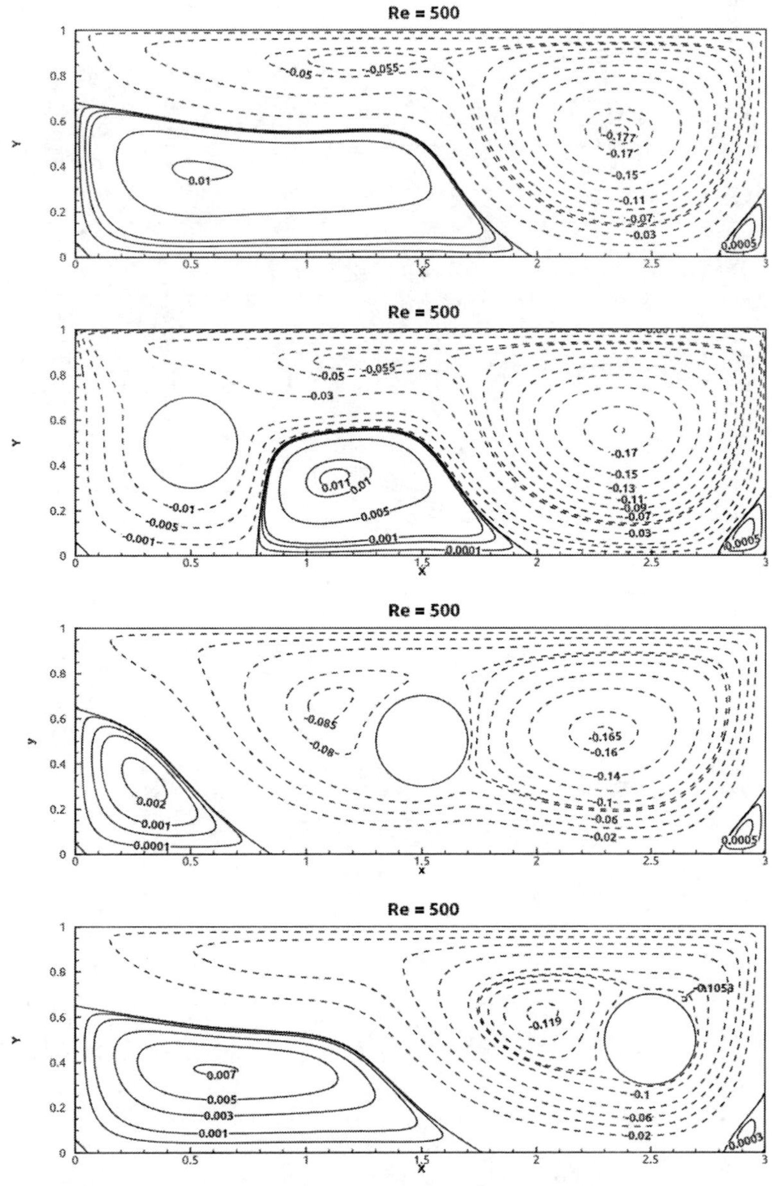

Figure 4. Streamfunctions within a steady lid-driven cavity of AR = 1/3 with and without a circular cylinder for Re = 500. The cylinder is positioned at (x, y) = (0.5, 0.5), (1.5, 0.5) and (2.5, 0.5).

a circular cylinder for $Re = 100$. The cylinder is located at (x, y) = (0.5, 0.5), (1.5, 0.5) and (2.5, 0.5). The cavity without the cylinder contains a clockwise primary vortex and two small anti-clockwise bottom corner vortices. These vortex structures qualitatively remain for the cylinder positioned at (0.5, 0.5) but with a slightly weaker primary vortex core and the nearly vanishing bottom left corner vortex. The bottom right corner vortex are weakly affected since it is far away from the cylinder. As the cylinder is positioned at (1.5, 0.5), the primary vortex present in the cavity without a cylinder appears to break into two clockwise vortices due to the presence of

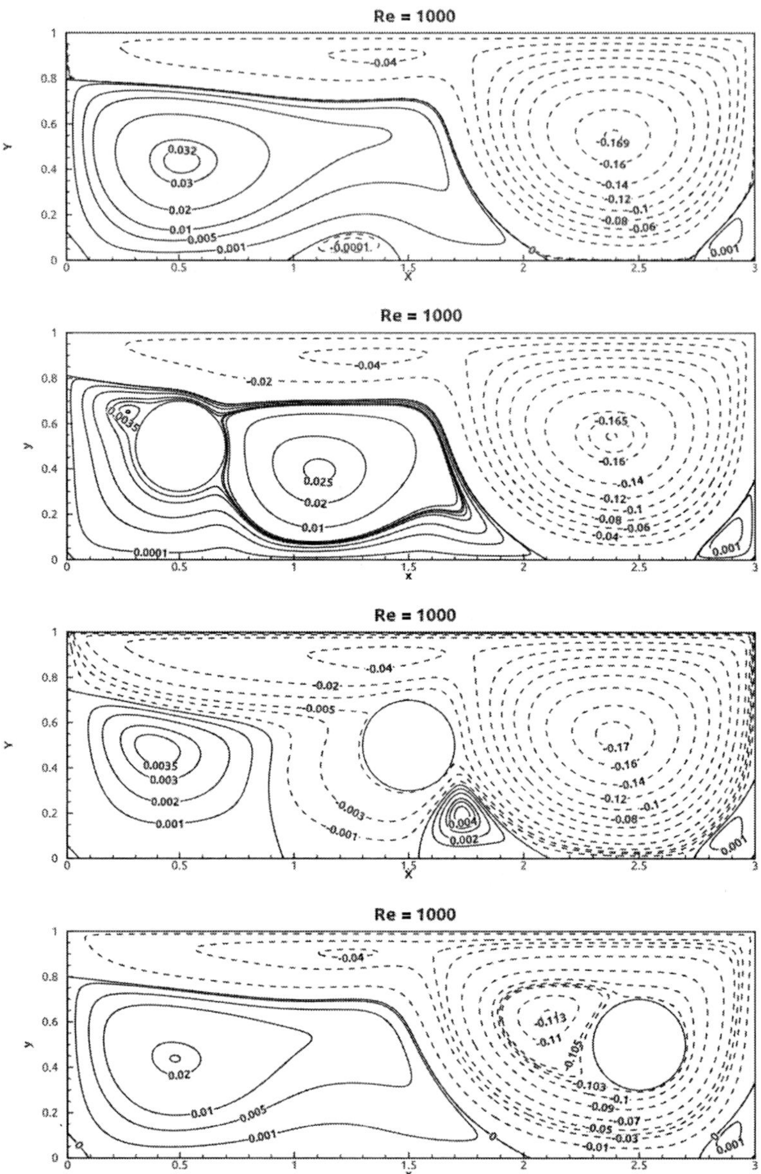

Figure 5. Streamfunctions within a steady lid-driven cavity of AR = 1/3 with and without a circular cylinder for Re = 1000. The cylinder is positioned at $(x, y) = (0.5, 0.5)$, $(1.5, 0.5)$ and $(2.5, 0.5)$.

the cylinder at the right side of which the flow driven by the lid rolls down. Similar qualitatively vortex structures were previously investigated by Young et al.[12] for creeping flow within a steady lid-driven rectangular cavity with a centered cylinder. They found two equal clockwise vortices attached to the upper left and right side of the cylinder, respectively. In present cases, the non-linear effect from the convective terms attributes to the asymmetry of the attached vortices where the left one is smaller and weaker than the right one. As the cylinder is shifted to (2.5, 0.5), a decrease in space between the cylinder and the right wall suppresses the growth of the right vortex, causing the left vortex becoming larger than the right one. This behavior qualitatively remains the same for $Re = 500$ as the cylinder moves rightwards from (1.5, 0.5) to (2.5, 0.5) shown in figure 4.

Figure 4 shows the streamfunctions for the same geometry shown in figure 3 but for $Re = 500$. In the cavity without the cylinder, a new clockwise vortex is formed beneath the lid, and flow separation occurs at $x = 2$, forming a larger bottom corner than that for $Re = 100$ (figure 3). As the cylinder is inserted at (0.5, 0.5), flow separation remains at $x = 2$ while the presence of the cylinder leads to the flow rolling down at the lower right side of the cylinder and reattaching the bottom wall, thus forming a large bottom vortex. For the cylinder located at (1.5, 0.5) and (2.5, 0.5), the clockwise vortex beneath the lid disappears, forming qualitatively similar vortex structures present for $Re = 100$ (figure 3) but with stronger vortex cores. Moreover, the attached vortices for $Re = 100$ (figure 3) is larger than those for 500 (figure 4) with the cylinder located at (2.5, 0.5). It appears that an increase in Re leads to a higher flow velocity around the cylinder, thus impeding the growth of the attached vortices.

Figure 5 shows the streamfunctions for the same geometry shown in figure 3 but for $Re = 1000$. In the cavity without the cylinder, an increase in Re forms a larger bottom left corner vortex and a small clockwise bottom vortex. This small bottom vortex disappears as the cylinder is inserted in the cavity. For the cylinder positioned at (0.5, 0.5), the cylinder is surrounded by a stronger anti-clockwise flow circulation due to a larger Re, such that the flow rolls down at the right side of the cylinder and up at the upper left side of the cylinder, forming two anti-clockwise attached vortices. The right one is much larger than the left one. As the cylinder is positioned at (1.5, 0.5), the bottom left corner vortex present in the cavity without the cylinder appears to break up into a large bottom left corner vortex and a bottom vortex. As the cylinder moves to (2.5, 0.5), the primary vortex in the cavity without the cylinder disappears (due to a higher Re) while the flow rolls up at the upper left side of the cylinder, forming the left attached vortex.

4. Summary and conclusion

The present paper provides a detailed investigation for flow in a steady lid-driven cavity of depth to width aspect ratio 1/3 containing a circular cylinder with a radius equals to 0.2. Three different Reynolds numbers and three different locations are considered. For a lower Reynolds number ($Re = 100$), the cylinder causes the flow rolling up and down, forming two clockwise vortices. As the cylinder moves rightwards, the left one grows while the right one decays. For a moderate Reynolds number ($Re = 500$), a large bottom vortex is formed for cylinder located near the left wall, and it disappears as the cylinder moves rightwards. For the largest Reynolds number ($Re = 1000$), two attached anti-clockwise vortices are formed around the cylinder located near the left wall. As the cylinder moves gradually closer the left wall, a large bottom vortex is formed and disappears together with the clockwise vortex to the right side of the cylinder.

Acknowledgement

We gratefully acknowledge the support for this research from the Department of Marine Technology, Norwegian University of Science and Technology and the China Scholarship Council (Grant no. 201506680058).

References

[1] Oztop H F, Zhao Z and Yu B 2009 *International Journal of Heat and Fluid Flow* **30** 886–901
[2] Khanafer K and Aithal S 2013 *International Journal of Heat and Mass Transfer* **66** 200–209
[3] Su S W, Lai M C and Lin C A 2007 *Computers & Fluids* **36** 313–324
[4] Ito K, Lai M C and Li Z 2009 *Journal of Computational Physics* **228** 2616–2628
[5] Ghaddar N K 1992 *International journal of Heat and Mass Transfer* **35** 2327–2334
[6] De A K and Dalal A 2006 *International Journal of Heat and Mass Transfer* **49** 4608–4623
[7] Lee J, Ha M and Yoon H 2010 *International Journal of Heat and Mass Transfer* **53** 5905–5919

[8] Fadlun E, Verzicco R, Orlandi P and Mohd-Yusof J 2000 *Journal of Computational Physics* **161** 35–60
[9] Peller N, Duc A L, Tremblay F and Manhart M 2006 *International Journal for Numerical Methods in Fluids* **52** 1175–1193
[10] Berthelsen P A and Faltinsen O M 2008 *Journal of Computational Physics* **227** 4354–4397
[11] Cai S G, Ouahsine A, Favier J and Hoarau Y 2017 *International Journal for Numerical Methods in Fluids* **85** 288–323
[12] Young D, Chen C, Fan C, Murugesan K and Tsai C 2005 *European Journal of Mechanics-B/Fluids* **24** 703–716

IOP Conf. Series: Materials Science and Engineering **700** (2019) 012009 doi:10.1088/1757-899X/700/1/012009

Numerical study on aerodynamic drag reduction and energy harvest for electric vehicle: a concept to extend driving range

A W Huluka[1] and C H Kim[2,*]

[1] Department of Automotive Engineering, Graduate School, Seoul National University of Science & Technology, 232 Gongneung-ro, Nowon-gu, Seoul 01811, Korea
[2] Department of Mechanical & Automotive Engineering, Seoul National University of Science & Technology, 232 Gongneung-ro, Nowon-gu, Seoul 01811, Korea

[*] Corresponding author: profchkim@seoultech.ac.kr

Abstract. Energy efficiency of road vehicles is highly attributed to its aerodynamic performance. Over the last decades, significant reduction of aerodynamic drag has been made. But, beyond the current achievement, further improvement become a challenge for car industries and professionals working in the area. This paper focuses on new approach to further find aerodynamic performance improvement particularly for electric vehicles. Therefore, a preliminary numerical study on ducting and concept of energy harvesting from aerodynamic resistance is introduced as an alternative to improve energy efficiency. Ducted and slightly modified Ahmed model is used to study aerodynamic characteristics of ducted models and how ducting would contribute to the reduction of energy consumption due to air resistance. Three-dimensional, incompressible, and steady flow governing equations are solved by CFD code (PHOENICS ver.2018) with extended turbulent model proposed by Chen-Kim (1987). From the study, it is found that the total drag reduction of over 15% on the ducted models compared to the base model. The kinetic energy preserved due to ducting is also significant. Harvesting this energy and utilizing it to extend driving range is believed to be a potential area to be dealt in detail. Therefore, the idea of duct application and energy harvesting introduced in this paper, either to reduce or use aerodynamic resistance is expected to have potential contribution to the development of future energy efficient road vehicles.

1. Introduction

Energy efficient vehicles are required to meet future emission and fuel consumption requirements. Environmental issues and public health security concerns have forced car industries and the public to look for alternatives to petroleum based means of transportation. In recent decades, the research and development activities related to transportation have emphasized the development of high-efficiency, clean and safe transportation. Electric vehicles, hybrid vehicles and fuel cell vehicles have been typically proposed to replace conventional vehicles in the future [1], [2].

Currently attention given to electric vehicle development is motivating and it is deemed to be a future car as it is energy efficient, no direct carbon emission, and low noise pollution. On the other hand, the limited driving range per charge is the challenge of current electric vehicles due to low energy density of the battery. To improve the driving range, one alternative is reducing the energy consumption of the moving vehicle. Therefore, identifying the energy consumption attributes of the moving vehicle and conducting a research on how to reduce consumption of energy would enhance the driving range of the vehicle. The largest portion of power driving the vehicle is consumed to overcome the air resistance [2], [3], [4], [5]. Aerodynamic drag force typically accounts for 65% of the total resistance for medium sized car at 100 km/h [2], [4], [5], [6]. Hence, reducing drag force contributes significantly to the energy efficiency of battery electric vehicles [7], [8].

The idea under consideration is harvesting energy from the air resistance without affecting the base drag. As the vehicle is moving in a sea of air, the kinetic energy of the fluid relative to the vehicle is significant particularly at higher vehicle speed. As a first step for a feasibility study, ducting and its aerodynamic effect on a moving vehicle is discussed using simple body. A numerical simulation was made as a preliminary study to either reduce or use aerodynamic resistance of road vehicle

2. Analytical theory

Theoretical background on road-load power of a moving vehicle, ducting and energy harvesting concept from the aerodynamic resistance is described here in short.

2.1. Road Load Power and Aerodynamic Drag

The total power (P_{batt}) required to overcome resistance forces for an electric vehicle of mass (M_v) cruising at a speed (V), can be expressed as:

$$P_{batt} = \frac{1}{\eta_t \eta_m}\left(M_v g f_r V \cos\theta + \frac{1}{2}\rho_a C_D A_f V^3 + M_v g V \sin\theta + M_v \delta V \frac{dV}{dt} \right) \tag{1}$$

Where η_t - efficiency of transmission, η_m - traction motor efficiency, g - acceleration due to gravity, δ - mass factor, θ - slope or grade, f_r - rolling resistance coefficient, C_D - drag coefficient, ρ_a - air density. The terms in the bracket represent rolling resistance, aerodynamic resistance, grade resistance and inertial resistance to the vehicle motion from left to right respectively. In a high speed driving situation, the power required to overcome the aerodynamic resistance is extremely significant consuming energy to keep its speed. This drag force (F_D) is a function of the frontal projected area (A_f) of the vehicle, drag coefficient (C_D) and the speed of the vehicle (V).

$$F_D = fun\,(C_D, A_f, V, \rho_a) \tag{2}$$

Therefore, the drag force is expressed as

$$F_D = C_D \frac{1}{2}\rho_a V^2 A_f \tag{3}$$

Two main components of the drag force are pressure and skin friction force. Aerodynamic resistance of a ground vehicle is dominated by pressure drag whilst skin-friction drag is dominant on aircraft and marine ships [9], [10], [11].

2.2. Ducting

The frontal area of internal combustion engine powered vehicles is opened with the meshed grill to induct incoming air into the engine room for engine cooling. To the contrary the frontal panel of most electric vehicles is closed by design and made up of single panel Figure 1. But the closed frontal panel of electric vehicle would have considerable effect on the increment of aerodynamic resistance of the vehicle.

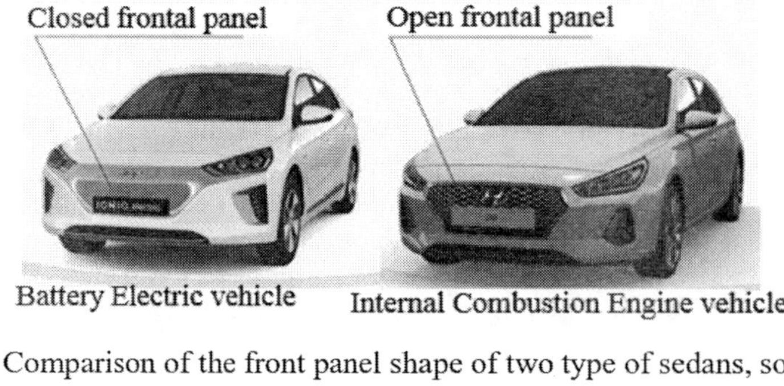

Figure 1. Comparison of the front panel shape of two type of sedans, source [12].

Ducting is not common in ground vehicle as a drag reduction technology. Here in this paper a deliberate duct for either drag reduction or energy harvesting system is made for feasibility study. This kind of approach is believed to be applicable in electric vehicles as mostly their front panel is closed by design.

2.3. . Aerodynamic Energy Harvest

When a vehicle moves at higher speed, serious air resistance which is called as drag exerts on the moving vehicle due to the relative speed. If this opposing aerodynamic force acting on a vehicle body due to the momentum loss is reduced or captured by any means, two very positive effects will be expected; fuel savings and emission reduction. If the momentum energy can be recovered, energy efficiency of road vehicle will increase. The energy harvesting concept can be modelled with a radial type power turbine as shown in Figure 2.

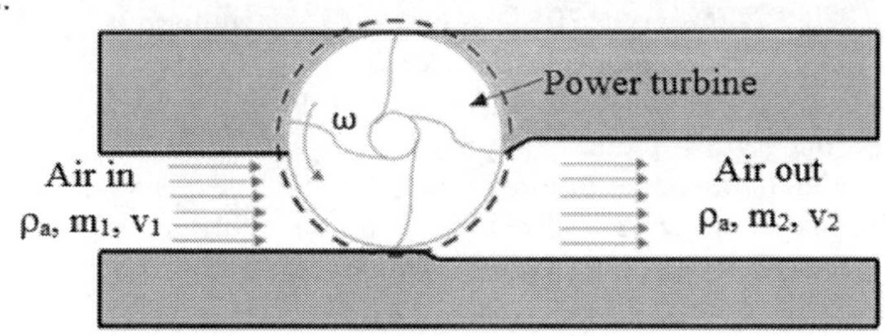

Figure 2. Schematic drawing of the wind energy harvesting system in an air duct.

General conservation of momentum for this model can be stated as the linear momentum equation and it is equal to external forces (F_{ext}) acting on the control volume (CV) indicated by red broken circle in Figure 2.

$$\sum \vec{F}_{ext,CV} = \frac{\partial}{\partial t} \int_{CV(t)} \vec{V} \rho_a d\forall + \int_{CS(t)} \vec{V} \rho_a (\vec{V} . \hat{n}) \mathbf{d}A \tag{4}$$

The force exerts on the power turbine due to the change in momentum at inlet and outlet will be used to calculate the torque (T) of the turbine ($dT = F_t \times dr$). The torque multiplied by the angular velocity of the turbine will give the power harvested from the aerodynamic resistance.

$$Power = \frac{2\pi NT}{60} \tag{5}$$

Where N is the rotational speed and T is the torque of the power turbine.

3. Description of the test model

Ahmed model [10] is used for this fundamental study with some modifications. A (4:1) scale model is used with the dimension 4.176 m x 1.556 m x 1.152 m (L x W x H), Figure 3. Five different models are defined with different size of duct as specified in Table 1.

Table 1. Specifications of the model body.

Model No.	Specification of the model	Area ratio[a]
M0	Base model with no duct	0%
M1	Model with 1072 mm x 160 mm (W`xH`) duct	9.6%
M2	Model with 1072 mm x 240 mm duct	14.4%
M3	Model with 1072 mm x 320 mm duct	19.1%
M4	Model with 1072 mm x 400 mm duct	23.9%

[a]Area Ratio(R_A); $R_A = {A_d}/{A_f}$ where A_f is the frontal projected area of the base model and A_d is the cross-sectional area of the duct (A_d =H`xW`, H` is duct height and W` is duct width)

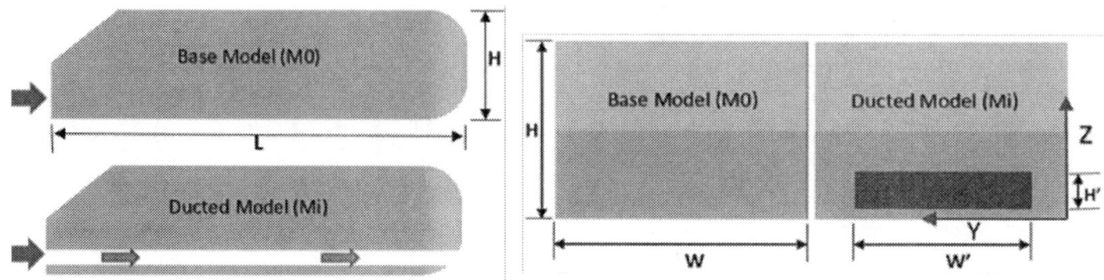

Figure 3. Side and font view of the base model (M0) and ducted model (Mi) for i = 1, 2, 3, 4

4. Numerical scheme and its conditions

In this study, Finite Volume Method (FVM) was used to perform the numerical calculation. The general-purpose CFD code, PHOENICS (ver. 2018) was used for a numerical simulation of the turbulent incompressible flow field. 3-dimensional Naiver-Stokes equations were solved with KECHEN turbulent model (Chen-Kim κ-ε model) which is a modified standard κ-ε turbulent model. The turbulent no slip condition near solid boundary was modelled by log law.

Chen-Kim κ-ε model believed to improve the dynamic response of the Epsilon equation by introducing an additional time scale and source term. In addition, several of the standard-model coefficients are adjusted so that the model maintains good agreement with experimental data on classical turbulent shear layers [13].

4.1. Numerical domain and its numerical grid

The numerical domain size is defined as shown in Figure 4. Especially 8L is given at the rear side of the model to have a stable convergence on the simulation result. The initial and boundary conditions are defined as given in the Table 2.

Table 2. Initial and boundary conditions.

Boundary surface	Boundary & initial conditions
Inlet	Velocity inlet (60, 80, 100, 120 km/h)
Outlet	Pressure outlet
Sides and top	No slip wall condition
Ground	moving ground type
Model exterior surface	No slip wall
Duct surface	No slip wall
Flow domain	Quasi-3D flow, Turbulent and incompressible flow

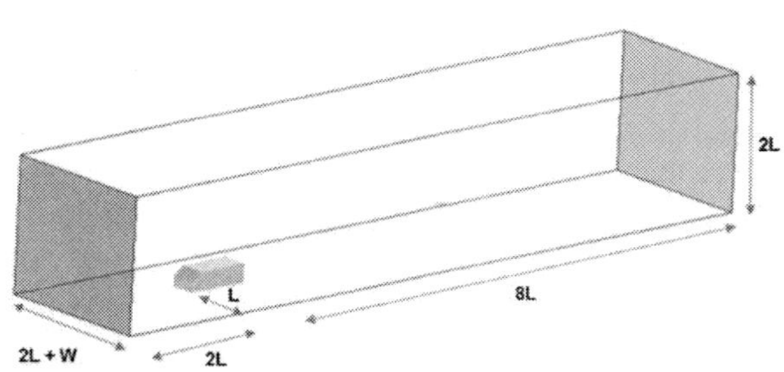

Figure 4. Numerical domain size for calculation.

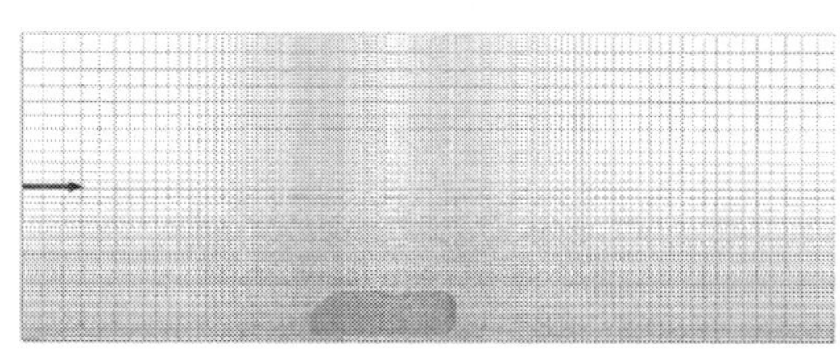

Figure 5. Optimized grid in the numerical domain (162x94x60)

With the aforementioned set up, numerical simulation has been done to study the duct effect on drag force, rear flow and related flow property variations. The calculation was conducted at optimized grid distribution Figure 5 which is obtained by thorough grid dependence evaluation, with different iterations and very nicely converged with an error cut off less than 0.01%. For more precise profile and boundary layer study very fine and highly dense grid distribution is also used in the study. In PHOENICS, the mesh generation does not need third party package when using Cartesian and polar coordinates and six sided rectangular brick element is used in this study.

4.2. Governing equations:

For steady incompressible turbulent flows, the flow field is characterized by the conservation laws (since the flow is assumed steady the time derivative component removed). Hence the governing equation for turbulent flow field is written:

1) Continuity equation:

$$\frac{\partial U_i}{\partial x_i}+\frac{\partial U_j}{\partial y_i}+\frac{\partial U_k}{\partial z_i}=0 \tag{6}$$

2) Momentum equation:

$$\frac{\partial}{\partial x_j}(U_iU_j)=-\frac{1}{\rho}\frac{\partial P}{\partial x_i}+\frac{\partial}{\partial x_j}\left[\nu\left(\frac{\partial U_i}{\partial x_j}+\frac{\partial U_j}{\partial x_i}\right)-\overline{u_iu_j}\right]-g_i \tag{7}$$

where, $-\overline{u_iu_j}=\nu_t\left(\frac{\partial U_i}{\partial x_j}+\frac{\partial U_j}{\partial x_i}\right)-\frac{2}{3}k\delta_{ij}$, $\delta_{ij}=1 \;\; if\; i=j \;\; and \;\; \delta_{ij}=0 \;\; if\; i\neq j$

3) Extended κ-ε closure turbulent model (KECHEN)

- Turbulent kinetic energy equation

$$\frac{\partial}{\partial x_i}(U_jk)=\frac{\partial}{\partial x_i}\left[\left(\nu+\frac{\nu_t}{\sigma_k}\right)\frac{\partial k}{\partial x_j}\right]+G-\varepsilon \tag{8}$$

$G=-\overline{u_iu_j}\frac{\partial U_i}{\partial x_j}$ (production rate due to deformation) [14],[15], $\nu_t=C_\mu\frac{k^2}{\varepsilon}$, $C_\mu=0.09$

- Energy dissipation equation $\frac{\partial}{\partial x_i}(U_j\varepsilon)=\frac{\partial}{\partial x_i}\left[\left(\nu+\frac{\nu_t}{\sigma_\varepsilon}\right)\frac{\partial \varepsilon}{\partial x_j}\right]+\frac{\varepsilon}{k}(C_{\varepsilon1}G)-C_{\varepsilon2}\frac{\varepsilon^2}{k})+C_{\varepsilon3}f_1\frac{P_k^2}{K}$ (9)

Where: ($C_\mu=0.09$, $C_{\varepsilon1}=1.15$, $C_{\varepsilon2}=1.9$, $C_{\varepsilon3}=0.25$ $\sigma_k=0.75$, $\sigma_\varepsilon=1.15$), k is turbulent kinetic energy, ε is energy dissipation rate, V_l & V_t laminar and turbulent kinematic viscosities, P_k is the volumetric production rate of kinetic energy by shear forces, σ_k, σ_ε Pradtl number connected to diffusivity of K and ε to eddy viscosity, f_1 is the Lam-Bremhorst (1981) damping function which tends to unity at high turbulence Reynolds numbers [16].

5. Result and discussion

The force distribution over the model is evaluated under 'integrated forces' indicating pressure force and friction forces. Based on this integrated force results, evaluation of the duct flow effect on each and every side is made possible. Special attention is given to the drag force being contributed from pressure force and skin friction force. The frontal pressure force and rear side drag force due to strong and counter rotating trailing vortex.

5.1. Aerodynamic flow Characteristics of the model

Aerodynamic characteristics of the flow around the model is demonstrated using the static pressure contour plot as depicted in Figure 6. The pressure distribution at front and rear of the un ducted and ducted model are different due to the duct effect. As shown in Figure 6, M1 has lower front stagnation pressure than the fully blocked body and higher rear pressure due to the flow momentum through the duct is injected to the rear wake resulting in reduced rear pressure drag. As the duct size increases, the decreasing trend of front stanation pressure continues. Similarly the rear pressure drag decreases with duct size.

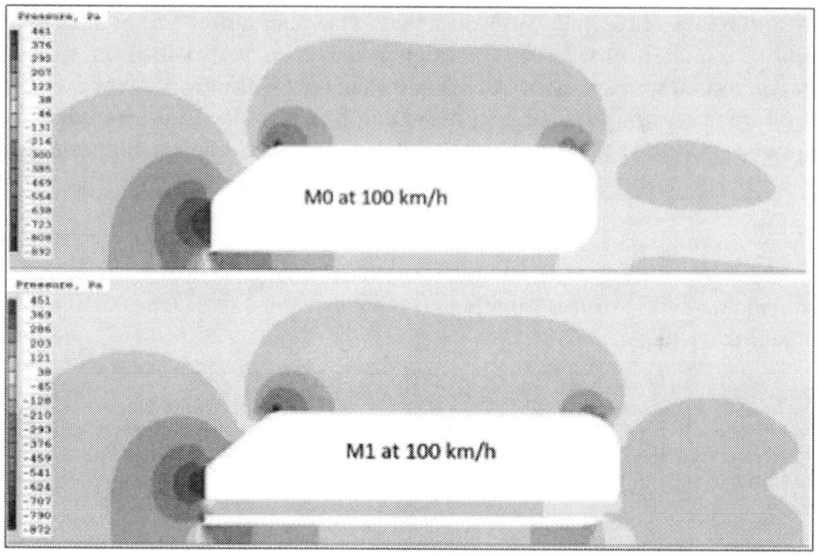

Figure 6. Pressure distribution on M0 (top) and M1(bottom) at 100 km/h.

5.2. Pressure Drag Force

Effect of the opened front panel on pressure drag force is evaluated in detail. The total pressure drag curve versus models with different duct size is shown in Figure 7. All pressure drags are nondimensionalized by the base model (M0) maximum pressure drag which is at 120 km/h. The highest pressure force is observed on the model M0. As the duct size increases the pressure force magnitude is decreasing. quantitatively, for model M1 at a speed of 120 km/h about 20.5% reduction of total pressure drag in reference to the base model is observed. This shows that the duct has a very significant effect on a pressure drag reduction of the bluff body under consideration.

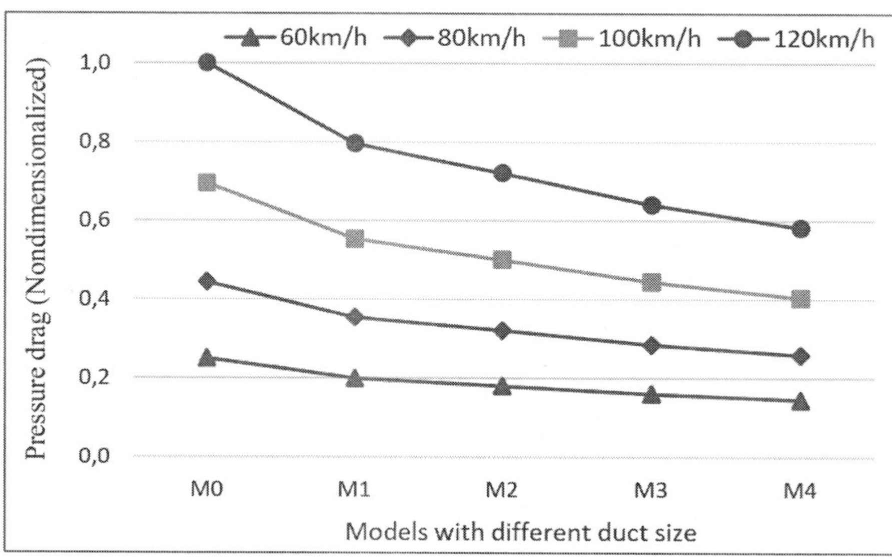

Figure 7. Variation of the pressure drag with the duct size at different speed.

5.3. Skin friction drag

The friction drag generated due to the skin friction shear stress at the boundary layer near the surfaces of the model solid surfaces is relatively small. During this study, contrary to the pressure drag force, the friction drag force of the ducted models revealed an increasing trend as duct size increases. Figure 8 demonstrates the friction drag trend in each model at four different test speeds. The model with no duct (base model) experienced relatively the smallest friction drag. This could be due to increased surface area and flow boundary layer created in each inner faces of the duct. Considering one of the ducted models with smallest duct size (M1) at speed of 100km/h a net total drag reduction of about 16.3% is observed quantitatively. The pressure drag force is reduced by 19.5% while the skin friction drag is increased by 3.2% of the total drag. As the duct size increases, the friction force increases as shown in Figure 8. This can be considered as side effect of ducting as far as drag reduction is concerned while the pressure force which is the dominant drag force in ground vehicle exhibit a decreasing trend. Since the contribution of skin friction to the total drag is very low, its increase with duct size did not affect much the decreasing trend of the total drag force.

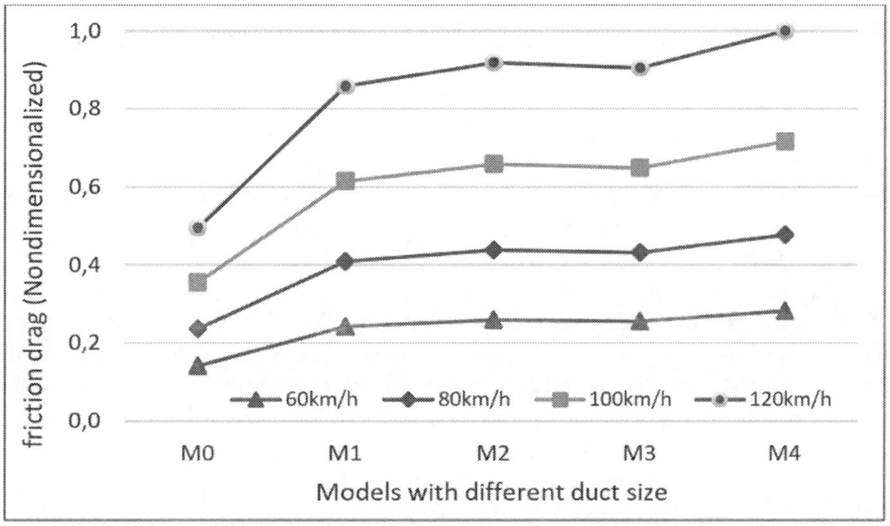

Figure 8. Variation of fiction drag with the duct size at different speed.

5.4. Drag coefficient

The coefficient of drag of the models also showed a reduction trend even though coefficient of drag is directly proportional to drag force and inversely proportional to reference area (A_f), reduction of both area and drag force by ducting does not assure a reduction in C_D. The relative change rate of area and change rate of drag force is a determining factor for reduction of the C_D of ducted model. As shown in Figure 9, about 15.3% of C_D was reduced in M4 compared to the base model (M0) while M1 C_D is reduced by about 7.8% in same speed range. The C_D on the Figure 9 is nondimensionalized by the C_{Do} of the base model. It should be noticed that an opened front panel on a moving body should have energy saving effect than the closed one and will contribute to driving range extension.

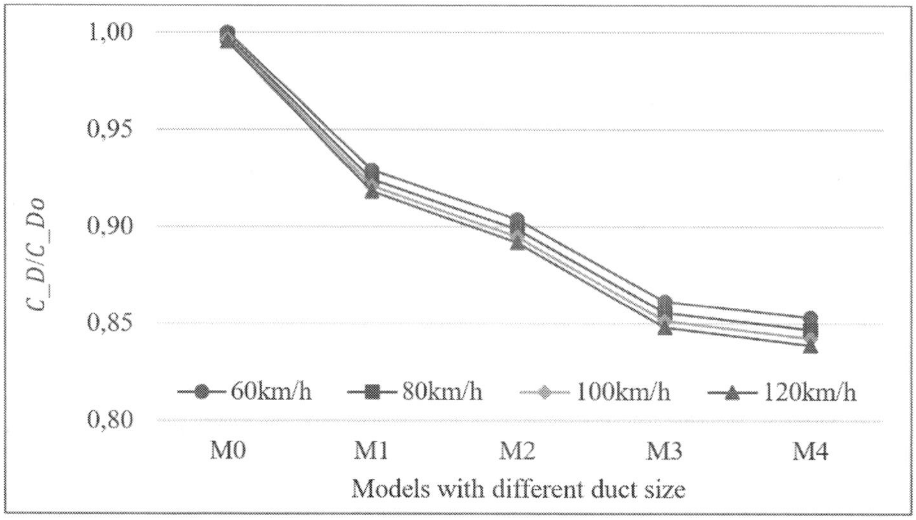

Figure 9. Variation of drag coefficient with the duct size at different speed.

5.5. Aerodynamic energy

The velocity profile just before the entry, at the entry, inside duct, at the exit and just after the exit of the duct gives some insight about the flow dynamics and energy level at the duct. To have a big picture of the flow domain along duct, a wider range velocity profile is plotted as dipicted on Figure 10. The profile is captured with in a 10 m long distance in the doman passing through the middle of the duct length to see the pattern of the flow speed.

At about 3 m infront of the duct, the velocity assumes a free stream velocity (27.78 m/s) which is a test speed of 100 km/h. This velocity declines as it approaches the object (model) front side and sharply drop to about 65% of the free stream velocity near the entry edge of the duct and then raise back to about 86% at the entry. Just after entry it slightly decelerates until it attain smooth flow inside the duct where it gradually accelerate. At the exit, it drastically decelerate again as it is exsposed to a chaotic tubulent flow at the rear of the object until it mix with and maintain the outside rear flow structure about 3 m behind the rear end. Without a duct, the kinetic energy of the whole bulk flow infront of the model is stagnated at the front exerting a resistance to the objec. But now it can be seen in the profile plot that some amount of the flow kinetics is maintaned and channeled to the rear through duct reducing the

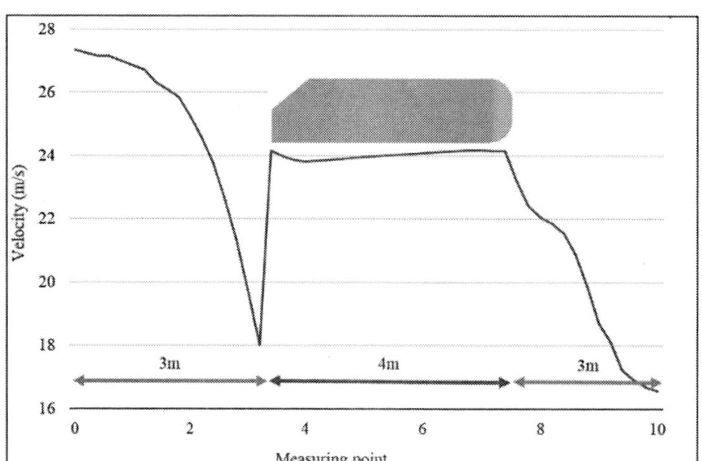

Figure 10. Longitudinal velocity profile through model duct center of M1.

aerodunamic resistance.This kinetic energy of the air can be harveted using the principle discussed in Subsection 2.3 of this article. Cosidering the typical senario power to be harvested is roughly estimated for model M1.

The calculation is performed with the following assumption. Air density is uniform throughout the duct. As the flow is guided to the turbine by duct, the turbine converts major portion of the kinetic energy in the air to mechanical energy turning the generator. Because it is a guided flow not in an open air turbine. Turbine is directly connected to the alternator mechanically. Therefore, the potential aerodynamic kinetic energy by the air flowing through a duct is estimated by:

$$\text{Kinetic Energy}= \frac{1}{2}\rho_a A_d \bar{U}^3 \tag{10}$$

The average air velocity $\bar{U}$ inside the duct:
$$U = \frac{\iint V_n dA_d}{A_d} \tag{11}$$

Available power harvested:
$$P_{available} = \frac{1}{2}\rho_a A_d \bar{U}^3 \eta_T \eta_g \tag{12}$$

Where η_T & η_g are efficiency of turbine and generator respectively. Table 3 shows the electric energy harvested by the kinetic energy of air passing through the duct of the model (M4) under reasonable assumptions.

Table 3. Rough estimation of the electric power harvested in M4.

Vehicle speed (km/h)	Averaged air velocity in the duct (m/s)	Turbine efficiency (%)	Generator efficiency (%)	Usable power (watt)	Contribution to total power in %
60	13.60	85%	88%	192	7.32%
80	18.18	85%	88%	458	7.39%
100	22.79	85%	88%	903	7.47%
120	27.39	85%	88%	1,567	7.51%

Power turbine and generator efficiency (assumed)

As shown in Table 3, the harvested electric energy (calculated) proportionally increases with the model speed. As the speed reaches 120km/h, the power harvested is up to 1.567kW. The ratio of usable harvested power to the total power consumed to overcome the aerodynamic resistance of the base model is over 7% as indicated in the last column of table 3. For the estimation of total driving range extension with this air duct installed at the front side of the model vehicle, both the aerodynamic drag reduction and the electric energy harvested should be considered. Experimentally supported detail analysis of the aerodynamic energy harvesting is a continuation of this work as a future plan.

6. Conclusion and future work

In this numerical study, a new approach to further reduce or use aerodynamic resistance is introduced. It is found that the air flow duct installed on the model vehicle show the positive effects and significant decrease on the total drag force. The pressure drag shows a decreasing trend as duct cross-sectional area increases although the skin friction drag increases with the duct size. It was also confirmed that skin friction drag increase due to ducting is insignificant as pressure drag force accounts over 90% of the total drag. Overall a drag reduction of over 15% is observed in a model with a minimum duct size among the test models in this study.

Velocity profile through the duct centre indicated the ducted flow characteristics particularly the fluid acceleration and deceleration before, at the duct entry, exit, and after exit respectively. The level of kinetic energy preserved through ducting is also clearly indicated on the velocity profile Figure 10. The electric energy harvested from this kinetic energy of the flow in the duct would definitely contribute to driving range extension of electric vehicle if effective and efficient system is designed to capture the energy. Effect of ducting on the stability of the model, feasibility of energy harvesting through ducting and front & rear flow control for better duct flow acceleration is part of the future work.

IOP Conf. Series: Materials Science and Engineering **700** (2019) 012009 doi:10.1088/1757-899X/700/1/012009

References

[1] Mehrdad Ehsani A E and Gao Y 2010 *Modern electric, hybrid electric and fuel cell vehicles*, 2nd Ed. CRC Press, 2010.
[2] Kim C 2011 A streamlined design of a high-speed coach for fuel savings and reduction of carbon dioxide, *Int. J. Automot. Eng.*, **2**, 101–7.
[3] Nalanagula S and Varadharajan G T 2016 Aerodynamics drag reductions methodology for the commercial vehicles using computational fluid dynamics, *SAE Tech. Pap. Ser.*, **1**, 2016.
[4] McCallen R C, Salari K and Ortega J M 2004 DOE's effort to reduce truck aerodynamic drag - joint experiments and computations lead to smart design," in *34th AIAA Fluid Dynamics Conference and Exhibit*, June 28 - July 1, Portland, USA.
[5] McCallen R, Couch, *et al* 2010 Progress in reducing aerodynamic drag for higher efficiency of heavy duty trucks (Class 7-8)," *SAE Tech. Pap. Ser.*, **1**(724), 2010.
[6] Altinisik A, Yemenici O and Umur H 2015 Aerodynamic analysis of a passenger car at yaw angle and two-vehicle platoon, *J. Fluids Eng.*, **137**(12) 121107.
[7] Hucho W and Sovran G 1993 Aerodynamics of road vehicles, *Annu. Rev. Fluid Mech.* **25**(1) 485 - 537.
[8] Johnson T V 2015 Review of vehicular emissions trends, *SAE Int. J. Engines*, **8**(3), 1152 - 67.
[9] Wood R M 2004 Impact of advanced aerodynamic technology on transportation energy consumption, *SAE World Congr.*, 113(21) 854 - 74.
[10] Ahmed S R, Ramm G and Faltin G 1984 Some Salient Features of the Time -Averaged Ground, *SAE Trans.* **93** 473–503.
[11] Kalyan D K and Paul A R 2013 Computational study of flow around a simplified 2D Ahmed Body," *Int. J. Eng. Sci. Innov. Technol.*, **2**(3) 148–153.
[12] www.hyundai.com, Posted on Oct. 21, 2017.
[13] Chen Y-S and Kim S-W 1987 *Computation of turbulent flows using an extended K-Epsilon turbulence closure interim report*, NASA-CR-179204, Alabama, USA.
[14] Tennekes H 1972 *A First course in turbulence*. Massachusetts: The MIT Press, 1972.
[15] Hinze J O 1975 *Turbulence*. McGraw-Hill, New York, 2nd Ed., 1975.
[16] Banks J 2007 Turbulence Modeling in PHOENICS [Online]. Available at: http://www.cham.co.uk/phoenics/d_polis/d_enc/turmod/enc_t342.htm.

Numerical simulation of flow around two 5:1 rectangular cylinders at a high Reynolds Number

G Yin[1,*], T Monaci[2] and M C Ong[1]

[1]Department of Mechanical and Structural Engineering and Materials Science, University of Stavanger, Stavanger, Norway
[2]CESI Engineering School, France

[*] Corresponding author: guang.yin@uis.no

Abstract. The unsteady flows around two stationary rectangular cylinders with a chord-to-thickness ratio of $B/D = 5$ at $Re_D = 2 \times 10^5$ (based on the free stream velocity and the width length D), are investigated using two-dimensional Unsteady Reynolds-Averaged Navier-Stokes equations with a standard $k - \omega$ SST turbulence model. The computational model is validated by carrying out simulation of the flow around a single rectangle and comparing the hydrodynamic quantities (time-averaged drag coefficient, root-mean-square of fluctuating lift coefficient) with the previous published data. The aim of the present study is to study the influence of the relative positions between the two rectangular cylinders on their hydrodynamic quantities and the surrounding flow fields.

1. Introduction

The developments of the offshore industries have increased the demanding for the investigation of the flow-structure interactions around marine structures. Flows around sharp-edged square and rectangular cylindrical structures are commonly seen in offshore and subsea engineering, such as submerged floating tunnels. These structures are often subject to high Reynolds number flows where strong currents are present. Since it is difficult to obtain the hydrodynamic quantities on the structures under the high Reynolds number flow conditions in experiments, Computational Fluid Dynamics (CFD) can serve as an efficient tool for the engineering design.

There have been extensively studies on the flows around rectangular cylinders with various aspect ratios. Early works includes the experimental measurements done by Okajima [12] on rectangular cylinders with aspect ratios from 1 to 4 in wind and water tunnels at Reynolds numbers ranging from 70 to 2×10^4. Nakamura & Yoshimura [11] used experiments to investigate the Strouhal number of the turbulent flow past rectangular cylinders with aspect ratio 0.2 to 5 at the Reynolds numbers (defined based on the height of the rectangle and the velocity of the inlet flow) of 5.5×10^3 to 1.4×10^5. Apart from experiments, numerical simulations have also been adopted to study the flow around rectangular cylinders with different aspect ratios. A benchmark case for the Benchmark on the Aerodynamics of a Rectangular 5:1 Cylinder (BARC) (Bartoli et al. [2]) was simulated using Large Eddy Simulation (LES) or Direct Numerical Simulation (DNS). The resulting force coefficients were in good agreement with the experimental data (Mannini et al. [8]). Mannini et al. [9] carried out two-dimensional (2D) Reynolds-averaged Navier-Stokes (RANS) simulations on the flow around a rectangular cylinder with an aspect ratio of 5:1 and compared the performance of different turbulence models.

Bruno et al. [3] performed 3D LES at $Re = 2 \times 10^5$ to study the fluctuating pressure field and it was found that the essential physics is 2D although the 3D flow features are observed. To the author's knowledge, most of the previous studies focused on the flows around a single rectangular cylinder. Only a few numerical studies have been conducted on the flow around two rectangular cylinders, which are frequently seen in many industries such as decks of two-tiered-bridges, offshore structures, heat exchangers, etc. The flow around two structures displays more complexity due to the interactions between shear layers, shedding vortices behind the structures. Therefore, it is important to study the effects of the arrangement of the two cylinders on the vortex dynamics behind the structures. Aboueian & Sohankar [1] carried out DNS of the flow over two square cylinders in staggered arrangement at a low Reynolds number of 150 for different gap spacing between cylinders to understand the flow structures. Lee et al. [7] used DNS to study the flow around a pair rectangular cylinders at a low Reynolds number of 100.

The goal of the present study is to extend the former studies to investigate the flow around two rectangular cylinders with different horizontal and vertical distances at a high Reynolds number, which has not been studied before. 2D RANS simulations are carried out. The $k-\omega$ SST model is used to resolve the turbulence stress. The paper is organized as follows. Section 2 gives a brief introduction to the mathematical formulation in the present study. The convergence study and the validation study are performed on the flow around a single rectangular cylinder to determine the grid and time-step resolution in Section 3. In Section 4, results of the flow around two rectangular cylinders are given. Finally, the conclusion is made in Section 5.

2. Mathematical Formulation

2.1. Flow Model

The unsteady Reynolds-averaged Navier-Stokes (URANS) equations for the conservation of mass and momentum are given by:

$$\frac{\partial u_i}{\partial x_i} = 0 \tag{1}$$

$$\frac{\partial u_i}{\partial t} + u_j \frac{\partial u_i}{\partial x_j} = -\frac{1}{\rho}\frac{\partial P}{\partial x_i} + \nu \frac{\partial^2 u_i}{\partial x_j^2} - \frac{\partial \overline{u_i' u_j'}}{\partial x_j} \tag{2}$$

where $i, j = 1,2$. Here x_1 and x_2 denote the streamwise (x) and cross-stream (y) directions respectively; u_1 and u_2 are the corresponding mean velocity components of the two direction (u, v); $\overline{u_i' u_j'}$ is the Reynolds stress component, where denotes the fluctuating part of the velocity; P is the pressure; ρ is the density of the fluid and t is the time.

The $k-\omega$ SST turbulence model (Menter [10]) is employed in the present study, which combines the $k-\omega$ and the $k-\varepsilon$ models. The original $k-\omega$ model of Wilcox [15] is used in the near-wall region and the standard $k-\varepsilon$ model of Jones & Launder [6] is adopted in the outer wake region and in the free shear layers. The detailed information of the $k-\omega$ SST turbulence model can be found in (Menter [10]; Tian et al. [14]; Jeong et al. [5]).

2.2. Numerical setup

The open source Computational Fluid Dynamics (CFD) code OpenFOAM (OpenFOAM v5) is used to carry out the simulations in the present study. The PISO (Pressure Implicit with Splitting of Operators) scheme (pisoFoam) is used. The spatial schemes for interpolation, gradient, Laplacian and divergence are linear, Gauss linear, Gauss linear corrected and Gauss linear schemes, respectively. All these schemes are in second order. The second order Crank-Nicholson scheme is employed for the time integration.

The computational domain is shown in Figure 1 for the single and the two rectangular cylinders. For the single rectangle cases, the origin of the coordinates is located at the centre of the rectangle. The distances from the rectangle centre to the inlet and outlet boundaries are $10D$ and $30D$, respectively. The top and bottom boundaries are located at a distance of $10D$ to the centre of the rectangle. For the cases with two rectangles, the origin of the coordinates is set to be located at the midpoint of the connection line of the centres of the two rectangles. The distance from the origin to the inlet is $18D$ and the distance from the origin to the outlet is $54D$. The top boundary is located at a distance of $10D$ from the centre of Rectangle 1 (marked in Figure 1) and bottom boundary is located at a distance of $10D$ to the centre of Rectangle 2 (marked in Figure 2).

At the inlet boundary, the flow is set to be a uniform flow with $u_1 = U_\infty$, $u_2 = 0$. At the outlet boundary, the velocities are set as zero normal gradient and the pressure is set as zero. The top and bottom boundaries are set as symmetry planes. On the surface of the rectangles, no-slip boundary condition is applied for the velocities, i.e. $u_i = 0(i = 1,2)$and zero normal gradient condition is applied for the pressure. The standard near-wall condition for k and ω is applied on the surfaces of the rectangles.

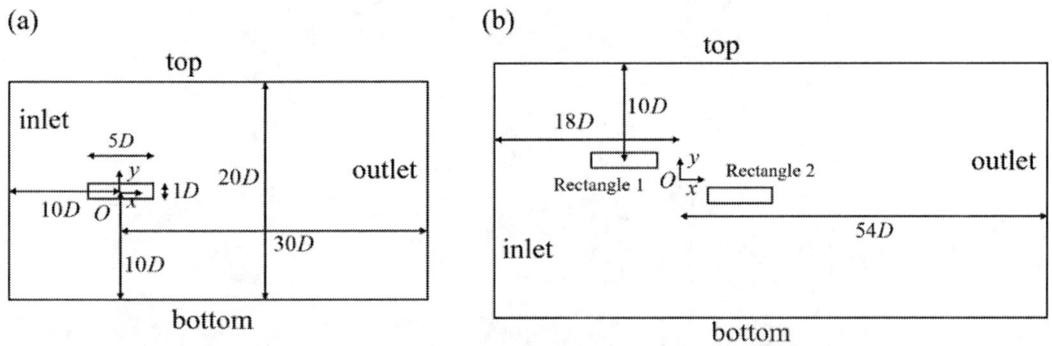

Figure 1. The computational domains for (a) the single rectangle case and (b) the two rectangles case.

3. Convergence studies and validation

The convergence studies are carried out for the single rectangle cases to determine the grid and time-step resolutions. The time-averaged drag coefficient ($\overline{C_d}$) and the root mean square of the lift coefficient (C_{lrms}) are considered, where the definitions of the two quantities are given as

$$\overline{C_d} = \frac{\overline{F_x}}{\frac{1}{2}\rho U_\infty^2 D} \tag{3}$$

$$C_l = \frac{F_y}{\frac{1}{2}\rho U_\infty^2 D} \tag{4}$$

$$C_{lrms} = \sqrt{\frac{1}{N}\sum_{i=1}^{N}\left(C_{l,i} - \overline{C_l}\right)^2} \tag{5}$$

where $\overline{F_x}$ is the time-averaged streamwise force acting on the rectangles per unit length and F_y is the cross-stream force acting on the rectangles per unit length, which are both computed from the integration of the pressure and viscous shear stress over the surface of the rectangles. N is the number of the sampling data.

Table 1 shows the results of the grid and time-step convergence studies of the single rectangle case. The differences of the numbers of the cells between cases are over 30%. The Courant numbers of all the simulations are kept below 0.4. It is shown that with the increasing number of cells, the maximum relative difference of $\overline{C_d}$ is 0.38% (denoted as $\%\overline{C_d}$ in Table 1) and the maximum relative difference of C_{lrms} is around 5% (denoted as $\%C_{lrms}$ in Table 1). With the smaller time-step, the relative differences of $\overline{C_d}$ and C_{lrms} are also negligible. As a result, it can be concluded the mesh and time-step of Case 3 can provide sufficient grid resolution. It is noteworthy that since the wall function is applied on the surface of the rectangle, a criterion of $y^+ > 30$ (defined as $y^+ = u_\tau \Delta y/\nu$ where Δy is the distance of the centre of the first cells away from the wall and u_τ is the friction velocity on the surface of the rectangle) is achieved in every case. An example of the grids of Case 3 is shown in Figure 2.

Table 1. Results of grid and time-step convergence studies.

Case	Elements	y^+	$\Delta t U_\infty/D$	$\overline{C_d}$	$\%\overline{C_d}$	C_{lrms}	$\%C_{lrms}$
1	125825	31.66	0.001	1.1013	-	0.2643	-
2	190427	33.43	0.001	1.1041	0.2542	0.2737	3.5566
3	257187	32.92	0.001	1.1105	0.5797	0.2882	5.2978
4	257187	32.92	0.0008	1.1111	0.0540	0.2906	0.8259

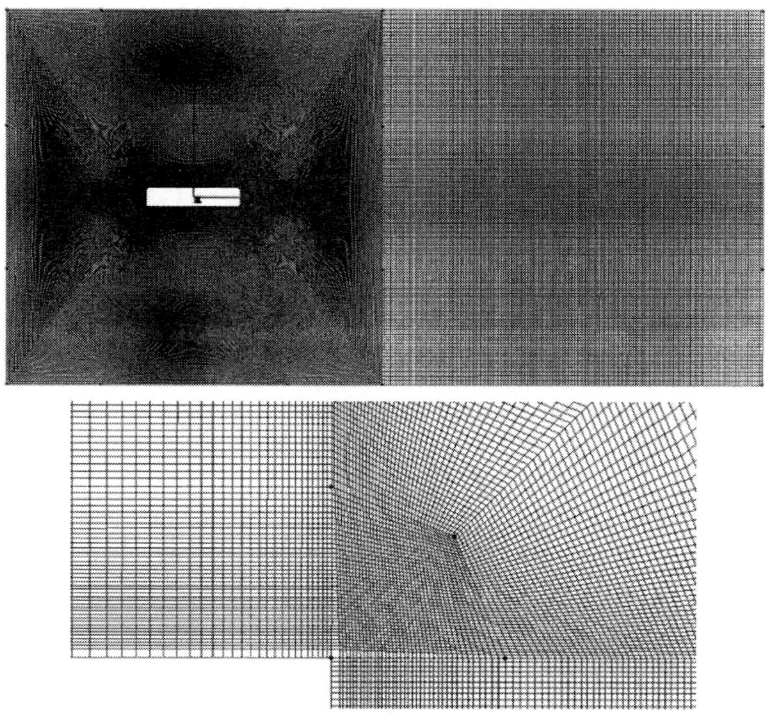

Figure 2. The computational mesh of case.

Validation studies are conducted by comparing the present results with the previous published data as shown in Table 2. It can be seen that $\overline{C_d}$ of the present study agrees well with the literatures while C_{lrms}, which is sensitive, is overpredicted compared with the numerical results at the same Reynolds number.

Table 2. Hydrodynamic quantities for rectangular cylinder with aspect ratio of 5:1.

Source / Author	Method	Re_D	$\overline{C_d}$	C_{lrms}
Present study	2D RANS $k-\omega$ SST	2×10^5	1.1105	0.2882
Dahl [4]	2D RANS $k-\varepsilon$	5×10^5	1.1008	0.2760
Bruno et al. [3]	3D LES	4×10^4	1.0500	0.7200
Mannini et al. [9]	3D DES	2.64×10^4	1.0290	0.4210
Schewe [13]	Exp.	2×10^4	1.0300	0.4000

4. Results and discussion

The grid resolution and time-step of the convergent results of Case 3 are used for the two rectangles cases. In the present study, four different combinations of (H, V) (H is the horizontal distance between the back face of Rectangle 1 and the front face of Rectangle 2 and V is the vertical distance between the bottom face of Rectangle 1 and the top face of Rectangle 2) are considered.

Time-averaged drag coefficients as well as the root-mean squares of the lift coefficients for the two rectangles (1 for Rectangle 1 and 2 for Rectangle 2) are displayed in Table 3. Figure 3 (a), (b) and Figure 4 (a), (b) show the time histories of the lift and drag coefficients of the two rectangles with different (H, V) in Table 3.

Table 3. Hydrodynamic quantities for two rectangular cylinders.

Cases	Elements	y^+(Rectangle 1/Rectangle 2)	$\overline{C_{d1}}$	C_{lrms1}	$\overline{C_{d2}}$	C_{lrms2}
$(H, V) = (0,1)D$	110000	35.12 / 35.18	1.391	0.032	1.388	0.030
$(H, V) = (0,3)D$	110000	31.90 / 32.73	1.244	0.121	1.248	0.129
$(H, V) = (3,1)D$	198000	30.38 / 32.59	1.042	0.111	1.196	0.529
$(H, V) = (6,1)D$	198000	30.50 / 31.61	1.047	0.100	1.142	0.389

For $(H, V) = (0,1)D$ and $(H, V) = (0,3)D$ with no horizontal distance between the two rectangles, $\overline{C_{d1}}$ and $\overline{C_{d2}}$ are almost the same. The sign of the lift coefficients of the two rectangles is opposite. This is due to the fact that the amplitudes of the low pressure regions above Rectangle 1 and below Rectangle 2 caused by the vortex are much stronger than the low pressure region between the two rectangles caused by high-velocity flow, as indicated in Figure 3 (c) and Figure 3 (d). With the increasing vertical distance between the two rectangles from $H = 1D$ to $H = 3D$, the root-mean-squares of the fluctuating lift coefficients of the two rectangles increase because of the increasing amplitude of the low pressure region between the two rectangles, as shown in Figure 3 (d). From the streamlines in Figure 3 (g) and Figure 3 (h), as the vertical distance increases, the small vortices below Rectangle 1 and above Rectangle 2 grow larger.

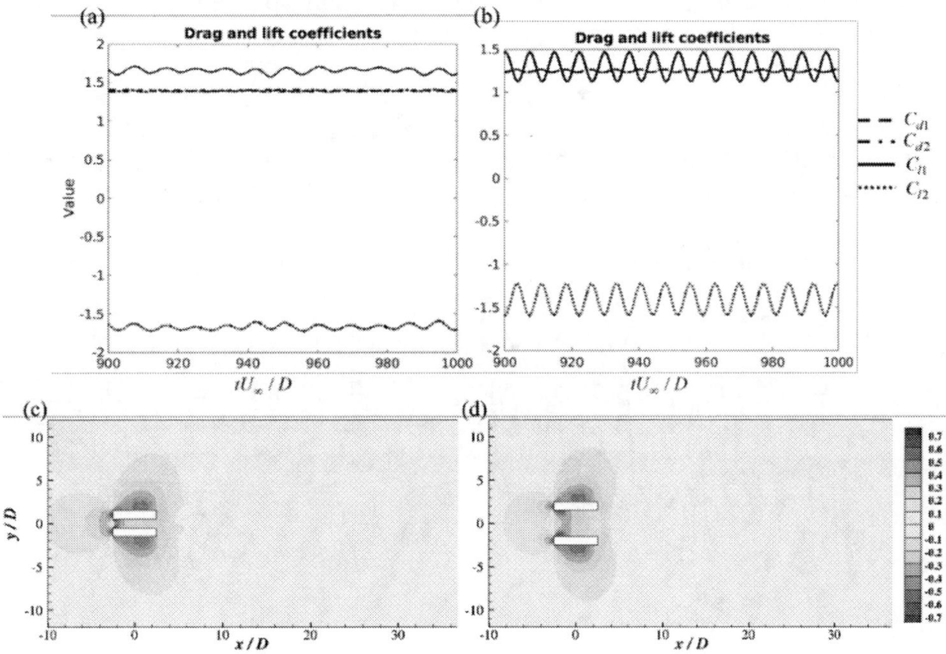

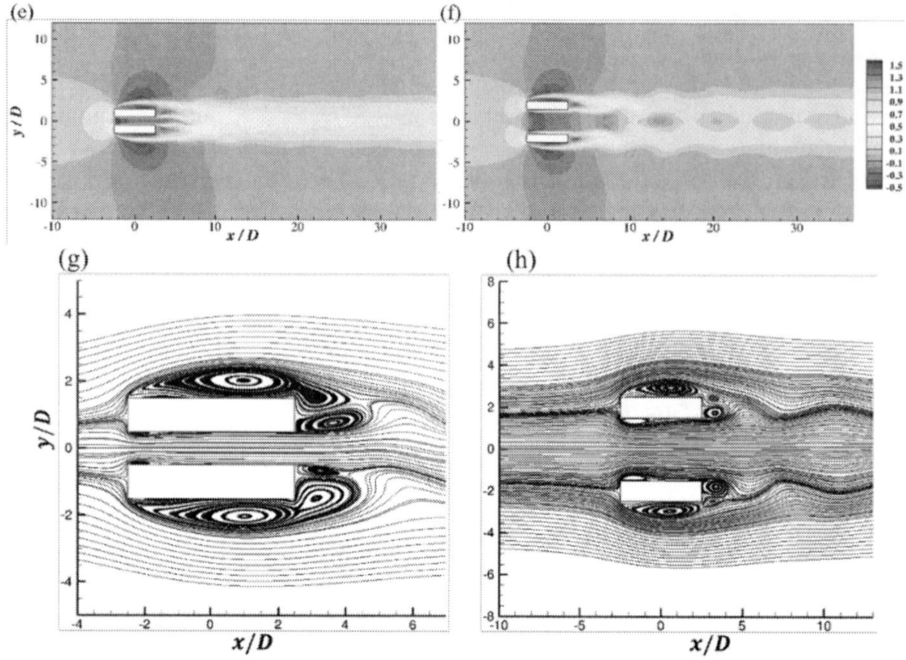

Figure 3. The Results for $(H, V) = (0,1)D$ (a), (c), (e), (g) and $(H, V) = (0,3)D$ (b), (d), (f), (h). (a),(b): time histories of C_{d1}, C_{l1} and C_{d2}, C_{l2}; (c),(d): instantaneous pressure $(2p/(\rho U_\infty^2))$ contours; (e),(f): instantaneous streamwise velocity (u/U_∞) contours; (g),(h): streamlines around the rectangles.

With a horizontal distance between the two rectangles, Rectangle 2 is influenced by the wake region behind Rectangle 1. As indicated in Figure 4 (a) and Figure 4 (b), the amplitudes of C_{lrms2} and $\overline{C_{d2}}$ are larger than those of Rectangle 1. With $(H, V) = (3,1)D$, the interaction between the surrounding flows of the two rectangles is strong, leading to an additional low-frequency oscillation modulating the high-frequency oscillation caused by the vortex shedding. This is more apparent for the lift coefficients than the drag coefficients as indicated in Figure 4(a). With the increasing H, this low-frequency oscillation disappears. This modulated periodic flow pattern has also been reported in Aboueian & Sohankar [1], where the flows around two staggered squares were studied. Instantaneous streamlines in Figure 4 (g) and Figure 4 (h) indicate that the vortices above Rectangle 2 is much larger than those in the cases with no horizontal distance where the vortices above Rectangle 2 are suppressed by the gap region between the two rectangles. This may be the contributor to the large amplitude of the lift coefficient of Rectangle 2.

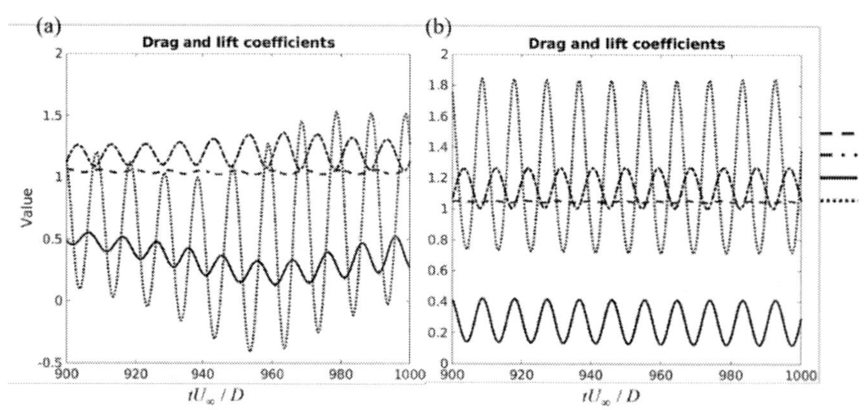

IOP Conf. Series: Materials Science and Engineering **700** (2019) 012010 doi:10.1088/1757-899X/700/1/012010

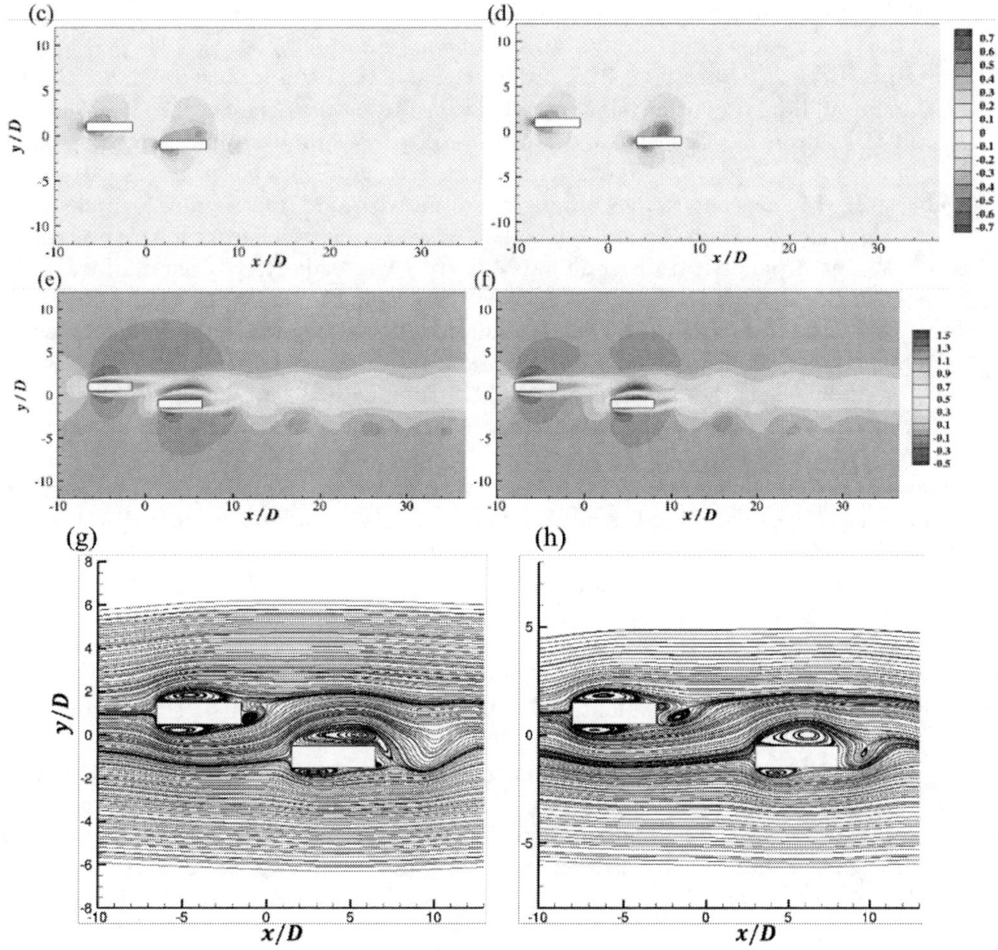

Figure 4. The Results for $(H,V) = (3,1)D$ (a), (c), (e), (g) and $(H,V) = (6,1)D$ (b), (d), (f), (h). (a),(b): time histories of C_{d1}, C_{l1} and C_{d2}, C_{l2}; (c),(d): instantaneous pressure $(2p/(\rho U_\infty^2))$ contours; (e),(f): instantaneous streamwise velocity (u/U_∞) contours; (g),(h): streamlines around the rectangles.

5. Conclusion

The flow around two rectangular cylinders with an aspect ratio of 5:1 has been simulated at a high Reynolds number of 2×10^5 using the 2D URANS combined with a standard $k - \omega$ SST and the wall function. The mesh and time-step convergence studies as well as the validation study have been carried out for the single rectangle case to prove the reliability of the present numerical model. The model is then applied to the two rectangle cases with different horizontal and vertical distances between them. It is found that with no horizontal distance, the drag coefficients of the two rectangles are almost the same. The lift coefficients of the two rectangles are opposite in sign with the same amplitude. As their vertical distance increases, the amplitude of the fluctuating lift coefficients of the two rectangles increases. With a horizontal distance, the lift coefficients display a modulated periodic flow pattern, which disappears with the increasing horizontal distance.

Acknowledgment

This study was supported with computational resources provided by the Norwegian Metacenter for Computational Science (NOTUR), under Project No: NN9372K.

IOP Conf. Series: Materials Science and Engineering **700** (2019) 012010 doi:10.1088/1757-899X/700/1/012010

References

[1] Aboueian J, and Sohankar A 2017. Identification of flow regimes around two staggered square cylinders by a numerical study, *Theor. Comput. Fluid Dyn.* **31**(3), 295-315.
[2] Bartoli G, Bruno L, Buresti G, Ricciardelli F, Salvetti M V and Zasso A 2008 BARC overview document. http://www.aniv-iawe.org/barc for further details (accessed 27/08/2019).
[3] Bruno L, Fransos D, Coste N and Bosco A 2010 3D flow around a rectangular cylinder: a computational study, *J. Wind Eng. Ind. Aerodyn.* **98**(6-7), 263-76.
[4] Dahl S M 2014 Unsteady RANS Simulation of Flow around Rectangular Cylinders with different Aspect Ratios at High Reynolds Number, Master's thesis, Institutt for marin teknikk, NTNU.
[5] Jeong W, Liu S, Jakobsen, J B and Ong M C 2019 Unsteady RANS Simulations of Flow around a Twin-Box Bridge Girder Cross Section, *Energies,* **12**(14) 2670.
[6] Jones W P and Launder B 1973 The calculation of low-Reynolds-number phenomena with a two-equation model of turbulence. *Int. J. Heat Mass Transfer*, **16**(6), 1119-30.
[7] Lee S J, Mun G S, Park Y G and Ha M Y 2019 A numerical study on fluid flow around two side-by-side rectangular cylinders with different arrangements, *J. Mech. Sci. Technol.* **33**(7), 3289-300.
[8] Mannini C, Šoda A and Schewe G 2010 Unsteady RANS modelling of flow past a rectangular cylinder: Investigation of Reynolds number effects, *Comput. Fluids,* **39**(9), 1609-24.
[9] Mannini C, Šoda A and Schewe G. 2011. Numerical investigation on the three-dimensional unsteady flow past a 5: 1 rectangular cylinder, *J. Wind Eng. Ind. Aerodyn.*, **99**(4), 469-82.
[10] Menter F R 1994 Two-equation eddy-viscosity turbulence models for engineering applications. *AIAA J.* **32**(8), 1598-1605.
[11] Nakamura Y and Yoshimura T 1982 Flutter and vortex excitation of rectangular prisms in pure torsion in smooth and turbulent flows, J. Sound Vib. **84**(3), 305-17.
[12] Okajima A 1982 Strouhal numbers of rectangular cylinders. *J. Fluid Mech.* **123**, 379-98.
[13] Schewe G 2013 Reynolds-number-effects in flow around a rectangular cylinder with aspect ratio 1: 5. J. Fluids Struct. **39** 15-26.
[14] Tian X, Ong M C, Yang J and Myrhaug D 2013 Unsteady RANS simulations of flow around rectangular cylinders with different aspect ratios, Ocean Eng. **58**, 208-16.
[15] Wilcox D C 1998 *Turbulence modeling for CFD* La Canada, CA: DCW industries.

IOP Conf. Series: Materials Science and Engineering **700** (2019) 012011 doi:10.1088/1757-899X/700/1/012011

Optimal sea floor placement of the oil/gas production equipment

S S Arsenyev-Obraztsov, A I Ermolaev and A M Kuvichko
Russian State Gubkin University of Oil and Gas, Russia

Corresponding author: arseniev@gubkin.ru

Abstract. In the construction designing of oil field surface facilities, exists a problem of the seabed production equipment optimal placement. As an initial approximation, we can represent this task in the form of the linear Boolean programming problem. An application of routine methods of discrete programming can theoretically give the desired solution. However, these methods do not take into account the specific nature of the problem. This problem belongs to the NP class. So we can run into significant computational difficulties. This situation is typical for construction of the ground surface or subsea located facilities for real oil/gas fields. To overcome it, we propose to replace a discrete programming model with a linear programming one, which takes into account problem-specific properties. Also, we present examples of the application of the proposed parallel optimization algorithms. Input data for them: the seabed profile, geometry, and space distribution of oil/gas reserves. Optimization problem objective function is the penalty for the irrational placement of seabed production equipment elements.

1. Introduction

The problems and tasks associated with the construction and placement of the seabed systems for the development of offshore oil/gas fields are of great importance for the economic effectiveness of production projects. There are several reasons for the subsea field development: proximity of deposit to the seacoast, deep enough seabed, the high total amount of initial reservoir energy and absence of need for processing and storing facilities preceding the transportation stage.

The development of offshore fields is carried out with the usage of expensive Seabed Production Complexes (SPC). Therefore, the losses from irrational facilities design and equipment placement dramatically increase. It is one of the main reasons for the usage of three-dimensional multi-phase/multi-component fluid flow simulators during the field development plan design. They are implanted into the computational comparison cycle using different optimization algorithms. In this paper, we analyze the procedures for optimal placement of the SPC units: wells, subsea production elements, and central gathering station. Under the term - subsea production element, we mean sub water platform with placed wellhead equipment and facilities for connecting wellheads to the oil/gas gathering network.

There is a significant number of articles devoted to well-placement optimization [1, 2, 3], and many others. For several years we develop an approach to the well-placement problem [4, 5, 6], which differs from similar research works in:

(1) We do not use technical and economic indicators: NPV, cumulative production, and some other effectiveness estimators. As optimization criteria, we select heuristic rules used for the reasonable deposit development, and tested by long-term field practice;

IOP Conf. Series: Materials Science and Engineering **700** (2019) 012011 doi:10.1088/1757-899X/700/1/012011

(2) This task can be set as the discrete programming problem which we further replace with the solution of a series of linear programming problems. reasonable

Both items increase the computational efficiency of optimization procedures, especially for the solution of high-dimensional problems common for designing development plans for oil and gas fields. Without fundamental changes, we can use proposed well-placement algorithms for the optimal placement of other elements of the SPC: subsea production facilities, a central gathering unit, and some other.

2. Optimization of the SPC elements placement

The following sequence of operations can approximately express the algorithm for SPC elements allocation on the seabed:

(a) determine a set of sections of the pay area where it is reasonable to locate bottom holes (the number of wells is predefined);

(b) determine the geometrical configuration (chart) of the horizontal wellbore (the length of the horizontal wellbore is preset);

(c) determine the allocation of the subsea production units (the number of units is pre-specified);

(d) determine central gathering unit allocation.

Initial information for the well-placement problem includes reservoir geometrical parameters, data about the deposit reserves spatial distribution, porous medium permeability, and reservoir fluids saturation. The generation of the initial information for optimization algorithms begins with the subdivision of the pay area into square blocks of equal size. Block sizes should be suitable for placing a full-sized horizontal wellbore in it. With the usage of geological modeling software for each block, we estimate the hydrocarbon reserves and other characteristics affecting well-placement. Then for each block, we evaluate the "expediency indicator" in terms of possible bottom hole placement in it.

Parameters for well-placement problem are:

- n and s - are numbers of blocks and wells, respectively, $n > s$, and n/s – is an integer (if this condition violated, then "dummy" blocks with zero reserves must be added so that ratio of the changed number of blocks to the number of wells equals to some integer);
- c_{ij} - "remote production fine" - a penalty for production of oil/gas from block j by well bottom hole placed in the block i, $(i, j = 1, 2, ..., n)$. It also can be interpreted as an "inclusion price for block j to be in the block i pay zone".

Problem solution is a set of variables $\{x_{ij}\}$ such that $x_{ij} = 1$ if block j included in the area of influence of the well located in the block i, and $x_{ij} = 0$ otherwise. The term "area of influence" means a set of blocks closest (adjacent) to the block with well. We assume that these blocks provide production well by the main part of the inflow of formation fluids. From the definition of solution it follows: if $x_{ii} = 1$, then the block i contains the bottom of the well i, if $x_{ii} = 0$, then there is no well in the block i.

Taking into account proposed assumptions and notations, the solution of this problem is the set of variables $\{x_{ij}\}$, satisfying the following system of terms and conditions [4]:

IOP Conf. Series: Materials Science and Engineering **700** (2019) 012011 doi:10.1088/1757-899X/700/1/012011

$$\sum_{i=1}^{n}\sum_{j=1}^{n} c_{ij}x_{ij} \to \min_{x}, \tag{1}$$

$$\sum_{i=1}^{n} x_{ii} = s, \tag{2}$$

$$\sum_{j=1}^{n} x_{ij} = \left(\frac{n}{s}\right) x_{ii}, \quad i = \overline{1,n}, \tag{3}$$

$$\sum_{i=1}^{n} x_{ij} = 1, \quad j = \overline{1,n}, \tag{4}$$

$$x_{ij} \in \{0,1\}, \quad i = \overline{1,n}, \quad j = \overline{1,n}. \tag{5}$$

The objective function (1) is a total penalty for the irrational well-placement. The equality (2) is a limitation for the number of wells placed on the oil field. Restrictions (3) are equivalent to the condition that the area of influence of each well contains the same number of blocks. The constraints (4) mean that any block can belong to only one area of influence. If (n/s) is an integer, then from the constraints (3) and (4) follows fulfillment of the constraint (2) [4]. If any block (for example, l-th) is forbidden for well placement, then the following conditions are added to the problem (1)-(5): $x_{lj} = 0$, $(j = 1, ..., n)$.

In [4, 5, 6] for the calculation of feasibility indicators, it is proposed to use the following heuristic rules for rational well-placement. The minimum size list of them looks like:

(a) it is desirable to guarantee the shortest possible distance from the well bottom to any point of the reservoir;

(b) also it is desirable to bring wells closer to the blocks with larger reserves.

Later this set of rules can be extended or changed.

The optimality criterion (1) expresses the desire to diminish the total penalty for violation of these rules. According to them, we can define values of c_{ij} as:

$$c_{ij} = \begin{cases} 0, & j = i \\ \lambda_j^{1-\gamma} \cdot r_{ij}^{\gamma} & j \neq i \end{cases}, \tag{6}$$

where

$$\lambda_j = \frac{V_j}{\max\limits_{1 \le k \le n}\{V_k\}}, \quad r_{ij} = \frac{R_{ij}}{\max\limits_{1 \le k \le n, 1 \le l \le n}\{R_{kl}\}}, \quad \gamma \in [0;1].$$

Let V_j - oil/gas reserves of the block j; R_{ij} - distance between the centers of blocks i and j; γ - expert estimation of the "distance" indicator importance ($0 \le \gamma \le 1$). From (1) it follows, that when we enlarge a distance from the well to the block with large reserves penalty increases.

To improve the adequacy of the model in [6] it is proposed to include into the set of input parameters, variables characterizing resistance of medium to the fluid flow: $\omega_i = \frac{\mu_i}{k_i h_i}$, where ω_i – coefficient of filtration resistance, μ_i – viscosity of fluid, k_i – permeability, h_i – fluid-saturated thickness of the block i. After that, we replace this set of blocks with a graph. Vertices of this graph are the centers of blocks, and arcs are lines connecting adjacent vertices. By adjacent vertices, we usually mean centers of adjacent blocks. Adjacent blocks are blocks that have a common boundary. The length of each arc is an average coefficient of flow resistance for two adjacent blocks. R_{ij} - the distance between blocks i and j, is the minimum total length of arcs connecting them. It means that for each pair of vertices; we needed to solve the shortest path problem [7].

After selecting blocks containing wells, we need to determine the profile of a horizontal wellbore. The mathematical definition of this problem is the following. After we selected blocks for a well and SPU (that the well connects to), we can define azimuth of a wellbore horizontal part as an azimuth of the vector directed from the SPU $(X_{SPU}; Y_{SPU})$ to the well (X_{well}, Y_{well}) location. We take as the initial point $(X_{in}; Y_{in})$ (formation entry) and as the terminal point $(X_{out}; Y_{out})$ (well bottom) of the horizontal part of the well. We now consider only X and Y point coordinates; Z-coordinate to be optimized further. Let us define the depth of the horizontal part of the wellbore as 0 at the formation entry point. and $L = \sqrt{(X_{in} - X_{out})^2 + (Y_{in} - Y_{out})^2}$ at well bottom point. We then define a set of points $(X_i; Y_i)$ for every $i = \overline{1, m}$ as:

$$X_i = X_{in} + X_{out}\frac{l_i}{L}, \quad (7)$$

$$Y_i = Y_{in} + Y_{out}\frac{l_i}{L}, \quad (8)$$

where l_i is the measured depth at point i:

$$l_i = L \cdot \frac{i-1}{m-1}, i = \overline{1, m}. \quad (9)$$

At every point $(X_i; Y_i)$ defined above we can get the limiting Z_i^{top} and Z_i^{bot} values from reservoir top and bottom surfaces, respectively, and get a Z-value Z_i between these limiting surfaces.We then take a desired $3D$-property (like remaining recoverable fluid-in-place, permeability, and some other), and we can calculate its value D at any $(X_i; Y_i; Z_i)$, but since the number of $3D$-grid blocks between top and bottom surfaces is limited, there could be only a limited number of Z-values at every layer i. We will sample Z_i at $3D$-property cell centers. The problem is to maximize the total sum of values at planned wellbore points, but to keep the inclination angle increase/decrease value not greater than a defined angle I (here we only vary inclination (vertical angle), but not azimuth). Thus, the following problem to be solved:

$$\sum_{i=1}^{m} D(X_i; Y_i; Z_i) \to \max_{Z_i}, \quad (10)$$

$$Z_i^{bot} \le Z_i \le Z_i^{top}, \quad i = \overline{1, m}, \quad (11)$$

$$\left| \arctan\left(\frac{Z_{i+1} - Z_i}{\sqrt{(X_{i+1} - X_i)^2 + (Y_{i+1} - Y_i)^2}} \right) \right| \le I, \quad i = \overline{1, m-1}, \quad (12)$$

The problem (10)-(12) can be solved using standard dynamic programming and a Bellman's Principle of Optimality [8], since we can split it to subproblems at layer i.

The definition of a mathematical model for optimal placement of seabed production units (SPU) begins with the replacement of deposit domain by a set of equal blocks. We can install SPU in any of them. Let m be the total number of SPUs. Other initial parameters are: M the number of SPU, and K the maximum allowable number of wells in SPU. We state that $KM \geq s$, where s is the number of wells (number of bottom holes). Let C_i be the cost of construction of SPU in the block i, $i = 1, 2, ..., m$; w_{ij} is the construction cost of well connecting the center of block i to bottom hole j, $j = 1, 2, ..., s$; R_{ij} is the distance between block j with a bottom hole in it and the center of block i; R is the maximum allowable distance between the bottom hole

IOP Conf. Series: Materials Science and Engineering **700** (2019) 012011 doi:10.1088/1757-899X/700/1/012011

and SPU. Let g_{ij} be coefficients such that if $R_{ij} \leq R$, then $g_{ij} = w_{ij}$; if $R_{ij} > R$, then $g_{ij} = W$, where, $W \gg \max\limits_{i,j}\{w_{ij}\}$.

Now problen solution can be defined as two groups of variables $\{x_{ij}\}$ and $\{y_i\}$, where $x_{ij} = 1$, if the bottom hole j is connected to the SPU located in the block i, and $x_{ij} = 0$, otherwise; $y_i = 1$, if there is a SPU in block i, otherwise $y_i = 0$.

Hence to solve the problem, we need to identify sets $\{y_i\}$, $\{x_{ij}\}$ such that:

$$\sum_{i=1}^{m}\sum_{j=1}^{s} g_{ij}x_{ij} + \sum_{i=1}^{m} C_i y_i \to \min_{x.y}, \tag{13}$$

$$\sum_{i=1}^{m} y_i = M, \tag{14}$$

$$\sum_{j=1}^{n} x_{ij} \leq K y_i, \quad i = \overline{1,m}, \tag{15}$$

$$\sum_{i=1}^{m} x_{ij} = 1, \quad j = \overline{1,s}, \tag{16}$$

$$x_{ij} \in \{0,1\},\ y_i \in \{0,1\},\ i = \overline{1,m},\ j = \overline{1,s}. \tag{17}$$

The problem of central gathering node (CGN) placement is defined by the mathematical model (13)-(17), in which s equals the number of SPU, $M = 1$, $K = s$, and inequality constraints (15) are replaced by equalities. In addition, g_{ij} is the cost of building of a gas flow-line between SPU j and the CGN, if CGN is placed into the block i, and C_i is the cost of building the CGN in the block i, $i = 1,2,...,m$, $j = 1,2,...,s$. To solve the problem, we need to inspect all acceptable options for the CGN placement. As the best, we choose one for which the total cost of building gas flow-lines and CGN will be minimal.

3. Method for solving problems (1)-(5) and (13)-(17)

The specifics of problem (1)-(5) are: First, we can represent the set of desired solution variables in the form of a square matrix; Second, if by some consideration values are assigned to the elements of the main diagonal of this matrix, then the original problem is converted to the classical transport problem with the cost criterion (T-problem)[9]. The proposed solution method uses these features. Below we present a description of this method applied for the problem (1)-(5).

Let consider some assigned of variables $\{x_{ii}\}$ satisfying constraints (2) and (5), i.e. $\{x_{ii}\}$ is a set of the one of several admissible well placement. Consider the sets A, B and C: $A = 1,2,...,n$, $B = \{i : x_{ii} = 1\}$, $C = \{i : x_{ii} = 0\}$, i.e.

$$A = B\bigcup C,\ B\bigcap C = \emptyset,\ |A| = n,\ |B| = s,\ |C| = n - s \tag{18}$$

Then, for a given set $\{x_{ii}\}$, it follows from (2) that $x_{ij} = 0$ for $i \in C$ & $j \in A$. Taking into account (18), problem (1) - (5) will be:

$$\sum_{i\in B}\sum_{j\in C} c_{ij}x_{ij} \to \min_x, \tag{19}$$

$$\sum_{j\in C} x_{ij} = \binom{n}{s} - 1, \quad i \in B, \tag{20}$$

$$\sum_{i\in B} x_{ij} = 1, \quad j \in C, \tag{21}$$

$$x_{ij} \in \{0,1\}, \quad i \in B, \quad j \in C. \tag{22}$$

Let's analyze the problem (19) - (22). First, the right-hand side parts of constraints (20) and (21) are positive integers. Therefore, all the support plans for this problem will be integer-valued [10]. Second, from (22) we obtain that $x_{ij} \geq 0$, and taking into account (21) we get the "automatic" fulfillment of $x_{ij} \leq 1$. Therefore, conditions (22) can be replaced by restrictions on the sign of the unknown variables:

$$x_{ij} \geq 0, \quad i \in B, \quad j \in C. \tag{23}$$

Third, $(n-s)$ - the sum of the right-hand sides of the constraints (20) equals to the sum of the right-hand side parts of constraints (21). Therefore, with a fixed set $\{x_{ii}\}$ satisfying constraints (2) and (5), problem (1) - (5) turns into model (19) - (21), (23), which is a classical transport problem. Therefore, one of the known methods, for example, the method of potentials [9], can solve this problem. Thus, any admissible set $\{x_{ii}\}$ uniquely defines some valid solution for the original problem (1) - (5). Therefore, the method for solving problem (1) - (5) is reduced to generating sets $\{x_{ii}\}$ satisfying constraints (2) and (5), solving T-problem (19) - (21), (23) for each and choosing the best in terms of criteria (1)..

Let us estimate z - the lower bound for the reasonable number of test sets $\{x_{ii}\}$ in the Monte-Carlo optimal solution search. By this, we mean that conducting $\#z$ tests we get an admissible result with a preset probability P. Lower bound for z we can obtain from the following considerations. Let N be the number of all admissible sets of $\{x_{ii}\}$, and $D(L)$ be best solutions subset of size L for the problem (1) - (5) ($|D(L)| = L$). We assume that if at least one of the test sets $\{x_{ii}\} \in D(L)$, then problem (1) - (5) is solved. Let Q be the probability that the test solution of the problem (1) - (5) belongs to the set $D(L)$. It is clear that $Q = L/N$. Let P be the predefined probability that at least one of the $\#z$ test sets belong to the set $D(L)$. For a known value of N and given values of P and L, an estimation of $\#z$ we can get as the solution of the equation:

$$P = 1 - (1 - L/N)^z$$

which gives:

$$z = \ln(1-P)/\ln(1-L/N)$$

We need to emphasize that the use of any enhanced random search algorithm or the application of additional information in the process of sets $\{x_{ii}\}$ generation could significantly reduce this estimation.

For the case of a large number of numerical computations, it is better to use high-performance computing systems. If the total amount of computation is relatively small, then workstation with general-purpose graphics accelerators (NVIDIA, AMD) or many-core computing systems based on Intel Xeon Phi, Sunway 26010, and some other processors will be a good option. From our point of view, the Sunway TaihuLight supernode is the best choice for conducting optimization with the usage of hydrodynamic simulators.

Since the total amount of memory required for the generation of the input sets and subsequent transport problem solution is not large, the calculations for one dataset can be entirely performed on the CUDA core or the fast PCE core in Scratch Pad Memory of the SW26010 processor [11]. This approach can significantly speed up calculations. The parallelization of the above algorithms is obvious.

The application of this method for the solution of the problem (13) - (17) requires preliminary transformation. By adding $s^* = KM - s$ fictitious wells to the model (13) - (17) inequality constraints (17) can be converted into equality constraints. For the model including fictitious wells coefficients should be set to zero, i.e. $c_{ij} = 0$. Then the above method can be applied in this case, only instead of the sets $\{x_{ii}\}$, it requires to generate sets $\{y_i\}$ satisfying constraints (14) and (17). The set $\{y_i\}$ represents one of the possible SPU (or CGN) placements.

4. Example

We apply the described algorithm to solve a test problem. We have created a sample (artificial) model similar to models of deepwater gas fields. We have divided this model into square blocks 1×1 km^2. The criterion we've used for (6) to calculate c_{ij} in (1) was a composite criterion $V_j = \overline{G_j}^{0.5} \times \overline{kH_j}^{0.4} \times \overline{P_j}^{0.1}$, where $\overline{G_j}$ is a normalized gas-in-place for the j-th cell, $\overline{kH_j}$ is a normalized kH property (cell permeability multiplied by cell height), and $\overline{P_j}$ is a normalized pressure. The map of the composite criterion for this test field is shown on the Figure 1.

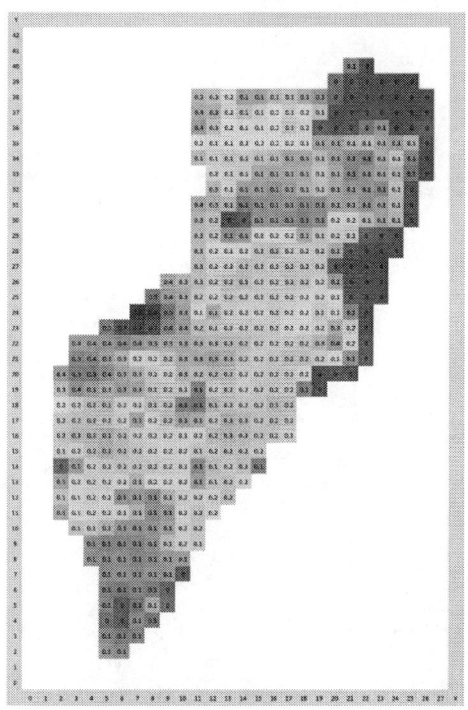

Figure 1. A representation of the used test field model: color scale represents the composite criterion value from the lowest (green) to the highest (red).

This model contains 492 blocks. Dividing total (recoverable) reserved to average produced gas estimate, we've decided to place 12 wells on this gasfield. We have tested different values of the parameter γ in equation (6). It is rational to select (as described in [6]) γ in the range from 0.05 to 0.25. We have selected 3 γ values: 0.05, 0.15 and 0.25. We have decided to pick well positions for $\gamma = 0.15$, as this case looks more realistic in terms of engineering criteria [4], however a question of selecting γ depends on the specific field. After running the iterative solver [12], we got the following results (Figure 2, (a)). The wells were located in blocks numbered 48, 63, 118, 125, 182, 209, 236, 259, 311, 362, 400 and 454. The resulting zones corresponded to cells are shown at Figure 2, (b). We count the number of times that a j-th block appears in the (local) optimal well placement (Figure 2, (c)). The distribution of (local) optima, i.e. the values of the objective function vs. the theoretical lognormal distribution (with an estimated mean (ξ) and standard deviation (σ)) is shown at Figure 2, (d).

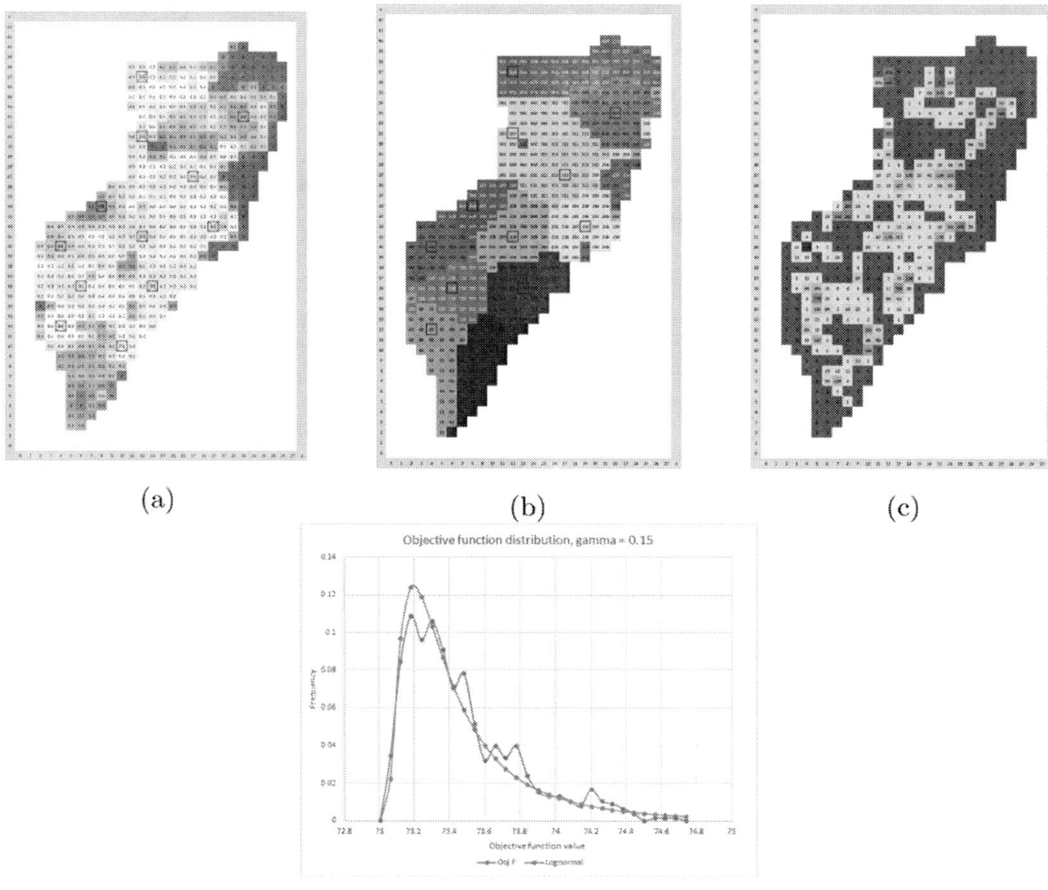

(a) (b) (c)

(d)

Figure 2. For $\gamma = 0.15$ (a) wells (crossed bold square block) optimal placement, color represents criterion V_j values from smallest (blue) to largest (red); (b) zones for wells (crossed block) distribution, color represents different zones; (c) the number of times that the block j has appeared in the (local) optimal solution, color represents values from smallest (green) to largest (red); (d) cumulative local optima distribution: solution vs. theoretical lognormal distribution with an estimated mean ξ and standard deviation σ.

For this location of wells, we then solve the problem of placing several SPU (13)-(17) having M set to 3 in (14). By doing this, we can drop the term $\sum_{i=1}^{m} C_i y_i$ in (13) since it turns into a constant and can be eliminated (since the cost values C_i are assumed the same for all i). Nevertheless, we use the constraint (15) in a way to connect any number of wells to an SPU. We also assume that the cost of a length unit of a well is equal for all wells. Using these assumptions, g_{ij} in (13) morphs into a simple distance from block i to j. The complexity of this problem for 492 cells, 3 SPUs and 12 wells is $C^3_{492} \times 12 \times 3 = 710221680$ cases, so it could be solved directly.

We have solved this problem for the test field model having wells placed into blocks defined above (Figure 2). The solution is shown at Figure 3, (a). The optimal result was unique, and SPUs were placed into blocks with numbers 118, 209, and 379. Note that tow first SPUs were placed in blocks with wells, while the thirds one between wells. We further solved the problem of placing a CGN for there three SPUs using the same approach. The solution is shown at Figure 3, (b). This position of the CGN guarantees optimal sum of distances; again, CGN was located at block number 209, where a well and an SPU were already placed.

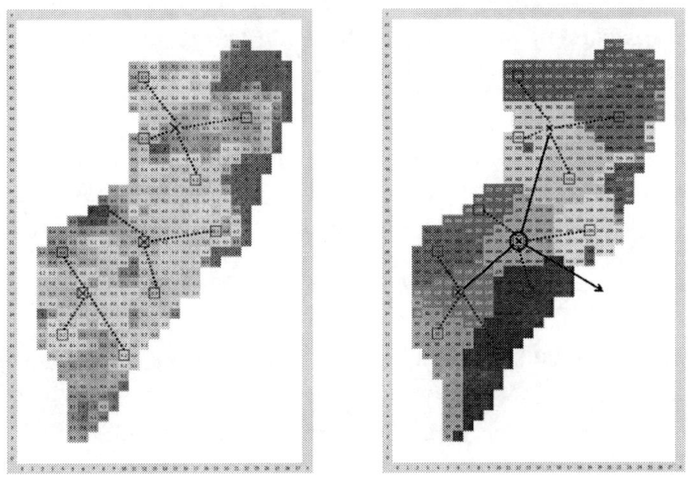

(a) (b)

Figure 3. For $\gamma = 0.15$ (a) SPU (crossed block) optimal placement; well (bold block) connections to SPUs are shown as dashed lines; color represents criterion V_j values from smallest (green) to largest (red), (b) zones for wells (bold block) distribution, SPUs (crossed blocks) and CGU (bold round); color represents different zones.

We also solve the problem of optimal well-placement inside the formation (10)-(12) for the found set of locations of wells and SPUs. As for the desired $3D$-property, we have selected kH above gas-water contact. The maximal inclination variation angle I is set to 2 per depth step. An example of the optimal position for the well in block number 259, presented in Figure 4.

5. Conclusion

The presented models and optimization algorithms allow taking into account the experience and knowledge of specialists in the design of oil and gas deposits development projects and using the capabilities of mathematical tools to support selected decisions. The primary purpose of models and algorithms is a creation of the preliminary set of optional projects for the development and reservoir surface facility construction for oil and gas deposits, from which subsequent choose of the best option needs to be conducted.

The advantages of the proposed approach for solving the problems of production technological elements placement include: First, a significant reduction in the number of calls to hydrodynamic

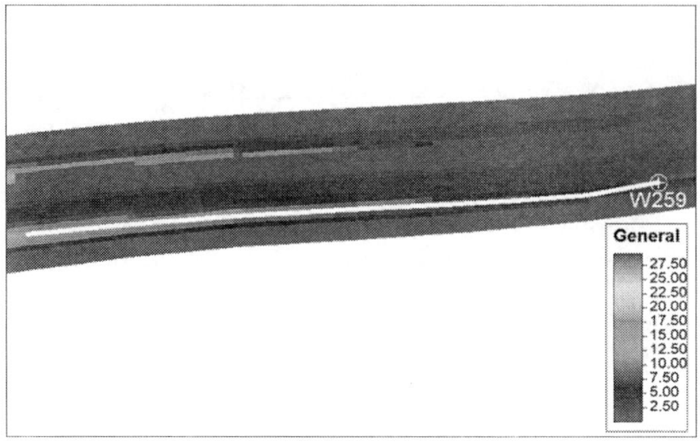

Figure 4. Optimal placement of the well 259 inside the formation with kH set as the target $3D$-property. The well is shown in white color.

simulators; Second, the possibility of using linear programming methods; Third, the proposed algorithms are well suited for different parallelization technologies. All this allows us to proceed for solving optimization problems of high dimensions, which is specific for the production project design for oil and gas fields.

References

[1] Badru O, Kabir C *et al.* 2003 Well placement optimization in field development *SPE Annual Technical Conference and Exhibition* (Society of Petroleum Engineers)

[2] Güyagüler B, Horne R N, Rogers L, Rosenzweig J J *et al.* 2002 *SPE Reservoir Evaluation & Engineering* **5** 229–236

[3] Rosenwald G W, Green D W *et al.* 1974 *Society of Petroleum Engineers Journal* **14** 44–54

[4] Ermolaev A and Ibragimov I 2006 *Proceedings of Trapeznikov Institute of Control Sciences of RAS* **27** 118–123

[5] Arsenyev-Obraztsov S, Ermolaev A, Kuvichko A, Naevdal G and Shafieirad A 2011 Improvement of oil and gas recovery by optimal well placement *IOR 2011-16th European Symposium on Improved Oil Recovery*

[6] Kuvichko A and Ermolaev A 2012 Hpc-based optimal well placement *ECMOR XIII-13th European Conference on the Mathematics of Oil Recovery*

[7] Christofides N 1975 *Graph theory: An algorithmic approach (Computer science and applied mathematics)* (Academic Press, Inc.)

[8] Bellman R 2003 *Dynamic Programming* (Dover Publications)

[9] Kantorovic L 1942 *Sci. URSS (NS)* **37** 199–201

[10] Sigal I K and Ivanova A P 2007 *Introduction to applied discrete programming* (M.: Fizmatlit)

[11] Arseniev-Obraztsov S 2018 *Automation, telemechanization and communication in oil industry* 52–61

[12] Kuvichko A, Andrianov N and Ermolaev A 2016 On solving large-scale well positioning problems *ECMOR XV-15th European Conference on the Mathematics of Oil Recovery*

Topology optimization through stiffness/weight ratio analysis for a three-point bending test of additive manufactured parts

A A Garcia-Granada[1,*], **J Catafal-Pedragosa**[1,2] **and H G Lemu**[2]

[1] GEPI-IQS Grup Enginyeria Producte Industrial, Universitat Ramon Llull, Via Augusta, 390. 08017, Barcelona, Spain.

[2] Department of Mechanical and Structural Engineering and Materials Science, University of Stavanger, Norway.

[*] Corresponding author: andres.garcia@iqs.url.edu

Abstract. Topology Optimization (TO) is a technique that allows for increasingly efficient designs and its objective is to maximize the performance of mechanical systems or structure in a variety of fields. Attempts to employ TO for parts manufactured with conventional methods such as casting, forging, injection moulding, CNC machining and the like could not lead to desired optimum results due to the existing manufacturing constraints regarding geometrical complexity. Currently, additive manufacturing (AM) techniques allow the fabrication of more complex shapes which in principle will lead to improved performances through application of the TO concept. This study focuses on structural optimization of additive manufactured parts of thermoplastic parts based on analysis of the stiffness/weight (mass) ratio for a beam subjected to a three-point bending load. The experimental work is done on optimization of parts manufactured by Fused Deposition Modelling (FDM) technology and finally compared with an identical model manufactured using Polyjet 3D printer. Different TO software are compared to conduct the optimization, and a module of SolidWorks 2018 from Dassault Systems is chosen for the topology optimization for the final experiment. The study focuses on the results on stiffness/mass ratios, paying attention to the influence of different printing parameters on the test results. An increase of stiffness/weight ratio of 31.7% was predicted by software while experiments showed an increase of just 27.04%.

1. Introduction

Topology optimization is a type of structural optimization that is used as a tool or technique in areas that need reduction of weight in a component by optimal distribution of mass (weight) and hence leads to improved stiffness to weight ratio. The optimization technique produces shapes by removal of materials from regions where the component shows low levels of stress under loading conditions. Such shapes can be complex and hence difficult to manufacture using the traditional manufacturing techniques. With the current opportunities provided by additive manufacturing, however, the difficulty of fabrication of complex shapes is not an issue, and hence topology optimized design can be realized. As additive manufacturing is a technology that is not yet fully mature, it requires detailed studies of several aspects to understand the material behavior under loading conditions. In this introduction, a brief literature review is reported to understand the previous work on material characterization of additive manufactured parts and also on topology optimization.

In order to perform a proper topology optimization, the optimization software needs to rely on material properties verified for each additive manufacturing strategy. In Akessa et al. [1], for instance, a study was conducted to characterize the mechanical properties of ABS-M30 material using rectangular samples that were subjected to 3-point bending tests. The samples were manufactured by varying the air gap, raster width and raster angle. A similar study was reported by Gebisa and Lemu [2] in which the effects of varying the FDM process parameters on the flexural properties of ULTEM 9085 were investigated. The objective of this study was to consider all possible combinations of parameters; air gap, raster width, raster angle, contour number and contour width.

According to other recent studies carried out by Domingo-Espin et al. [3], anisotropic material properties should be considered when using FEA simulation of FDM parts exceeding the elastic region limit. From this study, conclusions were achieved using tests and simulations of an "L" shaped cantilever beam with bending and torsion for polycarbonate materials. Furthermore, dynamic properties were studied as described by Domingo-Espin et al. [4], where a simple prismatic part was loaded using a dynamic mechanical analysis (DMA). Results showed that the building parameters, namely nozzle diameter, number of contours and distance between rasters can control the elastic behavior of the FDM manufactured part, being the number of contours the most influential parameter. Test parameters, such as amplitude, frequency and temperature, showed a great influence on the damping capacity of the part.

Creep behaviour of polycarbonate (PC) parts manufactured using FDM process were studied by Salazar-Martín et al. [5] using experimental method focusing on the effect of three process parameters: (1) part build orientation, (2) raster to raster air gap, and (3) number of contours. The study was conducted on the primary and secondary creep behaviour. It was found that increasing the density of the sample, by increasing number of countours and reducing air gap, causes creep strain to decrease. The study also shows the significance of arranging the deposited filaments in the same direction the sample is loaded. The influence of FDM manufacturing parameters on mode I fracture properties has been recently studied by Sedigi et al. [6] to explore how a part can hold deformation energy beyond elastic limits, taking into account plasticity and crack locations.

Other additive manufacturing techniques have been studied to understand the influence of manufacturing parameters on material properties. For example, Morales-Planas et al. [7] studied the influence of different manufacturing parameters such as part orientation on mechanical properties of Multi Jet Fusion PA12 focused on achieving the right design for watertightness, strength and tolerances. Once material properties are well studied, a topology optimization can be performed taking into account these values and a literature review on topology optimization is provided.

A review on topology optimization was performed by Hassani et al. [8] already in year 1998 and is continuously reviewed due to the growth of software and hardware developments. For example, Campbell et al. [9] provided a review of numerical optimization techniques for meta-device design for optical materials. In this paper, the literature review focused on lightweight design considering topology optimization needs for the best stiffness to weight ratio. Gebisa and Lemu [10] reported a case study on topology optimized design for additive manufacturing. An engine bracket was topologically redesigned to reduce its weight considering fabrication in AM. The study results show that topology optimization is a powerful technique to reduce the weight of a structural product while maintaining the design requirements if additive manufacturing is considered.

Faskhutdinov et al. [11] reported a study done on the topology optimization of a jet engine part with Selective Laser Melting (SLM) technology where the process of TO is described. The optimization is the process of choice of the best option imaginable. That decision is done based on some dependent values (design data) and a target function. The values of design data are found at which the target function has a minimum. While there may be a number of targets, one will have to have priority on others, as not all of them may be compatible. Then, topology optimization allows finding an optimal material distribution in a given design space under the certain loads and boundary conditions. Recently, Wang et al. [12] worked on lightweight design for robots by integrating topology optimization and parametric system optimization using TOSCA software. Their target was to maintain the deformations of the end-effector of a serial painting robot reducing the mass of components.

The objective of the work reported in this article is to conduct topology optimization of 3D printed parts of thermoplastic materials for improved stiffness to mass ratio using both experimental test (3-point bending test) and optimization software.

2. Materials and Methods

2.1. Three-point bending test

A three-point bending test is defined as a starting point for the topology optimization reported herein. This is because tensile tests are not adequate as they provide a uniform stress distribution across the section of the specimen and therefore the optimization is limited to a reduction of cross section. Three-point bending test is relatively simple to conduct in a common laboratory facility. It creates different stress values across the thickness of the specimen and provides room for topology optimization.

In this study, the stiffness to mass ratio is used as a parameter of optimization. Stiffness for a constant section specimen of a simple supported beam under transverse load is theoretically calculated as follows based on pure bending:

$$k = \frac{F}{y} = \frac{48EI}{L^3} \tag{1}$$

where k is stiffness, F is applied force, y is displacement in loading direction, E is Young modulus, I is section inertia and L is span length. For the same beam loading, maximum stress (σ) is obtained at the middle of the specimen and theoretically expressed as follows:

$$\sigma = \frac{FLh}{8I} \tag{2}$$

where h is section height of the specimen.

Theoretical equation for stress does not include stress concentrations near supports but test is defined to obtain fracture on lower parts of specimen. Therefore, in classical engineering optimization, increasing inertia will improve stiffness and at the same time reduce maximum stress avoiding plasticity and fracture with the exception of stress concentration effects if new shapes involve sharp edges.

2.2. Simulation software for topology optimization

Software for topology optimization is growing very fast together with the rapid increase in computational speed and hardware capacity. As a result, many modelling and simulation tools are incorporating topology optimization modules in their software package. In this study, a module of SolidWorks 2018 from Dassault Systems is chosen for the topology optimization. The software allows the definition of boundary conditions similar to the three-point bending test setup and provides design rules to select surfaces that should be defined as design features and non-design features. The design features are subjected to material removal if not contributing to load sharing while the non-design features should be kept as they are in initial design, regardless of the stress level acting on them. The software also allows the definition of other rules such as the optimization criteria. The optimization criteria in this case is defined based on the best stiffness to mass ratio with a load of 50 N, considering the material anywhere in the beam remains within the elastic region.

2.3. Additive fabrication and testing machine

Fortus 450mc machine from Stratasys (Figure 1(a)), FDM technology, is chosen to fabricate original and optimized parts using ABS – M30 material. According to the material data, the density is around 1040 kg/m3, Young's modulus between 2180 and 2230 MPa and yield strength between 26 and 31 MPa with elongation at break between 2 and 7%, depending on the orientation of the part (Table 1) [13]. For model slicing and machine control, the pre-processor software of the machine, Insight® 12.2, was used. Raster and contour width were set to 0.4064 mm with just one contour and 0 mm air gap between contours. Two orientations were chosen, flat and edge and for each orientation two angles 0° and 45° were chosen.

Optimized parts required support material for overhanging areas. The FDM process of the machine uses support material referred to as SR20, which dissolves at 70±3 °C and removed from the manufactured part. The solvent is a water solution with additions of sodium hydroxide and sodium carbonate. Two sample designs, i.e. rectangular samples and optimized samples were printed and tested. Four variations with different parameters where designed for each sample, and three specimens for each variation were tested. In other words, a total of 12 + 12 = 24 ABS specimens were tested.

Table 1. Properties for ABS- M30 and Verowhite (from [13] and [14]).

	Density [kg/m^3]	Young's modulus [MPa]	Yield stress [MPa]	Elongation at break [%]
ABS-M30	1040	2180-2230	26-31	2-7
Verowhite	1170	2000-3000	50-65	10-25

In order to compare the results with another printing technology, the identical designs were manufactured using an Objet30 Prime machine from Stratasys (Figure 1 (b)) which is based on Polyjet technology. This machine cures acrylic liquid by using ultraviolet lamps which can provide high accuracy with layers of 0.015 mm. The material used for this study was VeroWhite [14]. According to the material manufacturer, this material has the material properties given in Table 1. A total of 10 + 10 = 20 samples from Verowhite material were manufactured with the same manufacturing conditions to check repeatability. Finally specimens where tested using three-point bending test on an Instron 5985 (Figure 1 (c)) where displacement and force where recorded. Span length for the beam supports was set to 100 mm with a velocity of 1mm/min.

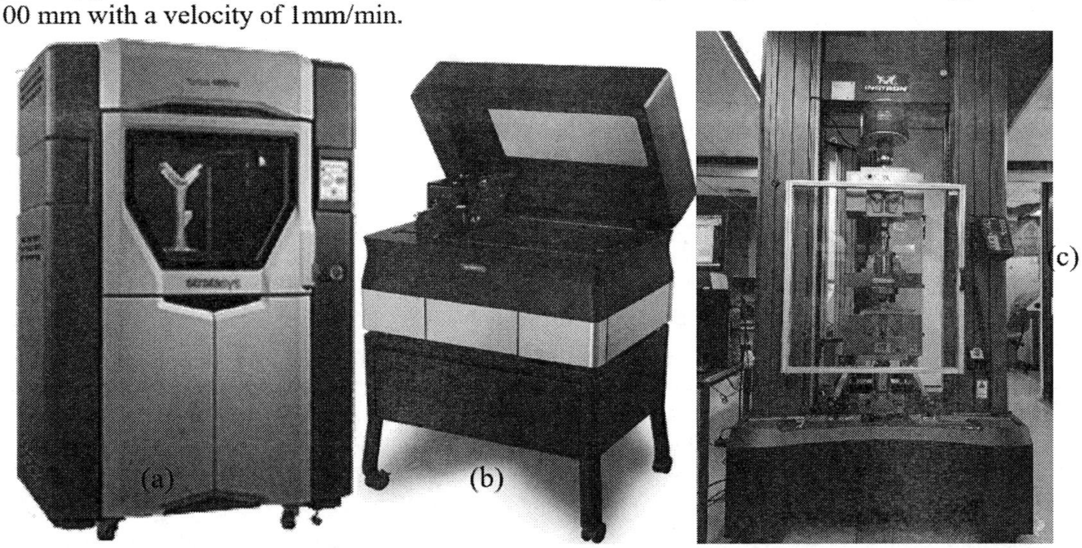

Figure 1. Machines used for additive manufacturing machine and testing (a) Fortus 450mc FDM machine (b) Objet 30 Prime Polyject machine and (c) Instron 5985 tensile test machine.

3. Results and discussion

The original geometry considered for this project is shown in Figure 2, where two small extensions were provided to a rectangular beam to avoid falling from the three-point bending supports. Theoretical calculation are provided with equation (1) and equation (2) ignoring these extensions. From theoretical calculations, the following values are obtained: mass, m = 20.8 g, stiffness, k = 640 N/mm (at E = 2000 MPa) and stress, σ = 1.875 MPa (for F = 50 N), which is below 26 MPa for the lowest yield stress and without considering stress concentration effects.

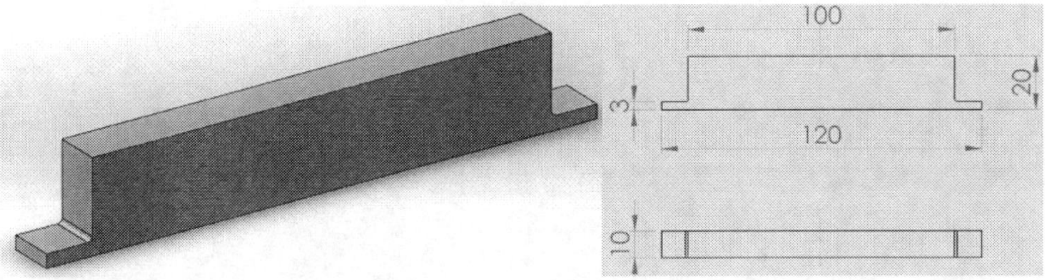

Figure 2. Original geometry to be optimized for a three point bend.

Once the samples are defined, topology optimization was performed to achieve an improvement in the stiffness/mass ratio. Figure 3 (a) shows the optimized model, i.e. after removing unwanted material but with sharp edges. Then, a smooth optimized geometry is generated (Figure 3 (b)) upon generating soft cure transitions. In this optimization process, the mass is reduced from 20.8 g down to 9.1 g. Since the section is not constant or of regular shape, theoretical calculations are not possible for stiffness and strength.

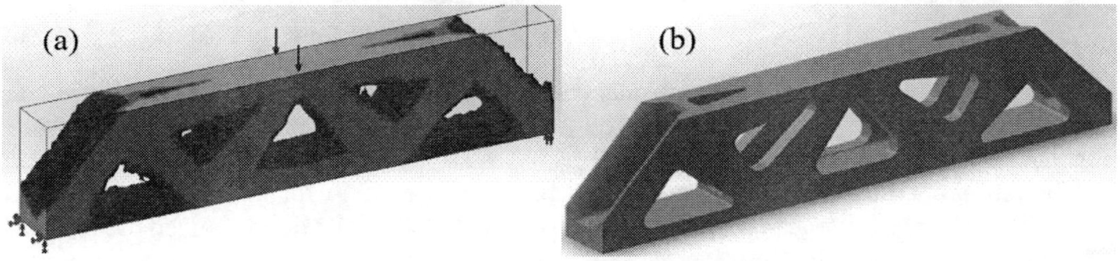

Figure 3. Optimized part (a) with rough surfaces as optimized and (b) smoothed surface geometry created base on optimized shape.

The original and the optimized parts are then manufactured with both machines. Figure 4 (a) shows the specimen manufactured using ABS-M30 material and SR20 support. After the support material is removed, the specimen is three-point bending tested as illustrated in Figure 4 (b).

Figure 4. Optimized part (a) as manufactured with material support material and (b) under three-point bending test.

The same procedure was repeated for the specimens fabricated from Verowhite material. Object30 Prime allows fabrication of parts which are closer to CAD geometry as it allows much smaller layer thickness. This means, it is possible to get a smoother surface in this case compared with that of Fortus 450mc machine. As shown in Figure 5(a), the fracture started from the top connection between front and bottom face but low friction led to sliding of the test sample (Figure 5(b)) with conditions dissimilar to the case in topology optimization, which is defined with ideal conditions where the load is placed in the middle and lower supports are always in the same place. Figures 6(a) and (b) show von Mises stress distribution for the original design and the optimized shape, respectively. In both figures, stress is much higher at the load point (middle of the beam) and at the support points (locations of boundary conditions), thus attention is paid to points in the lower part of the middle section of the specimen where fracture is expected to happen.

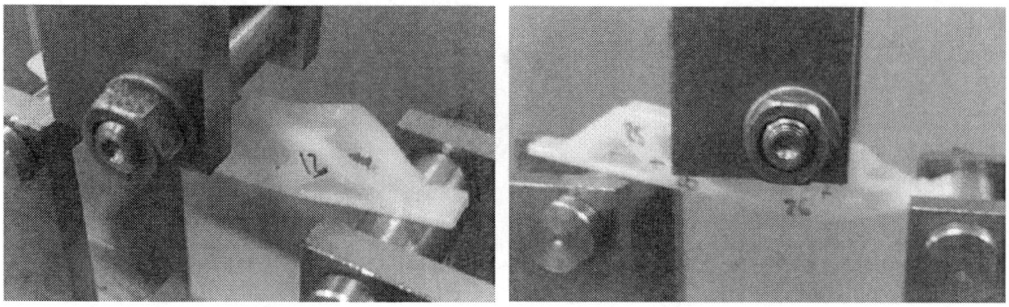

Figure 5. Optimized structure tested with unstable sliding support.

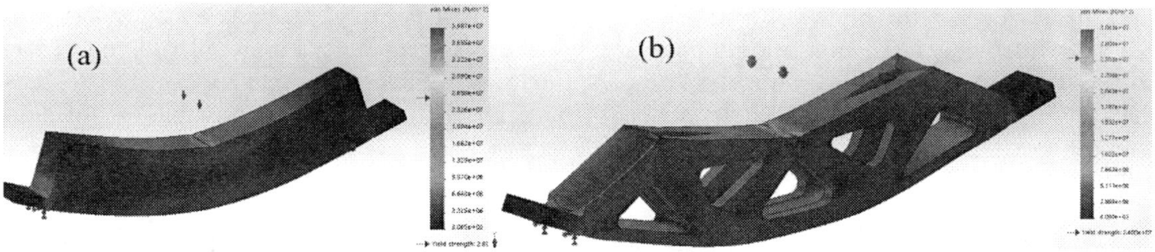

Figure 6. von Mises stress for 1 mm displacement for (a) original shape and (b) optimized shape.

Finally, Figure 7 shows results for all ABS-M30 tests for both original and optimized designs for each fabrication condition. The difference in stiffness to mass ratio for all orientations ranged from 18590 to 20830 N/m/g for original part and from 23460 to 26960 N/m/g for optimized parts. Comparison of all scenarios is provided in Table 2.

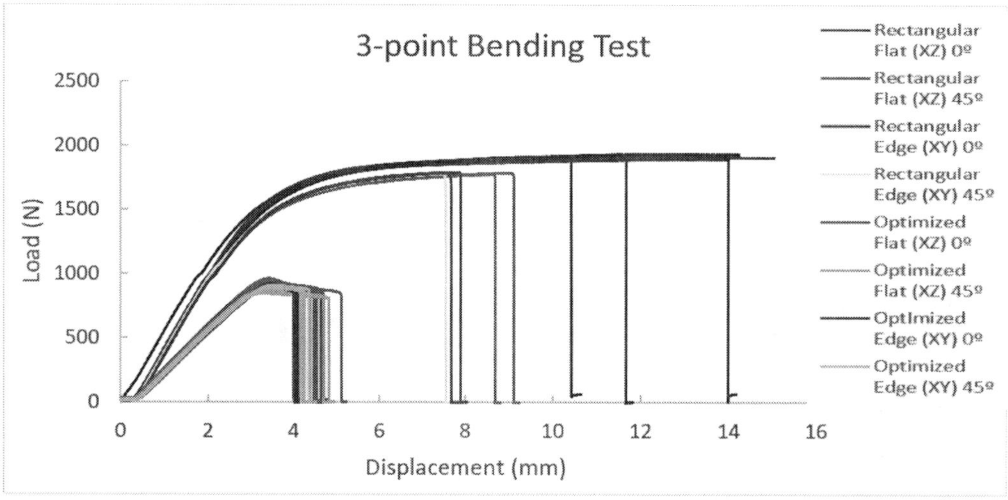

Figure 7. Test results for several printing directions for Fortus 450mc.

Analyzing all scenarios, it can be observed that the increase in stiffness to mass (k_y/m) ratio varies from 12.61% increase in the worst case to 45.03% in the best case. The achieved average increase in stiffness to mass ratio is 27.04%. This is very important since the layer orientation of real complex 3D components cannot be controlled for each location and therefore it is important to expect that improvement is achieved for any combination of orientations. Table 3 gives summarized values of the average mass, loads, displacements and stiffness.

Table 2. Stiffness comparison of ABS-M30 for different scenarios.

	Stiffness to mass (k_y/m) ratio (N/m/g)				
	Average value	Lowest value	Highest value	Worst case optimization	Best case optimization
Rectangular	1.975E+04	1.859E+04	2.083E+04	2.083E+04	1.859E+04
Optimized	2.509E+04	2.346E+04	2.696E+04	2.346E+04	2.696E+04
k_y/m increase after optimization	27.04%	26.19%	29.43%	12.61%	45.03%

Table 3. Average values for each sample design.

	Mass (g)	Max. load (N)	Displ. at max. load (mm)	Load at 1 mm displ. (N)	k_y (N/m)	k_y/m (N/m/g)
Rectangular	19.71	1847	10.06	389.2	3.892E+05	1.975E+04
Optimized	8.87	906	3.43	222.6	2.226E+05	2.509E+04

Furthermore, fracture is also analyzed for each specimen and fabrication orientation. Figure 8 shows fractured parts from edge fabrication ((a) and (b)) and flat fabrication ((c) and (d)). For each sample shown, the upper part is 0º and the lower part is 45º. Original parts are shown on the left and optimized parts on the right. Fracture is complete for flat orientations while fracture is produced on the bottom of the specimen (as expected) for edge manufactured components.

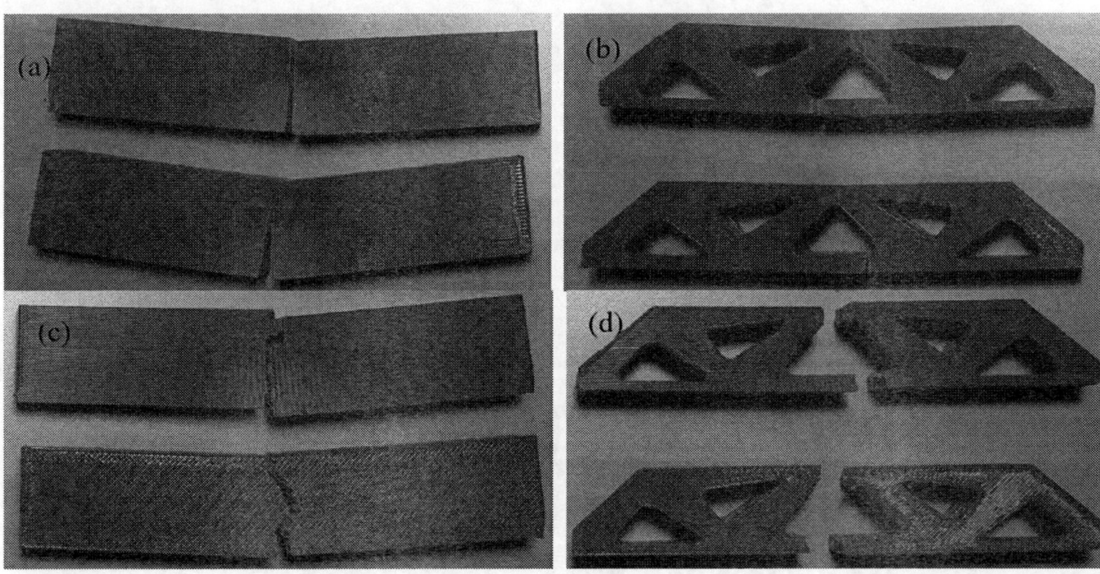

Figure 8. Fracture of original samples (left a and c) and optimized samples (right b and d).

Results are compared with predictions made by optimization software. In the simulations for optimization, a Young´s modulus of 1526 MPa was used for a better fitting to test data. 3D tetrahedral elements of side length 1 mm were used to obtain 12 elements per each mm^3 with 4 Jacobian points for each element. Therefore, for the original shape a total of 130087 nodes and 187218 elements were used to be eliminated during the optimization process and to define the light weight structure. However, the importance of optimization simulations is the reduction of weight using reliable material data. As given in Table 2, this varies depending on the material orientation. Comparison of the stress results of the optimized model with the original model is given in Table 4. This comparison shows that optimization software (FEA) resulted a 31.7% improvement in stiffness to mass ratio (with the same Young's modulus) while average results for the 3-point bending tests gave a 27.04% improvement. These percentage improvements between the simulation approach and experimental test approach have no significant difference.

Table 4. Comparison of FEA simulations with tests.

	Stiffness to mass (k_y/m) ratio (N/m/g)		
	FEA (E =1526 MPa)	3-point bending test (average)	Difference
Rectangular	2E+04	1.975E+04	1.25%
Optimized	2.634E+04	2.509E+04	4.75%
Improvement of k_y/m after optimization	31.7%	27.04%	

The work was repeated using another printer having better resolution. In this case, 10 samples for original design and 10 samples for optimized design were manufactured using Verowhite material and an Object30 Prime printer. For the last samples of Verowhite material, the support was not mechanically removed after solvent exposition for 24 hours in order to avoid possible damage to the structure. The weight of components is shown in Figure 9. Average values of each printing direction are taken for ABS-M30 samples. The weight of the Verowhite original design is very close to the computer aided design weight of 24.10 g considering density of 1170 kg/m^3 while ABS-M30 designs do not reach the expected density due to accuracy in layers and trajectories.

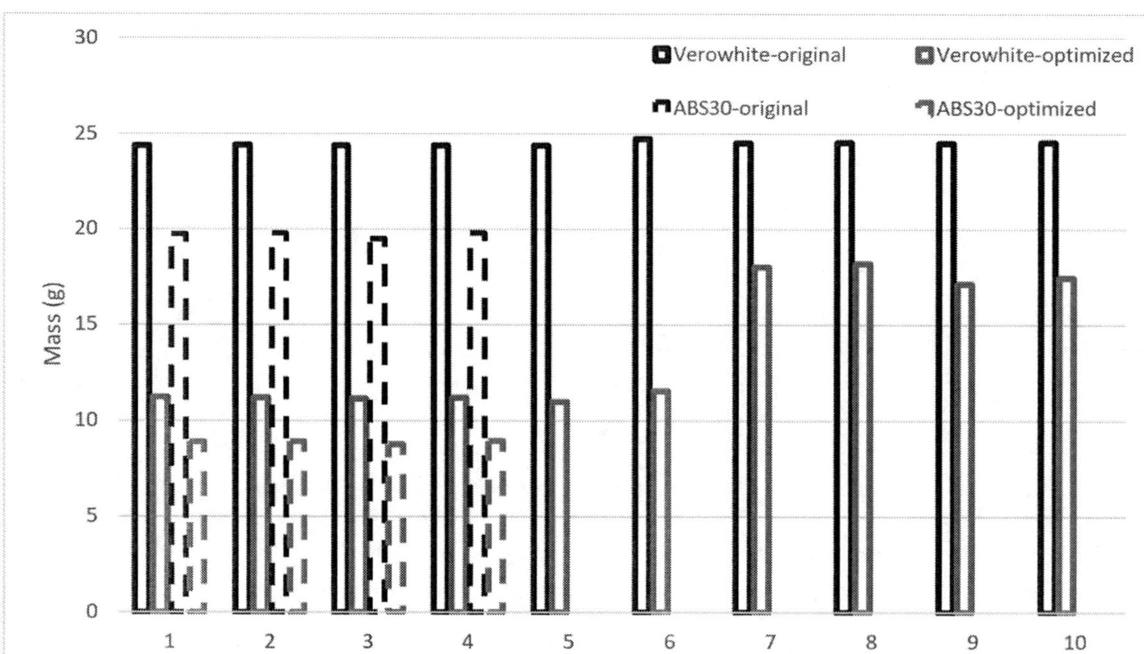

Figure 9. Mass comparison for Verowhite and ABS-M30.

On the other hand, Verowhite optimized shapes do not show any improvements in the stiffness to mass ratios as shown in Figure 10. A possible reason for this can be the sling of the Verowhite components that place the load in a point far from the center of the specimen due to a surface that looks like a polished surface. If focused on the original structures, the curves are very similar for ABS-M30 and Verowhite if the weights are normalized. The ABS 30 model, however, has shown improvement in stiffness to mass ratio. In both cases, for ABS-M30 and Verowhite, the optimized shape can take less deformation and less energy when compared to original designs.

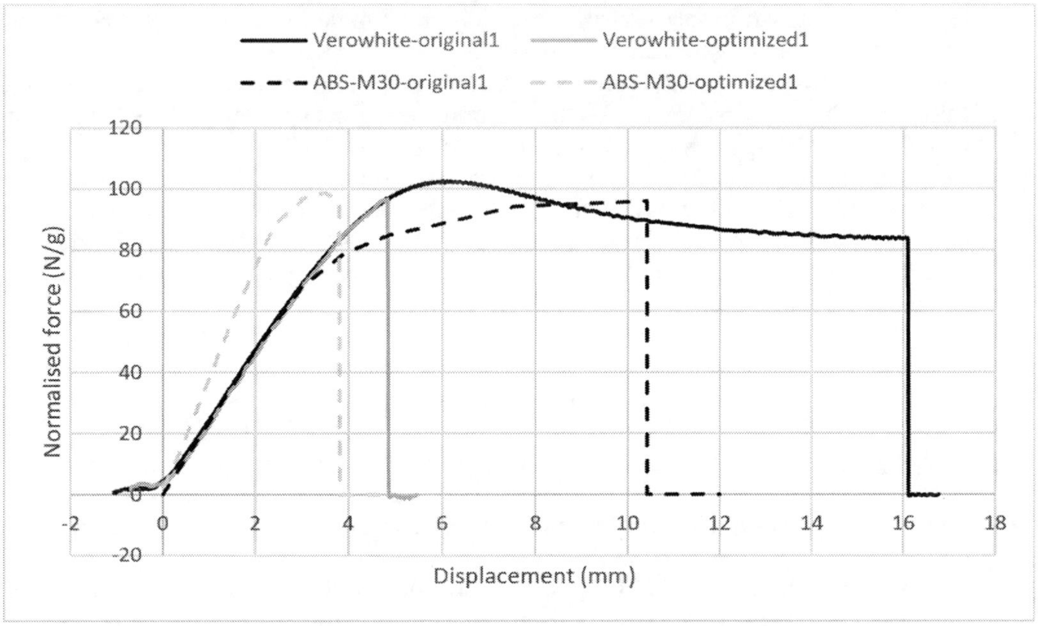

Figure 10. Comparison Verowhite to ABS-M30.

4. Conclusions

In this paper, a study focusing on topology optimization of additive manufactured part using Fortus 450mc (FDM technology) and Polyjet Object30 Primeis machines is reported. The selected part is tested using three-point bending test. The optimization targets maximum (improved) stiffness to mass ratio. It is observed that the obtained stiffness to mass ratio varies with printing orientation. For the ABS-M30 FDM manufactured parts, the topology optimized part under three-point bending test resulted in an average improvement in stiffness to mass ratio of 27.04%, worst case improvement of 12.61% and best case improvement of 45.03%. Similar design geometry was used to compare the optimized part performs by using Verowhite material printed in a Polyjet Objet30 Prime machine. It is observed that the part manufactured by Polyjet Objet30 Prime machine gives better accurate geometry compared with ABS-M30 using FDM Fortus 450mc. The part manufactured using Verowhite showed no improvement in stiffness to mass ratio. One reason for this lack of improvement is the glossy surface of the manufactured part when the support material is removed. The sliding of the test specimen under the three-point bending supports due to the glossy surface led to inaccuracies in the test results. It has also been observed that the optimized shapes provide a lower fracture displacement and absorb less energy when compared to original shapes.

Acknowledgements

This work has been performed within the Ris3Cat PlastFun (COMRDI16-1-0018) project (Plastic with functionalized surfaces), funded by ERDF through the Programa Operatiu de Catalunya 2014-2020.

References

[1] Akessa A D, Lemu H G and Gebisa A W 2017 Mechanical property characterization of additive manufactured ABS material using design of experiment approach, In: *Proc. ASME 2017 Inter. Mech. Eng. Congress and Exposition* (Nov. 3–9, 2017, Tampa, Florida, USA), IMECE2017-70144, V014T07A004.

[2] Gebisa A W and Lemu H G 2018 Investigating effects of Fused-Deposition Modeling (FDM) processing parameters on flexural properties of ULTEM 9085 using designed experiment, *Mater.* **11**(4), 500.

[3] Domingo-Espin M, Puigoriol-Forcada J M, Garcia-Granada A A, Llumà J, Borros S, and Reyes G 2015 Mechanical property characterization and simulation of fused deposition modeling Polycarbonate parts, *Mater. Des.* **83**, 670–77.

[4] Domingo-Espin M, Borros S, Agullo N, Amador A, Garcia-Granada A A, and Reyes G 2014 Influence of building parameters on the dynamic mechanical properties of polycarbonate fused deposition modeling parts, *3D Print. Addit. Manuf.* **1**(2), 70–7.

[5] Salazar-Martín A G, Pérez M A, García-Granada A A, Reyes G and Puigoriol-Forcada J M 2018 A study of creep in polycarbonate fused deposition modelling parts, *Mater. Des.* **141**, 414–25.

[6] Sedighi I, Ayatollahi M R, Bahrami B, Pérez-Martínez M A and Garcia-Granada A A 2019 Mechanical behavior of an additively manufactured polycarbonate specimen: tensile, flexural and mode I fracture properties, *Rapid Prototyp. J.*, **25**, 2019.

[7] Morales-Planas S, Minguella-Canela J, Lluma-Fuentes J, Travieso-Rodriguez J and García-Granada A A 2018 Multi Jet Fusion PA12 manufacturing parameters for watertightness, strength and tolerances, *Mater.* **11**(8), 1472.

[8] Hassani B and Hinton E 1998 A review of homogenization and topology optimization I—homogenization theory for media with periodic structure, *Comput. Struct.*, **69**(6), 707–17.

[9] Campbell S D, Sell D, Jenkins R P, Whiting E B, Fan J A and Werner D H 2019 Review of numerical optimization techniques for meta-device design, *Opt. Mater. Express*, **9**(4) 1842–63.

[10] Gebisa A W and Lemu H G 2017 A case study on topology optimized design for additive manufacturing, *Proc. of 1st COTech Conf.* (Stavanger: Nov 30 – Dec 1, 2017) Eds H G Lemu et al. (IOP Conference Series: Material Science and Engineering), **276**(1), 12026.

[11] Faskhutdinov R N, Dubrovskaya A S, Dongauzer K A, Maksimov P V and Trufanov N A 2017 Topology optimization of a gas-turbine engine part, In: Proc. Int. Conf. Mech. Eng., Autom. Control Syst. (27–29 Oct. 2016, Tomsk, Russian Federation): *IOP conf. series: Mat. Sci. Eng.* **177**(1), 12077.

[12] Wang X, Zhang D, Zhao C, Zhang P, Zhang Y and Cai Y 2019 Optimal design of lightweight serial robots by integrating topology optimization and parametric system optimization, *Mech. Mach. Theory*, **132**, 48–65.

[13] Croccolo D, De Agostinis M and Olmi G 2013 Experimental characterization and analytical modelling of the mechanical behaviour of fused deposition processed parts made of ABS-M30, *Comput. Mater. Sci.* **79** 506–18.

[14] Mueller J, Shea K and Daraio C 2015 Mechanical properties of parts fabricated with inkjet 3D printing through efficient experimental design, *Mater. Des.* **86** 902–12.

Optimization of well gas rates for offshore gas-condensate field

A I Ermolaev*, A A Nekrassov and I A Trubacheva

Russian State Gubkin University of Oil and Gas, Russia

* Corresponding author: ermolaev.a@gubkin.ru

Abstract. Condensate recovery enhancing on offshore gas-condensate fields cannot be realized using the cycling process due to significant financial costs and technical difficulties. Such a problem could be handled by the redistribution of well gas rates. This redistribution assumes the solution the following problem: maximize the total condensate production at a given time interval then gas production requirements are met. The results of solving the problem are functions, each representing the change in the well gas rate over the time. The problem is transformed to the linear optimization model so that the solution algorithms based on the linear programing theory can be applied. Their tests showed, that these algorithms easily find the solution of the optimization problem; and the condensate recovery is greater or equal than the one obtained by the optimization procedures implemented in Eclipse. The search time is an order of magnitude less than the one required by Eclipse.

1. Introduction

A way to enhance the condensate recovery efficiency for an offshore gas-condensate field is the production optimization — the allocation of well gas rates with the aim to maximize the condensate production under the required gas production. This is equivalent to the minimization of condensate loses in the reservoir, hence the loss of reservoir energy. Therefore, the problem is stated as following: to allocate the well gas rates so, that the reservoir energy loss is minimal, and the overall gas production requirements are fulfilled at any time interval.

The optimal rate allocation is influenced by the well interference. To take this effect into account, Charny [1] introduced the interaction coefficients for a homogenous oil reservoir with a far constant pressure external boundary. Meerov *et al.* [2] formulated the equation describing oil well interaction in the inhomogeneous reservoir, but didn't propose a method to evaluate the interaction coefficients. Zotov [3] proposed the system of equations for a homogeneous gas-condensate reservoir developing by a well pad and defined explicitly the interaction coefficients.

In contrast, in the current work a new approach is proposed allowing to take into account the well interference in the inhomogeneous gas-condensate reservoir. The interaction coefficients are interpreted as filtration resistance coefficients and expressed explicitly and then are used in the rate allocation algorithms. The solution algorithms are based on two hypotheses: the first one is analogous to Le Chatelier's principle [4]; the second one uses the relation between reservoir energy and well's bottom hole pressures. The obtained results confirm the validity of these hypotheses and showed better convergence rate compared with the numerical optimization methods implemented in the standard reservoir simulation software [5].

IOP Conf. Series: Materials Science and Engineering **700** (2019) 012013 doi:10.1088/1757-899X/700/1/012013

2. Definition of the filtration resistance coefficients

The key element of the problem statement is the relation between P_0 – the boundary pressure, P_i – the bottom-hole pressure (BHP) of the well i (i=1, 2,…, n) and $q_1,q_2,\dots,q_n$ – the well gas rates [3]:

$$\sum_{j=1}^{n} x_{ij}q_j = P_0^2 - P_i^2,\ i=\overline{1,n}, \tag{1}$$

where $n>1$, $q_i>0$, $P_0 \ge P_i > 0$, and $x_{ij} \ge 0$ are parameters that take into account the filtration resistance of gas flows from the drainage area of the j-th well to the drainage area of the well i.

As follows from equation (1), the increase of x_{ij} at constant flow rates and the boundary pressure leads to the decrease in BHPs. It means the increase of the reservoir energy required to move the gas-condensate mixture toward the wells. Therefore, the parameters x_{ij} are called the filtration resistance coefficients (FRC).

The approach to define FRC was presented earlier in [6]. The initial data for estimating the FRC are either the production data or the simulated data obtained by using a hydrodynamic simulator. Specifically, the $q_{it}>0$ and P_{it} are the well gas rate and the BHP of the well i, i=1,2,…,n, correspondently, as well as the P_{0t} is the boundary pressure at the time t.

According the equation (1), the search for the FRC is reduced to solving the system of linear equations:

$$\sum_{j=1}^{n} q_{jt}x_{ij} = P_{0t}^2 - P_{it}^2,\ i=\overline{1,n}, \tag{2}$$

The system (2) contains n equations and n^2 unknown variables – x_{ij} and, consequently, it has infinite number of solutions. Let's add to the problem statement the **Hypothesis I** allowing obtain the unique solution to the system of equations (2): *the formation provides maximum resistance to the filtration at any time.*

The Hypothesis I is a version of the Le Chatelier principle [4], whereby the system provides the maximum resistance to the action disturbing its equilibrium state. This formulation distinguish from the hypothesis used earlier for FRC evaluation [6] and is all the more valid.

Considering the equation (2) and the Hypothesis I, the search for x_{ij} is reduced to solving the problem:

$$\min_{i,j}\{x_{ij}\} \to \max_{x} \tag{3}$$

$$\sum_{j=1}^{n} q_{jt}x_{ij} = P_{0t}^2 - P_{it}^2,\ i=\overline{1,n}, \tag{4}$$

$$x_{ij} \ge 0,\ i=\overline{1,n},\ j=\overline{1,n}. \tag{5}$$

The problem (3) - (5) has an analytical solution – $\{x_{ij}^{(t)}\}$ [6]:

$$x_{ij}^{(t)} = \Delta_t / Q_t,\quad j \ne i, \tag{6}$$

$$x_{ij}^{(t)} = \Delta_t / Q_t + (\Delta_{it} - \Delta_t)/q_{it},\quad j = i, \tag{7}$$

where

$$Q_t \equiv \sum_{i=1}^{n} q_{it},\quad \Delta_t \equiv \min_{1\le i\le n}\{\Delta_{it}\},\quad \Delta_{it} \equiv P_{0t}^2 - P_{it}^2. \tag{8}$$

3. The mathematical formulation of the optimal distribution of the given overall gas production rate between the wells

After evaluating the FRC, it is possible to form the model of the optimal distribution of the given overall (summarized) gas production rate between the wells ($n > 1$) in accordance with the criterion of the minimal losses of the reservoir energy at the time t (or at the moment close to the t, while the values of FRC can be assumed as constants). In this case, the FRC are input data – $x_{ij}^{(t)}$, i = 1,2,…,n, j = 1, 2,…,n. The q_i and the P_i – the well gas rates and the BHPs at the time t, respectively — become the solution variables.

Let's introduce the parameters $\alpha_{it} \geq 0$:

$$\alpha_{it} = \left(P_{\max t}^2 - P_{it}^2\right) / q_{it}, \tag{9}$$

where $P_{\max t} \equiv \max_i \{P_{it}\}$.

The last notation shows that $\alpha_{it}=0$ for wells whose BHPs are equal to the P_{maxt}, i.e. they have the maximum values.

It follows from the equations (6) - (9) that for the time t, the system of equations (1) relating q_i and P_i can be reduced to the form:

$$P_i^2 = P_0^2 - \alpha_{it} q_i - (\Delta_t / Q_t) \sum_{j=1}^{n} q_j, \quad i = \overline{1,n}. \tag{10}$$

In accordance with the statement of the problem, let's introduce the parameter Q – the given overall gas production rate (summarized production rate) at some t, as well as the parameters a_i and b_i being respectively the minimal and the maximal limits for the gas production rate of the well i, i = 1,…,n. Assume that $0 \leq a_i < b_i < Q$.

The basis for the model of optimal distribution of the overall gas production rate between wells is the **Hypothesis 2:** *the minimization of reservoir energy losses is equal to the maximizing well's bottomhole pressures.*

Using our notations, the Hypothesis 2, and the equation (10), the problem takes the form:

$$\min_{i,j} \left[P_0^2 - \alpha_{it} q_i - (\Delta_t / Q_t) \sum_{j=1}^{n} q_j \right]^{\frac{1}{2}} \to \max_q \tag{11}$$

$$\sum_{i=1}^{n} q_i \geq Q \tag{12}$$

$$a_i \leq q_i \leq b_i, i = \overline{1,n}. \tag{13}$$

Note, that (i) it follows from equation (1) and the criterion (11) that the increase of the BHP of a producing well leads to the decrease of its flow rate, thus, the inequality (12) can be replaced by the equality constraint; (ii) neither P_0, nor the third summand depend on the index i in the radicand in the criterion (11); (iii) the radicand increases, its square root increases too; (iv) the product ($\alpha_{it} \cdot q_i$) in the radicand has the negative sign. The noted particularities of the problem (11)-(13) lead to simplification:

$$\max_{i,j} \{\alpha_{it} q_i\} \to \min_q \tag{14}$$

$$\sum_{i=1}^{n} q_i = Q \tag{15}$$

$$a_i \leq q_i \leq b_i, i = \overline{1,n}. \tag{16}$$

By introducing the additional variable (let's denote it by w) and the additional linear constraints, the problem (14)-(16) is replaced by the equivalent linear programming model:

$$w \to \min_{q,w} \tag{17}$$

$$\sum_{i=1}^{n} q_i = Q \tag{18}$$

$$\alpha_{it} q_i \leq w, i \in I_t \tag{19}$$

$$a_i \leq q_i \leq b_i, i = \overline{1,n}, \tag{20}$$

where $I_t=\{i:\alpha_{it}>0\}$.

Let's consider the case when all BHPs are equal to each other at the time t: $P_{it}=P_{maxt}$ to all $i=1,\ldots,n$. In this case, set $\alpha_{it}=1$, for $i=1,\ldots,n$.

4. Optimization of well gas rates at a given time point

Since the original problem (14)-(16) is equivalent to the problem (17)-(20), it suffices to find a solution to the problem (17)-(20). If the set $\{q_i^{ok}, w^{ok}\}$ is a permissible solution to the problem (17)-(20), then the set $\{q_i^{ok}\}$ is a feasible solution to the problem (14)-(16). If it is known that $\{q_i^{ok}\}$ is a feasible solution to the problem (14)-(16), then $\{q_i^{ok}, w^{ok}\}$ is a permissible solution to the problem (17)-(20), where w^{ok} is defined taking into account the criterion (17) and the constraints (19) as:

$$w^{ok}=\max_{1\le i\le n}\left\{\alpha_{it}q_i^{ok}\right\}. \tag{21}$$

Because the problem (17)-(20) is a linear programming model, its solution can be found using, e.g., the simplex method [7]. Nevertheless, in the following an algorithm is proposed that takes into consideration the peculiarities of the problem (17)-(20), and, thus, ensures the effective application even if the original problem (14)-(16) has a high dimension.

The results of the theoretical justification of the algorithm are presented below without the proof, which is caused by the limited volume of the article.

Before proceeding to the algorithm description, let's consider the necessary and sufficient conditions for the existence of permissible solutions to the problem (14)-(16) and the sufficient condition for their optimality.

The necessary and sufficient condition for the existence of the feasible solutions to the problem (14)-(16) is the satisfaction of the inequality:

$$\sum_{i=1}^{n}a_i\le Q\le\sum_{i=1}^{n}b_i.$$

Let's introduce w_t – the lower-bound estimate of the objective function (17) for permissible solutions to the problem (17)-(20):

$$w_t\equiv\max_{1\le i\le n}\left\{\alpha_{it}a_i\right\}. \tag{22}$$

The set of $\{q_i^{ok}\}$ is the optimal solution to the problem (14)-(16), if the following equality is satisfied:

$$w^{ok}\equiv w_t, \tag{23}$$

where w^{ok} is determined by the equation (21).

We now proceed to the investigation of conditions for obtaining an analytical solution to the problem (14)-(16). Let's introduce the notation: $\{q_i^{opt}\}$ – the optimal solution to the original problem (14)-(16), i.e. q_i^{opt} are optimal well gas rates for all $i=1,\ldots,n$. Let's denote $\gamma_i\equiv b_i-a_i$; $I_t\equiv\{i:\alpha_{it}>0\}$, $I_t=\{1,2,\ldots,m\}$; $I_0\equiv\{i:\alpha_{it}=0\}$, $I_0=\{m+1, m+2,\ldots,n\}$; additionally the variables G and Q_1 are calculated as

$$G\equiv Q-\sum_{i=1}^{n}a_i;\quad Q_1=Q-\sum_{i\in I_0}b_i. \tag{24}$$

Let's consider cases regarding the parameters G and γ_i for $i\in I_0$.

1. If the equality

$$G=\sum_{i\in I_0}\gamma_i. \tag{25}$$

is satisfied, then the optimal solution to the problem (14)-(16), $\{q_i^{opt}\}$, takes the form:

$$q_i^{opt}=\begin{cases}b_i, i\in I_0\\ a_i, i\in I_t.\end{cases} \tag{26}$$

2. If the inequality

$$G < \sum_{i \in I_0} \gamma_i . \tag{27}$$

is satisfied, we introduce n_0 – the number of elements in the ensemble I_0, i.e. n_0 is the number of α_{it}, that are equal to zero.

2.1. Let $\gamma_i \geq G/n_0$ be fulfilled for all $i \in I_0$ Then the optimal solution to the original problem (14)-(16) under the preceding condition (27) takes the form:

$$q_i^{opt} = \begin{cases} \dfrac{G}{n_0} + a_i, i \in I_0 \\ a_i, i \in I_t. \end{cases} \tag{28}$$

2.2. Let's consider the case than there exists $i \in I_0$ for which $\gamma_i < G/n_0$, and the condition (27) is satisfied. Let's introduce $I_1 \equiv \{i: \gamma_i < G/n_0, i \in I_0\}$, $I_2 \equiv \{i: \gamma_i \geq G/n_0, i \in I_0\}$, $n_1 \equiv |I_1|$, $n_1 \leq n_0$. Note, that $I_0 = I_1 \cup I_2$ and $I_2 \neq \varnothing$, because the inequality (27) is fulfilled.

Let's denote parameters:

$$H \equiv G - \sum_{i \in I_1} \gamma_i .$$

$$\beta_i \equiv H \cdot \gamma_i \cdot \left(\sum_{j \in I_2} \gamma_j \right)^{-1}, i \in I_2 .$$

Then, for this case, the optimal solution to the original problem (14)-(16) takes the form:

$$q_i^{opt} = \begin{cases} a_i, i \in I_t \\ b_i, i \in I_1 \\ \beta_i + a_i, i \in I_2 . \end{cases} \tag{29}$$

3. If the inequality

$$G > \sum_{i \in I_0} \gamma_i . \tag{30}$$

is satisfied, we introduce parameters:

$$\delta \equiv \sum_{i=1}^{m} \frac{1}{\alpha_{it}}; \; w^* = \frac{Q_1}{\delta}; \; q_i^* = \frac{w^*}{\alpha_{it}}, \; i \in I_t . \tag{31}$$

If the set $\{q_i^*\}$, $i \in I_t$ satisfies the constraints (16), then the optimal values of well gas rates in the problem (14)-(16) are determined by the following rule:

$$q_i^{opt} = \begin{cases} b_i, i \in I_0 \\ q_i^*, i \in I_t . \end{cases} \tag{32}$$

Therefore, the items 1÷3 define the conditions of existence an analytical solution to the original problem (14)-(16). These solutions are expressed by equations (26), (28), (29), and (32). The sufficient optimality condition (23) is fulfilled for all such solutions.

4. Let's consider the last possible case, when the inequality (30) is satisfied, but the set $\{q_i^*\}$, $i \in I_t$, does not satisfy the constraints (16).

Firstly, it should be noted, that and in this case, the optimal values of the well gas rates are equal to the maximal permitted values for $i \in I_0$:

$$q_i^{opt} = b_i, i \in I_0 .$$

The search for the optimal values of the problem variables q_i is an iterative procedure for $i \in I_t$. The iterations are aimed to obtain the variables q_i, $i \in I_t$, which should be close to q_i^*, $i \in I_t$, and satisfy the constraints (16) at the same time, whereby the $I_t = \{1,2,\ldots,m\}$ is the same as above.

The least squares criterion is proposed as the measure of proximity to the set $\{q_i^*\}$. Then the search of q_i, $i \in I_t$, is reduced to the solving the quadratic programming problem:

$$\sum_{i=1}^{m} (q_i - q_i^*)^2 \to \min_q \tag{33}$$

$$\sum_{i=1}^{m} q_i = Q_1 \tag{34}$$

$$a_i \le q_i \le b_i, i = \overline{1,m}, \tag{35}$$

where q_i^* are determined by equations (31), and Q_1 is determined by the equation (24).

The solution to the problem (33)-(35) is based on the use of Lagrangian relaxation [8]. This allows us to solve m minimization problems with one variable at each iteration instead of the solving the problem (33) - (35) with m unknowns, i.e. to find q_i for each $i \in \{1,2,\ldots,m\}$:

$$(q_i - q_i^*)^2 - \lambda^{(k)} q_i \to \min_{q_i} \tag{36}$$

$$a_i \le q_i \le b_i. \tag{37}$$

If $q_i(\lambda^{(k)})$ is used to denote the solution to the i-th problem (36), (37), then

$$q_i(\lambda^{(k)}) = \begin{cases} a_i, & c_i(\lambda^{(k)}) < a_i \\ c_i, & a_i \le c_i(\lambda^{(k)}) \le b_i \\ b_i, & c_i(\lambda^{(k)}) > b_i, \end{cases} \tag{38}$$

$$c_i(\lambda^{(k)}) = \frac{2q_i^* + \lambda^{(k)}}{2}, \tag{39}$$

where $i \in \{1,2,\ldots,m\}$.

The set of optimal solutions to the problem (36)-(37) – $\{q_i(\lambda^{(k)})\}$ is an approximation of the optimal solution to the problem (33)-(35) at iteration k.

Now it is necessary to check whether the constraint (34) is satisfied for the obtained set $\{q_i(\lambda^{(k)})\}$:

i) if

$$\sum_{i=1}^{m} q_i\left(\lambda^{(k)}\right) = Q_1, \tag{40}$$

then $\{q_i(\lambda^{(k)})\}$ is the optimal solution to the problem (33)-(35), which closeness to the solution of the original problem (14)-(16) of can be estimated; otherwise, the following cases should be checked;

ii) if

$$\sum_{i=1}^{m} q_i\left(\lambda^{(k)}\right) < Q_1,$$

then the next iteration is required, i.e. update $k=k+1$, set a new value of the penalty coefficient $\lambda^{(k)}>\lambda^{(k-1)}$ and proceed to the calculation of the new values $q_i(\lambda^{(k)})$ using the equations (38), (39);

iii) if

$$\sum_{i=1}^{m} q_i\left(\lambda^{(k)}\right) > Q_1,$$

then the next iteration is required, i.e. update $k=k+1$, set a new value of the penalty coefficient $\lambda^{(k)}<\lambda^{(k-1)}$ and proceed to the calculation of the new values $q_i(\lambda^{(k)})$ by the equations (38), (39).

Let's move on to the choice of penalty coefficients $\lambda^{(k)}$. It follows from equations (38) and (39) that $q_i(\lambda^{(k)})$ are the increasing functions of $\lambda^{(k)}$, $i=1,2,\ldots,m$. Therefore, the left-hand side of the equation (40) is also the increasing function of $\lambda^{(k)}$. The satisfiability of the equation (40) is equivalent to the optimality condition for solutions to the problem (33)-(35). Therefore, the search for $\lambda^{(k)}$ is reduced to the finding the root of the equation:

$$\sum_{i=1}^{m} q_i\left(\lambda^{(k)}\right) - Q_1 = 0. \tag{41}$$

Some numerical methods for finding the roots of a nonlinear equation with one variable can be applied to solve the equation (41) with respect to $\lambda^{(k)}$. For instance, the dichotomy method [9] can be used. Iterations continue until the equality (40) will be fulfilled with the required accuracy:

$$\left|\sum_{i=1}^{m} q_i\left(\lambda^{(k)}\right) - Q_1\right| / Q_1 < \varepsilon, \tag{42}$$

where $\varepsilon>0$ is the required accuracy.

The set $\{q_i(\lambda^{(k)})\}$, $i=1,2,\ldots,m$, becomes an approximate solution to the problem (33)-(35), when the inequality (42) is realized. In this case, the set $\{q_i^{ok}\}$

$$q_i^{ok} = \begin{cases} b_i, i \in I_0 \\ q_i(\lambda^{(k)}), i \in I_t \end{cases}$$

is accepted as the final solution of the problem (14)-(16).

The set $\{q_i^{ok}, w^{ok}\}$ will become the final solution to the equivalent problem (17)-(20), where w^{ok} is found by the equation (21). Let w^{opt} be the value of the objective function (17) of the solution to the problem (17)-(20). Then, the inequality $w_t \leq w^{opt} \leq w^{ok}$ has to be satisfied, where w_t is calculated by the equation (22). Therefore, the relative residual of the solution $\{q_i^{ok}, w^{ok}\}$ can be evaluated by the parameter θ:

$$\theta = \left(w^{ok} - w_t\right) / w_t.$$

Because the problem (17)-(20) and the problem (14)-(16) are equivalent, the parameter θ can serve as the measure of the solution accuracy to the original problem (14)-(16).

5. Well rate optimization for the long-time production period

The described in the previous section algorithm may be applied for the prediction of the gas well rate changes in the time. Let's consider a case, where at the initial time step $t = 0$ the real well rates, BHPs, and the boundary pressure are known. Let's assume that at the time $t \in \{0,1,\ldots,T\}$ the overall production rates are given: $Q^{(t)}$, $t = 0,1,\ldots,T$. One has to find the well gas production rates: $q_i^{(t)}$, which ensure the minimal loses of the reservoir energy during the period $[0, T]$ under the fulfillment of the constrains:

$$\sum_{i=1}^{n} q_i^{(t)} = Q^{(t)}, t = \overline{0,T} \tag{43}$$

$$a_i^{(t)} \leq q_i^{(t)} \leq b_i^{(t)},\ i = \overline{1,n},\ t = \overline{0,T}, \tag{44}$$

where $a_i^{(t)}$ and $b_i^{(t)}$ are the minimal and the maximal limits of the well gas rates for the well i, $i=1,\ldots,n$, at the time $t \in \{1,2,\ldots,T\}$, respectively.

The solution of the stated problem is separated into $T+1$ stages. At the first stage ($t=0$), the parameters α_{i0} are estimated on the basis of the real production data, and the problem (14), (43), (44) is solved at $t=0$. Using our notations, the set $\{q_i^{(0)}\}$ is this solution. Then, during the time period [0, 1], the development process is simulated using hydrodynamic modeling with the well production rates $\{q_i^{(0)}\}$.

The new well production rates and BHPs for all wells will be estimated as result of simulation at the time step $t=1$. Well production rates $q_i^{(0)}$ and the simulated BHPs become analogous real data. The new values of α_{i1} are found, and, then, the problem (14), (43), (44) is solved at $t = 1$, etc. Such calculations are repeated until the time $t = T$. By this mean, the solutions obtained at the previous stage become the input data for the next one.

Let's consider the cases where the number of wells is changed during the production period. Let the well n will be shut at the moment $t=s$. Then, at the time step s all calculation procedures are carried out taking into account the well n. Based on these results, α_{is}, well production rates, and BHPs are evaluated and will be further considered as the real data. Then, the time step s is recalculated taking into account the "real" data but without modeling the well n.

Let's consider a case then at the time step s the well $(n+1)$ comes into operation. Then the production rates and BHPs can be found using the standard group control option including the new well. The obtained data are used as the "real" ones. Taking these data into account the α_{is}, $i=1,\ldots,(n+1)$ are estimated and then the problem (14),(43),(44) is solved at $t=s$.

6. Case study

To evaluate the capabilities of the developed method the results were compared with methods of the overall production rate distribution implemented into the standard reservoir simulation software. Two cases were considered: the results of the implemented optimization algorithm [5] and the rate distribution described above. The final goal of optimization was to maximize the ultimate condensate recovery.

Algorithms were applied to the stochastically created synthetic model of an offshore gas production facility with six production horizontal wells. The wells surround the production unit with equal production sectors of 60°. The length of well's production interval is 500 m. The offset from the production unit is also 500 m (see Figure 1). Porosity is normally distributed.

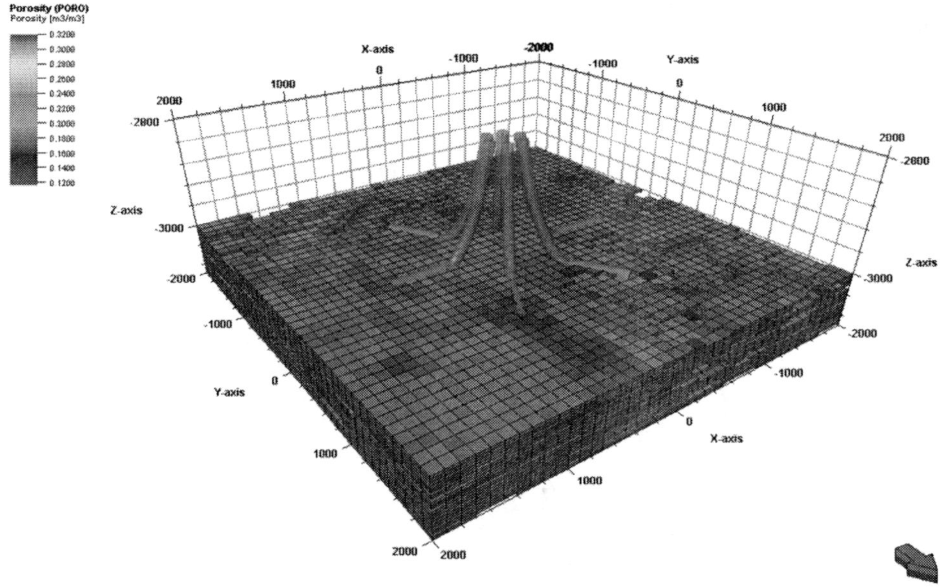

Figure 1. Porosity distribution in the synthetic model.

The considered production area is 4×4 km with the total thickness of 100 m. Parameters of the model were generated stochastically: the mean net to gross ratio is 0.5, the average permeability is 250 mD and the average porosity is 0.23. The reservoir was saturated with the gas-condensate mixture and the residual water. The gas saturation is 80%. The initial pressure was 32 MPa. The gas-condensate characteristic is shown in the Figure 2.

IOP Conf. Series: Materials Science and Engineering **700** (2019) 012013 doi:10.1088/1757-899X/700/1/012013

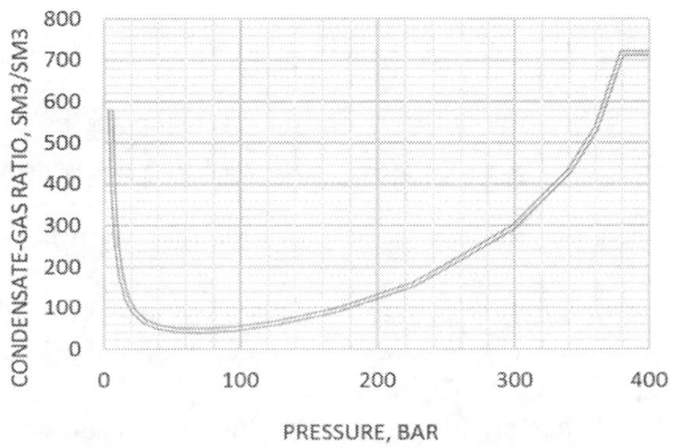

Figure 2. Condensate-gas ratio of the reservoir mixture.

The development strategy consisted of two stages. Firstly, the wells come into operation one by one with the same gas production rate of 1.34 mln. st. m^3/day. The interval between the well production starts was six months. After 2 years the swinging gas production begins: for 6 months the overall gas production rate was set to 7.6 mln. st. m^3/d (simulating the summer period) and for 6 months the overall gas production rate was set at 8.4 mln. st. m^3/d (simulating the winter period). The swinging production period continued 5 years. The upper and the lower limits for well's gas rates were set as b_i=2.0 mln. st. m^3/d and a_i=0.5 mln. st. m^3/d, i=1,…,n , respectively. Results are shown in the Figure 3. The cumulative condensate recovery for ECLIPSE OPTIMIZE option was 6.09 mln. m^3, for the proposed method – 6.12 mln. m^3.

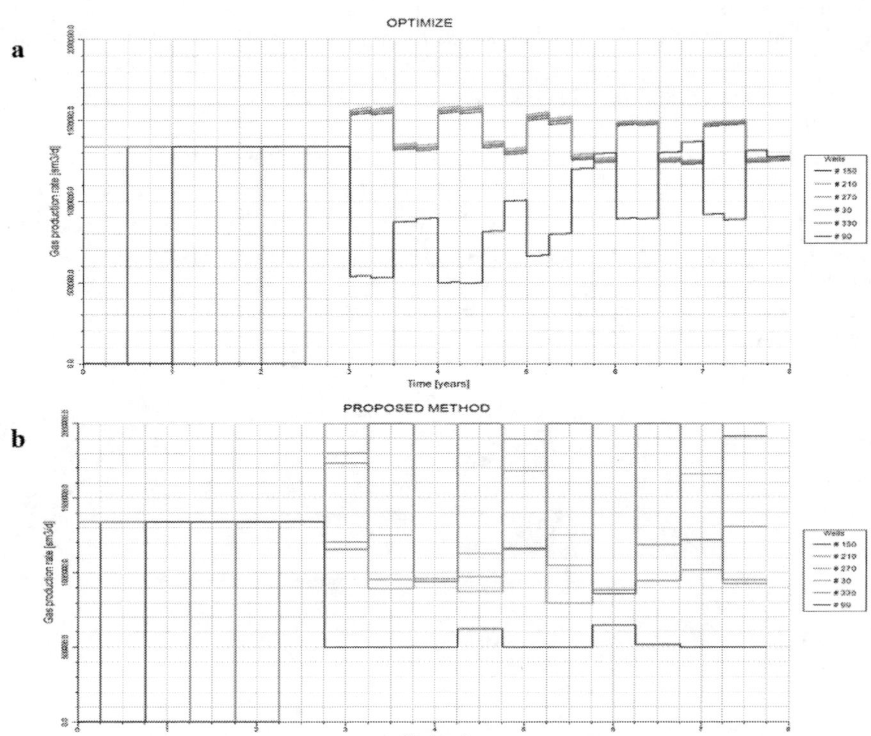

Figure 3. Well's gas rate changing in time: a – OPTIMIZE (Eclipse); b – proposed method.

IOP Conf. Series: Materials Science and Engineering **700** (2019) 012013 doi:10.1088/1757-899X/700/1/012013

7. Conclusions

1. The developed method allows to obtain either the optimal or an approximate solution, the error of which can be evaluated. The conditions required to obtain an analytical solution to optimization problem are defined. If these conditions are not met, the method leads to the iterative procedure, whereby the m parameter task is replaced by m one-dimensional optimization problems. This allows to avoid computational difficulties characteristic to the large-dimensional problems. Thus, the method can be recommended for solving problems of optimizing the development of real gas-condensate fields.
2. The given examples of the numerical study of the developed models and methods have confirmed the validity of the proposed Hypotheses and the computational capabilities of these optimization models and algorithms.
3. The case study showed that at the practically the same efficiency with respect to the objective function, the solving time required by the proposed method is significantly less than the one needed by the Eclipse internal optimization features.
4. The main application area of the developed algorithms is the forming the starting point for further search of the optimal field development strategy.

References

[1] Charny I A 1949 *Podzemnaja gidrogazodynamica* (Moscow: Gostoptekhizdat) p 397 (in Russian)

[2] Berschansky Y M, Kulibanov V N, Meerov M V and Pershin O J 1983 Upravlenie razrabotkoj neftjanykh mestorozhdenij Meerov M V (Moscow: Nedra) p 309 (in Russian)

[3] Zotov G A 1998 Produktivnost i dobyvnye vozmozhnosti kusta gazovykh skvazhin *Voprosy metodologii I novykh technologiy razrabotki mestorozhdenij prirodnogo gaza* (Moscow: Vniigaz) p 116-134 (in Russian)

[4] Landau L D and Lifshitz E M 1980 *Statistical physics* Vol. 5 (Oxford: Butterworth-Heinemann) p 544.

[5] Sokolnikoff I S, Redheffer R M 1967 *Mathematics of Physics and Modern Engineering* (New York: McGraw-Hill) p 828

[6] Ermolaev A, Nekrassov A and Trubacheva I 2018 Optimization of well rates for condensate field development *ECMOR XVI - 16th European Conf. on the Mathematics of Oil Recovery.*

[7] Chvatal V 1983 *Linear programming* (New York: W. H. Freeman and Company) p 425.

[8] Minoux M 1986 *Mathematical Programming Theory and Algorithms* (New York: John Wiley & Sons) 489 p.

[9] Hamming R W 1962 *Numerical methods for scientists and engineers* (New York: MC Graw-Hill Book Company) p 411.

Optimal thermobaric parameters determination of natural gas dehydration in LNG production

E B Fedorova*, V B Mel'nikov, E B Gafarova and V A Fedorova

Gubkin Russian State University of Oil and Gas (National Research University), Russia

* Corresponding author: fedorova.e@gubkin.ru

Abstract. Gas dehydration is one of the main technological processes in natural gas treatment for pipeline transportation and LNG production. The main task of this research is to increase the efficiency of adsorption gas dehydration. This report presents the results of a comparative assessment and determination of the optimal thermobaric parameters of the process of natural gas dehydration with KSMG type of silica gel and NaA-BS type of zeolite-containing adsorbent. It has been shown that reducing the temperature of the natural gas before dehydration reduces the mass of loaded adsorbent, the mass of adsorber, the consumption of regeneration gas, cooling gases, and fuel gas for heating regeneration gas. Recommendations on the dehydration unit placement after pre-cooling stage in order to reduce capital and operating costs are given at the end of the article. The results are important in the design and modernization of technological processes of the LNG production.

1. Introduction

An adsorption gas dehydration plays an important role in the natural gas treatment for natural gas liquefaction [1, 2]. When liquefying natural gas, the important technological parameter is the moisture dew point temperature - the moisture content in natural gas before liquefaction must be less than $1 \cdot 10^{-6}$ m^3/nm^3. Gas dehydration is carried out to prevent the formation of hydrates in process piping and vessels when producing liquefied natural gas (LNG). Hydrate deposits inside piping and vessels reduce their inner size, increase the hydraulic resistance, which leads to additional energy consumption, and operational costs, and sometimes to emergency shutdowns [3].

In general, such adsorbents as silica gel and aluminum silicates of alkaline elements are used in industrial processes of natural gas adsorption drying. Adsorbents in the form of beads having a diameter of from 2 to 8 mm, and in the form of extrudates (thin cylinders) having a diameter of from 1 to 6 mm are used [3-6].

2. A comparative evaluation of silica gel and zeolite for natural gas dehydration

The purpose of this study is a comparative evaluation of silica gel KSMG type and zeolite NaA-BS type (4A type) adsorbents and determination of the optimal temperature and pressure of adsorption natural gas dehydration process in order to increase its effectiveness in the natural gas treatment for liquefaction. Properties of the adsorbents produced by the Salavat catalyst plant are presented in Tables 1 and 2.

IOP Conf. Series: Materials Science and Engineering **700** (2019) 012014 doi:10.1088/1757-899X/700/1/012014

Table 1. Physical and chemical properties of silica gel KSMG with diameter 4.5mm

Specific surface (m^2/g)	Pore volume (cm^3/ g)	Bulk density (g/cm^3)
720	0.4	0.85

Table 2. Physical and chemical properties of granulated zeolite NaA-BS

Size of the granules on the average diameter (mm)	Pore volume (cm^3/ g)	Bulk density (g/cm^3)
4.5 ± 0.5	0.38	0.82

The dehydration process pressure doesn't impact on an adsorption activity of adsorbents, so the pressure is determined by a liquefaction unit pressure. However, the adsorption process temperature has a significant effect on the process efficiency. A drying unit with inlet gas pressure of 4.5 MPa is considered for comparison of adsorbents properties. The temperature of the incoming gas stream is 20°C. At these thermobaric conditions the moisture content in natural gas is 30.44 mg/m^3.

An experiment was carried out to study the temperature influence on the equilibrium activity of the selected adsorbents. An adsorbent sample was placed in a U- tube, and the tube was installed in a thermostat. The temperature inside the thermostat was changed from plus 20 to minus 20 ° C with the step of 5 ° C. Air saturated with water vapor was passed through the adsorbent sample at each step of the temperature change. The tube with the adsorbent sample was weighed before the experiment and after getting the equilibrium. The adsorption activity of the sample was calculated by the equation

$$a_t = (G_S - G_0) \cdot G_0^{-1} \tag{1}$$

where a_t is an equilibrium adsorption activity of the sample at temperature t; G_S is a weight of the sample at the equilibrium state (g); G_0 is a sample weight before passing saturated air through adsorbent (g).

The dependence of the adsorbents KSMG and NaA-BS equilibrium adsorption activity on temperature at a pressure of 4.5 MPa is shown in Figure 1. It can be noticed that, with a decreasing in temperature from 20°C to minus 20°C, silica gel activity increases from 4 to 6.7% mass, while zeolite activity increases from 16 to 27.8% mass.

The results obtained allow us to note that lowering the temperature of the adsorption gas dehydration process from 20°C to minus 20°C reduces the amount of adsorbent to provide the required quality of gas dehydration. For example, the calculations of adsorbent charge for two types of adsorbents were made, using a simplified equation [6]:

$$G_a = \Delta W \cdot G_g \cdot \tau_c \cdot (A_t^{\infty})^{-1} \tag{2}$$

where G_a is the amount of adsorbent (kg); ΔW is the amount of moisture that must be removed (kg/Nm^3); G_g is installation capacity (Nm3/h); τ_c is time of adsorption stage (h); A_t^{∞} is the equilibrium activity of the adsorbent (% mass).

Taking the dehydration unit capacity - 300 t/h, the amount of moisture that must be removed – 30.44 mg/Nm3, time of adsorption stage – 24 h, results of adsorbent amount calculations are presented in Table 3.

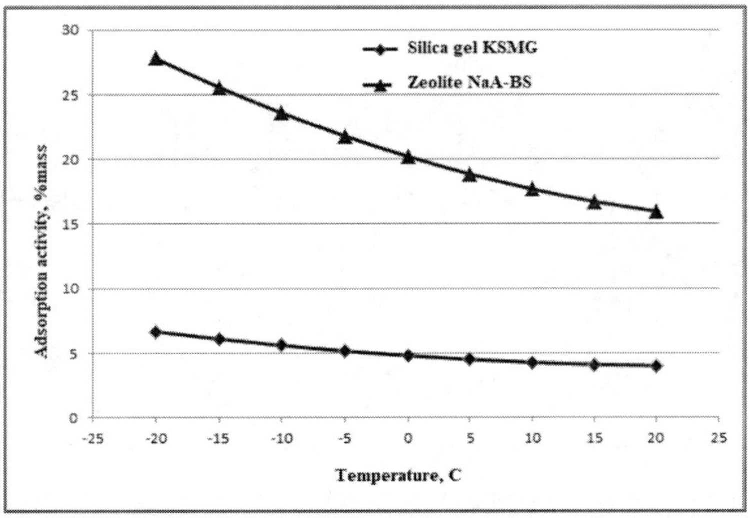

Figure 1. Dependence of the adsorbent's equilibrium adsorption activity on temperature at a pressure of 4.5 MPa.

Table 3. Adsorbent charge, t

Temperature, C	20	15	10	5	0	-5	-10	-15	-20
KSMG	18.65	18.22	17.52	16.62	15.58	14.47	13.35	12.25	11.20
NaA-BS	4.68	4.47	4.23	3.97	3.70	3.43	3.17	2.92	2.69

According to calculation results, at temperature of minus 20°C the amount of KSMG silica gel is 1.66 time less than at temperature of 20°C, the amount of NaA-BS zeolite is 1.74 time less for the same temperature change.

At the same time, the amount of NaA-BS zeolite in comparison with KSMG silica gel is approximately 4 times less. The reduction in the required mass of the adsorbent charge for a given degree of gas drying allows the size of the adsorbers and, consequently, the steel intensity and the cost of the adsorber to be reduced. It means that low-temperature adsorption process with zeolite could increase the efficiency of gas dehydration.

3. Research on low-temperature adsorption for LNG production and optimal thermobaric parameters determination

The low-temperature adsorption unit can be accomplished into the LNG train in different ways. For example, two of them are:

1) by cooling the gas stream entering the drying unit due to heat exchange with the boiloff gases from the end-flash process (Figure 2);
2) by placing the drying unit in the train at a pre-cooling stage of liquefaction unit (Figure 3).

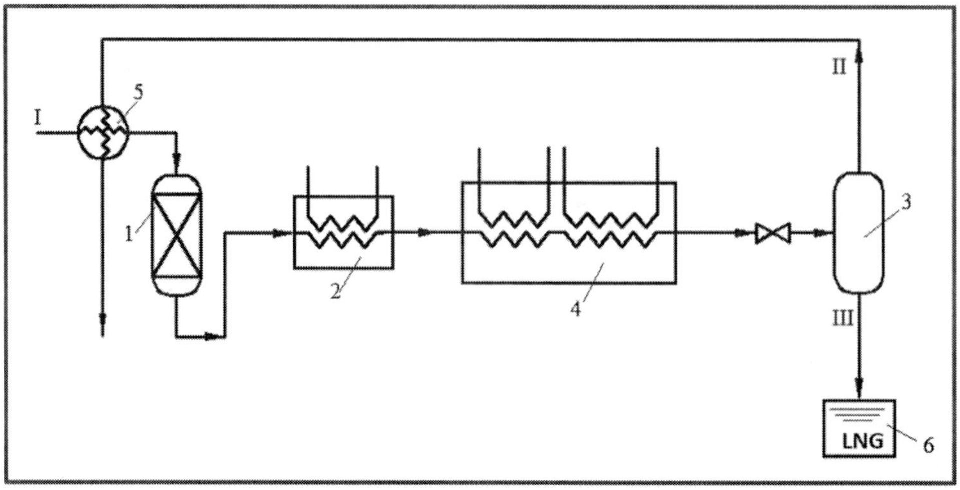

Figure 2. Scheme of natural gas cooling with boiloff gas upstream the dehydration unit (option 1)
(I - Natural gas; II - boiloff gas; III - LNG; 1 - adsorption drying unit; 2 - pre-cooling unit; 3 – separator; 4 – liquefaction unit; 5 - recuperative heat exchanger; 6 - LNG storage tank)

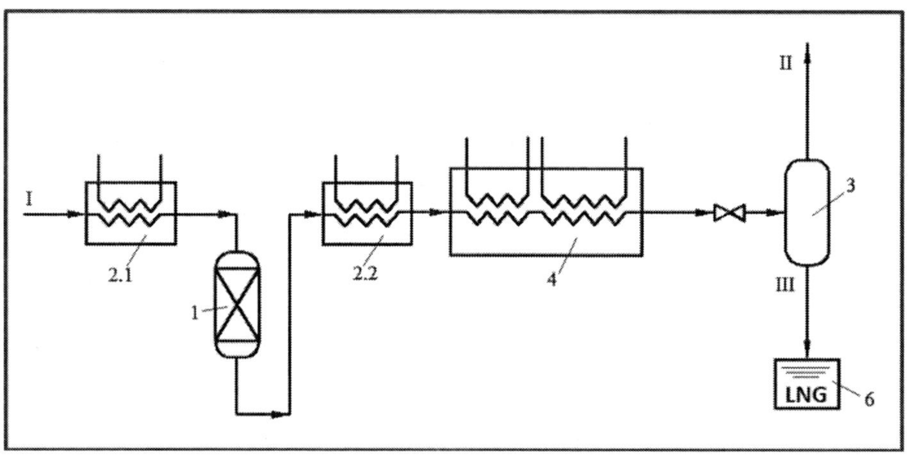

Figure 3. Scheme of natural gas cooling upstream the dehydration unit (option 2)
*(I - Natural gas; II - boiloff gas; III - LNG; 1 - adsorption drying unit;
2 - pre-cooling unit; 3 – separator; 4 – liquefaction unit; 6 - LNG storage tank)*

Feed gas from the main gas pipeline corresponds to STO Gazprom 089-2010 (the temperature of the moisture dew point varies from minus 10 to minus 20 ° C, depending on the season and the macroclimatic region). To prevent possible liquid formation from gas stream, which may include moisture and hydrocarbons, the temperature to which gas flow should be cooled at the inlet of dehydration unit, should not be below the moisture and hydrocarbons dew point.

Thermodynamic calculations of both options were carried out with the Aspen HYSYS. To simplify the calculations, the following assumptions were made:

–pure methane was taken as a feed stream;

–the hydraulic resistance of heat exchangers and heat exchange with the environment weren't taken into account.

Calculation results of the first option (cooling gas stream with boiloff gas) showed an insignificant decrease in the temperature of natural gas stream: from 20 to 14°C (stream I downstream the heat

IOP Conf. Series: Materials Science and Engineering **700** (2019) 012014 doi:10.1088/1757-899X/700/1/012014

exchanger 5, Figure 2). This is due to small amount and low heat transfer coefficient of the boiloff gas. Thus, when natural gas flow is 300 t/h, the boil-off gas flow is 13 t/h. Downstream heat exchanger 5, the temperature of boiloff gas rises from minus 159.5 to 11°C.

Second technological option of low-temperature dehydration provides gas cooling in the pre-cooling refrigeration cycle (Figure 3) and has some features. Placing the dehydration unit between the pre-cooling cycle and liquefaction cycle in the train can cause the formation of hydrates, because the gas flow temperature at this stage might vary from minus 30°C (for example, for the *APCI* process *C3MR*) down to minus 80°C (for example, in winter for the *Shell* process *DMR*).

As a rule, pre-cooling cycles, regardless of refrigerant composition, are multi-stage, so that energy efficiency increases. In order to select the optimum temperature of the low-temperature adsorption process, it is possible to place the dehydration unit between separate stages of the pre-cooling cycle.

The effect of the low-temperature adsorption process with decreasing temperature is shown in Figure 4. The changes in the main parameters of the unit are presented as a function

$$\varphi(t) = 1 - (A_t \cdot A_{20}^{-1}) \quad (3)$$

where A_t is parameter value at temperature t; A_{20} is parameter value at 20°C.

The following parameters were chosen as parameters of the dehydration unit:

- mass of loaded adsorbent;
- mass of an adsorber;
- regeneration gas flow rate;
- gas flow rate for adsorber cooling;
- fuel gas flow rate for regeneration gas heating in the furnace.

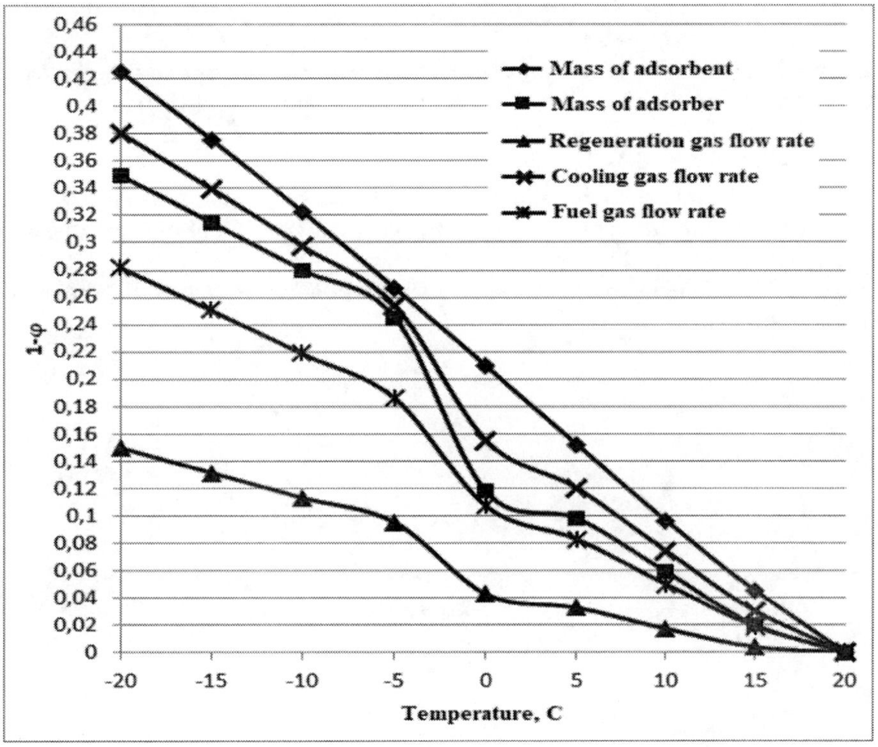

Figure 4. Dependence of dehydration unit parameters on the temperature

IOP Conf. Series: Materials Science and Engineering **700** (2019) 012014 doi:10.1088/1757-899X/700/1/012014

Thus, reducing temperature of natural gas entering the adsorption drying unit allows to reduce mass of loaded adsorbent, mass of adsorber, regeneration gas flow rate and cooling gas flow rate, as well as to reduce fuel gas flow rate. Therefore, placing an adsorption drying unit between the stages of the pre-cooling cycle, without additional energy consumption for cooling the natural gas stream, both capital and operating costs for gas dehydration at LNG production can be significantly reduced.

4. Conclusions

The comparative evaluation of silica gel KSMG and zeolite NaA-BS adsorbents was carried out. It was shown that with a decrease in temperature, the activity of silica gel and zeolite increases, therefore, a decrease in temperature of the adsorption gas dehydration leads to an increase in the energy efficiency of the process. A comparison of two types of adsorbents has shown that adsorption activity of zeolite is sufficiently more than the one of silica gel at any temperature. The amount of NaA-BS zeolite for adsorber loading in comparison with KSMG silica gel is approximately 4 times less. The reduction in the required mass of the adsorbent charge for a given degree of gas drying allows the size of the adsorbers and, consequently, the steel intensity and the cost of the adsorber to be reduced. Two ways to reduce the temperature of the drying process were considered, it is more efficient to place the drying unit between the stages of the pre-cooling cycle at the liquefaction train. As a result, of calculations, the dependence of dehydration unit parameters on gas temperature was obtained: reducing the process temperature leads to decrease in both capital (adsorbent charge mass, adsorber steel intensity) and operating costs for gas dehydration (regeneration, cooling and fuel gas flow rates).

References

[1]. Fedorova E B 2011 *The state-of-the-art and development of the global LNG industry: technology and equipment*. (Moscow, Gubkin Russian State University of Oil and Gas (NRU)), 159.

[2]. Fedorova E B and Mel'nikov V B 2015 The peculiarity of natural gas treatment in LNG production. *Proc. of the Gubkin Russian State University of Oil and Gas (NRU)*. **4** (281), 100-114.

[3]. Melnikov V B 2017 *Commercial collection and processing of gas and gas condensate*. Textbook. (Moscow, Gubkin Russian State University of Oil and Gas (NRU)), 464.

[4]. *Technology of natural gas and gas condensate processing 2002: Handbook in 2 volumes*. (Moscow, Nedra-Business Center), **1**, 517.

[5]. Skoblo A I, Molokanov Yu K, Vladimirov A I and Schelkunov V A 2012 *Processes and equipment for oil refining and petrochemistry*. (Moscow, Gubkin Russian State University of Oil and Gas (NRU)), 725.

[6]. Bekirov T M and Lanchakov G A 1999 *Gas and condensate treatment technology*. (Moscow, Nedra), 596.

IOP Conf. Series: Materials Science and Engineering **700** (2019) 012015 doi:10.1088/1757-899X/700/1/012015

Fretting fatigue and wear of mechanical joints: Literature study

Ø Karlsen* **and H G Lemu**
Department of Mechanical and Structural Engineering and Materials Science, University of Stavanger, Stavanger, Norway

*Corresponding author: oyvind.karlsen@uis.no

Abstract. Most mechanical connections are in one way or another exposed to some form of tear, wear, corrosion and fatigue, and likely to fail over time. If the contacting surfaces in a connection are exposed to tangential loading due to vibrations, small amplitude displacements called fretting can be induced at the surface, and might result in crack nucleation and possible propagation. The review and analysis reported herein are based on review of a wide range of studies reported in the literature during the last 25 years, which look into earlier studies around fretting fatigue in interference fit connections. In addition, a new method on how to obtain an interference fit is being mentioned as a possible way of increasing the joints fretting fatigue life. Previous investigations show clearly positive effects of finding and operating with the correct combinations of interference fit levels, palliative treatments and material choices to increase the fretting fatigue life.

1. Introduction

In most mechanical systems, there are needs for power transmission and connection techniques using various shapes and dimensions, and the most familiar transmission systems might be of electrical or mechanical type. Among those, spline, dovetail, riveted, interference fitted, pre-stressed and open-hole pin/bolt connections are the typical mechanical joint types [1].

In general, with an increasing transmitted power, the mechanical connections are exposed to increasing loads and stresses at contact areas, and the probability for damages at the contact areas is therefore increasing. Already early in the Nineteenth century, when the railway still had a very important role and position, it was known that repetitive loads caused failure in the axles and connected components. At the time, the number of axles in England was around 200 000, and in the middle of the 19th century, hundreds of people lost their lives in accidents all over Europe due to axle failures [2].

Many studies have been conducted during the last decades to analyse and increase the understanding and perception of the mechanisms that work in the contact areas and surfaces of different mechanical connections, like train wheels and axles (wheel seats, gear seats, brake disk seat) [3, 4]. The press-fitted parts on a railway vehicle are extremely important and critical because of the easily initiation of cracks in those areas and spots [2, 5, 6].

This literature study aims to look deeper into previous studies and investigations on the mechanisms of damages that occur in interference fitted connections of mechanical systems with relative movements between the connected parts. A special focus will be on fretting fatigue, what the causes are, and what can be done to prevent or to avoid factors that reduce the life of the component and the system. In addition, this review aims to have a critical approach to the previous studies and identify research issues and ways to approach problematics around these connection types. The review will focus on a few specific mechanical systems, but still mainly on interference fit challenges that could occur in mechanical systems with interference fitted couplings.

2. Mechanical connection techniques

There are a variety of different mechanical connections in use, depending on the type of industry, type of machine, equipment or joint, type and level of loads, materials involved, etc. Different connection types transfer forces and loads between different parts in the connection in different ways, and with an increasing power, the stress level increases, with damage and reduced fatigue life as a possible result. The damages in a contact surface can range from minimal surface or topographic changes to severe crack initiation, which again can lead to crack propagation if exposed to cyclic and/or dynamic forces. Such dynamic forces can originate from the type of operation and type of connection the system is designed for, or from external sources not within the original operational scope, like vibrations. It is well known that crack initiation at a contact surface, with crack propagation due to dynamic loading results easily in fretting fatigue and reduced fatigue life [3, 4, 7-10].

2.1. Interference fit connection

Interference fit, also known as press fit or friction fit, is a fastening method where two parts, typically circular cross section shape, are pushed together (one into the other) and kept in place due to the pressure between them. Typically, the inner part will have a nominal outer diameter slightly greater than the nominal inner diameter of the outer part, and when pushed or forced together the contact friction forces keep the parts from moving relative to each other (nominally). The level of interference depends on the negative difference in nominal diameters.

Figure 1 shows a principle sketch for an interference fit. The interference fit causes the bore to expand and induces the tensile tangential (tangent to the bore surface) stresses, i.e. Hoop pre-stresses at the bore edge. This increases the mean stress level but this pre-stress considerably reduces the magnitude of the local cyclic stress, meaning the alternating stress, which has a much more fatigue life reducing effect [11]. An advantage of this kind of connection is the superior level of possible torque transmission between two assembled parts, but the pressure distribution is the source responsible for fatigue failure.

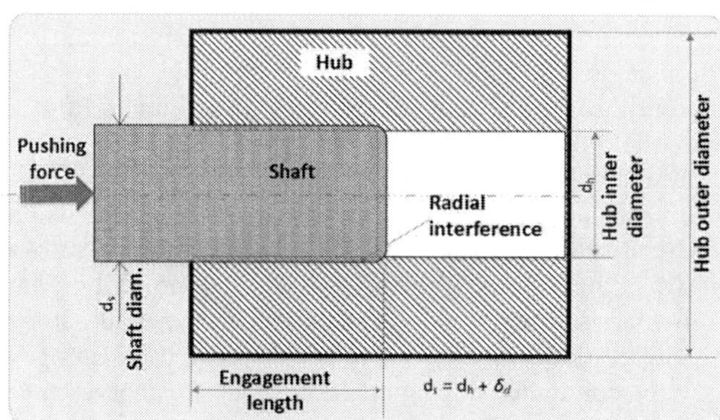

Figure 1. Principle sketch of an interference fit

The classical analytical pressure in the connection, under the assumption of solid shaft, hollow hub, axis symmetry and plane stress, is given by [12]:

$$p_f = \frac{E\delta_d}{2D_f}\left(1 - \left(\frac{D_f}{D_h}\right)^2\right) \tag{1}$$

where p_f is interference pressure (MPa); E is the elasticity modulus (MPa); δ_d is the interference fit level (mm); D_f is nominal shaft diameter and hub inner diameter (mm) and D_h hub outer diameter (mm).

There are other equations, known as Lame's equations (elasticity theory) for thick-wall cylinder, commonly used to calculate stresses and deformations at interference fit connections [13-16]. Although the shapes typically used in Lame's analysis could differ from the real shape of interference fit machine components, and they give approximate results, they are often used due to their practical applicability.

Lame's equations are based on certain assumptions, among others:

- The thickness of the cylinder wall is constant.
- Thick-wall cylinder.
- Material of the cylinder is homogeneous and isotropic.
- The two mating parts (hub and cylinder) have the same axial length.
- Plane sections of the cylinder perpendicular to the longitudinal axis remain plane under the pressure.

The Lame's equations indicate that Hoop stress and radial stress in an interference fit joint, and in addition axial stress in a closed cylinder, are subjected to a defined outer and inner pressure.

2.2. Shrink fitted connection

Shrink fit is an interference fit technique where typically the inner part is being cooled down (shrinks), or the inner surface of the outer part is being heated up, to make the fit possible. After assembly, the temperature-affected part will adjust its size back to original size, and there will be an interference fitted connection. The dimension-change of the temperature affected part will decide the final interference fit level, and contact pressure. Induction heating is a typical technique for heating and liquid nitrogen is often used for cooling.

The basic formula for calculating the required temperature (T) for the expansion of the metal is:

$$T = \frac{d}{c * \alpha} \tag{2}$$

where d is the required radial expansion (of the radius), c is the nominal hub inner radius and α is the coefficient of thermal expansion.

2.3. Open hole pin/bolt connection

A typically open-hole connection is a cylindrical pin in a joint with an operation tolerance equal to the installation tolerance, which means that the pin is "loose" within the bore and having the possibility to move relative to the bore during operation, as moveable joints in cranes or heavy machinery.

For instance, Bondura PIN technology (Figure 2 [17]) utilizes such open hole pin connection or expanding pin that is a form of an interference fit technique, where it is possible to expand and shrink the pin, mechanically, without need for any heating, cooling or axial force to push the pin into the bore, neither before nor during or after the expanding process. The pin has tapered ends, and end-sleeves with a conical inner surface that fits to the tapered pin. The sleeves are pushed into the pin by torqueing the tightening end screws or nut and the sleeves climb the tapered pin ends, lock the pin to the bore as a wedge, and prevent any relative movement between the pin and the supports. The interference fit level depends on the torqueing level of the tightening screws, and/or nut. By reversing the process, the complete expanding pin assembly can easily be removed, and the technique works well on all pin/hole sizes.

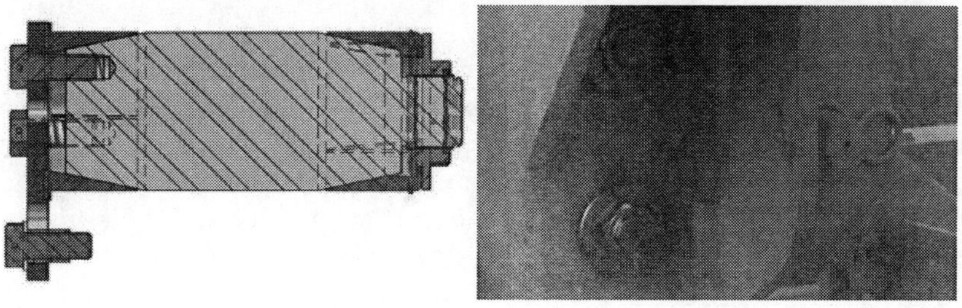

Figure 2. Bondura PIN technology (a) Expanding pin solution (b) Expanding pin in a joint

3. Fretting fatigue in different types of connections

3.1. Fatigue - general

The American Society for Testing and Materials (ASMT) defines the concept fatigue as [18]: "The process of progressive localized permanent structural change occurring in a material subjected to conditions that produce fluctuating stresses and strains at some point or points and that may culminate in cracks or complete fracture after a sufficient number of fluctuations".

It is well known and experienced during many years that many mechanical joints subjected to cyclical loads will develop fatigue cracks at the hole or bore where the bolt or pin is inserted. Many studies have been performed in the last decades to investigate problems with fatigue in mechanical connections [1, 3, 4, 7-9, 11, 13, 19, 20].

The process of fatigue involves three main stages:

1. An initial fatigue damage, which leads to a crack nucleation and possibly crack initiation.
2. Cyclic growth of a crack (propagation) until the remaining cross-section becomes too weak to withstand the loading.
3. A sudden fracture of the remaining cross section.

Among the various types of fatigue failure, the typical once include: mechanical fatigue, creep fatigue, thermo-mechanical fatigue, corrosion fatigue, and fretting fatigue. Most of these fatigue forms and related failures are caused by cyclic loads combined with frictional sliding, or micro movements.

3.2. Fretting - general

Fretting as indicated by ASTM [21] is: "A small-amplitude oscillatory motion, usually tangential, between the solid surfaces in contact" and: "The process of crack formation at a fretting damage site, progressive crack growth, possibly culminating in complete fracture, occurring in a material subjected to concomitantly fretting and fluctuating stresses and strains".

Another term related with fretting wear arises as a result of a combination of fretting and corrosion. This is a form of fretting wear in which different types of corrosion play a significant role and could reduce fretting fatigue life substantially [22-25]. The risk of fretting arises when two or more material surfaces are repeatedly moved against each other under load. Such loads can be cyclic loads from the normal operation, or maybe unwanted vibrations. The fretting process from first mechanical wear to a breakdown, Figure 3 (adapted from [26]), can be explained as: The loss of material due to fretting may not be so significant, but it might be enough to degrade the structural strength and to reduce the fatigue life. If the combination of amplitude and normal force is "right" the debris from the wear process could be trapped between the contact surfaces, and play an important role in the further wear process, and the topographies of the wear scars become complicated [24].

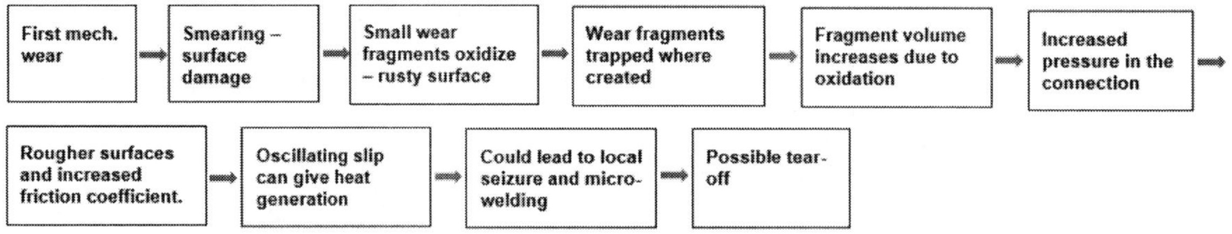

Figure 3. The fretting process

Fretting wear is one of the main reasons for failure of key components and can reduce the fatigue life up to 20 – 50% [27]. It is a type of wear damage induced by a short amplitude sliding motion between two loaded surfaces [24, 28]. It is a microscopic dynamic process involving deformed structures, cracks, oxidation process and subsequent of behaviour of oxide debris between contact surfaces [29-32].

In general, fretting can be treated as three interrelated problems: (1) Fretting wear, (2) fretting corrosion and (3) fretting fatigue. Fretting wear is typically wear damage due to a fretting problem, whose prediction is a critical aspect for prediction of fretting fatigue [33-38]. Fretting corrosion, on the

IOP Conf. Series: Materials Science and Engineering **700** (2019) 012015 doi:10.1088/1757-899X/700/1/012015

other hand, is a chemical reaction problem in combination with a fretting problem [22, 23, 28, 39-44], while fretting fatigue can occur when both fretting and fatigue conditions are working at the same time. Investigations have shown that by increasing the normal force, the fretting regime transforms from typically gross slip towards partial slip [24, 45, 46]. The main parameters affecting fretting wear normally include: Normal load, slip amplitude, cyclic frequency, surface roughness, material properties and contact geometry.

3.3. Contact surface zones and geometrical shapes
Different bodies in a joint will have different geometric shapes, and therefore often different contact surface shapes, and they could act differently under loading. Figure 4 [1] shows examples for different contact geometries. The contact zone will vary with loading, and the contact status can be defined as sticking, slip/sliding and opening area, as shown in Figure 5 [27]. The stick zone is the area with no relative movement between the two surfaces, and therefore wear will not occur. In the slip-zone, there will be relative movement between the surfaces, and fatigue and abrasive wear will occur. The fretting corrosion behaviour depends strongly on the fretting regime, being a microscopic dynamic process involving deformed structures, cracks, oxidation process and subsequent behaviour of oxide debris between contact surfaces [28-32].

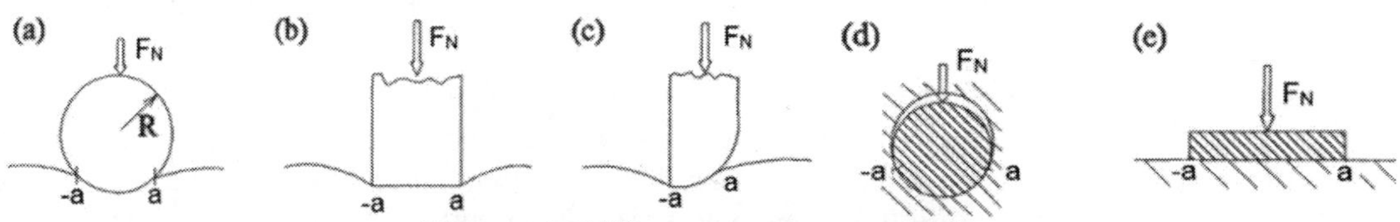

Figure 4. Different contact geometries

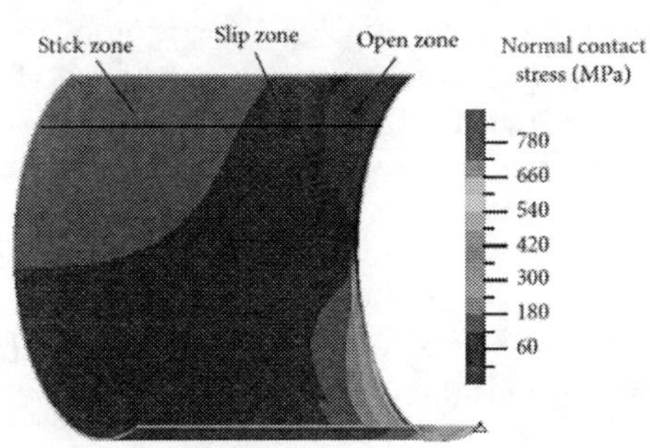

Figure 5. Stick, slip and open zone

3.4. Fretting fatigue in railway interference fit connections, and aircraft fuselage connections
Fretting problems such as fretting wear, fretting corrosion and fretting fatigue have been studied for many years, with the latest studies adding new knowledge to previous knowledge. Fretting fatigue, which is a typical and frequently appearing type of fatigue, results from infinitesimal relative surface motion between two contact materials when cyclic loads are applied. In pressurized water reactors [23, 24] for instance, fretting fatigue is a common corrosion fatigue degradation phenomenon between the steam generator tubes and anti-vibration bars/tubes supports [22, 47, 48].

A number of studies have been performed to investigate more about the different fretting and fatigue problems that occur in different industries and mechanical systems. Some of the studies were aimed to find possible solutions to decrease the negative effects of fretting, and increase the parts′ and the systems′ fatigue life, among them, the following can be mentioned as examples: Railway axles with interference fit connections, aircraft structure and fuselage with riveted or pinned connections, steam generators in nuclear energy production environment, threaded tube connections, and high power excavator tracks.

3.4.1. Railway axles with interference fit connections

From a safety point of view, the axle with its connections is the most important component in railway vehicles. The axle, gear seats and wheel seats are joined by interference / press fit, and they are critical parts, due to the long-time exposure to crack initiation because of fretting fatigue [2, 5, 6]. Figure 6 [3] shows a typical interference fitted railway wheel axle with wheel hub and seat. Many actions have been taken and a number of investigations have been reported with an objective of reducing the problems related to fretting fatigue and improve the fretting fatigue life [5, 6, 49]. Examples can be; optimizing the contact pressure/interference level, surface treatments, stress relief grooves, strength, and more. For rotary bending axles the wear mechanism of fretting damage is typically combinations of abrasive wear, oxidative wear and delamination [50].

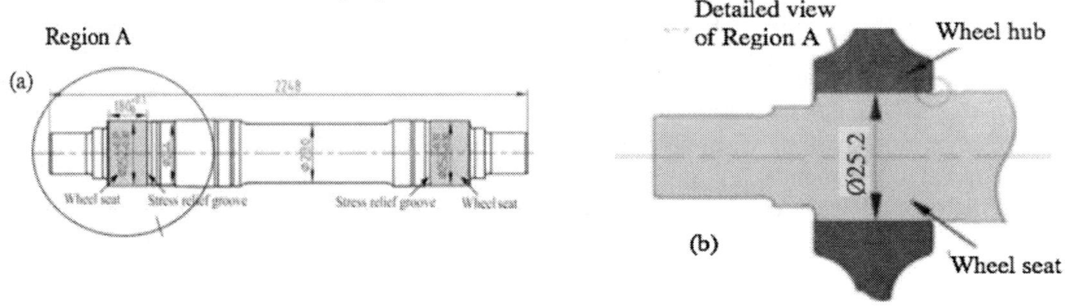

Figure 6. Railway axle (a) with hub and seat (b)

(a) Optimizing the contact pressure / interference level: Studies conducted to investigate the fretting fatigue life consequences of changing interference fit level, or contact pressure between the two contact surfaces shown that the transmission from high wear regime to low wear regime occurs when the contact pressure drops below a critical mean pressure of approx. 150 MPa (Figure 7 (a) [51]). As it is observed in the figure, the evolution of the wear kinetics as a function of the mean pressure identifies a critical pressure between mild and severe wear regimes.

These results are based on the fretting rig test (Figure 7 (b)) by imposing displacement independent of the normal force. As a result, one can read that with an imposed displacement combined with an increased normal pressure (> 150 MPa), the contact area will suffer severe wear. In many cases, and maybe in an interference fitted railway axle with hub and seat, it can be expected that there exists no relation between the normal force (interference fit level) and the actual displacement (relative movements between the hub and the axle). It could be expected that an increased normal force level would decrease the displacement, and therefore possibly decrease of the wear, or at least change the wear pattern.

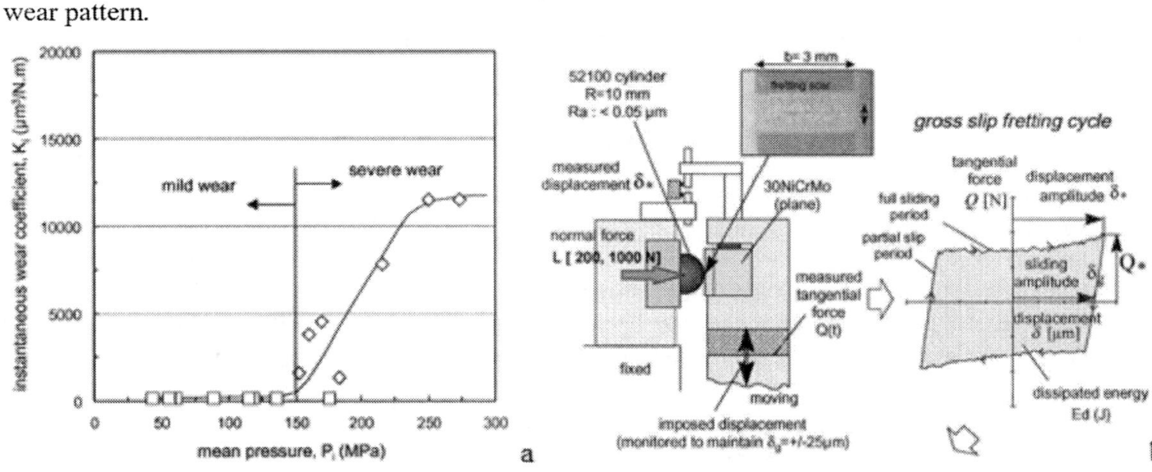

Figure 7. (a) Evolution of the wear kinetics as a function of the mean pressure, and (b) schematic of the fretting rig test set-up.

(b) Surface treatment and fretting fatigue palliatives: One way to avoid fretting fatigue and its negative consequences is to avoid fretting fatigue contacts; though not always possible. There exist palliative mechanisms for fretting fatigue that could reduce the problem, including [1];

- reduction of friction (expected reduction of the tangential contact force) [52]
- introduction of residual compressive stresses (reduction of fretting wear) [7, 11, 53, 54]
- increasing surface hardness (by different peening processes) [55-61]
- altering surface chemistry and modifying the micro-topography [55]

(c) *Stress relief grooves and fillets*: For many years, it has been investigated on how to achieve important improvements of fatigue life [2, 3, 5, 62-64]. One way is to optimize the place, size and form of the stress relief grooves and fillets, to maximize their effect and increase the fatigue strength and life [62, 63]. The European Norm EN13104 [65] states the requirements for calculations for powered axles.

The other method is use of finite elements methods such as the one conducted by Zeng, et al. [3] who used the method for prediction of fretting fatigue crack initiation in a full-scale railway axle, powered railway axles used in China, taking into consideration the stress redistribution and surface change due to fretting wear. Thereafter, the influence of stress relief groove on the fretting wear and fatigue was investigated.

3.4.2. Aircraft structure and fuselage with riveted or pinned connections

The normal method how to connect skin plates, or fuselage, for an aircraft construction is by rivets, as shown in Figure 8 [1]. Fatigue of the pressurized fuselages of transport aircraft is a significant problem all designers of aircraft have to take into consideration for assuring a sufficient lifetime and safety. Such joints in a fuselage require holes that could easily weakens the strength, increase the risk of fatigue, and reduce lifetime. Due to the cyclic nature of the service loading of the joints in aircraft structures, fatigue failures are most common failure types. Both pins, bolts, and rivets are commonly used in these structures, often with an interference interface, or an open-hole connection, often cold expansion treated.

Cold expansion: Cold expansion does not include a cooling, or any other temperature dependent processes, so "cold" is referring to the absence of "introduction of heat" into the process, which is the normal way when forming metal. Cold expansion is a cold forming of the bore wall to make the material more resistant, before introducing a pin or bolt for operationally purposes [7, 11, 53, 54].

Interference fit: At the junction of the main load-bearing components in the aircraft, such as wing beams, fuselage strengthening frames, and other important structural connected parts, interference fit bolted joints are widely used to improve the structural fatigue performance. Connections like press-fit, shrink-fit and mechanical radial expansion can be seen as variations within the definition of interference fit.

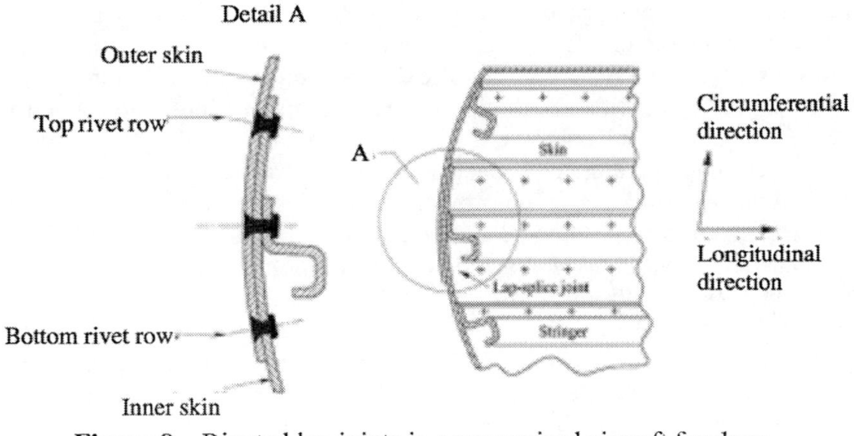

Figure 8. Riveted lap joints in a pressurized aircraft fuselage

Iyer, et al [66] concluded in their study that an interference fit of 2% tended to inhibit fretting fatigue. In a related work Iver et al. [67] studied the influence of interference and clamping on fretting fatigue in single rivet-row lap joints, and the study was concerned with the determination of the three-dimensional contact stress and displacement fields in the panels of realistic riveted lap joints.

Moreover, Chakherlou, et al. [9] investigated both experimentally and numerically the effect of interference fit on holed plate with interference fit levels of 1%, 1.5%, 2% and 4%. The results showed that by increasing the interference fit level from 1% to 1.5% and further to 2%, the fatigue life also increased, but by increasing the level further up to 4%, it did not give any more increase in fatigue life.

To improve fatigue life of holed plates made of aluminum, in double shear lap joints, Chakherlou, et al. [11] conducted experimental and numerical studies of cold expansion (CE) and interference fit (IF) techniques. Both the cold expansion and the interference fit test levels were set to 1.5% and 4.7%. With increasing IF level, the amplitude of the alternating stress reduces, but the mean stress increases. At CE and IF level of 1.5%, the interference fit connection gives a longer fatigue life, and at 4.7% level the differences between cold expansion and interference fit are small.

Bi, et al. [68] have also performed experiments and FE analysis and concluded, in their case, that the maximum hoop tensile stress was at its minimum at an interference fit size of 1.5%. The interference fit size of 1.5% was therefore defined as the optimal value in aircraft interference assembly. Mirzajanzadeh, et al. [20] also conducted similar studies by considering different friction factors for different interference levels.

4. Discussions and outlooks

Fretting problems and challenges are well known and still be a continuous issue in many industries, although numerous investigations have been conducted for many years. Within the railway industry, especially when it comes to the axle/hub connections, many actions have been taken, and a number of investigations are conducted in order to reduce the problems related to fretting fatigue and improving the fretting fatigue life [5, 6, 49]. It has been focused on optimizing the contact pressure / interference level, palliatives, stress relief grooves and more, with the aim to prevent fretting related problems, and consequently prolong the fatigue life of the components and equipment. In general, the newer investigations are building their results on results from earlier investigations combined with new and often better techniques, test equipment and improved FE tools and methods, in addition to improved materials and palliatives / surface protection techniques.

When it comes to interference / press fit, and other treatments like cold expansion, it seems that all try to use similar techniques. An interference fit is made by forcing an axle axially into a hub, where the nominal axle diameter is slightly greater than the nominal hub diameter. The difference in nominal diameters defines the interference fit level. An alternative is to cool down the axle before inserting into the hub, and let it expand back towards nominal diameter, and create the interference fit level required.

The mechanical radial expansion, or expanding PIN technology [17] mentioned earlier, i.e. by companies like Bondura Technology and Expander, is not mentioned in any of the investigations found in this literature study. The mechanical radial expansion creates the interference fit slightly different from both the "normal" press fit and shrink fit techniques, and it is possible to control and adjust the fit level over time. The cold shrink energy issue is not present, neither the press force for reaching the fit level, nor the effects of temperature differences and friction between the connecting surfaces of the joint. This indicates future possibilities for improving press-fitted connection solutions of powered axles, typically like the axle/hub connection in a railway wheel. Further research and investigations will be necessary to clarify how efficient an expanding PIN solution can be, compared to existing solutions, like the traditional press-fit and shrink-fit.

Fatigue, in different forms, has caused many very serious accidents for many years, also within the aircraft and offshore oil & gas industries. A world-wide survey of aircraft accidents covering both fixed-wing aircraft and helicopters for the period 1934 – 1979, having metal fatigue as a related cause, showed 306 fatal accidents with 1803 fatalities [69]. Two-thirds of the accidents involved fixed-wing aircraft, with wing failure as the most common cause followed by engine failure. For the helicopter accidents, a third were due to failure in the main-rotor system, and a quarter due to tail-rotor failure. Some of the most serious accidents at the Norwegian shelf would be the Alexander L. Kielland hotel platform fatal accident in March 1980, with 123 lost lives [70]. A hydrophone was fillet welded to a tube, and due to low mechanical strength combined with poor welding and high residual stresses, the fillet weld partially cracked, which again resulted in a fatigue fracture of the tube.

The Eurocopter AS332 Super Puma helicopter accident in the North-Sea in 1997, also called the Norne accident, had 13 fatalities. The official report [71] from 2001 stated that the initial and direct reason for the accident was due to fatigue in the spline-sleeve of the right gear box. The Eurocopter EC225 Super Puma helicopter had a fatal accident with 16 lives lost, close to Turøy (Norway) in 2016, when coming in from the North-Sea. The AIBN (Norwegian Accident Investigation Board) released their final report [72] in 2018 stating that the accident was a result of a fatigue fracture in a second stage planet gear in the main rotor gearbox which initiated from a surface micro-pit in the upper outer race of the bearing, propagating subsurface while producing a limited quantity of particles from spalling, before turning towards the gear teeth and fracturing the rim of the gear without being detected.

In general terms, the influence of fretting fatigue can be reduced or eliminated by means of the methods mentioned previously in this study; optimizing contact pressure, correct use of palliatives, correct use of stress relief grooves, improved test techniques, methods and numeric analysis tools and methods, and maybe also by investigating the use of expanding pin technology. The previously mentioned accidents [70-72] prove that the influence or consequence of fatigue can be fatal, and often a result of a highly complex chain of reactions. The investigation of the Turøy accident [72] has shown that the combination of material properties, surface treatment, design, operational loading environment and debris gave rise to a failure mode which was not previously anticipated or assessed. The investigators at the Norne accident [71] could not determine the exact cause why the fatigue fracture occurred, but it was discovered serious problems with the hard-metal coating of the spline-sleeve, with over-sized carbide grains and out-of-standard thickness variations of the coating, in combination with some undefined scratches on the coating surface from before the accident, which all together could be (partly) the reason for breakage. The Alexander L. Kielland platform accident investigation [70] revealed quality problems with material, weld, heat treatment and design. All these accidents could probably have been avoided, or had reduced consequences, with improved quality control of design and construction (platform), and of parts and materials (Norne and Turøy) and a higher general knowledge among the engineers and designers about the consequences of combinations of various individual fatigue-risk mechanisms (platform, Norne, Turøy).

5. Conclusion

Fretting problems and challenges are sources for concerns not only in most mechanical industries, but also in other industries that experience vibrations and micro movements relatively between adjacent material surfaces. In addition to continued investigations, it is imperative to improve material qualities, palliative methods, test techniques and the way of how to connect the parts of the joint. It is imperative to understand how combinations of mechanisms can result in fatigue problems, although each and one of them may not result in such problems. In practical life, minimizing the possible damage effects of a fatigue problem could be as important as avoiding having the fatigue problem itself. New methods, as the mechanical radial expansion technique have been in the marked for many years, but are possibly not well enough known among the researches. The mechanical radial expansion technique should be applied more in future investigations regarding how to reduce fretting fatigue and other related issues, in interference fitted joints. Its advantages when it comes to installation, retrieval, elimination of the need for heat/cold, and possibility to quickly adjust the interference fit level, makes this solution worth investigate further.

References

[1] De Pauw J 2016 *Experimental research on the influence of palliatives in fretting fatigue*, Ghent University, Faculty of Engineering and Architecture, Ghent, Belgium.
[2] Hirakawa K and Kubota M 2001 *Proc. Instit. Mech. Eng.*, Part F, **215**(2) 73-82.
[3] Zeng D, Zhang Y, Lu L, Lang Z and Zhu S 2019, *Int. J. Fatigue*, **118** 225-236.
[4] Luke M, Burdack M, Moroz S and Varfolomeev I 2016 *Int. J. Fatigue,* **86** 24-33.
[5] Hirakawa K, Toyama K and Kubota M 1998 *Int. J. Fatigue,* **20**(2) 135-144.
[6] Makino T, Kato T and Hirakawa K 2011 *Eng. Fract. Mech*, **78**(5) 810-825.
[7] Sun Y, Hu W, Shen F, Meng Q and Xu Y 2016 *Int. J. Mech. Sci.*, **107** 188-200.
[8] Croccolo D, De Agostinis M, Ceschini L, Morri A and Marconi A 2013 *Fatigue Fract. Eng. Mater. Struct.*, **36**(7) 689-698.
[9] Chakherlou T N, Mirzajanzadeh M, Abazadeh B and Saeedi K 2010 *Eur. J. Mech., A/Solids*, **29**(4) 675-682.
[10] Liu J, Liu D, Liu Y and Tang C 2005 *J. Mech. Strength*, **27**(4) 504-509.
[11] Chakherlou T N, Taghizadeh H and Aghdam A B 2013 *Aerosp. Sci. Technol.*, **29**(1) 351-362.
[12] Pedersen N L 2016 *Struct. Multidiscip. Optim.*, **54**(2) 349-359.
[13] Alfredsson B 2009 *Int. J. Fatigue*, **31**(10) 1559-1570.
[14] Gutkin R and Alfredsson B 2008 *Eng. Fail. Anal.*, **15**(5) 582-596.
[15] Truman C E and Booker J D 2007 *Eng. Fail. Anal.*, **14**(4) 557-572.
[16] Özel A, Temiz S, Aydin M D and Şen S 2005 *Mater. Des.*, **26**(4) 281-289.
[17] Bondura, "Bondura," [Online]. Available: www.bondura.no. [Accessed 28th January 2019].
[18] ASTM, Standard Terminology Relating to Fatigue and Fracture Testing, 2013.
[19] Oku Y, Sugino M, Ando Y, Takazaki D and Kubota M 2017 *Tribol. Int.* **108** 111-120.
[20] Mirzajanzadeh M, Chakherlou T N and Vogwell J 2011 *Eng. Fract. Mech.*, **78**(6) 1233-1246.
[21] ASTM, E2789: Standard guide for fretting fatigue testing, 2015.
[22] Liao J, Wu X, Tan J, Qian H and Xie Y 2018 *Corros. Sci.*, **133**, 423-431.
[23] Zinkle S J and Was G S 2013 *Acta Materialia*, **61**(3) 735-758.
[24] Chung I and Lee M 2011*Nuclear Eng. Des.*, **241**(10) 4103-4110.
[25] Abbas F and Majzoobi G H 2018 *Theor. Appl. Fract. Mech.*, **93** 144-154.
[26] Klebanov B M 2008 *Machine Elements Life and Design,* 2008.
[27] Han C and Zhang J 2014 *Sci. World J.*, **2014**.
[28] Li J, Yang B, Lu Y, Wang Z and Shoji T 2017 *Mater. Charact.*, **131** 224-233.
[29] Xin L, Wang Z, Li J, Lu Y and Shoji T 2016 *Mater. Charact.,* **115** 32-38.
[30] Lee Y H and Kim I S 2002 *Wear*, **253**(3 to 4) 438-447.
[31] Lee Y, Kim H K, Kim H D, Park C Y and Kim I S 2003 *Wear*, **255**(7 to 12) 1198-1208.
[32] Diomidis N and Mischler S 2011 *Tribol. Int.*, **44**(11) 1452-1460.
[33] Madge J J, Leen S B and Shipway P H 2007 *Wear*, **263**(1 to 6 SPEC. ISS.) 542-551.
[34] Jin O and Mall S 2004 *Wear*, **256**(7 to 8) 671-684.
[35] McColl I R, Ding J and Leen S B 2004 *Wear*, **256**(11 to 12) 1114-1127.
[36] Nakazawa K, Maruyama N and Hanawa T 2003 *Tribol. Int.*, **36**(2) 79-85.
[37] Hattori T and Watanabe T 2006 *Tribol. Int.,* **39**(10) 1100-1105.
[38] Sabelkin V and Mall S 2005 *Fatigue Fract. Eng. Mater. and Struct.,* **28**(9) 809-824.
[39] Wang Z H, Lu Y H, Li J and Shoji T 2016 *Tribol. Int.*, **95** 162-169.
[40] Lgried M, Liskiewicz T and Neville A 2012 *Wear*, **282 - 283**, 52-58.
[41] Kwok C T, Cheng F T and Man H C 2000 *Mater. Sci. Eng.,* A **290** (1 to 2) 145-154.
[42] Mi X, Cai Z B, Xiong X M, Peng J F and Zhu M-H 2016 *Tribol. Int.*, **100** 400-409.
[43] Xin L, Yang B, Li J, Lu Y and Shoji T 2017 *Corros. Sci.,* **123** 116-128.
[44] Zhong X, Han E-H and Wu X 2013 *Corros. Sci.,* **66** 369-379.
[45] Zhang X and Liu D 2009 *Acta Metall. Sinica*, **22**(2) 131-137.
[46] Xin L, Yang B B, Wang Z H, Lu Y H and Shoji T 2016 *Wear,* **368 - 369** 210-218.
[47] Young B A, Gao X, Srivatsan T S and King P J 2007 *Mater. Des.,* **28**(2) 373-379.

[48] Hong S M and Kim I S 2005 *Wear*, **259**(1- 6) 356-360.
[49] Yamamoto M and Ishiduka H 2017 *Int. J. Fatigue*, **97** 48-55.
[50] Song C, Shen M X, Lin X F, Liu D W and Zhu M H 2014 *Fatigue Fract. Eng. Mater. Struct,* **37**(1) 72-84.
[51] Fouvry S 2001 *Wear*, **250 - 251** (PART 2) 1320-1331.
[52] Zhou Z R and Vincent L, 1999 *Wear,* **225 - 229** (PART II) 962-967.
[53] Chakherlou T N and Vogwell J 2003 *Eng. Fail. Anal.,* **10**(1) 13-24.
[54] Chakherlou T N and Vogwell J 2004 *Fatigue Fract. Eng. Mater. Struct.,* **27**(5) 343-351.
[55] Kasarekar A T, Sadeghi F and Tseregounis S 2008 *Wear,* **264**(7 - 8) 719-730.
[56] Menezes M R, Godoy C, Buono V T, Schvartzman M M and Wilson J C 2017 *Surf. and Coat. Technol.,* **309** 651-662.
[57] Donzella G, Gerosa R, Pertogalli C, Silva G and Beretta M 2011 *Procedia Eng.,* **10** 3399-3404.
[58] Cseh D, Mertinger V and Lukács J 2013 *Mater. Sci. Forum,* **752**, 95-104.
[59] Mordyuk B N and Prokopenko G I 2007 *J. Sound Vib.,* **308**(3 - 5) 855-866.
[60] Kumar S A, Raman S G S, Narayanan T S S and Gnanamoorthy R 2012 *Surf. Coat. Technol.,* **206**(21) 4425-4432.
[61] Montross C S, Wei T, Ye L, Clark G and Mai Y-W 2002 *Int. J Fatigue,* **24**(10) 1021-1036.
[62] Kubota M, Kataoka S and Kondo Y 2009 *Int. J. Fatigue,* **31**(3) 439-446.
[63] Kataoka S, Sakae C, Kubota M and Kondo Y 2007 *Key Eng. Mater.,* **353 - 358** (PART 2) 856-859.
[64] Cervello S 2016 *Int. J. Fatigue,* **86** 2-12.
[65] European Norm 2009, *EN 13104 Railway applications. Wheeksets and bogies. Powered axles. Design method.*
[66] Iyer K, Hahn G T, Bastias P C and Rubin C A 1995 *Wear,* **181 - 183** (PART 2) 524-530.
[67] Iyer K, Rubin C A and Hahn G T 2001 *J. Tribol.,* **123**(4) 686-698.
[68] Bi Y, Jiang J and Ke Y 2015 *Adv. Mech. Eng.,* **7**(6) 1-12.
[69] Campbell G S 1981 *Int. J. Fatigue,* **3**(4) 181-185.
[70] Almar-Naess A, Haagensen P J, Lian B, Moan T and Simonsen T, 1984 *J. Energy Res. Technol.,* **106**(1) 24-30.
[71] Havarikommisjonen for Sivil Luftfart, 2001, RAP.:47/2001.
[72] AIBN, 2018, Report SL 2018/04.

Numerical simulation of FDM manufactured parts by adopting approaches in composite material simulation

A D Akessa [*], **A W Gebisa and H G Lemu**

Department of Mechanical and Structural Engineering and Materials Science, University of Stavanger, Norway

[*] Corresponding author: adugna.d.akessa@uis.no

Abstract. Additive manufacturing (AM) process is a promising manufacturing method that can replace conventional manufacturing methods, particularly for parts with complex geometry. It is the process of joining materials layer by layer to make object directly from 3D model data. Due to the inherent manufacturing process characteristics, such products experience material anisotropy with mechanical properties not easy to calculate analytically. On the other hand, numerical simulation is becoming increasingly important to solve complex problems such as in composite materials. However, no significant work of numerical simulation is reported for additive manufactured parts. This paper reports an approach to numerical simulation of additive manufactured parts by adopting methodologies developed for composite materials. To fill the existing gap in this area, the mechanical properties of FDM parts are experimentally studied. The experimental samples are produced from ULTEM9085 material with different printing parameters. The mechanical properties of the samples are then analyzed and numerical simulation using finite element method is done to compare the results with experimental results and verify the simulation model. The main aim of the study is to devise a numerical simulation method for additive manufactured parts by adopting existing methods for composite materials.

1. Introduction

Additive manufacturing (AM) also knows as 3D printing is a manufacturing technology that is showing a promising progress for the future manufacturing sector. To advance the technology to the next level, the mechanical properties of the materials produced by the technology should be well documented, organized and standardized. In the conventional manufacturing industry, the mechanical properties of materials are well organized and compiled on plenty of material handbooks. Furthermore, the mechanical properties of the materials can be predicted using diverse tools and methods including finite element analysis (FEA) techniques. When it comes to additive manufacturing, on the other hand, since the technology is still immature, the use of FEA for the prediction of their mechanical properties are very limited [1]. This can be attributed partly to the fact that the AM of load carrying structural parts is an emerging approach from the application of 3D printing in rapid prototyping. In addition, the AM process involves a combination of, among others, different materials, material processing energy, process parameters and fabrication (printing) methods. Thus, computational tools that are developed so far are dedicated to specific material type [2, 3] or processing technology, for example laser sintering [4 - 6] and there exists no sufficiently developed approach that can address each combination.

Fused deposition modelling (FDM), one of additive manufacturing techniques of thermoplastic materials, builds parts by heating, extruding and depositing thermoplastic materials in layer by layer fashion as illustrated in Figure 1. As a result of lack of numerical computation tools to predict the behavior of structural components by this technology, most of the reported studies focus on experimental methods [7, 8], theoretical analysis [8] and analytical/empirical methods [9, 10].

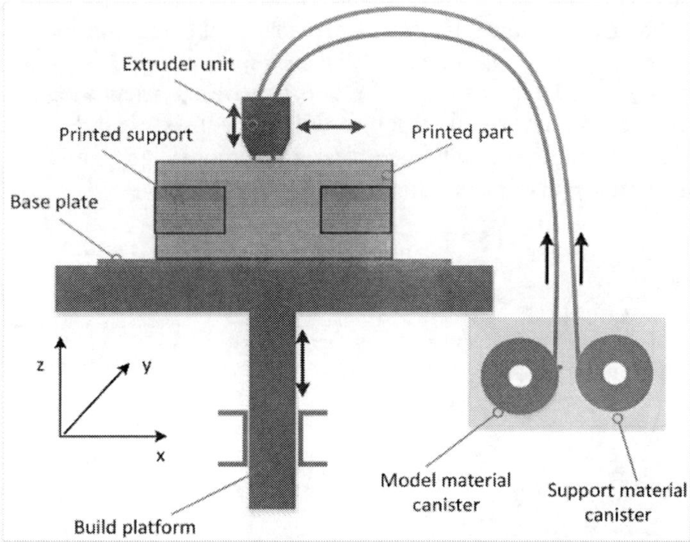

Figure 1. Illustration of fused deposition modelling (FDM) technique

The layer-wise deposition in fused deposition modelling resembles the structure of laminated composites in which the fabrication process builds by piling up layers of laminates. Laminated composites consist thin layers of different materials bonded together, such as bimetals, clad metals, plywood, formica and the like. As the FEA of composites is relatively developed and employed in different FEA software, the current research attempts to adopt FEA approach in ANSYS software that is developed for laminated composite materials to the study of the mechanical property (tensile) of ULTEM 9085 produced by fused deposition modelling and verify the approach through experimental investigation. In other words, the aim of the study reported herein is to investigate the possibility of applying the already developed FEA technique for composite material structures for the parts produced by fused deposition modelling.

The remaining part of the article is divided into 4 main sections. Section 2 explains the basic theories and mechanics of composite materials with focus on the theory that gives the rationale to use the numerical simulation approaches developed for composite laminates to additive fabricated parts. Then, the materials and methods used to conduct the study are presented in Section 3 followed by discussion of results in Section 4. Finally, the conclusions drawn from the study are summarized in Section 5.

2. Basic composite material theory and analogy with additive fabrication

2.1. Mechanics of composite materials and structures

A composite material is combination of two or more materials at macroscopic scale, which are not soluble in each other to form a third material. The combined material constituent is called the reinforcing phase and the one in which it is embedded is called the matrix [11]. Due to properties of composite structure, mechanical characterization is more complex compared to metal structures which have isotropic properties. To conduct mechanical analysis, more material properties are needed compared with monolithic materials like steel. There by, structural analysis computational and experimentally more complicated and intensive due to complexity of its nature. The properties of composite material depends on the properties of the constituents, their geometry and the distribution of the phases [12].

Lamina, or ply, is a plane (curved) layer of unidirectional fiber or woven fabric in a matrix. It is called unidirectional lamina in case of unidirectional fiber. Lamina is an orthotropic material with principal material axes in the direction of the fibers (longitudinal), normal to the fibers in the plane of the lamina (in-plane transverse) and normal to the plane of the lamina. These principal axes are designated as 1, 2 and 3 respectively as shown in Figure 2.

Laminate is a combination of two or more unidirectional laminae or plies stacked together at various orientations. The laminae can be of different thicknesses and consists of different materials, based on the orientation of the principal axes, laminates are analyzed using common coordinates (x, y, z). Lamina is a single flat layer of unidirectional fibers or woven fibers arranged in a matrix, whereas a laminate is a stack of plies of composites. Each layer can be laid at various orientations and can be made up of different material systems. However, a real structure will not consist a single lamina, which has approximately a thickness of 0.125 mm, but a laminate consisting of more than one lamina bonded together through their thickness will be required to take realistic loads [13].

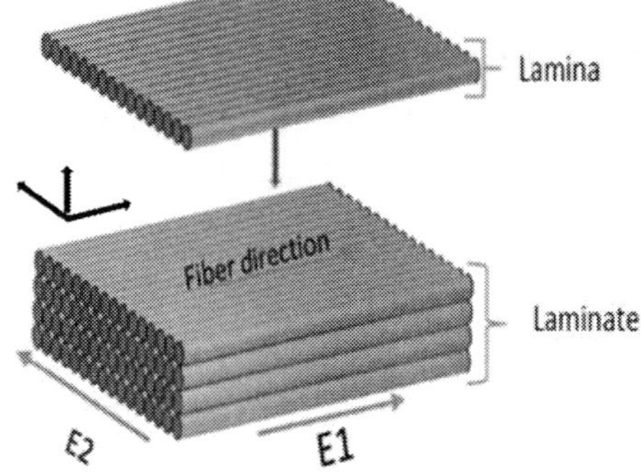

Figure 2. Illustration of composite stacking and their principal axes

2.2. Mechanical behaviour of composite materials

The mechanical behaviors of composite materials are different from conventional engineering materials. While some properties of these materials are just modifications of conventional behaviors, there are also some completely new characteristics that require new analytical and experimental approaches.

For instance, most ordinary engineering materials are both homogeneous and isotropic. For homogeneous bodies, the material properties are uniform throughout the part, i.e., the properties are independent of position in the body. Whereas isotropic bodies have material properties that are identical in all the three directions at a point in the body. Composite materials, on the other hand, are often both inhomogeneous and non-isotropic (either orthotropic or anisotropic). Thus, they have non-uniform properties throughout the body. The material properties of orthotropic bodies are different in the three mutually perpendicular directions at a point in the body i.e. the body has three mutually perpendicular planes of material property symmetry, that the properties a point in the body are orientation dependent.

Some composites like laminated safety glass have three layers, each of which is homogeneous. Composites with fibres placed at any angle are more complex. Due to the inhomogeneity nature of composite structures, they are studied from two different points of view. One is based on micromechanics point of view, where the interaction of the constituent materials is investigated at the microscopic scale to see their effects on the performance of the material. The second is micromechanics point of view, where the properties of the constituent materials are examined at the macroscopic scale assuming that the composite material is homogeneous [11].

Barbero [13] states that the single lamina of fiber reinforced composite behaves as an orthotropic material, i.e. the material has three mutually perpendicular planes of symmetry. Further, when force is applied on a composite structure, every point in the body may rotate and translate as a rigid body as well as deform to occupy a new space. However, random arrays of aligned fibers are considered as transversely isotropic having five independent elastic constants including two poison's ratio values which are caused by axial stress and transverse stress. These Poison's ratios are not independent [14].

2.3. Numerical simulation of FDM parts

Simulation in engineering is conducted to evaluate the designed performance of mechanical parts or systems using relevant computation tools. In addition, it is used in the product development process throughout the concept development, design and manufacturing phases. As a result, recent research has focused on application of simulation-driven design in order to help designers get immediate feedback on design performance by embedding simulation into the CAD environment and designers' work [15].

FDM is new and promising technology in the field of manufacturing in which one can produce complex parts from 3D model (CAD). There is limited work of numerical simulation of additive manufacturing using FDM process. Therefore, this work is done by adopting the numerical simulation methodology already developed for composite materials for FDM produced parts. This approach is adopted because the internal structure of FDM produced part is assumed analogous to the fiber layout in composite materials and as a result, classical laminate theory in composite simulation is adopted in the study. In the literature [13, 15], it is also stated that a single lamina of fiber reinforced composite behaves as an orthotropic material. In macro-mechanics, it is also assumed that the lamina, which is a building block of laminate, is macroscopically homogeneous, linear elastic, macroscopically orthotropic and initially stress free. In this simulation of FDM parts, each layer of a component is considered as a lamina and the total layers of the component is taken as a laminate.

3. Materials and methods

3.1. Sample preparation

To conduct experimental study, the samples were prepared based using ISO standard (ISO 172). The dimensions of the samples are shown in Figure 4. The models were developed in Autodesk Inventor® and saved in STL file format. Then, the STL file was loaded to a pre-processing software called Insight, which is provided with the FDM 3D printer to be used for slicing and process parameter settings. The software, then saves the data in Chromeleon Backup Archive (CBM) file format which is ready for printing.

Three printing orientations were considered: (1) along the planned loading direction (0°), (2) transverse to loading direction (90°) and (3) 45° to the loading direction. Identical geometries were used for both 0° and 90° samples (Figure 3(a)) and another geometry, shown in Figure 3(b), was used for the 45° samples. Five replicas were used for both orientations (0° and 90°) samples and four samples were tested for 45° samples. The samples were fabricated using the FDM machine, Fortus 450 mc.

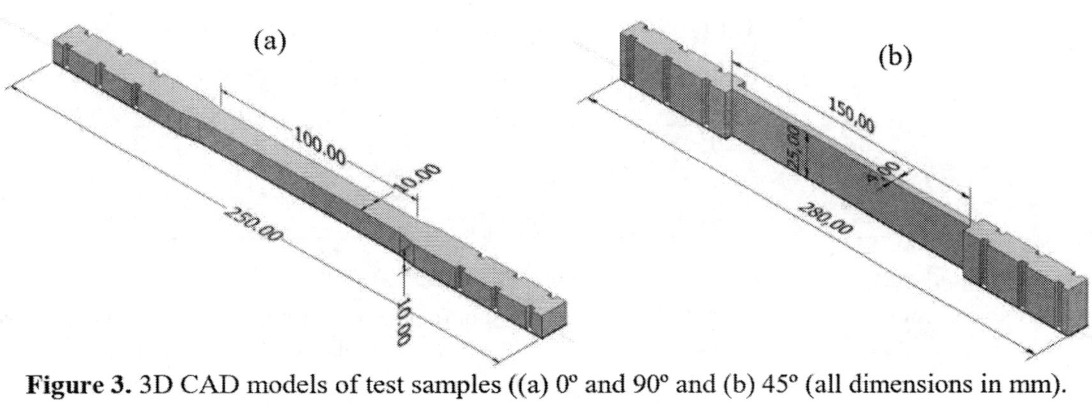

Figure 3. 3D CAD models of test samples ((a) 0° and 90° and (b) 45° (all dimensions in mm).

3.2. Experimental setup and procedure

The experiment was performed on Instron 5895 using a tensile load of 250 kN. The tensile test was performed by placing the specimen into the grips of the test machine, as illustrated in Figure 4. The grips were tightened and then the tension load was applied at a constant recommended strain rate of 0.01 mm/mm/min. The data were then collected and analyzed.

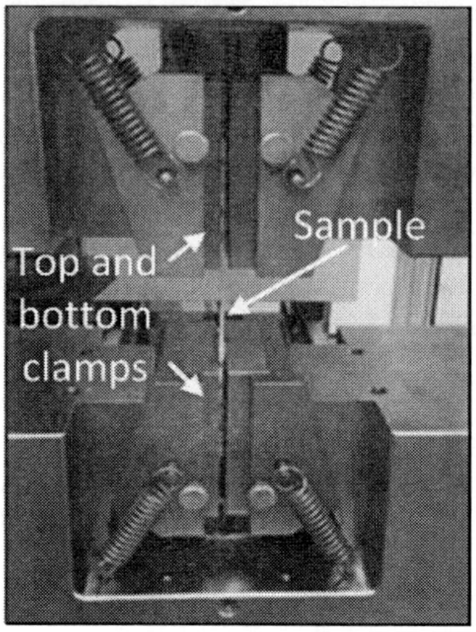

Figure 4. Experimental setup on INSTRON 5895 Universal test machine.

The tensile tests were conducted by slowly pulling the test pieces until they break. The test machine measures stress as the ratio of force applied to its original area, and the strain is the ratio of change in length to the original length. The calculated maximum stress and strain values are given in Table 1.The values of the principal elastic moduli, the shear modulus and the Poison's ratio values were obtained from experimental tests. These values for the material used (ULTEM 9085) are evaluated in elastic region and the results are given in Table 2.

Table 1. Values of maximum stress and strain for each orientation.

Raster angle	Maximum stress (MPa)	Strain 1 (mm/mm)	Strain 2 (mm/mm)
0^0	87	0.064	0.274
45^0	68	0.043	0.07
90^0	37	0.0197	0.073

Table 2. Elastic properties of ULTEM 9085 material.

Elastic modulus	Values (MPa)	Shear modulus	Values (MPa)	Poison's ratios	Values
E_{11}	2206	G_{12}	2055	v_{12}	0.39
E_{22}	1826	G_{13}	2055	v_{13}	0.39
E_{23}	1826	G_{23}	793	v_{23}	0.4

3.3. Numerical simulation approach

As stated above, the elastic properties of the test samples (i.e. modulus of elasticity along principle axes, shear modulus and Poison's ratios) are obtained from experimental tests. Properties along the principal axes are assumed from specimens printed upright in Z-direction in 0° raster orientation, whose specimen is similar with those printed in 90° orientation.

The numerical analysis in the study is performed using a FEA software, ANSYS Workbench. Finite element analysis of laminated composites is done at different levels depending on the level of post-processing required. Where detailing is necessary, the stress and strain are computed at level of fiber and

IOP Conf. Series: Materials Science and Engineering **700** (2019) 012016 doi:10.1088/1757-899X/700/1/012016

matrix. In this case, it is found necessary to describe the microstructure, including the fiber shape and geometrical distribution, and the material properties of the constituents. When the composite material is a woven fabric, or the laminate is very thick, the composite should be analyzed as solid. However, most of laminate structures can be analyzed using simplified models of plates and shells. On the other hand, the composite material can be considered as a homogeneous equivalent material. In this case, its structural behavior can be analyzed using orthotropic properties. To calculate stress and strain, the actual laminate stacking sequence (LSS), the elastic properties of each lamina, as well as thickness and fiber orientation of every lamina must be given [4]. Therefore, in this study, finite element analysis is done assuming the FDM fabricated parts as linear orthotropic composite material, in which stacking sequence, elastic properties of each laminate, thickness and fiber orientation is defined from experimental tests.

As stated, test specimens for this study were printed in the following orientations; 0° where load is applied along filament orientation, 90° in which filament orientation is perpendicular to load application direction and 45° in which the filament is oriented at 45° to the applied load direction. The specimens were tested to obtain values of the nine independent constants (Table 2). Using these constants, numerical analysis was done in ANSYS workbench. The number of lamina was found by dividing specimen gauge length to layer thickness and each lamina is stacked one up on the other to form the laminate as shown in Figure 5 (a) – (c).

As stated above stiffness matrix for the orthotropic constitutive model was used to simulate the mechanical response of the same parts physically tested and data is obtained from the experimental test. In this research, three different simulations were made, one for each experimentally tested orientation. Material property data (i.e. elastic modulus, poison's ratio and shear modulus in direction 1, 2, and 3).

The models were developed using Design Modeler, through modeling each layer with thickness of 0.254 mm which is equal to layer thickness adding each layer as 'Add Frozen' for every orientation. Engineering data is defined from experimental test. ANSYS Workbench 17.0 software was used to simulate the part considering the building orientation by assuming the material is orthotropic. In ANSYS Mechanical, material is assigned for each layer as ULTEM 9085. Finally the model is meshed by setting the model properties and analysis was conducted by defining the boundary conditions according to experimental tests.

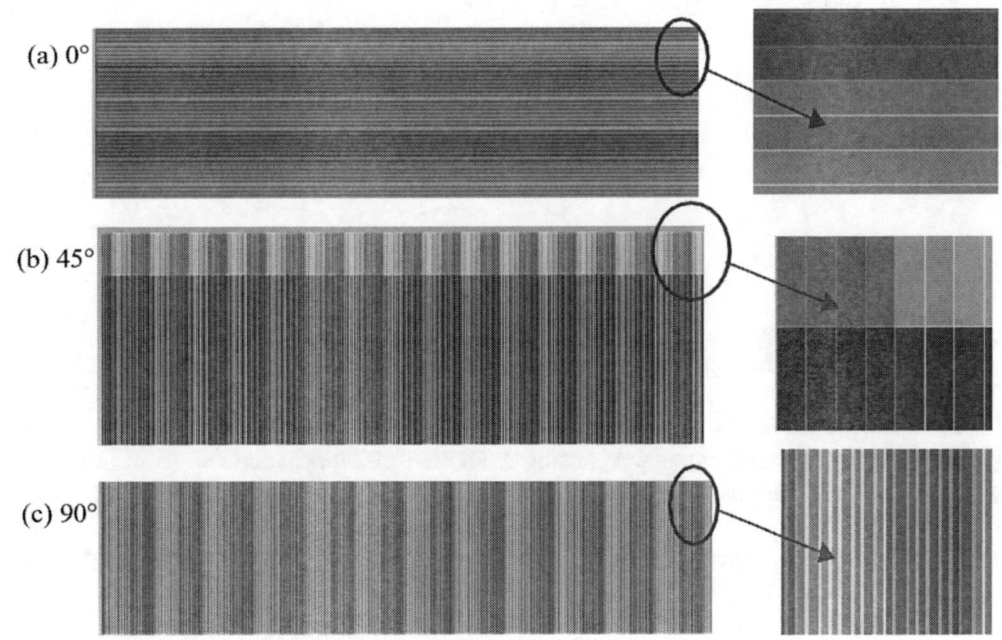

Figure 5. Illustration showing models of different orientations in ANSYS.

4. Result and Discussion

4.1. Experimental result

Tensile stress-strain graphs from experimental test for each printing orientation (i.e., 0°, 45° and 90°) are given in Figure 6(a) – (c). The maximal tensile stress obtained from the experimental results are about 78 MPa, 63 MPa and 37 MPa for 0°, 45° and 90° respectively. In the experimental test, 5 specimens were tested for each orientation, except the case in Figure 6(b), in which test 5 was observed unjustified.

In all the considered cases, the results show the same trend. As normally expected the maximum tensile stress is achieved for loading along printing direction (0°), while the lowest maximum tensile stress is obtained for loading in transverse direction (90°). The corresponding strain values at maximum tensile stress are 0,04, 0,035 and 0,018 for 0°, 45° and 90° respectively.

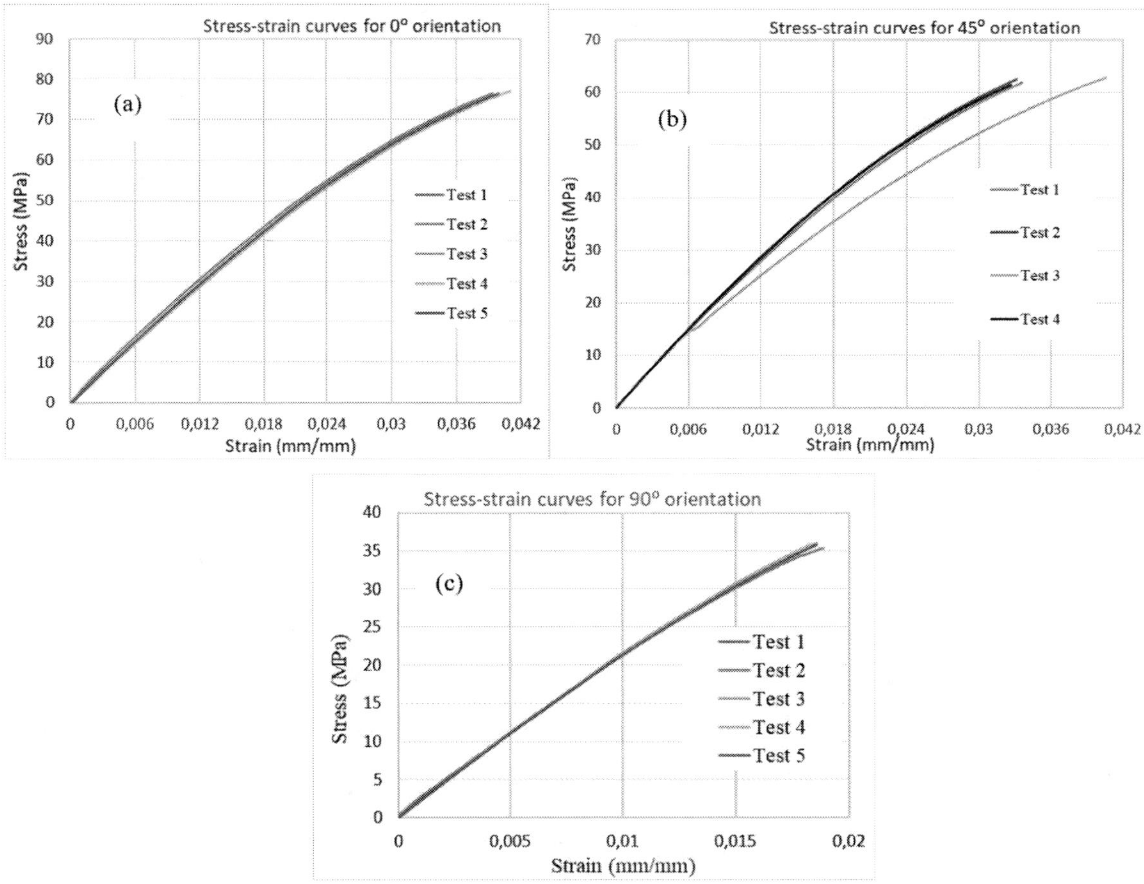

Figure 6. Tensile stress-strain curve for (a) 0° orientation (b) 45° orientation and (c) 90° orientation.

4.2. Finite element analysis results

The results of the finite element analysis using ANSYS 17.0 workbench are given in Figure 7 (a) – (c) for 0°, 45° and 90° orientations respectively. In the simulation of parts produced by 3D printing, which are identical with those produced for experimental test, orthotropic material properties are considered. The corresponding material property for FEA modeling is taken from previous study conducted by El-Gizawy et al. [16]. The boundary conditions for the FEA are set in similar way as the experimental setup, i.e. one end of the specimen is fixed and the load is applied at the other end. Default mesh sizes are used to generate the mesh model.

IOP Conf. Series: Materials Science and Engineering **700** (2019) 012016 doi:10.1088/1757-899X/700/1/012016

The result from FEA shows maximum stress at boundary conditions due to stress concentration as a result of specimen geometry and boundary conditions. However, the result taken in the study is the average result. The result from FEA shows maximum value in 0^0 orientation where the load is applied along filament orientations and minimum in 90^0 orientation where load is applied transvers to filament orientations. In the latter case the applied load try to break the bond between filaments.

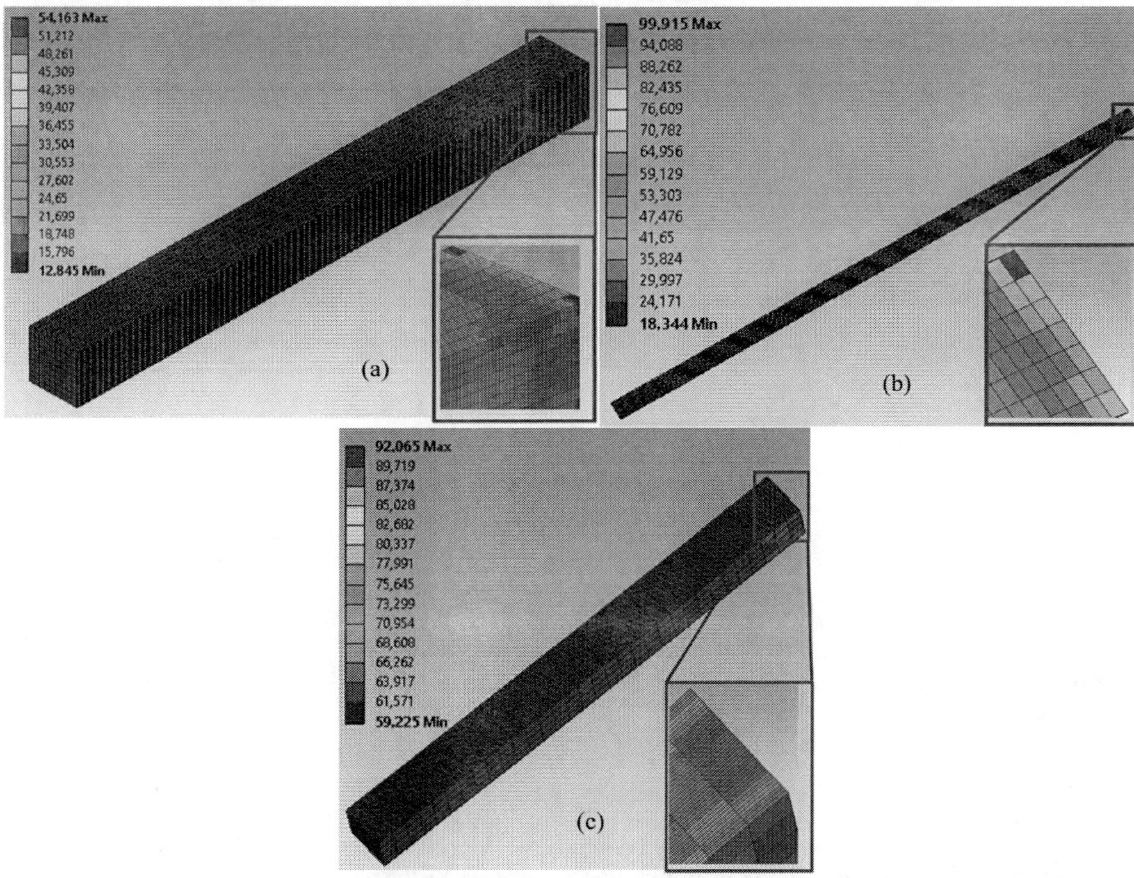

Figure 7. Illustration of stress distribution for orientations in (a) 0° (b) 45° and (c) 90° direction.

4.3. Comparison of experimental and numerical results

The results from the experiment and FEA are compared as plotted in Figure 8(a) – (c). As previously stated, the results taken from FEA are the average value and experimental results are raw data from INSTRON 5895. In the case of loading along the printing direction (0^0 orientation), the experimental and the numerical results almost overlap until the strain value reaches 0,02. For higher strain values, however, the numerical simulation (average value) proceeds linearly, while the experimental results indicate that the material gradually gives up for higher stresses. This trend is observed for the other two cases (45^0 and 90^0 orientation), though the results, in general, show that the numerical simulations underestimate the stress values for low range of strain values.

IOP Conf. Series: Materials Science and Engineering **700** (2019) 012016 doi:10.1088/1757-899X/700/1/012016

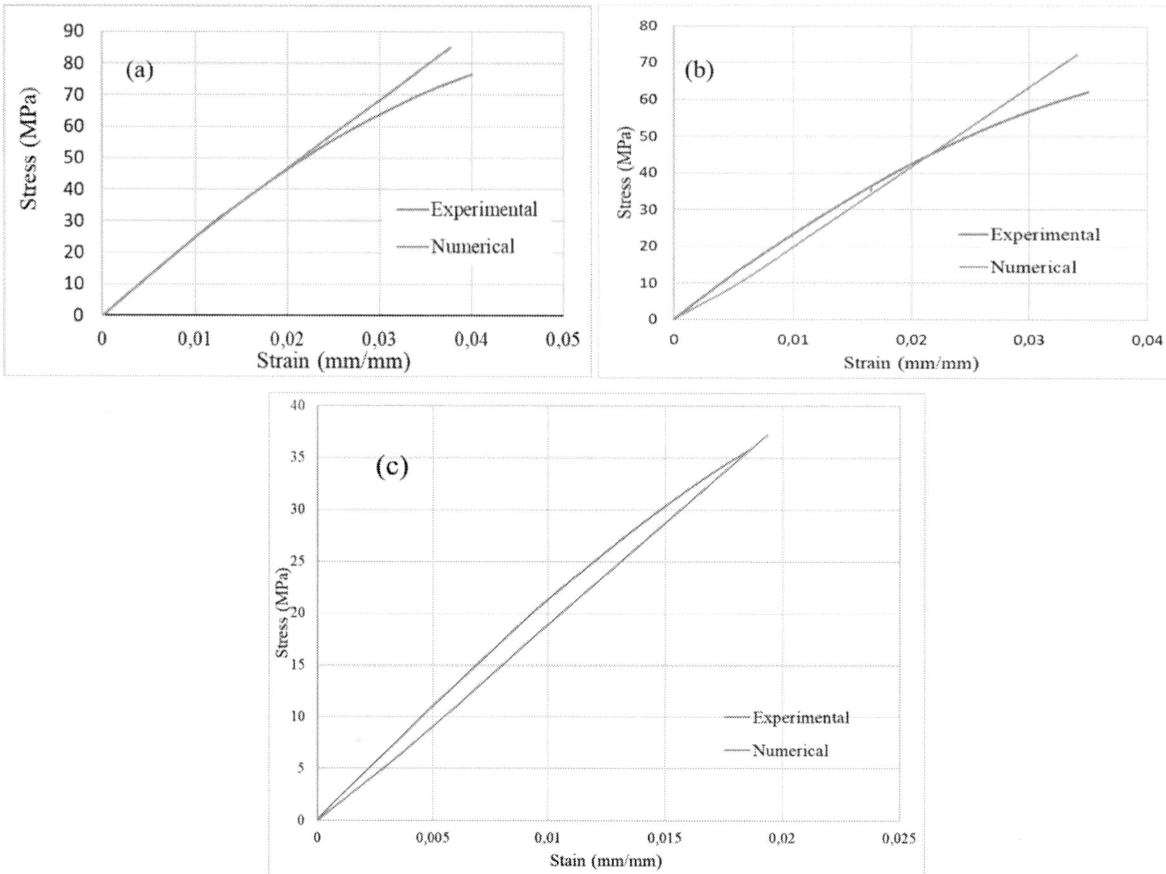

Figure 8. Comparison of results from experimental and FEA for (a) 0° (b) 45° and (c) 90° orientations.

5. Conclusion

The study of additive manufactured parts using both experimental and numerical methods are reported in this article. The numerical simulation approach is based on the approach available in ANSYS workbench for simulation and analysis of laminated composite materials. The reason to adopt this simulation technique is that the fabrication process of both cases uses layer by layer arrangement of materials that will have impact on the mechanical behavior.

In order to verify the employed numerical simulation technique, experimental tests were conducted on additive manufactured samples of the same size and material property. The results show that both results follow the same trend, though some deviations are observed for 45^0 and 90^0 printing orientations. In general, the adoption of the simulation technique of laminated composites for additive manufactured parts gives acceptable results particularly when the load acts in the printing direction. For higher strain levels than about 0,02, however, the simulation remains linearly varying and overestimates the stress value.

References

[1] Bikas H, Stavropoulos P and Chryssolouris G 2016 Additive manufacturing methods and modelling approaches: a critical review, *Int. J. Adv. Manuf. Technol.* **83**, 389–405.

[2] Zargarian A, Esfahanian M, Kadkhodapour J and Ziaei-Rada S 2016 Numerical simulation of the fatigue behavior of additive manufactured titanium porous lattice structures, *Mater. Sci. Eng.*: C, **60**(1), 339 – 347.

[3] Lindgren L-E, Lundbäck A, , Fisk M and , Draxler J 2019 Modelling additive manufacturing of superalloys, *Procedia Manuf.* **35**, 252–258.

[4] Hu D and Kovacevic R 2003 Sensing, modeling and control for laser-based additive manufacturing, *Int. J. Mach. Tools Manuf.* **43**(1), 51 – 60.
[5] Nie P, O.A.Ojo and Lia Z 2014 Numerical modeling of microstructure evolution during laser additive manufacturing of a nickel-based superalloy, *Acta Mater.*, **77**, 85 – 95.
[6] Zhang Z, Tan Z J, Yao X X, Hu C P Ge P, Wan Z Y, Li J Y and Wu Q 2019 Numerical methods for microstructural evolutions in laser additive manufacturing, *Comput. Math. Appl.* **78**(7), 2296 – 2307.
[7] Byberg K I, Gebisa A W and Lemu H G 2018 Mechanical properties of ULTEM 9085 material processed by fused deposition modeling, *Polymer Testing,* **72**, 335-347.
[8] Zhao Y, Chen Y and Zhou Y 2019 Novel mechanical models of tensile strength and elastic property of FDMAM PLA materials: Experimental and theoretical analyses, *Mater. Des.* **181**, 108089
[9] El Moumen A, Tarfaoui M and Lafdi K 2019 Additive manufacturing of polymer composites: Processing and modeling approaches, *Composites Part B*, **171**, 166 – 182.
[10] Gebisa A W and Lemu H G 2018 Investigating effects of Fused-Deposition Modeling (FDM) processing parameters on flexural properties of ULTEM 9085 using designed experiment, *Mater.* **11**(4), 500.
[11] Kaw A K 2006 *Mechanics of composite materials*. 2nd Ed. CRC press, NY, USA.
[12] Daniel I M and Ishai O 2006 *Engineering mechanics of composite materials*, 2nd Ed. Oxford University Press, Oxford, UK.
[13] Barbero, E.J., 2013. *Finite element analysis of composite materials using ANSYS.* 2nd Ed. CRC press, Taylor & Francis Group, NW, USA.
[14] Chawla K K 1987 *Composue materials: Science and Engineering*, Springer-Verlag, NY, USA
[15] Siemens home page: https://www.plm.automation.siemens.com/global/en/ (Last accessed: 15.09.2019)
[16] El-Gizawy A S, Corl S and Graybill B 2011 *Process-induced properties of FDM products*. In Proc. of the ICMET, Int. Conf. on Mech. Eng. Technol. Congress and Exposition, 2011.

IOP Conf. Series: Materials Science and Engineering **700** (2019) 012017 doi:10.1088/1757-899X/700/1/012017

Mechanical behavior of aluminum honeycomb sandwich structures under extreme low temperature conditions

S P Zaoutsos

Laboratory of Advanced Materials and Constructions, University of Thessaly, Gaiopolis, 41110, Larissa, Greece

E-mail: zaoutsos@teilar.gr

Abstract. The current study is an effort to characterize the mechanical response of panels that are used as structural elements at low temperature applications. For the above reason, plaques of aluminum honeycomb sandwich were exposed for a certain period of time at different low temperature profiles ranging from 23°C to -70°C so that possible variation of their mechanical properties due to temperature exposure to be justified. Four different temperatures namely 23°C, 0°C, -40°C and -70°C were applied to these plaques prior mechanical testing that was performed at different loading modes. Specimens were obtained from these plaques in order to study the effect of low temperature on mechanical behavior and damage response of the specific sandwich structures. Extensive mechanical testing was carried out by means of tensile, compression and bending loading and the respective mechanical properties occurring from these tests were determined. From the results occurring from experimental data obtained from mechanical testing it is evident that there is a moderate decrease of ultimate strength and yield strength of the material as service temperature decreases in tensile, edgewise and flatwise compression as well as in three point bending loading modes. Moreover a drop in temperature results in a decrease in strain at rupture at tensile and compression loading.

1. Introduction

Sandwich constructions are lightweight structures that are used in many industrial and engineering applications like automotive industry, offshore and marine applications or aero-space. As the demand for low weight in combination with excellent mechanical properties increases in the above applications, extensive studies for the mechanical behavior of sandwich constructions are needed to be carried out in order to assure their structural integrity and reduce the possibility of failures.

Usually sandwich components consist of two facing layers (skins) separated by a core material. In most cases the skins are aluminium [1-3], reinforced polymers [4,5] or metals [6,7] depending on the application. In addition the core material can be a lightweight polymer foam or even a geometrically more complex material such as aluminium, sheet or polymer honeycomb. Generally the sandwich is an attractive structural design concept since, by the proper choice of materials, constructions having high ratios of stiffness-to-weight can be achieved.

However sandwich constructions usually suffer from debonding and delamination between skin and core occurring either from impact, bending, tensile or compressive loading. Other common kind of failures are buckling of the skins or intra-cell wrinkling mainly occurring from compressive or bending loading resulting once more in delamination between core and skin.

IOP Conf. Series: Materials Science and Engineering **700** (2019) 012017 doi:10.1088/1757-899X/700/1/012017

When these materials are used in offshore, marine or aerospace applications undergo important temperature variations in service. Due to difference between the coefficient of thermal expansion (CTE) of the skin and the core, as well as mechanical degradation occurring from low temperature exposure significant stresses are induced in the overall construction. Those stresses can produce microcracking and cause deterioration of the mechanical properties of the sandwich structure.

As the use of aluminum alloys is expanding in a variety of applications that are used in offshore infrastructures such as jacket structures, liquid natural gas carriers or drillships, these structures demand a detailed knowledge of the mechanical behavior of the materials that are used in the service temperature. Thus, the issue of the mechanical characterization of these materials and the understanding of their response under loading and fracture in extreme low temperatures is of great significance for their use as structural elements.

As far as the aluminum sandwich structures are concerned although their use is very popular in many engineering structures, a few researchers have studied the specific behavior that they exhibit in low temperature conditions [8, 9]. Usually aluminum preserves its structural efficiency in low temperature conditions although may lose its moderate ductility thus temperature may have an impact on the type of failure [10-12]. On the contrary, geometrical complexity of a sandwich structure may result to lower values of strength and rigidity due to the core or the skin. Moreover, interface between the core and the skin, which is usually an epoxy resin, suffers from temperature variations due to change in ductility and the resulting embrittlement of the polymeric nature of the polymeric system.

In the current study aluminum honeycomb sandwich plaques were exposed at low temperatures profiles and possible variations of their mechanical properties were experimentally investigated at different loading modes. Changes in these properties at the above mentioned conditions will justify the use of these materials in demanding applications such as offshore or aeronautics structures.

2. Materials and Methods

Sandwich panel used in the study consisted of two aluminum sheets of 3000 series type and a lightweight thicker aluminum honeycomb core material of 3003 type, of a mean weight 6,3 kg/m^2.

The nominal dimensions of each panel were a standard width of 1000 mm with a width tolerance 0/2.0 mm and a standard length 1000 mm with a length tolerance 0/3.0 mm. The thickness of each panel was measured at 20 mm with a tolerance of ± 0.2 mm. Aluminum thickness top and back cover sheet had 1 mm thickness while typical mechanical properties of the sandwich panel aluminum elements are shown in Table 1.

Table 1. Aluminum honeycomb panel technical data

	Aluminum skin	Core
Aluminum type	Aluminum 3000 series	Aluminum alloy 3003
Ultimate tensile strength Rm [N/mm^2]	159,31	
Modulus of elasticity, E [N/mm^2]	35288	
Compressive strength [MPa]		195
Core density [kg/m^3]	38,4	38,4
Cell size [mm]		9,5

3. Experimental Procedure

3.1. Conditioning of aluminium sandwich specimens

Specimens in the desired dimensions were cut from the aluminum honeycomb sandwich plaques, set in an environmental chamber Votch VC^3 4060 and exposed for a certain period of time of 10 months at low temperatures profiles ranging from 23°C to -70°C so that possible variations of their mechanical properties to be experimentally investigated. Four different temperatures namely 23°C, 0°C, -40°C and -70°C were applied to these plaques in order to study the effect of low temperature on the mechanical response and damage of aluminum sandwich structures. Temperature was measured in each series of

IOP Conf. Series: Materials Science and Engineering **700** (2019) 012017 doi:10.1088/1757-899X/700/1/012017

specimens for all the exposure period in order to assure temperature was kept constant and that no temperature variation was observed during the temperature exposure period.

3.2. Quasi-Static Mechanical Test

A test program was applied in all specimens through a series of mechanical tests that were performed in order to characterize the aluminum honeycomb sandwich material at different quasi static loading modes. Mechanical testing in the desired loading modes was performed in an ISTRON 3382 Universal Testing Machine of a load capability of 100 kN (22,400 lbf). The device is equipped with a data acquisition system controlled from Instron Bluehill® Lite Software to record load and displacement.

3.2.1. *Tensile tests.* Tensile tests were conducted in order to determine tensile mechanical properties of the material at different temperature profiles so that the possible mechanical degradation of the structure to be determined. The tests were performed according to the ASTM standard D 638-02A [13] with nominal dimension of the specimens were 10 mm X 20 mm X 200 mm. and a crosshead speed rate of 0.5 mm/min.

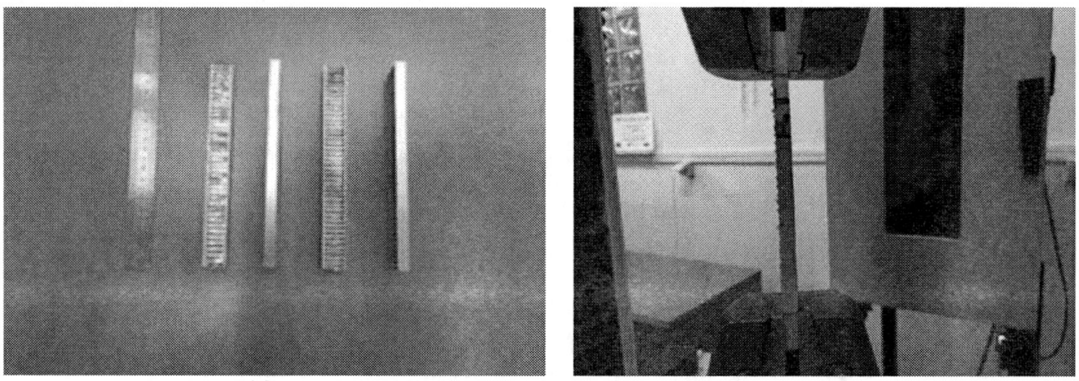

(a) (b)

Figure 1. Specimens a) before and b) during tensile tests.

3.2.2. *Compressive tests.* Flatwise compressive tests were performed on specimens that were cut for this purpose from the aluminum panels according to ASTM standard ASTM C-365-94 [14]. Testing was performed at crosshead speed rate of 0.5 mm/min on specimens with nominal dimensions of 100mm X 100 mm X 20 mm (Figure 1a). The edgewise compressive tests allowed evaluating the in-plane bending behavior of the sandwich panels. Edgewise compressive tests were performed according to ASTM standard C-364-94 [15] on three specimens of the same temperature exposure, with dimensions of 100 mm X 200 mm x 20 mm each (Figure 2a), at a crosshead speed rate of 6 mm/min.

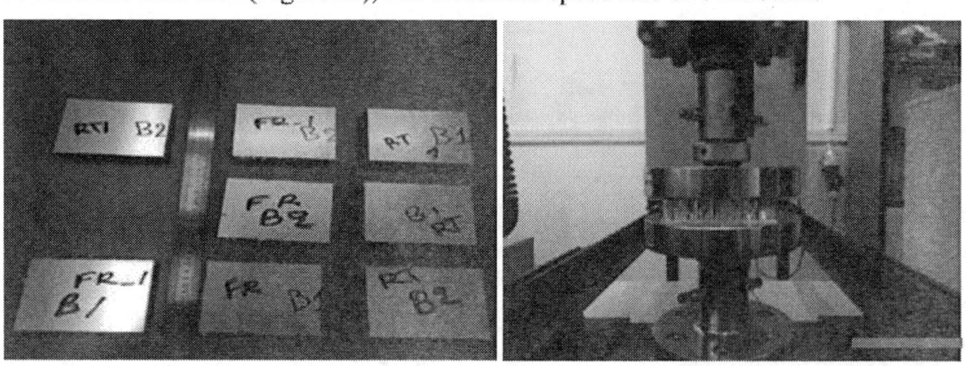

Figure 2. Specimens a) before and b) during flatwise compression tests.

Figure 3. a) Specimens for edgewise compression b) Edgewise compression testing in aluminum sandwich specimen.

3.2.3. Flexural tests. Quasi-static three-point bending tests were carried out on specimens from aluminum sandwich panels according to ASTM C-393-94 [16] in order to study the bending behavior of the sandwich panel. The nominal dimensions of each specimen were 75 mm x 200 mm x 20 mm (Figure 4) while the crosshead speed rate of the test was set as 2 mm/min. A group of three specimens at each temperature condition was tested in order to assure that the repeatability of the experimental procedure.

Figure 4. Specimen tested at three point bending loading mode.

4. Results and Discussion

Experimental stress-strain curves of the tested sandwich specimens occurring from respective tensile tests are shown in Figure 5. Additionally from Figure 6a it can be seen that the tested material preserves its ultimate tensile strength as the temperature drops from 23 °C to -70 °C. The same observation is also evident for the yield tensile strength. Tensile failure in the aluminum sandwich specimens loaded in tension took place in form of breakage of the skin as it can be seen if Figure 7. By comparing the tensile failure points for all temperatures, we can observe that at low temperature tensile failure occurs at almost the same value of stress but in lower values of strain implying that the material becomes less ductile as temperature decreases (Figure 6b).

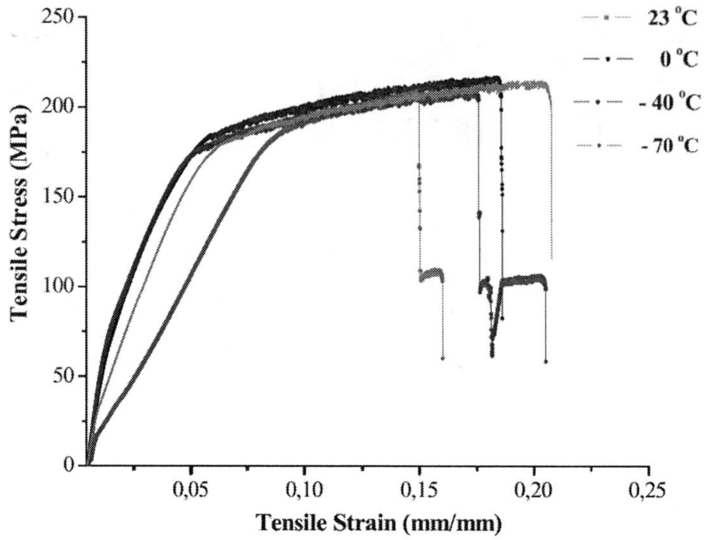

Figure 5. Variation of tensile stress versus tensile strain at different test temperatures.

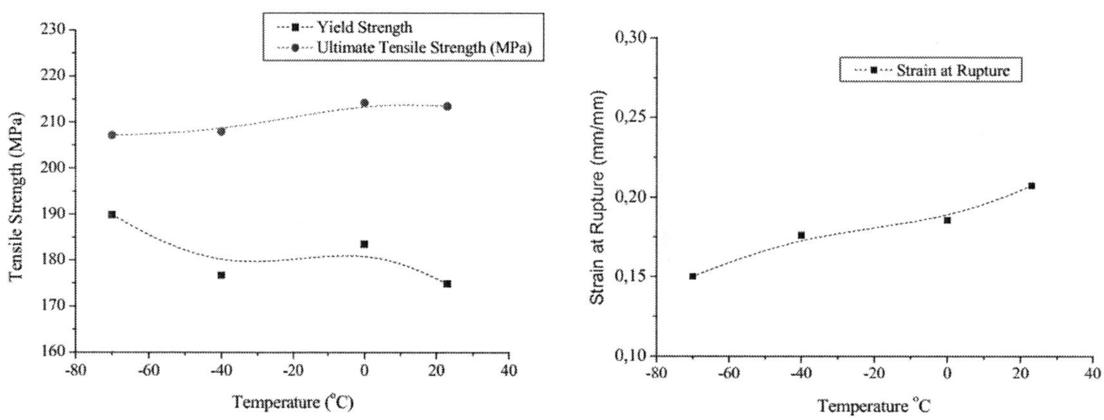

Figure 6. a) Variation of a) Yield tensile strength and ultimate tensile strength and b) Strain at rupture as a function of temperature.

Figure 7. View of a specimen after a tensile test at -70 °C.

Compression stress versus compression strain curves in the edgewise direction are depicted in Figure 8 respectively. The ultimate and yield compressive strength is decreased especially in the case of -40 and -70 °C as it can be seen in Figure 9a. In that case strain at rupture also exhibits lower values as temperature a decrease denoting that the material loses in ductility (Figure 9b).

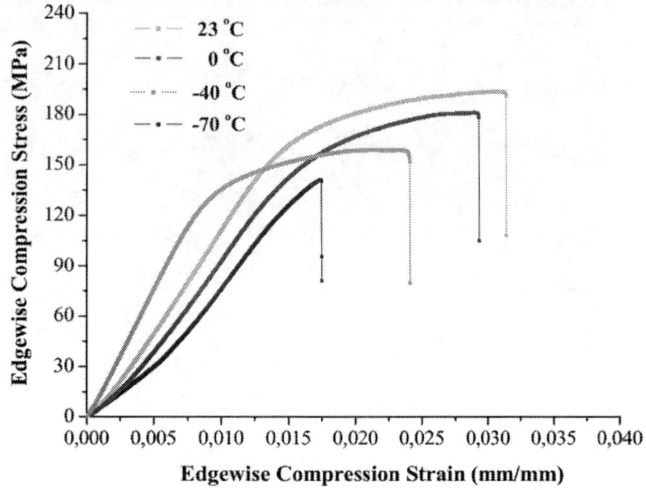

Figure 8. Stress-strain curves occurring from edge wise compression for specimens that were exposed at different temperature

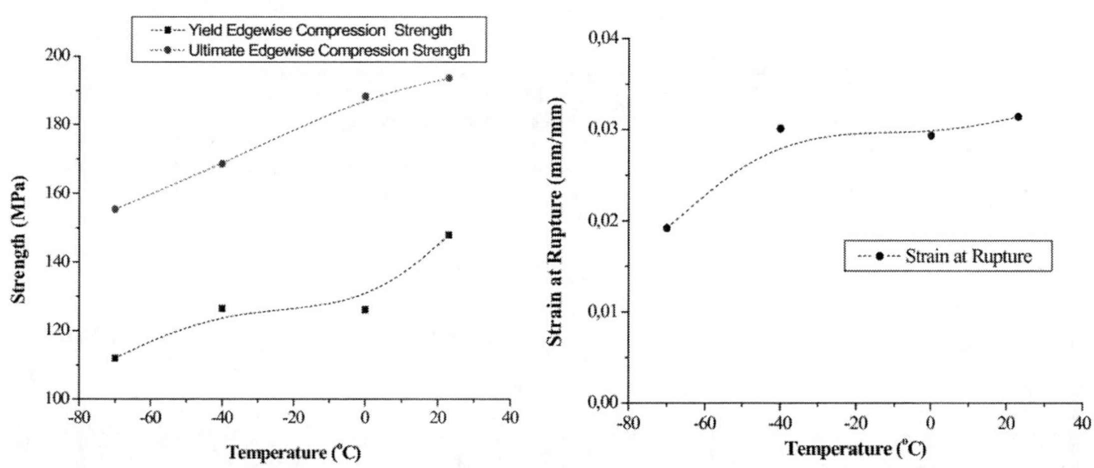

Figure 9. Graphical representation of a) ultimate edgewise compression strength and yield edgewise compressive strength and b) strain at rupture for compression test, as a function of temperature exposure.

The fracture mode of the specimens subjected to edge wise compression is characterized by buckling instability of the skins, followed by delamination and by crushing of the skins near the hydraulic machine's plates as it can be seen Figure 10. A complete delamination between the skins and the core was observed in all temperature levels.

The stress strain curves of aluminum sandwich material subjected to flatwise compression at different test temperatures are graphically represented in Figure 11. As the temperature decreased the flatwise compression strength slightly decreases, and the flat wise compression strain at rupture drops as represented in Figures 12a and 12b respectively. This is probably due to the fact that aluminum does not change its properties in brittleness radically even at low temperature. The specimens showed core failure at any temperature. The fracture pattern of the test specimens of the sandwiches subjected to flatwise compression is shown in Figure 13. The damage induced in the specimen mainly concerns the aluminum core leading to collapse of its cells and the failure of the sandwich panel.

IOP Conf. Series: Materials Science and Engineering **700** (2019) 012017 doi:10.1088/1757-899X/700/1/012017

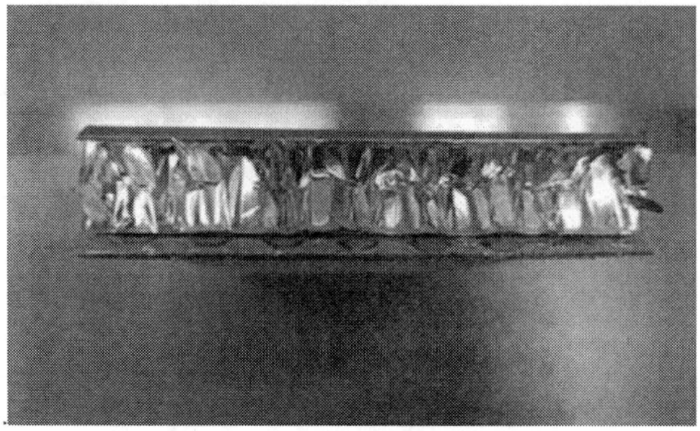

Figure 10. View of a specimen after an edgewise compression test at -70 °C.

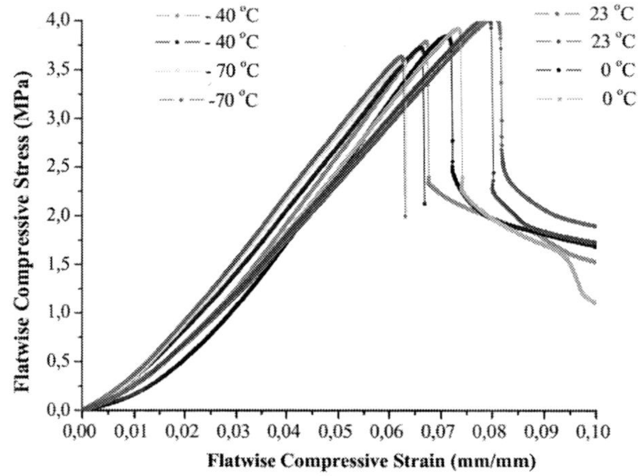

Figure 11. Stress-strain curves occurring from flatwise compression tests at different temperature profiles.

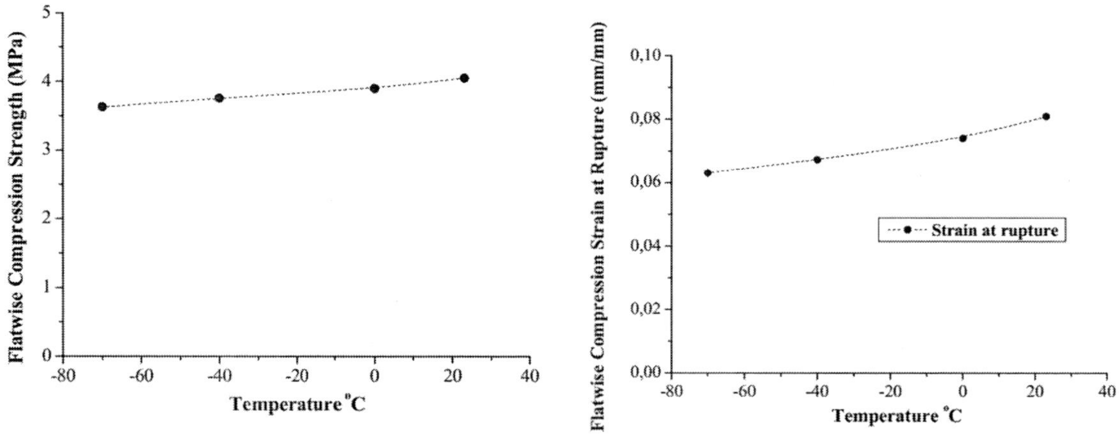

Figure 12. Graphical representation of a) ultimate flatwise compression strength and b) strain at rupture for compression flatwise test, as a function of temperature exposure.

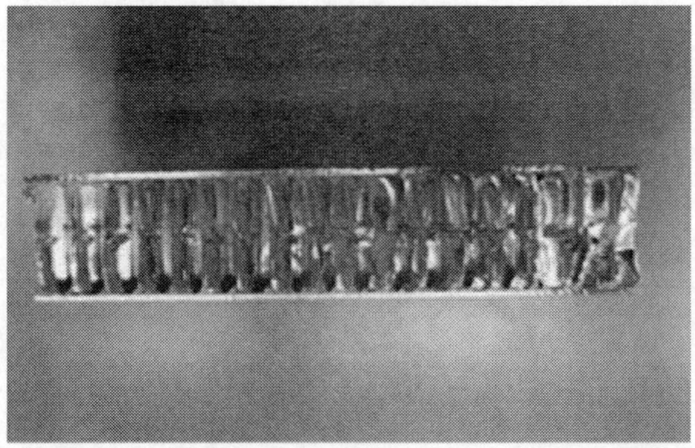

Figure 13. View of a specimen after a flatwise compression test at -70 °C.

The stress versus strain curves of the sandwich panels with aluminum honeycomb core and skins under three-point bending testing for different tested temperature profiles are shown in Figure 14. These curves can be divided into two groups, the former with an inelastic evolution at 23 °C and 0 °C and the later with an elastic evolution at -40 °C and -70 °C (Figure 15). However both ultimate bending strength and yield bending strength exhibit a moderate sensitivity in the changes of temperature as it can be seen in Figure 15. The ultimate bending strength of aluminum sandwich panels is governed by the instability of the top skins under compressive forces as it can be seen in Figure 16.

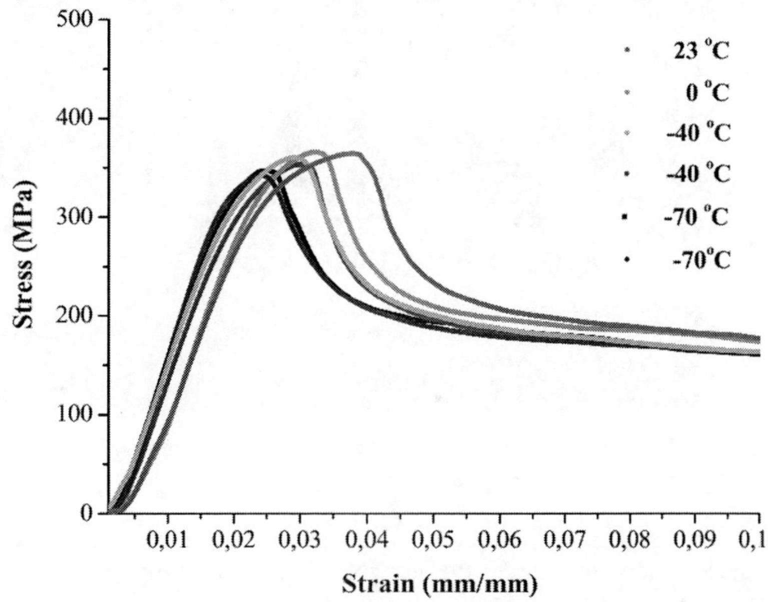

Figure 14. Stress-strain curves occurring from three point bending tests for specimens that were exposed at different temperature.

IOP Conf. Series: Materials Science and Engineering **700** (2019) 012017 doi:10.1088/1757-899X/700/1/012017

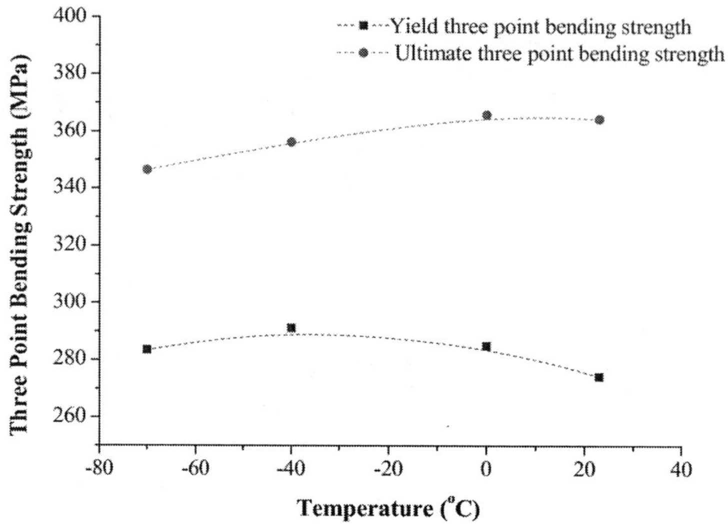

Figure 15. Graphical representation of ultimate flatwise compression strength and yield flatwise compression strength, as a function of temperature exposure.

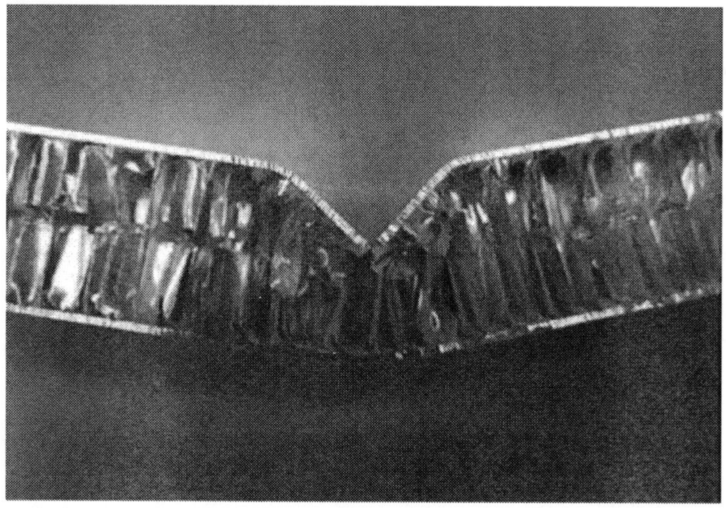

Figure 16. View of a specimen after a three point bending test at -70 °C.

5. Conclusions

Mechanical tests were performed in aluminum honeycomb specimens that were prior exposed to different temperature conditions in order to study the effect of this type exposure in the strength of and the overall mechanical behavior of the tested materials.

Ultimate tensile strength and yield strength are not affected from changes of temperature; while in the case of compression loading the respective values of strength seem to be decreased as temperature decreases. However, in all cases strain at failure is decreasingly affected by temperature conditions a fact that is attributed to the embrittlement of the material as temperature drop in value. As far as the bending strength is concerned the values of ultimate strength and yield strength are not affected by the temperature.The experimental data can be used for the selection of lighter and stiffer combination of sandwich panels according to the temperatures and loading conditions.

References

1. Khan M K 2006 *Proceedings of the Institution of Mechanical Engineers, Part G Journal of Aerospace Engineering*, Vol. 220 ISSN: 0954-4100 (Sage) DOI: 10.1243/09544100JAERO76
2. Ibrahim N A, Al Hazzal M H F, Adestal E Y T, Atiah Bt. Abdullah Sidekl and Endutl N A 2017 *ICAMME 2017: IOP Conf. Series: Materials Science and Engineering* **290** 012083, (*IOP Publishing*) DOI:10.1088/1757-899X/290/1/012083.
3. Barbieri G, Cognini F and Lari Gm Vivio F 2015, *Procedia Eng.* **109** 427
4. Mohammed Elamin, Bing Li and Tan K L 2017 *Proc. of the 21st Int. Conf. on Composite Materials* (Chinese Society for Composite Materials, China)
5. Birmana V and Kardomateas G A 2018 *Composites Part B* **142** 221
6. Remennikov A M and Kong S Y, 2012 *Compos. Struct.* **94** (12) 3546
7. Valdevit L, Wei Z, Mercer C, Zok F W and Evans A G 2006 *Int. J. Solid Struct.* **43** (16) 4888
8. Rajkumar S, Arulmurugan B, Manikandan M, Karthick R and Kaviprasath S 2017 *Applied Mechanics and Materials* **867** *245 DOI:10.4028/www.scientific.net/AMM.867.245*
9. Khashaba U A and Ramzi Othman 2017 *Int. J. Impact Eng.* **108** 191
10. Doo-Hwan Park, Sung-Woong Choi, Jeong-Hyeon Kim, Jae-Myung Lee 2015 *Cryogenics* **68** 44 DOI: 10.1016/j.cryogenics.2015.02.001
11. Senkov O N Bhat R B and Senkova S V 2004 in *Metallic Materials with High Structural Efficiency* High Strength Aluminum Alloys for Cryogenic Applications Senkov et al (eds.) 2004 (Kluwer Academic Publishers Netherlands) p 151
12. Kaufman J G 1999 *Properties of aluminum alloys: tensile, creep, and fatigue data at high and low temperatures* (Materials Park, Ohio: The Aluminum Association Inc. and ASM International)
13. ASTM D638–14, Standard Test Method for Tensile Properties of Plastics, 2014, (American Society for Testing and Materials, Philadelphia)
14. ASTM C-365-94. , Standard test method for flatwise compressive properties of sandwich cores, 1994 (American Society for Testing and Materials, Philadelphia) 1994.
15. ASTM C-364-94. Standard test method for edgewise compressive strength of sandwich constructions, 1994 (American Society for Testing and Materials, Philadelphia).
16. ASTM C-393-94. Standard test method for flexural properties of sandwich constructions, 1994 (American Society for Testing and Materials, Philadelphia).

Assessment of jacket-type platform stress state in corrosive environment: Case study

I Starokon and A Ermakov

Russian State Gubkin University of Oil and Gas, Russia

Corresponding authors: Starokon79@mail.ru, AlexanderIE@yandex.ru

Abstract. The structural tubular elements of the offshore platform substructure work in combined conditions of bending, tension, compression and torsion loading. Evaluation of the platform stress state is an important task. Effective solution of which can be received by a CAD modelling. The real stress state of the structure depends on the characteristics of corrosion processes. Therefore, corrosion defects should be taking into account for numerical and analytical modelling to determine/predict actual state of stress of bottom fixed offshore platforms. The stress concentration factor is the real characteristic of stress value for the structural elements of the platform in the zone of corrosion. The paper deals with the methods of the actual stress state calculation for the structural elements and welded joints of stationary platform under corrosion. The features of corrosion defects are also a subject of analysis. The proposed a calculation method of stress concentration factors for the primary tubular members of platform substructure in condition of corrosion.

1. Introduction

Offshore fixed platforms play an important role in offshore hydrocarbon production [1-5]. To ensure reliable and safe operation of the offshore structures, it is important to determine the real stresses acting both in structural elements and in welded joints of the offshore fixed platform (OFP), in due time preventing overstress of structures and possible accidental situation.

The considered offshore Jacket-type platforms are the most common steel structures welded from tubular members of different diameters. Methods for calculating the values of the loads and stresses are described in the normative documents and specialized literature. Most elements of the support block (SB) for the OFP are in a complex-stressed state under the action of wind, wave, vibration, temperature and other influences [4-6]. They often subjected to the common action of bending, tension, compression and torsion. Therefore, the assessment of the actual stress state, taking into account the mutual interaction of elements, is a difficult and time-consuming task. An effective solution of task can be found by using a computer aided FE models in the software environment of StructureCAD, ANSYS, SolidWorks, etc.

2. Considered platform and stress evaluation

The authors of the report have analyzed the inspection reports of offshore platforms located on the Black Sea shelf. This data showed that the greatest damage of the platform is in the splash zone located from (below) -10 meters from sea level and (above) +14 meters from sea level. According to the normative documents, the fatigue damage occurs in this zone at a wave height of 13.9 meters with 1%

probability. However, in fact, as shown by the authors' calculations according to [5, 6], the wave height directly causing fatigue failure does not exceed 6.3 meters. Only the impact of extreme winds at a speed of more than 49 m/s can cause a wave height of 13.9 m.

The authors carried out an analytical study, which showed that in the elements of the underwater part of the platform there are large stresses due to both the weight characteristics of the platform and the action of bending and torque from wind-wave load (WWL). In the absence of WWL, these stresses are static and operate in the platform columns along their axes. Studies of platforms a computer aided FE models prove that the maximum total values of stresses from static and dynamic loads for columns will be achieved both in the areas of fixation to the ground and in the points of the wave load action. For horizontal elements, maximum stresses arise due to the action of wave, ice and other dynamic loads, which reach their maximum values in the splash zone. For braces, both of these provisions are true, since in addition to the dynamic loads on them, part of the load is redistributed from the weight of the structure. Therefore, the zone of greatest stress for braces can be identified only during the calculation.

Figure 1. Offshore platform Support Block of the Subbotinskoye field in the Black Sea.

Figure 2. The model of the platform SB in the software package StructureCAD.

This paper presents results of the stressed state study of the supporting block platform installed at the Subbotinskoye field in the Black Sea (Figure 1 and Figure 2). To assess the stresses in the platform elements by SCAD (Rev. 2017) software package, a platform model was developed. The structure was subjected to the following loads: equipment weight and marine fouling, buoyancy force of the aquatic environment (combination of loads K1) and wind wave load action (WWL). Moreover, to take into account the mutual interaction of elements, the direction of the wind-wave load was set both along the X axis (combination of K2 loads) and in the direction at an angle of 45^0 to the X axis (combination of K3 loads). The magnitude of the wave load was chosen to correspond to the wave action at a wind speed of 25 m/s and a wave height of 6.8 m. Structural architecture of the platform is a truss constructed by tubular members. It has five sections, each of which is approximately ten meters high. The platform is installed at a depth of 52 meters. The following results were obtained (Table 1).

Table 1. Values of equivalent stresses for the elements of platform support block for various combinations of loading and various design solutions

Load combinations	Section numbers of the platform support block														
	1	2	3	4	5	1	2	3	4	5	1	2	3	4	5
	Values of equivalent stress, MPa														
	Columns					Horizontal elements					Bracing				
K_1	45	43	18	32	51	6	7	32	17	20	77	50	104	112	114
K_2	104	99	39	58	68	19	25	54	62	23	136	116	133	138	131
K_3	101	100	61	93	105	35	37	59	121	92	135	110	138	132	127

IOP Conf. Series: Materials Science and Engineering **700** (2019) 012018 doi:10.1088/1757-899X/700/1/012018

3. Analysis of factors affecting the platform elements stress state

3.1 Influence of cracks on the stress state of the platform elements

It is important to conduct a stress study of the platform structural elements with the cracks. Depending on the orientation and type of crack, a concentration of annular, radial or longitudinal stresses occurs. The increase of stress is described using the Stress Concentration Factor (SCF). For a longitudinal surface crack, the stress concentration is calculated primarily from the values of ring stresses, and for a circumferential crack by the values of longitudinal stresses.

Change of stress state due to presence of various sizes of cracks were observed by a parametric study. As an example, the influence of crack on the stress state of structure legs is evaluated. The initial low static stress of 45 MPa was considered (see Table 2).

Table 2. Values of stress concentration in the presence of circumferential cracks in the elements of offshore platform, on the example of a leg, MPa

Depth of crack, mm	Crack length, mm					
	5	10	20	30	40	50
1	68	74	79	81	84	81
2	72	97	106	114	118	119
3	74	108	131	141	147	151

From the Table 2 it becomes obvious that in the presence of a 3 mm deep crack, even insignificant static stresses in a column of 45 MPa can significantly increase and exceed the permissible level.

3.2. The effect of corrosion defects on state of stress of the members

The aggressive marine aquatic environment causes the metal corrosion of the operating platforms. In the presence of stresses, caused by various loads, the corrosive effect creates a concentration of stresses, including those exceeding the normative level.

Various theories of stress corrosion have been analyzed. However, in this process of studying the corrosive effect on the platforms stress state, the authors refer to the classical theory of fracture mechanics. It is obvious that various forms of corrosion on the surface of the platform structural elements have a significant impact on the stress state. It is considered that continuous surface corrosion leads to a uniform thinning of the wall thickness of the pipes with a uniform decrease in the total bearing capacity. Corrosion defects (CD), extending deep into the base metal and having a small radius at the end of the defect, have a much more serious increasing effect on stress. Such CD dramatically change the surface shape of the platform structural members, leading to a change in the overall , and become potentially overstressed of the member, known in classical fracture mechanics as stress concentration (SC). This state is characterized by Stress Concentration Factor (SCF). Provided studies show that the magnitude of the stress concentration is the greater when corrosion defect is sharper [5, 6].

A corrosive defect in the form of a cavity is characterized by the parameters of length, width, depth, and radius of curvature at the end of CD. The method for determining the parameters of corrosion defects for offshore platforms is described in detail in [8]. In order to control the changes in the stress state of a platform, it is necessary to determine the corrosion rate (CR).

The studies do not consider the influence of electrochemical protection systems, insulating coatings, sacrificial protection, and other options. But the intensity of the corrosion process is different and depends largely on the locations of the elements, their design features and the quality of structure manufacturing. The elements of platform structures in the splash zones have the mostly surface and through corrosion damages.

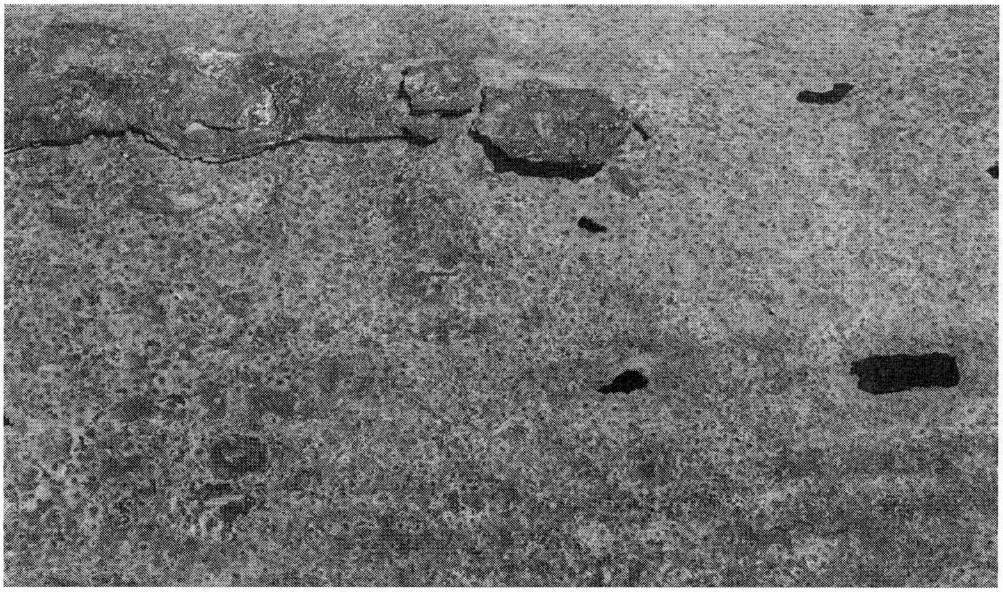

Figure 3. The actual state of the platform elements after 10 years of operation.

The analysis of diagnostic examinations data shows that the entire surface of a structural element is damaged by corrosion. It is about 500-600 and more ulcerative CD with a complex spatial form were identified. This makes the calculation process difficult enough. Assessment of joint effect of CD on the platform stress state is a task that can be resolve by schematization and assembling of separate corrosion defects. To solve this problem, the entire surface of the studied structural element was divided into sections with ulcerative CD, allocated against the background of general surface corrosion.

Then a single CD of a cylindrical structural element is analyzed [5, 6]. The tube element has outer diameter DH and wall thickness δ. It was considered the a cavity has the height H, length L along the tube axis, angular size is Θ, linear width is W. Based on this schematization, stress concentration factors for the most common defect sizes was calculated (Table 3). Analysis of the table showed that the stress concentration is growing with increasing of depth and decreasing of the length and angle of the corrosion defect.

3.3. The effect of uniform surface corrosion on state of stress of the members

The authors have analyzed the effect of surface corrosion defects of tube elements on the stress concentration. Based on the data of corrosion rates for various elements of the platform [2, 5, 6], the stress concentration effect under the action of uniform surface corrosion was investigated. Consider that all types of loads can be classified as compression-tension, bending and torsion. The stresses arising under these loads will depend on the cross-sectional areas of structural elements (SE), values of axial moments of sections resistance under bending and torsion.

The analysis was performed for the tube legs with diameter from 720 to 1020 mm with wall thicknesses from 16 mm to 30 mm. Analysis have shown that the values of the rate of corrosion processes obtained as a result of examination of platforms at various fields of the Black Sea differ from the rate (Table 4), recommended by norms [3, 4].

IOP Conf. Series: Materials Science and Engineering **700** (2019) 012018 doi:10.1088/1757-899X/700/1/012018

Table 3. Stress concentration factors of typical platform structural elements with different parameters of corrosion defects

Depth corrosive defect H,	Corrosion defect length L, mm											Corner size of corrosion defect Θ, degrees							
	1	**5**	**10**	**15**	**20**	**25**	**30**	**35**	**40**	**50**	**60**	**1**	**5**	**10**	**20**	**45**	**135**	**160**	**170**
	1) Structural element with diameter of 478 mm and wall thickness of 11 mm																		
0.04	1.01	1.01	1.01	1.01	1.01	1.01	1.01	1.01	1.00	1.00	1.00	1.01	1.01	1.01	1.01	1.01	1.01	1.01	1.01
0.52	1.11	1.10	1.10	1.09	1.09	1.9	1.08	1.07	1.07	1.05	1.03	1.17	1.17	1.12	1.13	1.12	1.08	1.07	1.10
1	1.21	1.20	1.19	1.19	1.18	1.17	1.16	1.15	1.14	1.12	1.09	1.31	1.31	1.21	1.23	1.22	1.14	1.13	1.18
1.52	1.32	1.31	1.30	1.29	1.28	1.27	1.26	1.25	1.23	1.21	1.18	1.44	1.45	1.29	1.32	1.31	1.19	1.18	1.26
2	1.42	1.41	1.40	1.38	1.37	1.36	1.35	1.33	1.32	1.29	1.26	1.56	1.57	1.34	1.39	1.38	1.23	1.21	1.32
2.52	1.54	1.52	1.50	1.49	1.47	1.46	1.45	1.43	1.42	1.39	1.36	1.67	1.69	1.39	1.46	1.44	1.25	1.23	1.39
3	1.64	1.62	1.60	1.58	1.56	1.55	1.53	1.2	1.51	1.48	1.45	1.77	1.9	1.41	1.50	1.9	1.26	1.3	1.44
	2) Structural element with diameter of 530 mm and wall thickness of 12 mm																		
0.04	1.01	1.01	1.01	1.01	1.01	1.01	1.01	1.01	1.00	1.00	1.00	1.01	1.01	1.01	1.01	1.01	1.01	1.01	1.01
0.52	1.10	1.10	1.09	1.09	1.08	1.08	1.08	1.07	1.07	1.06	1.04	1.15	1.16	1.11	1.12	1.11	1.14	1.08	1.07
1	1.19	1.19	1.18	1.17	1.17	1.16	1.15	1.14	1.14	1.12	1.10	1.28	1.29	1.20	1.22	1.21	1.26	1.13	1.12
1.52	1.29	1.29	1.28	1.27	1.26	1.25	1.24	1.23	1.22	1.20	1.17	1.41	1.42	1.27	1.30	1.29	1.39	1.18	1.17
2	1.39	1.38	1.37	1.36	1.34	1.33	1.32	1.31	1.30	1.28	1.25	1.52	1.53	1.33	1.37	1.36	1.50	1.22	1.20
2.52	1.49	1.48	1.46	1.45	1.44	1.42	1.41	1.40	1.39	1.36	1.34	1.63	1.64	1.37	1.43	1.42	1.62	1.24	1.22
3	1.58	1.57	1.55	1.54	1.52	1.51	1.50	1.48	1.47	1.45	1.42	1.72	1.74	1.40	1.48	1.46	1.25	1.23	1.42
	3) Structural element with diameter of 1420 mm and wall thickness of 15.7 mm																		
0.04	1.01	1.01	1.01	1.01	1.01	1.01	1.00	1.00	1.00	1.00	1.00	1.01	1.01	1.01	1.01	1.01	1.01	1.00	1.00
0.52	1.07	1.07	1.07	1.07	1.07	1.07	1.07	1.06	1.06	1.06	1.06	1.12	1.12	1.09	1.09	1.09	1.11	1.06	1.05
1	1.15	1.14	1.14	1.14	1.13	1.13	1.13	1.13	1.12	1.12	1.11	1.22	1.23	1.16	1.17	1.16	1.20	1.11	1.10
1.52	1.22	1.22	1.22	1.21	1.21	1.20	1.20	1.20	1.19	1.18	1.18	1.32	1.33	1.22	1.24	1.23	1.30	1.15	1.14
2	1.30	1.29	1.29	1.28	1.28	1.27	1.27	1.26	1.26	1.25	1.24	1.41	1.42	1.27	1.30	1.29	1.39	1.18	1.17
2.52	1.38	1.37	1.36	1.36	1.35	1.35	1.34	1.33	1.33	1.32	1.31	1.50	1.51	1.32	1.36	1.35	1.48	1.21	1.19
3	1.45	1.44	1.43	1.43	1.42	1.41	1.41	1.40	1.39	1.38	1.37	1.58	1.59	1.35	1.41	1.39	1.57	1.23	1.21
3.52	1.53	1.52	1.51	1.50	1.49	1.49	1.48	1.47	1.46	1.45	1.44	1.66	1.68	1.38	1.45	1.44	1.65	1.25	1.22
4	1.60	1.59	1.58	1.57	1.56	1.55	1.54	1.54	1.53	1.51	1.50	1.73	1.75	1.40	1.48	1.47	1.26	1.23	1,42

Table 4. Corrosion rates of tubular structural elements, taking into account the zone of their location.

Name of elements	Location zone	Field		
		Golitsinskoe	Strelkovoe	Shtormovoe
		Corrosion rate of tubular members, calculated according to diagnostic examinations, mm/year		
Columns	Underwater zone	0.0531	0.0641	0.0804
	Splash zone	0.2468	0.1957	0.2993
	Overwater zone	0.1029	0.0467	0.0436
Braces	Underwater zone	0.08053	0.0896	0.0701
	Splash zone	0.2532	0.1886	0.269
	Overwater zone	0.0103	0.0182	0.0024
Horizontal elements	Underwater zone	0.0654	0.0668	0.0780
	Splash zone	0.167	0.172	0.185
	Overwater zone	0.00813	0.0063	0.0175

The table shows that, for example, the corrosion rate of the columns located in the splash zone is 0.2993 mm / year, which is as much as 1.89 times higher than recommended by norms [3, 4]. As well known, the stress concentration coefficient is determined from the ratio of actual and nominal voltages. Compressive or tensile stresses are calculated as the ratio of the force to the cross-sectional area of the tube element of the platform. Since surface corrosion in an offshore field leads to a thinning of the outer surface of the structural elements of the platform, reducing the value of the cross-sectional area while the magnitude of the action of the loads is constant, the actual stress increases.

Knowing the corrosion rates, which depend on the location of the platform elements, it is possible to predict a decrease in the cross-sectional areas of the elements. Based on the ratio of design and reduced as a result of the corrosion effects of the transverse areas, it becomes possible to determine the stress concentration factor.

By a fifth-degree polynomial approximation, which gives a result with an accuracy of 90%, the authors has proposed the following formulas for calculating the SCF for columns located in different zones under the action of longitudinal forces, bending and torsional moments.

For underwater zone:

$$\mathrm{K} = 2{,}7 - (2628\rho + 67{,}06t - 132{,}7\rho^2 - 10{,}45\rho t) \cdot 10^{-4} \\ -(219{,}1\rho^3 - 6{,}02t^2 + 21{,}12\rho^2 t - 1{,}983\rho t^2) \cdot 10^{-6} - 3{,}153 \cdot 10^{-19} t^3, \quad (1)$$

For splash zone:

$$K = 3{,}429 - (3770\rho + 95{,}61t - 191{,}1\rho^2 - 13{,}68\rho t) \cdot 10^{-4} \\ -(317{,}3\rho^3 - 116{,}1t^2 + 22{,}56\rho^2 t + 3{,}997\rho t^2) \cdot 10^{-6} + 24{,}68 \cdot 10^{-8} t^3, \quad (2)$$

For surface zones:

$$\mathrm{K} = 2{,}441 - (2261\rho + 9{,}106t - 115{,}9\rho^2 - 3{,}773\rho t) \cdot 10^{-4} \\ -(193{,}7\rho^3 - 17{,}16t^2 - 7{,}069\rho^2 t - 3{,}44\rho t^2) \cdot 10^{-6} - 98{,}77 \cdot 10^{-8} t^3, \quad (3)$$

Where, (for Eqs. 1-3): *K*-value of stress concentration factor, ρ-relative coordinate, determined from the ratio of the radius of the studied structural element to its wall thickness. Taking into account the values of corrosion rates in surface, splash and underwater zones, the authors carried out numerical-analysis of the structure stress state. As a result, the values of stress concentration factors were

IOP Conf. Series: Materials Science and Engineering **700** (2019) 012018 doi:10.1088/1757-899X/700/1/012018

calculated depending on the operation time of the reference platform and the dimensions of the test element. The obtained stress concentration factors (SCF) indicate that their values in cases of compression-tension forces and bending or torsional moments differ from each other no more than 2%. This allows calculate them by using a common formulation.

In the considered structures, all horizontal, transverse and diagonal members have diameters from 325 mm to 530 mm with wall thicknesses from 10 mm to 18 mm. They are connected with columns or with other elements, like inclined transverse and longitudinal elements. For analysis in accordance with [2], the corrosion rate is 0.18 mm/year for underwater, splash and surface zones. As in the previous case, CCS values are equal for various forms of loading. On the base of the CCS value calculation and approximation of the obtained results, the following formula was performed:

$$K = 1{,}035 + 0{,}07193t - 0{,}2372\,\rho - 0{,}00592\,t\rho + 0{,}2512\rho^2 - 0{,}05864t\rho^2$$
$$+0{,}6685\rho^3 + 0{,}02765t\rho^3 - 0{,}2376\rho^4 + 0{,}03548t\rho^4 - 0{,}3197\rho^5 \quad (4)$$

Where, K is the value of the stress concentration factor, ρ is the relative coordinate determined from the ratio of the radius of the studied structural element to the thickness of its wall; t is the time of operation.

As a result of the analysis, it was revealed that the corrosion rate of 0.2993 mm/year for columns located in the splash zone is 1.89 times greater was indicated (0.16 mm/year). Therefore, when using formulas for calculation (1-4), it is advisable to multiply the value of t by 1.89 for columns and 1.49 for diagonal lines. At the same time, for all elements located in the underwater zone, the corrosion rate is significantly lower (approximately 0.02 mm/year), which is explained by significant marine fouling.

To determine the SCS values of welded joints (WJ) depending on the operation time, the WJ models of the platform were built using SolidWorks (Rev. 2018) software. The values were calculated by points with the maximum stress concentration taking into account the form of loading. The results are shown in Table 5.

The significant effective amplitudes of alternating stresses lead to early exhaustion of the structural elements resource. This is especially important when assessing the resource of restored (repaired) welded joints. The authors [7] have proposed the structural elements that allow reducing the amplitudes of alternating stresses acting in a welded joint. This will increase their remaining life.

Table 5. The stress concentration coefficients of WJ under corrosive effects, taking into account the shape of the applied load at the points with the maximum stress concentration

Tube structural elements	Duration of corrosion exposure, years					
	3	9	15	21	27	30
Action of a bending moment in the plane						
Horizontal transverse in the area of connection with columns (CR - 0.18 mm / year)						
325x10	1,04	1,16	1,34	1,52	1,88	2,00
426x12	1,05	1,16	1,25	1,44	1,72	1,86
530x14	1,04	1,16	1,26	1,41	1,72	1,87
Inclined longitudinal in the area of connection with columns (CR - 0.15 mm / year)						
325x10	1,04	1,12	1,26	1,41	1,55	1,73
426x12	1,04	1,13	1,20	1,30	1,54	1,68
530x14	1,03	1,14	1,25	1,34	1,48	1,53
Underwater parts of columns (CR - 0.12 mm / year)						
325x10	1,03	1,08	1,17	1,30	1,41	1,51
426x12	1,03	1,10	1,14	1,23	1,31	1,40
530x14	1,03	1,12	1,20	1,24	1,31	1,47
Action of axial force on the waist tube and bending moment outside the plane						
Horizontal transverse in the area of connection with columns (CR - 0.18 mm / year)						
325x10	1,10	1,32	1,52	1,56	2,13	2,37

Tube structural elements	Duration of corrosion exposure, years					
	3	9	15	21	27	30
426x12	1,05	1,15	1,38	1,58	1,79	1,91
530x14	1,05	1,19	1,30	1,38	1,63	1,72
Inclined longitudinal in the area of connection with columns (CR- 0.15 mm / year)						
325x10	1,08	1,15	1,32	1,44	1,65	1,83
426x12	1,03	1,14	1,29	1,47	1,57	1,62
530x14	1,04	1,16	1,25	1,32	1,43	1,49
Underwater parts of columns (CR - 0.12 mm / year)						
325x10	1,18	1,20	1,31	1,34	1,48	1,50
426x12	1,02	1,09	1,18	1,34	1,48	1,53
530x14	1,02	1,10	1,20	1,24	1,33	1,37

4. Conclusions

The stress state of the platform elements in marine conditions is affected by cracks, caverns, and general metal degradation resulting from corrosion. The real rate of corrosion processes for the structural elements of the platform in the marine environment, determining its stress state and stress concentration, differs from the values recommended by the normative design documents.

References

[1]. Borodavkin P P 2006 *Offshore oil and gas facilities: Textbook for universities.* Part 1. Design. – M.: OOO «Nedra-Biznestsentr», 555 (in Russian)

[2]. Gudmectad O T, Zolotukhin A B, Ermakov A I et al. 1999 *Basics of Offshore petroleum engineering and development of marine facilities: with emphasis on the arctic offshore-oil and gas*, Neft'i Gaz, Moscow.

[3]. SP 38.13330.2010 "SNiP 2.06.04-82 2011 Loads and impacts on hydraulic structures (wave, ice and from ships) "Ministry of Regional Development of Russia. 116 p. (in Russian)

[4]. Rules for the classification, construction and equipment of floating drilling rigs and offshore fixed platforms - Russian Maritime Register of Shipping, St. Petersburg., 2014. -484p. (in Russian)

[5]. Starokon I V 2015 Assessment of the fatigue life of welded joints and the main structural elements of the supporting blocks of offshore stationary platforms, *Basic Res. J.*, 2015 (7-4) 691-696.

[6]. Starokon I V 2012 *Fundamentals of the theory and practice of the formation of fatigue cracks in offshore oil and gas facilities*, Modern problems of science and education. - 2012. - № 4; URL: www.science-education.ru/104-6605

[7]. Starokon I V, Golovachev A O and Nadyrov R I 2019 Methods for increasing the fatigue life of repaired welded joints of offshore oil and gas facilities, *IOP Conf. Series: Earth and Environ. Sci.*, **272**(3) 8.

[8]. Starokon I V, Golovachev A O 2019 Method of determining the sizes of corrosion defects of elements of marine oil and gas industrial constructions on the basis of data on temperature contrasts, *IOP Conf. Series: Earth and Environ. Sci.*, **272**(3) 8.

IOP Conf. Series: Materials Science and Engineering **700** (2019) 012019 doi:10.1088/1757-899X/700/1/012019

Numerical analysis of the influence of friction conditions on the pile-up effect in Vickers hardness measurements

T Trzepiecinski[1] and H G Lemu[2*]
[1] Department of Materials Forming and Processing, Rzeszow University of Technology, al. Powst. Warszawy 12, 35-959 Rzeszów, Poland
[2] Department of Mechanical and Structural Engineering, University of Stavanger, N-4036 Stavanger, Norway

* Corresponding author: Hirpa.g.lemu@uis.no

Abstract. Vickers indentation load-depth curve may be used to determine basic mechanical properties of metallic materials such as hardness, yield stress, modulus of elasticity, and elastic and plastic work. The pile-up phenomenon observed during indentation causes underestimation of the projected contact area and the diagonal dimensions of the impression in different hardness measurement scales (nano-, micro-, and macro-). The aim of this paper is to conduct a numerical analysis of the effect of friction conditions on the pile-up phenomenon during testing of DC04 steel sheet. The mechanical properties of the sheet metal used for the modelling purpose were first characterized by tensile tests on samples cut along the rolling direction (0°), transverse to the rolling direction (90°) and at an angle of 45° to the rolling direction. The numerical computation was conducted using ABAQUS, which is one of the powerful finite element-based programs. A wide range of variation of friction coefficients, i.e. 0-0.3 has been used in the analysis. It has been observed that the results of indentation of anisotropic materials are significantly affected by friction. The difference in the pile-up height measured at rolling direction of sheet metal and transverse to the rolling direction decreases with the reduction of the maximum displacement of the indenter. For higher values of the coefficient of friction, the higher the value of the indenter displacement, the lower becomes the increase in the pile-up height value.

1. Introduction

Mechanical properties of metallic materials are important features that determine their applicability to the specific bulk metal forming process or to the construction of a particular element type. For this reason, metals are subjected to different types of tests which aim is to establish the ability to transfer the load without damage [1, 2]. Due to the static, dynamic or cyclical nature of various loads, a number of test methods have been developed to determine the mechanical properties of materials, such as yield stress, ultimate tensile strength, modulus of elasticity, hardness, and strain hardening parameters. Hardness testing is a non-destructive test that can be carried out directly on structural elements and can be carried out in nano-, micro- and macro- scales according to the specimen sizes. The measurements in static conditions, especially in the nano- and microscale give reasonable results. The use of dynamic-plastic or dynamically elastic devices for measuring hardness causes significant errors to the results of testing thin-walled elements. Indentation hardness significantly depends on the elastic-plastic properties

IOP Conf. Series: Materials Science and Engineering **700** (2019) 012019 doi:10.1088/1757-899X/700/1/012019

of material which on elastic recovery of the bulk deformed material [3]. Furthermore, the hardness measurement accuracy may be disturbed by surface roughness of the material tested [4, 5].

Hardness may be defined as a measure of the resistance of the metal when pressing the indenter tip or a measure of the resistance of the metal against the permanent deformation resulting from the indentation. A hardness test consists of indenting the material for a specified time and at a given load on a smooth metal surface, which results in a localized permanent deformation in the shape of a dimple impression. The prevalence of static hardness measurements may be easily explained by establishing the relationship between hardness test results and the results of other tests of mechanical properties carried out in a very small volume of material. In order to study the mechanical properties of materials, i.e. Berkovich, Brinell, Rockwell and Vickers hardness, many indentation based studies have been conducted using diverse parameters such as the load applied, material dimensions and shape of the indenter.

Pile-up or sink-in of the material (Figure 1) around the indenter-tip is a phenomenon which leads to a significant error in the determination of material properties [6, 7]. These phenomena cause significant underestimation during indentation of the impression diagonals and the true contact area of impression. As shown by Bolshakov and Pharr [8], the true contact area may be overestimated by 60%.

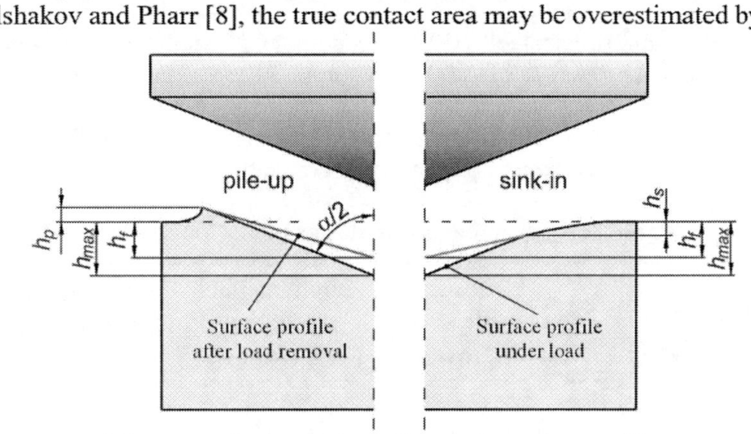

Figure 1. Cross section of the indentation profile.

Definition of abbreviations in the figure: h_f - final displacement of the indenter after complete unloading, h_{max} - maximum displacement of the indenter, h_p - pile-up height, h_s - sink-in depth, α - angle between the two opposite faces of indenter (α = 136°±0.5°)

Many studies have devoted to the formation of pile-up/sink-in. For instance, several studies [9-11] have shown that these phenomena can be correlated to the elastic and plastic regimes of deformation, such as the ratio of the final indentation depth h_f to maximum indentation depth h_{max} or the strain hardening exponent n. An empirical method of approximating the pile-up contact perimeter has been developed by Kese and Li [12] who concluded that correlating pile-up to various material and loading conditions can be invaluable in appropriately accounting the effect of pile-up on hardness. Zhou and Yao [13] suggested accounting the plastic work and plastic deformation volume to accounting the pile-up. It is well known that the Oliver and Pharr method [14] can overestimate the hardness value of metallic materials that plastically deform due to piling up around the indentation. Some of authors proposed a correction to this standard Oliver and Pharr method to account for errors in hardness measurement due to pile-up effect. In general, imaging-based hardness corrections for pile-up or sink-in formation are based on the consideration that the contact area depends on the tip-area function.

Cabibbo et al. [15] used scanning probe microscopy (SPM) measurements and analytical methods to analyse indentation hardness. As it has been concluded, the SPM-based method showed a quite good quantitative and qualitative literature data agreement. In the work of Burik et al. [16] pile-up correction of mechanical properties was based on the ratio of pile-up height measured by atomic force microscopy (AFM) and contact depth. Bandara et al. [17] conducted 2D and 3D numerical simulations of nano-indentations to understand the behaviour of pile-up formation. They concluded, however, that the pile-up height increased linearly relative to the load, while the rate (slope) increases with the amount of strain hardening. Zhang et al. [18] found numerically that high temperature hardness does not change with the change of load when pile-up of material during indentation is ignored. Tang et al. [19] conducted the finite element (FE) analysis to examine the pile-up effect on the accuracy of sharp indentation testing. They concluded that the effect of pile-up becomes significant when $h_f/h_{max} > 0.85$. The pile-up effect depends on the anisotropy of the material tested and, in general, the anisotropic pile-up or sink-in patterns were, so far, studied in relation to single crystals [20-22].

The aim of this study is to build a finite element based model and investigate the relationship between the frictional conditions and the amount of the pile-up effect during testing of anisotropic material. Three dimensional (3D) numerical models have considered to account for real strain hardening properties and anisotropic yield criterion of the sheet metal tested. The remaining part of the article is divided into 4 Sections. Section 2 presents the material used for the study and the mechanical properties of the steel sheet determined through tensile test. The numerical modelling work is reported in Section 3 followed by the discussion of the results in Section 4. Finally, Section 5 presents the conclusions drawn from the study.

2. Material

The test material was a 2-mm-thick steel DC04 sheet commonly used in the automotive industry. Table 1 presents basic mechanical properties of the sheet determined according to the EN ISO 6892-1 standard. The samples for the tensile tests were cut along the rolling direction (0°), transverse to the rolling direction (90°) and at an angle of 45° to the rolling direction. Three specimens were tested for each sample orientation and basic mechanical parameters were determined.

Table 1. Mechanical properties of the DC04 steel sheet.

Orientation	Yield stress $R_{p0.2}$ (MPa)	Ultimate tensile stress R_m (MPa)	Strengthening coefficient K (MPa)	Strain hardening exponent n	Lankford coefficient r
0°	182.1	322.5	549.3	0.214	1.751
45°	196	336.2	564.9	0.205	1.124
90°	190	320.9	541.6	0.209	1.846
Average value	191.2	328.95	555.17	0.208	1.461

The values of the strain hardening parameters, i.e. the strength coefficient *K* and the strain hardening exponent *n*, were determined by approximation of the experimental stress-strain curve by the Hollomon power law relationship:

$$\sigma_p = K \cdot \varphi_i^n \tag{1}$$

where σ_p is the yield stress and φ_i is equivalent plastic strain.

Although the values of the determination coefficient R2 of the experimental data in equation (1) were above 0.985, it is well known that Hollomon equation does not fit accurately the experimental data in the range of small amounts of plastic deformations. So, in the numerical models, the real strain hardening curve has been included. Figure 2 shows exemplary stress-strain relations. It is observed in this figure that the stress-strain relation curves for rolling direction (0°) and transverse direction (90°) overlap.

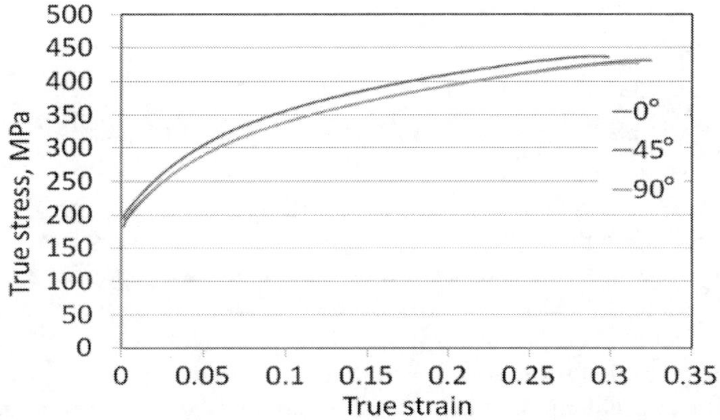

Figure 2. Exemplary true stress-true strain curves for the DC04 sheet.

3. Numerical modelling

The numerical modelling of Vickers pyramid indentation test has been performed in FE-based ABAQUS software. 3D model of the indenter and workpiece were analysed. Due to the plane symmetry of the indentation process, a quarter model with appropriate symmetric boundary conditions was used(Figure 3). The hardness of diamond indenter is much greater that the hardness of the workpiece. So, the geometry of the indenter is represented by plane that was meshed using rigid elements. The maximum displacement of the indenter h_{max} into workpiece was set to four values: 0.0025 mm, 0.005 mm, 0.075 mm and 0.1 mm.

To select the appropriate element size, especially in the area of contact of the indenter with workpiece, the mesh sensitivity analysis was conducted. The indenter surface has been meshed by 8375 4-node 3-D bilinear rigid quadrilateral elements since ABAQUS allows only these options for rigid elements. All rotations and displacements of the indenter surface in X and Y direction (Figure 3) were fixed, while the displacement of nodes laying on the base of the sheet metal model were fixed only in Y direction. Furthermore, the displacement of nodes laying on the YZ plane and XY plane were fixed in X and Z direction respectively. The workpiece was modelled with a 46202 8-node linear brick with the capability of representing large deformations and material and geometrical non-linearity. A finer mesh at the contact region and a gradually coarser mesh further away from the contact area were used to reduce the computational time and increase the computational accuracy.

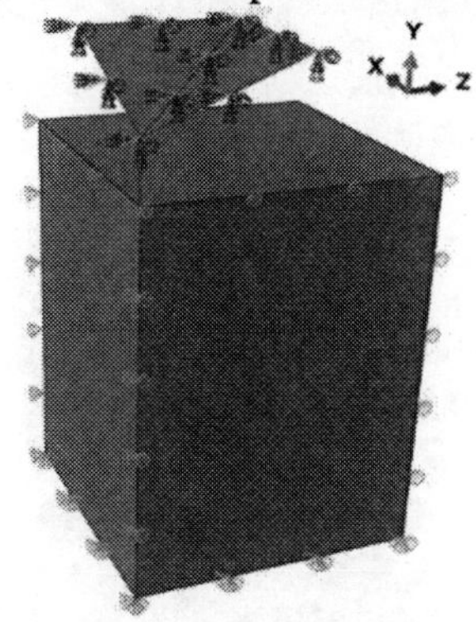

Figure 3. Finite element mesh in the numerical simulation.

The workpiece material is treated as elastic-plastic. The elastic behaviour of sheet material has been specified using the following parameters: Young's modulus $E = 210$ GPa and Poisson's ratio $\nu = 0.3$. The plastic behaviour of the steel sheet was described by the Hill(1948) [23] quadratic yield function, commonly used to describe the yielding of anisotropic steel sheets [24].

The Hill(1948) formulation is an extension of the isotropic von Mises function, and can be expressed in terms of rectangular Cartesian stress components as:

$$\overline{\sigma} = \sqrt{F(\sigma_{22} - \sigma_{33})^2 + G(\sigma_{33} - \sigma_{11})^2 + H(\sigma_{11} - \sigma_{22})^2 + 2L\sigma_{23}^2 + 2M\sigma_{31}^2 + 2N\sigma_{12}^2} \quad (2)$$

where $\overline{\sigma}$ is the equivalent stress, and indices 1, 2, 3 represent the rolling, transverse and normal directions to the sheet surface. Constants F, G, H, L, M and N define the anisotropy state of material and they are determined based on the Lankford's coefficients and yield stresses in specific directions of the sheet metal according to the formulas which have been described, for example in [25].

The contact properties have been described by Coulomb friction model using the following friction coefficients: 0.1, 0.2 and 0.3. Moreover, the frictionless conditions were assumed for the purpose of studying only the effect of material anisotropy on the pile-up effect. ABAQUS integration implicit code has been successfully applied for the static simulation of the indentation process. To solve geometrical and material non-linear problem, the incremental-iterative Newton-Raphson method is used.

4. Results and discussion

The pile-up height values were measured along the rolling direction (RD) and transverse (TD) to the rolling direction starting from the impression tip (Figure 4). The variations of the displacement of the nodes laying in the RD (Figure 5) and TD directions may be divided into three areas: (i) straight line corresponded to the edge of indenter, (ii) pile-up effect and (iii) the horizontal straight line where the material is not subjected to the plastic deformation.

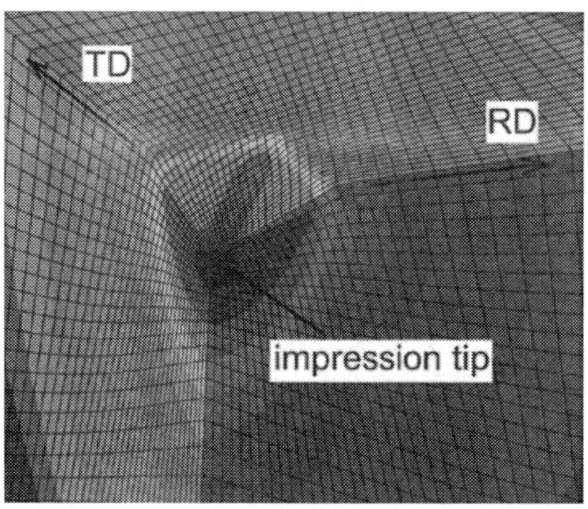

Figure 4. Schematic of measurement of the pile-up height in two orthogonal directions

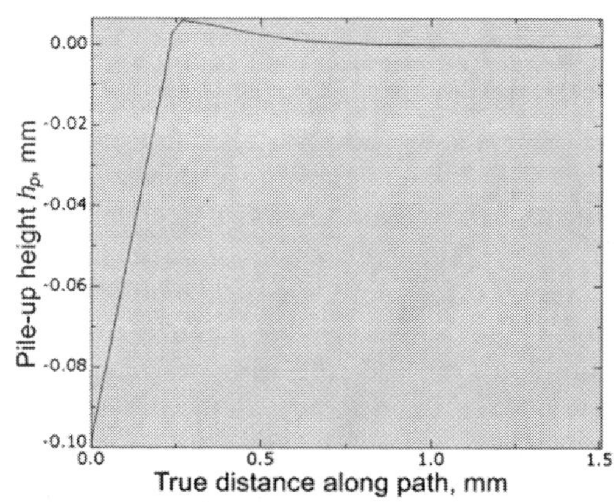

Figure 5. Exemplary variation of the pile-up height along the RD ($\mu = 0.3$, $h_{max} = 0.1$ mm)

The increase in the friction coefficient μ value causes a reduction of the pile-up height h_p for all analysed impression depths (Figure 6). In the case of frictionless conditions, a pile-up height increases nearly linear with the maximum displacement of the indenter h_{max}. For higher values of the coefficient of friction, the higher the value of the indenter displacement, the slower is the increase in the pile-up height value.

In the RD direction, greater values of pile-up height were observed (Figure 6a) compared to the TD direction (Figure 6b). This phenomenon may be linked to the differences in the values of Lankford's coefficients of the material and in the yield stress. The value of the yield stress in the RD direction is lower than in the TD, thus the material is more susceptible to flow in the RD direction.

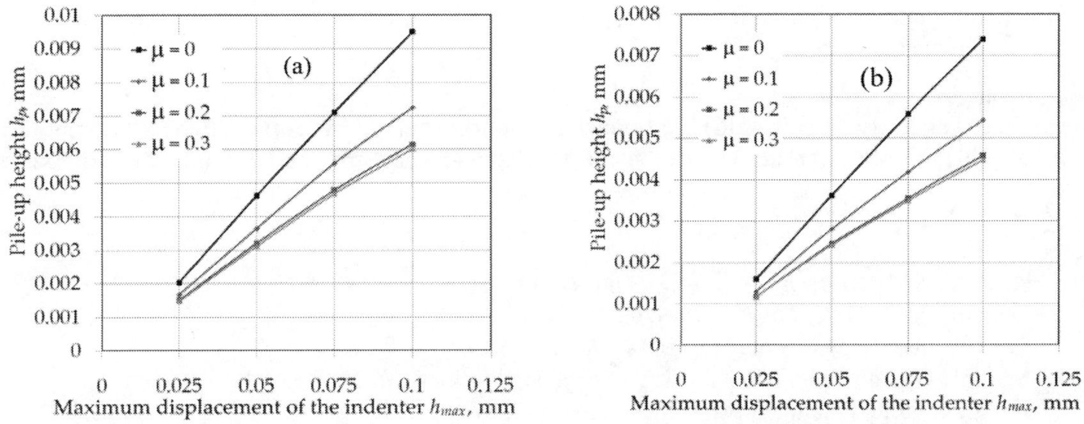

Figure 6. Variation of the pile-up height along (a) rolling direction (b) transverse direction

The difference in the pile-up height measured at RD and TD decreases with the reduction of the maximum displacement of the indenter h_{max} (Figure 7). A similar effect has been observed for the indentation depths h_{max} = 0.05 mm and 0.075 mm (not presented in Figure 7). Anisotropic pile-ups have important effects on the diagonal dimensions of the impression and evaluation of the contact area from indentation load–displacement curves. Thus, the area is always underestimated while the hardness is always overestimated. Moreover, elastic properties of the anisotropic material cause the non-uniform stress recovery of the material near the pile-up.

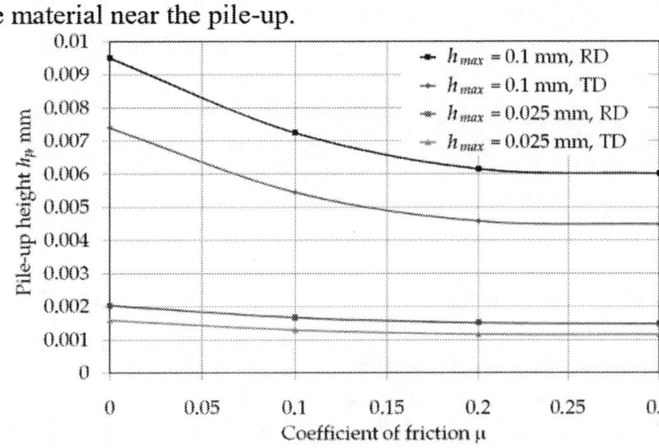

Figure 7. Effect of the friction coefficient on the pile-up height

5. Conclusions

In this paper, the frictional conditions and anisotropic material properties were addressed through determining a pile-up height. The results of 3D numerical simulations of the indentation that have been accounted for the material anisotropy and real strain hardening properties, allow to draw the following conclusions:

- In the case of anisotropic elastic-plastic sheets exhibiting plane anisotropy, the flow of the material around the tip, and formation of the pile-up around the tip, is non-uniform.
- The height, and as a consequence, the volume of the pile-up around the indenter depends on the friction coefficient value and the maximum indentation depth.
- The difference in the pile-up height measured at RD and TD, decreases with the reduction of the maximum displacement of the indenter h_{max}.

- The friction between indenter and specimen can significantly affect the contact area and piling-up or sinking-in which is in good agreement with the results of comprehensive computations conducted by Guo et al. [26].

Acknowledgement

The authors gratefully acknowledge the financial support provided by University of Stavanger through the COTech Program Area at Department of Mechanical and Structural Engineering and Materials Science.

References

[1] Kowalik M 2010 *Arch. Civ. Mech. Eng.* **10** 45.
[2] Kowalik M 2011 *Mater. Sci.* **46** 679.
[3] Trzepieciński T and Lemu H G 2017 *Metals* **7** 380.
[4] Lemu H G and Trzepieciński T 2013 *Stroj. Vestn.-J. Mech. Eng.* **59** 41.
[5] Jaworski J and Trzepieciński T 2016 *Kovove Mater.* **54** 17.
[6] Gale J D and Achuthan A 2014 *J. Mater. Sci.* **49** 5066.
[7] Miyake K, Fujisawa S, Korenaga A, Ishida T and Sasaki S 2004 *Jpn. J. Appl. Phys.* **43** 4602.
[8] Bolshakov A and Pharr G M 1998 *J. Mater. Res.* **13** 1049.
[9] Wang J, Hodgson P D and Yang C 2012 *J. Mater. Res.* **27** 313.
[10] Rodriguez J and Garrido-Maneiro M A 2007 *Mech. Mater.* **39** 987.
[11] Taljat B and Pharr G M 2004 *Int. J. Solids Struct.* **41** 3891.
[12] Kese K and Li Z C 2006 *Scripta. Mater.* **55** 699.
[13] Zhou L and Yao Y 2007 *Mater. Sci.Eng.* A **460-461** 95.
[14] Oliver W C and Pharr G M 1992 *J. Mater. Res.* **7** 1564.
[15] Cabibbo M, Ciccarelli D and Spigarelli S 2013 *Phys. Procedia* **40** 100.
[16] Burik P, Zubko P, Pešek L and Voleský L 2017 *Mater. Sci. Forum* **891** 78.
[17] Bandara R M D C N, Srilal D U, Sivaram S, Jayasinghe S C and Bandara C S. *Proc. Int. Conf. on Sustainable Built Environment (At Kandy)* p. 197.
[18] Zhang T, Jiang F, Yan L and Xu X 2018 *Materials* **11** 34.
[19] Tang Z, Guo Y, Jia Z, Li Y and Wei Q 2015 *Adv. Mater. Sci. Eng.* article id 528729.
[20] Casals O and Forest S 2009 *Comp. Mater. Sci.* **45** 774.
[21] Liu M, Tieu A K, Peng C T and Zhou K 2015 *Mater. Lett.* **161** 227.
[22] Kearney C, Zhao Z, Bruet B J, Radovitzky R, Boyce M C and Ortiz C 2006 *Phys. Rev. Lett.* **30** 255505.
[23] Hill R 1948 *Proc. Royal Soc. A Math., Phys. Eng. Sci.***193**, 281.
[24] Trzepieciński T 2010 *Arch. Civ. Mech. Eng.* **10** 95.
[25] *Analysis User's Manual.* 2007 *Volume III: Materials* (Providence: Dassault Systèmes).
[26] Guo W C, Rauchs G, Zhang W H and Ponthot J 2010 *J. Comput. Appl. Math.* **234** 2183.

IOP Conf. Series: Materials Science and Engineering **700** (2019) 012020 doi:10.1088/1757-899X/700/1/012020

Contact stresses associated with the wedge-lock mechanism in a prototype subsea pipeline recovery tool

Y Xing[1], N B Eriksson[1,2], M C Ong[1] and C Knutsen[2]

[1]Department of Mechanical and Structural Engineering and Material Science, University of Stavanger, Kjell Arholmsgate 41, 4036 Stavanger, Norway
[2]IK-Norway, Christian August Thorings veg 9, 4033 Stavanger, Norway

Corresponding author: yihan.xing@uis.no

Abstract. This paper examines the contact stresses associated with the wedge-lock mechanism in a prototype subsea pipeline recovery tool (PRT) through numerical methods and experimental testing. The prototype PRT studied combines IK-Norway subsea plug technology in a subsea PRT. This allows the PRT to seal the subsea pipeline while it is being retrieved. The PRT can then flush the pipeline before retrieval and therefore providing for operational cost savings. This paper presents the first stage of the study which the sealing mechanism is not studied. The current wedge-lock mechanism uses a cone to press steel balls against the insides of the subsea pipeline thereby locking the PRT to the pipeline. Designing for sufficient contact stresses is critical to ensure that the PRT can grab the subsea pipeline tightly in order to ensure a successful recovery. The contact pressure distribution between the ball and cone in the wedge-lock mechanism is modelled using finite element method in Ansys. The resulting indentation depths occurring at the pipeline are compared against physical tests. It was found that the contact problem associated with the wedge-lock mechanism is complex and the actual contact stress/final gripping force is sensitive to the input parameters. Second, the results show that the Ansys numerical model is able to represent the contract stresses fairly accurately if the friction coefficient is well-controlled. Third, it was observed that a low friction coefficient ensures low indentation depths. Last, the results highlight that an increase in the cone-angle leads to a decrease in the indentation depth.

1. Introduction and background

Pipeline recovery tools (PRTs) are used for the retrieval of subsea pipelines. During the next years, the decommissioning market for offshore and subsea pipeline is projected to grow significantly [1]. The reason is that many producing fields will reach the end of their lives. There has been a large production in subsea pipelines since 2003 [2]. These pipelines will eventually be decommissioned, and this will potentially lead to increased market demand for PRT's in the near future. There are many different solutions available in the market used for subsea pipeline recovery. An example of a PRT is presented in Figure 1. PRTs often use a mechanical system to insert/wrap over and lock itself to the subsea pipeline.

IOP Conf. Series: Materials Science and Engineering **700** (2019) 012020 doi:10.1088/1757-899X/700/1/012020

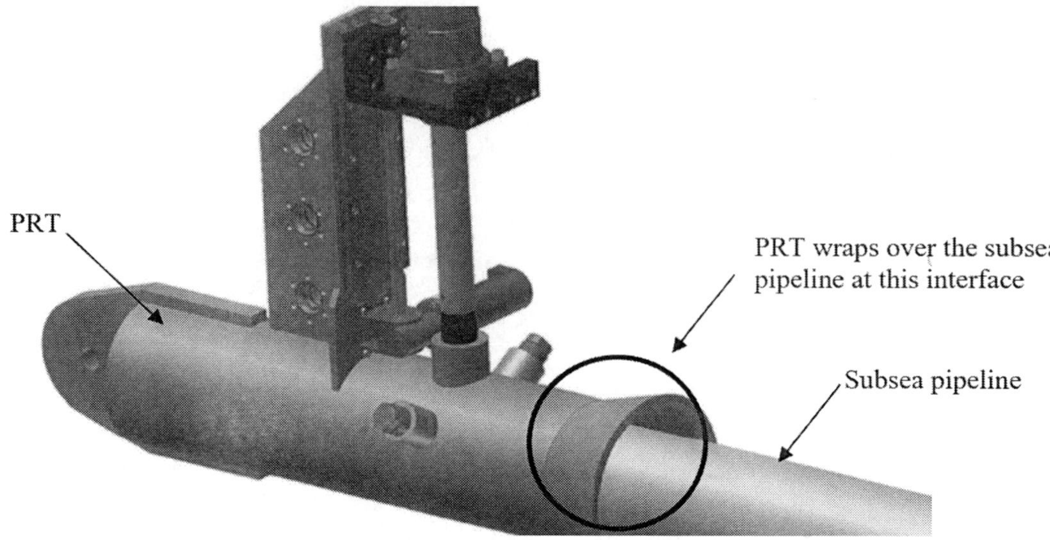

Figure 1. IK-Norway pipeline recovery tool, Ref. IK-Norway [3].

A unique wedge-lock and sealing technology that utilizes steel balls casted in polyurethane (PUR) is presented in this paper. The steel balls form a ball-grip mechanism together with a cone and presses steel balls against the inner diameter of the subsea pipeline to generate large contact stress. This in turns locks the PRT onto the subsea pipeline. The polyurethane presses itself on the inner diameter of the pipeline and this forms a sealing mechanism sealing off the pipeline during the recovery operation. This sealing mechanism that has its origins from IK-Norway's plug technology. Details of the prototype PRT and its unique wedge-lock and sealing technology is presented in Section 3. The ability to seal the subsea pipeline allows the pipeline to be flushed before retrieval. This leads to reduction in operational costs.

This paper studies the contact stresses associated the wedge-lock mechanism using numerical and physical testing methods. The purpose is to characterize the design of the wedge-lock mechanism. This paper presents the first stage of the study in which the sealing mechanism is not studied. This means that the steel balls without the PUR is modelled and physically tested. There are plans to study the steel balls casted in PUR in the second stage of the study in the near future.

2. Subsea pipeline recovery operation

The subsea pipeline recovery operation is briefly described in this section. The operation typically follows the steps below:

- The PRT is inserted at the end of the subsea pipeline using a remotely-operated vehicle.
- A hydraulic load is applied to the PRT to activate the locking mechanism. At this stage, the PRT is locked onto the subsea pipeline.
- The PRT is then pulled up to the surface thereby recovering the subsea pipeline.

The PRT can grip the pipeline on the inner or outer diameter. In the case where the PRT grips the pipeline on its outer diameter i.e., the PRT would warp over the pipeline as illustrated in Figure 1,.

3. Prototype pipeline recovery tool (PRT)

The prototype PRT is presented in Figure 2. This tool is designed to recover a 12-inch subsea pipeline. It is designed to recover pipelines manufactured in accordance with the ASME B36.10M [4] dimensions. Item number 8 in Figure 2 is a bearing steel ball embedded in polyurethane (PUR). Illustrative details of this can be found in Figure 4.

IOP Conf. Series: Materials Science and Engineering **700** (2019) 012020 doi:10.1088/1757-899X/700/1/012020

Part List	
Item	**Title**
1	Main Body PRT
2	Set-Force Body
3	PUR-Pipeline Packer
4	PUR-sealing packer with bearing steel balls
5	ROV-Handle
6	Two-Way Cylinder
7	Studbolt
8	Bearing Steel Ball

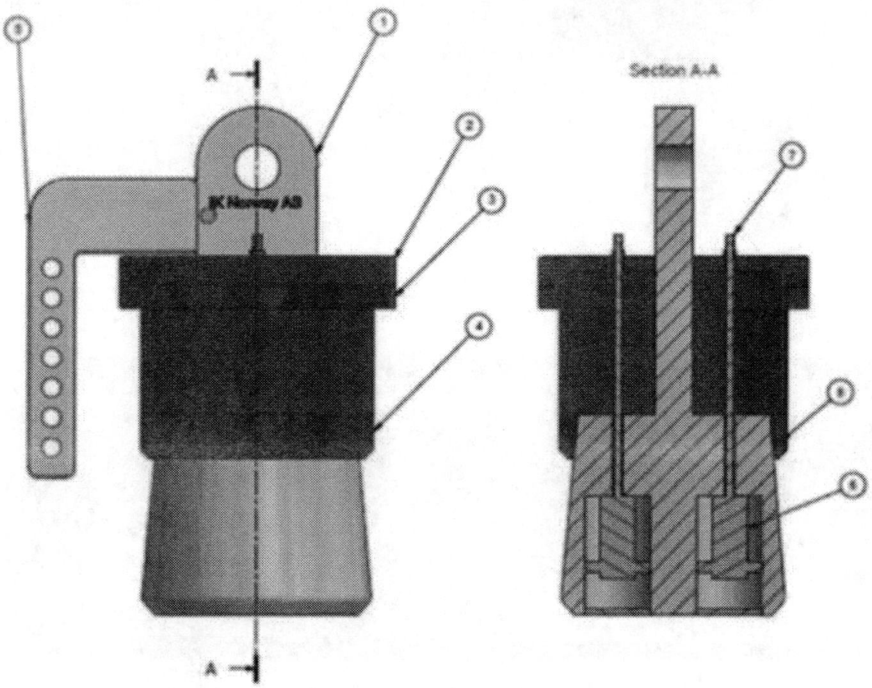

Figure 2. Prototype pipeline recovery tool.

3.1. The wedge-lock and sealing mechanism

The mechanism functions in the following manner as illustrated in Figure 3:

- The tool is inserted into one end of the subsea pipeline; Step 1 in Figure 3
- The cone is pulled towards the force-set pipe using hydraulics. This causes the force-set pipe to press onto the polyurethane (PUR); Step 2 in Figure 3
- The PUR then presses onto the steel balls (casted in the PUR); Step 3 in Figure 3
- The steel balls press onto the cone and the pipeline. This generates the large contact stresses and therefore the gripping force required to lock the PRT to the pipeline.
- The PUR also presses against the cone and pipeline. This seals the pipeline with the PRT.
- The PRT can now pull the pipeline; Step 4 in Figure 3

The above steps are also illustrated in Figure 3. An illustration of the wedge-lock and sealing mechanism is presented in Figure 4.

IOP Conf. Series: Materials Science and Engineering **700** (2019) 012020 doi:10.1088/1757-899X/700/1/012020

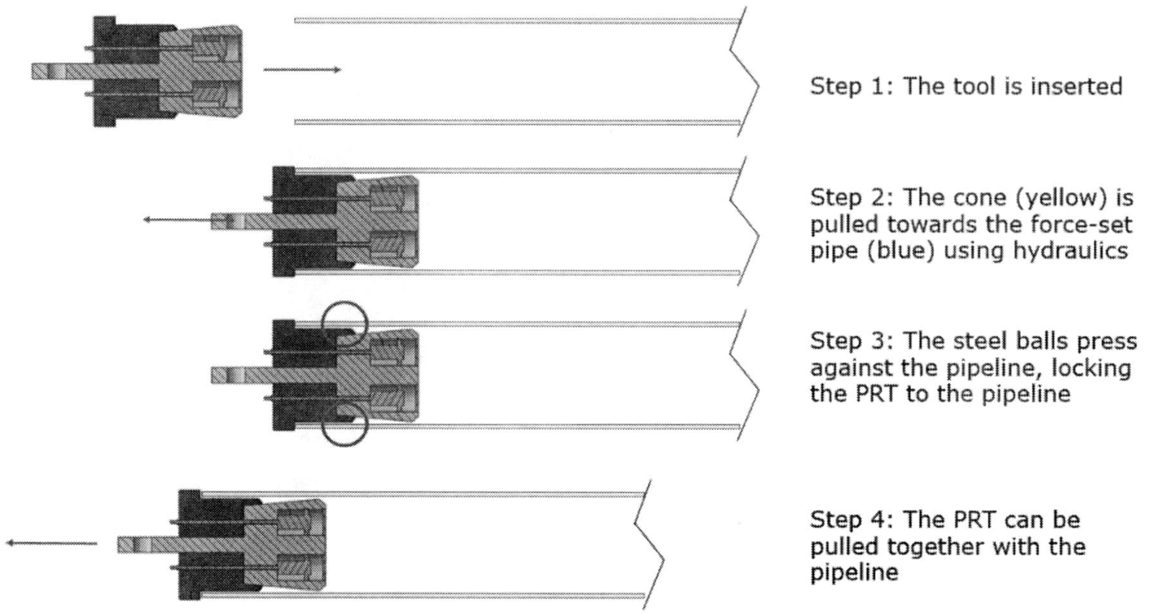

Figure 3. Step-by-step illustration of the PRT in use.

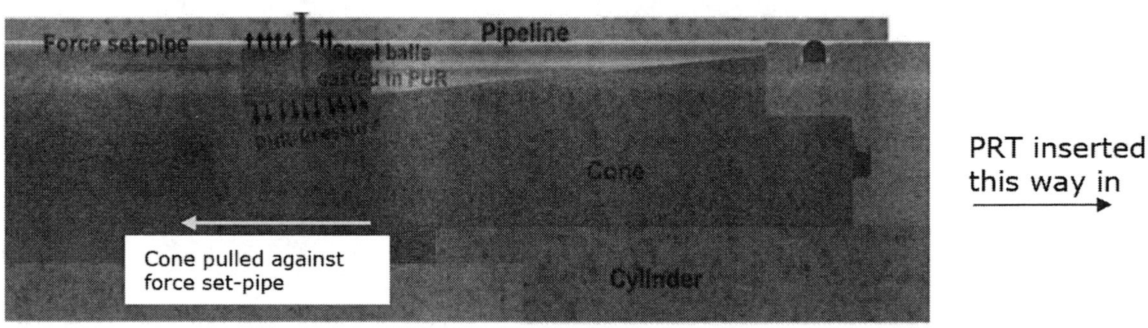

Figure 4. Wedge-lock and sealing mechanism.

4. Finite element modelling

The load and stress patterns are symmetrical about the mid-plane of the steel ball therefore a 1/32 model is used; there are a total of 32 steel balls.

4.1. FE model details

This section presents the FE model details.

4.1.1. Loads and boundary conditions. The loads and boundary conditions applied on the model are illustrated in Figure 5. Note that frictionless support is applied at all out-of-plane surfaces to take into account of the symmetrical condition about the middle of the steel ball. A 3D view of the model showing the loads and boundary conditions is presented in Figure 6.

4.1.2. Material model. A bilinear material curve is used in the FE model for the pipeline. The material properties are taken from the properties for S355 steels in DNVGL-RP-C208 [5]. The elastic modulus is 210 000 Mpa and the yield strength is 335 MPa. The cone and steel ball are modelled as elastic material with elastic modulus of 210 000 MPa. The cone and steel ball are prescribed to be high strength

IOP Conf. Series: Materials Science and Engineering **700** (2019) 012020 doi:10.1088/1757-899X/700/1/012020

steel with much higher yield strength than the pipeline. They are not meant to plastically deform in real life.

4.1.3. Mesh details. The mesh details are presented in Figure 5. The contact element size used is 0.2 mm. There are a total of 162 960 10-node tetrahedral elements with 236 880 nodes. The mesh size is determined via a mesh refinement study presented in Section 4.2.

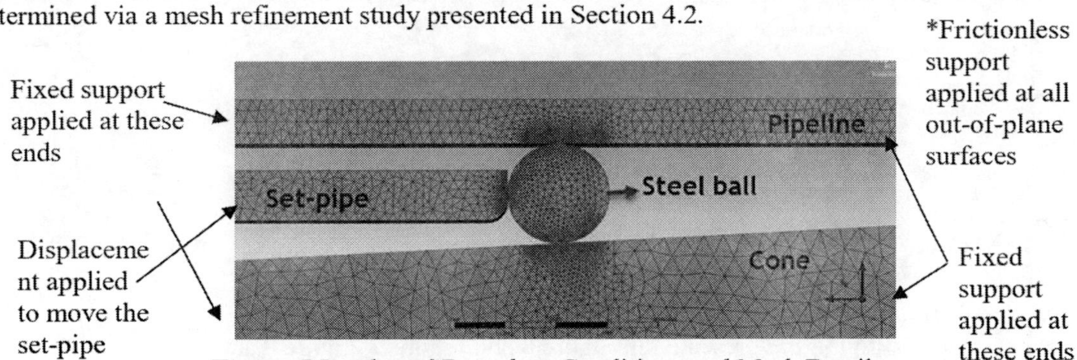

Figure 5 Loads and Boundary Conditions and Mesh Details

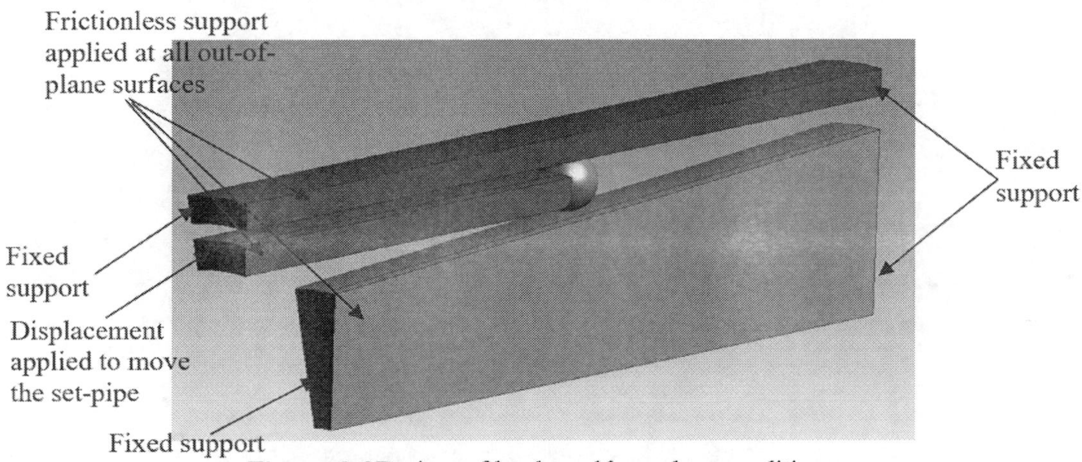

Figure 6. 3D view of loads and boundary conditions.

4.2. Mesh refinement study

A mesh refinement study focusing on the contact area with the following cases as presented in Table 1 investigated. A mesh refinement study was not performed for the other parts of the model. In these parts, there are at least five layers of elements in any given direction for these parts. This is considered to be sufficiently fine.

Table 1. Cases studied for mesh refinement study.

Contact element size (mm)	No. of elements	No. of nodes
0.15	337 523	478 814
0.2	162 960	236 880
0.3	73 879	112 109
0.4	51 437	79 632
0.5	42 879	67 229

IOP Conf. Series: Materials Science and Engineering **700** (2019) 012020 doi:10.1088/1757-899X/700/1/012020

The results of the mesh refinement study are presented in Figure 7 and show that a contact element sizing of 0.2 mm produces sufficiently converged results. Therefore this is used in the paper.

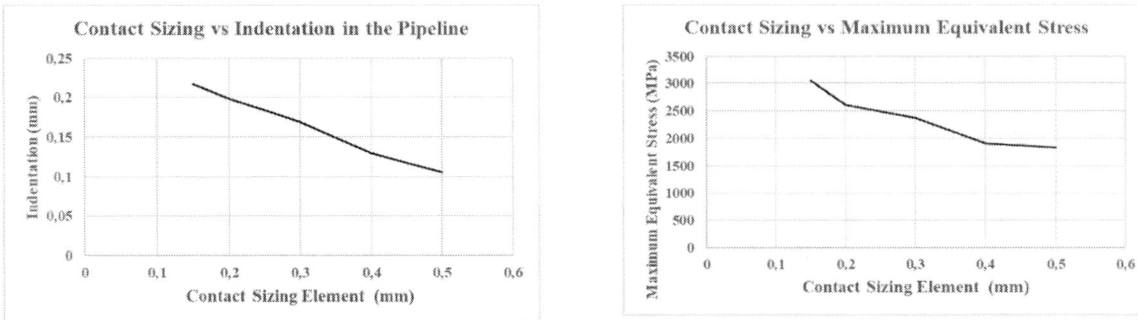

Figure 7. Results of mesh refinement study.

4.3. Example results

Some example result plots are presented in Figure 8 and Figure 9. Figure 8 shows the von Mises stress patterns.

Figure 8. Von-Mises stresses pattern.

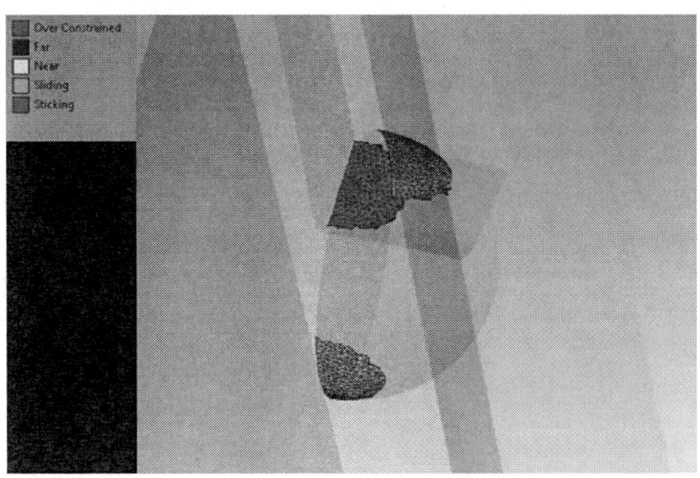

Figure 9. Overview of sticking/sliding status at the steel ball.

5. Experimental test set-up

A test rig was fabricated to model the wedge-lock and sealing mechanism. The cross-sectional view of the test rig is presented in Figure 10. In this test rig, the set-pipe is pushed towards the cone using a hydraulic press as presented in Figure 11. This is in the opposite direction of what will occur in the actual design where the set-pipe is pulled towards cone. However, the physical stresses/forces generated are the same for both cases. Note that Figure 10 shows the polyurethane part. In the actual tests, the steel balls were tested without the polyurethane part, i.e., only the wedge-lock mechanism is investigated.

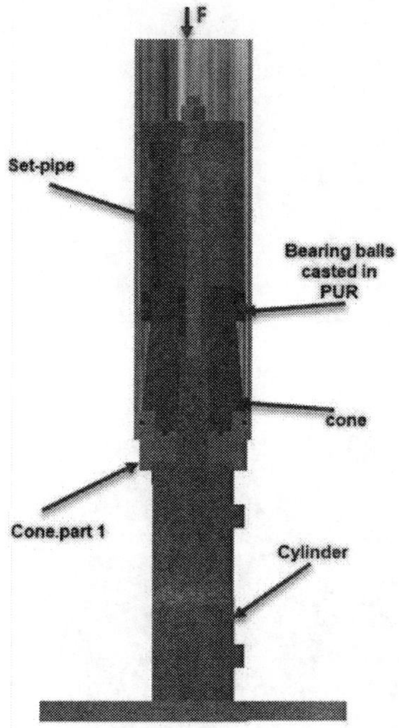

Figure 10. Cross-sectional view of test jig

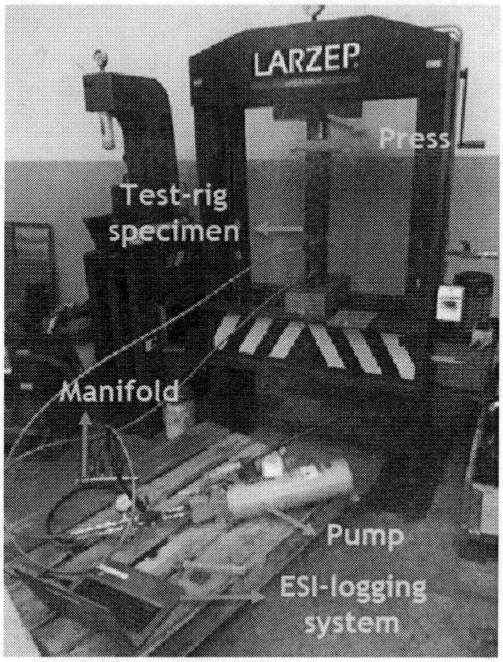

Figure 11. Test jig on the hydraulic press.

5.1. Test rig components

The components in the test rig and their corresponding numbering are presented in Table 2 and Figure 12, respectively.

Table 2. Components used in test rig.

Part number	Component	Comment
1	Steel ball	Ref. Section 5.1.1.
2	Cone part 2	Ref. Section 5.1.2.
3	Pipe	Ref. Section 5.1.3.
4	Cone part 1	Ref. Section 5.1.2.
5	Set-pipe	Ref. Section 5.1.4.
6	Polyurethane	Not tested

IOP Conf. Series: Materials Science and Engineering **700** (2019) 012020 doi:10.1088/1757-899X/700/1/012020

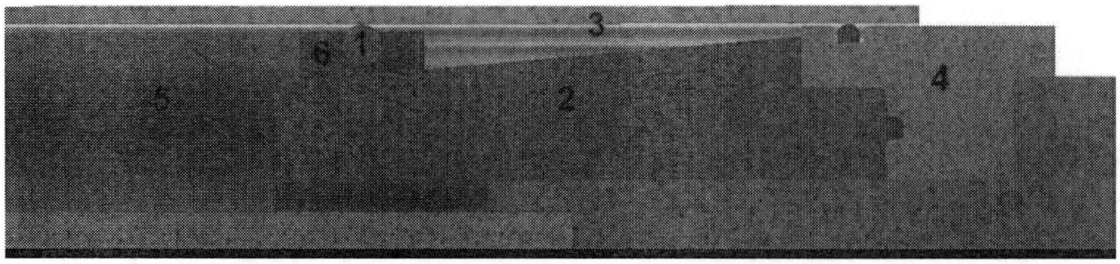

Figure 12. Numbering of components in test rig.

5.1.1. Steel ball. The steel balls used have a diameter of 10 mm and are made in accordance with the standard DIN 17230 [5] and have a hardness of at least 740 HV10, which is between 57 and 58 HRC. There are a total of 32 steel balls in the test rig.

5.1.2. Cone. The cone consist of two parts, part 1 and part 2. Part 2 is the part that presses against the steel balls. Part 1 is an interfacing component that is used to hold Part 2 in place. A total of six cones as presented in Figure 13 were manufactured. However, only the 3 °cone were completely tested in the test. The definition of the cone angle is presented in Figure 14. The cones are made from Uddeholm Calmax steel for high toughness, high wear resistance and high through hardening properties [7].

Figure 13. Six cones (part 2) with angles of 10°, 8°, 6°, 5°, 4° and 3° were manufactured. However, only the cone with 3°angle was completely tested in the test.

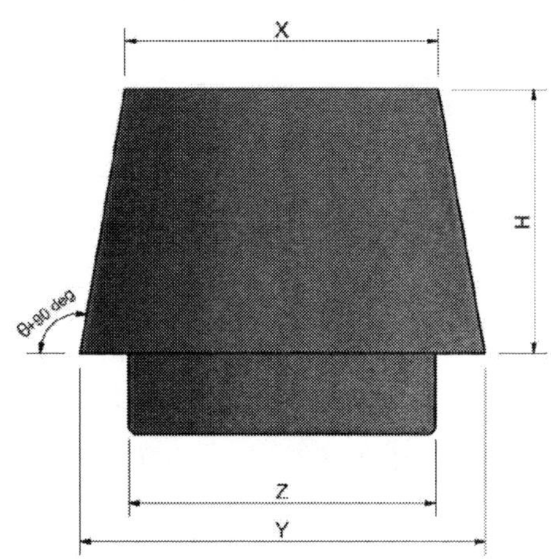

Figure 14. Definition of cone angle, θ

5.1.3. Pipe. The pipe is a steel tube with a tensile strength of 450 MPa and a yield strength of 388 MPa at the measured section. This is the component with the lowest yield strength that is in contact with the steel balls.

5.1.4. Set-pipe. This is a machined pipe. A hydraulic press is used to press this against the cone and pipe.

5.2. Indentation at test components

Figure 15 and Figure 16 show the indentations observed at the test components after the physical tests.

Figure 15. Indentation at the pipe and cone.

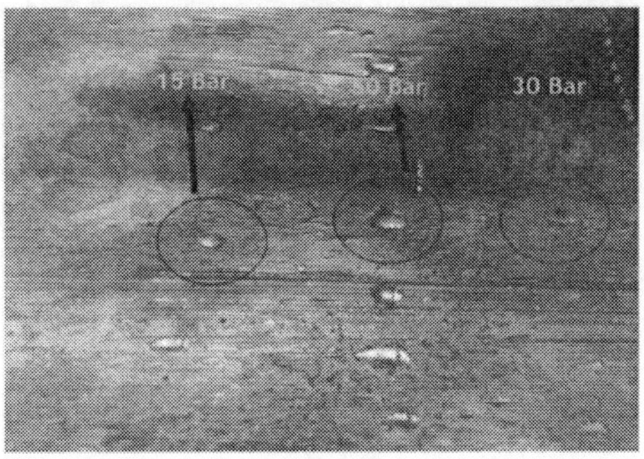

Figure 16. Zoom-in at indentation at the pipe

6. Results and discussions

6.1. FE results and discussions

The results from the linear-elastic and elastic-plastic model shows that friction coefficient has a major impact on reaction force, equivalent stress and indentation depth. This can be observed from the results presented in Figure 17 and Figure 18. Another parameter that affects the results significantly is yield stress of the pipe. This is shown in Figure 19.

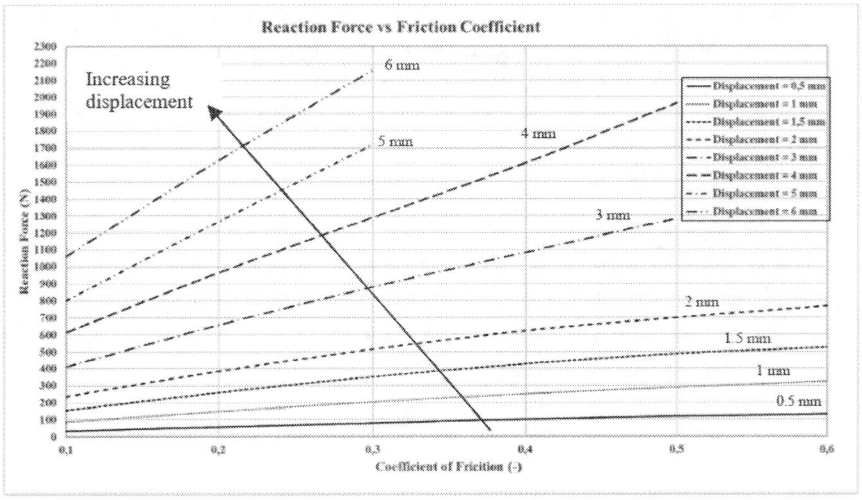

Figure 17. Reaction force vs. friction coefficient, FE model tesults, 3° cone.

IOP Conf. Series: Materials Science and Engineering **700** (2019) 012020 doi:10.1088/1757-899X/700/1/012020

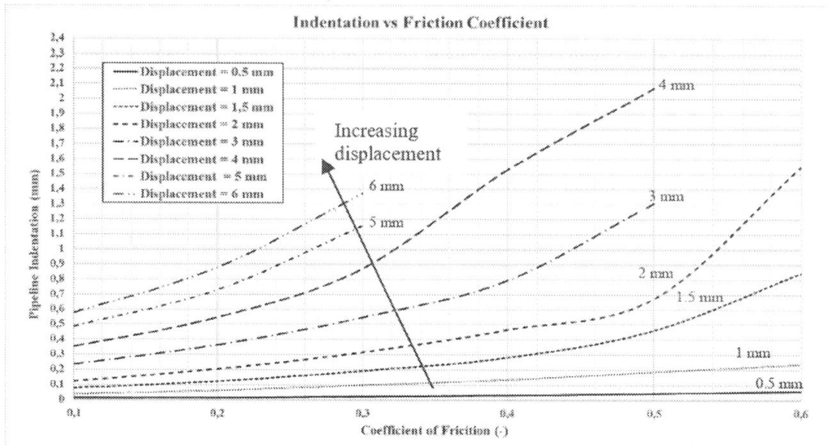

Figure 18. Indentation vs. friction coefficient, FE model results, 3° cone

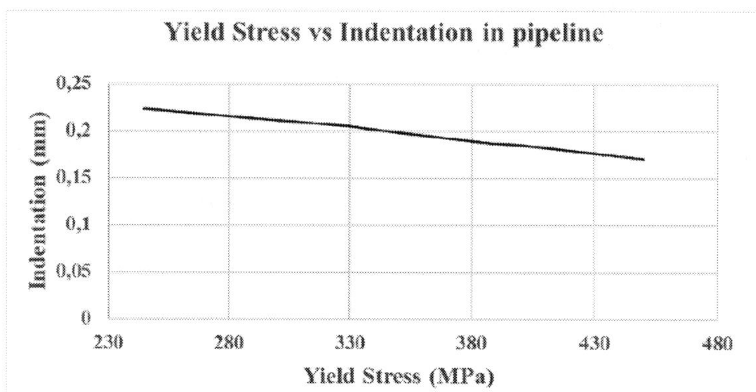

Figure 19. Indentation vs. yield stress, FE model results, 3° cone

6.2. Physical testing results and discussions

The displacement of the steel balls caused by the applied force from the set-pipe causes the steel balls to deform both cone and steel pipe as shown in Figure 15 and Figure 16. As expected, the results as presented in Figure 20 show that there is a correlation between the angle of the cone, set-pressure and indentation. The figure shows that indentation has an approximately linear relationship with the set-pressure. However, the results with the cone-angle of 4 ° does not have the same slope on the results as the other cone-angles. This implies that there are irregularities in these results. A possible explanation is that the cone is misaligned during the test.

IOP Conf. Series: Materials Science and Engineering **700** (2019) 012020 doi:10.1088/1757-899X/700/1/012020

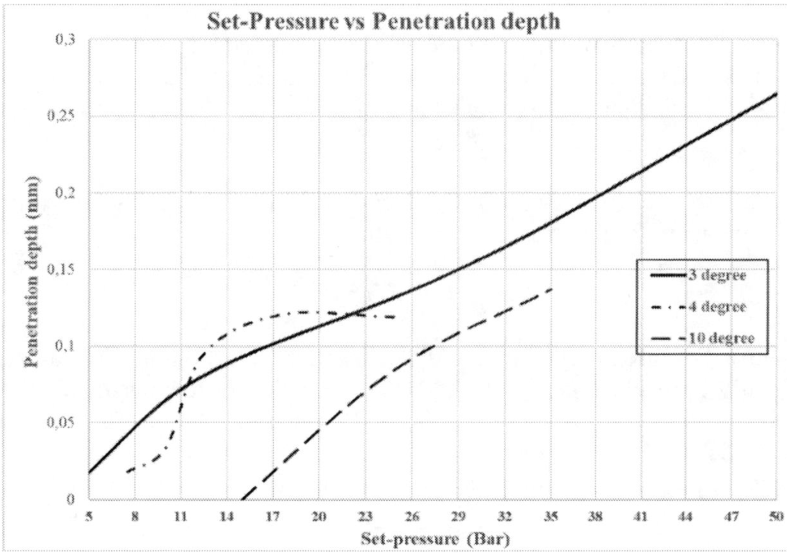

Figure 20. Indentation depth vs set-pressure, physical testing.

6.3. Comparison between FE model and physical test results

The comparison between the FE model and physical test results is presented in Table 3. The table is read in the following manner:

- The indentation depth is measured from the physical tests and corresponds to a given set-pressure. In this case, three set-pressures of 10, 15 and 30 bar were investigated.
- The set-pressure is converted to the corresponding reaction force and displacement at the set-pipe and the corresponding friction coefficient range is read from the FE model results as presented in Figure 17, Figure 18 and Figure 19.

The results compare well for a certain range of friction coefficients which in fact corresponds to the typical range of friction coefficients founds in engineering application for lubricated metal surfaces.

Table 3. Comparison between FE model and physical test results, 3° cone.

Set-pressure – physical test (bar)	Indentation depth – physical test (mm)	Friction coefficient range – FE model
30	0.18	0.15 to 0.3
15	0.085	0.1 to 0.4
10	0.06	0.15 to 0.4

7. Conclusions

The following conclusions are made:

- The size and simplicity of the prototype PRT suggests savings in maintenance and operations can be made.
- Uncertainties in the coefficient of friction and yield stress lead to significant variations in the results.
- The FE model gives a good representation of the test-rig if the friction coefficient is known.
- A low friction coefficient ensures a minimum indentation depth. Low friction coefficients can be achieved by lubricating the steel balls.
- The results show that an increase in the cone-angle will lead to a decrease in the indentation depth when the same magnitude of force is applied on the steel balls.

IOP Conf. Series: Materials Science and Engineering **700** (2019) 012020 doi:10.1088/1757-899X/700/1/012020

8. Future Work

The sealing mechanism of the PRT is planned to be studied in future work. The planned approach is to investigate the sealing capacity of the polyurethane using numerical modelling and experimental tests. Polyurethane is a polymer which can be very sensitive to input conditions such as strain rates and temperature and therefore requires special considerations.

Acknowledgments

The authors would like to express thanks to IK-Norway for the support in the financing of the test rig and providing the facilities for the experimental testing.

References

[1] Manoucherhri, S., Subsea Pipelines and Flowlines Decomissioning – What We Should Know for a Rational Approach, Paper No. OMAE2017-61239, Proceedings of ASME 2017 36th Internal Conference on Ocean, Offshore and Arctic Engineering, Trondheim, Norway, 2017

[2] Venas, A., Offshore Pipelines Outlook, DNVGL Report, 2016

[3] IK-Norway, Special Design, Lifting Tools, Retrieved 18.07.2019 from https://ik-worldwide.com/product/products-and-services/subsea/studies/consept-selection/special-design-lifting-tools/

[4] ASME B36.10M -2018 Welded and Seamless Wrought Steel Pipe

[5] DNVGL-RP-C208, Determination of Structural Capacity by Non-Linear FE Analysis Methods, 2013

[6] DIN 17230 Beall and Roller Bearing Steels – Technical Condition of Delivery, 1980

[7] Uddeholm Calmax, Assab, Retrieved 18.07.2019 from http://www.assab-singapore.com/media/CALMAX-D20140711.pdf

The correlation and determination matrices associated with the burst design of a subsea carbon-fibre-epoxy composite flow-line

Y Xing[1], W Xu[1], V Buratti[2]

[1]Department of Mechanical and Structural Engineering and Material Science, University of Stavanger, Kjell Arholmsgate 41, 4036 Stavanger, Norway
[2]Aibel AS, Vestre Svanholmen 14, 4313 Sandnes, Norway

Corresponding author: yihan.xing@uis.no

Abstract. This paper examines the correlation and determination matrices of the burst design in a subsea carbon-fibre-epoxy composite flow-line. A case study of a 4" flow-line is investigated. The correlation and determination matrices are calculated and compared using Pearson and Spearman correlation methods. A comprehensive suite of failure modes that comprises of Maximum Stress, Tsai-Wu, Hashin and Puck failure criteria is used to quantify the burst design. The results reveal that the failure criteria are strongly correlated to each other. The applied pressure and ply thickness are moderately correlated to the failure criteria. It is observed that due to the nature of burst loads, most material failure parameters with the exception of the Tsai-Wu constant, F_{12} do not exhibit correlations with the failure criteria. Second, both Pearson and Spearman correlation methods identified the same set of major design parameters. Third, it was found that the identification of the major design parameters is not affected by the sample size. This paper provides an analysis framework to aid in the identification of the major design parameters which is the initial and crucial step of a design optimisation exercise.

1. Introduction and background

Carbon-fibre-epoxy composite (CFEC) materials consist of an epoxy matrix reinforced by carbon fibres. The principal advantage of using CFEC materials is the high strength to weight ratio; carbon fibres are characterized by a strength-to-weight ratio 50 times higher than steel. Hence, structures made from CFEC materials are typically 20 % lighter than steel and 30% to 50% stronger. However, unlike isotropic materials, CFEC materials have directional strength properties. The orthotropic property is both an advantage and a disadvantage. CFEC materials can be aligned in the strong direction optimally to achieve higher strength-to-weight ratios in the structure. However, it requires careful design to ensure that the loads transverse to the fibre directions are not overly large; this is the weak direction. Another advantage of CFEC materials is their high fatigue strength. The fatigue limit of CFEC materials is as much as 70 % of the ultimate strength compared to only 30 % in the case of steel materials. A third advantage of using CFEC materials is their corrosion resistant properties. The above-mentioned properties make CFEC flow-lines attractive in high demanding applications.

The stress analysis and strength evaluation of CEFC flowlines have been well-studied. Some studies are briefly described here. Azar et al. [1] calculated the optimum angle of filament-wound pipelines used for natural gas transmission using approximation methods. Theotokoglou [2] used finite element analyses to quantity the effect of delamination on the loss in load carrying capacity. Xia et al. [3] analysed multi-layered filament-wound composite pipes under internal pressure using a simplified elastic solution. Chouchaoui et al. [4] studied the stresses and displacements of a laminated cylindrical tube using analytical models. Guz et al. [5] analysed the stress distributions through the pipe thickness for various lay-ups when the pipe is subjected to different outer pressure loads. These papers presented some form of parametric studies. However, there has not been any published studies applying the parametric correlation method to composite flowlines to identify the major design parameters.

This paper examines the correlation and determination matrices associated with the burst design of a 4" CFEC flow-line. Four failure criteria based on different failure theories are investigated. These are the Maximum Stress, Tsai-Wu [6], Hashin [7] and Puck [8-10] failure criteria. Furthermore, two correlation methods, namely Pearson [11, 12] and Spearman [13-15] correlation are studied. The purpose is to investigate the identification of the major design parameters associated with each individual failure criterion. The identification of the major design parameters is the initial and crucial step in a design optimization exercise [16].

Nomenclature

$\rho_{P_1P_2}$	Pearson linear correlation coefficient
$cov(P_1, P_2)$	Covariance
σ_{P_1}	Standard deviation of variable P_1
σ_{P_2}	Standard deviation of variable P_2
$\rho_{P_1{}^2,P_2{}^2}$	Pearson quadratic correlation coefficient
$\rho_{rg_{P_1},rg_{P_2}}$	Spearman linear correlation coefficient
$cov(rg_{P_1}, rg_{P_2})$	Covariance of the rank variable
$\sigma_{rg_{P_1}}$	Standard deviation of the rank variable P_1
$\sigma_{rg_{P_2}}$	Standard deviation of the rank variable P_2
$\rho_{rg_{P_1}{}^2,rg_{P_2}{}^2}$	Spearman quadratic correlation coefficient
σ_A	Axial stress
σ_H	Hoop stress
σ_1	Stress in the x-direction
σ_2	Stress in the y-direction
σ_3	Stress in the z-direction
τ_{12}	Shear stress in the xy-plane
τ_{23}	Shear stress in the yz-plane
τ_{13}	Shear stress in the xz-plane
σ_{ut1}	Ultimate tensile strength in the x-direction
σ_{uc1}	Ultimate compressive strength in the x-direction
σ_{ut2}	Ultimate tensile strength in the y-direction
σ_{uc2}	Ultimate compressive strength in the y-direction
σ_{ut3}	Ultimate tensile strength in the z-direction
σ_{uc3}	Ultimate compressive strength in the z-direction
τ_{u12}	Ultimate shear strength the in the xy-plane
τ_{u13}	Ultimate shear strength the in the xz-plane
τ_{u23}	Ultimate shear strength the in the yz-plane

IOP Conf. Series: Materials Science and Engineering **700** (2019) 012021 doi:10.1088/1757-899X/700/1/012021

2. Correlation and determination matrix

The correlation matrix measures the linear correlation between the identified design input parameters and the output variables using correlation coefficients. The correlative coefficient is a number between -1 and 1. A positive value indicates that P_2 increases with P_1, while a negative value indicates that P_2 decreases with P_1. The closer the value of correlation is to +1 or -1, the stronger the correlation between the input and the output variables. The objective of using a correlation matrix is to identify the major design parameters. The following correlation methods are used in this paper:

- Pearson correlation (Ref. Section 2.1.)
- Spearman correlation (Ref. Section 2.2.)

The determination matrix measures the quadratic correlations between P_1 and P_2. It indicates how close the points are to a quadratic curve. The determination matrix provide a measure of how well future outcomes are likely to be predicted. The correlation and determination matrices can be used to reduce the number of input parameters to a selection of major parameters in the design of experiments, or to reduce the number of output variables considered in a design optimization study.

2.1. *Pearson correlation*

The Pearson correlation [11, 12] is a measure of the linear correlation between two variable P_1 and P_2. The Pearson correlation coefficient is defined as:

$$\rho_{P_1 P_2} = \frac{cov(P_1, P_2)}{\sigma_{P_1}\sigma_{P_2}} \tag{1}$$

Equation (1) is used in the correlation matrix. For the determination matrix, i.e., quadratic correlation, the Pearson correlation coefficient is defined as:

$$\rho_{P_1{}^2, P_2{}^2} = \frac{cov\left(P_1{}^2, P_2{}^2\right)}{\sigma_{P_1{}^2}\sigma_{P_2{}^2}} \tag{2}$$

2.2. *Spearman correlation*

The Spearman correlation [13-15] is defined as the Pearson correlation between the rank variables. It is given as:

$$\rho_{rg_{P_1}, rg_{P_2}} = \frac{cov\left(rg_{P_1}, rg_{P_2}\right)}{\sigma_{rg_{P_1}}\sigma_{rg_{P_2}}} \tag{3}$$

Similar to the Pearson correlation, Equation (3) is used in the correlation matrix. For the determination matrix, the Spearman correlation coefficient is defined as:

$$\rho_{rg_{P_1{}^2}, rg_{P_2{}^2}} = \frac{cov\left(rg_{P_1{}^2}, rg_{P_2{}^2}\right)}{\sigma_{rg_{P_1{}^2}}\sigma_{rg_{P_2{}^2}}} \tag{4}$$

Note that the symbol σ in Equation (1) to (4) is used to represent standard deviations. The symbol σ is used to represent stresses in the rest of the paper.

3. Burst design of a subsea carbon-fibre-epoxy composite flow-line

3.1. *Thin wall cylinder*

The flow-line is a long thin wall cylinder subjected to internal pressure. Under this load condition, an element in the cylinder wall will experience two distinct stresses, axial (σ_A) and hoop stresses (σ_H) as illustrated in Figure 1.

IOP Conf. Series: Materials Science and Engineering **700** (2019) 012021 doi:10.1088/1757-899X/700/1/012021

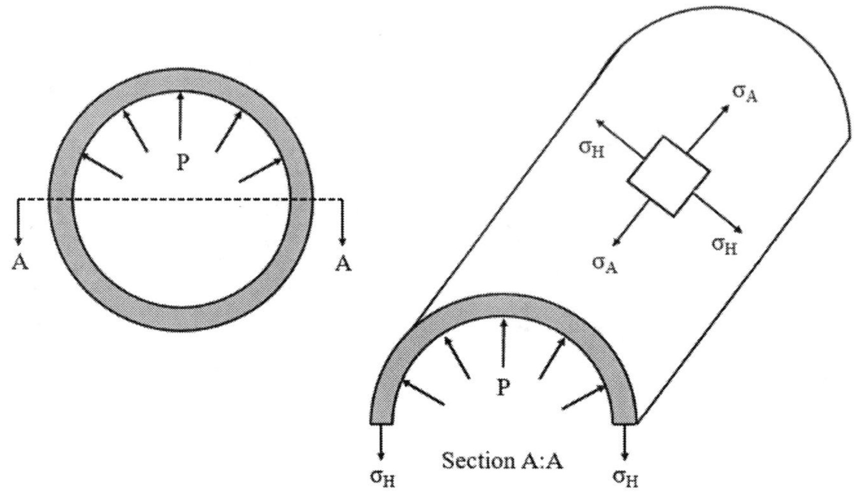

Figure 1. Stresses in a thin wall cylinder.

3.2. Classical laminate theory
The classical laminate theory (CLT) [17-19] is used in this paper. This theory extends the classical plate theory for isotropic and homogeneous materials proposed by Kirchhoff [20, 21]. The CLT assumes the following:

- There is no slip between the adjacent layers. This means that the laminate is perfectly bonded.
- Each lamina is a homogenous layer with its effective properties known and uniform throughout the lamina.
- Each lamina is in a state of plane stress
- Each lamina can be isotropic, orthotropic or inversely isotropic.
- The laminate deforms in accordance with Kirchhoff's theory. This means the normal to the mid-plane remain straight and normal to the mid-plane after deformation. In addition, the normal to the mid-plane does not change lengths.

3.3. Failure criteria
The four failure criteria listed in Table 1 are investigated in this paper. The corresponding failure theories and their failure functions are described in some details in the following sub-sections. It is mentioned that failure is defined to occur when the failure function returns a value equal or greater than 1.0.

Note that since CLT is used, the inter-laminar normal and shear stresses are zero and not computed, i.e., $\sigma_3 = \tau_{23} = \tau_{13} = 0$. However, for completeness, the generalized 3D failure criteria are presented in Section 3.3.1. to 3.3.4.

3.3.1. Maximum stress failure criterion. This failure criterion compares the ratios of the actual stresses to the failure stresses in the ply principal coordinate system. The generalized failure function is defined as:

$$f = max\left(\left|\frac{\sigma_1}{X}\right|, \left|\frac{\sigma_2}{Y}\right|, \left|\frac{\sigma_3}{Z}\right|, \left|\frac{\tau_{12}}{S}\right|, \left|\frac{\tau_{13}}{R}\right|, \left|\frac{\tau_{23}}{Q}\right|\right) \quad (5)$$

where:

$$X = \begin{cases}\sigma_{uc1}, \sigma_1 < 0 \\ \sigma_{ut1}, \sigma_1 \geq 0\end{cases}, \quad Y = \begin{cases}\sigma_{uc2}, \sigma_2 < 0 \\ \sigma_{ut2}, \sigma_2 \geq 0\end{cases}, \quad Z = \begin{cases}\sigma_{uc3}, \sigma_3 < 0 \\ \sigma_{ut3}, \sigma_3 \geq 0\end{cases}$$

$$S = \tau_{u12}, \quad R = \tau_{u13}, \quad Q = \tau_{u23}$$

Table 1. Failure criteria studied.

Failure Criterion	Physical Basis	Usage Convenience	Ref. Section
Maximum Stress	Tensile behaviour of brittle material	Requires only few parameters by testing	3.3.1.
Tsai-Wu	Interactive tensor polynomial fitted to the failure behaviour of the material	Requires numerous parameters by a comprehensive testing program	3.3.2.
Hashin	Interactive criterion considering failure in fibre, transverse and delamination separately.	Requires only few parameters by testing	3.3.3.
Puck	Complex interactive criterion considering fibre and inter-fibre failures separately.	Requires numerous parameters by a comprehensive testing program	3.3.4.

3.3.2. Tsai-Wu failure criterion. The Tsai-Wu failure criterion [6] uses a quadratic failure function and is a simplification of Gol'denblat and Kapnov's generalised failure theory for anisotropic materials [22]. It is expressed as:

$$f = f_i\sigma_i + f_{ij}\sigma_i\sigma_j \tag{6}$$

where i, j = 1, 2, 3, 4, 5, 6. In the plane stress condition, Equation (6) reduces to:

$$f = F_{11}\sigma_1^2 + F_{22}\sigma_2^2 + F_{66}\tau_{12}^2 + 2F_{12}\sigma_1\sigma_2 + F_1\sigma_1 + F_2\sigma_2 \tag{7}$$

where $F_{11} = \frac{1}{\sigma_{ut1}\sigma_{uc1}}$, $F_{22} = \frac{1}{\sigma_{ut2}\sigma_{uc2}}$, $F_{66} = \frac{1}{S^2}$, $F_1 = \frac{1}{\sigma_{ut1}} - \frac{1}{\sigma_{uc1}}$, $F_2 = \frac{1}{\sigma_{ut2}} - \frac{1}{\sigma_{uc2}}$

and the nominal value of F_{12} is defined in Table 3.

3.3.3. Hashin failure criterion. This Hashin failure criterion [7] distinguishes the various different failure modes and model each of the modes separately. This is in contrast to the Tsai-Wu failure criterion (Ref. Section 3.3.2.) which uses a quadratic failure function that 'mixes' all the failures modes together. The Hashin failure function is defined as:

$$f = max(f_f, f_m, f_d) \tag{8}$$

where ff, fm and fd are the failure functions for failures in the fibre direction, transverse direction and delamination, respectively. For failure in the fibre direction, f_f is defined as:

$$f_f = \begin{cases}\left(\frac{\sigma_1}{\sigma_{ut1}}\right)^2 + \left(\frac{\tau_{12}}{\tau_{u12}}\right)^2 + \left(\frac{\tau_{13}}{\tau_{u13}}\right)^2, \sigma_1 \geq 0 \\ -\frac{\sigma_1}{\sigma_{uc1}}, \sigma_1 < 0\end{cases} \tag{9}$$

IOP Conf. Series: Materials Science and Engineering **700** (2019) 012021 doi:10.1088/1757-899X/700/1/012021

For failure in the transverse direction, f_m is defined as:

$$f_m = \begin{cases} \left(\frac{\sigma_2}{\sigma_{ut2}}\right)^2 + \left(\frac{\tau_{23}}{\tau_{u23}}\right)^2 + \left(\frac{\tau_{12}}{\tau_{u12}}\right)^2 + \left(\frac{\tau_{13}}{\tau_{u13}}\right)^2, \sigma_2 \geq 0 \\ \left(\frac{\sigma_2}{2\tau_{u23}}\right)^2 + \left(\frac{\tau_{23}}{\tau_{u23}}\right)^2 + \left(\frac{\tau_{12}}{\tau_{u12}}\right)^2 + \left[\left(\frac{Y_c}{2\tau_{u23}}\right)^2 - 1\right]\frac{\sigma_2}{\sigma_{uc2}}, \sigma_2 < 0 \end{cases} \tag{10}$$

For delamination failure, f_d is defined as:

$$f_d = \begin{cases} \left(\frac{\sigma_3}{\sigma_{uc3}}\right)^2 + \left(\frac{\tau_{13}}{\tau_{u13}}\right)^2 + \left(\frac{\tau_{23}}{\tau_{u23}}\right)^2, \sigma_3 < 0 \\ \left(\frac{\sigma_3}{\sigma_{ut3}}\right)^2 + \left(\frac{\tau_{13}}{\tau_{u13}}\right)^2 + \left(\frac{\tau_{23}}{\tau_{u23}}\right)^2, \sigma_3 \geq 0 \end{cases} \tag{11}$$

3.3.4. Puck failure criteria. The Puck's action plane strength criterion is used in this paper. The criterion considers two separate failures, namely fibre failure (FF) and inter-fibre failure (IFF). For FF, the Maximum Stress criterion as defined below is used [8-10]:

$$f_{FF} = \begin{cases} \frac{\sigma_1}{\sigma_{uc1}}, \sigma_1 < 0 \\ \frac{\sigma_1}{\sigma_{ut1}}, \sigma_1 \geq 0 \end{cases} \tag{12}$$

The failure function for IFF is complex and will not be presented here. Briefly, the IFF failure function considers a 3D stress state using the definition of fracture resistances, R and slope parameters of the fracture curves, p. The Puck IFF failure function also takes into account that some fibres might already break under uniaxial loads much lower than loads which cause ultimate factors using degradation factors. Details of the IFF failure function can be found in Puck et al. [8, 9]

4. Case study

The flow-line studied is a 4" pipe with an applied pressure of 2500 psi.

4.1. Base case

The general base case properties are presented in Table 2.

Table 2. General properties of flow-line, Base case.

Property	Symbol	Value	Unit
Outer Diameter	OD	114.3	mm
Wall Thickness	WT	6	mm
Total Number of Ply	-	30	
Ply Thickness	t_{ply}	0.2	mm
Fibre orientation	-	+/- 45	o
Applied Pressure	$P_{applied}$	17.25	MPa

Table 3. Material data - ply (prepreg epoxy carbon UD)

Material Property	Symbol	Value	Unit
Elastic Modulus	E_1, E_2, E_3	121000, 8600, 8600	MPa
Shear Modulus	G_{12}, G_{23}, G_{13}	4700, 3100, 4700	MPa
Poisson's Ratio	v_{12}, v_{23}, v_{13}	0.27, 0.4, 0.27	-
Tensile Strength	σ_{ut1}, σ_{ut2}, σ_{ut3}	2231, 29, 29	MPa
Compressive Strength	σ_{uc1}, σ_{uc2}, σ_{uc3}	-1082, -100, -100	MPa
Shear Strength	τ_{u12}, τ_{u23}, τ_{u13}	60, 32, 60	MPa
Tsai-Wu Constants	F_{12}, F_{23}, F_{13}	-1, -1, -1	-
Puck Constants	p_{21+}, p_{22+}, p_{21-}, p_{22-}	0.35, 0.25, 0.3, 0.25	-
Additional Puck Constants	s, M, F_{IW}	0.5, 0.5, 0.8	-

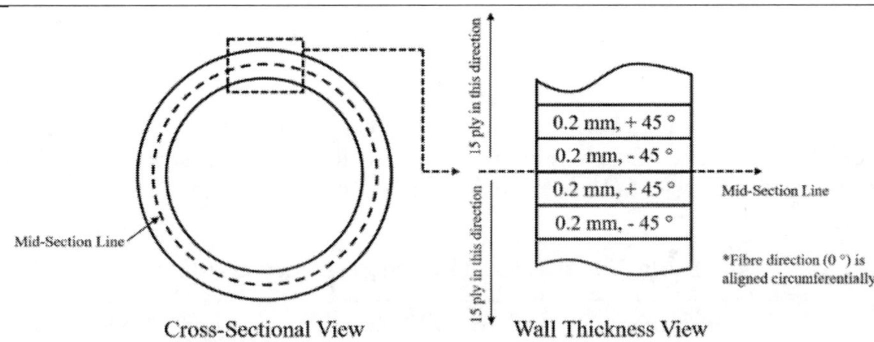

Figure 2. Stacking sequence.

The base case failure criteria values, corresponding to the base case design talked into this sub-section, are presented in Table 4. These are calculated using the finite element model presented in Section 4.2. The results are calculated using the mesh size of 10 mm as discussed in Section 4.2.2. The failure modes are mainly caused by the matrix, i.e., the material fails in shear. The Tsai-Wu failure criterion does not distinguish failure modes. One can also notice the large difference in failure values calculated using Maximum Stress failure criterion vs. the rest of the criteria. This is because Maximum Stress failure criterion does not consider the strength contribution from the fibre when evaluating matrix failure. In this particular case, the strength contribution from the fibres would be significant as they are aligned 45° to the direction of hoop stress and therefore help to share some of the load away from the matrix.

Table 4. Base case failure criteria values.

	Maximum Stress	Tsai-Wu	Hashin	Puck
Failure Criterion Value	0.642	0.905	0.899	0.934
Failure Mode	τ_{u12} exceeded	-	Matrix failure	Matrix tension failure

4.2. Finite element model

A 2 000 mm section of the flow-line is modeled. This is considered to be sufficiently long by engineering judgement in order to avoid end effects due to loads and boundary conditions in the finite element model. The results are obtained at the middle part of the section, i.e., at the 1 000 mm point. ANSYS R17.0 is used for the finite element modelling.

IOP Conf. Series: Materials Science and Engineering **700** (2019) 012021 doi:10.1088/1757-899X/700/1/012021

4.2.1. Loads and boundary conditions. The loads and boundary conditions applied on the model are illustrated in Figure 3. Pressure (Load A) is applied in the interior of the flow-line. Fixed support (Load B) is applied on the right edge of the flow-line. An end cap force (Load C) due to the internal pressure is applied on the left edge of the flow-line.

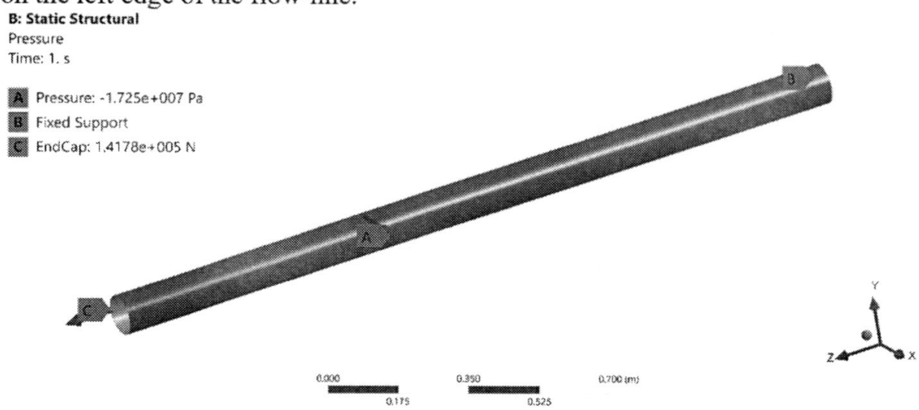

Figure 3. Loads and boundary conditions

4.2.2. Mesh refinement study. The cases studied and the corresponding results obtained for the mesh refinement study are presented in Table 5 and Figure 4 respectively. The element used is the 4-node SHELL181 element [23]. The results show that a 10 mm element size is sufficient to produce converged solutions for the four failure criteria investigated in this paper. The mesh details used in this paper are presented in Figure 5.

Table 5. Cases studied for mesh refinement study

Element Size (mm)	No. of Elements	No. of Nodes
30	888	900
20	900	1 552
15	2 904	2 926
10	6 386	6 400
5	24 896	24 960

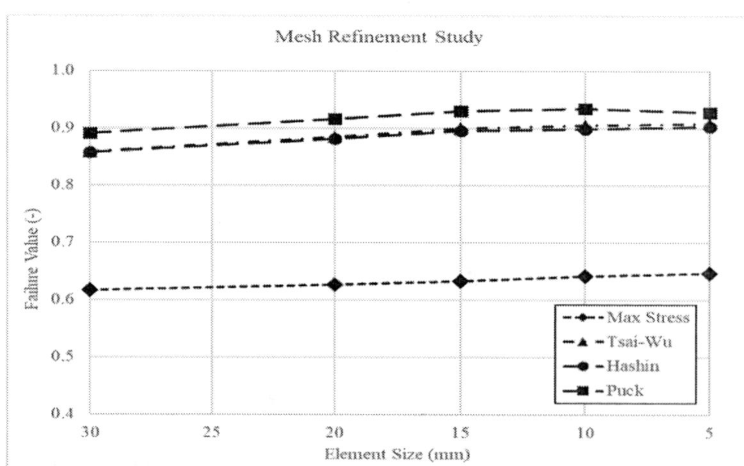

Figure 4. Results of mesh refinement study.

IOP Conf. Series: Materials Science and Engineering **700** (2019) 012021 doi:10.1088/1757-899X/700/1/012021

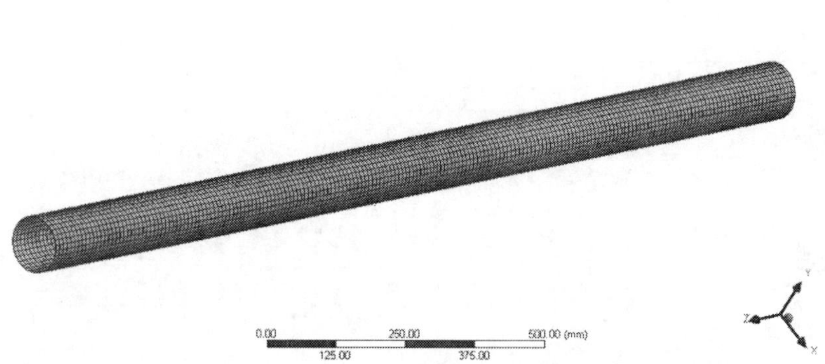

Figure 5. Mesh details, 10 mm element size, 6 386 SHELL181 elements with 6 400 nodes.

4.2.3. Example result plots. An example plot of the maximum stress failure is presented in Figure 6. The maximum value generated by the failure function in the middle of the 2 000 mm flow-line is extracted and used in this paper.

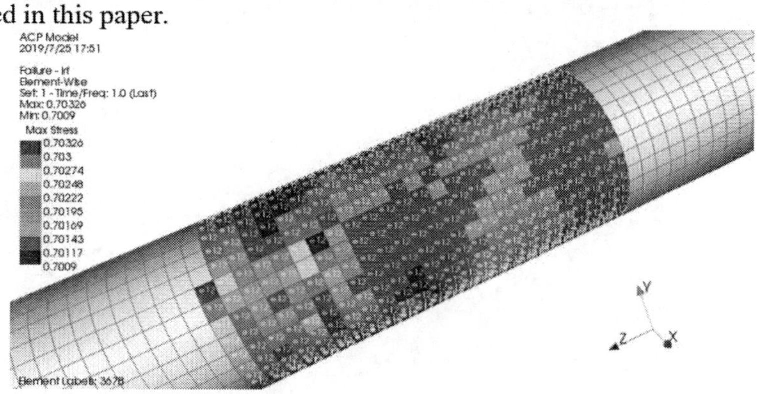

Figure 6. Example plot of maximum stress failure.

5. Results and discussions

5.1. Correlation and determination matrices

The correlation and determination matrices are presented in Figure 7 and Figure 8, respectively. The matrices are calculated using Spearman correlation method and a sample size of 200. Notice that the correlation matrix is symmetrical. However, the determination matrix is not symmetrical. This is because $Cov(P_1^2,P_2^2) \neq Cov(P_2^2,P_1^2)$ The matrix calculated using the Pearson correlation identified the same set of major design parameters and therefore not presented here. The comparison of Pearson and Spearman correlation methods is presented in Section 5.2. The observations made are presented in the following sub-sections.

5.1.1. Strong correlations between failure criteria. The failure criteria have strong linear and quadratic correlations with one and other. The values of these correlations are all above 0.8. This is intuitive, an increase in the failure value calculated from one failure criterion will imply that the failure values calculated from other failure criteria would also increase. A detailed investigation into the correlation values presented in Table 6 and Table 7 revealed that the correlation is stronger between the Tsai-Wu, Hashin and Puck criteria. The Maximum Stress criterion is not as correlated, particularly for the quadratic correlations. This could be explained that Tsai-Wu, Hashin and Puck criteria are non-linear criteria while the Maximum Stress criterion is a linear criterion. In addition, Tsai-Wu and Hashin have failure functions of a quadratic nature.

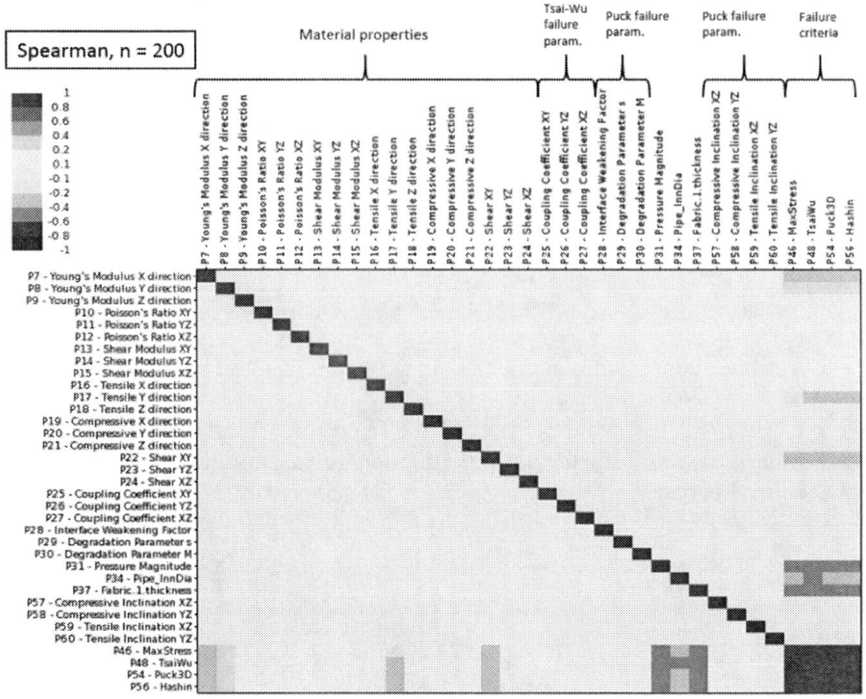

Figure 7. Correlation matrix, Spearman correlation, n = 200.

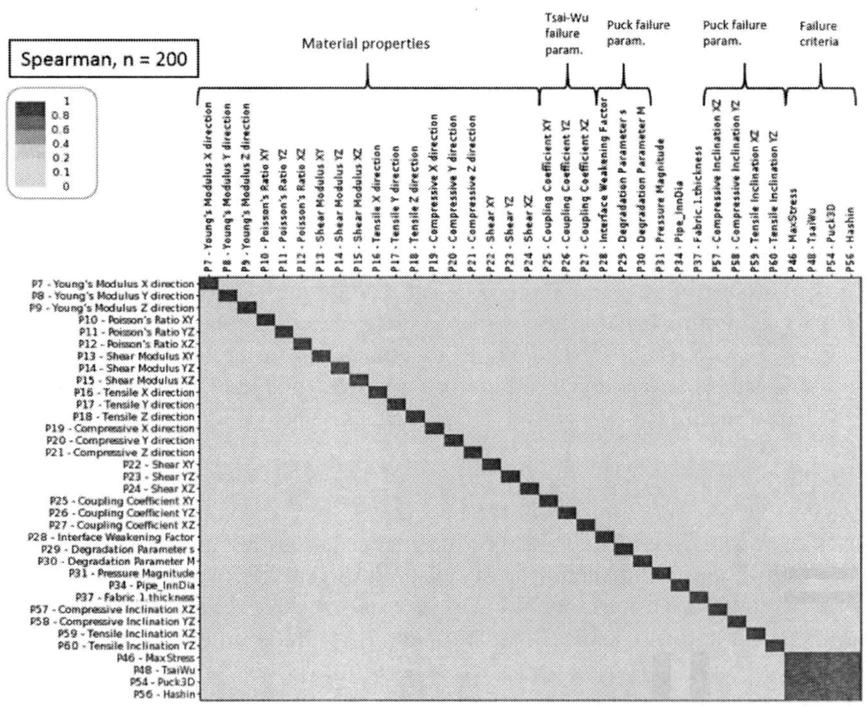

Figure 8. Determination matrix, Spearman correlation, n = 200.

IOP Conf. Series: Materials Science and Engineering **700** (2019) 012021 doi:10.1088/1757-899X/700/1/012021

Table 6. Spearman linear correlation, n = 200, failure criteria

	Maximum Stress	Tsai-Wu	Hashin	Puck
Maximum Stress	1.00	0.90	0.91	0.91
Tsai-Wu	0.96	1.00	0.96	0.97
Hashin	0.91	0.96	1.00	0.99
Puck	0.91	0.97	0.99	1.00

Table 7. Spearman quadratic correlation, n = 200, failure criteria

	Maximum Stress	Tsai-Wu	Hashin	Puck
Maximum Stress	1.00	0.81	0.83	0.84
Tsai-Wu	0.81	1.00	0.94	0.96
Hashin	0.82	0.94	1.00	0.98
Puck	0.84	0.96	0.98	1.00

5.1.2. Moderate correlations between applied pressure & ply thickness and failure criteria. The applied pressure, $P_{applied}$ and ply thickness, t_{ply} are moderately linearly correlated (0.4 to 0.6 or -0.6 to -0.4) with the failure criteria. These parameters are also slightly quadratically correlated (0.1 to 0.4 or -0.4 to -0.1). It is intuitive to understand that applied pressure and ply thickness would be naturally correlated with the failure criteria. The counter-intuitive thing is that the correlation is not strong. It can be natural to think that since hoop stress is directly proportional to the applied pressure and inversely proportional to wall thickness, therefore the correlation must be very strong. The moderate correlations can be understood by taking a closer examination at the correlation scatter diagram of Puck failure criteria vs. applied pressure and ply thickness as presented in Figure 9. Note the applied pressure (horizontal axis) is plotted in decreasing values, i.e., the right side of the horizontal axis denotes smaller values of applied pressure. From Figure 9, it is clear that the scatter is quite large; this is due to the fact that failure of the flow-line is not just fully dependent on the values of applied pressure and/or ply thickness but a result of many design parameters at play.

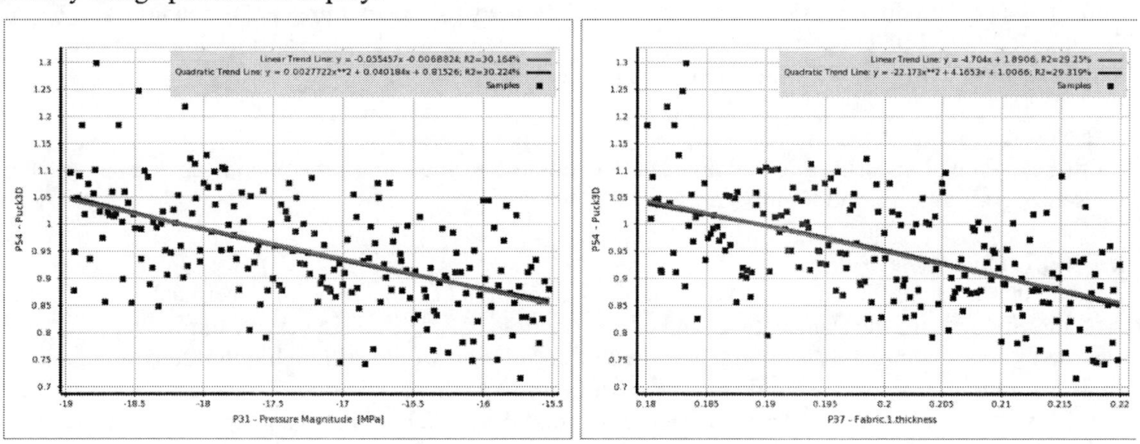

Figure 9. Correlation scatter diagram, Puck failure criterion vs. applied pressure and ply thickness, Spearman correlation, n = 200.

5.1.3. Other parameters with slight correlation with failure criteria. The parameters that have slight linear correlations (0.1 to 0.4 or -0.4 to -0.1) with the failure criteria are elastic moduli, E_1, E_2, tensile strength, σ_{ut1}, shear strength, τ_{u12} and outer diameter, OD. In addition, Tsai-Wu constant, F_{12}, has a slight correlation with the Tsai-Wu failure criterion. The other parameters that have slight quadratic correlations (0.1 to 0.4 or -0.4 to -0.1) with the failure criteria are shear strength, τ_{u12} and outer diameter, OD. The slight correlations of tensile strength σ_{ut1} and shear strength τ_{u12} can be surprising to the reader. However, this can be understood by examining their correlation scatter diagrams. Figure 10 presents the correlation scatter diagrams of Maximum Stress criterion vs tensile strength σ_{ut1} and shear strength τ_{u12}. It becomes clear why these two design parameters are only slightly correlated to the Maximum Stress criterion. As previously mentioned in Section 5.1.2. , the failure of the flow-line is not just dependent ontensile strength σ_{ut1} and shear strength τ_{u12} , but it is a result of many design parameters at play.

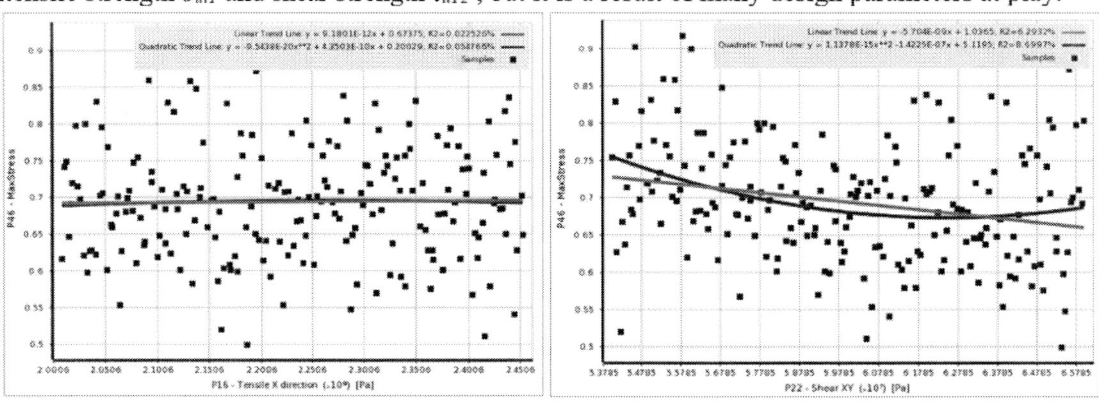

Figure 10. Correlation scatter diagram, maximum stress criterion vs. tensile strength σ_{ut1} and shear strength τ_{u12}, Spearman correlation, n = 200

5.1.4. Most material-related failure parameters insignificantly correlated with failure criteria. Most material-related failure parameters with the exception of Tsai-Wu constant, F_{12}, do not exhibit correlation (below |0.1|) to the failure criteria. This is due to the nature of the current loading considered which predominantly result in membrane stress along the pipe walls. This causes the load to be predominantly in the fibre direction. The failure parameters in the failure criteria, i.e., Tsai-Wu constants and all Puck constants considered are primarily defined to model transverse, bending and/or inter-laminar loading. It is expected that these failure parameters would play a more important role in the failure criteria when more complex loadings are considered.

5.2. Pearson vs. Spearman

Both Pearson and Spearman correlation methods identified the same set of major design parameters as presented in Figure 11 and Figure 12.

5.3. Effect of sample size

In general, the larger the sample size, the more likely design parameters with lower values of correlation get highlighted in the correlation and determination matrices. However, the governing design parameters, i.e., parameters with higher values of correlation would still be highlighted on the matrices. This is highlighted by the example presented in Figure 13 where the Spearman correlation matrix from n = 50 is compared against n = 200. It is observed that the darker coloured squares, i.e., high correlation values are practically the same for n = 50 and n = 200. The difference is in the lighter coloured squares, i.e., low correlation values. For practical engineering purposes, a sample size of 100 seemed to be sufficient in this paper.

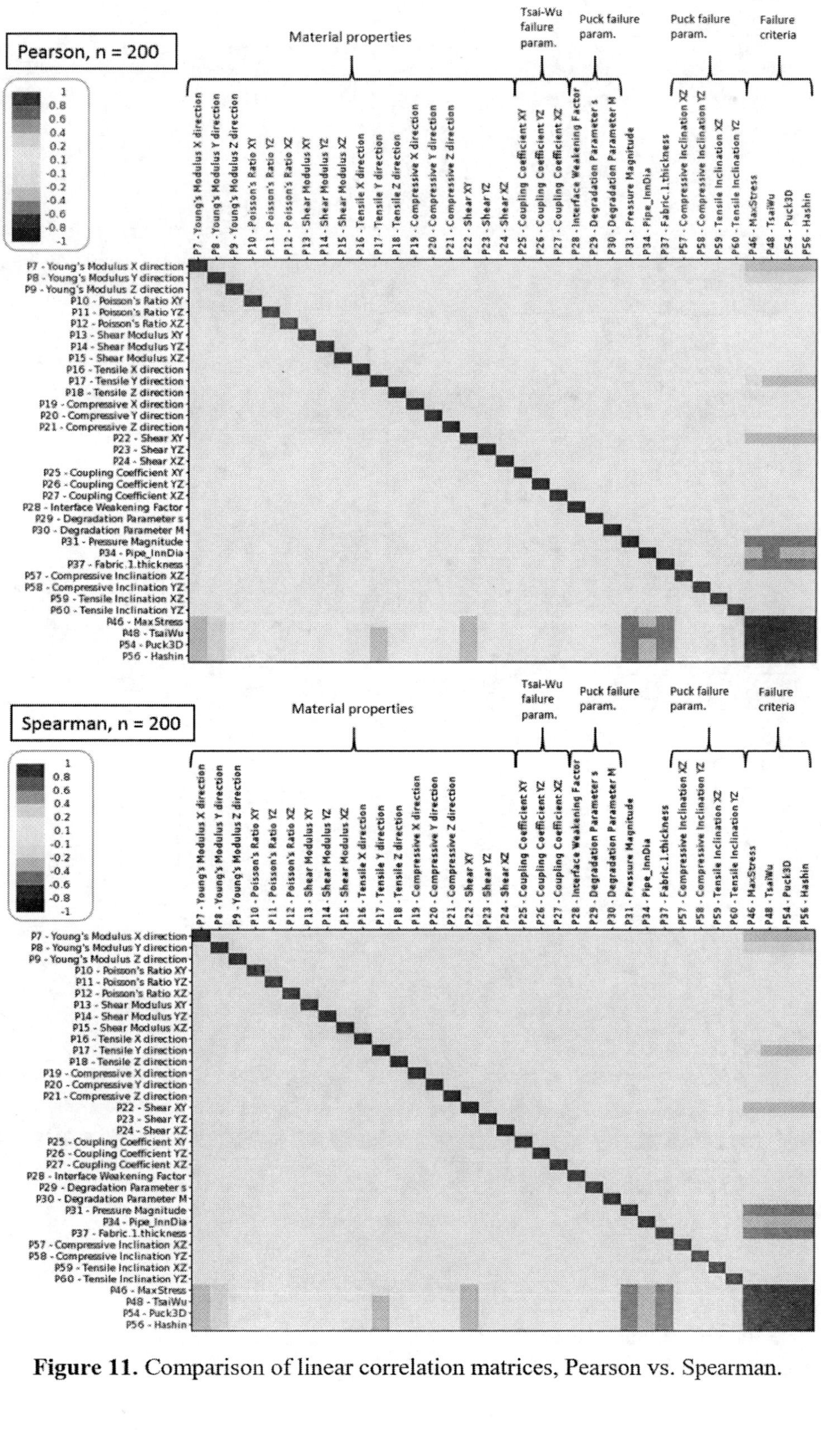

Figure 11. Comparison of linear correlation matrices, Pearson vs. Spearman.

IOP Conf. Series: Materials Science and Engineering **700** (2019) 012021 doi:10.1088/1757-899X/700/1/012021

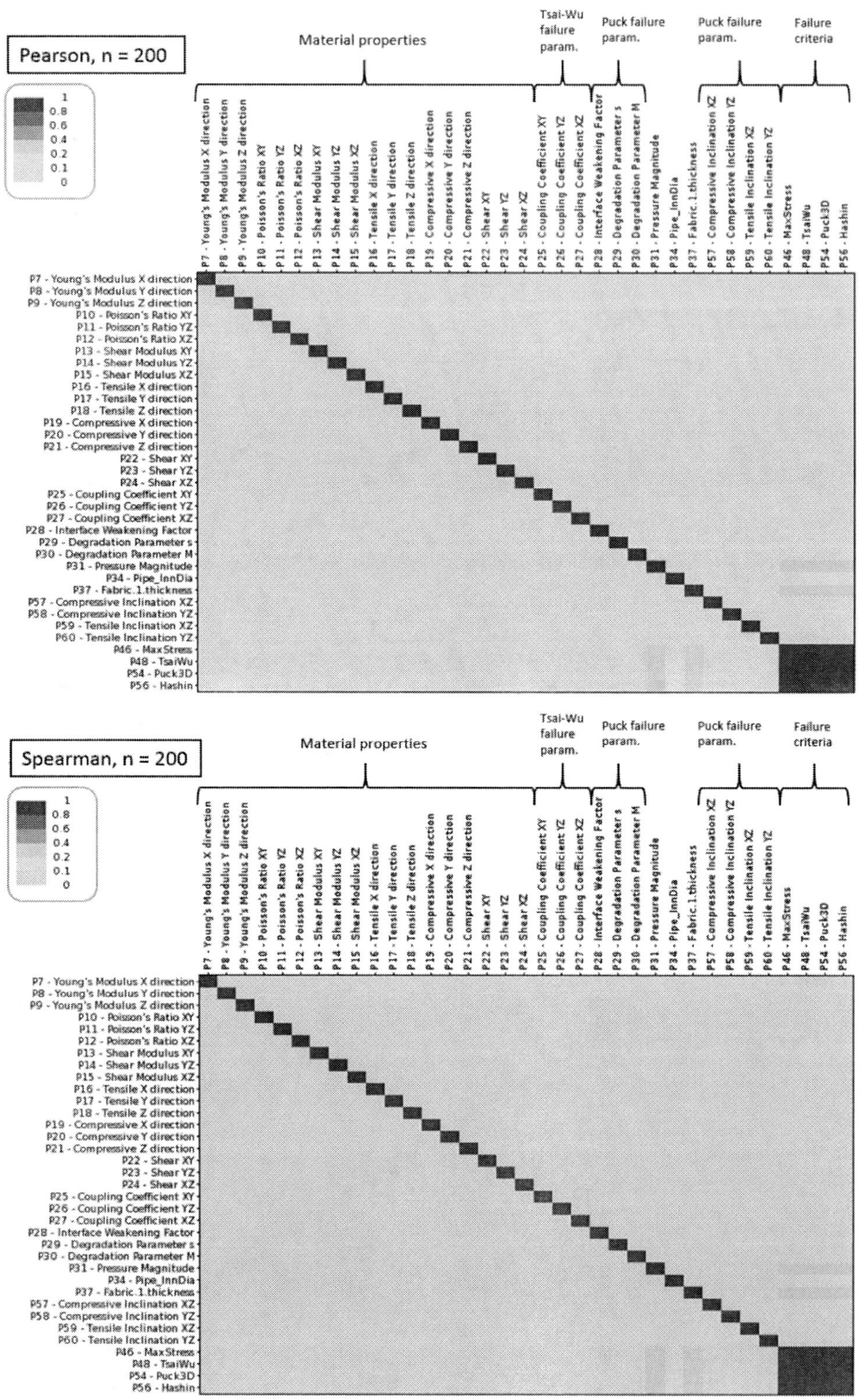

Figure 12. Comparison of quadratic correlation matrices, Pearson vs. Spearman

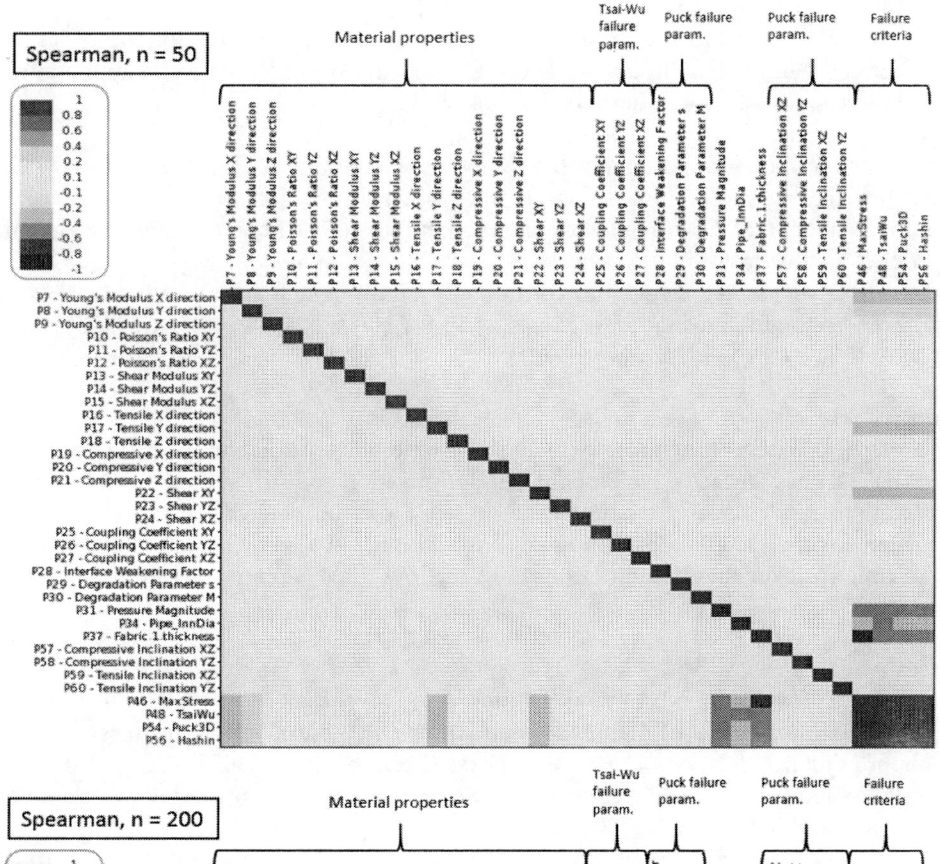

Figure 13. Comparison of correlation matrices, Spearman correlation, n = 50 vs. n = 200.

6. Conclusions

The following conclusions are made:

- Strong linear and quadratic correlations are observed between the failure criteria.
- Moderate linear and slight quadratic correlations are observed between applied pressure & ply thickness and the failure criteria.
- Most material failure parameters with the exception of Tsai-Wu constant, F_{12}, are not strongly correlated to the failure criteria. This can be explained by the fact that the loads experienced by the flow-line are pre-dominantly in the fibre direction.
- Both Pearson and Spearman correlation methods identified the same set of major design parameters for the correlation matrix.
- The effect of sample size is not profoundly significant to the results. There were no large differences found in using sample sizes of 50 vs. 100 vs. 200. For practical engineering purpose a sample size of 100 is recommended.

7. Future work

There are plans to extend the study to include (i) external tension/compression forces and bending moments, (ii) response surface development and optimisation and (iii) six-sigma analysis. Furthermore, additional failure modes such as fatigue and collapse can be studied in future work. Other design scenarios such as pipeline global buckling and installation can also be investigated. The objective is to provide a framework for the design optimisation of flow-lines when non-traditional materials such as CFEC are used.

References

[1] Reza Khoshravan Azar M, Emami Satellou A A, Shishesaz M and Salavati B 2013 Calculating the optimum angle of filament-wound pipes in natural gas transmission pipelines using approximation methods, *J. Pressure Vessel Technol.* **135**(2), 021702.

[2] Theotokoglou E E 2006 Behaviour of thick composite tubes considering of delamination, *Theor. Appl. Fract. Mech.* **46**(3), 276-85.

[3] Xia M, Takayanagi H and Kemmochi K 2001 Analysis of multi-layered filament-wound composite pipes under internal pressure, *Compos. Struct.* **53**(4), 483-91.

[4] Chouchaoui C S and Ochoa O O 1999 Similitude study for a laminated cylindrical tube under tensile, torsion, bending, internal and external pressure. Part I: governing equations, *Compos. Struct.* **44**(4), 221-29.

[5] Guz I A, Menshykova M and Paik J K 2017 Thick-walled composite tubes for offshore applications: an example of stress and failure analysis for filament-wound multi-layered pipes, *Ships and Offshore Structures.* **12**(3), 304-22.

[6] Tsai S W and Wu E M 1971 A general theory of strength for anisotropic materials, *J. Compos. Mater.* **5**(1), 58-80.

[7] Hashin Z 1980 Failure criteria for unidirectional fiber composites, *J. Appl. Mech.* **47**(2), 329.

[8] Puck A 1996 Festigkeitsanalyse von faser-matrix-laminaten. (Carl Hanser).

[9] Puck A 1998 Failure analysis of FRP laminates by means of physically based phenomenological models, *Compos. Sci. Technol.* **58**(7), 1045-67.

[10] Puck A 2002 Schürmann H. Failure analysis of FRP laminates by means of physically based phenomenological models, *Compos. Sci. Technol.* **62**(12-13), 1633-62.

[11] Pearson K 1920 Notes on the history of correlation, *Biometrika.* **13**(1), 25-45.

[12] Pearson K 1986 Mathematical contributions to the theory of evolution. III. Regression, Heredity, and Panmixia, *Philos. Trans. R. Soc. London, Ser. A, Containing Papers of a Mathematical or Physical Character.* **187**, 253-318.

[13] Spearman C E 1904 The proof and measurement of association between two things, *Am. J. Psychol.* **15**, 72-101.

[14] Spearman C E 1904 General intelligence, objectively determined and measured, *Am. J. Psychol.* **15**, 201-93.
[15] Spearman C E 1910 Correlation calculated from faculty data, *Br. J. Psychol*, 1904-1920. **3**(3), 271-95.
[16] Vanderplaats G N 2007 Multidiscipline design optimization (Colorado Springs)
[17] Nettles A T 1994 Basic mechanics of laminated composite plates. (Marshall Space Flight Center, NASA).
[18] Palantera M 1998 ESAComp 4.1 - theoretical background of ESAComp analyses.
[19] Jones R M 1975 Mechanics of composite material (Hemisphere).
[20] Kirchhoff G 1850 Über das gleichgewicht und die bewegung einer elastischen scheibe, *J. Reine und Angewante Mathematik* (Crelle). **40**, 51-88.
[21] Love A E H 1906 A treatise on the mathematical theory of elasticity (Cambridge).
[22] Gol'denblat II, Kopnov V A 1966 Strength of glass-reinforced plastics in the complex stress state - english translation, *Polymer Mechanics*. **1**, 54.
[23] ANSYS®. Academic research mechanical, release 17.0, help system, element reference, chapter I, SHELL181, ANSYS, Inc.

Flexural dynamic behavior of submerged cylindrical structures under wave loads

T Giotis [1,2,*], D Pavlou [1] and K A Belibassakis [2]

[1]Dept. of Mechanical and Structural Engineering and Materials Science, University of Stavanger, Norway
[2]School of Naval Architecture and Marine Engineering, National Technical University of Athens, Greece

* Corresponding author: fanisgiotis@gmail.com

Abstract. An analytical solution for the dynamic response of submerged slender circular cylindrical structure subjected to linear wave loads is presented. A double Laplace transform with respect to temporal and spatial variables is applied both to motion equation and boundary conditions. The dynamic deflection of the beam is obtained by inversion of the Laplace transform. The latter with respect to spatial variable is obtained analytically, while the one concerning the temporal variable is numerically calculated using Durbin numerical scheme. Results in the case of a representative example for a monopile foundation subjected to Airy waves are presented and discussed, and the analytical result is compared against numerical dynamic and static solutions.

1. Introduction

The analysis of dynamic response of circular cylindrical structures under wave loads has attracted attention by many investigators. The focus is mainly on the determination of dynamic deflection obtained through numerical methods, and experimental studies in wave tank, e.g., [1]. In the present work an analytical solution for the dynamic response of slender cylindrical structures under Airy waves is presented, which is obtained by using Laplace transform techniques. The proposed solution has the advantage that the solution of the PDE is analytic and could be easily extended to include the additional effect of current. Formulation of the problem is valid for rocky or stiff seabed. The waves encountered in the installation area are characterized by a small steepness ratio H/λ and as a result they can be considered as linear. The mono-pile foundation is assumed fixed on the seabed. This assumption The geometric characteristics of the pile remain constant throughout its length and the outer diameter (D_{out}) of the foundation is small with respect to the wavelength (λ) of the installation area. The small ratio H/λ allows the assumption that the flow disturbance is not significant due to the presence of the structure and as a result, the monopile can be considered as a slender structure. The origin (O) of the system is located on the center of the foundation base (Figure 1).

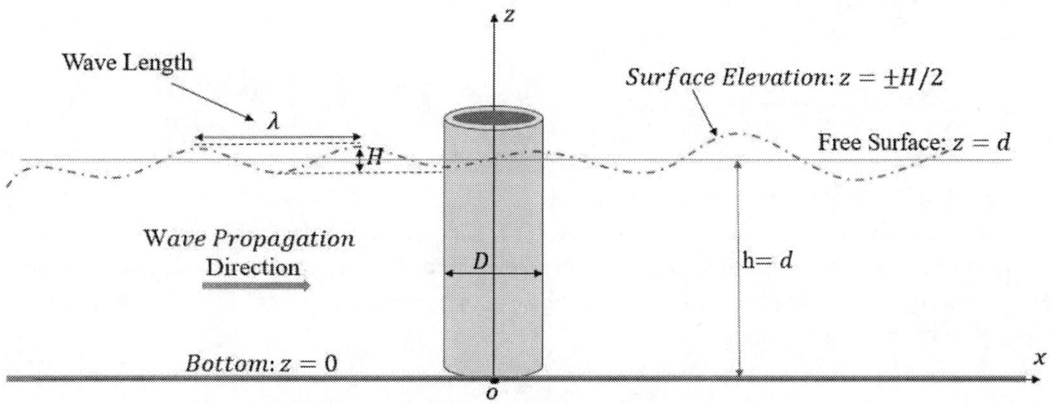

Figure 1. The origin system of the monopile.

2. Analytical solution of the problem

2.1 Boundary conditions

The fixed end which is located on the seabed z = 0. Thus, slope and deflection are zero at z = 0.

$$w(0,t) = \left.\frac{\partial w(z,t)}{\partial z}\right|_{z=0} = 0 \tag{1}$$

The upper end at z = L is free and as a result, both bending moment and shear force are zero at z = L.

$$E \cdot I(z) \cdot \left.\frac{\partial^2 w(z,t)}{\partial z^2}\right|_{z=L} = 0 \tag{2}$$

$$\left.\frac{\partial}{\partial z}\left[E \cdot I(z) \cdot \frac{\partial^2 w(z,t)}{\partial z^2}\right]\right|_{z=L} = 0 \tag{3}$$

2.2 Equation of motion

The dynamic displacement field is expressed by means of Euler-Bernoulli beam model [2]:

$$E \cdot I \cdot \frac{\partial^4 w(z,t)}{\partial z^4} + P \cdot A \cdot \frac{\partial^2 w(z,t)}{\partial t^2} + C \cdot \frac{\partial w(z,t)}{\partial t} = q(z,t) \tag{4}$$

The right hand side part of (4) is the hydrodynamic load is calculated by using potential wave theory [3, 4] and Morison Equation [5]. The circular cylinder is considered as a moving structure in waves. Thus, the relative formulation of Morison Equation is used. However, the structure represents a monopile offshore wind turbine, which has to be rigid enough for the safety of the rotor. The rigidity of the structure ensures that the maximum deflection amplitude is much smaller than the structure diameter (D_{out}), and as a result, the velocity of the structure could be neglected. The final form of the hydrodynamic load is given by (5).

$$q(z,t) = -\rho \cdot C_A \cdot A \cdot \frac{\partial^2 w(z,t)}{\partial t^2} + \rho \cdot C_M \cdot A \cdot \frac{Du(t)}{Dt} + \frac{1}{2} \cdot \rho \cdot C_D \cdot D_{out} \cdot u(t) \cdot |u(t)| \tag{5}$$

The drag term of Eq. (5) is linearized and the final form of equation of motion is:

$$E\cdot I\cdot\frac{\partial^4 w(z,t)}{\partial z^4}+\left[P\cdot A+\rho\cdot C_A\cdot A\right]\cdot\frac{\partial^2 w(z,t)}{\partial t^2}+C\cdot\frac{\partial w(z,t)}{\partial t}=$$

$$-C_M\cdot\rho\cdot\frac{\pi\cdot D_{out}^2}{4}\cdot\omega^2\cdot\frac{H}{2}\cdot\frac{\cosh(k\cdot z)}{\sinh(k\cdot d)}\cdot\sin(\omega\cdot t)+\frac{1}{2}\cdot\rho\cdot C_D\cdot D\cdot\left[\frac{H}{2}\cdot\omega\cdot\frac{\cosh(k\cdot z)}{\sinh(k\cdot d)}\right]^2\cdot\cos(\omega\cdot t) \quad (6)$$

2.3 General solution of the motion equation

For the solution of the boundary value problem, Laplace transform will be applied to the spatial and temporal variables. The definition of these transforms and their inverse forms used on the solution is given in the Appendix. Application of the Laplace transform with respect to time and introduction of $w^{(n)}(z,0)=0$ for n=1,2,3,4 yield:

$$E\cdot I\cdot\frac{\partial^4 w^*(z,s)}{\partial z^4}+w^*(z,s)\cdot\left[\left(P\cdot A_C+\rho\cdot C_A\cdot A\right)\cdot s^2+C\cdot s\right]=$$

$$-C_M\cdot\rho\cdot\frac{\pi\cdot D_{out}^2}{4}\cdot\omega^3\cdot\frac{H}{2}\cdot\frac{\cosh(k\cdot z)}{\sinh(k\cdot d)}\cdot\left(\frac{1}{s^2+\omega^2}\right)+\frac{1}{2}\cdot\rho\cdot C_D\cdot D\cdot\left[\frac{H}{2}\cdot\omega\cdot\frac{\cosh(k\cdot z)}{\sinh(k\cdot d)}\right]^2\cdot\left(\frac{s}{s^2+\omega^2}\right) \quad (7)$$

where

$$w^*(z,s)=L\left\{w(z,t);t\to s\right\}=\int_0^\infty e^{-s\cdot t}\cdot w(z,t)dt \quad (8)$$

Application of the Laplace transform with respect to z and introduction of zero initial conditions into (7) yield:

$$E\cdot I\cdot\left[-\bar{w}'''(0,s)-q\cdot\bar{w}''(0,s)-q^2\cdot\bar{w}'(0,s)-q^3\cdot\bar{w}(0,s)+q^4\cdot\bar{w}^*(q,s)\right]+$$

$$\left[\left(P\cdot A+\rho\cdot C_A\cdot A\right)\cdot s^2+C\cdot s\right]\cdot\bar{w}^*(q,s)=$$

$$-C_M\cdot\rho\cdot\frac{\pi\cdot D_{out}^2}{4}\cdot\omega^3\cdot\frac{H}{2}\cdot\frac{1}{\sinh(k\cdot d)}\cdot\left(\frac{1}{s^2+\omega^2}\right)\cdot\left(\frac{q}{q^2-k^2}\right)+$$

$$\frac{1}{8}\cdot C_D\cdot\rho\cdot D_{out}\cdot H^2\cdot\omega^2\cdot\frac{1}{\sinh^2(k\cdot d)}\cdot\left(\frac{s}{s^2+\omega^2}\right)\cdot\frac{q^2-2\cdot k^2}{q\cdot\left(q^2-4\cdot k^2\right)} \quad (9)$$

where

$$\bar{w}^*(q,s)=L\left\{w^*(z,s);z\to q\right\}=\int_0^\infty e^{-q\cdot z}\cdot w^*(z,s)\cdot dz \quad (10)$$

For the simplification of (9), the Laplace transform is applied into the boundary conditions (2). Thus, the conditions at z=0 takes the following form:

$$\bar{w}(0,s)=L\left\{w(0,t);t\to s\right\}=0 \quad (11)$$

$$\bar{w}'(0,s)=L\left\{w'(0,t);t\to s\right\}=0 \quad (12)$$

IOP Conf. Series: Materials Science and Engineering **700** (2019) 012022 doi:10.1088/1757-899X/700/1/012022

Introducing Eqs. (11), (12) into Eq. (9) yields:

$$\bar{w}^*(q,s)\cdot\left[E\cdot I\cdot q^4+\left(P\cdot A+\rho\cdot C_A\cdot A\right)\cdot s^2+C\cdot s\right]=E\cdot I\cdot q\cdot \bar{w}''(0,s)+E\cdot I\cdot \bar{w}'''(0,s)-$$
$$C_M\cdot\rho\cdot\frac{\pi\cdot D_{out}^2}{4}\cdot\omega^3\cdot\frac{H}{2}\cdot\frac{1}{\sinh(k\cdot d)}\cdot\left(\frac{1}{s^2+\omega^2}\right)\cdot\left(\frac{q}{q^2-k^2}\right)+$$
$$\frac{1}{8}\cdot C_D\cdot\rho\cdot D_{out}\cdot H^2\cdot\omega^2\cdot\frac{1}{\sinh^2(k\cdot d)}\cdot\left(\frac{s}{s^2+\omega^2}\right)\cdot\frac{q^2-2\cdot k^2}{q\cdot\left(q^2-4\cdot k^2\right)} \tag{13}$$

Solving (13) with respect to $\bar{w}^*(q,s)$, yields:

$$\bar{w}^*(q,s)=\bar{M}_1^*(q,s)\cdot\bar{w}''(0,s)+\bar{M}_2^*(q,s)\cdot\bar{w}'''(0,s)+\bar{M}_3^*(q,s)+\bar{M}_4^*(q,s) \tag{14}$$

where

$$\bar{M}_1^*(q,s)=\frac{q}{q^4+\dfrac{\left(P\cdot A+\rho\cdot C_A\cdot A\right)\cdot s^2+C\cdot s}{E\cdot I}} \tag{15}$$

$$\bar{M}_2^*(q,s)=\frac{1}{q^4+\dfrac{\left(P\cdot A+\rho\cdot C_A\cdot A\right)\cdot s^2+C\cdot s}{E\cdot I}} \tag{16}$$

$$\bar{M}_3^*(q,s)=-\frac{C_M\cdot\rho\cdot\dfrac{\pi\cdot D_{out}^2}{4}\cdot\omega^3\cdot\dfrac{H}{2}\cdot\dfrac{1}{\sinh(k\cdot d)}\cdot\left(\dfrac{1}{s^2+\omega^2}\right)\cdot\left(\dfrac{q}{q^2-k^2}\right)}{E\cdot I\cdot q^2+\left(P\cdot A+\rho\cdot C_A\cdot A\right)\cdot s^2+C\cdot s} \tag{17}$$

$$\bar{M}_4^*(q,s)=\frac{\dfrac{1}{8}\cdot C_D\cdot\rho\cdot D_{out}\cdot H^2\cdot\omega^2\cdot\dfrac{1}{\sinh^2(k\cdot d)}\cdot\left(\dfrac{s}{s^2+\omega^2}\right)\cdot\left(\dfrac{q^2-2\cdot k^2}{q\cdot\left(q^2-4\cdot k^2\right)}\right)}{E\cdot I\cdot q^4+\left(P\cdot A_C+\rho\cdot C_A\cdot A\right)\cdot s^2+C\cdot s} \tag{18}$$

Application of the inverse Laplace transform with respect to q into Eq. (14) yields:

$$\bar{w}(z,s)=\bar{M}_1(z,s)\cdot\bar{w}''(0,s)+\bar{M}_2(z,s)\cdot\bar{w}'''(0,s)+\bar{M}_3(z,s)+\bar{M}_4(z,s) \tag{19}$$

where

$$\bar{w}(q,s)=L^{-1}\left\{\bar{w}^*(q,s);q\rightarrow z\right\} \tag{20}$$

$$\bar{M}_i(z,s)=L^{-1}\left\{\bar{M}_i^*(q,s);q\rightarrow z\right\},\quad i=1,2,3,4 \tag{21}$$

Equation (19) contains two unknown variables $\bar{w}''(0,s)$ and $\bar{w}'''(0,s)$. The first expresses the bending moment at the position z=0 and the second is equal with the shear force at the base of the foundation (z=0). The determination of those variables is achieved by the introduction of the boundary conditions of the beam at the position z=L. Application of the Laplace transform with respect to time into Eqs. (2) and (3) yields:

$$\bar{w}''(L,s) = \mathsf{L}\left\{w''(L,t); t \to s\right\} = 0 \tag{22}$$

$$\bar{w}'''(L,s) = \mathsf{L}\left\{w'''(L,t); t \to s\right\} = 0 \tag{23}$$

Introduction of Eq. (25) and (26) into Eq. (19) yields:

$$\bar{w}''(0,s) = \frac{\left[\bar{M}_3'''(L,s) + \bar{M}_4'''(L,s)\right] \cdot \bar{M}_2''(L,s) - \left[\bar{M}_3''(L,s) + \bar{M}_4''(L,s)\right] \cdot \bar{M}_2'''(L,s)}{\bar{M}_2'''(L,s) \cdot \bar{M}_1''(L,s) \cdot \left[1 - \dfrac{\bar{M}_2''(L,s) \cdot \bar{M}_1'''(L,s)}{\bar{M}_2'''(L,s) \cdot \bar{M}_1''(L,s)}\right]} \tag{24}$$

$$\bar{w}'''(0,s) = -\frac{\left[\bar{M}_3'''(L,s) + \bar{M}_4'''(L,s)\right] \cdot \bar{M}_2''(L,s) - \left[\bar{M}_3''(L,s) + \bar{M}_4''(L,s)\right] \cdot \bar{M}_2'''(L,s)}{\bar{M}_2'''(L,s) \cdot \bar{M}_1''(L,s) \cdot \left[1 - \dfrac{\bar{M}_2''(L,s) \cdot \bar{M}_1'''(L,s)}{\bar{M}_2'''(L,s) \cdot \bar{M}_1''(L,s)}\right]} \times$$

$$\frac{\bar{M}_1'''(L,s)}{\bar{M}_2'''(L,s)} - \frac{\bar{M}_3'''(L,s) + \bar{M}_4'''(L,s)}{\bar{M}_2'''(L,s)} \tag{25}$$

Application of the inverse Laplace transform with respect to s into (19) finally yields:

$$w(z,t) = \mathsf{L}^{-1}\left\{M_1(z,s) \cdot \bar{w}''(0,s) + \bar{M}_2(z,s) \cdot \bar{w}'''(0,s) + \bar{M}_3(z,s) + \bar{M}_4(z,s); s \to t\right\} \tag{26}$$

where

$$w(z,t) = \mathsf{L}^{-1}\left\{\bar{w}(z,s); s \to t\right\} \tag{27}$$

2.4 Linearization of the hydrodynamic load

It is important to mention that the equation which describes the horizontal Drag load on strip is a non-linear equation. The analytical calculation of the pylon response under the horizontal wave loads requires the formulation of the load equation as the addition of linear harmonic terms. It can be achieved by developing the Drag load into a Fourier series. The process of converting the Drag load into a sum of linear harmonic terms is described in detail as follows. Non-linear term could be written as an independent function named N.

$$N = C_2 \times \left|\cos(k \times x_c - \omega \times t)\right| \times \cos(k \times x_c - \omega \times t) \tag{28}$$

where

$$C_2 = \frac{1}{2} \cdot \rho \cdot C_D \cdot D_{out} \cdot \left[\frac{H}{2} \cdot \omega \cdot \frac{\cosh[k \cdot (z+d)]}{\sinh(k \cdot d)} \right]^2$$

By setting $\theta = k \times x_c - \omega \times t$, (28) takes the following form:

$$N(\theta) = C_2 \cdot |\cos(\theta)| \cdot \cos(\theta) \Rightarrow N(\theta) = \begin{cases} C_2 \cdot \cos^2(\theta), -\frac{\pi}{2} \le \theta \le \frac{\pi}{2} \\ -C_2 \cdot \cos^2(\theta), \frac{\pi}{2} \le \theta \le \frac{3 \cdot \pi}{2} \end{cases} \quad (29)$$

Let as now consider the representation of N(θ) in a Fourier series as follows:

$$\frac{N(\theta)}{C_2} = \frac{1}{2} \cdot a_0 + \sum_{j=1}^{\infty} a_j \cdot \cos(j \cdot \theta) + \sum_{j=1}^{\infty} b_j \cdot \sin(j \cdot \theta) \quad , \quad \text{where } a_j = \frac{2}{\pi} \cdot \int_0^T N(\theta) \cdot \cos(j \cdot \theta) \cdot d\theta,$$

$$\text{and } b_j = \frac{2}{\pi} \cdot \int_0^T N(\theta) \cdot \sin(j \cdot \theta) \cdot d\theta.$$

In the case when $a_0=0$ and $|a_j| << |a_1|$, and $|b_j| << |b_1|$, for $j>1$, as happens to be the ones considered here, the function N(θ) is approximated by using only the first term in the series as follows:

$$\frac{N(\theta)}{C_2} \approx a_1 \times \cos(\theta), \qquad \text{with } a_1 = \max\left|\frac{N(\theta)}{C_2}\right| \quad (30)$$

Using the above approximation in Eq. (5) we finally obtain:

$$\frac{dF_{N_D}(t)}{dz} \approx \frac{1}{2} \cdot \rho \cdot C_D \cdot D_{out} \cdot \left[\frac{H}{2} \cdot \omega \cdot \frac{\cosh[k \cdot (z+d)]}{\sinh(k \cdot d)} \right]^2 \cdot \cos(k \cdot x_c - \omega \cdot t) \quad (31)$$

3. Numerical example of the analytical solution

A vertical circular cylindrical structure, fixed in the seabed is considered, as described by Table 1.

Table 1. Structural characteristics of the pile.

Magnitude	Name	Value
Height (m)	h	30
Outer Diameter (m)	D_{out}	6
Shell Thickness (m)	thik	0.05
Modulus of Elasticity (GPa)	E	210
Poisson's Ratio	v	0.30
Density of Material (kg/m^3)	P	7820
Damping Coefficient	C	4398
Added Mass Coefficient	C_A	1
Drag Coefficient	C_D	0.65

The wave characteristics of the installation area of the structure are given in Table 2.

IOP Conf. Series: Materials Science and Engineering **700** (2019) 012022 doi:10.1088/1757-899X/700/1/012022

Table 2. Wave characteristics of the installation area.

Magnitude	Name	Value
Wave Height (m)	H	3.5
Wave Period (sec)	T	6
Water Depth (m)	d	30
Wavenumber	k	0.1121
Wave Frequency (rad/sec)	w	1.0472
Density of Water (kg/m^3)	ρ	1024.7

The comparison of the non-linear drag load at the top of the structure (z = h) and its linearized approximation by means of (31) is presented in Figure 2.

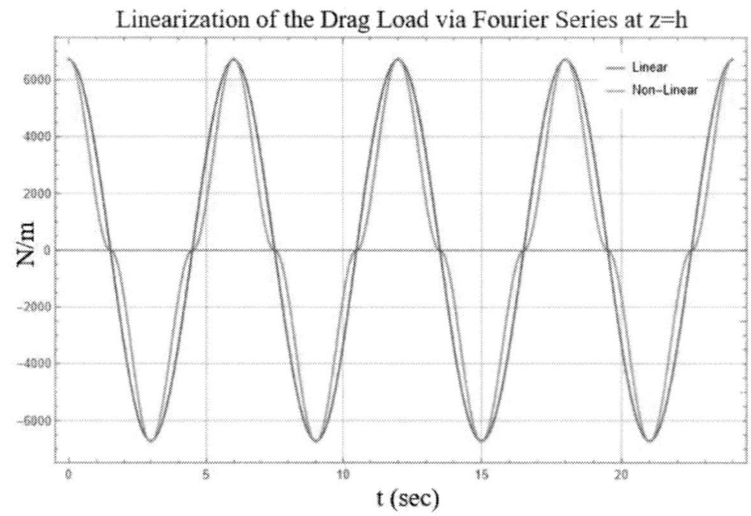

Figure 2. Linear and non-linear drag load at z = h.

Applying the Morison Equation for both linear and non-linear drag load, the total hydrodynamic load on the top of the structure (z = h) is presented in Figure 3.

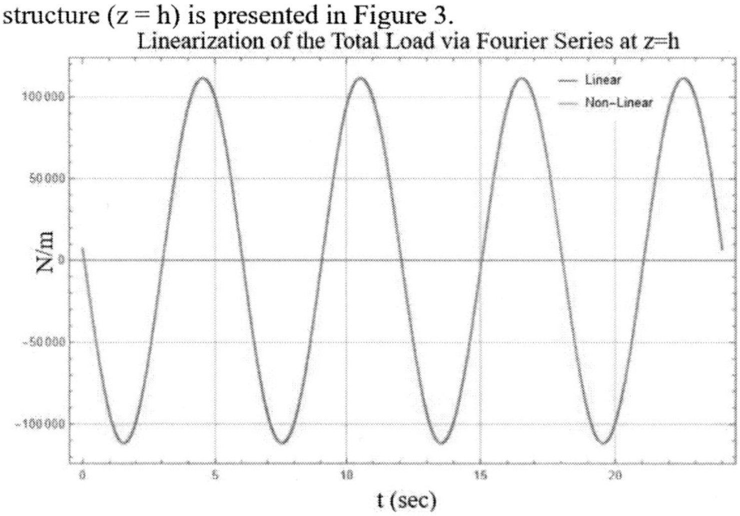

Figure 3. Linear and non-linear total load at z = h.

Applying (23), the deflection of the structure at z = h, during four periods, is given in Figure 4.

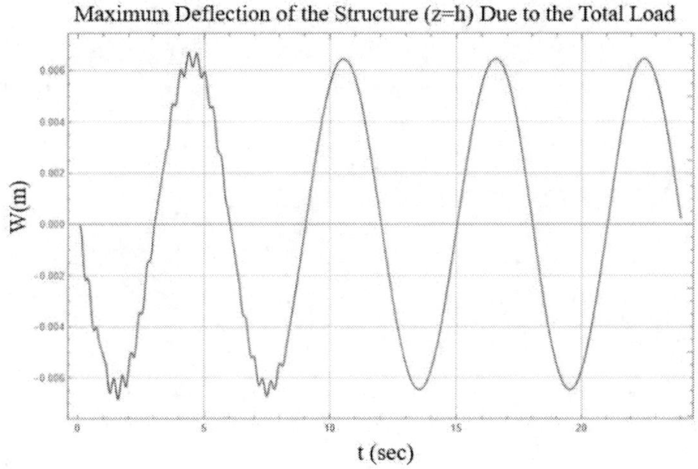

Figure 4. Deflection of the structure at z = h.

The response of the structure at z = h follows the form of the total hydrodynamic load. Deflection is equal with zero at t = 0 due to the zero initial conditions used during the analytical solution. Thus, the maximum dynamic deflection of the beam is equal to $w(L)_{Dynamic} = 0.0064$ m. The calculated deflection of the pillar due to the hydrodynamic load is presented in Figure 5 for one period (T).

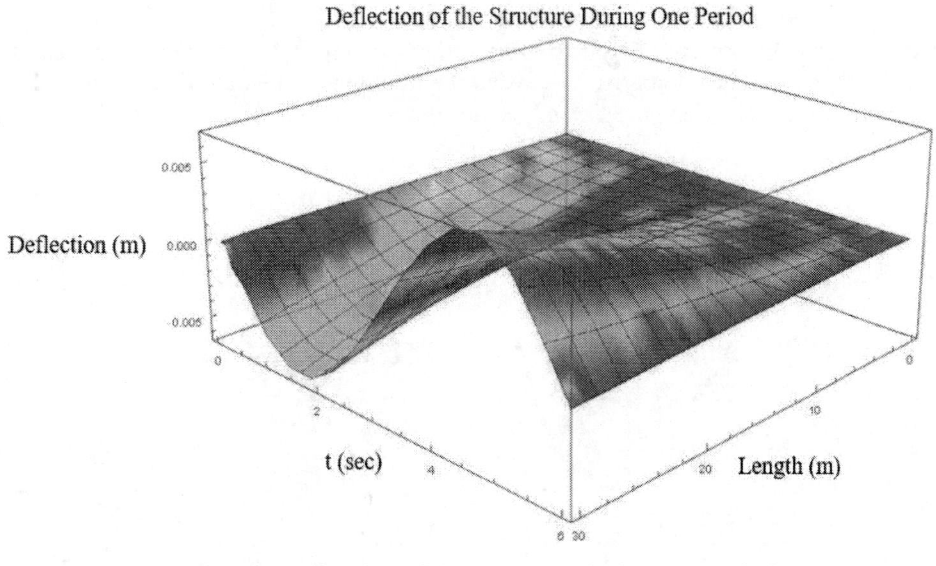

Figure 5. Deflection of the structure during one period.

4. Evaluation of the analytical solution

4.1 Comparison between analytical dynamic and static solution

To verify the reliability of the results, the dynamic maximum deflection of the structure due to the hydrodynamic load, is compared with the maximum static deflection. The solution of the static problem is achieved by using the finite element method (FEM). The maximum static deflection is $w(L)_{Static} = 0.0064$ m, and compared with the static maximum deflection, they are the equal. The dynamic deflection of the beam with respect to the maximum static deflection due to the hydrodynamic load, is presented in Figure 6.

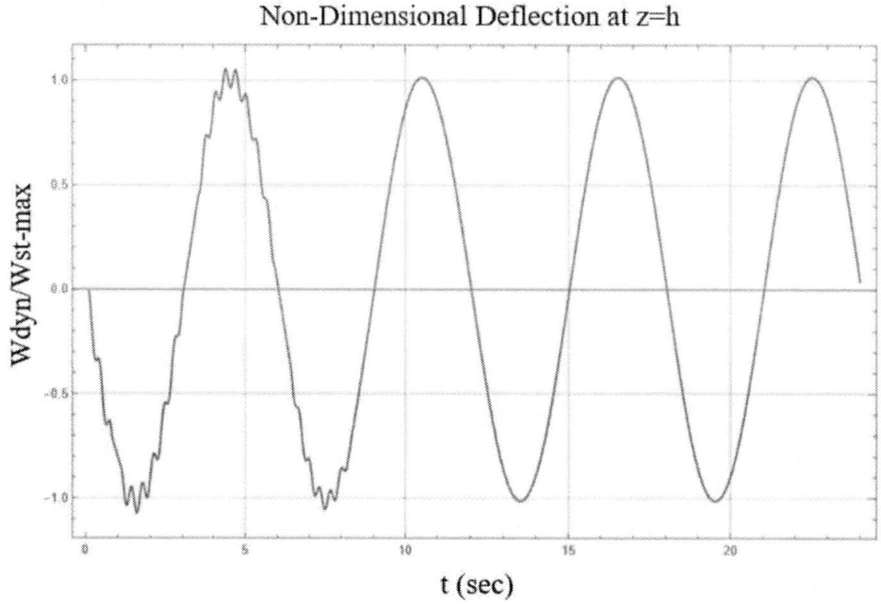

Figure 6. Comparison between maximum static and the dynamic deflection.

4.2 Comparison between analytical and numerical solution

For the validation of the present analytical model, the problem is also solved numerically by a Finite Differential Method, and the comparison between the numerical and analytical solution is presented in Figure 7. The numerical solution with FDM, verifies that the analytical solution is characterized by high accuracy.

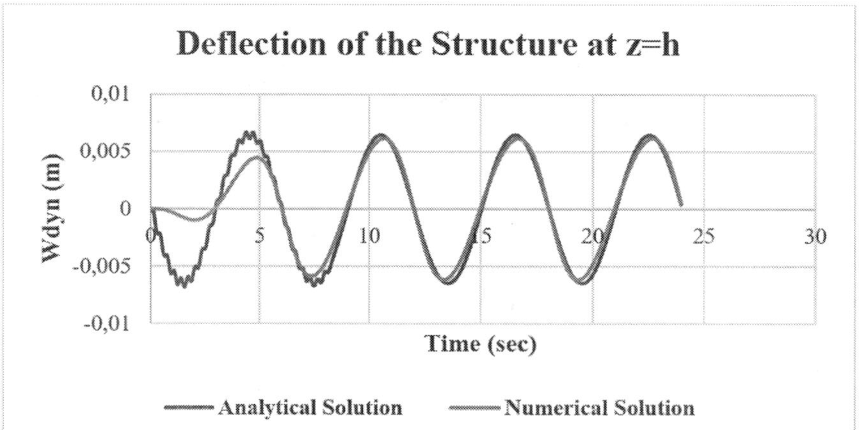

Figure 7. Comparison between numerical and analytical solution.

IOP Conf. Series: Materials Science and Engineering **700** (2019) 012022 doi:10.1088/1757-899X/700/1/012022

5. Conclusions

In this work the calculation of wave loads on slender vertical cylindrical structures is presented, as obtained by application of Morison Equation, in conjunction with potential flow theory. Results of representative examples for a monopile foundation subjected to short waves are presented and discussed. Subsequently, the elastic response of the vertical pillar in waves is studied and an analytical solution concerning the dynamic response is also derived based on linearization. For the validation of the present method, a numerical solution for the dynamic response of the monopile foundation subjected to general wave loads is derived, based on finite difference method (FDM). The numerical and the analytical solution give approximately the same result, verifying the accuracy of the analytical solution for waves of relatively small amplitude. Moreover, the calculation of the maximum static deflection of the structure subjected to the maximum wave loads is obtained by the finite element method (FEM) and is found compatible with the amplitude of analytical dynamic deflection.

Future work includes the introduction of current effects and wind loads into the analytical and numerical model which is important for a more efficient and realistic estimation of the dynamic response of such structures. Also, the developed analytical solution is realistic in the case of a seabed characterized by high stiffness due to the choice of fixed boundary condition at the base of the structure. For more realistic modelling of seabed material and foundation, enhanced boundary condition can be used.

References

[1]. El Kaddah N H 1983 *Water wave-structure interaction for small amplitude structural oscillations*. Doctoral thesis, City University London.

[2]. Graff K F 1975 *Wave motions in elastic solids*, Clarendon Press.

[3]. Massel S.R, (1989). *Hydrodynamics of Coastal Zones*, Elsevier.

[4]. Dean R G, Dalrymple R A 1991 *Water wave mechanics for engineers and scientists*, World Scientific.

[5]. Det Norske Veritas Recommended Practice DNV-RP-C205 (2010), *Environmental Conditions and Environmental Loads*, DNV-RP-C205, October 2010.

[6]. Pavlou D G 2015 *Essentials of the Finite Element Method for Mechanical and Structural Engineers*, Elsevier.

[7]. Oberhettinger F and Badii L 1973 *Tables of Laplace Transforms*. Berlin Heidelberg, New York.

[8]. Cheng H-D, Sidauruk P 1994 Approximate Inversion of the Laplace Transform, *The MATHEMATICA J.*, **4**(2) 76 - 82.

IOP Conf. Series: Materials Science and Engineering **700** (2019) 012022 doi:10.1088/1757-899X/700/1/012022

Appendix

The Laplace transform (with respect to z) of a function f(z) is defined as

$$\mathsf{L}\{f(z); z \to p\} = \bar{f}(p) = \int_0^{\infty} f(z) \cdot e^{-p \cdot t} \cdot dp \tag{A.1}$$

and its inverse is

$$f(z) = \mathsf{L}^{-1}\left\{\bar{f}(p); p \to z\right\} = \frac{1}{2 \cdot \pi \cdot i} \int_{\gamma - i\infty}^{\gamma + i\infty} \bar{f}(p) \cdot e^{p \cdot t} \cdot dp \tag{A.2}$$

The Durbin method [8] for numerical inversion of Laplace transform approximates the time function f(t) by a periodic function of period T. In the present case the following representation is used

$$f(t) \approx \frac{e^{at}}{T} \cdot \left\{ \frac{F(a)}{2} + \sum_{k=1}^{N} \mathrm{Re}\left[F\left(a + \frac{i\pi k}{T}\right)\cos\left(\frac{i\pi k}{T}\right)\right] - \mathrm{Im}\left[F\left(a + \frac{i\pi k}{T}\right)\sin\left(\frac{i\pi k}{T}\right)\right]\right\}, \tag{A.3}$$

where $a = a - \frac{\ln E}{2T}$, E is the error tolerance, for which the default value $E = 1 \times 10^{-8}$ has been assigned, and a is the real part of the leading pole of the function F(s). For functions without poles, the default value a = 0 has been set.
The variables used in the paper are described in Table A1.

Table A1. Table of variables.

Magnitude	Symbol	Units
Outer Diameter	D_{out}	m
Shell Thickness	thick	m
Modulus of Elasticity	E	GPa
Poisson Ratio	v	-
Density of Material	P	kg/m^3
Damping Coefficient	C	
Deflection of the Structure	w	m
Added Mass Coefficient	C_A	-
Drag Coefficient	C_D	-
Wave Length	λ	m
Wave Frequency	ω	rad/sec
Wave Amplitude	H	m
Wave Period	T	sec
Water Depth	d	m
Density of Water	ρ	kg/m^3
Wavenumber	k	-

IOP Conf. Series: Materials Science and Engineering **700** (2019) 012023 doi:10.1088/1757-899X/700/1/012023

Machine learning in reservoir permeability prediction and modelling of fluid flow in porous media

A B Zolotukhin[1,2,3*] and A T Gayubov[1]

[1] Gubkin Russian State University of Oil and Gas (National Research University), Leninsky prospect 65, Moscow, 119991, Russia

[2] Northern Arctic Federal University, Severnaya Dvina Emb. 17, Arkhangelsk, 163002, Russia

[3] The University of Stavanger, P.O. Box 8600 Forus N-4036 Stavanger, Norway

* Corresponding author: anatoly.zolotukhin@gmail.com

Abstract. Reliable data on the properties of the porous medium are necessary for the correct description of the process of displacing hydrocarbons from the reservoirs and forecasting reservoir performance. The true permeability of the reservoir is one of the most important parameters which determination is time-consuming, costly and require skilled labor. The paper describes the methodology for determining the permeability of a porous medium, based on machine learning. The results of laboratory experiments, available in the database (terrigenous reservoirs with permeability in the range from 12 to 1132 md), are used to train the neural network, and then to predict the reservoir permeability. Comparison of the predicted and calculated permeability values showed a fairly good match between them with the determination coefficient of 0.92. The last task considered in this paper is to obtain an analytical expression describing a fluid flow in a porous medium using machine learning. This procedure enabled to obtain a resultant equation of fluid flow in a wide range of reservoir parameters and pressure gradients, which can be used in reservoir simulators.

1. Introduction

The determination of the absolute permeability of a rock sample based on standard laboratory studies requires special attention at the stage of processing experimental data. The traditional technique is based on the use of the Forchheimer equation [1], which takes into account the deviation from the Darcy law at high flow velocities caused by inertial forces:

$$|grad\, p| = \frac{\mu}{k}u + \beta\rho u^2 \tag{1}$$

where $|grad\, p|$ – pressure difference, μ – dynamic viscosity coefficient, k – permeability coefficient, u – velocity, β – non-Darcy coefficient, ρ – fluid density. Although equation (1) seems to be a simple relation for determining permeability, estimating the inertial (high velocity) coefficient β is not a simple task [2,3]. To evaluate it, data from experiments conducted at high pressure gradients, i.e. under conditions when inertial forces become comparable with viscous forces and when the deviation from the (linear) Darcy law becomes obvious and can be fixed.

It has been established by numerous experiments (for example, [2,3]) that the coefficient β is not constant and can change its value thousands of times depending on the type of porous medium. It has also been noted in numerous literature [4-8] that β depends on porosity and permeability and that its determination is not an easy task.

At low pressure gradients, the gas exhibits a slip property (superconductivity of the porous medium), and Darcy's law also ceases to correctly describe the fluid flow. This phenomenon was first described and investigated by Klinkenberg [9]. Thus, the flow of fluids in a porous medium is quite complex and exhibits nonlinearity properties over the entire range of pressure gradients.

In [10,11], an attempt was made to describe the law of flow of fluids in porous media using dimension theory, which enabled to write it in the following compact form:

$$\Pi = 1 + C \cdot Re - exp(-b \cdot Re^{\gamma}), \tag{2}$$

$$\Pi = \frac{k\Delta p}{\mu u l}, \quad Re = \frac{u\rho\sqrt{k}}{\mu\phi^{1.5}}. \tag{3}$$

where l – characteristic length, ϕ – porosity. Let us note that parameter Re is a macroscopic Reynolds number in porous media. Although equation (2) captures the main characteristics of flow in the entire range of pressure gradients. Coefficients C, b and γ are to be determined. However, this task is not a simple one.

In this paper, an approach based on Machine Learning (ML) is used in order to identify these coefficients, compose a methodology of determining the permeability of porous media and obtain a universal equation of fluid flow through porous media.

2. Methodology

The solution of the above task based on the use of ML will be carried out in stages, namely:

1. Determination of the permeability of samples of a porous medium based on available experimental data using artificial neural networks.
2. Obtaining the universal law of fluid flow in a porous medium by normalizing dimensionless parameters obtained using dimensional analysis.
3. Using the machine learning method - a multiple regression model to test and refine the universal law of fluid flow in a porous medium.

2.1. Artificial Neural Network Model

The significant benefit of ML predictive models is that one may not have to perform core analysis (CA) measurements for all the core samples, but can conduct prediction of the results [12]. Once a predictive model of any CA results is trained it could be effectively used for future forecasting. There are quite a number of ML algorithms to build a predictive model [13]. In our work, we used the neural network to predict the permeability of porous media.

The choice of algorithm is highly dependent on the problem at hand, data quality and dataset size. For example, it would be unnecessary to build recursive or convolutional neural networks in our problem due to the small dataset size and its structure.

Artificial neural network (ANN) is a mathematical representation of the biological neural network (Figure 1) [14]. It consists of several layers with units that are connected by links. Each connection has its own weight and level of activation. Each node has an input value, an activation function and an output. In ANN information propagates (forward pass) from first (not hidden) layer with inputs to the next hidden layer and then to further hidden layers until the output layer would be achieved. The value in each node of the first hidden layer obtained after calculation of activation function for the dot product of inputs and weights. Next hidden layer receives the output of the previous one and puts its dot product with weights to the activation, and so on. Initially, all weights for all nodes are assigned randomly. ANN calculates first output with random weights. Then compare it to real values, calculate the error and adopts weights to obtain smaller error on the next iteration via backpropagation (ANN training algorithm). After training all weights are tuned, and one may make a prediction for a new data by passing them into an ANN, which will calculate output via forward propagation throw all activations [15]. Scikit-learn Python library provides ANN representation in MLPRegressor method [16]. This implementation allows to indicate the number of hidden layers, number of nodes in each layer, activation function, learning rate and some other parameters.

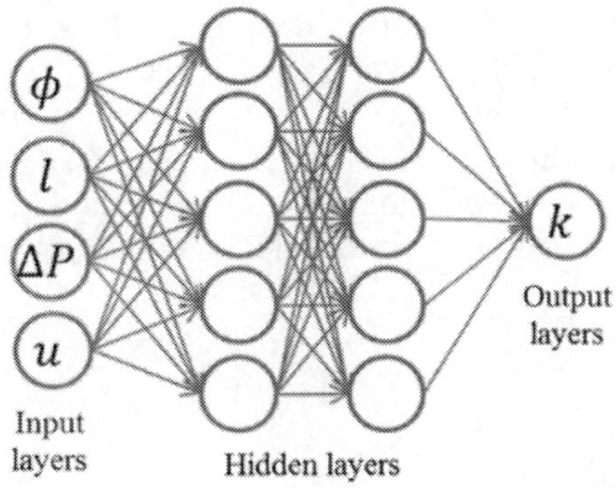

Figure 1. Artificial neural network with two hidden layers.

The objective of this research is to use ANN techniques to build a model to predict reservoir permeability. In the proposed model, the physical characteristics of the rock and fluid sample are used as input parameters, which fully describe the gas flows in the porous medium and are determined prior to the simulation with a high degree of reliability from laboratory studies. As the output (predicted) parameter of the model, the permeability characteristic of the rock sample was used. The choice of this predicted characteristic was based on the fact that when determining the coefficient C in the Forchheimer equation (2), it is necessary to preset the approximate permeability from the obtained physical parameters of the sample and fluid. Sometimes databases contain low-quality data on laboratory measured permeability values. This may be due to the lack of competence of the stuff, inaccuracy of instruments, and other reasons. In this case, the use of a neural network can serve as an additional tool that evaluates the quality of the obtained permeability values. In some cases, inaccurate data maybe replaced with the values obtained by ANN. For permeability predictive models we have used 4 features. All 4 features accounting in machine learning algorithm are presented in Table 1. In our model, we did not consider such input parameters as sample diameter, gas viscosity and density, and ambient temperature, because in all rock samples they had the same value.

Table 1. All features used in predictive model.

No.	Feature	Unit
1	Sample porosity, ϕ	–
2	Length of the sample, l	m
3	Pressure difference, ΔP	Pa
4	Flow rate, u	m/s

The creation of an ANN model was implemented in the programming language Python. Ninety percent of the total data were used for the training of ANN models while the remaining data points were used for testing. Data were randomly mixed for training and testing. The training sample consisted of 203 experimental measurements on four physical parameters of the rock sample (terrigenous reservoirs with permeability in the range from 12 to 1132 md). The network architecture was chosen on the basis of the dependence of the number of hidden layers and the number of neurons in each hidden layer shown in Figure 2. The figure shows various models from 1 to 4 hidden layers, and in each layer consisting from 5 to 20 neurons. For training ANN, we chose a model consisting of 4 hidden layers. The neurons in the hidden layer were 11. Rectified Linear Unit activation function was used as a transfer function between input and hidden layer and linear type activation functions were used between hidden and output layer.

As follows from Figure 2, the best (most reliable) results are coming from the ANN with a limited number of hidden layers (up to four) a number of neurons in each not exceeding high numbers. This is a well-known thesis (not yet formally proven): The number of hidden layers and neurons in each layer should be somehow proportional to the available dataset.

Performance of the ANN model could be demonstrated by plotting predicted values (via cross-validation) of permeability after ablation versus the actual values (Figure 3). The permeability values are set in a normalized form since the values of low-permeability rock samples are not visually visible against the background of other high-permeability samples. The determination coefficient (R^2) between the estimated permeability using the developed ANN model and the actual core permeability was 0.92. Based on these results, we decided to select the ANN model for permeability prediction.

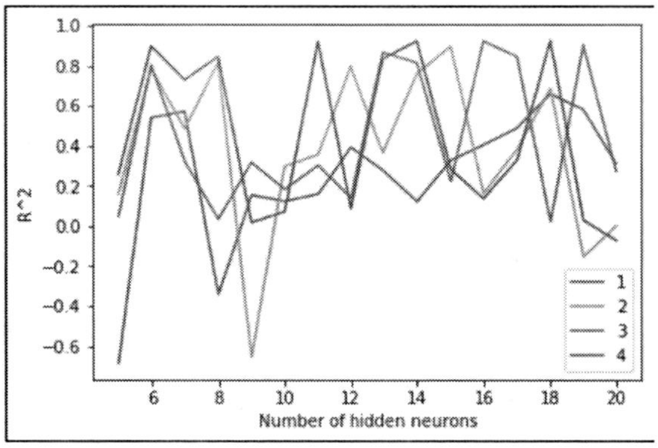

Figure 2. Neural network architecture graph.

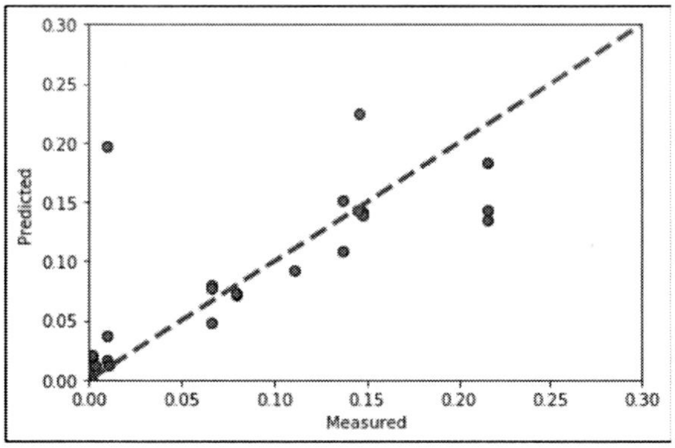

Figure 3. Comparison of real and predicted core permeability by ANN model.

2.2. Method of data normalization

By data normalization in this work, we mean the translation (mapping) of a series of curves into one universal straight line. In accordance with the results of laboratory experiments and previous studies, for further analysis, we select dimensionless criteria described in (2). Using the research results [10,11,17], it is possible to determine two areas in the $\mathbf{\Pi} - \boldsymbol{Re}$ diagram (Figure 4), where the dependence has non-linear (interval 1) and linear (interval 2) parts.

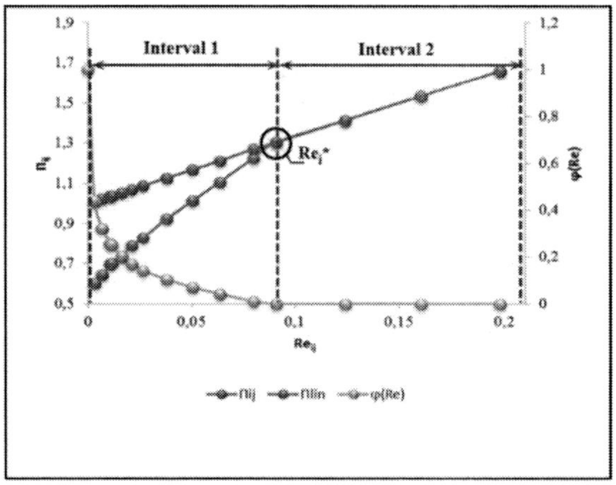

Figure 4. Typical characteristics of the dependency $\Pi = f(Re)$.

2.2.1. Linear area. Let us start with the interval 2 as the initial study area. As follows from earlier works, the fluid flow in the linear interval 2 of the $\Pi - Re$ diagram can be described using the following equation [11,17]:

$$\Pi = 1 + C \cdot Re \quad (4)$$

In accordance with the accepted research algorithm the following dependency graph based on laboratory experiments can be built:

$$\Pi_{ij} = 1 + C_j \cdot Re_{ij} \quad (5)$$

for the linear interval 2 of all laboratory experiments in expiremental database. Here index j indicates experiment number, and index i – measurement number in the j-th experiment. On Figure 5 experimental points of the same color belong to the same experiment.

Analysis of the data obtained from processing each of the experimental straight lines (Figure 6) showed that the value of the coefficient C_j depends on the permeability, porosity and length of a rock sample measured in the j-th experiment and can be approximated by the following dependency:

$$ln\left(\frac{k_j}{\phi_j \cdot l_j^2}\right) = 1.9679\, ln\left(\frac{C_n}{C_j}\right) - 20.109 \quad (6)$$

where $C_n = 1.2085$, coefficient obtained for a rock sample with a permeability of 1132 mD.

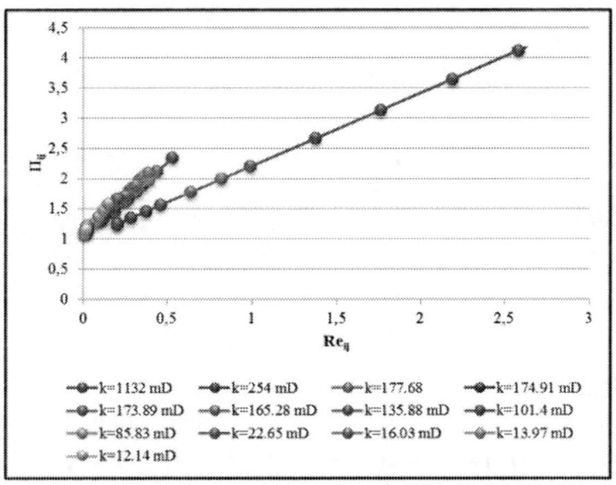

Figure 5. Building the dependency $\Pi = f(Re)$ on experimental data.

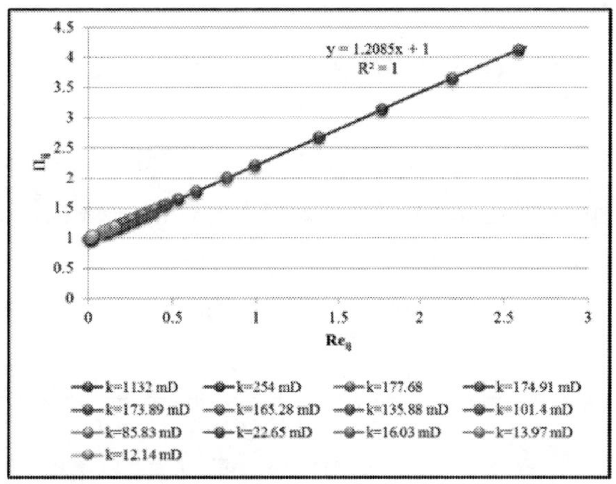

Figure 6. Normalization of experimental data and their translation into a universal dependence.

A dimensionless group $\frac{\phi \cdot l^2}{k}$ can be interpreted as tortuosity τ of porous channels, i.e.

$$\tau_j = \frac{\phi_j \cdot l_j^2}{k_j} \quad (7)$$

This definition of tortuosity is referred to [18], where it is stated that "Tortuosity of the streamline S is given by $\tau(S) = {}^{l}/_{l_S}$ for the length of the porous medium l divided by the length of the streamline l_S.

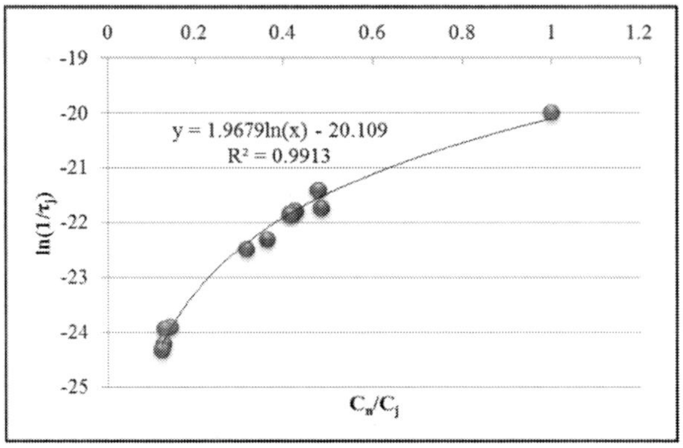

Figure 7. Dependence of $\frac{C_n}{C_j}$ on τ_j.

We believe that more natural and similar to the definition of the term tortuosity will be the ratio of the inverse of the above, i.e. $= {}^{l_s}/_{l}$, which is further converted to (7). With the last result dependency of $\frac{C_n}{C_j}$ on τ_j can be illustrated as shown on Figure 7. Taking into account the last relation, dependence (6) can be rewritten as follows:

$$C_j = B_1 \cdot (\tau_j)^{\alpha} \tag{8}$$

where $B_1 = 4.41 \cdot 10^{-5}$; $\alpha = 0.5$. In view of the last result, the universal dependence takes the following form:

$$\Pi_{ij} = 1 + B_1 \cdot (\tau_j)^{\alpha} \cdot Re_{ij} \tag{9}$$

Using the new equation (9) the new values of the normalized coefficient $C_{j,\mathrm{n}}$ were obtained (Table 2) and the dependence $\Pi_{ij}(Re_{ij})$ was built (Figure 8).

Table 2. The results of calculating the coefficients $C_{j,\mathrm{n}}$ and $b_{j,\mathrm{n}}$ based on the normalization method

No.	k_j, mD	ϕ_j	l_j, m^2	$1/\tau$	$ln(1/\tau)$	Re_j^*	C_j	$C_{j,\mathrm{n}}$	b_j	$b_{j,\mathrm{n}}$
1	1132	0.204	0.051	2.10E-09	-19.983	0.20	1.209	1.133	5.000	3.043
2	254	0.188	0.051	5.11E-10	-21.395	0.18	2.531	2.323	7.177	5.657
3	165.28	0.173	0.051	3.65E-10	-21.730	0.19	2.494	2.754	6.326	6.553
4	177.68	0.192	0.051	3.50E-10	-21.774	0.17	2.832	2.816	7.076	6.680
5	135.88	0.147	0.051	3.49E-10	-21.775	0.18	2.832	2.818	4.970	6.683
6	173.89	0.192	0.052	3.37E-10	-21.811	0.11	2.900	2.870	7.617	6.790
7	174.91	0.202	0.051	3.27E-10	-21.840	0.11	2.914	2.913	6.137	6.878
8	101.4	0.184	0.051	2.09E-10	-22.288	0.09	3.331	3.657	6.620	8.372
9	85.83	0.201	0.049	1.73E-10	-22.475	0.07	3.830	4.023	7.309	9.090
10	22.65	0.202	0.051	4.25E-11	-23.880	0.02	8.444	8.214	14.714	16.843
11	16.03	0.149	0.051	4.07E-11	-23.926	0.01	9.300	8.405	17.956	17.181
12	13.97	0.169	0.051	3.12E-11	-24.189	0.01	9.450	9.610	20.439	19.288
13	12.14	0.162	0.052	2.76E-11	-24.315	0.01	9.854	10.242	20.819	20.379

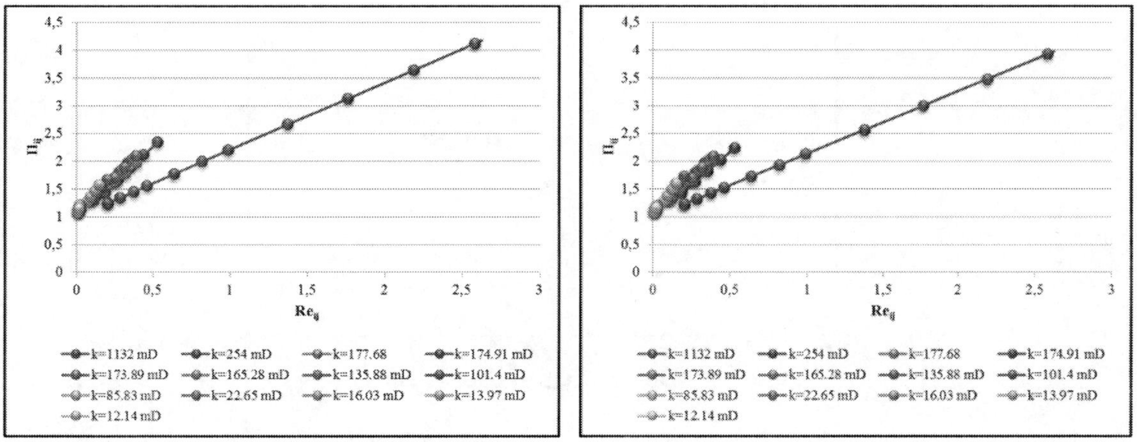

a) Dependence $\Pi_{ij}(Re_{ij})$ on the experimental data

b) Dependence $\Pi_{ij}(Re_{ij})$, calculated according to equation (9)

Figure 8. Graphs $\Pi_{ij}(Re_{ij})$, obtained by using the data normalization method.

2.2.2. Nonlinear area. Now let us consider interval 1 (see Figure 4), in which the effect of nonlinearity of the gas flow is observed. This effect is due to a so-called gas slippage effect, first described by Klinkenberg [9]. To take into account the slippage effect, we can use the following approximation, in which the function $\varphi(Re)$ describes interval 1:

$$\Pi_{ij} = 1 + C_j \cdot Re_{ij} - \varphi(Re_{ij}) \tag{10}$$

where $\varphi(Re)$ can be approximated by the following function used in equation (2) (Figure 9) [11]:

$$\varphi(Re_{ij}) = e^{-b_j \cdot Re_{ij}^{\gamma}} \tag{11}$$

In accordance with the above algorithm of the data normalization, we build dependency graphs. The generalized straight line takes the following form (Figure 10):

$$\ln(\Pi_{ij} - 1 - C_j Re) = b_j \cdot Re_{ij}^{\gamma} \tag{12}$$

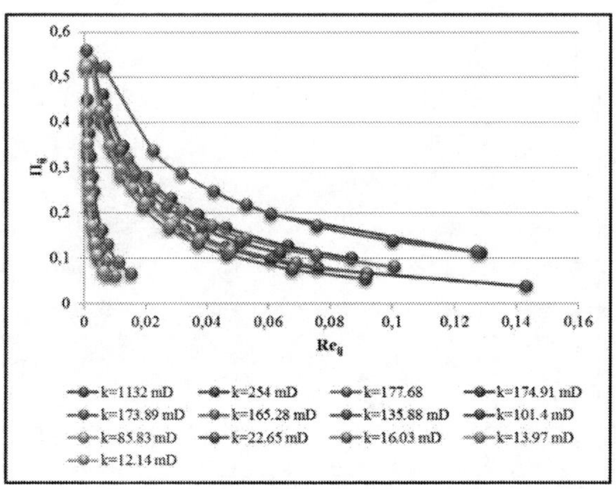

Figure 9. Dependence $\Pi_{ij}(Re_{ij})$, calculated according to equation (11).

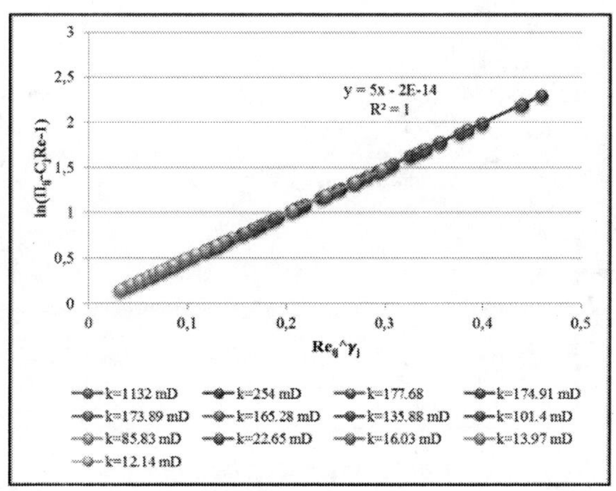

Figure 10. Normalization of experimental data and their translation into a universal dependence $\Pi_{ij}(Re_{ij}{}^{\gamma})$.

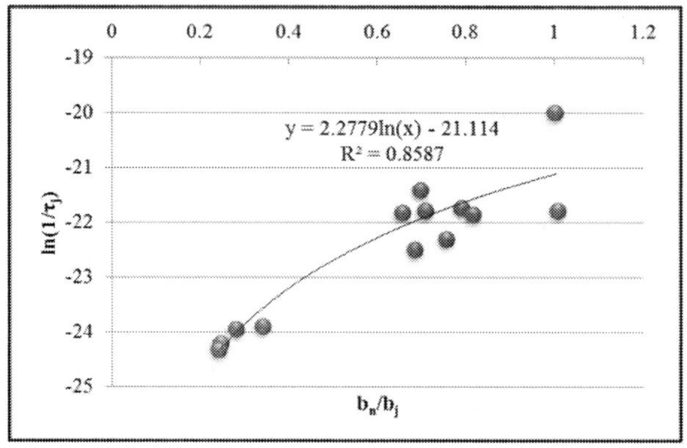

Figure 11. Dependence of coefficient $\frac{b_n}{b_{ij}}$ on τ_j.

Let us now determine the value of the universal constant (or functional) b_j in equation (11). Analysis of the available experimental data showed that the coefficient b_j depends on permeability and porosity and the length of the rock sample, as well as the coefficient C_{ij}, and can be approximated by the following dependency (Figure 11):

$$\ln\left(\frac{1}{\tau}\right) = 2.2779 \ln\left(\frac{b_n}{b_j}\right) - 21.114 \quad (13)$$

where $b_n = 5$, coefficient obtained for a rock sample with a permeability of 1132 mD.
Equation 13 can be rewritten as:

$$b_j = 4.72 \cdot 10^{-4}(\tau_j)^{0.44} = B_2 \cdot \tau_j^{\gamma_1} \quad (14)$$

where $B_2 = 4.72 \cdot 10^{-4}$, $\gamma_1 = 0.44$. With the last result (11) can be written in the following form:

$$\varphi(Re_{ij}) = e^{-B_2 \cdot \tau_j^{\gamma_1} \cdot Re_{ij}^{\gamma_2}} \quad (15)$$

Using the new equation (15) we obtained the values of the normalized coefficient $b_{j,n}$ (Table 2) and constructed the dependence $\varphi(Re_{ij})$ (Figure 12).

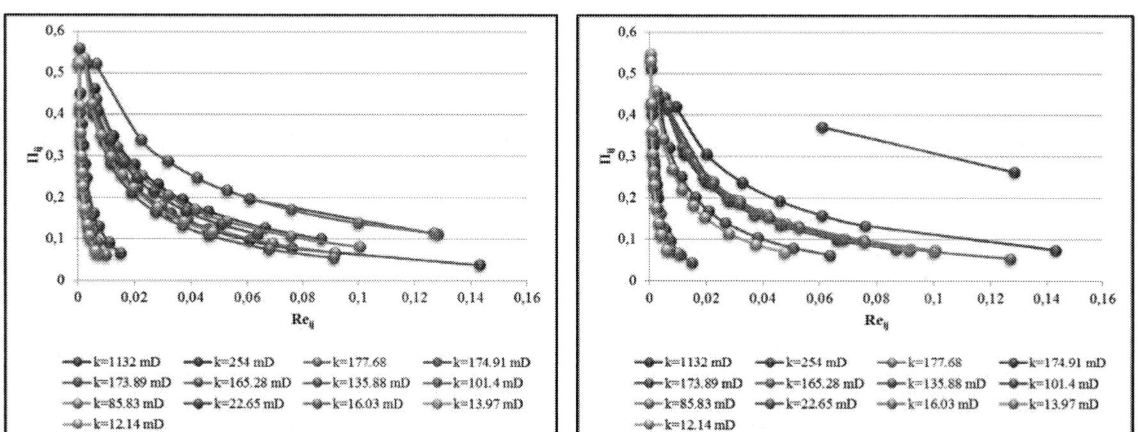

a) Dependence $\varphi_{ij}(Re_{ij})$ based on the experimental method

b) Dependence $\varphi_{ij}(Re_{ij})$, calculated according to the equation (15)

Figure 12. Graphs $\varphi_{ij}(Re_{ij})$, obtained using the data normalization method.

2.2.3. Combining linear and nonlinear areas. Based on the newly obtained universal coefficients C, b and γ, we can rewrite equation (10) as follows:

$$\Pi = 1 + B_1 \tau^{\alpha} \cdot Re_{ij} - e^{-B_2 \cdot \tau_j^{\gamma_1} \cdot Re^{\gamma_2}} \quad (16)$$

where $B_1 = 4.41 \cdot 10^{-5}$, $B_2 = 4.72 \cdot 10^{-4}$, $\alpha = 0.5$, $\gamma_1 = 0.44$, $\gamma_2 = 0.4$

2.2.4 The border Reynolds number.
Let us note that the border Reynolds number Re_j^* defined as:

$$\varphi(Re_j^*) \leq \varepsilon , \quad (17)$$

where ε is a small number, is the number dividing intervals 1 and 2 on the dependency $\Pi = f(Re)$ (Figure 1). Dependency of Re_j^* on tortuosity τ is shown on Figure 13.

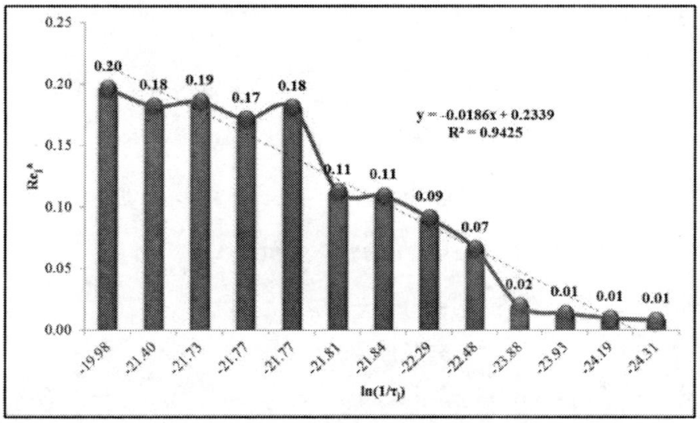

Figure 13. Dependency of the border Reynolds number Re_j^* on tortuosity τ.

2.3. Linear regression model
As was shown above, the dependence of Π_{ij} values on the parameters Re and τ (tortuosity) is generally nonlinear, approaching the linear one only asymptotically (for large Re values). To simplify the subsequent analysis, we divide the entire area of variation of the parameter Re into two, namely, the interval 1 of small Re values (non-linear part) and the interval 2 of large Re values (linear part). We first consider the linear part of the equation, leaving the border number Re_j^* separating both zones for further analysis in more detail.

2.3.1. Linear part of the equation. For the linear part, as was shown above, the following equation holds:

$$\Pi_{ij} = 1 + C_j \cdot Re_{ij}$$

or, in terms of the tortuosity parameter τ introduced above:

$$\Pi_{ij} = 1 + B_1 \cdot \tau_j^{\alpha} \cdot Re_{ij} \quad (18)$$

where $\tau_j = \frac{\phi_j l_j^2}{k_j}$. We rewrite equation (18) in a more convenient form for subsequent analysis:

$$\ln(\Pi_{ij} - 1) = \alpha \cdot \ln(\tau_j) + \ln(Re_{ij}) + \ln(B_1) \quad (19)$$

or in terms commonly used in regression analysis:

$$y = w_0 \cdot x_0 + w_1 \cdot x_1 + b_0 \quad (20)$$

where the following notation is used: $w_0 = \alpha$, $w_1 = \lambda$, $b_0 = \ln(B_1)$. The input data for the regression model: $x_0 = \ln(\tau_j)$, $x_1 = \ln(Re_{ij})$, the output is: $y = \ln(\Pi_{ij} - 1)$. The regression model produces the following parameters:

$$\ln(\Pi_{ij} - 1) = 0.471 \cdot \ln(\tau_j) + 0.977 \cdot \ln(Re_{ij}) - 9.244,$$

or

$$\Pi_{ij} = 1 + B_1(\tau_j)^{\alpha} Re_{ij}{}^{\lambda}, \tag{21}$$

where $B_1 = 9.67 \cdot 10^{-5}$, $\alpha = 0.47$, $\lambda = 0.98$.

2.3.2. Linear and nonlinear parts of the equation. The generalized equation for a linear and non-linear zone has the following form:

$$\Pi_{ij} = 1 + B_1 \cdot \tau_j{}^{\alpha} \cdot Re_{ij}{}^{\lambda} - e^{-B_2 \cdot \tau_j^{\gamma_1} \cdot Re_{ij}{}^{\gamma_2}}, \tag{22}$$

which can be rewritten as:

$$e^{-B_2 \cdot \tau_j^{\gamma_1} \cdot Re_{ij}{}^{\gamma_2}} = 1 + B_1 \cdot \tau_j{}^{\alpha} \cdot Re_{ij}{}^{\lambda} - \Pi_{ij}$$

$$B_2 \cdot \tau_j^{\gamma_1} \cdot Re_{ij}{}^{\gamma_2} = -ln(1 + B_1 \cdot \tau_j{}^{\alpha} \cdot Re_{ij}{}^{\lambda} - \Pi_{ij}).$$

In the final form, convenient for using the regression model, equation (22) will take the form:

$$\ln(B_2) + \gamma_1 ln(\tau_j) + \gamma_2 ln(Re_{ij}) = ln[-ln(1 + B_1 \cdot \tau_j{}^{\alpha} \cdot Re_{ij}{}^{\lambda} - \Pi_{ij})], \tag{23}$$

or in a form convenient for regression analysis:

$$y = w_0 \cdot x_0 + w_1 \cdot x_1 + b_0$$

The input data for the regression model: $x_0 = ln(\tau_j)$, $x_1 = \ln(Re_{ij})$, the output data is: $y = ln[-ln(1 - \Pi_{ij} + B_1 \cdot \tau_j{}^{\alpha} \cdot Re_{ij}{}^{\lambda})]$. The regression model produces the following parameters: $w_0 = \gamma_1$, $w_1 = \gamma_2$, $b_0 = \ln(B_2)$.

$$ln[-ln(1 - \Pi_{ij} + B_1 \cdot \tau_j{}^{\alpha} \cdot Re_{ij}{}^{\lambda})] = \gamma_1 \cdot \ln(\tau_j) + \gamma_2 \cdot \ln(Re_{ij}) + \ln(B_2)$$

where $\gamma_1 = 0.34$, $\gamma_2 = 0.36$, $B_2 = 3.82 \cdot 10^{-3}$. Finally, the general formula generated by the method of the regression model is:

$$\Pi_{ij} = 1 + B_1 \cdot \tau_j{}^{\alpha} \cdot Re_{ij}{}^{\lambda} - e^{-B_2 \cdot \tau_j^{\gamma_1} \cdot Re_{ij}{}^{\gamma_2}}$$

where $B_1 = 9.67 \cdot 10^{-5}$, $B_2 = 3.82 \cdot 10^{-3}$, $\alpha = 0.47$, $\lambda = 0.98$, $\gamma_1 = 0.34$, $\gamma_2 = 0.36$

3. Analysis of the results

This section provides a brief analysis of the results of two methods employed for obtaining the universal equation of fluid flow in a porous medium, namely: 1) Normalization method and 2) Regression model. Table 3 lists the formulas obtained using these approaches. As the determination coefficients show, the normalization method showed a slightly better result than the regression model of machine learning. The obtained high values of the determination coefficient confirm the correct choice of the initial parameters and their presentation in a dimensionless form. Note that during the data processing using the normalization method it was found that some data fall far beyond the general trend. The response to the request to the laboratory where the tests were carried out confirmed that the initial porosity of the two samples instead of the correct 0.149 and 0.127 was recorded as 0.008 and 0.010, the correction of which put them back into the general trend.

IOP Conf. Series: Materials Science and Engineering **700** (2019) 012023 doi:10.1088/1757-899X/700/1/012023

When obtaining the universal gas flow equation using the two above methods, the corrected values were used, which ensured high determination coefficients between the experimental and calculated data. This result is illustrated by Figure 14, where experimental (blue points) and forecasted data (red points for normalization method and green points for regression method, respectively) are perfectly matched it a wide range of reservoir permeability and Reynolds number values. Due to the lack of space, we managed to place in Figure 14 only a part of the results.

Table 3. Results of using the normalization method and regression model

Method for determining the coefficients.	Formula for calculating	R^2
Normalization method	$\Pi_{ij} = 1 + B_1 \cdot \tau_j{}^{\alpha} \cdot Re_{ij} - e^{-B_2 \cdot \tau_j^{\gamma_1} \cdot Re^{\gamma_2}}$ $B_1 = 4.41 \cdot 10^{-5}$, $B_2 = 4.72 \cdot 10^{-4}$, $\alpha = 0.5$, $\gamma_1 = 0.44$, $\gamma_2 = 0.4$	0.98
Regression model	$\Pi_{ij} = 1 + B_1 \cdot \tau_j{}^{\alpha} \cdot Re_{ij}{}^{\lambda} - e^{-B_2 \cdot \tau_j^{\gamma_1} \cdot Re_{ij}{}^{\gamma_2}}$ $B_1 = 9.67 \cdot 10^{-5}$, $B_2 = 3.82 \cdot 10^{-3}$, $\alpha = 0.47, \lambda = 0.98, \gamma_1 = 0.34, \gamma_2 = 0.36$	0.97

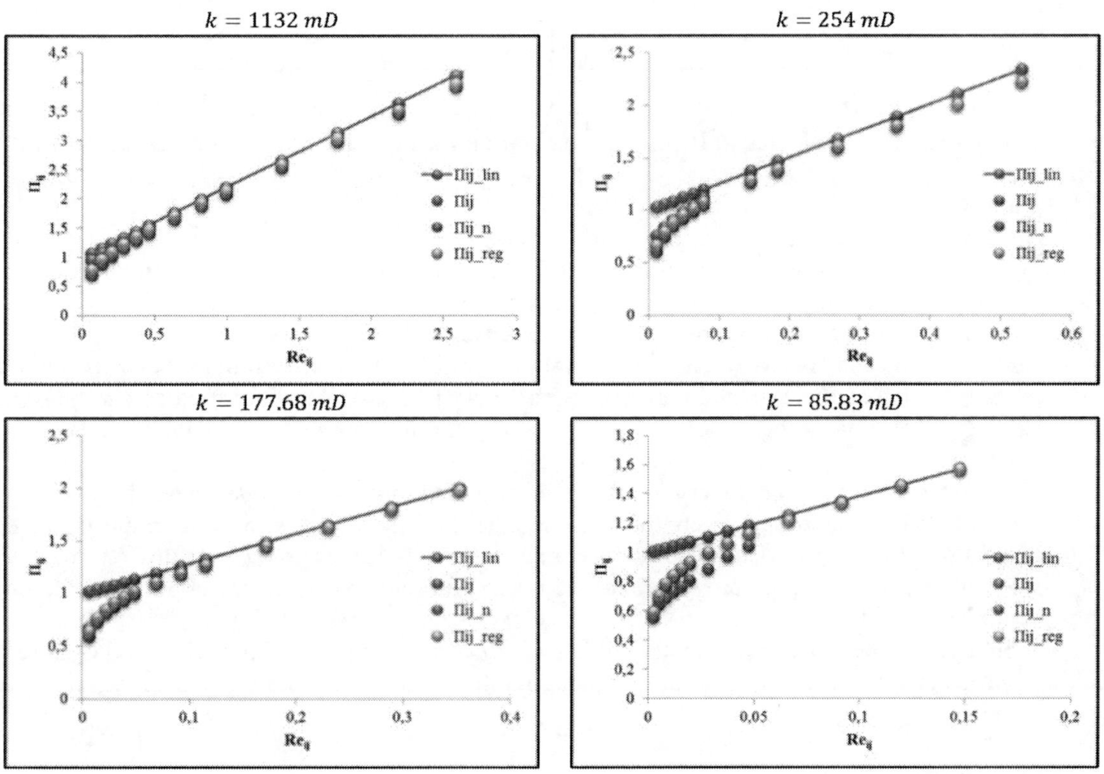

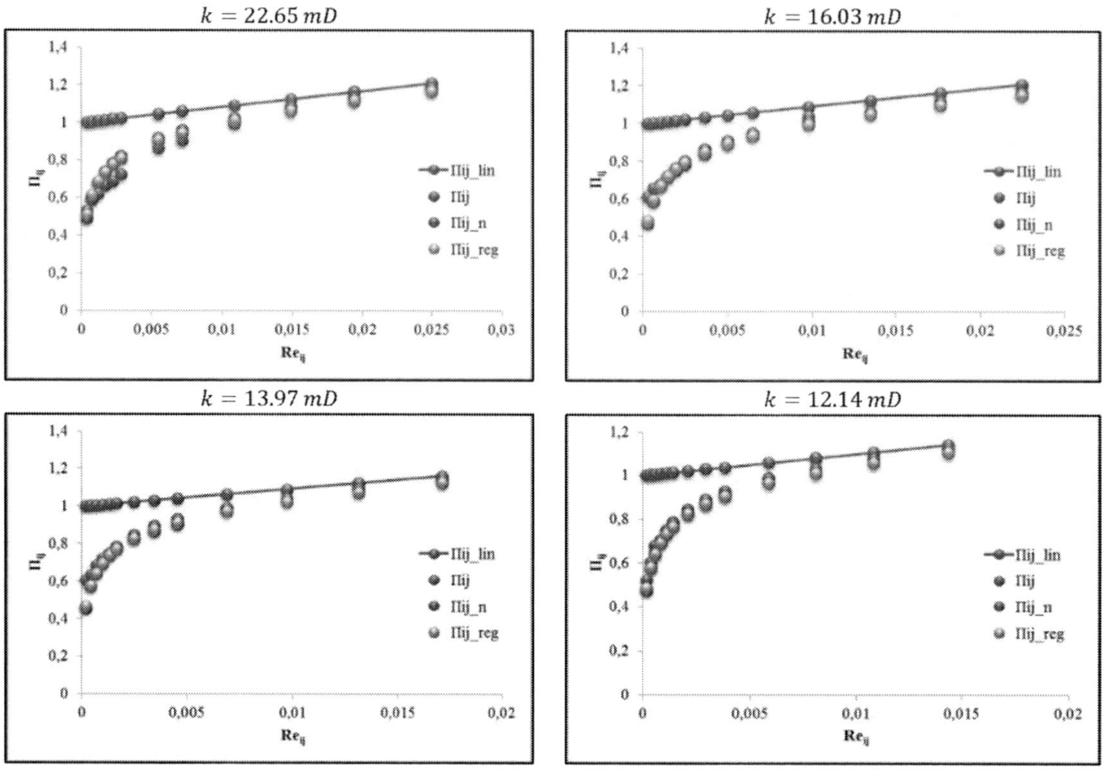

Figure 14. Dependence plots $\Pi - Re$ obtained using: 1. Π_{ij_lin} – the linear part of the equation; 2. Π_{ij} – experimental data; 3. Π_{ij_n} – normalization method; 4. Π_{ij_reg} – regression model.

4. Conclusions

The following general conclusions can be derived from the previous analysis:

- The use of the dimensional analysis method can significantly reduce the number of variables involved in the analysis, improve the quality of the resulting solutions and at the same time simplify the obtained equations and increase the reliability of the analysis.
- In particular, this approach made it possible to evaluate the permeability of porous medium samples using artificial neural networks with a very high correlation of predicted values with the available data in the sample, even with a very limited number of available experimental data.
- The use of dimensional theory methods in combination with regression analysis made it possible to define more precisely the constants included in the universal semi-empirical law of fluid flow. In its turn, it enables high accuracy and reliability of predictive flow regimes over the entire range of pressure gradients at reservoirs and experimental conditions for core samples.
- The use of the universal equation of fluid flow in porous media makes it possible to significantly simplify and refine computer modeling of the oil and gas field development forecast.

Acknowledgments
We are grateful to Dr. Abbakumov V. for valuable advice given in the preparation of this work.

IOP Conf. Series: Materials Science and Engineering **700** (2019) 012023 doi:10.1088/1757-899X/700/1/012023

References

[1] Forchheimer P 1901 *Wasserbewegung durch Boden* Zeit ver Deutsch Ing.
[2] Tessem R 1980 *High Velocity Coefficient's Dependence on Rock Properties: A Laboratory Study*, Norway: Thesis Pet. Inst. NTH. pp 26-30.
[3] Torsæter O, Tessem R. and Berge B 1981 *High Velocity Coefficient's Dependence on Rock Properties*" SINTEF Report (Norway: NTH Trondheim).
[4] Cornell D and Katz D L 1953 Flow of gases through consolidated porous media, *Ind. Eng. Chem.* **45**(10), 2145–52.
[5] Geertsma J 1974 Estimating the coefficient of inertial resistance in fluid flow through porous media *SPE J.* **14**(05), 445–50.
[6] Janicek J D and Katz D L 1955 *Applications of unsteady state gas flow calculations,* paper presented at Research Conference on Flow of Natural Gas Reservoirs, Proc. U. of Mich. Res. Conf. (June 30, 1955).
[7] Jones S C 1987 *Using the inertial coefficient, β, to characterize heterogeneity in reservoir rock.* Paper presented at 62nd SPE Annual Tech. Conf. and Exhibit. (Texas: Dallas).
[8] Liu C Y, Civan F and Evans R D 1995 Correlation of the non-Darcy flow coefficient, *J. Can. Petrol. Technol.*, **34**(10), pp 50–53.
[9] Klinkenberg L J 1941 *Permeability of Porous Media to Liquids and Gases*, API Drill. and Prod. Practice.
[10] Zolotukhin A B 2002 *A new gas flow equation and permeability determination technique* Reg. SPE offshore Bergen Conf. on Drilling, Completion and Reservoir Management (Norway: Bergen).
[11] Zolotukhin A B 2006 *Equation of gas flow in porous media and ghe technoque of determining the permeability of a porous medium* Report presented at an Int. Conf. dedicated to the 50th anniversary of TATNIPIneft (Russia: Bugulma).
[12] Unsal E, Dane J and Dozier G V 2005 A genetic algorithm for predicting pore geometry based on air permeability measurements, *Vadose Zone J.* **4**(2), 389–97.
[13] Hastie T, Tibshirani R, Friedman J 2001 *The Elements of Statistical Learning: Data Mining, Inference, and Prediction* (New York: Springer–Verlag).
[14] Haykin S 1994 *Neural networks: a comprehensive foundation* (New York: Prentice Hall).
[15] Rumelhart D E, Hinton G E and Williams R J 1986 Learning representations by back-propagating errors, *Nature* **323**(6088), 533–36.
[16] Pedregosa F et al. 2012 Scikit-learn: Machine learning in python, *J. Mach. Learn. Res.* **12,** 2825–30.
[17] Zolotukhin A B, Shulev V E 2018 *Fluid Flow in Porous Media, Dimensional Analysis and Permeability Determination*. Proc. of Int. scientific-technical conf. "Development of exploration technique and exploitation of onshore and offshore hydrocarbon fields" ("Geopetrol-2018") (Poland: Zakopane-Kostelisco) pp 573–80.
[18] Berg C F 2014 Permeability Description by Characteristic Length, Tortuosity, Constriction and Porosity, Transp. Porous Media **103** pp 381–400.

IOP Conf. Series: Materials Science and Engineering **700** (2019) 012024 doi:10.1088/1757-899X/700/1/012024

Strength analysis of 3D printed carbon fibre reinforced thermoplastic using experimental and numerical methods

F Ghebretinsae*, O Mikkelsen and A D Akessa

Department of Mechanical and Structural Engineering and Materials Science, University of Stavanger, Stavanger, Norway

* Corresponding author: f.ghebretinsae@stud.uis.no

Abstract. A study of strength of composite materials produced by 3D printing technology is presented. The samples fabrication and tests to determine the strength both in bending and in tension of the composite materials have been carried out. The composite samples were additively manufactured using Markforged® 3D printer of type Mark-Two. The fabricated composite samples were of carbon fiber filament combined with a thermoset plastic matrix, by its producer named "Onyx". The tests provided sample mean value for the ultimate tensile strength of 560 MPa and the tensile modulus of 25 GPa. Based on the three point bending tests the ultimate flexural strength of 271 MPa and flexural modulus of 16 GPa were estimated. The tests are reported and discussed in view of stress analysis modeling the layered composite with finite element models.

1. Introduction

Additive Manufacturing (AM) is defined as manufacturing of a component by laying a material layer upon a layer [1]. It is one of the currently developing method of manufacturing process [2]. Fast fabrication of prototypes using AM reduces product development time significantly. Prototypes make communication easier by "touch and feel" type of communication. Specially 3D printing has a great advantage due to 3D models can be fabricated without geometrical limitations and no additional machines/tools are required [2]. However, products fabricated with 3D printing have size limitations due to the capability of the 3D printer. 3D printed products have anisotropic mechanical properties and are more porous than traditionally manufactured products [3, 4]. In this study the terms such as '3D printer' referrers specifically to the 3D AM machine and '3D printing' to the AM processes by 3D printers.

Laying continuous reinforcing fiber within a matrix has been commonly practiced in traditional manufacturing of parts from composite materials in aerospace and automobile industries [5, 6]. However, the use of composite materials in AM occurs to be relatively new and it significantly increases the structural applicability of parts fabricated by AM [2]. The introduction of composite materials to AM takes the technology from prototyping stage further to the fabrication of strong functional parts. However, there is limited understanding of mechanical property of parts produced by AM and their mechanical property depends on; the building direction [7], layer-thickness [3, 8], bonding strength between layers [8], formation of voids [9], type of materials [8], etc. A unique advantage of AM when used for composite materials is that the orientation and alignment of continuous fiber can be located accurately in complex geometries, which is extremely challenging in traditional molding fabrication.

Composite materials have high strength to weight ratio, high corrosion and wear resistance, good fatigue life and thermal conductivity [10]. These properties make composite materials attractive in engineering applications. Composites have anisotropic material properties ($E_{11}≠E_{22}≠E_{33}$). However, by applying fiber in different orientation at each layer a "quasi-isotropic" property can be achieved [5, 11]. Furthermore, the fiber provides maximum strength when load is applied in the direction of the fiber, whereas the loads applied perpendicular to the fiber mostly depend on the weak matrix phase [12]. The term "fiber" is used in synonym to the strengthening material in this study. The strength and stiffness of the composite materials is mainly dependent on the type and volume fraction of the reinforcing phase, angle of orientation and fabrication method [3, 9, 10, 13]. Basically, higher fiber volume fraction provides higher strength and stiffness. However, to obtain adequate matrix support to the fiber, the volume percentage of the reinforcement should be less than 70% of the total volume of a component [14]. Elastic modulus in the longitudinal direction for a continuous and unidirectional (UD) fiber reinforced composite can be expressed by the 'Rule of Mixtures' (ROM). The rule shows that "the stiffness of the composite material is a weight-mean of the modulus of the two phases and simply depends on the volume fraction of fibers" [1].

$$E_c = E_m V_m + E_f V_f \quad (1)$$

Where E stands for the E-modulus, V stands for the volume fraction. Whereas the subscripts c stands for the composite part, m for the matrix and f for the fiber (strengthening) phase. A similar equation to Equation (1) can be used to predict the stress, density, Poisson's ratio and shear strength.

Lozada, J.N. et al. studied effect of density and type of infill patterns on the stiffness and strength of Nylon and Onyx and found that both properties increased with increasing density. However, among all types of patterns, the triangular pattern provide best strength per weight result [13]. Moreover, Ning et al, studied the effect of printing parameters such as infill-speed, nozzle temperature and layer thickness and found that thicker layers led to largest average strength, while high nozzle temperature increased porosity [3]. Generally, both the AM parameters and fiber orientation influence the mechanical property of finished composite part [15].

For numerical simulation, composite material can be modelled using layer elements. After creating a model using layered elements, structural analysis including large deflection and stress can be performed [16]. It is important to take special attention when modeling composite materials due to finite element analysis (FEA) of composites require several orthotropic material properties [16]. At least two materials and nine orthotropic properties of each material are required. The nine parameters are elastic moduli (E_1, E_2, E_3), shear moduli (G_{12}, G_{23}, G_{31}), and Poisson's ratios (v_{12}, v_{23}, v_{31}) [16]. Subscripts 1, 2, and 3 represents x-, y- and z- directions, respectively.

2. Methodology

To characterize the mechanical properties of 3D printed composite materials, tensile and flexural experimental tests were performed following the ASTM D3039 and ASTM D7264-07 standards. Moreover, a finite element model was developed.

2.1. Markforged® Mark-Two 3D Printer

Mark-Two is the second generation desktop 3D-Printer for composite materials, introduced in 2016 by Markforged® [17]. Mark-two is a compact 3D printer which benefits from both the Fused Filament Fabrication (FFF) and Continuous Filament Fabrication (CFF) technology. Test samples were fabricated from carbon fiber (CF) filament and a thermoset plastic matrix, named "Onyx" [18], where both materials were delivered by Markforged®. In this report, "Onyx" is used as synonym to the matrix material. The 3D printer can also use Nylon as a matrix and fibers such as fiberglass (FG), Kevlar and high-strength high-temperature fiberglass (HSHT) [19].

IOP Conf. Series: Materials Science and Engineering **700** (2019) 012024 doi:10.1088/1757-899X/700/1/012024

Table 1. Slicing software (Eiger®) setting used when 3D printing test samples using Mark-Two.

Description	Tensile Onyx/CF	Flexural Onyx/CF
Nozzle temperature	272 °C	272 °C
Heat bed temperature	No heat	No heat
Fill pattern	Solid	Solid
Fill density	100%	100%
Sample dimension [mm]	250x15x1.75	154x13x4
Floor layers	2	4
Roof layers	2	4
Layer thickness [mm]	0.125	0.125
Wall layers	1	2
Total fiber layers	10	20
Total matrix layers	4	12
Fiber fill type	Isotropic	Isotropic
Concentric fiber rings	1	1
Fiber angle (degrees)	0	0
Print time per pcs	2h 05min	1h 44min

The maximum size that can be built by mark Two is 320 mm x 132 mm x 154 mm (x-y-z respectively). Mark-Two has two nozzles where one nozzle is used for the matrix material and the second nozzle is used for the reinforcing fiber. The filament material is heated to its glass transition inside the nozzle and extrudes while the printer head move in the XY-plane. Only one nozzle at a time extrudes material and the continuous fiber is terminated at every layer. After one layer is deposited the working bed moves one-layer thickness along the z-direction as specified in the slicing software, which is 0.125 mm if CF is used. This thickness is pre-defined when printing parts with reinforcing fiber. Unlike other AM machines, the working bed of Mark-two do not need heating and parts can be removed right after fabrication is finished. The 3D printer settings used in this study are summarized in Table 1.

2.2. Tensile Test and sample preparation

Hart et al. [7] performed tensile test on dog-bone shaped CF reinforced composite fabricated by 3D printing and failure occurred at the grip section in all the samples. It was later recommended rectangular specimen with bonded tabs of standard type ASTM D3039. The tensile test in this study was performed following this recommendation and ASTM D3039 standard for tensile test of composite materials was used.

a) Tensile setup using Instron 5895.

b) 3-point flexural setup using Zwick/Z020.

Figure 1. Tensile and flexural setups using corresponding universal testing machine.

IOP Conf. Series: Materials Science and Engineering **700** (2019) 012024 doi:10.1088/1757-899X/700/1/012024

Five rectangular CF reinforced tensile specimen with 250 mm in length,15 mm in width and 1.75 mm in depth was loaded in tension by Instron-5895 testing machine at a constant strain rate of 0.008 min^{-1}. The standard states that the strain rate should be selected so as failure to occur within 1 to 10 min and 0.01 min^{-1} is suggested [20]. However, during the trial test, the suggested strain rate led to fast and sudden failure. To overcome this, a reduced strain rate of 0.008 min^{-1} was used. The sample was clamped to a fixed part of the testing machine at the bottom and to the movable head at the other end as shown on Figure 1(a). Load was applied parallel to the UD CF and the samples were loaded until complete failure. The extension of the specimen were measured by an extensometer with gauge length of 100 mm. The test was carried at room temperature and normal humidity. It was important to consider the thickness of a single layer when selecting the thickness of the test sample because it is impossible to fabricate layers less than the minimum thickness accepted by the 3D printer [20]. Detailed information of each layer and applied material for the tensile sample is provided on Table 2. Moreover, the tensile sample had total 14 layers at the gauge section, where ten of them was made from CF and the rest was Onyx. The continuous CF was UD, oriented at 0° (along the x-axis) inside Onyx. To avoid damage and residual stress from the griper, the specimens were provided with 15 mm wide, 56 mm long and 1.5 mm thick tabs, beveled at 5° towards the gauge section as shown on Figure 2 (a). The tabs were simply printed together with the rectangular samples only with the matrix material. About 62% fiber volume fraction was estimated at the gauge section. Moreover, a solid infill with 100% density was used to reduce formation of voids and achieve enough fiber volume fraction. Each sample took approximately 2 hours printing time to complete.

2.3. Flexural Test and sample preparation

The three point flexural test was performed by Zwick/Z020 testing machine following the ASTM standards D- 7264M – 07, procedure-A. Five specimens with 154 mm length, 13 mm width and 4 mm thick (Figure 3) was fabricated by Mark-two 3D printer from UD continuous CF and Onyx matrix. A span to thickness ratio of 32:1 was used when selecting dimensions from the standard. The testing sample was simply supported at two ends, 124 mm apart and a concentrated load, uniformly distributed along the width of the beam was applied at the mid-span (Figure 3 (b)) [21]. Samples had been tested until failure at a head speed of 1 mm/min.

The distribution of the CF layers in the flexural test samples were decided by considering the loading conditions of a classical beam from the general beam theory, wherein the neutral axis of the beam is theoretically unstressed. Therefore, no reinforcing fiber was provided at the center of the beam (Figure 3(b)). Detailed information about each layer and applied materials type are provided on Table 4. For a single specimen of 4 mm thick, a total of 32 layers each 0.125 mm thick were required. Furthermore, the flexural samples had an approximately 42% fiber volume fraction and the CF were symmetrically located with reference to neutral axis of the beam (Figure 3 (b)).

$$\sigma = \frac{3PL}{2bh^2} \tag{2}$$

Where σ is stress at the mid-span, P, L, b and h represent the applied force, support span, initial beam width and initial thickness, respectively [21].

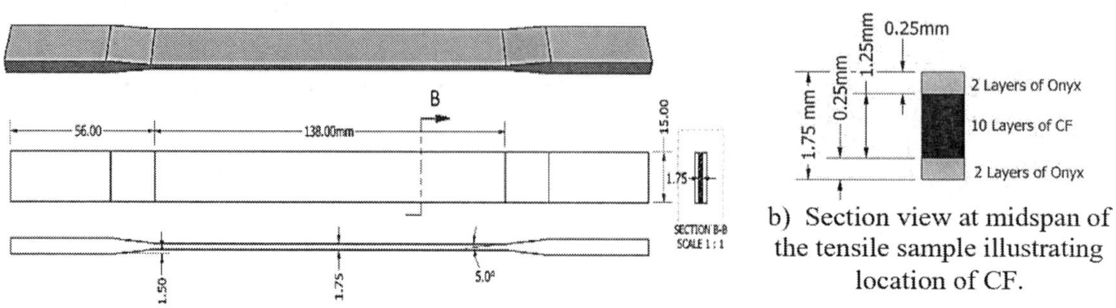

a) Dimensions of tensile specimen with tabs

b) Section view at midspan of the tensile sample illustrating location of CF.

Figure 2. Tabbed tensile test specimen.

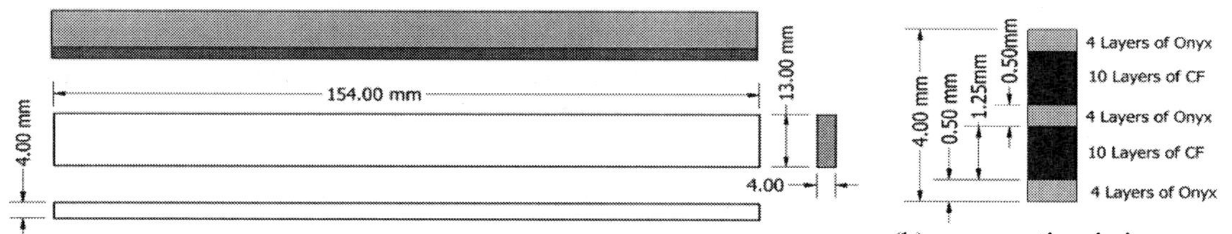

(a) Flexural specimen

(b) cross-sectional view

Figure 3. Flexural specimen and detail section view of layers and assigned material.

2.4. Fiber volume fraction estimation

The total cross-sectional area was calculated by summing up the individual cross-sectional area covered by each material at the gauge section. The cross-sectional area was taken from a location within the gauge section. To find the fiber volume fraction the "rule of mixtures" were used. ROM uses the material properties of the individual composite materials and their volume ratio to predict the property of the composite part. Considering the matrix and fiber layer cover same length and width of the part, the volume fraction can be represented by area fraction equation (3).

$$V_f = \frac{A_f}{A}, \quad V_m = \frac{A_m}{A} \tag{3}$$

$$E_c = E_m \frac{A_m}{A} + E_f \frac{A_f}{A} \tag{4}$$

Where A represents the total cross sectional area, while A_f and A_m stands for cross sectional area occupied by the fiber and matrix, respectively.

Considering Figure 3 (b) & 2 (b), and the approximated width of a single printed carbon filament, the area covered by CF was estimated .The area calculated was assumed to be at the middle of gauge length and the average dimension values of the printed sample were used on both tests. Moreover, the thickness of the layers was obtained by dividing the final thickness by the total number of layers in the sample. In similarity to the FE models, the presence of voids was ignored when estimating the fiber percent. From the 3D model and final printed sample dimensions the tensile and the flexural samples had a cross-sectional area increase of 7% and 1%, respectively. However, porosity and distribution of voids within the CF and Onyx was not studied.

IOP Conf. Series: Materials Science and Engineering **700** (2019) 012024 doi:10.1088/1757-899X/700/1/012024

2.5. Modeling tensile and flexural samples in Ansys

ANSYS Mechanical APDL 17.0 were used as a tool to model and analyze the specimen models of composite materials. In this study element type SHELL181 4-node structural shell was used and the samples were considered as thin walled. Shell181 was selected due to shell elements allow to define layered composite of thin-walled structures [22]. The CF and Onyx composite specimens were modelled as a thin layered lamina. Since the layers fabricated in the 3D printer were made of either the reinforcing or the matrix material, the model was also designed to contain only one type of material per lamina. This means one layer was made of either only CF or Onyx. Material properties presented on Table 3 were used for the simulation of both models. During developing FE model there have been made several assumptions. The assumptions made were; Constituents show linear elastic behavior, the matrix (Onyx) has isotropic material properties, fibers are transversely isotropic, fiber-matrix has perfect bonding and no voids or defects present in the test samples. Moreover, a Laminate Stacking Sequence (LSS) was used to create the 0.125 mm thick laminas and material properties and its orientation were defined. The lay-up was made in similar way as in the 3D printer settings. Details LSS are provided on Table 2 for tensile and Table 4 for flexural.

2.5.1. Tensile model: A finite element model of 1.75 mm thick, 15 mm wide and 138 mm in length (represents only the gauge section) was modeled in ANSYS for the tensile analysis. The left end of the model had a fixed boundary condition with zero displacement in all possible movements, whereas negative 559.9 MPa uniformly distributed pressure was applied at the right end of the model. All boundary conditions (BC) were applied on a line. Furthermore, the model was meshed with an element size of 0.75 producing 3680 total elements. The tensile model has 20 elements along its width and it contains only one element throughout its thickness.

Table 2. Layer orientation and material for each lamina for the tensile specimens.

Layer number	Total number of layers	Material type	Layer thickness [mm]	Orientation Angle, [°]
13, 14	2 layers	Onyx	0.125	(+45°,-45°)2
3-12	10 layers	CF	0.125	(0°)10
1, 2	2 layers	Onyx	0.125	(+45°,-45°)2

2.5.2. Flexural model: A finite element model of 4 mm thick, 13 mm wide and 186 mm length was modeled in ANSYS. The FE-model had a zero displacement only in z-direction at the two supports and only in y-direction at the left end of the model (Figure 4). Whereas the right end was constrained with zero DOF in x- and y- direction. The supports was 124 mm apart and the model was extended from the supports by 31 mm to each side. A total force of 303 N, with 33.7 N to each nodes was applied at 9 nodes located at the mid-span of the beam. The model was meshed into 8 elements along its width, with refined mesh at the mid-span of the beam and contains one element throughout the thickness. Furthermore, the output of the stress analysis are based on settings for full integration with incompatible modes of the shell elements.

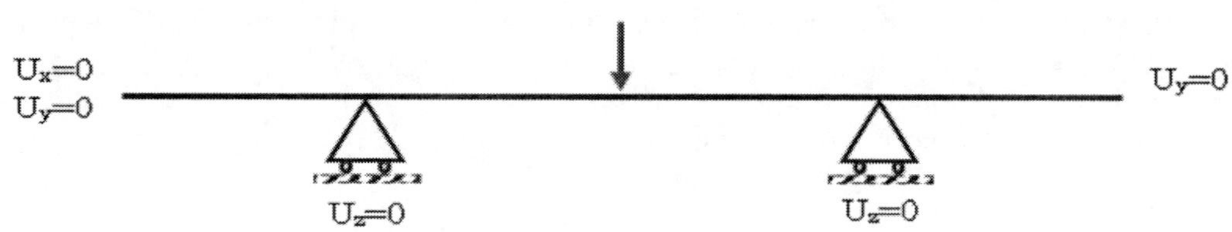

Figure 4. Boundary conditions for the flexural model.

2.5.3. Orthotropic material property assumption: van der Klift et al. [2] estimated the amount of fiber volume fraction on a single bundle of Markforged® carbon fiber filament by evaporating the matrix according to JIK K7075 and found a fiber volume fraction of 34.5% and the rest 65.5% was other coating and adhesive polymers. This estimation was used to calculate the assumed orthotropic material properties CF filament provided by Markforged® in this study. For making assumptions on the Poison's ratio of Onyx, five samples was tested and it is assumed be 0.43.

Table 3. Assumed material properties of Markforged® carbon fiber and Onyx used for simulation in Ansys. Typical AS4 CF yarn properties are adapted from Meddad, 2002 [23].

Description	AS4 carbon fiber yarn [23]	Markforged® CF		Data provided by
		Tensile	Flexural	
E-Modulus Onyx, [GPa]	-	1.4	2.9	Markforged®
Axial E-Modulus CF E11, [GPa]	231	54	51	Markforged®
Transverse E-Modulus CF E22= E33, [GPa]	22.4	7.728*	7.728*	Assumed
Axial Shear Modulus CF, G12= G13, [GPa]	22.1	7.625*	7.625*	Assumed
Transverse Shear Modulus, CF, G23, [GPa]	8.3	2.864*	2.864*	Assumed
Axial Poisson's Ratio , CF, v12= v13	0.3	0.104*	0.104*	Assumed
Transverse Poisson's Ratio, CF, v23	0.35	0.121*	0.121*	Assumed

Note: That values marked with * on Table 3 were obtained by considering only 34.5% of the AS4 CF values and unmarked values are provided by Markforged®.

Table 4. Layer orientation and material of each layer lamina of the flexural specimens. [1 is bottom layer and 32 is top layer].

Layer number	Total number of Layers	Material type	Layer Thickness [mm]	Orientation Angle, [°]
29 - 32	4 layers	Onyx	0.125	(+45°,-45°)2
19 - 28	10 layers	CF	0.125	(0°)10
15 - 18	4 layers	Onyx	0.125	(+45°,-45°)2
5 - 14	10 layers	CF	0.125	(0°)10
1 - 4	4 layers	Onyx	0.125	(+45°,-45°)2

3. Results

The stress on experimental tensile test results is calculated by dividing the instantaneous force by the average initial cross-sectional area of the sample measured at the gauge section. Whereas the flexural stress is calculated using equation (2).

3.1. Results from tensile experimental test

The average maximum tensile strength obtained from the tensile samples was 560 MPa with a standard deviation (SD) of 17.70 MPa and tensile elastic modulus of 25 GPa with 2.7 GPa SD. The average yield strain was about 0.026 mm/mm. A stress-strain curve was generated from the individual tensile-test-sample results data and it is presented on Figure 5.

Table 5. Experimental results from tensile samples fabricated from CF and Onyx.

Description	Average Width [mm]	Average Thickness [mm]	Max. Load [kN]	Max Tensile Strength [MPa]	Tensile Modulus [GPa]	Tensile Strain [mm/mm]
S1	15.06	1.87	14.99	558.87	24.63	0.027
S2	14.98	1.86	15.07	537.75	20.59	0.032
S3	14.96	1.87	16.05	567.61	26.81	0.024
S4	14.98	1.88	15.40	550.37	26.27	0.025
S5	14.96	1.86	16.47	584.68	26.89	0.023
Mean	14.99	1.87	15.6	559.9	25	0.026
SD	0.04	0.01	0.64	17.72	2.65	0.004
CV [%]	0.3%	0.4%	4%	3%	11%	14%

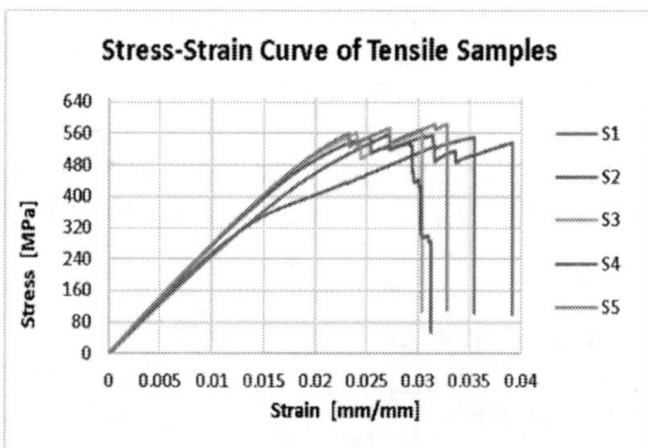

Figure 5. Stress-Strain curve of tensile specimens.

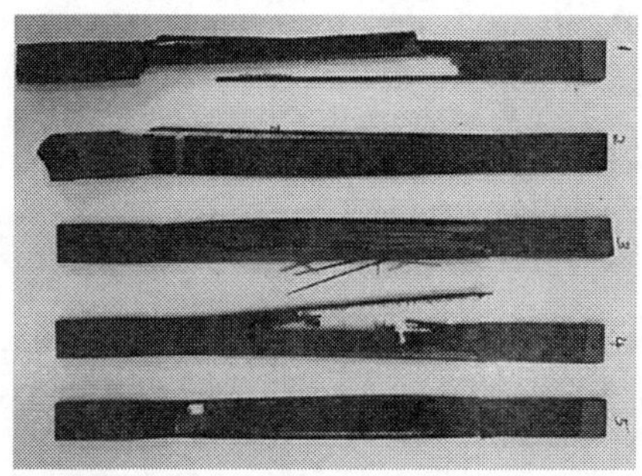

Figure 6. Failed UD tensile samples.

3.2. Experimental results from the 3-point flexural test

From the experiment, a maximum mean flexural strength of 271 MPa was obtained. Equation (2) was used to estimate the flexural stress and a stress strain curve is provided on Figure 7. Results from the flexural experimental test are provided on Table 6.

Table 6. Flexural 3-point test experimental results.

Description	Average width [mm]	Average Thickness [mm]	Max. Load [N]	Flexural Strength [MPa]	Flexural Modulus [GPa]	Flexural Strain [%]
T1	12.98	4.01	328.27	293.55	15.53	2.30
T2	12.97	4.04	271.58	242.86	16.17	4.19
T3	13.01	4.03	339.05	303.19	16.73	2.28
T4	12.96	4.04	270.26	241.68	18.56	2.64
T5	12.98	4.03	304.47	272.27	15.11	2.63
Mean	12.98	4.03	302.7	270.7	16.4	2.8
SD	0.02	0.01	31.62	28.27	1.35	0.79
CV [%]	0.14%	0.30%	10%	10%	8%	28%

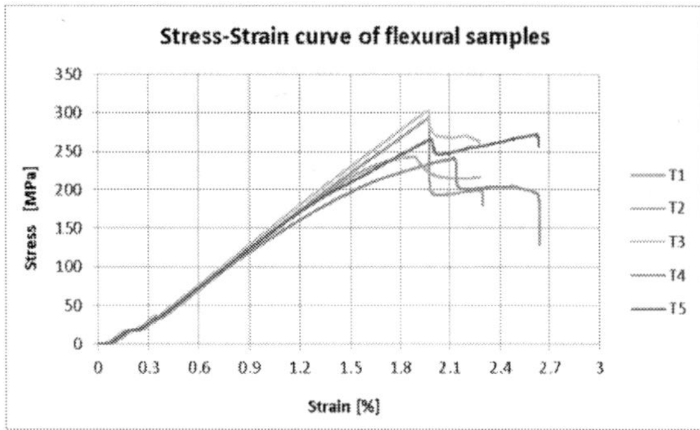

Figure 7. Stress-Strain Curves of flexural Samples.

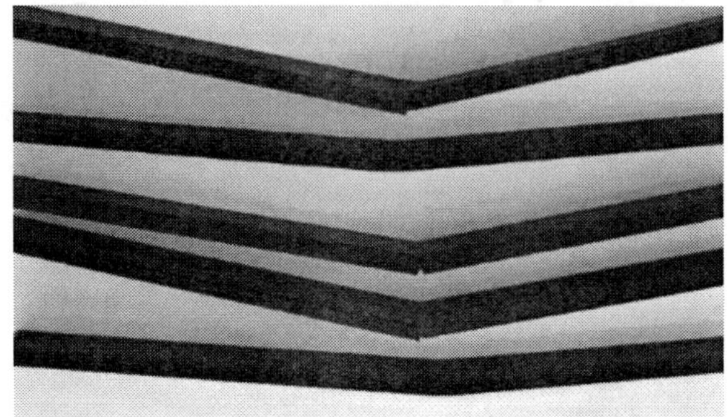

Figure 8. Failed three-point flexural sample.

3.3. FEA results from tensile model

From the FE-model, a nodal longitudinal displacement of 1.13 mm was determined. The longitudinal distribution of tensile stress in CF layer is shown in Figure 9, representing layer 7 of the finite elements. Apart from the steep tensile stress gradient close to the fixed boundary as presented in Figure 10 a uniform distributed axial stress of 443 MPa is estimated in CF layer. The uniform stress in Onyx layers is 13 MPa along the main length of the model. It is to be noted that the different values in tensile stress are within their respective individual limits of strength. Marforged® reports 36 MPa stress at yield for the pure Onyx matrix and tensile strength of 700 MPa for CF. The 0.9% strain is also well below the 1.5% strain at break reported by Marforged®.

Table 7. FE model results from tensile model.

Description	Nodal solution	Units
Displacement UX	1.13	mm
CF: SX, Tensile stress	443	MPa
Onyx: SX, Tensile stress	13.2	MPa
Max longitudinal strain	0.009	mm/mm

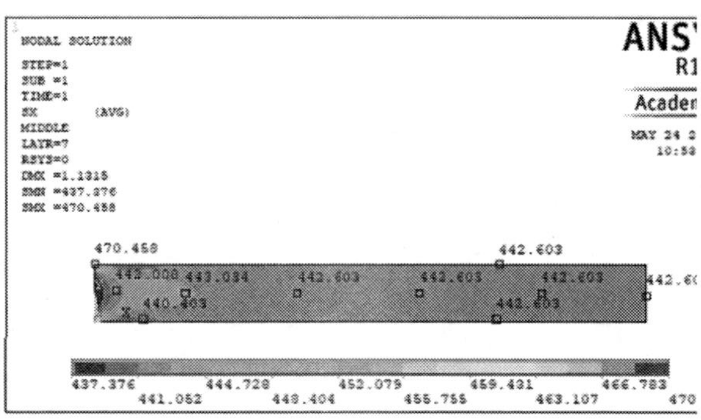

Figure 9. Tensile stress in CF layer

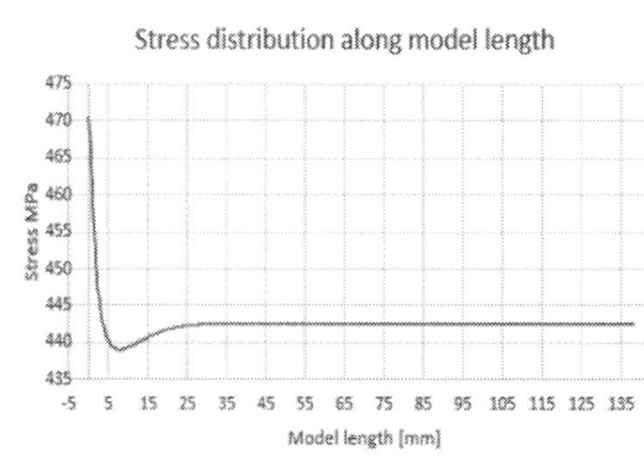

Figure 10. Tensile stress distribution in CF layer.

3.4. FEA results from flexural model

The total load at mid-span was 303 N generating a maximum deflection of 7.6 mm. From the lowest CF-layer (layer 5) at mid-span of the beam, a maximum flexural stress of 445 MPa and equivalent stress

IOP Conf. Series: Materials Science and Engineering **700** (2019) 012024 doi:10.1088/1757-899X/700/1/012024

of 450 MPa was estimated. Moreover, the surface of the first Onyx layer (1st layer from bottom) was subjected to a flexural stress of 37 MPa and equivalent stress of 34 MPa. Marforged® reported flexural strength of 81 MPa for pure Onyx and 470 MPa for pure CF, thus no yield is to be expected in the Onyx layers [18]. The distribution of the flexural stress (SX) at the mid-span of the FE-model is presented in Figure 11. Furthermore, maximum stress values for selected layers at the mid-span are presented in Table 9.

Table 8. Results from flexural FE model at the mid-span of the beam.

Description	Nodal solution	Units
Max. disp.,UZ	7.6	mm
CF: SX, Flexural stress	445	MPa
CF: SEQV, Von Mises Stress	450	MPa
Onyx: SX, Flexural Stress	37	MPa
Onyx: SEQV, Von Mises Stress	34	MPa
Max. strain, EPELX	0.012	mm/mm

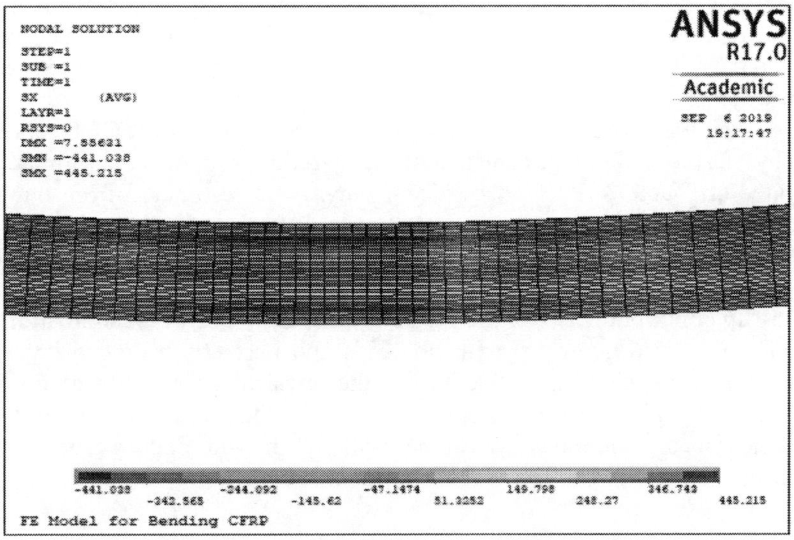

Figure 11. Max Stress at the mid-span of flexural Model.

A summary of results from experiment, ROM and FEA are presented in Table 10, both for tensile and flexural samples. The fiber volume fraction estimated from experimental samples is used in ROM calculations. In the FEA the values are taken from the mid-span of the models. The FE result for the tensile and flexural was obtained from layer 7 and 5, respectively. Both layers are made from CF.

Table 9. Flexural stress from some layers of the FE model at the mid-span.

Layers →	L-1	L-2	L-3	L-4	L-5	L-6	L-12	L-13	L-14	L-17
Material	Onyx	Onyx	Onyx	Onyx	CF	CF	CF	CF	CF	onyx
Stress: SX [MPa]	37	34	31	29	445	390	170	134	97	0
Node nr.	1391	1391	1391	1391	1391	1391	1391	1391	1391	1445
Deflection, [mm]	7.6	-	-	-	-	-	-	-	-	-
Strain: EPLX	0.011	0.011	0.010	0.009	0.008	0.008	0.003	0.003	0.002	0.0

Table 10. Results of the composite samples from ROM, experimental test and FEA.

Description	Tensile stress [MPa]	Flexural Stress [MPa]
Experimental	560	271
ROM	453	244
FEA of CF: SX	443	445
FEA of Onyx: SX	13.2	37

4. Discussion

In the experiment, delamination failure was observed in both flexural (Figure 8) and tensile test (Figure 6) samples. The delamination was observed between the matrix and fiber layers and the matrix material was easy to peel off from the fiber. Also, it had been observed formation of voids between layers on both matrix- and fiber- materials during fabrication of samples. The delamination was probably due to the nature of AM, specifically weak bonding between successive layers. Referring to the tensile tests, all the tensile specimens fail within the gauge length close to the grip in CF layers (Figure 6). After successive longitudinal delamination of the samples, the failures at final stage were characterized in all cases as typical brittle failures with a sudden and strong energy release. The estimated tensile strength of 560 MPa and an E-modulus of 25 GPa, was obtained from the experimental tensile test. From flexural test results, the flexural strength of 271 MPa and E-modulus of 16 GPa was obtained. The bending samples failed at the mid-span and a delamination between fiber and matrix layers were observed in the samples (Figure 8).

In the FE-model it was assumed orthotropic material properties for the CF layers: Some material data was provided by Markforged®, while the rest of the material properties are estimated from AS4 carbon yarn as presented on Table 3. Poisson's ratio of Onyx were estimated from our tests based on five samples, fabricated in our lab.

FE- model with layered shell elements based on its idealized assumptions predict accurate values of stress in each composite layer. A significant deviation from the physical samples at critical levels of strain is the no-slip condition between the layers in the FE-model vs. delamination experienced in the test samples. The idealized geometry with no voids and perfectly arranged layers in the FE-model, compared to assumed voids and imperfections in the physical samples introduce further errors when predicting failure with FE-models. FEM is versatile tool for the understanding of the elastic properties of composite materials and the modeling can be further improved. Failure criteria e.g. Tsai-Wu criteria can be implemented but should be based on experimental data.

5. Conclusion

A delamination failure was the common cause of failure mainly in the tensile test. Bonding between successive non-identical material layers and adjacent layers of CF was weak. In contrast to all, the composite material provided a promising property with about 560 MPa in tensile strength and about 271 MPa in flexural strength, respectively. Furthermore, the UD composite material achieved an elastic modulus of 25 GPa and 16 GPa in tensile and bending, respectively.

IOP Conf. Series: Materials Science and Engineering **700** (2019) 012024 doi:10.1088/1757-899X/700/1/012024

References

[1]. Hull, D and Clyne T W 1996 *An introduction to composite materials* 2nd Ed. (Cambridge University Press) p 344.

[2]. Van Der Klift, Koga, F Y, Todoroki A, Ueda M, Hirano Y, and Matsuzaki R 2016 *Open J. Compos. Mater.* **6** (1) 18-27.

[3]. Ning F, Cong, W, Qiu J, Wei J, and Wang S 2015 *Composites Part B* **80** 369-78.

[4]. Jiang D and Smith D E 2017 *Addit. Manuf.*, **18** 84-94.

[5]. Strong, A B 2008 *Fundamentals of composites manufacturing: materials, methods and applications* 2nd Ed. (Society of Manufacturing Engineers) p 599.

[6]. Miracle D B, Donaldson S L et al. 2001 *ASM handbook Volume 21: Composites* 21 Ed. (ASM international) p 1201.

[7]. Hart R J, Patton E G and Sapunkov O 2018 *Characterization of continuous fiber-reinforced composite materials manufactured via fused filament fabrication*. Technical report, available at: https://apps.dtic.mil/dtic/tr/fulltext/u2/1062410.pdf. (Accessed: 20.10.2019].

[8]. Caminero M, Chacón J, García-Moreno I and Reverte J 2018 *Polym. Test.*, **68** 415-23.

[9]. Dickson A N, Barry J N, McDonnell K A and Dowling D P 2017 *Addit. Manuf.*, **16** 146-52.

[10]. Schwartz M M 1984 *Composite materials handbook* Ed. (McGraw-Hill) p 651.

[11]. Gay D and Hoa S V 2007 *Composite materials: design and applications* 2nd Ed. (CRC press) p 545.

[12]. Nicolais L, Meo M and Milella E 2011 *Composite materials: A Vision for the Future* 1st Ed. (Springer Science & Business Media) p 218.

[13]. Lozada J N, Ahuett-Garza H, Castañón P O, Verbeeten W M and González D S 2019 *Addit. Manuf.*, **26** 227-41.

[14]. Campbell F C 2010 *Structural composite materials* Ed. (ASM international) p 612.

[15]. Lee C, Kim S, Kim H and Ahn S 2007 *J. Mater. Process. Technol.*, **187** 627-30.

[16]. Barbero E J 2013 *Finite element analysis of composite materials using ANSYS®* 2nd Ed. (CRC press) p 366.

[17]. Markforged, <*Mark Two product specification.pdf*>. (Last accessed: 20.10.2019)

[18]. Markforged, *Onyx, material specification*. Available at https://www.3axis.us/matetials/markforged-materials.pdf (Last accessed: 20.10.2019)

[19]. Markforg3D, <*Markforged-Products-and-Applications.pdf*>. 2016.

[20]. Standard A, *ASTM D3039-Standard test method for tensile properties of polymer matrix composite materials*. 2008, ASTM International. p. 13.

[21]. Standard A J A S fT and Materials P: Philadelphia, USA 2007 D 7264/D 7264M-07 11.

[22]. Mac Donald, B.J. 2007 *Practical Stress Analysis with Finite Elements* 1st Ed. (Glasnevin Publishing) p 350.

[23]. Meddad, A, Azaiez J, Ait-Kadi A and Guenette R 2002 *J. Compos. Mater.*, **36** (4) 423-41.

Implementation of linear, double-linear, and nonlinear fatigue damage accumulation rules for fatigue life prediction of offshore drilling top-drive tie-rods

J E Inoma [1,*], D G Pavlou [1] and J Zec [2]

[1] Department of Mechanical and Structural Engineering, University of Stavanger
[2] NOV Rig Technologies Buraasen 24, 4604 Kristiansand S, Norway

* Corresponding author: jesseinoma@gmail.com

Abstract: The offshore oil and gas industry has been exposed to major challenges over the last decade, particularly demanding cost cuts and more effective technical solutions. Predictive systems and remaining life assessments for both machine and structural components are known to be one of the core areas that has gathered much attention lately. This paper focuses on multiaxial fatigue that is a problem in number of engineering structures and equipment. The ability to properly assess and quantify multiaxial fatigue of offshore equipment and structures has major benefits for owners of engineering assets both in terms of technical and safety integrity. Traditionally, Palmgren-Miner's damage rule is used for life estimation involving multiaxial fatigue due to its ease of use. There are however some known shortcomings with Palmgren-Miner's rule namely: it does not consider the loading sequences. This for instance can result in overestimation of fatigue life in scenarios where stress amplitudes are decreasing, and moreover underestimation of fatigue life in scenarios where stress amplitudes increase. This research work thus involves closely studying the application of Palmgren-Miner's linear damage rule, Manson's double linear damage rule, and Subramanyan's non-linear damage rule for the purpose of enhancing the predictability of damage as well as accuracy of fatigue life assessments. Each of these techniques were applied for tie-rods of an offshore drilling top-drive having a known stress history. The loading histories were provided by a drilling company. The torque and axial force values are transformed to stress components for the corresponding critical spots. The "rainflow" counting technique is applied to the obtained stress histories and the mean stress effect is considered for the damage accumulation calculations. The fatigue life prediction of the three models are justified and discussed. This research work was aimed at contributing towards the utilization of more robust fatigue life estimation techniques such as Manson's double linear damage rule and simplified non-linear damage rules which capture the true nature of the fatigue that equipment's are subjected to, and thus to improve the reliability of the fatigue life predictions.

1. Introduction

Condition monitoring and equipment maintenance are essential activities which must be carried out by managers of modern engineering assets, equipment and infrastructure. These activities are important to ensure that the structural integrity of an asset is maintained as well as to ensure an asset performs its function and at the same time can maintain resilience when it is exposed to sudden shocks or changes. In a nutshell, condition monitoring of critical equipment is essential for companies to stay competitive and meet the strict HSE regulations set by the Petroleum Safety Authority (PSA) and other regulators for operating on the Norwegian Continental Shelf (NCS). To fully harness the advantages of condition monitoring, correct models need to be used to determine the structural integrity of equipment and infrastructure that are being utilised for operations. This is so important because using wrong models and assumptions will result in scheduled maintenance or replacement being initiated for equipment which could have been used for many more years without failure or can underestimate the fatigue damage increasing the risk when using such equipment.

Structural integrity of assets is an important issue and a key cost factor in the oil and gas industry. Several time-scheduled non-destructive testing (NDT) techniques are used to inspect integrity (fatigue, wear, corrosion) of structural members (for example every 5 years), this has some significant drawbacks such as: it is time consuming, requiring often full disassembly of equipment, while few failures are found in these structures. There usually is a great need for competence outside the crew to carry out this analysis – which drive high costs and results in significant downtime for structural engineers to access these equipment's.

It has been observed that these periodical inspections usually add more failures to the system (due to coupling errors). Due to this reason the industry has moved into transferring inspection scheduling from 'time-based' inspections to 'condition-based monitoring' of equipment usage by applying 'physics based' fatigue or wear mathematical condition focused (machine learning & AI) data driven models.

The challenge with such models especially physics and statistics based) are as follows:

- These involve high level of intellectual property (IP) and this is seldomly shared.
- These are often hard-coded and difficult to evolve, when data contradicts conservative model assumptions.
- It is difficult to document such model changes and foundation for improvements.
- It is difficult to get consensus on what to improve, when and by using what quality assurance process, involving who.
- Access to results require access to special systems, without direct coaction to what these data are used for.
- Relation between data/results is manually compared and translated to value, KPI or certification process.
- Physics based models require a lot of infrastructure / hardware power to be utilized (several layers of differential equations, hard coded according to system taxonomy).
- On other hand data driven approaches require equality of configurations, data quality and datasets to achieve complete validation.

This paper provides insight on the current approach of using standards (API 8C *"Specification of Drilling and Production Hoisting Equipment"* and F.E.M. *"Rules for the Design of Hoisting Appliances"*) by a drilling company to assess fatigue damage and the structural integrity of their top-drive members with special focus on weakest link (i.e. the tie rods of the top-drive). The current approach being used by the drilling company to access fatigue damage will be compared to two other models (Manson's Double Linear Damage model and Subramanyan's Non-Linear Damage model). The F.E.M 1.001 standard is based on the use of Palmgren-Miner's model and S-N curves when calculating fatigue damage on equipment.

Generally, Miner's rule is widely accepted as the method to access the fatigue damage due to its simplicity and ease of use in design of equipment where loading history is not available. The drawback of using this method is that it does not consider the load sequence for example an equipment might be exposed to a higher loading cycle of let's say 900 MPa followed by 600 MPa or vice versa but the Miner approach will still predict the same damage even if the smaller load cycle comes before the higher cycle. The result of this is that the life of a structure or equipment will be underestimated when the cycles are reversed low - high load sequence and overestimated in a high - low load sequence. For random loading sequences Miner's rule is conservative in fatigue life predictions and this is one of the main reasons why no faults are discovered in the top drives after they are dismantled.

This paper will utilize the same approach adopted by the drilling company as prescribed by the F.E.M 1.001 standard when reviewing Palmgren Miner's approach to fatigue damage. The results of this approach will be compared to those obtained by using Manson's Double Linear Damage model and Subramanyan's Non-linear damage model and conclusions will be drawn from these approaches to estimating fatigue damage.

2. Objectives of the study

This paper's focus is centred on improving process of control and validation of structural integrity of the drilling company's top-drive tie-rods. This will be achieved by applying Palmgren Miner's rule, Manson's double linear damage rule and Subramanyan's non-linear damage rule to improve damage predictions and fatigue life assessments. The results from these tests using the selected fatigue damage models will serve as a basis for optimizing the current methodology of fatigue analysis employed at the drilling company. Fatigue calculation results from using the selected fatigue damage models will be compared. Von-Mises' equivalent stress approach will be applied to transform torque and axial forces into an equivalent stress. This approach is utilised for a variety of multiaxial fatigue problems and it is suitable for materials that exhibit ductile behaviour as is the case in this research work. Von Mises' equivalent stress approach was utilized to get the equivalent stresses for each of the stress combinations. This approach was adopted because the axial and torsional loads obtained from the loading history of the top drive were considered to be proportional and in the same phases.

According to [1], "In situations in which we can reasonably expect an overloaded part in service to fail in the same manner as the standard tensile test bar made of the same material, it is recommended that the maximum-distortion-energy theory be used to predict ductile yielding". The material of construction of the top-drive is high strength steel which exhibits ductile behaviour, and this is additional justification why Von-Mises equivalent stress approach was used. It should be noted that Von-Mises equivalent stress approach is an approximation and there are more robust models such as those proposed by Fatemi and Papadopoulos [2-4], however these more robust approaches are more complicated for implementation in engineering applications. Details on Von- Mises' equivalent stress equation can be seen in the equation (1) below.

$$\sigma_{Eq.m} = \frac{1}{\sqrt{2}}\sqrt{[\sigma_{m1} - \sigma_{m2}]^2 + [\sigma_{m2} - \sigma_{m3}]^2 + [\sigma_{m3} - \sigma_{m1}]^2} \quad (1)$$

where,

- $\sigma_{Eq.m}$ is the equivalent nominal stresses amplitude
- σ_{m1}, σ_{m2} and σ_{m3} are the principal mean nominal stresses acting on the material

The selected models considered in this research work will be built and coded in Python [5] to attain a shift from physics dominated differential equation-based component models to hybrid models adapted from linear algebra and vector calculus. The aim of this approach is to reduce internal IT infrastructure capacity dependence (now in cloud or big data centres, depending again on good connectivity), speed up calculations and implement machine learning techniques to improve results. Such simplification will enable execution on edge devices or within control devices.

Focus will be placed on building these models in a way that it is user-friendly and easily adaptable to other equipment apart from the tie-rods on the top drives. This will be necessary to ensure that calculations are quick and allow for the use of machine learning techniques to further optimize the results from this model.

IOP Conf. Series: Materials Science and Engineering **700** (2019) 012025 doi:10.1088/1757-899X/700/1/012025

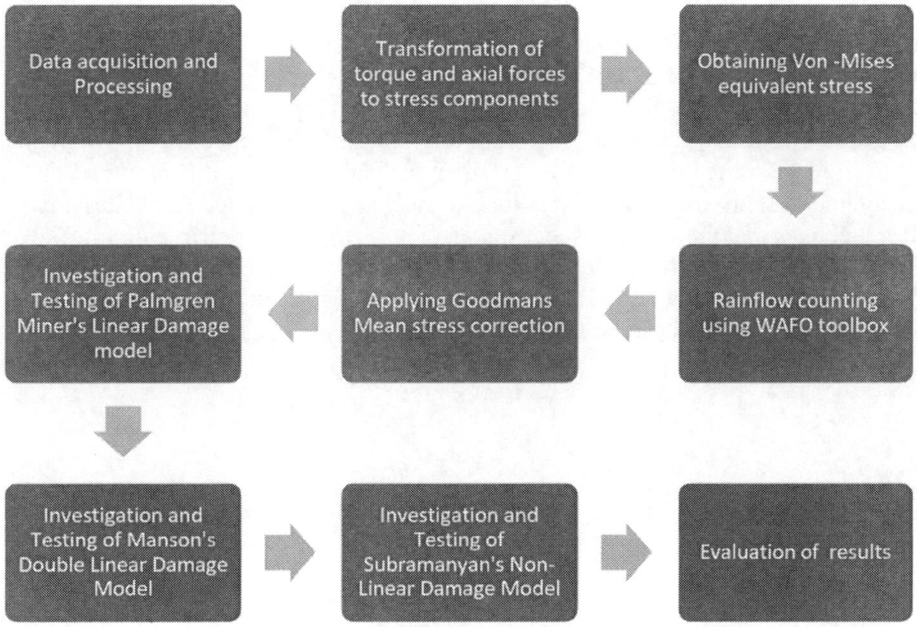

Figure 1. Research methodology.

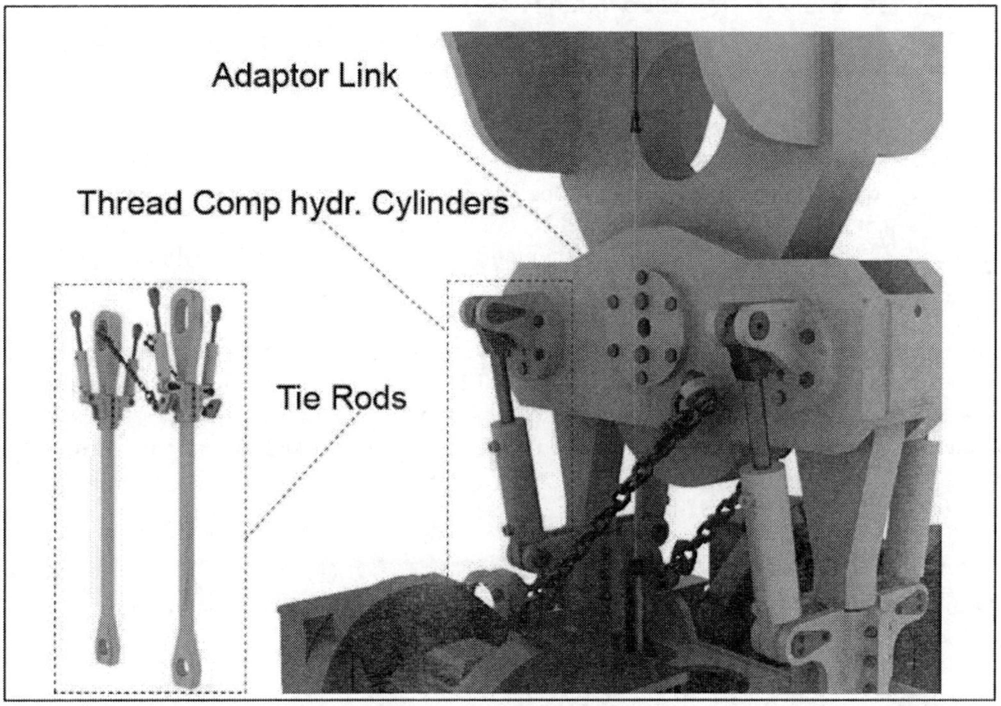

Figure 2. Illustration showing the suspension system of the top drive with the tie-rods which are analysed in this research work.

3. Review of selected fatigue damage models

3.1 Palmgren-Miner's linear damage model

Miner's rule was popularized in 1945 by Miner [6] in his work on fatigue involving tension-tension axial fatigue data for aircraft skin material, this was a development of an earlier cumulative damage model proposed by Palmgren in 1924 [7]. Palmgren Miner's model operates on two major assumptions: the load spectra is assumed to be fully reversed sinusoidal cycle and it is assumed that the total work absorbed by the system will result in failure occurring. This model postulates that "where there are k different stress magnitudes in a spectrum, S_i ($1 \leq i \leq k$), each contributing n_i (S_i) cycles, then if N_i (S_i) is the number of cycles to failure of a constant stress reversal S_i (determined by uniaxial fatigue tests), failure occurs when damage(D) i.e. the ratio of the applied cycles to the number of cycles to failure is equal to 1" [6]. This can be further explained with the equations below.

$$D = \sum_{i=1}^{k} \frac{n_i}{N_i} \tag{2}$$

$$D = \left(1 - \sum_{i=1}^{k} \frac{n_i}{N_i}\right) \tag{3}$$

Where:

- n_i is the number of cycles accumulated at stress Si.
- N_i is the number of cycles a material can take until failure at the given stress
- D is the fraction of life consumed by exposure to the cycles at the different stress levels.

Palmgren-Miner's linear damage model is the most popular model used in various industries for analysing fatigue life and expressing the damage that a material is being subjected to, this is due to the simplicity in utilizing this model. Also, most of the stress-life curves available in standards such as DNV-RP-C203 were developed from experimental data based on Miners rule.

To overcome Miner's rule shortcomings various probabilistic methods have been used to counteract the load sequence effects thus resulting in acceptable predictions for fatigue life under random loading [8-11]

3.2 Manson's double linear damage model

This model was developed by Manson in 1967, where he considered fatigue to be occurring in two major stages namely: crack initiation and crack propagation. The main assumption in this initial postulate by Manson, was that the crack initiation period N_p could be used to express the total life of a material by using (4) and (5) .Various revisions have been made to this model until 1981 where Manson abandoned the use of the terminologies of crack initiation and crack propagation, rather he chose to call this Phase I and Phase II, he also presented equations to implement the double linear damage rule. This was done to achieve simplicity in the application of this principle in comparison to the damage curve approach that he had suggested in his previous publications [12, 13]. Manson's double linear damage rule can be regarded as Miner's rule applied to two phases of fatigue damage. According to Manson, [14] when block loading exceeds two levels the following equations can be used to compute damage in the two phases.

Total fatigue life is expressed as:

$$N_f = N_I + N_{II} \tag{4}$$

The relationship between Phase I damage and total fatigue life is then expressed as

$$N_I = N_f \exp(ZN_f\Phi) \tag{5}$$

Z and Φ are constants and can be determined from the knee points of the curve for Phase I damage. The knee points are the same for all materials and can be determined from the maximum and minimum lives present in the loading cycle.

IOP Conf. Series: Materials Science and Engineering **700** (2019) 012025 doi:10.1088/1757-899X/700/1/012025

$$N_{I,N1,f} = N_{1,f}\left(\frac{n_1}{N_{1,f}}\right)_{knee} = 0.35N_{1,f}\left(\frac{N_{1,f}}{N_{2,f}}\right)^{0.25} \quad (6)$$

$$N_{I,N2,f} = N_{2,f}\left(1-\left(\frac{n_{2,f}}{N_{2,f}}\right)_{knee}\right) = N_{2,f}\left(1-0.65\left(\frac{N_{1,f}}{N_{2,f}}\right)^{0.25}\right) \quad (7)$$

By substituting equations (6) and (7) into equations (4) and (5) a solution can be obtained for the constants Z and Φ as shown by the equations below.

$$\Phi = \frac{1}{ln\left(\frac{N_{1f}}{N_{2f}}\right)}\ln\left[\frac{ln\left(0.35\left(\frac{N_{1f}}{N_{2f}}\right)^{0.25}\right)}{ln\left(1-0.65\left(\frac{N_{1f}}{N_{2f}}\right)^{0.25}\right)}\right] \quad (8)$$

$$Z = \frac{ln\left(0.35\left(\frac{N_{1f}}{N_{2f}}\right)^{0.25}\right)}{N_{1,f}^{\Phi}} \quad (9)$$

Therefore, N_{II} can be expressed as:

$$N_{II} = N_f - N_I = N_f\left(1-exp\left(ZN_f^{\Phi}\right)\right) \quad (10)$$

Where:

- N_f is the total number of cycles to failure
- N_I is the number of cycles to failure in phase one
- N_{II} is the number of cycles to failure in phase two

Manson's double linear damage model has been shown to conform with experimental results as demonstrated by the research work carried out by Manson et al for NASA [14] and other publications from notable researchers within the fatigue subject area [15, 16]. By utilizing two linear damage phases for fatigue, the ease in the use of Miner's rule can be carried over into this model, like it was stated earlier the double linear damage rule is similar to applying Miner's rule in two phases of damage. It eliminates the deficiencies that are present in Miner's rule by having the co-ordinates for the knee-point of the S-N-curve and due to its linear nature, application of this model for designing components or analysing fatigue is easily done.

The drawbacks when utilizing this model is that the knee-point of the S-N curve has to be determined to properly implement this model, also Manson's model does not take into account the retardation mechanisms for crack growth [17, 18] and mixed mode cracks [19-21].

3.3 Subramanyan's non-linear damage model

This non-linear model was developed by S. Subramanyan in 1976, in this model the concept of iso-damage lines that converge at the knee-point of an S-N curve is utilized when analysing fatigue damage. This postulate of iso-damage lines is where Subramanyan's model deviates from Miner's rule, because in Miner's rule it is assumed that a constant damage line lies on an S-N curve and this constant damage line is parallel to the S-N curve for all the stress and number of cycles to failure combinations [22]. This cumulative damage model operates under the assumption of 100% damage existing on the S-N curve of a material. When stress and equivalent number of cycles to failure below the endurance limit of a material are read off an S-N curve their combinations will result in no damage (0% damage). The interval between the S-N curve (100 % damage) and stress and cycle combinations below the endurance limit (0% damage) will have a set of straight iso-damage lines which will converge at the knee-point of the S-N curve. Revisions were made to this model in 1978 to account for the reduction in the endurance limit of a material at various stress levels [23].

According to Subramanyan's model, the condition for fatigue failure is:

$$C_i + \{C_{i-1} + [C_{i-2} + \cdots + (C_2 + C_1^{\alpha_1})^{\alpha_2} \ldots]^{\alpha_{i-2}}\}^{\alpha_{i-1}} = 1 \quad (11)$$

The cycle ratio of this stress loading combinations is the ratio between the loading cycles and the number of cycles to failure which is obtained from the S-N curve.

$$C_i = \frac{n_i}{N_i} \quad (12)$$

$$\alpha_i = \frac{\log N_e - \log N_{i+1}}{\log N_e - \log N_i} \quad (13)$$

Where:

- n_i is the number of counted load cycles at the given stress level n
- N_i is the number of cycles to failure at a given stress level n on the S-N curve
- C_i is the cycle ratio at a given stress level i
- N_e is the number of cycles at the knee point of the S-N curve

Subramanyan's model has shown slightly non-conservativeness when compared to actual experimental results for SAE 4130 [16] which makes its usage acceptable. However, application of this model to analyse fatigue must be carried out with caution. According to Fatemi and Yang [24] Subramanyan's model is not applicable in cases where the stress amplitudes are near the endurance limit of the material, because S-N curves exhibit non-linearity close to their knee-point and because there is singularity at the knee point since all iso-damage lines will converge at this point.

4. Results and discussion

This section contains the results obtained from using Miner's linear damage rule, Manson's double linear damage rule and Subramanyan's non-linear damage rule. The comparison of results obtained from the various fatigue damage models will focus on two (2) key areas, damage results, and number of loading blocks until failure results. Here, emphasis will be placed on how the fatigue damage models compare to each other on the Main well and Auxiliary well. The key difference between these two is that the top-drive for the Main well was used for more operations than the top-drive on the Auxiliary well.

4.1 Comparison of Fatigue Damage Results

In Figure 3(a) the fatigue damage results from top drive on the Main well of Rig 1 shows slight increases in damage from Manson's DLDR and Subramanyan's NLDR. Damage increases by 50 percent from 0.065 to 0.13 in Manson's DLDR model, while in Subramanyan's model there is an 8 percent increase in damage from 0.096 to 0.104. In Miner's linear damage rule there is a 60 percent increase in damage from 3.81E-6 to 9.08E-6.

From these damage results it can be observed that Subramanyan's non-linear model and Manson's DLDR are quite close in their estimates for accumulated damage. Subramanyan's model gives the most damage accumulation of 0.2 which is slightly higher than damage accumulation in Manson's DLDR model (0.195). Meanwhile, the accumulated damage from Miner's rule (1.289E-5) is quite low in comparison to Manson and Subramanyan's models. This has significant implications on the operational life of this top-drive, because according to the damage accumulation obtained from Miner's rule this equipment can go on in operation without experiencing fatigue failure for a much longer time in comparison to the damage accumulation results from Manson's double linear damage rule and Subramanyan's non-linear damage which predict a shorter operational life for the top-drive. These deviations in the damage accumulation results from Subramanyan's model, Manson's and Miner's model are in agreement with variable amplitude tests that have been carried out by [8-11] that show

Miner's rule overestimates fatigue life when there is a decrease in applied stresses in a high-low loading sequence.

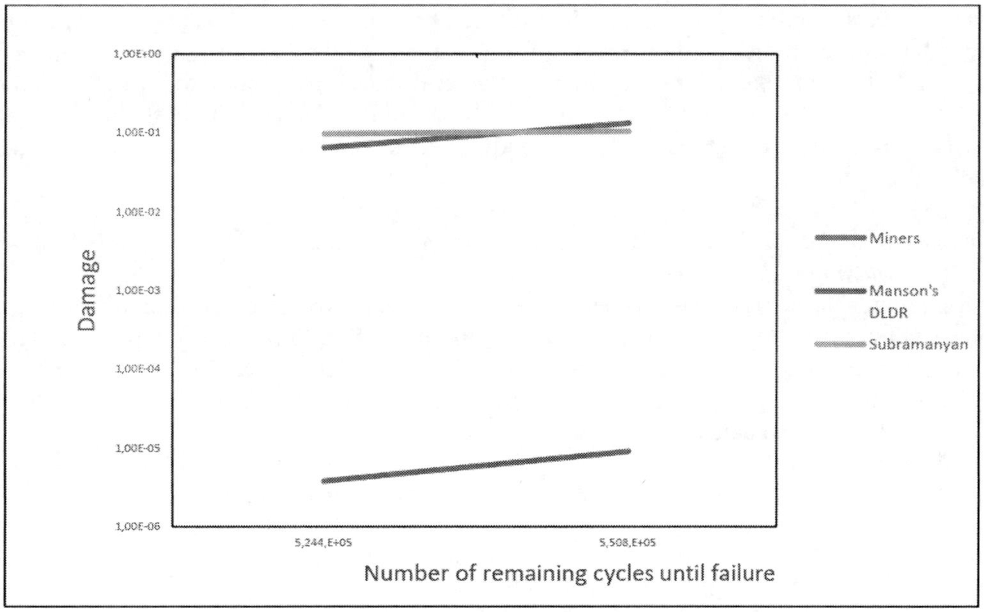

(a)

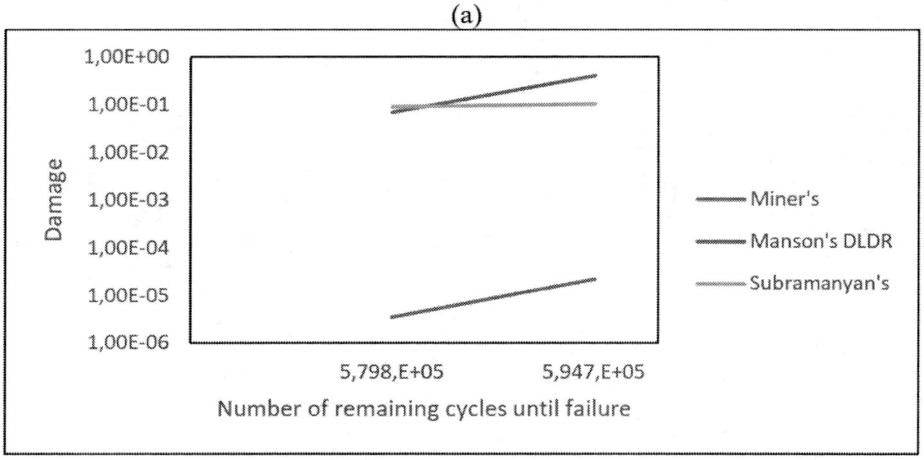

(b)

Figure 3. Estimates of damage versus the number of remaining cycles until failure from the selected fatigue damage models for Rig 1 Main Well (a) and Auxiliary Well (b) in 2014.

From the fatigue damage results shown in figure 3(b) for the top drive on the Auxiliary well on Rig 1, it can be observed that the accumulated damage results from Manson's DLDR and Subramanyan's model are more conservative than the accumulated damage predicted by Miner's rule. Here, the accumulated damage from Subramanyan's model is 0.208 while that from Manson's DLDR is 0.47. Meanwhile, the accumulated damage results obtained from using Miner's rule is 9.71E-6 which is significantly lower than the accumulated damage results obtained from Manson's DLDR or Subramanyan's model, this continues in the trend of overestimation of the remaining fatigue life by Miner's rule that has been seen in the previous results.

In summary, the damage results obtained from the rigs for the year 2014 shows close correlations between the damage results obtained using Manson's double linear damage rule and Subramanyan's non-linear damage rule with Subramanyan's model being slightly less conservative than Manson's model. In the results obtained, Miner's rule consistently overestimated the fatigue life for the analysed top-drive members in comparison to the fatigue life estimates gotten from Manson's and Subramanyan's models.

Given that these three models and consider different mechanisms and assumptions when measuring damage this brings into question how much information can be derived from their damage estimates. It is because of this that the number of loading blocks until failure was computed using each model. It is believed that this will be more informative when comparing the results from these three fatigue damage models. This comparison using the number of loading blocks until failure will be shown the next section of this discussion.

4.2 Comparison of the number of loading blocks until failure results for Miners linear damage rule and Manson's double linear damage rule models

In this comparison the analysed top-drive members are subjected to a stress loading history which results in damage when these applied stresses are above the endurance limit of the top-drive. This loading history is assumed to represent one (1) loading block of applied stresses on the top drive, from this analogy the number of loading blocks until failure is computed as the number of applied stress loading blocks that will result in a damage value of unity (1).

For Miner's rule and Manson's DLDR, the number of loading blocks until failure was computed as the inverse of the total damage accumulation, where in Manson's model this was computed for the two damage phases considered in this model. While in Subramanyan's model the stress block iterations were done until damage was equal to unity (1), the stress block where damage was equal to (1) was taken as the predicted number of loading blocks until failure.

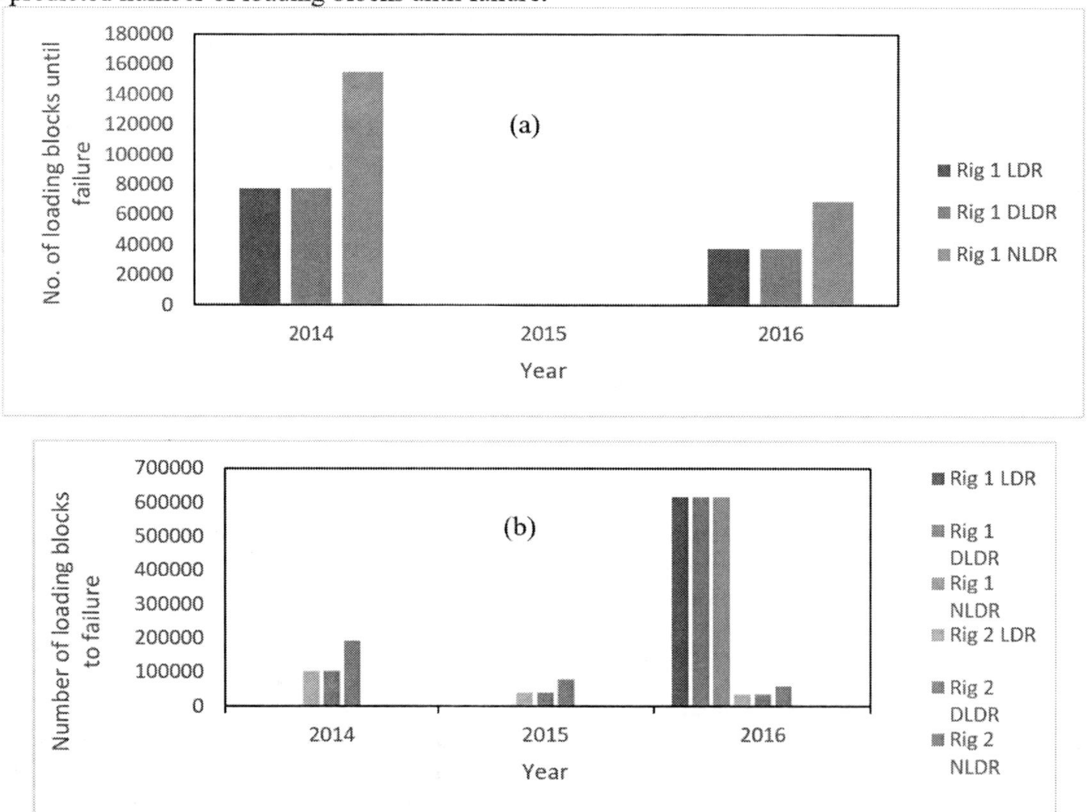

Figure 4. Estimates of number of loading blocks until failure from the selected fatigue damage models for Rig 1 Main Well (a) and Auxiliary Well (b).

Table 1. Estimates of number of loading blocks until failure from the selected fatigue damage models for Rig 1 Main Well (a) and Auxiliary Well (b).

Rig	Loading Blocks to Failure	Year		
		2014	2015	2016
	Miners Rule	77563.87101	No Damage	37562.5057
Rig 1	Manson's DLDR	77563.84533	No Damage	37558.85643
	Subramanyan's NLDR	155080	No Damage	68848
	Miners Rule	No Damage	No Damage	No Damage
Rig 2	Manson's DLDR	No Damage	No Damage	No Damage
	Subramanyan's NLDR	No Damage	No Damage	No Damage
	Miners Rule	No Damage	No Damage	No Damage
Rig 3	Manson's DLDR	No Damage	No Damage	No Damage
	Subramanyan's NLDR	No Damage	No Damage	No Damage

(a)

Rig	Loading Blocks to Failure	Year		
		2014	2015	2016
	Miners Rule	No Damage	No Damage	615923.7589
	Manson's DLDR	No Damage	No Damage	615923.7589
Rig 1	Subramanyan's NLDR	No Damage	No Damage	615922
	Miners Rule	102999.654	39509.11615	35290.93691
	Manson's DLDR	102993.7873	39509.11448	35287.01475
Rig 2	Subramanyan's NLDR	191989	79013	59224
	Miners Rule	No Damage	No Damage	No Damage
	Manson's DLDR	No Damage	No Damage	No Damage
Rig 3	Subramanyan's NLDR	No Damage	No Damage	No Damage

(b)

The results shown in figure 4 (a) covers the predicted number of loading blocks until failure for the Main Well on the analysed rigs. It can be observed that Miner's rule and Manson's DLDR show close similar predictions for the number of loading blocks until failure. This can also be seen in table 1(a), there is very little between Miner's rule and Manson's DLDR. Meanwhile, Subramanyan's model shows big deviations in its predictions of number of loading blocks until failure in comparison to Miner's rule and Manson's DLDR. These results show a trend of Miner's rule and Manson's double linear damage rule having more conservative estimates for the number of loading blocks of the applied stresses that the top drive can take before it experiences fatigue failure, Subramanyan's non-linear damage rule shows less conservative estimates. The implication of this is that the top drive will have to be recertified or replaced much earlier if the estimates from Miner's rule and Manson's DLDR are used in comparison to Subramanyan's model.

The results for the Auxiliary well on the analysed rigs can be seen in figure 4 (b) and table 1(b) below. The same trend observed in the Main Well continued here except in 2016 where the predictions for the number of loading blocks until failure estimates from Miner's rule and Manson's double linear damage rule and Subramanyan's non-linear damage rule are fairly equivalent. This can be due to the very low stress loading for the top drive on the Auxiliary well and its infrequency of use, from the rainflow count there was just one counted cycle (1) of a damaging stress that was applied to this top drive.

In summary, the results from the non-conservative estimates for the three (3) Rigs show that Subramanyan's non-linear damage rule is less conservative in its assessment of the number of loading blocks until failure in comparison to Miner's rule and Manson's double linear damage rule which are more conservative in their estimates. This observation is in agreement with experimental results from tests carried out Lee et al on SAE 4130 [16] where it was shown that Subramanyan's model was less conservative in its predictions of number of loading blocks until failure.

5. Conclusions

The damage accumulation results obtained from the selected models show that Miner's rule was consistently less conservative in its prediction for the accumulated damage in comparison to Manson's double linear damage rule and Subramanyan's non-linear damage rule. The selected double-linear and non-linear damage accumulation rules showed close correlations in their predictions of the amount of damage that had been accumulated by the tie-rods on the drilling top-drives which was in most cases more than the damage accumulation that was predicted by Palmgren Miner's rule. It can be concluded from these results that Palmgren Miner's rule underestimates the amount of accumulated damage which will result in an overestimation of the number of cycles remaining before the component experiences a failure. This finding is consistent with the research carried out by [15, 24-27] where it was proven that Palmgren Miner's rule predicts longer life for components when the stress amplitude is decreasing. Manson's and Subramanyan's damage accumulation rules should be utilised to provide more reliable damage accumulation estimates.

It can be concluded that the results from this project will serve as an enabler for the utilization of more comprehensive fatigue life estimation techniques such as Manson's double linear damage rule and simplified non-linear damage rules which capture the true nature of the fatigue that equipment's are subjected to, and thus to improve the reliability of the fatigue life predictions.

Acknowledgements

Appreciation goes to National Oilwell Varco for providing a workspace and resources in this research project.

References

[1] Juvinall R C and Marshek K M 2006 *Fundamentals of Machine Component Design*. 4th ed. ed. New York: Wiley.

[2] Papadopoulos I V, Davoli P et al. 1997 A comparative study of multiaxial high-cycle fatigue criteria for metals. *Int.J. Fatigue.* **19**, 219-235.

[3] Papadopoulos I. V 2001 Long life fatigue under multiaxial loading, *Int.J. Fatigue,* **23**, 839-49.

[4] Fatemi, A and Shamsaei N 2011 Multiaxial fatigue: An overview and some approximation models for life estimation *Int. J. Fatigue.* **33**(8) 948-58

[5] Miner M A 1945 Cumulative damage in fatigue. *J. Appl.Mech.* **12** 149–64.

[6] Palmgren A G 1924 Die Lebensdauer von Kugellagern [Life Length of Roller Bearings]. *Zeitschrift des Vereines Deutscher Ingenieure (in German).* **68**(14) 339–41

[7] Blason S, Correia J et al. 2016 A probabilistic analysis of Miner's law for different loading conditions. *Struct. Eng. Mech.* **60**(1) 71–90.

[8] Fernández-Canteli A, Blasón S et al. 2014 A probabilistic interpretation of the Miner number for fatigue life prediction. *Frattura. integr. strutt,* **8**(30) 327-39

[9] Pavlou D G 2002 A phenomenological fatigue damage accumulation rule based on hardness increasing, for the 2024-T42 aluminum. *Eng. Struct.* **24**(11) 1363-68.

[10] Pavlou D G 2018 The theory of the S-N fatigue damage envelope: Generalization of linear, double-linear, and non-linear fatigue damage models. *Int. J. Fatigue*, **110** 204-14.

[11] Ensign C R, Freche J C and Manson S S 1966 Application of a double linear damage rule to cumulative fatigue - *NASA-TM-X-52226.*

[12] Manson S S 1966 Interfaces between fatigue, creep, and fracture. *Int. J. Fracture* **2**(1) 327-63.

[13] Manson, S S and Halford G 1981 Practical implementation of the double linear damage rule and damage curve approach for treating cumulative fatigue damage. *Int. J. Fracture,* **17**(2) 169-92.

[14] Fissolo A, Gourdin C, Chen Y, 2015 Investigations into the cumulative fatigue life of an AISI 304L austenitic stainless steel used for pressure water reactors: Application of a double linear damage rule. *Int. J. Fatigue,* **77** 199-215.

[15] Yung-Li L 2005 *Fatigue Testing and Analysis*, ed. Pan J, Hathaway R B et al. Butterworth-

Heinemann 57-76
[16] Pavlou D G 2000 Prediction of fatigue crack growth under real stress histories. *Eng. Struct.* **22**(12) 1707-13.
[17] Pavlou D G and Vlachakis N V 2004 Estimation of fatigue crack growth retardation due to crack branching. *Comput. Mater.* **29**, 446-452.
[18] Mikkelsen O, Rege K, et al. 2017 Numerical estimation of the stop holes-induced fatigue crack growth retardation in offshore structures taking into account the corrosion effect. in *The 27th Int. Ocean and Polar Eng. Conf.* International Society of Offshore and Polar Engineers.
[19] Pavlou D G 2015 Fatigue crack deflection-induced retardation based on the principle of the minimum potential energy. *Int. Review Mech. Eng.* **9**(3) 324-30.
[20] Pavlou D G, Vlachakis N V et al. 2003 Fatigue crack propagation trajectories under mixed-mode cyclic loading. *Eng. Struct.* **25**(7) 869-75.
[21] Subramanyan S 1976 A Cumulative Damage Rule Based on the Knee Point of the S-N Curve. *J. Eng. Mater-T.* **98**(4) 316-21.
[22] Srivatsavan R and Subramanyan S 1978 A Cumulative Damage Rule Based on Successive Reduction in Fatigue Limit. *J. Eng. Mater-T.* **100**(2) 212-214.
[23] Fatemi A and Yang L 1998 Cumulative fatigue damage and life prediction theories: a survey of the state of the art for homogeneous materials. *Int. J. Fatigue.* **20**(1) 9-34.
[24] Mesmacque G, Garcia S et al. 2005 Sequential law in multiaxial fatigue, a new damage indicator. *Int. J. Fatigue,* **27**(4) 461-67.
[25] Schoenborn S, Sonsino C M et al. 2015 Cumulative Damage of High-strength Cast Iron Alloys for Automotive Applications. *Procedia Eng.* **101** 440-49.
[26] Theil N 2016 Fatigue life prediction method for the practical engineering use taking in account the effect of the overload blocks. *Int. J. Fatigue.* **90** 23-35.

IOP Conf. Series: Materials Science and Engineering **700** (2019) 012026 doi:10.1088/1757-899X/700/1/012026

Numerical study on the ventilation performance of a livestock house built in porous panels in Cold Regions

Y Xu*, M Y Mustafa, R K Calay and B R Sørensen

Department of Building, Energy and Materials Technology, UiT/ the Arctic University of Norway

* Corresponding author: Yizhong.xu@uit.no

Abstract. This paper presents a numerical prediction on the assessment of natural ventilation on a livestock house built in porous panels, aided by Computational Fluid Dynamics techniques. A typical climatic environment in Cold Regions was taken into consideration. The ventilation performances have been assessed by the distribution of indoor air velocity, turbulence intensity and local mean age of air through the cases: the house without porous panels, with wind erosion porous panels, and with snow erosion porous panels. The study has found that the porous panels improve the indoor environment in term of the reduction of air velocity and turbulence, and the wind erosion panel performs better than the snow erosion panel. However, applying porous panels may have negative effects on air exchange effectiveness. The presented numerical model is easy to be constructed, flexible to investigate different case scenarios, and quick and cheap to get solutions.

1. Introduction

Cold Regions are part of the Earth system characterized by the presence of snow and ice at least part of a year. A broadly accepted definition of Cold Region is to use the 0°C isotherm in the coldest month of the year [1]. Climate in Cold Regions during warm and cold seasons is significantly different. In warm seasons, the climate could be mild. In cold seasons, the climate however goes to extreme, with the presence of low temperature, severe snow & ice, and strong wind. Tromsø is located over 300 kilometres north of the Arctic Circle at 69°40'33"N 18°55'10"E that is a typical city in Cold Regions. The warm season in Tromsø lasts less than 3 months, and the cold season about 5 month. The snowy period of a year lasts for 8 months, from October, the last year to May, the next year.

Porous panel systems like fences are widely applied to create a sustainable environment for human being, livestock and machinery under the harsh climate conditions in wintry Cold Regions. A fence is an artificial structure with openings of certain shape, size and distribution to effectively mitigate the damages caused by wind and sediments transported (i.e. drifting snow and accumulating ice). The significant feature of fences is in the form of windbreaks allowing the natural climate to be controlled [2-5]. Fence Porosity β, defined as the ratio between the perforated area and the total area of the fence screen, is widely recognized as the most influential factor determining the characteristics of windbreak and the performance of the fence [2] [4] [6].

An optimal fence should offer the possible maximum wind reduction over the possible longest shelter distance. These are two contradictory factors that maximizing wind reduction will sacrifice the length of the shelter distance. The concept of optimal porosity β_{opt} is to identify the porosity where the best balance between the wind reduction and the shelter distance is achieved. β_{opt} varies depending on the applications. For wind erosion fences, β_{opt} is in the range from 0.25 to 0.35 [2] [4] [6]. For snow erosion fences, β_{opt} is in the range around 0.5 [5]. The characteristics of windbreak induced by the porous panel system essentially affects the ventilation conditions in its confined space.

Many countries and societies have developed their standards related to ventilation for buildings. Norway follows the European standards. A recent movement of the European Committee for Standardization (CEN) standards is the publication of EN 19798:2017, dealing with the designs of natural ventilation in buildings (including non-residential buildings). A clear guidance of assessing the quality of air in a purely fence-confined environment is still absent today. Nevertheless, the utilization of porous materials in buildings to improve natural ventilation is increasing nowadays. In particular, for those natural ventilation dominant non-residential buildings. Porous materials like porous panels, ventilated clads and space boards are built in the buildings, together with walls and roofs to form an indoor environment, which ventilation and Indoor Air Quality (IAQ) can be assessed by the relevant standards.

Applying porous materials in livestock houses to improve the well-being of the animals is common practice in Cold Regions. These livestock houses are usually naturally ventilated. Attentions have been paid on the issues of the effect of the different porous materials and configurations, wind velocity distribution, the mean age of air (MAA), and thermal comfort, etc. [7-9]. These research works assessed the effectiveness of porous materials on natural ventilations, mainly focusing on the porous materials with the porosity of 0.5. This paper presents a study focusing on the assessment of ventilation conditions of a livestock house built in porous panels with different porosities, bearing in mind the significance of porosity influencing the performance of porous panels.

2. Methods

2.1. Feasibility of CFD technique

CFD technique has attracted increasing research and industrial interest today. Blocken B [10] described a bright future of the application of CFD technique in Wind Engineering, based on the research works done over the last 50 years. The feasibility of CFD technique on the performance of porous systems and indoor environment assessments have been proven by considerable researches with successes [8] [11-13]. Compared with physical experiments, CFD solutions create virtual testing models and virtual environments that require no cost on true model preparations and measurement equipment. Additionally, CFD solution provides detailed information in the calculated domain without similarity constraints, which is a big challenge for physical experimental tests. However, physical experiments are not dispensable, it is imperative that the accuracy and reliability of CFD solutions must be validated and verified against the related physical experimental tests.

2.2. Study case

Figure 1 shows the 2D domain for the study case. The livestock house was in the middle of the domain. The cross section of the house was the width of 9.6 m, the wall height of 3 m, and the roof ridge height of 5.2 m. There was a gap of 0.6 m between the wall and the eave, where the porous panels were situated. The size of the entire 2D domain was 60m high and 39.6 m long. Symmetry wall was applied to the top of the domain, which made half height of the domain needed to be modelled. The blockage ratio for the domain was less than 5%.

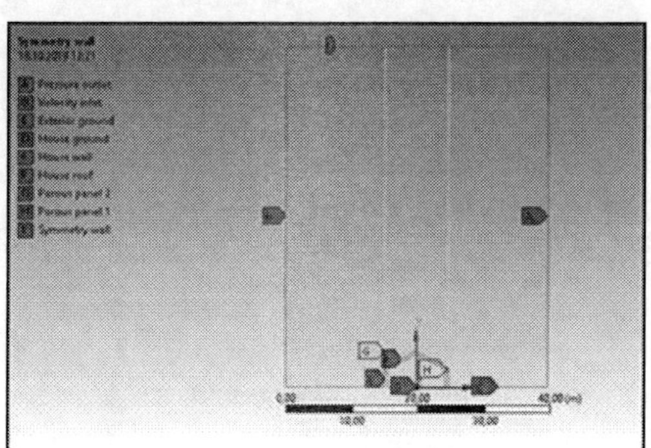

Figure 1. The creation of 2D domain.

IOP Conf. Series: Materials Science and Engineering **700** (2019) 012026 doi:10.1088/1757-899X/700/1/012026

For the convenience of study, the cell zone enclosed by the house was named as the interior zone, and the cell zone outside the house was the exterior zone.

Three case scenarios were investigated in this paper as follows:

- Case A: The house without porous panel ($\beta = 1.00$) that the gap is totally open
- Case B: The house with porous panel of $\beta = 0.50$ that the porosity is in the optimal range for snow erosion
- Case C: The house with porous panel of $\beta = 0.27$ that the porosity is in the optimal range for wind erosion

ANSYS FLUENT workbench v.18.2 was employed in the simulations.

2.3. Boundary condition

Yr is the joint online weather service from the Norwegian Meteorological institute and the Norwegian Broadcasting Corporation. According to its report, the maximum average wind speed in Tromsø was 4.3 m/s from September 2018 to September 2019, and the average temperature was -3.4°C [14]. Neutral stability condition is assumed to apply the atmospheric boundary layer (ABL). Hence, the wind profile in this modelling follows the power law relationship as the following:

$$u = u_r(y/y_r)^{\alpha} \tag{1}$$

where: u is the wind speed at the height y, and u_r is the known wind speed (4.3 m/s) at a reference height y_r (10m). The exponent α is 0.143.

The inlet velocity profile in this study was described in equation (1). The inlet boundary condition, including wind velocity, turbulent kinetic energy, turbulent dissipation rate and specific turbulent dissipation rate, was written in programming C language and interpreted into the model.

At the pressure out, the Gauge pressure was set to zero. Both of the backflow turbulence intensity and the backflow turbulent viscosity ratio were set as 5% in the simulation. A no-slip condition was applied onto all of the walls in the domain.

2.4. Element arrangement and mesh independence

The arrangement of finite elements in CFD simulation is vital to the accuracy and computational efficiency. Figure 2 is the meshed domain. Quadrilateral elements were applied in the entire domain. Cares have been taken as follows:

- Symmetry wall was applied to the top of the domain that made half height of the domain needed to be meshed;
- Since the house interior zone was of interest for this study, dense elements were arranged in this area, and coarse elements were applied to the house exterior zone;
- In the ground wall region, elements were inflated from the ground-adjacent layer up to ten layers, with the total thickness of 0.06m at the growth rate of 0.2, which ensured each ground-adjacent cell's centroid within the viscous sublayer $y^+ < 5$. Such treatment was essential to the fidelity of simulating wall-bounded turbulent flow by $k - \varepsilon$ turbulence model [15];
- Apart from the ground wall, the elements near the regions of the rest walls were refined.

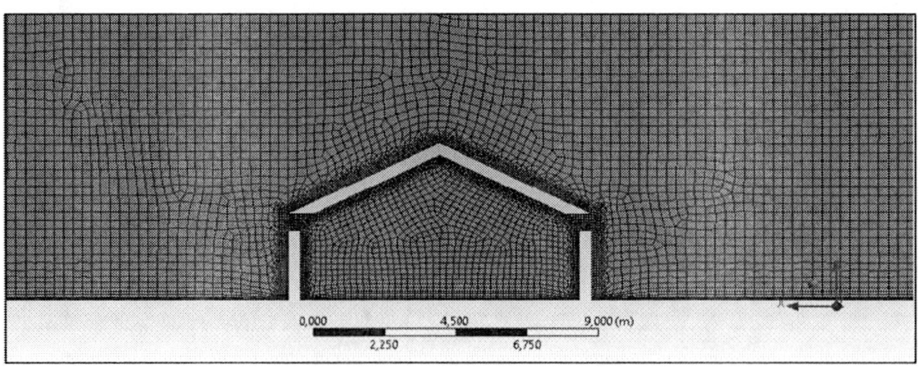

Figure 2. The arrangement of elements.

In the present CFD simulations, the mathematical model was governed by the RANS equations, describing time-averaged motion of fluid flow. The additional unknown variables (called Reynolds stresses) exhibited from the RANS were solved by the Standard $k-\varepsilon$ turbulence model. The model sensitivity analysis in this paper focused on the assessment of mesh independence. Mesh independence is that the numerical solution must be ensured to be independent of the mesh resolution. A good quality of mesh should be efficient and economic which is as close to the ideal mesh as possible. During the mesh sensitivity analysis, the velocity magnitudes and turbulence intensities at the positions of comparing vertical lines (y-axial) were selected as the monitoring variables. These positions of comparing lines were carefully chosen to reflect the sensitive regions in the domain, such as the porous panel influencing zones. The initial mesh guess was used by ANSYS default mesh, then increasing the quality of mesh by a factor of 1.5 as the second guess, and so on. When the solution was close to the one solved by the last mesh, decreasing the quality of mesh by a factor of 0.5 until the mesh independence was identified. Mesh independence was achieved at the number of elements of 33973.

2.5. Porous jump

The characteristics of the porous panel itself was not physically modelled in the simulations. It was treated as a pressure discontinuity surface by applying a porous jump condition, where momentum was absorbed from the flow as a momentum sink and added to the Reynolds-averaged Navier-Stokes (RANS) fluid flow equations.

For turbulent flow around a perforated plate like planar porous panels, the viscous loss term can be neglected [11]. Hence, the face permeability α for the porous panel was assigned as $1e+20m^2$. The inertial resistance C_2 was associated with the resistance coefficient or pressure loss coefficient k_r as follows:

$$C_2 = k_r/\Delta m \tag{2}$$

$$k_r = 0.52\,(1-\beta^2)/\beta^2 \tag{3}$$

where Δm is the thickness of the porous panel, and β is the porosity.

Porous jump must be constructed as an interface in the modelling, which porous cells are treated as 100% open to the domain that differs the fact where the porous panel is only partially open to the airflow. Presuming the flow rate unchanged with or without the fence of $\beta = 0.27$, $u_{27\%open} = 3.7u_{100\%open}$. Therefore, k_r needs to be adjusted to k_r':

$$k_r' = k_r\left(u_{27\%open}^2/u_{100\%open}^2\right) \tag{4}$$

Table 1 lists the calculated variables for the porous jump (the porous panel), where: α, Δm and C_2 are the input parameters for modelling the porous panels.

Table 1. The variables for the porous jump (the porous panel).

Item	Unit	Case A	Case B	Case C
Panel height $\boldsymbol{H}$	m	N/A	0.6	0.6
Porosity $\boldsymbol{\beta}$		N/A	0.50	0.27
Panel thickness$\boldsymbol{\Delta m}$	m	N/A	0.005	0.005
Pressure loss coefficient $\boldsymbol{k_r}$		N/A	1.56	6.61
Adjusted pressure loss coefficient $\boldsymbol{k_r'}$		N/A	6.24	90.67
Inertial resistance $\boldsymbol{C_2}$	m^{-1}	N/A	1248	18134
Face permeability $\boldsymbol{\alpha}$	m^2	N/A	1e+20	1e+20

2.6. Local MAA

In building physics, local MAA is defined as the average time for a parcel of air from a supply inlet are to any location in a ventilated room. It is used to define an air change effectiveness parameter named as the air change ratio (ACR).

MAA is not automatically included as a predefined variable in ANSYS FLUENT. It can however be defined in the pre-processing stage as an additional variable by the method in (5). MAA, denoted as T_{mean}, is obtained by solving the following advection equation as:

$$d\,T_{mean}/dt = 1 \tag{5}$$

In ANSYS FLUENT Solvers, scalar quantities are multiplied by the density ρ in the transport equation. Therefore, the advection equation is written as:

$$d(\rho T_{mean})/dt = \rho \tag{6}$$

The derivative with respect of time of the source is ρ that the integral itself is time. When integrated, the age scalar will equal to the residence time. Equation (6) was written in programming C language and interpreted into Fluent during the simulations. The term on the right hand side of equation (6) is the source term that should be assigned to the cell zones. A diffusivity term should be assigned to the properties of the air.

2.7. Other considerations

No-slip condition was applied to all of the walls. Standard $k - \varepsilon$ turbulence model was selected to address the 'closure problem' of the governing equation- RANS equations in the study.

The convergence criteria were assessed by monitoring the scaled continuity, x-velocity, y-velocity and turbulence kinetic energy residuals, and the convergence absolute criteria was 0.001. The solution initialization run the hybrid method at 10 of iterations. The ultimate solution was obtained on the Second Order Upwind discretization.

3. Results and discussions

3.1. Indoor air velocity

When indoor air velocity above 0.2 - 0.3 m/s, it may cause draughty indoor environments for livestock houses [16-17]. It has been reported that the risk of infection from diarrhoea, respiratory diseases increased when young calves (0-90 days old) were kept in draughty regions of a calf house [18]. Therefore, understanding the relationship between outdoor wind speed and indoor air velocity is important to livestock house design.

Figure 3 displays the contours of velocity magnitudes in the clip range from 0-0.86 m/s that was 80% reduction of the outdoor wind velocity of 4.3 m/s. From Figure 3, it can be observed that the indoor air velocities were reduced with reducing the porosity of the fence. Considering the breather height of 1m for the animals, only Case C offers the covering region of the air velocity less or equal to 0.86 m/s at such a height.

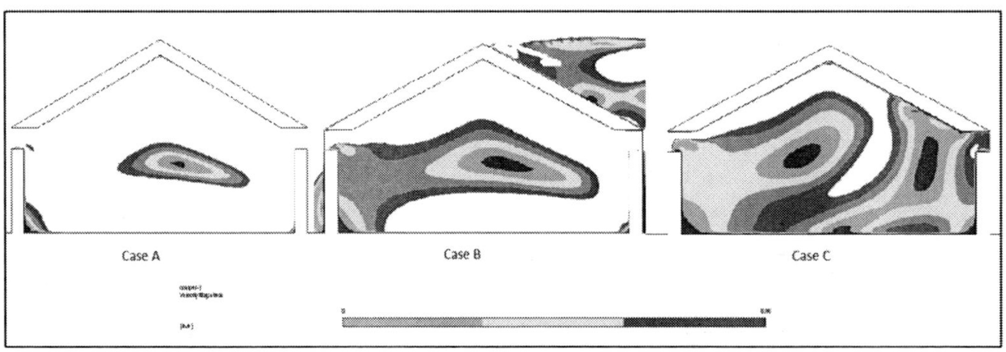

Figure 3. Contours of velocity magnitude in the clip range from 0 to 0.82 m/s.

Tromsø often experiences heavy wind in the cold seasons. The study found that this livestock house cannot create an indoor environment with the indoor air velocity less than 0.2 - 0.3 m/s, which may cause draughty indoor environment that is a potential risk for animal diseases.

3.2. Indoor turbulence

Turbulence is another factor to induce draughty indoor environment. High-level turbulence in livestock houses may cause animal diseases and discomfort. Fanger's draught model is commonly used and accepted in Europe that proposed a DR (draught rate) number to qualify the indoor environment. This model use turbulence intensity to evaluate the level of turbulence. CEN standards suggest that an average

turbulence intensity 40% for draught comfort design in mixing ventilation. Figure 4 shows the contours of turbulence intensity in the clip range less or equal to 40% for the livestock house.
The turbulence level in the livestock was reduced with the reduction of the porosity of the fence. Case C create an almost entire indoor environment with the turbulence intensity no more than 40%, while Case A suffered the strongest turbulence level that its turbulence intensities were almost all above 40%.

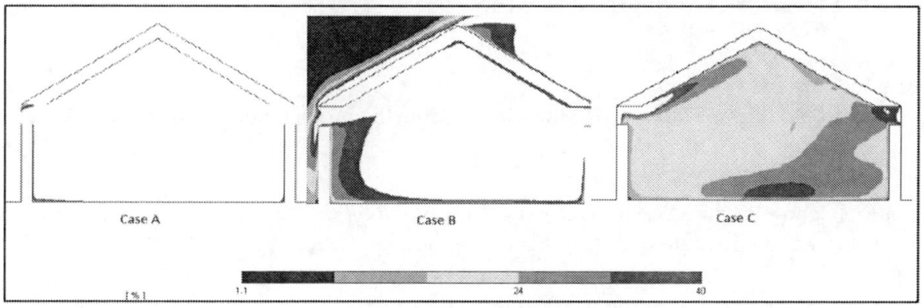

Figure 4. Contours of turbulence intensity in the clip range no more than 40%.

3.3. Indoor MAA
By adding equation (6) as a user defined scalar (UDS) into the simulations, the information of local MAA in the entire domain can be obtained. Figure 5 is the contour of MAA for Case C. According to the simulation results, the maximum MAA value is 24 seconds, 67 seconds, and 110 seconds for Case A. Case B and Case C, respectively. The presence of the fences has retarded the air travel in the house with significance.

Figure 6 is a comparison of the distribution of local MAA at the breath height of 1m for each case. The local MAA was increased with the reduction of the fence porosity. Wind erosion fence (with the porosity $\leq$ 0.35) creates a more complex structure of flow than snow erosion fence (with the porosity around 0.50) does, resulting in a region of recirculating air leeward of the fence [12]. It is the reason to explain why there was fluctuations on the pattern of Local MAA for Case C in Figure 6.

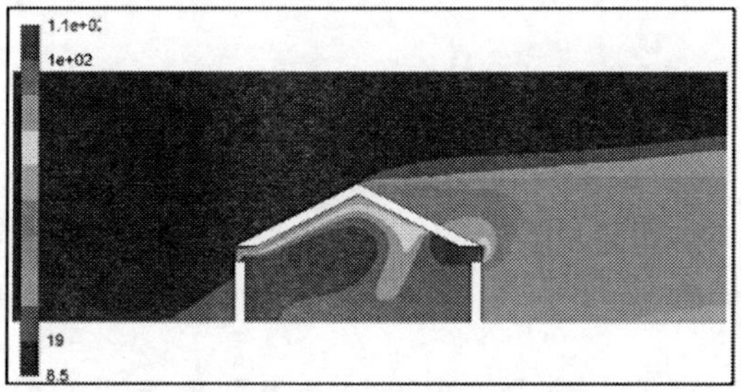

Figure 5. Contour of MMA for Case C.

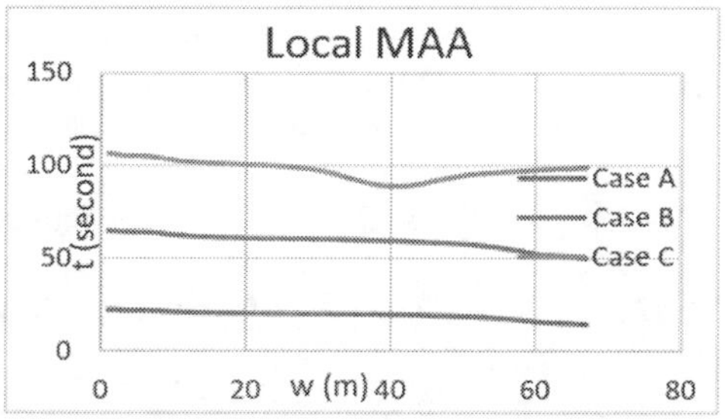

Figure 6. Comparison of the local MAA at the breath height of 1m.

4. Conclusions

This paper presents a numerical study through the analyses of indoor air velocity, turbulence and local MAA for a livestock house in Cold Regions. The conclusions are as follows:

1. The porous panel is numerically modelled as a porous jump, instead of creating a true physical panel model. By this mean, the CFD model is easy to be constructed in terms of domain creation and element arrangement, flexible to different scenarios, and quick and cheap to get solutions. By adding an additional transport equation into the modelling, the model can explicitly present the information of MAA in full resolution.
2. The built in porous panels significantly improve the indoor environment in terms of air velocity and turbulence if subjected strong outdoor wind. The study has found that the wind erosion panel performs better than the snow erosion panel. It is useful information that most porous materials used in livestock houses are the porosity around 0.5 today.
3. The use of porous materials in livestock house may have negative impact on indoor air change effectiveness (ACE).

4. The studied livestock house may cause draughty indoor environment under the wintry climate in Tromsø. Improvement of the livestock house in design and other measures is of necessity for the welfare of the animals.

The limitations of this study mainly lie on as follows:

1. The current work is informative, since the model has not been fully validated. The validation will be done in the future work.
2. Two equation based turbulence models are weak to predict airflow turbulence. As such, the results of the indoor turbulence in this work may suffer in question.

Acknowledgement
The authors would like to acknowledge the support from the project number A74234 (BA Center Nord).

References

[1] Bates R and Bilello M 1966 *Defining the Cold Regions of the Northern Hemisphere* (New Hanpshire:Cold regions Research and Engineering Laboratory).

[2] Hagen L 1976 Windbreak design for optimum wind erosion control *Proc. of the Sympos.: Sheltbelts on the Great Plains* (Denver) 31-34.

[3] Bradley E and Mulhearn P 1983 Development of velocity and shear stress distributions in the wake of a porous shelter fence *J. Wind Eng. Ind. Aerod.* **15**,145-56.

[4] Dong Z, Luo W, Qian G and Wang H 2011 Evaluating the optimal porosity of fences for reducing wind erosion *J. Science in Cold and Arid regions.* **3**,1-12.

[5] Tabler R 2003 *Controlled Blowing and Drifting Snow with Snow Fences and Road Design* (USA:Transportation Research Board of the National Academies)

[6] Cornelis W and Gabriels D 2005 Optimal windbreak design for wind-erosion control J. *Arid Environ.* **61**, 315-32.

[7] Bruce J 1981 *Ventilation and Temperature Control Criteria for Pigs* in *Environmental* (London: Butterworths) 197-216.

[8] Norton T S, Grant J, Fallon R and Sun D 2010 Assessing the ventilation performance of a naturally ventilated livestock building with different eave opening conditions *J.Comput. Electron. Agr.* **71**, 7-21.

[9] Lago A, McGuirk S M, Bennett T B, Cook N B and Nordlund K V 2006 Calf respiratory disease and pen microenvironments in maturally ventilated calf barns in winter *J. Dairy Sci.* **89**,4014-25.

[10] Blocken B 2014 50 years of computational wind engineering: past, present and future *J.Wind Eng. Ind. Aerod.* **129**, 69-102.

[11] Bouidin P and Wilson J 2008 Windbreak aerodynamics: is computational fluid dynamics reliable? *J.Boundary-Layer Meteorol.* **126**,181-208.

[12] Xu Y, Calay R K and Mustafa M Y 2018 Feasibility of computational fluid dynamics for analyzing airflow around porous fences *J. Heat Transfer Eng.*

[13] Buratti C, Mariani R and Moretti E 2011 Mean age of air in a natually ventilated office: Experimental data and simulations *J.Energ. Buildings.* **43**, 2021-27.

[14] Yr, "Weather statistics for Tromsø (Troms)," Yr, 18 10 2019. [Online]. Available: https://www.yr.no/place/Norway/Troms/Troms%C3%B8/Troms%C3%B8/statistics.html.

[15] Chmielewski M and Gieras M 2013 Three-zonal wall function for k-e turbulence models *J. CMST*,**19**, 107-14.

[16] Webster A 1984 *Calf Husbandary, Health and Welfare* (London: Granada Technical Books)

[17] Martig J, Boss P H, Nicolet J and Steck F 1976 Etiology and predisposing factors in respiratory disease of milk-fattened veal calves *J. Livest. Prod. Sci.* **3**, 285-95.

[18] Lundborg G K, Svensson E C and Oltenacu P A 2005 Herd-level risk factor for infectious diseases in Swedish dairy calves aged 0-90 days *J. Prev. Vet. Med.* **68**, 123-43.

Brief review on the limit state function of dynamic scour protections

T Fazeres-Ferradosa[1,2]*, M Welzel[3], F Taveira-Pinto[1,2], P Rosa-Santos[1,2] and J Chambel[1,2]

[1] CIIMAR - Interdisciplinary Centre of Marine and Environmental Research of the University of Porto, Terminal de Cruzeiros do Porto de Leixões, Av. General Norton de Matos, S/N, 4450-208 Matosinhos, Portugal

[2] FEUP - Faculty of Engineering of the University of Porto, Department of Civil Engineering, rua Dr. Roberto Frias, s/n, 4200-465, Porto, Portugal

[3] Leichtweiß-Institute for Hydraulic Engineering and Water Resources, Technische Universität Braunschweig, Beethovenstraße 51a, 38106 Braunschweig, Germany

* Corresponding author: tferradosa@fe.up.pt

Abstract. Offshore foundations, namely for offshore wind, wave and tidal applications, often require the use of scour protections. Rip-rap scour protections are an important element of the foundation to ensure that the natural frequency stays within the design limits. Scour protection design still presents a remarkable empirical nature, which typically leads to uncertainty on their behaviour under extreme met-ocean conditions. Therefore, reliability assessment of scour protections has been seen as a possibility to account for design uncertainty and to optimise the scour protections. However, the definition of a suitable limit state function is still a matter of research focus, namely, regarding the proper definition of the acceptable damage level for dynamic scour protections. This research provides a brief review on the recent studies related to both the limit state function and the calculation of damage numbers through bathymetric data. A discussion is raised on how the methodologies for calculating the damage number may influence the limit state function and a theoretical example is given to assess the effects on the probability of failure. Results have shown that the acceptable damage number requires a clearer definition, which should be based on the number of layers of rock material and the area of filter exposure. In addition, this research highlights the need for alternative ways to assess damage.

1. Introduction

Offshore foundations, namely for offshore wind, wave and tidal applications, often use rubble-mound material as the main component of the scour protection. Rip-rap scour protections are an important element of the foundation to ensure that the natural frequency stays within the design limits. However, scour protection design still presents a remarkable empirical nature, which typically leads to uncertainty under extreme met-ocean conditions [1].

Aiming at uncertainty reduction and a better perception of the protection's failure, recent research has been developed on the topic of reliability based design of scour protections, *e.g.* [1], [2] and [3]. These works follow the trend of reliability and risk analysis applied to other maritime and coastal structures, commonly made of rock material, *e.g.* breakwaters [4]. The application of these techniques to rubble-mound structures is still far from the mature development registered in other types of known structures, *e.g.* structural design of tubular elements for an offshore platform or mooring lines for floating foundations. In [5], a review of scientific and technical challenges of reliability analysis of scour protections, for offshore wind foundations, is extensively given. A key aspect identified in [5], which requires further insight and discussion for a proper application of reliability techniques, is the definition of an accurate limit state function, which might be suitable to define the protections' failure. While [3] extensively addresses the failure mode known as the "erosion of the top layer" for both static scour protections, without movement of rock material, and dynamic scour protections, where movement is allowed, there is still a lack of studies considering other failure modes (*e.g.* [6]).

In order to understand the failure mechanism of the rock armour material at these protections, more knowledge is required in terms of the physical scour pattern and damage severity at the top layer. Therefore, physical modelling studies have been extensively dedicated to the analysis of damage in scour protections, namely, under waves and current combined (*e.g.* [7, 8]). In [8], a physical model is used to discuss the design limits of the damage number (S_{3D}), introduced by [9], to describe the failure of dynamic scour protections. The results showed that further research on the damage definition and range is required. It was latter seen in [5] that the methodology used to derive the damage number from bathymetric profiles influences the damage numbers obtained, thus ultimately influencing the reliability assessment of a scour protection, by means of its limit state function.

In addition, currently existing methods to define the protection's failure in physical models are often based on visual criteria (e.g. [9, 10]). Despite the importance of visual damage criteria, they may contribute to cases where damage is sometimes undetected or miss-evaluated. Even though several works have been performed on alternative ways of evaluating damage (e.g. [11, 12]), the damage assessment based on the damage number combined with visual observation remains as the most used criteria for rip-rap scour protections at offshore wind foundations.

The present research aims at highlighting the importance of the acceptable damage number and its calculation for a proper reliability assessment. The effects of this parameter in the limit state function are shown and discussed, by means of a physical model and the computation of damage according to different methodologies. It is proven that the sub-area arrangement used to analyse the bathymetric profiles influences the damage number significantly. An example is used to identify the potential effects that damage assessment methodologies may have on the limit state function of a scour protection.

2. Methodology

2.1. Damage number based design

In [9], the design of dynamic scour protections for monopile foundations is based on a predicted damage number (S_{3D}), whose predictive equation was derived from an extensive set of physical model tests. The predictive equation (1) is a regression based equation obtained from a set of damage numbers calculated directly from the bathymetric profiles taken after a certain number of waves (N).

$$\frac{S_{3D}}{N^{b_0}} = a_0 \frac{U_m^3 T_{m-1,0}^2}{\sqrt{gd}\,(s-1)^{\frac{3}{2}} D_{n50}^2} + a_1 \left(a_2 + a_3 \frac{\left(\frac{U_c}{w_s}\right)^2 (U_c + a_4 U_m)^2 \sqrt{d}}{g D_{n50}^{\frac{3}{2}}} \right) \quad (1)$$

where U_c is the average current velocity, s is the ratio between sediment's density (ρ_s) and water density (ρ_w), g is the gravitational acceleration, d is the water depth, U_m is the orbital bottom velocity and w_s is the sediments' fall velocity. $T_{m-1,0}$ is the energy spectral wave period, which for a JONSWAP spectrum, with $\gamma = 3.3$ can be obtained from the peak period (T_p) as $T_{m-1,0} = m_{-1}/m_0 = 1.107T_p$. In (1) b_0, a_0, a_2 and a_3 are equal to 0.243, 0.00076, -0.022 and 0.0079, respectively. The constants a_1 and a_4 depend on the wave-current relative direction, the current velocity and the Ursell number. A detailed calculation of these parameters is given in [9].

The damage numbers derived from the bathymetric profiles, *e.g.* those which are not predicted but, instead, obtained through physical modelling of prototype conditions, are obtained as a function of the eroded volume (V_e) over a specified sub-area (A_{sub}) of the scour protection itself, (2). In the original formulation, [9] defines the sub-area as being equal to the cross-section of the monopile foundation, with D_p being the pile diameter (Figure 1). Then the representative damage number of the protection is equal to the maximum value for all the S_{3D} calculated from Eq. (2), *i.e.* S_{3Dmax}.

$$S_{3D} = \frac{V_e}{D_{n50}A_{sub}} = \frac{V_e}{D_{n50}\pi \frac{D_P^2}{4}} \quad (2)$$

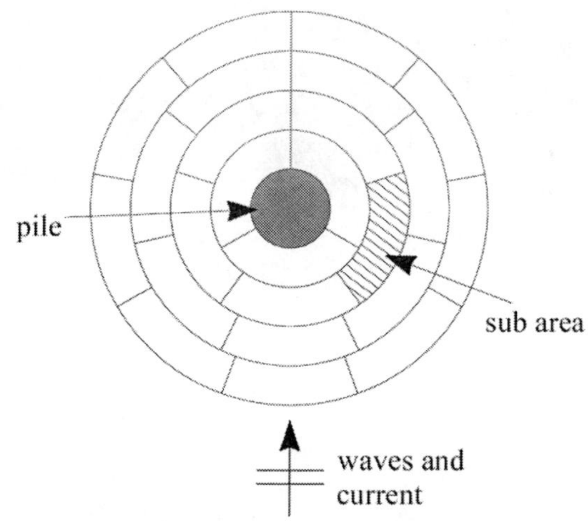

Figure 1. Scheme of a scour protection divided in sub-areas equal to the pile cross-section.

De Vos et al. [9] proposes a design procedure for dynamic scour protections by means of Eq. (1), where local sea-state conditions, the rock material characteristics and a predefined damage level (*e.g.* $S_{3D} = 0.5$) serve as input. Then the equation can be solved in order to calculate D_{n50}, thus providing the nominal median diameter of the armour layer. For further details and discussion of this methodology, [3] and [9] are recommended. As defined in [9], the damage number provides a measure on the number of layers of rock material, which have been removed over the specified sub-area. [9] also defines failure visually as in [10], *i.e.* the scour protection fails when the filter layer is exposed over and area equal or larger than $4D_{50}^2$.

The underlying design assumption of this methodology is the proper definition of an acceptable damage level. In [9], a value of S_{3D} smaller or equal to 0.25 is suggested for scour protections without movement of the armour material (static scour protection), whereas a value of S_{3D} between 0.25 and 1.00 is suggested for dynamic scour protections. S_{3D} values above 1.00 were identified through physical modelling as being within the failure domain of the scour protection.

The reference values presented by [9] were identified as conservative since dynamically stable scour protections were found for larger values of S_{3D}. Research performed by [3, 8] concluded that the division of the physical model into sub-areas has an important effect on the maximum damage number. Alternative divisions are suggested in [5], which concludes that further discussion on the definition of the acceptable damage is required. However, this implies a deeper study on the definition of the sub-areas used in (2), which ultimately affect the regression leading to (1), thus having an effect on the design methodology of dynamic scour protections.

Figure 2 shows a sub-area arrangement alternative to [9], which was applied in [13]. In this arrangement, the sub-areas vary between each other and are divided between concentric circular rings. These sub-areas vary depending on the ring and sector on which the sub-area is located. The areas of the outer rings are larger than the ones near the pile, but the number of sectors remains the same. Since the eroded volume is being evaluated in different sub-areas, differences are expected when (2) is applied, thus invalidating a direct comparison with the limits for the acceptable damage discussed in [9]. In [8], a comparison between the data presented in [9] and [13] is made and it is concluded that the damage numbers obtained for dynamic scour protections can largely exceed the value of S_{3D} equal to 1.00 (see [3, 5]). [5] discusses the use of an arrangement of overlapped sub-areas that may potentially provide a more accurate knowledge on the damage pattern occurring at the protection. However, the literature shows that such concept is yet to be systematically applied to a large set of bathymetric data, as in [9] and in a shorter extent as in [13]. The present application demonstrates that the use of different sub-areas not only impacts directly the assessment of the damage number, but it also affects the assessment of the reliability of a scour protection (Section 3.1).

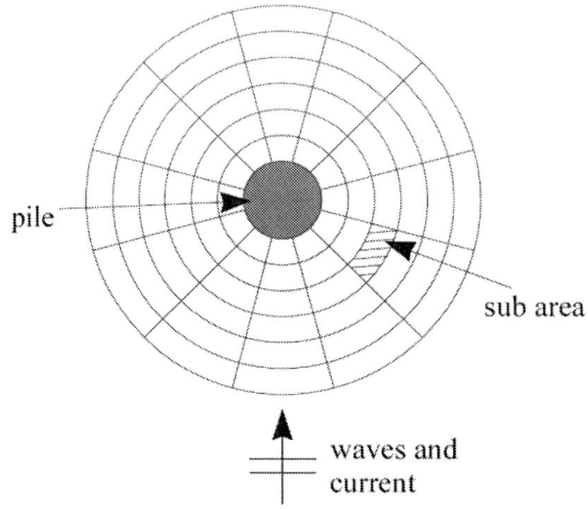

Figure 2. Scheme of sub-areas arranged in concentric circular rings.

2.2. Reliability of dynamic scour protections

The definition of the ultimate limit state of a dynamic scour protection, for the erosion of the top layer failure mode, can be defined as in [1], (3) where the reference level of $S_{3D} = 1.00$ is used as the resistance parcel of the scour protection (R) and the expected damage for the local loading conditions is calculated from (1), which defines the limit state function g(X). A full description on how to perform reliability analysis for dynamic scour protections is given in [1, 2 and 3].

$$g(X) = R - S_{3D} = 1.00 - S_{3D} \tag{3}$$

where X represents the vector of random variables included in the calculation of S_{3D}, *e.g.* U_m, Uc, T_p. While (3) seems to be linear, the design procedure is dependent on relation given in (1), which is highly non-linear, in particular, when considering the design variable D_{n50}. Therefore, and considering the relatively low computation effort required to simulate (3), it was shown by [3] that Monte-Carlo simulation method can be applied to assess the domain of g(X). In order to obtain a measure of the protection's reliability, the probability of failure (P_f) can be calculated, as in (4).

$$Pf = \frac{\#\left(g(X) \leq 0\right)}{n} = \frac{\sum_{1}^{n} I(g(X))}{n} \tag{4}$$

where n is the number of simulations, *i.e.* the number of times g(X) is calculated for random hydrodynamic conditions and remaining variables of (1). I(g(X)) is an indicator function equal to 1 if $g(X) \leq 0$ or 0 if $g(X) > 0$. If $g(X) \leq 0$, the scour protection fails, i.e. the top layer is eroded, because the acceptable level of damage, *i.e.* the $S_{3D} = 1.00$, has been exceeded for certain random conditions.

Here the definition of failure is linked to the non-compliance with the design criteria. In practical situations, the actual physical failure (exposure of the filter) may not have occurred yet. However, as the design criterion is not being met the protection is considered to be in eminent failure. Taking this definition into consideration, it is noted that if R changes from 1.00 to any other value, then the probability of failure will also change.

Therefore, it is important to understand, which value of R should be used. The changes in the sub-areas definition may lead to changes in the acceptable damage limits as defined in [9]. Moreover, the calculation of the S_{3D} from the bathymetric data has a direct impact on the regression fit used to develop the predictive design formula of the S_{3D}, also included in g(X).

2.3. Physical model

The aforementioned changes, influence the probability of failure. In this research, a minor example of a physical model test is provided to show that the damage number may vary depending on the sub-area arrangement and to provide a prospective discussion on the potential effects on the probability of failure.

The physical model consists in a protected monopile foundation, with rip-rap scour protection with an extent of $5D_p$ and an outer slope of 1:3. The model had a geometrical scale of 1:50 with Froude's similitude. The test included the effect of waves and current combined, with a D_{n50} of 4.2 mm and a protection thickness of $3D_{50}$, which is within the tested range in [9]. The rock material density was 2630 kg/m^3. A ratio of 0.84 between the median diameter (D_{50}) and the nominal median diameter (D_{n50}) was considered.

A total of 9000 waves were performed and no visual exposure of the filter was noticed. The wave-current flume had a total length of 32 m with a cross-section, 1 m wide by 1 m height. Full details of the experimental setup, equipment, measurements accuracy and testing conditions are given in [13]. The scour protection was subjected to a mean significant wave height of 0.11 m (5 m in prototype) and a peak period of 1.60 s (11 s in prototype). A depth averaged current velocity of 0.15 m/s was measured at 40% of the mean water level, counting from the sand bed to surface. Figure 3 shows the drained tested model before testing and after 9000 waves. The scour protection showed very minor movements of the scour protection, being between a static stability and a dynamically stability with small movement of the armour material.

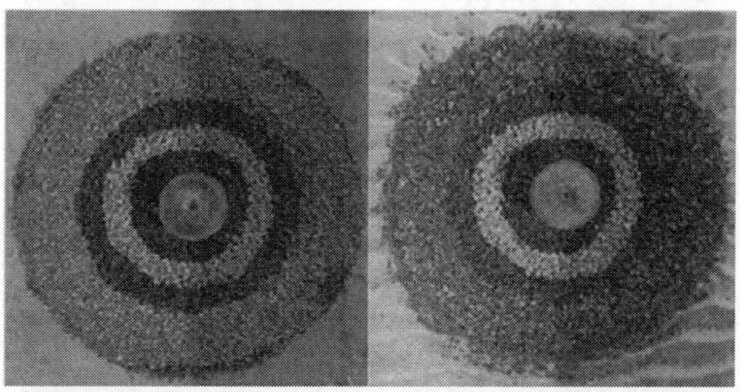

Figure 3. Top view of monopile foundation and scour protection before testing (left) and after 9000 waves (right) – current and waves from bottom to top.

3. Results and discussion

3.1. Deriving the damage number from different sub-areas arrangements

Based on the physical model described, the bathymetric data was analysed under the sub-areas arrangement from [9] and [13] (Figure 4 and Figure 5). Despite the reasonable agreement in the damage pattern, where the majority of the scour occurs near the pile and at the edge region, the difference in the maximum damage number is significant and varies roughly 68% from the methodology presented in Figure 4 to Figure 5. The methodology [13] provides a larger value near the pile. This was expected due to the smaller area size considered. It can be noted that the method presented in [13] is concentrating the damage at the pile centre more than at the edges. The scour pattern is different from the typical shapes

IOP Conf. Series: Materials Science and Engineering **700** (2019) 012027 doi:10.1088/1757-899X/700/1/012027

reported in [3, 8, 9]. At the upstream region more edge scour should be expected. The present model shows a larger edge scour at the lee-wake region. The remaining damage at the upstream side of the pile is consistent with typical damage under waves and current combined. Nevertheless, the physical model enables to prove the main assumption, i.e. the damage number is influenced by the sub-area size and the analysis applied to the bathymetric data. Results show that the damage number varies considerably depending on the methodology used, thus the reliability assessment is affected in the sense that the limits for static and dynamic stability, formerly introduced by [9], may require a re-evaluation if different sub-areas are used. This is particularly important as it can affect the limit state function, g(X), particularly if the failure of the protection (failure criterion) is defined in alternative ways, *e.g.* different areas of filter exposure.

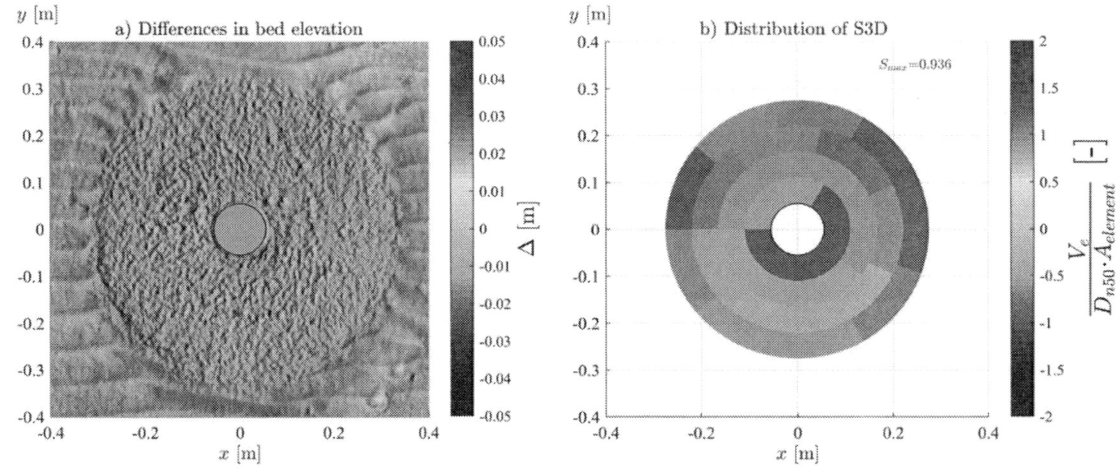

Figure 4. (a) Bathymetry differences between 0 and 9000 waves, (b) sub-areas analysis according to [9], S_{3Dmax}=0.936 for 24 sub-areas equal to the pile cross-section (waves from bottom to top).

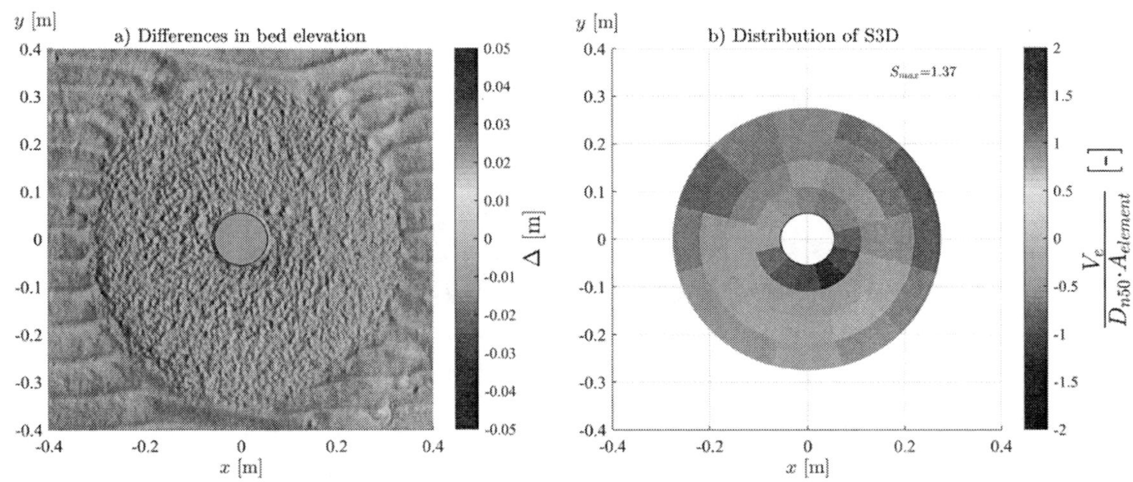

Figure 5. (a) Bathymetry differences between 0 and 9000 waves, (b) sub-areas analysis according to [13], S_{3Dmax}=1.37 for 48 sub-areas with radial division (waves from bottom to top).

3.2. Theoretical example of the influence in the probability of failure

Considering different values of acceptable damage leads to a direct influence on the probability of failure in the sense that the design criterion is being more or less restrictive depending on the chosen R (Eq. 3). As discussed previously these may change depending on the methodology used to calculate the S3D value from bathymetric data, whereas such methodology may also lead to alternatives to Eq. (1).

In this section, the data from Horns Rev 3 offshore wind farm, fully reported in [1] and [3], is used to show how the probability of failure is affected by the S_{3D} value chosen for Eq. (3). Monte-Carlo method was used to evaluate the limit state function g(X) based on a bi-variate version of the kernel density estimation method for n = 300 000 simulations, which has been shown in [1] to be a sufficient number to obtain a stabilized value of the probability. The input of the random variables that form vector X is also the same as defined in [1]. In brief, the significant wave height ranged from 0.14 m to 6.11 m and the peak periods from 1.60 s to 22 s (see *e.g.* [1, 3]).

For the present case three curves are presented in Figure 6 for different values of D_{50} ($0.84D_{n50}$). The first curve shows the probability of failure for R = 0.25, the limit for static stability, the second curve for R=1.00, the limit for dynamic stability (both as defined in [9]). The third curve is calculated for R=2.00. The use of larger values of R has a physical meaning, which corresponds to considering a less restrictive design criterion. The meaning is that the designer is assuming that more layers of stones, *i.e.* rock armour material, can be removed from the scour protection before the filter is exposed (*i.e.* before failure).

Figure 6 shows that increasing the acceptable damage in the design criteria leads to smaller probabilities of failure, thus meaning that the protection for the same random loading conditions fulfils the design criterion more often. As the scour protection is designed to be dynamic, applying an acceptable damage of R = 0.25, compatible with static stability, is leading to large probabilities of failure, thus meaning that, for the same random loading conditions, the protection is often having an estimated S_{3D} larger than 0.25. The effects on the probability of failure shown in Figure 6 are in agreement with those given in [2], where a set of copula-based models is used. As the D_{50} increases there is no damage on the protection for the considered conditions.

As the definition of a proper acceptable damage is a key aspect to estimate the reliability of the scour protection, the underlying question, is which value of R should be used. The answer to this relies on the failure definition, the method used to analyse the bathymetry and finally on the predictive equation used.

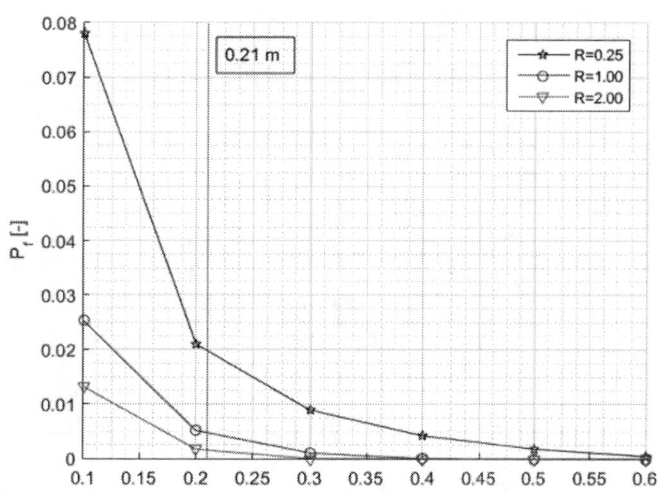

Figure 6. Example of the influence of the acceptable damage number as introduced by [9] in the probability of failure.

Thinking that the S_{3D} can represent the number of layers removed over the analysed sub-area, then the probabilities derived with R = 0.25 could be analysed under a static stability criterion, *e.g.* [14] which is based on the threshold of motion and the critical shear-stress. A comparison between probabilities derived from statically stable and dynamically stable criteria can be seen in [1]. Although the present calculation serves its purpose of showing the influence of the acceptable damage, for statically stable scour protections, the use of g(X) based on [14] is recommended by [3].

Considering, for example, a scour protection with a thickness of rock material equal to $3D_{50}$. It seems appealing to state that depending on the sub-area being considered the exposure of the filter occurs for $R = S_{3D} = 3$. A theoretical reliability analysis for the physical model of section 3.1, *i.e.* $D_{50} = 0.21$ m and R = 3, yields a probability of failure of 1.9×10^{-4} (see Table 1), which is lower than the ones presented for the remaining values of R (Table 1). In this sense, the values shown in Figure 6 may be seen as conservative assessments of the protection's reliability.

Nevertheless, an important problem is that the limit state function from (3) does not directly consider the area of filter layer exposure, namely the area of $4D_{n50}^2$ used to identify the failure visually, as in [9] and [10]. Therefore, present research highlights the importance of developing new methodologies to calculate the damage number that might be directly related to a meaningful area of filter exposure. Thus contributing to improving the definition of g(X), both to define a proper parcel R and, eventually, a more accurate way of estimating the S_{3D}.

Table 1. Probabilities of failure for $D_{50} = 0.21$ m.

R (acceptable S_{3D})	P_f
0.25	2.0×10^{-2}
1.00	4.6×10^{-3}
2.00	1.3×10^{-3}
3.00	1.9×10^{-4}

4. Conclusions

Reliability analysis remains as a rather novel but important topic to scour protections. Despite the recent research aiming at understanding the failure frequency at scour protections for offshore wind foundations, there are still aspects to be clarified.

One of the most important relies on the accuracy of the limit state function, which largely depends on the definition of an acceptable damage level (R). The definition of this parameter is yet to be fully understood and is currently progressing based on the knowledge derived from physical model studies and different methodologies to calculate the damage at the top layer.

The present research addressed the influence of the acceptable damage number on the reliability assessment of a scour protection by means of a theoretical example. It was concluded that varying this parameter leads to considerable changes in the probability failure, which agrees with recent findings from [1, 2 and 3]. Moreover, an example is used to show that a value of S_{3D}=1 may lead to a conservative estimation of the protection's failure, in scour protections with thicknesses that exceed one times the median stone size ($1D_{50}$). However, regardless of this influence's magnitude, it was discussed that defining an acceptable damage number largely depends on the methodology employed to calculate the S_{3D}. A brief analysis of the physical model was performed with the different areas, based on [9] and [13], being concluded that the methodology applied influences the maximum damage number. This fact formerly introduced by [5] is hereby proved. This implies that the predictive formula, Eq. (1), and the associated acceptable damage is re-evaluated if new methodologies are used. The model showed a variation in the maximum damage number close to 68%. In fact, while 24 sub-areas are used in [9], 48 are used in [13]. Thus, not only the size and shape of each sub-area differ, but the sample of damage numbers, from which the maximum is chosen, is also different. This may lead to an influence when addressing the damage distribution in the scour protection. Such aspect requires further research in order to understand how the number of sub-areas contributes to a better insight on the damage pattern and the definition of new threshold parameters for the limit state function.

Different arrangements of the sub-areas influence the limit state function, not only in terms of the failure threshold, but also on the potential alternatives for damage prediction, *i.e.* alternatives to Eq. 1. These findings highlight the need for further research on bathymetric analysis and damage estimation as a way to provide more accurate reliability assessments. In addition, it was seen that a clear relationship between the damage limits proposed in [9] and discussed in [8] and the visual area of exposed filter [9, 10] is yet to be determined. The use of more flexible sub-areas arrangements, as formerly suggested in [5] may provide a better insight on how to define the acceptable damage for common scour protections, allowing for limit state functions that simulate the actual exposure of the filter.

Acknowledgments
This work is supported by the project POCI-01-0145-FEDER-032170 (Oracle project), funded by the European Fund for Regional Development (FEDER), through the COMPETE2020, the Programa Operacional Competitividade e Internacionalização (POCI) and FCT/MCTES through national funds (PIDDAC).

References
[1] Fazeres-Ferradosa T, Taveira-Pinto F, Romão X, Vanem E, Reis T and das Neves L 2018 Probabilistic design and reliability analysis of scour protections for Offshore Windfarms, *Eng. Fail. Anal.* **91**, 291–305.
[2] Fazeres-Ferradosa T, Taveira-Pinto F, Romão X, Reis M T, das Neves L 2019 Reliability Assessment of Offshore Dynamic Scour Protections using Copulas, *Wind Engineering* **43**(5), 506-538.
[3] Fazeres-Ferradosa T 2018 Reliability Analysis Applied to the Optimization of Dynamic Scour Protections for Offshore Windfarm Foundations. PhD Thesis. Faculty of Engineering, University of Porto, Porto.
[4] Galiatsatou P and Makris C 2018 Optimized reliability based upgrading of rubble mound breakwters in a changing climate, *Journal of Marine Science and Engineering* **6**(3), 92.
[5] Fazeres-Ferradosa T, Taveira-Pinto F, Rosa-Santos P and Chambel J 2019 Reliability Analysis of Offshore Scour Protections – Review of scientific and technical challenges, *Proc. of the ICE Mar. Eng.* (*in press*).
[6] Sumer M and Nielsen A 2013 Sinking failure of scour protection at wind turbine foundation, *Proc. of the ICE Energy* **166**, no. EN4, 170-188.
[7] Chavez C, Stratigaki V, Wu M, Troch P, Schendel A, Welzel M, Villanueva R, SchulrmannT, De Vos L, Kisacik D, Taveira-Pinto F, Fazeres-Ferradosa T, Rosa Santos P, Baelus L, Szengel V, Bolle A, Whitehouse R and Todd R 2019 Large-Scale Experiments to Improve Monopile Scour Protection Design Adapted to Climate Change—The PROTEUS Project, *Energies* **12**(9), art. n. 1709, 2019.
[8] Fazeres-Ferradosa T, Taveira-Pinto F, Reis M T and das Neves L 2018 Physical modelling of dynamic scour protections: Analysis of the Damage Number, *Proc. of the ICE Mar. Eng.* **171**(1),1-24.
[9] De Vos L, De Rouck J, Troch P and Frigaard P 2012 Empirical design of scour protections around monopile foundations - Part 2 - Dynamic approach, *Coast. Eng*, **60**, 286-298.
[10] den Boon J H, Sutherland J, Whitehouse R, Soulsby R, Stam C J M, Verhoeven K, Høgedal M and Hald T 2004 Scour behaviour and scour protection for offshore Monopile Foundations of Offshore Wind Turbines, *in (EWEC) European Wind Energy Conference & Exhibition*, London.
[11] B. Hofland, M. Disco and M. V. Gent 2014 Damage characterization of rubble mound roundheads,

IOP Conf. Series: Materials Science and Engineering **700** (2019) 012027 doi:10.1088/1757-899X/700/1/012027

in Proc. 5th COASTLAB, Varna.
[12] de Almeida E, van Gent M R A and B. Hofland B 2019 Damage characterization of rock slopes," *J. of Mar. Sc. and Eng.* **7**(1), art. n. 10,1-15.
[13] de Schoesitter P, Audenart S, Baelus L, Bolle A, Brown A, das Neves L, Fazeres-Ferradosa T, Haerens P, Taveira-Pinto F, Troch P and Whitehouse R 2014 Feasibility of a dynamically stable rock armour layer scour protection for offshore wind farms, *in Proc. of the Int. Conf. OMAE,* San Francisco, California.
[14] Soulsby R 1997 *Dynamics of marine sands: a manual for practical applications*, Thomas Telford, London.

IOP Conf. Series: Materials Science and Engineering **700** (2019) 012028 doi:10.1088/1757-899X/700/1/012028

Fatigue design challenges: Recent linear and nonlinear models

D G Pavlou

Dept. of Mechanical and Structural Engineering and Materials Science, University of Stavanger, Norway.

Corresponding author: dimitrios.g.pavlou@uis.no

Abstract. Reliable fatigue design rules affect the proactive identification of safety parameters in engineering structures. Numerous fatigue crack initiation and propagation models, linear and nonlinear, have been developed for designing purposes or estimation of the remaining life of aging structures. Depending on the adopted assumptions, the accuracy varies for different loading histories, loading types, and materials. Semi empirical models are simple but yield significant inaccuracies. Models with better theoretical basis provide better accuracy, but implementation in real conditions is problematic. In the present work, a review of author's recent fatigue crack initiation and propagation models based on physical mechanisms is presented and improvements are proposed. Verification of the models on test results is provided and discussed.

1. Introduction

Successful structural design and data driven safety assessment require reliable fracture mechanics tools for fatigue damage accumulation modelling. Palmgren and Miner were the first who attempted to calculate the fatigue damage [1, 2]. Due to its simplicity, their fatigue damage accumulation rule is widely used. However, the Palmgren-Miner's rule is inconsistent with the nonlinear material damage mechanisms [3, 4]. Depending on the type of loading history, the Miner's rule can overestimate or underestimate the fatigue damage accumulation prediction [5]. Its accuracy is better for loading spectra containing high stress levels. The above rule is linear and does not take into account the loading sequence effect and the material damage accumulation theory. The Miner's rule can be used for crack initiation only. Manson and Halford [6, 7] have improved the linear fatigue damage rule. Their double-linear damage model (DLDR) preserves (to some extend) the simplicity of Palmgren-Miner's rule and takes into account two different fatigue mechanisms. They have proposed two fatigue stages in order to model both the crack initiation and crack propagation phases.

However, the load interaction effects [8, 9] and the mixed mode fatigue crack propagation [10-12] are not accounted for. Known nonlinear models are based on continuum damage theory [13], isodamage straight lines [e.g. 14, 15], and Manson-Halford's concept [16]. Further literature review on nonlinear damage accumulation models has been published recently [17]. For crack propagation modelling the Paris rule [18] is well known. However, this old rule is linear, and therefore it does not take into account the load interaction effects [19, 20]. Modification of this rule has been attempted by Wheeler [21] and Willenborg [22] in order to improve the fatigue crack growth prediction for loading histories containing overloads. However, the modelling of the retardation effects has semi empirical basis. Elber [23-25] is the first who modelled the crack closure as a predominant mechanism for overload-induced fatigue crack growth retardation. The proposed model yields accurate results for structural elements subjected to conditions where the crack closure mechanism is predominant (large overload-induced plastic zones).

However, apart from the crack closure mechanism due to residual plastic deformation, more than four other material mechanisms affecting the fatigue crack growth are known, i.e. strain hardening of the material within the overload crack-tip plastic zone [e.g. 26, 27], plastic blunting and re-sharpening [28], fracture surface roughness which causes contact between the crack faces at non-zero loads reducing the effective ΔK [29-31], and crack branching [9, 10].

In the present work, the new crack initiation theory of the S-N fatigue damage envelope [32], developed by the author, will be implemented on Al-2024 specimens subjected to high-low (H-L) and low-high (L-H) two-stage loading. Moreover, the overload-induced crack growth retardation model [26, 27, 33, 34] based on strain hardening, will be improved to take into account both overload and underload effects.

2. Fatigue crack initiation

A recent advance in fatigue crack initiation modelling is based on the proposal that the area bounded by the S and N axes and the S-N curve can provide a damage map for the material [32].

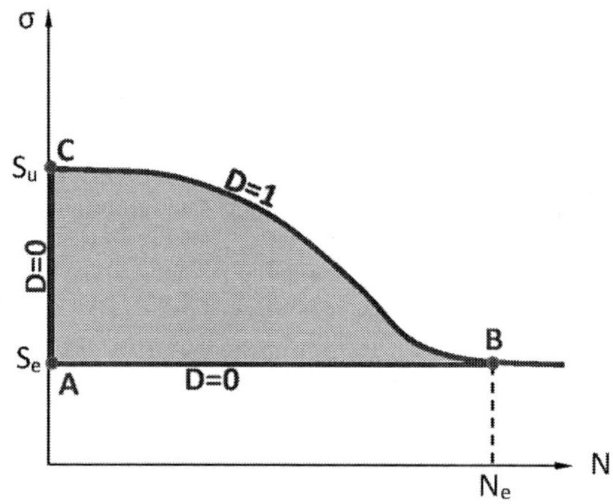

Figure 1. The S-N damage envelope.

The boundary AB of the envelope (Fig. 1) corresponding to the endurance limit S_e should be considered as an isodamage line with damage D=0. The boundary AC (Fig. 1) corresponds to number of loading cycles N=0. Therefore, it should also be considered as an isodamage line with D=0. Moreover, the boundary CB is the S-N curve corresponds to crack initiation, i.e. D=1. Finite element analysis performed in [32] has derived a damage map and isodamage lines. The damage zones for Al-2024 have been derived with the aid of ANSYS [32, 35] using the dimensionless parameters in (1) and (2) for stress and loading cycles and are demonstrated in Figure 2.

$$\sigma_i^* = \frac{\sigma_i - S_e}{S_u - S_e} \qquad 0 \le \sigma_i^* \le 1 \tag{1}$$

$$n_i^* = \frac{n}{N_e} \qquad 0 \le n_i^* \le 1 \tag{2}$$

The colours of the damage zones correspond to $Log(D)$. Therefore, taking into account the vertical legend of colors for $Log(D)$, the isodamage lines correspond to values $D = 10^{-10}, 10^{-8.88889}, ..., 10^0$. Using the results of Fig. 2, damage curves D vs n / N_f have been derived in Fig. 3 for stress levels $\sigma^* = 0.088,\ 0.248,\ 0.408,\ 0.888$.

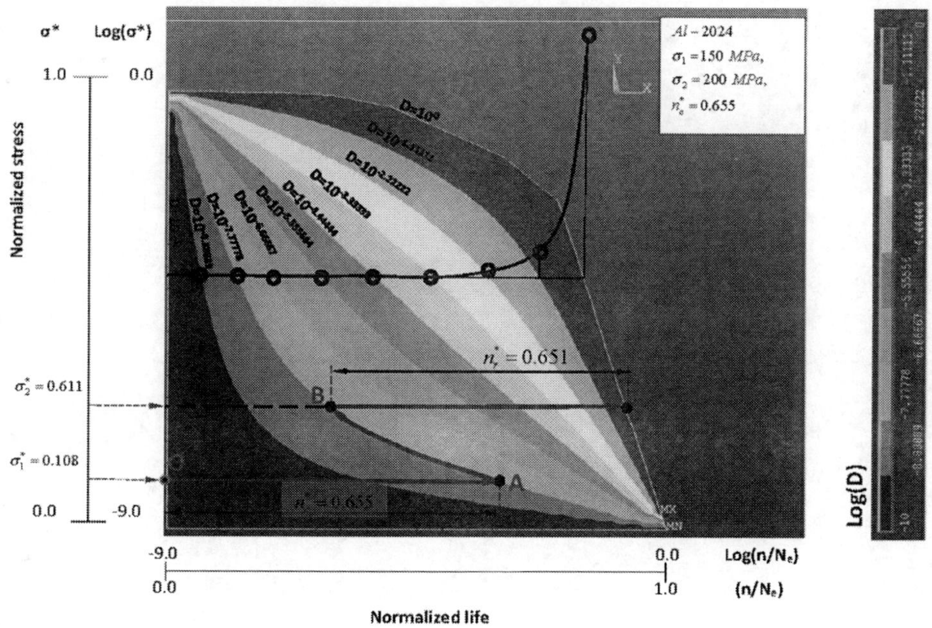

Figure 2. Damage zones for Al-2024.

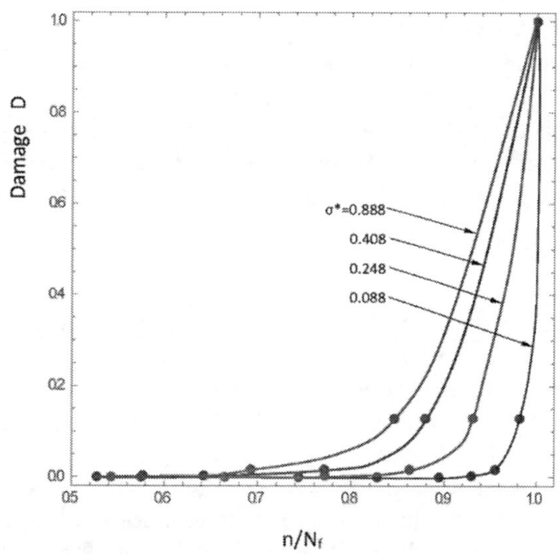

Figure 3. Damage curves for Al-2024.

The damage envelope in Figure 2 can be used for the calculation of the remaining life for two-stage loading (Figure 4). For a loading block with stress level $\sigma_1 = 150MPa$ (i.e. $\sigma_1^* = 0.108$) and loading period $n_c^* = 0.655$ (consumed life), the accumulated damage is $D = 10^{-7.77778}$ (point A in Figure 2). The continuation of the loading for a higher stress $\sigma_2 = 200MPa$ (i.e. $\sigma_2^* = 0.611$) should start from the already accumulated damage $D = 10^{-7.77778}$ corresponding to point B (same isodamage line with point A). The remaining life is demonstrated by the segment BC, i.e. $n_r^* = 0.651$. This procedure is followed by a number of loading data H/L and L/H borrowed by the ref. [36] and the results are shown in Figure 4.

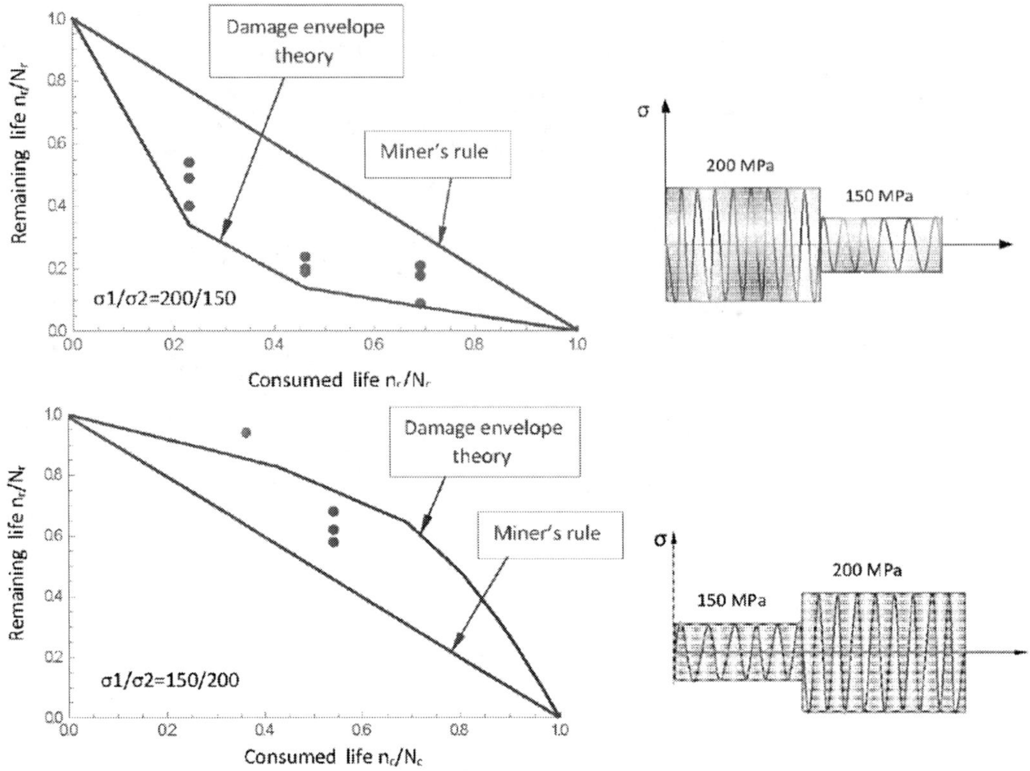

Figure 4. Remaining life predictions for Al-2024 specimens subjected to H/L and L/H two-stage loading

The results indicate that the S-N damage envelope theory provides successful predictions of the remaining life for these loading cases. Unlike the above nonlinear predictions, Miner's rule has overestimated the remaining life for H/L loads, and has underestimated it for the inverse loading sequence.

3. Fatigue crack propagation

Variable amplitude loading sequences are associated with transient effects in fatigue crack growth. Overloads yield significant retardation in the fatigue crack growth rate, and underloads reduce the above effect. Among the proposed tools for fatigue crack growth prediction for variable amplitude loading, Wheeler's [21], Whillenborg's [22], and Elber's [23] models are the most known, and have been used for further improvements [37-40]. Wheeler's and Whillenborg's models have semi empirical basis. Elber's model and its improvements take into account the overload retardation effect due to the crack closure mechanism [41-44] and provide accurate predictions for loading cases where the crack closure mechanism is dominating the fatigue crack growth. The first attempt for modelling the effect of the material hardening within the overload plastic zone took place in 1995 [26, 27]. In the original version of this research, the crack growth rate after an overload is correlated to the actual yield stress of the hardened material within the overload plastic zone. The main idea is based on the assumption that the baseline plastic zone in the crack tip (Figure 5) is subjected to low-cycle-fatigue conditions due to high strain amplitude [26, 27, 33, 34].

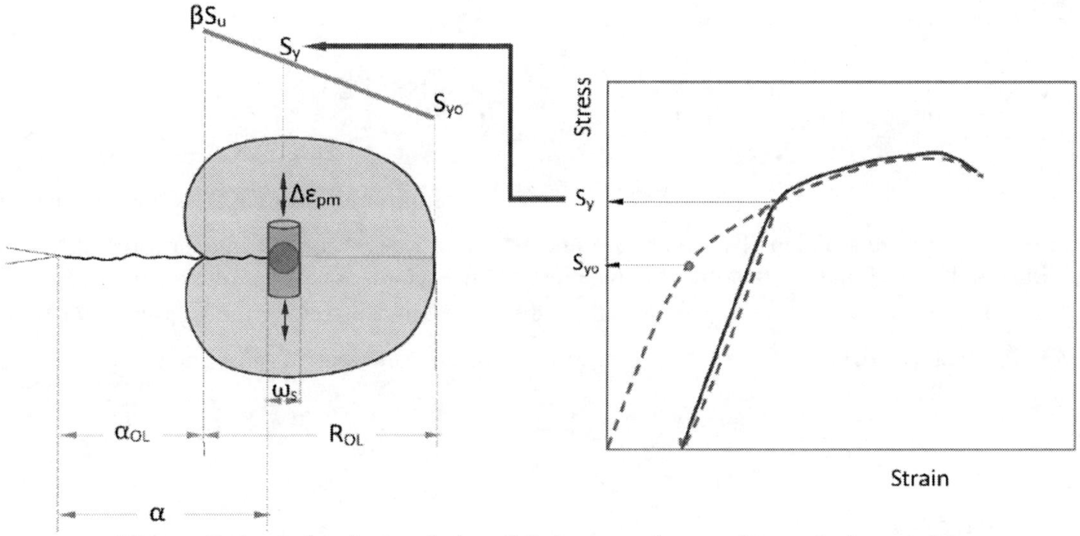

Figure 5. Strain hardening-induced fatigue crack growth retardation model.

The fatigue crack growth rate da / dN is approximated by the formula

$$\frac{da}{dN} = \frac{\omega_s}{\Delta N} \tag{3}$$

where ω_s is the plastic zone size (Figure 5) of the constant amplitude loading $\Delta\sigma$ (Figure 6), and ΔN is the required number of cycles for crack growth by a length ω_s. With the aid of Coffin-Manson rule for low cycle, and following fracture mechanics analysis, the following crack growth retardation model [26, 27] is proposed:

$$\frac{da}{dN} = \frac{\pi}{32}\left(\frac{1}{2K_{cr}^2}\right)^{1/m} \frac{\Delta K^{2(1+1/m)}}{S_y^2} \tag{4}$$

where S_y is the actual yield stress of the hardened material within the overload plastic zone (Fig. 5). Depending on the location of the crack tip, the S_y of the hardened material takes values $S_y = S_u$ for $a = a_{OL}$ (immediately after the overload), and $S_y = S_{yo}$ for $a = a_{OL} + R_{OL}$ when the retardation effect ends.

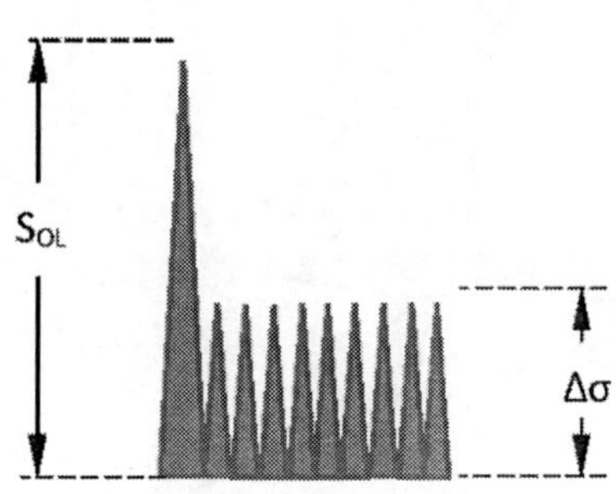

Figure 6. Constant amplitude loading containing an overload.

A simple linear distribution of the S_y between the boundary values S_u, S_{yo} is assumed, and a retardation factor is derived, as given in (5).

$$\lambda = \left(\frac{S_{yo}}{S_y}\right)^2 \qquad \left(S_{yo}/S_u\right)^2 \le \lambda \le 1 \tag{5}$$

Since the power of the ratio $\left(S_{yo}/S_y\right)$ is 2, the effect of the strain hardening seems to dominate the overload-induced retardation. However, the above model does not take into account the effect of underloads. It is well known that underloads eliminate the retardation effect of overloads [e.g. 45]. An underload with stress $\sigma_{UL} = -\sigma_{OL}$ can cancel the retardation effect of the overload. Therefore, a correction factor in (6) is proposed to adjust the value of the parameter S_u in Figure 5.

$$\beta = \left(\frac{S_{yo}}{S_u} - 1\right)\frac{|\sigma_{UL}|}{|\sigma_{OL}|} + 1 \tag{6}$$

Therefore,

$$\beta S_u = \begin{cases} S_{yo} & for\ \sigma_{UL} = -\sigma_{OL} \\ S_u & for\ \sigma_{UL} = 0 \end{cases} \tag{7}$$

For $\sigma_{UL} = -\sigma_{OL}$ the value of the maximum yield stress of the hardened material (Figure 5) is $\beta S_u = S_{yo}$. In this case $\lambda = \left(S_{yo}/\beta S_u\right)^2 = 1$. Therefore, no retardation effect takes place. For $\sigma_{UL} = 0$, eq. (6) yields $\beta = 1$. Therefore, $\lambda = \left(S_{yo}/S_u\right)^2$ meaning that considerable retardation occurs. Taking into account the above improvement, the model is applied to the random loading history M90 (0.27) borrowed by ref. [46]. The above loading contains overloads and underloads. The modified rainflow counting method described in [26, 27] is implemented in order to transform the irregular history to a loading history containing full cycles. The theoretical predictions are correlated to test results [46], indicating good agreement (Figure 7).

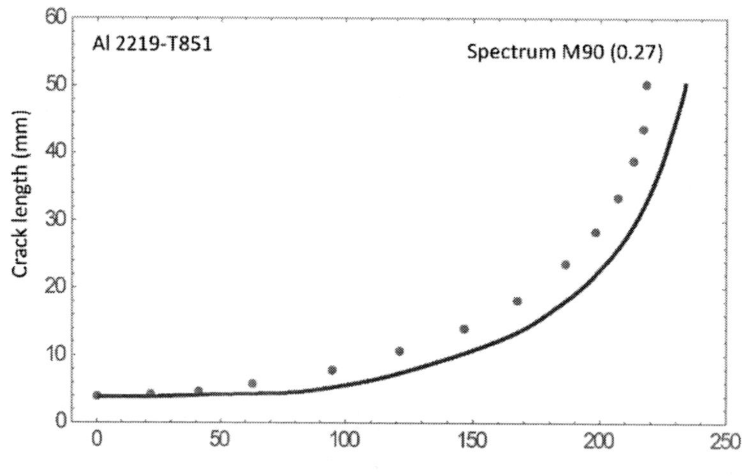

Figure 7. Correlation of the prediction of the proposed method to test results borrowed from ref. [46].

4. Conclusions

1. Fatigue design challenges in aerospace industry have been discussed.
2. Recent advances in fatigue crack initiation and propagation prediction have been presented.
3. Author's recent models and their improvements have been implemented to predict the fatigue crack initiation and propagation in variable amplitude loading cases.
4. Verification of the S-N damage envelope theory for crack initiation, and the strain hardening model for crack propagation on test results for Al-2024 and Al-2219 has been provided.

References

[1] Palmgren A and Die Lebensdauer von Kugellagern 1924 *Z Ver Dtsch Ing.* **68**(14) p 339
[2] Miner M A 1945 *J. Appl. Mech.* **12**(3) A159
[3] Pavlou D G 2002 *Eng. Struct.* **24** p 1363
[4] Shun-Peng Zhu, Yong-Zhen Hao, José A F O Correia, Grzegorz Lesiuk, Abilio M P, De Jesus, 2018 *Fatigue Fract. Eng. Mater. Struct.* **1** 12,
[5] Rege K and Pavlou D G 2017 *Int. J. Fatigue* **98** 234
[6] Manson S S, Halford G R 1981 *Int. J. Fract.* **17**(2)169
[7] Lee Y-L et al. 2011 *Fatigue testing and analysis theory and practice* (Saint louis: Butterworth-Heinemann)
[8] Pavlou D G 2000 *Eng. Struct.* **22** 1707
[9] Pavlou D G, Vlachakis N V, Pavlou M G, Vlachakis V N 2004 *Comput. Mat.Sc.* **29** 446
[10] Pavlou D G 2015 *Int. Review Mech. Eng. (IREME)* **9**(3) 324
[11] Pavlou D G 2018 *J. Aerosp. Technol. Manage.* **10** 1
[12] Pavlou D G, Labeas G N, Vlachakis N V and Pavlou F G 2003 *Eng. Struct.* **25** 869
[13] Shang D-G and Yao W-X 1999 *Int. J. Fatigue* **21**(2) 187
[14] Subramanyan S 1976 *ASME J. Mater. Technol.* **98**(4) 316
[15] Hashin Z and Rotem A 1978 *Mater. Sci. Eng.* **34**(2) 147
[16] Gao H et al. 2014 *Sci. World J.*
[17] Santecchia E et al. 2016 *Adv Mater Sci Eng*
[18] Paris P C, Tada H and Donald J K 1999 *Int. J. Fatigue* **21** 35
[19] Schijve J 2009 *Fatigue of Materials* Springer
[20] Suresh S 2003 *Fatigue of Materials* Cambridge University Press
[21] Wheeler O E 1972 *J. Basic Eng. Trans ASME, Ser D* **94**(1) 181
[22] Willenborg J D, Engle Jr R M, Wood H A 1971 *AFDL-TM-71-1-FBR*
[23] Elber W 1970 *Engineering Fracture Mechanics* **2** p 37
[24] Bathias C 2013 *Fatigue of Materials and Structures: Fundamentals* John Wiley and Sons
[25] Yang B, Zhao Y.-X, Liang H-Q, Wu P-B, Zeng J 2005 *Eng. Mech.* **22**(5) 99
[26] Pavlou D G 2000 *Eng. Struct.* **22** p 1707
[27] Παυλου Δ 1994 Προβλεψη της διαδοσης των ρωγμων σε τυχαια μεταβαλλομενα φορτια *PhD Thesis* (University of Patras)
[28] Ward-Close C M, Blom A F, Ritcie R O 1989 *Eng. Fract. Mech.* **32** (4) 613
[29] Parry M R, Syngelakis S, Sinclair I 2000 *Proc. of the 6th Int. Conf. of Damage and Fracture Mechanics*, (Southampton: WIT press) p 313
[30] Mayer H R, Stanzl S E, Tschegg E K, Schijve J 1992 *Fatigue Fract. Eng. Mater. Struct.* **15** (3) 265
[31] Louat N, Sadananda K, Duesbery M, Vasudevan A K 1993 *J. Metall. Trans.* **24A** p 2225
[32] Pavlou D G 2018 *Int. J. Fatigue* **110** 204
[33] Pavlou D G 1998 *Proc. of the 5th Int. Conf. on Structures under Shock and Impact (SUSI 98)* (Southampton: Computational Mechanics Publications) p 633
[34] Pantelakis S, Kermanidis Th, Pavlou D. 1995 *J. Theoret. Appl. Fract. Mech.* **22** p 35
[35] ANSYS website: https://www.ansys.com/
[36] Pavlou D G 2002 *Eng. Struct.* **24** p 1363
[37] Shan J, Wei Z, Xiaoyang L and Fuqiand S 2014 *Math. Prob. Eng.* ID 713678, 1

[38] Jiang S, Zhang W, He J and Wang Z 2016 *Chin. J. Aeronaut.* **29**(6)1618-25
[39] Marcos V S P, Fathi A I D, Arnaldo F Camarao and Sergio H M 2007 *Mater. Res.* **10**(2) 101
[40] Michal J, Dorota K and Henryk T 2010 *Fatigue Aircraft Struct.* **1** 37
[41] Khan S U, Alderliesten R C, Scjijve J, Benedictus R 2007 *Computational and Experimental Analysis of Damaged Materials* (Kerala: Research Signpost) p 77
[42] Luiz C H R and Carlos A J M 2016 *Fratturaed Integrita Strutturale* **35** 456
[43] Ricardo L C H, Andrade A H P and Topper T H 2008 *Proc. of SAE Brazil Congress Society of Automotive Engineers* Article 2008-36-0007.
[44] Ribeiro A S, Jesus A P, Costa J M, Borrego L P, Maeiro J C 2010 *8th Congresso Nacional de Mecanica Experimental* (Guimaraes) p 21
[45] Huang X, Torgeir M and Cui W 2008 *Int. J. Fatigue* **30** 2
[46] Newman Jr J C 1981 *Methods and Models for predicting Fatigue Crack Growth under Random Loading* ASTM STP 748

Buckling capacity of simulated patch corroded tubular columns – laboratory tests

T Vo[1], K Hestholm[1], G Ersdal[1,2,*], N Oma[2] and M Sivertsvik[3]

[1] University of Stavanger, Norway
[2] Petroleum Safety Authority, Stavanger, Norway
[3] Norwegian University for Science, Trondheim, Norway

[*] Corresponding author: gerhard.ersdal@uis.no

Abstract. This paper presents the experimental test results for the buckling capacity of tubular columns exposed to simulated unsymmetrical patch corrosion damage. These tests were performed to investigate how the results of these experiments compare with the capacities obtained for the formulae provided in the 2004 revision of NORSOK N-004. Formulae for patch corroded tubular columns are needed in standards and as such it is unfortunate that these formulae are removed from the present revision of the NORSOK N-004 standard. Prior to suggesting introducing these previous formulae in standards, experimental tests to confirm good correlation between experiments and formulae are needed. The tests performed in this work are presented in this paper for 100% (intact columns), 69%, 34% and 0% remaining wall thickness in the patch.

1. Introduction

The vast majority of the infrastructure (offshore platforms, buildings, bridges, etc) in the world is existing and now ageing. This large number of existing structures is of vital importance for society, but they are in many cases degrading and need significant investments for maintenance. Replacement with new structures might be both economically and environmentally unsound. Still, most universities educate students in civil and structural engineering merely in the design of new structures. Fortunately, the University of Stavanger has established a course and research work into the continued use of ageing structures.

However, using ageing structures beyond their original design life also raises several problematic issues. Structures in operation are exposed to conditions of stress and environment that ultimately will degrade them from their initial state and damage will accumulate until the structures may be judged to be no longer fit-for-service. If not withdrawn from further service or being repaired, failure of some kind will eventually occur. Also, the cost of the required maintenance, inspection and repair needed to cope with this deterioration and damage will at some stage become unacceptable compared to the revenue that can be gained from the use of these structures. Hence, it becomes vital to know:

1. how structures change with age,
2. how their condition and other aspects influencing their safety can be determined,
3. how their capacity (strength and fatigue life) can be determined as a result of ageing mechanisms,
4. how any anomalies in the structures should be evaluated,
5. how anomalies found in an existing structure can be repaired and mitigated, and
6. how the integrity management of ageing structures should be performed.

IOP Conf. Series: Materials Science and Engineering **700** (2019) 012029 doi:10.1088/1757-899X/700/1/012029

This paper is work performed under item 3 as listed above and is specifically aimed at providing insight into the ultimate capacity of tubular columns in compression with patches representing localized corrosion.

Earlier work on ultimate capacity of tubular columns with patch corrosion includes Ricles et al. and Hebor [1, 2]. These works are the result of laboratory work testing damaged and repaired tubular-profile steel columns under axial compression load. These studies included larger scale tests as previous experiments mostly included diameters below 76 mm. A comparison between the damaged and un-damaged columns showed that the capacity was greatly reduced due to the corrosion. The reduction was reinforced by the fact that an eccentricity arose due to the geometric change as a result of unsymmetrical metal loss. In addition, high stresses occurred in the corroded area.

As we will show in this paper the comparison between documented laboratory tests of ultimate capacity of dented tubular columns compared to NORSOK N-004 [3] formulae is quite promising. In contrast, the comparison between the tests on corroded tubular columns as performed by Hebor [1] was rather disappointing. Hence, it was decided to perform more tests of tubular columns with corrosion damage to provide more data to validate the method provided by NORSOK N-004 [3]. The first stage in this test programme was documented by Hestholm and Vo [4] and the main findings of this programme are given in this paper.

2. Basic buckling theory

In engineering, buckling is a well-known phenomenon. Buckling is an instability by a sudden lateral deflection of a structural element when subjected to compressive loads and can lead to structural failure. Historically, several structural accidents and disasters have occurred due to buckling and instability. The solution to the stability problem is often attributed to Leonard Euler (1707-1783). The solution was based on the following assumptions (Timoshenko and Gere [5]):

- The column material is homogeneous and linearly elastic (that is, it follows Hooke's law)
- The column is perfectly straight and has no imperfections
- The load is applied centrically
- The cross section is constant throughout the length
- The length of the column is considerably longer as opposed to the cross-sectional dimensions

The Euler-formula for buckling capacity is expressed as:

$$N_E = \frac{\pi^2 \cdot EI}{(k \cdot L)^2} \tag{1}$$

where:

N_E is the Euler buckling capacity,
E is the modulus of elasticity of the material,
I is the second moment of inertia of the column,
k is a factor correcting for the effective length of the column (buckling factor),
L is the length of the column.

When using the Euler formula, the maximum compressive load capacity the column will withstand in the elastic regime before becoming unstable is obtained. Being unstable means that the smallest external impact will create an imbalance and the column will experience large lateral deflections and will buckle.

The Euler formula will in practice be of little use to ordinary designs as a column will not be perfectly straight and will have initial imperfections. Further, material will not behave linearly elastic but will rather experience plasticity prior to buckling and especially after buckling. Therefore, in order to determine the capacity of a column both the Euler buckling capacity and the material strength should be taken into account. This means that for columns with low slenderness (L/r where L is the length of the column and r is the radius of gyration), the material strength will be dominant, while for slimmer columns, the Euler formula will apply.

Several empirical formulae for the buckling capacity of columns have been developed. One such formula is the Perry Robertson-formula [6] that is used frequently in civil engineering standards such as Eurocode:

$$\frac{N}{N_d} + \frac{N \cdot (w_0 + e) \cdot \frac{1}{1 - \frac{N}{N_E}}}{M_d} \leq 1.0 \tag{2}$$

where:

N is axial compression loading,
N_d is column buckling capacity,
w_0 is the initial deformation (fabrication allowance),
e is the eccentricity of the axial compression load,
N_E is the Euler buckling capacity,
M_d is moment capacity.

Another commonly used approach to buckling capacity of columns is the Johnson-Ostenfelt correction. This is an empirically based formula that relates the slenderness ratio to the stress that provides the ultimate load required to buckle a column. The formula is based on results from J.B. Johnson from around 1900 [7]:

$$f_{cr} = \phi \cdot f_y$$
$$\phi = \begin{cases} 1 - \lambda^2/2, & \lambda^2 \leq 2 \\ 1/\lambda^2, & \lambda^2 > 2 \end{cases} \tag{3}$$

where $\lambda^2 = f_y/f_E$ is the reduced slenderness, $f_E = N_E/A$ is the Euler buckling stress and f_y is the yield stress.

In addition to global buckling of a column, local buckling of web, flanges or the pipe wall needs to be accounted for in certain circumstances. Global buckling will induce a curvature of the entire length of the element. Local buckling will cause parts of the profile to deform and will weaken the capacity of the local section and thereby the global buckling capacity. Local buckling capacity will in some cases dominate the column's ultimate compressive capacity. Diameter-to-thickness ratio, D/t for tubular sections, is an important factor in classifying the column with regards to its susceptibility to local buckling. A large D/t ratio indicates thin walls and susceptibility for local buckling (slender column), and this needs to be accounted for in the buckling capacity. A lower ratio means that the column is more massive and will not be prone to local instability (compact column). D/t in the mid-range is called a non-compact column.

Steel members used in structures are continuously exposed to various degradation mechanisms and to events that may inflict damage. The typical degradation mechanisms and events are corrosion, fatigue and impacts. In general, degradation mechanisms and events will lead to four different types of effects that need to be accounted for in the calculation of the capacity of deteriorated steel structural members:

- Metal loss and wall thinning (corrosion, erosion, wear, tear, etc.)
- Cracking (fatigue, creep, hydrogen intrusion, etc.)
- Changes to material properties (hydrogen embrittlement, hardening, temperature, etc.)
- Geometric changes (dents from impacts, bowing from impacts, permanent plastic deformation, corrosion, etc.)

This paper intends to study the effect of patch corrosion that will give both metal loss and possible geometrical changes. The metal loss and geometrical changes will reduce the buckling capacity of a damaged column compared to an intact column. The metal loss will result in a reduction in axial and moment capacity in a reduced cross-section. The geometric changes will introduce an eccentricity to the profile if an unsymmetrical change occurs as a result of the patch corrosion as shown in Figure 1. For a simple rectangular section this can be illustrated by three possible corrosion patterns as shown in Figure 2.

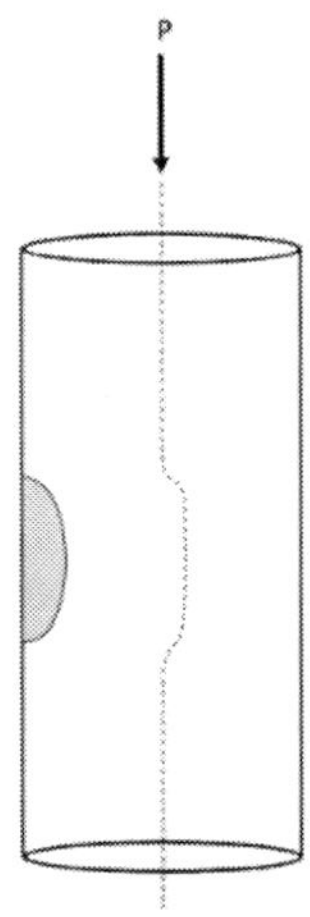

Figure 1. Eccentricity due to unsymmetrical patch corrosion.

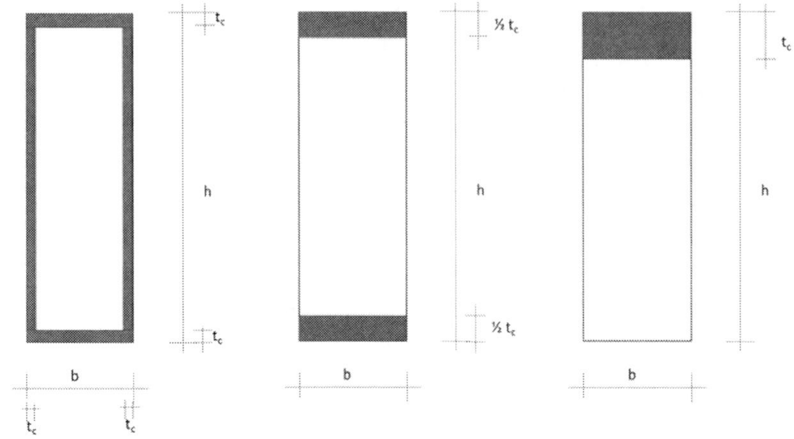

Figure 2. Possible corrosion patterns in a rectangular section.

In the first two corrosion patterns shown in Figure 2, no eccentricity is introduced. However, in the last, unsymmetrical corrosion pattern an eccentricity of $e = t_c/2$ is introduced. Hence, a moment equal to $M = N \cdot e$, where N is the axial force on the column, has to be accounted for in the strength calculations.

In Norway the standards used for design and assessment of structures are Eurocode for land-based structures and NORSOK for offshore structures. In this paper tubular members typical of offshore structures with patch corrosion are studied and, hence, the NORSOK N-004 standard has been used as the basis for capacity calculations.

NORSOK N-004 Rev. 2, October 2004 [3] provides formulae for calculating the capacity of undamaged tubular columns and tubular columns with unsymmetrical corrosion damage based on the formulae for a dented tubular section. In NORSOK N-004 the corrosion is illustrated by an equivalent dent (section 10.6.2.2).

The buckling capacity of an undamaged tubular column in NORSOK N-004 is given by:

$$N_{Sd} = \frac{f_c \cdot A}{\gamma_m} \tag{4}$$

where:

f_c is the compressive strength given by:

$$f_c = \begin{cases} \left(1{,}0 - 0.28\bar{\lambda}_0^2\right) \cdot f_y & for\ \bar{\lambda}_0 \leq 1{,}34 \\ \dfrac{0.9}{\bar{\lambda}_0^2} \cdot f_y & for\ \bar{\lambda}_0 > 1{,}34 \end{cases}$$

λ_0 is the column slenderness parameter given by: $\lambda_0 = \frac{k \cdot L}{\pi \cdot r_{gyr}} \sqrt{\frac{f_{cl}}{E}}$

where:

k is the buckling coefficient
L is the length of the column
r_{gyr} is the radius of gyration for the section given by $\sqrt{I/A}$
f_{cl} is the characteristic local buckling capacity

$$f_c = \begin{cases} f_y & \frac{f_y}{f_{cle}} \leq 0.170 \\ \left[\left(1.047 - 0.274 \cdot \frac{f_y}{f_{cle}}\right) \cdot f_y\right] & 0.170 < \frac{f_y}{f_{cle}} \leq 1.911 \\ f_{cle} & \frac{f_y}{f_{cle}} > 1.911 \end{cases}$$

$f_{cle}$is the characteristic elastic local buckling capacity given by:

$$f_{cle} = 2 \cdot C_e \cdot E \cdot \frac{t}{D}$$

Where C_e is the characteristic buckling coefficient.

The characteristic buckling coefficient C_e is 0.6 if initial deformations are not included and 0.3 if initial deformations are included. Hence, in this paper $C_e = 0.3$ is used.

NORSOK N-004 provides the following formulae for the ultimate buckling capacity of a dented tubular column:

$$N_{Sd} \leq N_{dent,c,Rd} = \frac{N_{dent,c}}{\gamma_M} \tag{5}$$

where:

$N_{dent,c,Rd}$ is the design capacity of the damaged column,
$N_{dent,c}$ is the characteristic capacity of the damaged column, and
γ_M is a partial material factor.

$N_{dent,c}$ can be calculated as:

$$N_{dent,c} = \begin{cases} (1{,}0 - 0.28\bar{\lambda}_d^2) * \xi_c * f_y * A_0 & for\ \bar{\lambda}_d \leq 1{,}34 \\ \frac{0.9}{\bar{\lambda}_d^2} * \xi_c * f_y * A_0 & for\ \bar{\lambda}_d > 1{,}34 \end{cases} \tag{6}$$

where:

A_0 is the un-damaged section area
$\bar{\lambda}_d$ is the reduced slenderness of the damaged column given by:

$$\bar{\lambda}_d = \sqrt{\frac{\xi_c}{\xi_M}} * \bar{\lambda}$$

where:

$\bar{\lambda}$ is the relative slenderness of the column
$\xi_c \quad = e^{-0{,}08*\left(\frac{\delta}{t}\right)}, \quad for\ ^{\delta}/_{t} < 10$
$\xi_M \quad = e^{-0{,}06*\left(\frac{\delta}{t}\right)}, \quad for\ ^{\delta}/_{t} < 10$

where:

δ is the dent depth
t is the thickness of the tubular member

The equivalent dent depth that is intended to illustrate the corrosion damage is given by:

$$\delta' = \frac{1}{2}\left(1 - cos\left(\pi \frac{A_{corr}}{A}\right)\right) D \tag{7}$$

where:

δ' is the equivalent dent depth
D is the diameter of the tubular column
A is the cross-section area of the tubular column
A_{corr} is the cross-section area of the corroded tubular column

As a patch corrosion may introduce eccentricities causing a local moment the formulae for moment capacity of a damaged tubular also has to be included. According to NORSOK N-004 the ultimate moment capacity of dented tubular member can be written as:

$$M_{sd} \leq M_{dent,Rd} = \begin{matrix} \xi_M \cdot M_{Rd} & \textit{if dented area is in compression} \\ M_{Rd} & \textit{otherwise} \end{matrix} \tag{8}$$

where:

$M_{Rd} = \frac{f_m W}{\gamma_M}$ is the design moment capacity of the undented tubular member,
$M_{dent,Rd}$ is the design moment capacity of dented tubular member,
M_{sd} is the design moment capacity,
f_m is the characteristic stress in bending,
W is the elastic moment of inertia of the section,
γ_M is the material partial factor.

where:

$$f_m = \frac{Z}{W} f_y \qquad for\ \frac{f_y D}{Et} \leq 0.0517$$
$$f_m = \left(1.13 - 2.58\left(\frac{f_y D}{Et}\right)\right)\left(\frac{Z}{W}\right) f_y \qquad for\ 0.0517 < \frac{f_y D}{Et} \leq 0.1034$$
$$f_m = \left(0.94 - 0.76\left(\frac{f_y D}{Et}\right)\right)\left(\frac{Z}{W}\right) f_y \qquad for\ 0.1034 < \frac{f_y D}{Et} \leq 120\frac{f_y}{E}$$

where:

Z is the plastic moment of inertia of the section

The combination of axial compression and bending moment that the tubular column will experience when exposed to un-symmetric patch corrosion needs to be combined when checking the capacity. NORSOK N-004 gives an interaction formula based on the Perry Robertson formula presented earlier. The capacity of the tubular column is seen as acceptable if the following formula is satisfied:

$$\frac{N_{Sd}}{N_{dent,c,Rd}} + \sqrt{\left(\frac{N_{Sd}\Delta y_2 + C_{m1}M_{1,Sd}}{\left(1 - \frac{N_{Sd}}{N_{E,dent}}\right) M_{dent,Rd}}\right)^{\alpha} + \left(\frac{N_{Sd}\Delta y_1 + C_{m2}M_{2,Sd}}{\left(1 - \frac{N_{Sd}}{N_E}\right) M_{Rd}}\right)^{2}} \leq 1.0 \tag{9}$$

where:

$$\alpha = \begin{cases} 2 - 3\frac{\delta}{D} & \textit{... if the damaged area is in compression} \\ 2 & \textit{... if the damaged area is in tension} \end{cases}$$

N_{Sd} is the design axial force exposed to the tubular column,
$M_{1,Sd}$ is the design moment about the axis parallel to the damage,
$M_{2,Sd}$ is the design moment about the axis perpendicular to the damage.
$N_{E,dent} = \pi^2 \frac{EI_{dent}}{(kl)^2}$
k is the effective length factor,
I_{dent} is the second moment of inertia for the damaged cross-section $= \xi_M I$
I is the second moment of inertia for the undamaged cross section,
Δy_1 is the member eccentricity perpendicular to the dent,
Δy_2 is the member eccentricity in-line to the dent,
C_{M1}, C_{M2} is moment reduction factors provided by NORSOK N-004

IOP Conf. Series: Materials Science and Engineering **700** (2019) 012029 doi:10.1088/1757-899X/700/1/012029

In the tests the columns are subject to axial compression exclusively. Any resulting moment is due to the eccentricity due to patch corrosion. Hence, the interaction formulae by setting the moment reduction factor equal to 1.0 can be simplified to $M_{1,Sd} = N_{Sd}\Delta y_2$ providing the following formula for capacity check:

$$\frac{N_{Sd}}{N_{dent,c,Rd}} + \sqrt{\left(\frac{N_{Sd}\Delta y_2}{\left(1-\frac{N_{Sd}}{N_{E,dent}}\right)M_{dent,Rd}}\right)^{\alpha}} \leq 1.0 \qquad (10)$$

3. Comparison of existing tests and NORSOK N-004 formulae

A comparison of the results from Hebor [1] and the calculated capacities according to NORSOK N-004 [3] for patch corroded columns in compression are shown in Figure 3. The tests performed by Hebor [1] included specimens with a diameter of 218 mm and a thickness ranging from 3.4 to 6.5 mm with thickness loss ranging from 0% to 100%. As can be seen in Figure 3 the results from Hebor [1] indicate significantly larger capacity compared to the calculation formulae in NORSOK N-004.

A similar comparison was performed where results of experiments documented in the existing literature of tests of dented tubulars were compared to the NORSOK N-004 formulas [8-19]. In order to achieve reasonably similar results from the two sources (the NORSOK N-004 formulas and the experimental results) the added moment due to the eccentricity introduced by the deformation of the dented section had to be included from NORSOK. This fact is not clearly stated in the standard. After the inclusion of the moment due to the eccentricity introduced by the dented section the comparison between the experimental results and capacity based on NORSOK N-004 was as shown in Figure 4, indicating a reasonable compliance.

Due to the lack of similarity in the comparison with the tests performed by Hebor [9] and NORSOK N-004 calculations for patch corroded columns it was decided to perform more tests of tubular columns with simulated corrosion damage to provide more data for the validation of the NORSOK N-004 formulae.

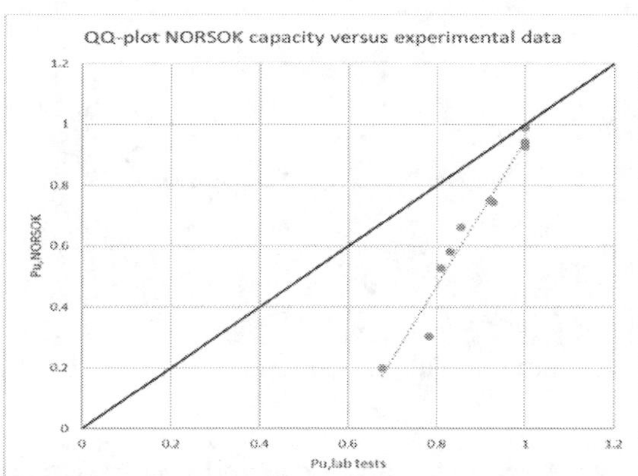

Figure 3. A comparison of the results of the tests performed by Hebor [9] and calculations according to NORSOK N-004 [3].

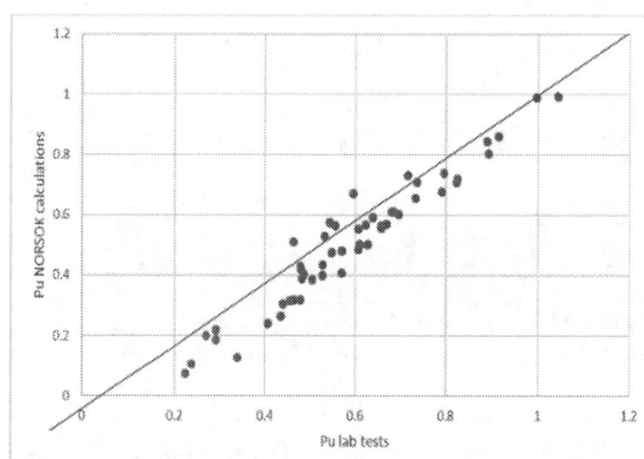

Figure 4. Comparison between capacity of dented tubular columns according to experiments (x-axis) and calculations according to NORSOK N-004 [3] (y-axis).

IOP Conf. Series: Materials Science and Engineering **700** (2019) 012029 doi:10.1088/1757-899X/700/1/012029

4. Overview of test programme

The test equipment used was limited by a 400 kN maximum compression force and a maximum specimen length of 2 m. The strength of the steel was selected to S235 and the test specimens used were 70 mm x 2.9 mm steel pipes with a length of 1.5 m. The slenderness of the selected columns was 63 and the outer diameter to thickness ratio D/t was 24.1 which is a reasonable value compared to tubular columns used in several structural applications.

The material certificate of the longitudinal welded steel tubes specified a Re min. of 235 MPa, mechanical test value Re of 280.4 MPa and a corresponding tensile strength Rm of 391.7 MPa. Five (5) independent mechanical tension tests were performed in the UiS test laboratory from test specimens taken directly from the received tubulars as shown in Figure 5. These tests indicated a mean value of the yield strength of 340 MPa.

The test machine for the buckling test is the TONI TECHNIK Baustoffprüfsysteme GmbH D-1000. The machine is designed to test bending capacity for beams and hence the machine is not optimized for the testing of columns. However, some modifications were introduced to enable buckling tests of aluminium columns with this machine previously and the same type of setup was used for these tests.

In order to position the tubular column in the test machine two fasteners cups were machined to hold the tubular column in position at each end as shown in Figure 6.

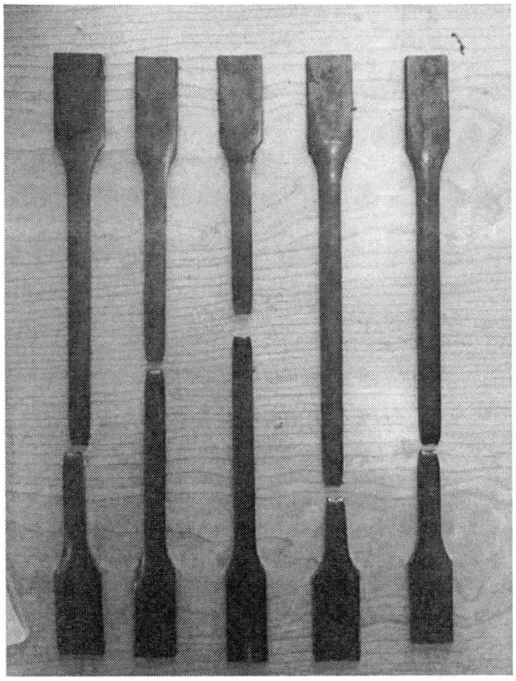

Figure 5. Test specimens from the yield strength test.

Figure 6. Tubular column mounted in test machine.

The test setup was intended to give a column that was fixed in the lower end and simply supported (pinned) in the upper end. Buckling tests of undamaged cross-sections were performed as a calibration test. This calibration test was intended to provide:

- A basis for the ultimate capacity of an undamaged column,
- A check of the buckling mode and hence the correctness of the assumed support condition.

The tests indicated a buckling mode with maximum deflection and buckling failure at L/2 as opposed to the expected 0.7 L if the assumptions for the support conditions were true. This indicated that the column either experienced both supports as pinned or as fixed. Hence, calculations according to NORSOK N-004 have been performed with pinned, fixed and fixed-pinned boundary conditions.

IOP Conf. Series: Materials Science and Engineering **700** (2019) 012029 doi:10.1088/1757-899X/700/1/012029

Initial deformation was measured prior to testing by use of a straight-edge. With the available equipment at the UiS laboratory no global initial deformation could be detected. Some small local deformation (ovalities and changes in diameters) was indicated in some measurements, but these were of minor nature and not included in further evaluations.

The damage area simulating patch corrosion was selected to be 70mm in height and 70mm along the arc length of the pipe giving an angle of damaged area to be 114.6 degrees. The background for the chosen amount of corrosion damage (wastage) inflicted on the tubulars was based on an attempt to achieve results for a range as wide as possible of capacity reductions (ratio between damaged and un-damaged columns). A range from 1.0 down to approximately 0.4 was achieved in the tests. This was achieved by removing 0.9mm, 1.9mm and 2.9mm by grinding in order to simulate corrosion damage (wastage).

In order to distinguish between the different columns, they were named with the following particulars: *Pipe diameter, wall thickness, damage extent as a percentage of thickness, chronological test number within their group*. As an example, a tube with 1mm remaining thickness is marked as 70-29-34-4, where 70 is outer diameter, 2.9 is material thickness, 34 is percent of remaining thickness and 4 would be the test number within the group. In addition, the tubulars with a simulated through thickness patch corrosion has an "h" or "r" at the end of the identification that indicates whether the corners of the damage are hard or round. Hard corners were cut with angle grinders and round corners were drilled.

The reasoning behind the simulated patch corrosion area was rather pragmatic and an area of DxD where D is the diameter of the pipe seems a reasonable size to start the investigation. The attempt was to have a uniform thickness reduction in this area. However, as can be seen in Figure 7 the grinding process may have left the specimens with some minor variations in thickness.

Grinding was chosen over other possible methods in order to reduce heat and possible changes to the material properties during the process. However, there is still a possibility that some residual stresses were introduced during the grinding process.

Figure 7. Example of tubular column with simulated corrosion damage by grinding.

5. Results of test programme and comparison with NORSOK N-004 calculations

Figure 8 shows the resulting failure mode for 3 of the tests performed. In the cases with 34% and 69% thickness loss it seems like local buckling has occurred. NORSOK N-004 does not indicate the same amount of local buckling as observed in the patch corroded area for the test specimens. However, the local buckling observed may be more prominent for these rather small-scale tests than a more realistic size specimen relevant for the NORSOK N-004 (minimum thickness for NORSOK N-004 formulas is t = 6 mm).

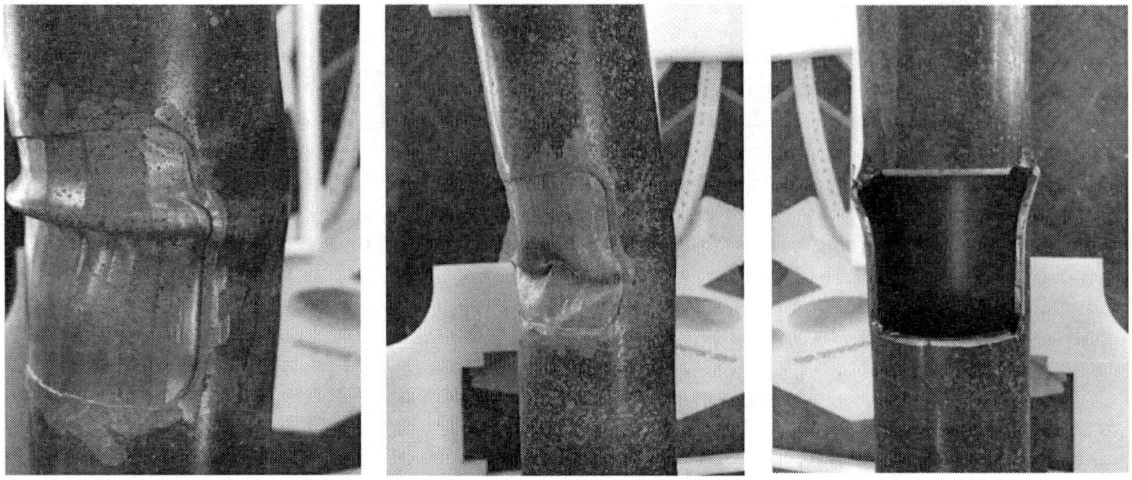

Figure 8. Example of the failure mode of the columns with 34%, 69% and 100% of the thickness "corroded" respectively.

Control measurements of the remaining thickness of the simulated corrosion patches were performed for two specimens. These two specimens were selected for control measurements due to their rather large deviation in their test series. Some deviations from the intended thicknesses were registered and new calculations were performed according to NORSOK N-004 with updated remaining thickness for the specimens with deviations (marked with * in Table 1).

In addition, there were some doubt regarding the straightness of the support cups for the first runs of each series (marked with ** in Table 1). This may have introduced an additional moment in the column that reduced the ultimate capacity slightly. However, a proper estimate of the influence of this effect has not been established.

Figure 9 shows a q-q-plot of measured capacities and capacities according to NORSOK calculations where all values are normalized using the largest measured value of the undamaged column (192.31 kN). Apart from the two results giving significantly lower capacities compared to NORSOK N-004, the estimated capacities of the tubular columns compares reasonably well (bias 1.09 and CoV 17%). For columns with low remaining material thickness (~1 mm) the calculations consistently provide higher capacity than the experiment. This indicates that the calculation method in NORSOK N-004 may underestimate the effect of local buckling. The two largest deviations are found for case 70-29-34-1 and 70-29-34-2 obtaining only 81% and 85% of the calculated capacity respectively.

Table 1. Corrected values for tests and calculations according to NORSOK N-004.

Sample	Damage position	t'	f_u	f_y	P_u NORSOK (partial factors =1.0)			P_u Test
					k=0.5	k=0.7	k=1.0	
70-29-100-1	0	2.9	388.7	343	190.7	180.9	159.9	184.12**
70-29-100-2	0	2.9	388.7	343	190.7	180.9	159.9	191.05
70-29-100-3	0	2.9	388.7	343	190.7	180.9	159.9	192.31
70-29-69-1	975	2 (1.5)	388.7	343		124.0*		145.15**
70-29-69-2	975	2	388.7	343	154.4	145.5	125.9	164.14
70-29-69-3	750	2	388.7	343	154.4	145.5	125.9	158.92
70-29-69-4	750	2	388.7	343	154.4	145.5	125.9	155.34
70-29-69-5	750	2	388.7	343	154.4	145.5	125.9	162.67
70-29-34-1	975	1 (0.4)	388.7	343	78.7*	78.4*	65.9*	63.79**
70-29-34-2	975	1	388.7	343	105.9	101.9	86.7	82.91
70-29-34-3	750	1	388.7	343	105.9	101.9	86.7	99.69
70-29-34-4	750	1	388.7	343	105.9	101.9	86.7	98.08
70-29-34-5	750	1	388.7	343	105.9	101.9	86.7	98.89
70-29-0-1h	975	0	388.7	343	62.7	59.5	53.2	75.43**
70-29-0-2r	975	0	388.7	343	62.7	59.5	53.2	92.97
70-29-0-3r	975	0	388.7	343	62.7	59.5	53.2	81.25

1. Capacities calculated based on NORSOK N-004 is performed by eliminating the safety factors (all safety factors set to 1.0) and applying yield stress values of 340MPa from the tensile tests of the steel used in the tests. Eccentricity of the column due to fabrication misalignment is set to L / 2000 in these calculations.

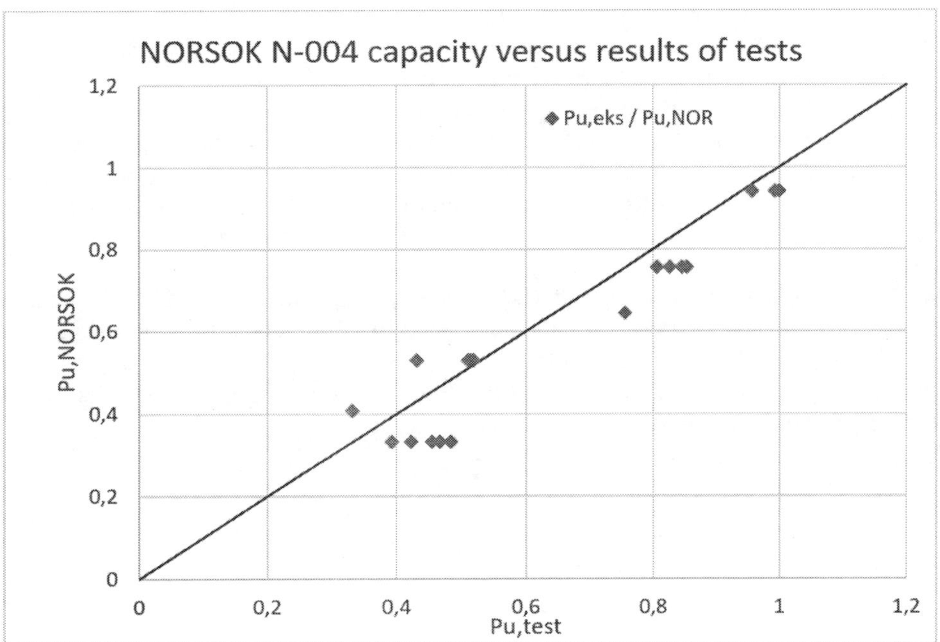

Figure 9. Q-Q plot of results of laboratory tests versus calculated capacities according to NORSOK N-004 [3].

6. Conclusions and recommendation for further work

A set of tests of tubular columns with varying degrees of corrosion thickness (simulated by grinding) has been performed with reasonable results compared to the calculation formulae given in NORSOK N-004 [3]. Several of the tests showed signs of local buckling in the area of patch corrosion. However, the calculation formulae in NORSOK N-004 did not indicate local buckling as being a major factor in the capacity of these specimens. For columns with low remaining material thickness (~1 mm) the calculations consistently provided higher capacity than the experiment. This indicates that the calculation method in NORSOK N-004 may underestimates the effect of local buckling. The two largest deviations were found for case 70-29-34-1 and 70-29-34-2 obtaining only 81% and 85% of the calculated capacity respectively. The dimensions of the specimens used in these tests are rather small (thickness of 2.9 mm) and may be a reason for local buckling being more prominent compared to the prediction by the equations in NORSOK N-004.

Further work in this area is needed, especially more realistic ways of introducing realistic patch corrosion to the specimens. Preferably this should be introduced by means of local accelerated corrosion. This will introduce some uncertainty into the tests as the remaining thickness has to be measured after the corrosion process and most likely after the buckling tests have been performed.

Other vital issues that need further investigation include:

1. The effect of the position of damage (L/2 or 2*L/3) showed no significant difference in the capacity in the tests performed in this work. This seems unrealistic but may be further investigated by more FEM-analysis or by a new series of tests with this as the main purpose to investigate.
2. Boundary conditions are not sufficiently well established in the laboratory experiments. Some of the results indicated an effective buckling length of approximately 0.5 times L (fixed-fixed condition). A test setup that is more in line with the intended support conditions should be used for future tests. A FEM-analysis with the full test rig applying contact elements (between column and "shoe" and between "shoe" and end-plate connected to rig) is a possible way to model and investigate the actual boundary conditions obtained by the actual test set-up.
3. The test specimens are rather small compared to what is normally used in the offshore and building industry. The effect of local buckling may therefore be affected by these rather small-scale test. Test with specimens of more realistic dimensions and various reduced slenderness ratios should be performed at a later stage.
4. All tests in this work were performed with the same rather low slenderness ratio. For future tests specimens with a higher slenderness ratio should be performed.
5. The test setup used in these tests introduced a small rotation to the lower end support as the "cup" was not 100% levelled. After some 4-6 tests the cup loosened and seemed to introduce less rotation, which was also seen in the resulting capacities and load-displacement curves. A test setup that is less sensitive to such irregularities should be used for future tests.

Acknowledgement

The authors would like to acknowledge the assistance from Sudath Siriwardane, Samdar Kakay, Jarle Berge and Emil Surnevik Kristiansen for the assistance in laboratory testing, simulation of corrosion, technical assistance and materials for testing.

The opinions expressed herein are those of the authors and they should not be construed as reflecting the views of the Petroleum Safety Authority Norway.

References

[1] Hebor M and Ricles J M 1994 *Residual strength and repair of corroded marine steel tubulars* ATLSS Eng. Research Center (Lehig University Betlehem, Pennsylvania, USA)

[2] Ricles J M, Bruin W M, Sooi T K, Hebor M F and Schonwetter P C 1995 Residual strength assessment and repair of damaged offshore tubulars *Offshore Techn. Conf.* (Houston, Texas)

[3] Standard Norge 2004 *Design of Steel Structures*, Norsok Standard N-004, Rev. 2 (Lysaker, Norway)

[4] Hestholm K and Tri V 2019 *Knekningskapasitet av korroderte stålsøyler-rørprofiler / Buckling capacity of corroded steel tubular members* (University of Stavanger Stavanger, Norway)

[5] Timoshenko S P and Gere J M 1963 *Theory of elastic stability* McGraw-Hill International Edition (Singapore)

[6] Robertson A 1925 The strength of struts, Institution of Civil Engineers. Selected Engineering Papers. no. 28.

[7] DNV 1995 *Hull structural rules – development, background, motives* (with reference to DNV Rules for classification for ships 1999 for actual formula) Det Norske Veritas (Høvik, Norge)

[8] Ostapenko A, Berger T, Chambers S and Hebor M 1996 *Residual strength and repair of damaged and deteriorated offshore structures: effect of corrosion damage on strength of tubular columns with patch corrosion* ATLSS Engineering Research Centre (Lehigh University, Bethlehem, Pennsylvania, USA)

[9] Ricles J M, Sooi T K and Bruin W M 1994 Experimental and analytical studies on the residual strength and repair of dent-damaged platform bracing *Offshore Techn. Conf.* (Houston, Texas)

[10] Ostapenko A, Wood B, Chowdhury A and Hebor M 1993 *Residual strength of damaged and deteriorated offshore structures: Vol 1; residual strength of damaged and deteriorated tubular members* ATLSS Engineering Research Centre (Lehigh University, Bethlehem, Pennsylvania, USA)

[11] Loh J T 1993 Ultimate strength of dented tubular steel members *Int. Offshore Polar Eng. Conf.* (Singapore)

[12] Loh J T, Kahlich J L and Broekers D L 1992 *Dented tubular steel members* Exxon Production Co. (Houston, Texas)

[13] Landet E and Lotsberg I 1992 Laboratory testing of ultimate capacity of dented tubular members, *J. Struct. Eng.* **118** 1071-89

[14] Yao T, Taby J and Moan T 1988 Ultimate strength and post-ultimate strength behavior of damaged tubular members in offshore structures *J. Offshore Mech. Arct. Eng.* **110** 254-62

[15] Taby J and Moan T 1987 Ultimate behaviour of circular tubular members with large initial imperfections *Annu. Tech. Session, Struct. Stab. Res. Counc.*

[16] Smith C S 1986 Residual strength of tubulars containing combined bending and dent damage *Proc. Offshore Oper. Symp. 9th Annu. Energy Sour. Tech. Conf. Exhb.* (New Orleans, Louisiana, USA)

[17] Ueda Y and Rashed S M H 1985 Behavior of damaged tubular structural members *J. Energy Res. Technol.* **107** 342-9

[18] Taby J and Moan T 1985 Collapse and residual strength of damaged tubular members *4th Int. Conf. on Behav. of Offshore Struct.* 395 - 409 (Delft, Netherlands)

[19] Smith C S 1983 Assessment of damage in offshore steel platforms *Proc. Int. Conf. Offshore Mar. Safe.* (Glasgow, UK)

IOP Conf. Series: Materials Science and Engineering **700** (2019) 012030 doi:10.1088/1757-899X/700/1/012030

Inspection planning for fatigue cracks in offshore structures by Monte Carlo simulations

H Neeraas[1], G Ersdal[1,2,*], N Oma[2] and N-Ch Hellevig[3]

[1] University of Stavanger, Norway
[2] Petroleum Safety Authority, Stavanger, Norway
[3] Aker Solutions, Bergen, Norway

[*] Corresponding author: gerhard.ersdal@uis.no

Abstract. Probability-informed inspection planning has traditionally a tendency to produce longer inspection intervals for older structures in contrast with the intuitive assumption that shorter inspection intervals are needed. In this paper an alternative method for probability-informed inspection planning is suggested based on a parameter updating rather than the general updating used in most cases. The proposed method produces constant inspection intervals after the theoretical first inspection. Further, the method produces reasonable inspection intervals more or less in line with what is used in practise in the offshore industry at present.

1. Introduction

The vast majority of the infrastructure (offshore platforms, buildings, bridges, etc) in the world is existing and now ageing. This large number of existing structures is of vital importance for society, but they are in many cases degrading and need significant investments for maintenance. Replacing such structures with new ones might be both economically and environmentally unsound. Still, most universities educate students in civil and structural engineering merely in the design of new structures. The University of Stavanger, however, has established a course and research work into the continued use of ageing structures.

Using ageing structures beyond their original design life can raise several concerns. Structures in operation are exposed to conditions of stress and environment that ultimately will degrade them from their initial state and damage will accumulate until the structures may be judged to be no longer fit-for-service. If not withdrawn from further service or being repaired, failure of some kind will eventually occur. Also, the cost of the required maintenance, inspection and repair needed to cope with this deterioration and damage will at some stage become unacceptable compared to the revenue that can be gained from the use of these structures. Hence, it becomes vital to know:

1. how structures change with age,
2. how their condition and other aspects influencing their safety can be determined,
3. how their capacity (strength and fatigue life) can be determined as a result of ageing mechanisms,
4. how any anomalies in the structures should be evaluated,
5. how anomalies found in an existing structure can be repaired and mitigated, and
6. how the integrity management of ageing structures should be performed.

This paper is work performed under item 6 as listed above and is specifically aimed at providing insight into the inspection planning for old structures. The work is based on the master thesis performed by Neeraas [1].

The integrity management of existing structures requires that changes to all aspects that influence the safety of structures are identified, registered, evaluated and if needed mitigated or repaired. One such change that will influence the safety of a structure in operation is the development of fatigue cracks and an inspection regime to identify such fatigue cracks are needed for any structure that is experiencing cyclic loading. Historically the inspection for fatigue cracks typically have followed a calendar-based programme with inspection intervals in the range of every 4-5th year for critical parts of the structure. Probability-based inspection planning (sometimes called risk-based inspection planning or RBI) was developed in the 1980'ties, see e.g. Madsen et al [2]. In general, the probability-based inspection planning was introduced to reduce the amount of inspections by evaluating the probability of the presence of a crack at different positions and evaluating the consequence of having the crack in these positions. These evaluations would then be used to prioritize where to inspect by choosing to inspect the positions where the combination of probability and consequence of having a crack has the largest influence on safety level.

In addition to prioritization of where to inspect, the probability-based inspection analysis also allowed for taking into account the results of inspections performed. Most importantly, if an inspection were performed and no cracks were determined, the probability of finding a fatigue crack in the following years could be reduced by probabilistic updating methods (Bayesian updating).

An example of such updating is shown in Figure 1. In the example, inspection has been determined to be needed after approximately 18 years. After the first inspection, which is modelled as "no cracks found", the probability is updated and the next inspection is determined to be needed after an additional 12 years. The inspection intervals indicated in this example figure, assuming that all inspection will turn out negative, is then by reading the figure approximately 18, 12, 20 and 25 years.

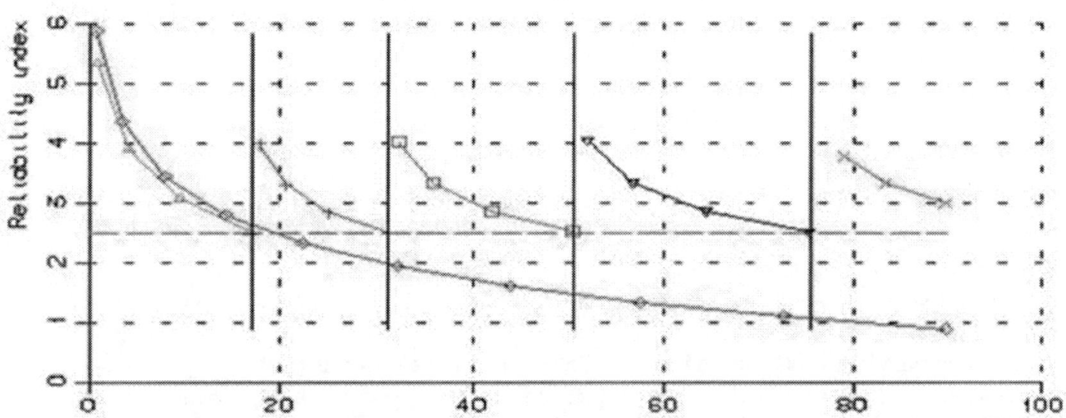

Figure 1. Updated probability of fatigue crack in structure (www.dnvgl.com).

Although the example in Figure 1 is purely an illustration it indicates the trend that has motivated this study, namely that the inspection intervals for ageing structures seems to increase with age rather than to decrease as one would intuitively expect.

2. Overview of simulations performed

The simulations presented in this paper studies the crack growth and the inspection need for a node in an offshore jacket. The study includes evaluating the effect of these parameters:

- three different calculated fatigue lives (20, 40 and 60 years)
- uncertainties in the initial crack size (a_0)

- uncertainties in the applied load
- level of acceptable failure probability corresponding to 0.01 and 0.001 respectively.

An overview of the simulation process is shown in Figure 2.

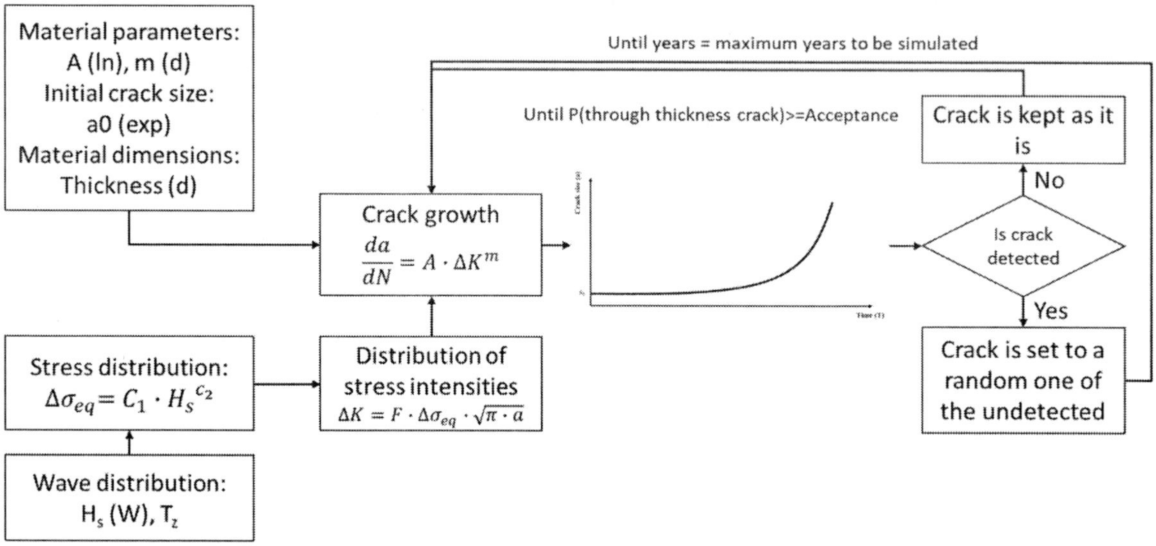

Figure 2. A proposed process for probability-based inspection planning based on Ersdal and Oma [12]. In the figure ln indicates a lognormal distributed parameter, exp indicates an exponential distributed parameter, W indicates a Weibull distributed parameter and d indicates a deterministic parameter.

The first step in this process is to define the input parameters: material parameters, initial crack size, wave height and period distribution. The initial crack size in the welded detail is assumed to follow and exponential distribution with an mean initial crack size of $\mu_{a0} = 0.11mm$ [3] and the probability distribution function and cumulative distribution function described by:

$$f(x) = \lambda \cdot e^{-\lambda \cdot x} \tag{1}$$

$$F(x) = 1 - e^{-\lambda \cdot x} \tag{2}$$

where $\lambda = \frac{1}{\mu_{ao}}$.

The slope of the Paris equation m (or m_a) is in these analyses fixed to 3.0. The material parameters lnA is determined by calibration of the crack growth curve to provide similar probability of failure as a SN-fatigue analysis as shown in Figure 3.

In the SN-fatigue analysis the design value of log(a) is 11.764 and the slope of the SN curve is set to $m_{SN} = 3$. Mean value of log(a) is assumed to be 2 standard deviations larger than the design value and the standard deviation is assumed to be 0.25. Hence, the mean value of log(a) is 12.264. Based on the calibration performed the material parameters of the Paris equation has been modelled by a mean value of lnA is 29.1 and a standard deviation of lnA is 0.64.

The long-term distribution of significant wave heights is given by:

$$F_{H_s}(h) = 1 - exp\left[-\left(\frac{h - H_0}{H_C - H_0}\right)^{\gamma}\right] \tag{3}$$

where $H_C = 2.895m$, $H_0 = 0.198m$ and $\gamma = 1.499$.

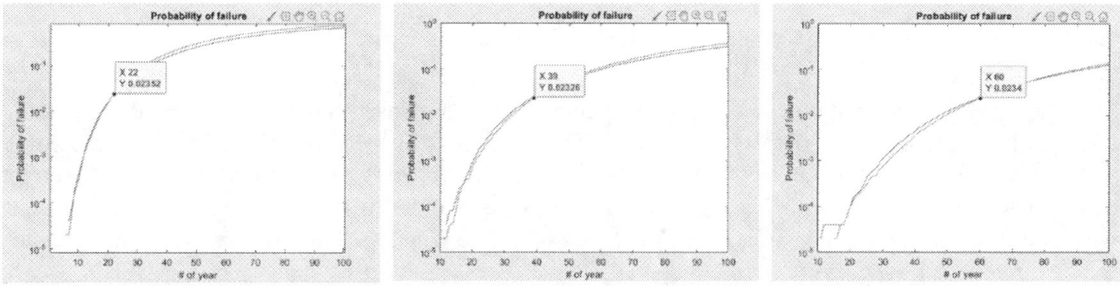

Figure 3. Result of calibration the fracture mechanics parameters to the SN-curve.

Mean zero up-crossing period of the sea-state is given by:

$$T_Z = 3.3 \cdot \sqrt{H_s} \qquad \text{with a minimum value of } T_Z = 5.72 \text{ s.} \tag{4}$$

Based on the wave height and period the width of the stress intensity, ΔK, is defined by the equivalent stress range $\Delta\sigma_{eq}$ implemented as described in Ersdal [4], given by the wave distribution and the geometric function of a tubular jacket node.

$$\Delta\sigma_{eq} = \frac{C_1}{1.702 - 0.138 \cdot m_a} \cdot (H_s)^{C_2 - 0.03} \tag{5}$$

where C_1 and C_2 are coefficients, H_s is the significant wave height and m_a is the slope of the crack-growth curve in the Paris equation.

Crack-growth is modelled by the Paris equation:

$$\frac{da}{dN} = A \cdot \Delta K^{m_a} \tag{6}$$

where A and m_a are material parameters and:

$$\Delta K = \Delta\sigma \cdot F(a,t) \cdot \sqrt{\pi \cdot a} \tag{7}$$

where ΔK is the stress intensity range, $\Delta\sigma$ is the stress range of each cycle, $F(a,t)$ is the geometric stress function for the detail and a is the present crack depth. For a tubular member detail, the geometric stress function may be set to (Dalane, 1993):

$$F(a,t) = \left(1.08 - 0.7 \cdot \frac{a}{t}\right) \cdot \left(1.0 + 1.24 \cdot e^{-22.1\frac{a}{t}} + 3.17 \cdot e^{-357\frac{a}{t}}\right) \tag{8}$$

where a is the present crack depth and t is the thickness of the material. For each sea-state the crack growth can be calculated as [5]:

$$da_i = A \cdot \left[\Delta\sigma_{eq} \cdot F(a_{i-1},t) \cdot \sqrt{\pi \cdot a_{i-1}}\right]^{m_a} \cdot N_w \tag{9}$$

where N_w is the number of cycles in the sea-state.

As sea-states are simulated by a Monte-Carlo simulation the crack size increases. At the end of each year the number of cracks calculated to be larger than the thickness (through-thickness cracks) is counted and if the number of through thickness cracks divided by the total number of simulations exceeds the acceptance requirements (0.01 and 0.001) an inspection is performed.

The inspection is simulated by drawing a random number and checking if the probability of detection for the size of the simulated crack is larger than this random number. The probability of detection (PoD) curves are given by DNVGL [6]:

$$PoD(d) = 1 - \frac{1}{1 + \left(\frac{d}{X_0}\right)^b} \tag{10}$$

where d is the dimension of the crack (depth or length), X_0 and b is parameters for individual inspection methods.

IOP Conf. Series: Materials Science and Engineering **700** (2019) 012030 doi:10.1088/1757-899X/700/1/012030

The ratio between crack depth and crack length $\frac{a}{c}$ is assumed to be 0.15 [5]. In the simulations the crack depth is simulated and where crack length is used this is transferred to crack depth by $a = \frac{1}{2} \cdot 0.15 \cdot (2 \cdot c)$.

PoD-curves is dependent on the NDT method used, the environment they are used in and the accessibility of inspection. DNVGL-RP-C210 (DNVGL 2015) gives the values for X_0 and b for MPI and CVI respectively for various environments and accessibilities. In these simulations under water conditions and low accessibility are selected for both cases. For MPI $X_0 = 1.16$ and $b = 0.90$ are used with depth of crack as dimension. For CVI $X_0 = 83.03$ and $b = 1.079$ are used with length of crack as dimension in the PoD curve.

If the simulations indicate that a crack is detected by the procedure described above it is set to the largest of 1) a random one of the undetected ones and 2) a random value of the initial distribution of cracks. The undiscovered cracks are retained as they are. Both discovered and undiscovered cracks are then simulated further until the maximum life cycle is achieved.

In order to obtain a reasonable accuracy in the simulations the number of Monte-Carlo simulations is set according to the formula [7]:

$$s = \sqrt{\frac{\hat{P}_f(1 - \hat{P}_f)}{N}} \tag{11}$$

where s is the standard error of the failure probability, $\hat{P}_f$ is the failure probability, and N is the number of simulations. In order to obtain a 5% accuracy for a problem with a $\hat{P}_f \sim 10^{-4}$ the number of simulations need to be:

$$N = \frac{P_f(1 - P_f)}{s^2} = \frac{P_f(1 - P_f)}{(0{,}05 \cdot P_f)^2} = \frac{(1 - P_f)}{P_f \cdot (0{,}05)^2} \approx 4\ 000\ 000$$

In these simulations the acceptable failure probability is set to 0.01 and 0.001 respectively. This implies that 40 000 and 400 000 simulations are needed for each case respectively.

The parameter sensitivity study includes the model uncertainty (load and crack growth model) and the initial crack size. The model uncertainty is included by stochastic variables with mean value of 1.0 and a coefficient of variation (CoV) in the crack growth model as shown in Equation 12.

$$\frac{da}{dN} = A \cdot \left[\alpha_G \cdot \alpha_Y \cdot \Delta\sigma_{eq} \cdot F \cdot \sqrt{\pi \cdot a}\right]^{m_a} \tag{12}$$

The model uncertainty in the load model is modelled by a stochastic variable that is kept for the full life of the simulation (GCoV) and a stochastic variable that is changed every year (YCoV). Each of these variables are modelled with a CoV in the range of 0.0-0.3. The model uncertainty in the initial crack size is modelled by a stochastic variable with a CoV in the range of 0.0-0.2.

In addition, the effect of two probability levels for when inspections should be performed is studied ($P_{acceptance}$ equal to 0.01 or 0.001). The $P_{acceptance}$ values are based on normally used acceptance criteria for important and medium important structural nodes and members.

3. Results of the simulations

Figure 4 and Figure 5 show the simulated initial crack sizes and the crack growth parameter lnA respectively. Figure 6 shows the crack growth of 20 random simulations of the detail. The updates after inspection can be detected by a discrete reduction of the crack size. Also, some examples of deeper cracks can be seen. These are a result of that the updated crack size is selected as a random value between the undetected cracks, and there will always be a likelihood that a deeper crack is un-detected and can be selected as the new crack size of a detected crack.

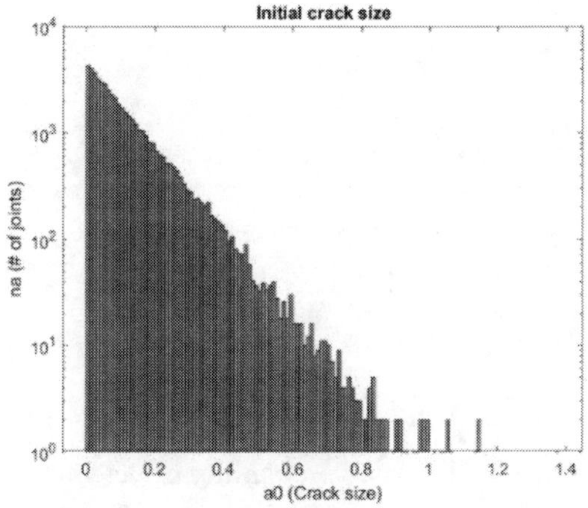

Figure 4. Example of simulated initial crack size.

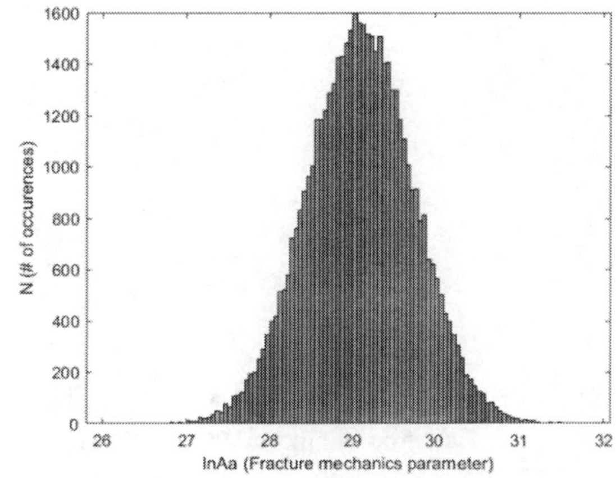

Figure 5. Example of simulated fracture mechanics material parameter.

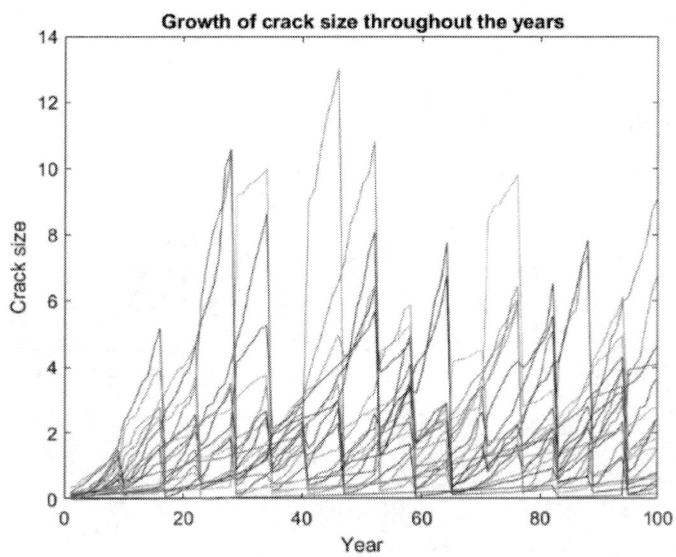

Figure 6. Example of crack growth and updating after inspection for 20 random simulations.

Figure 7 gives the probability of failure calculated for one of the simulations. As indicated by the Figure 7 this method of updating the failure probability after inspection provides a constant inspection intervals after the first inspection. Hence, we have in the main result part of this paper looked exclusively on the number of years until first inspection is needed and the inspection interval after the first inspection.

As an additional benefit the likelihood of a crack being detected (in Figure 8 called "repair required") can be calculated using this simulation method. The trend indicated in Figure 8 is quite similar in all cases simulated. For CVI the curve flattens out at a plateau of 0.4 and for MPI on a plateau value of 0.6-0.67 as also shown in the example in Figure 8.

IOP Conf. Series: Materials Science and Engineering **700** (2019) 012030 doi:10.1088/1757-899X/700/1/012030

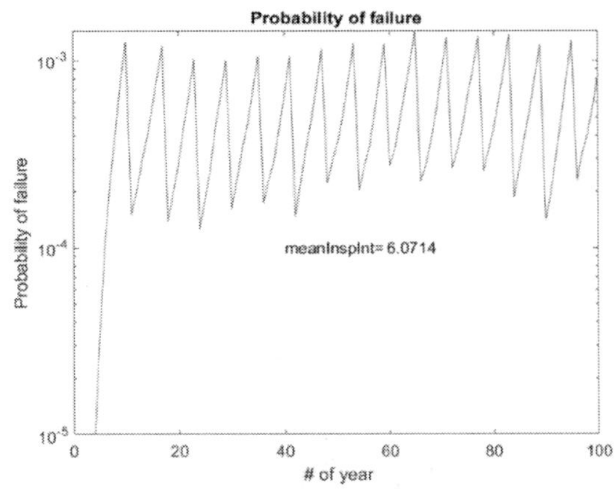

Figure 7. Example of simulated probability of failure.

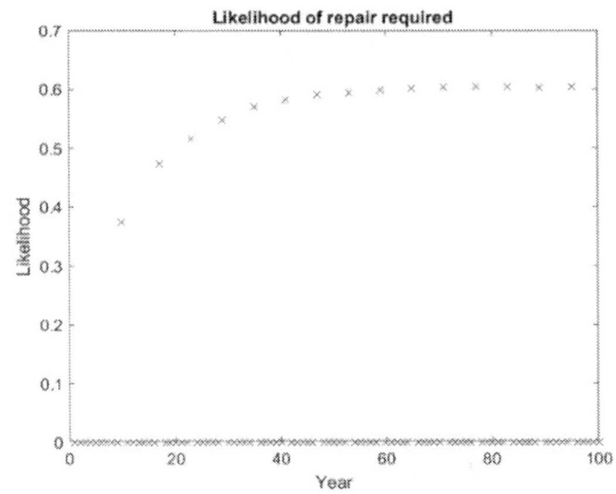

Figure 8. Simulated probability of detection of crack.

The core of the updating is the distribution of cracks prior to and after inspections. An example of two such distributions are shown in Figure 9 and Figure 10. The distribution of cracks prior to updating is shown in red, and the distribution of cracks after updating is shown in blue (note that blue over red becomes purple in these figures). It can easily be seen that the number of large cracks is reduced, and it is also possible to see that the number of small cracks is increased after the updating.

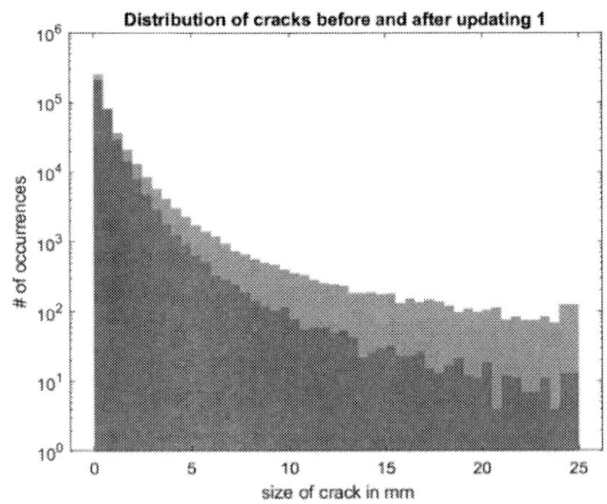

Figure 9. Distribution of simulated crack sizes before and after first inspection.

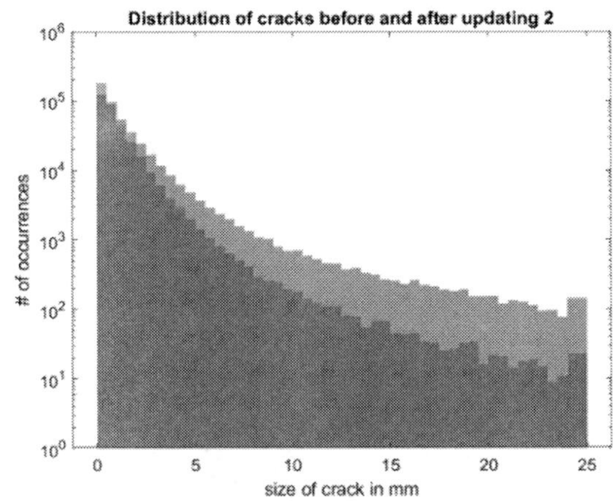

Figure 10. Distribution of simulated crack sizes before and after second inspection.

4. Results of the simulations

A key part of this study was to investigate the effect of the parameters representing the model uncertainties and uncertainty in the initial crack growth. As expected, an increase in the model uncertainty both for the GCoV and YCoV decreases the inspection intervals as shown in Figure 11 and Figure 12. The decrease in calculated inspection interval is more prominent for the GCoV case.

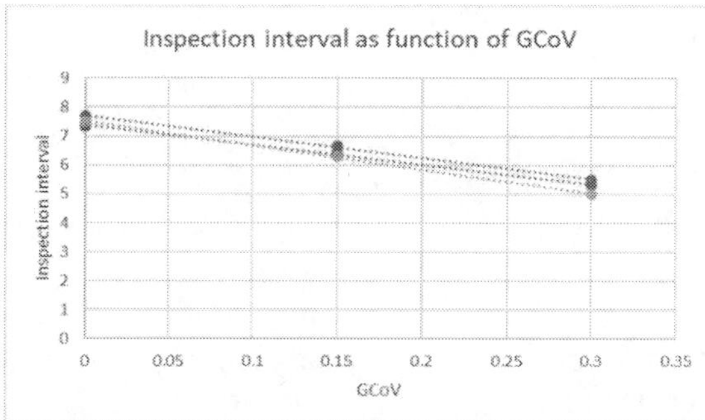

Figure 11. Inspection interval as a function of GCoV.

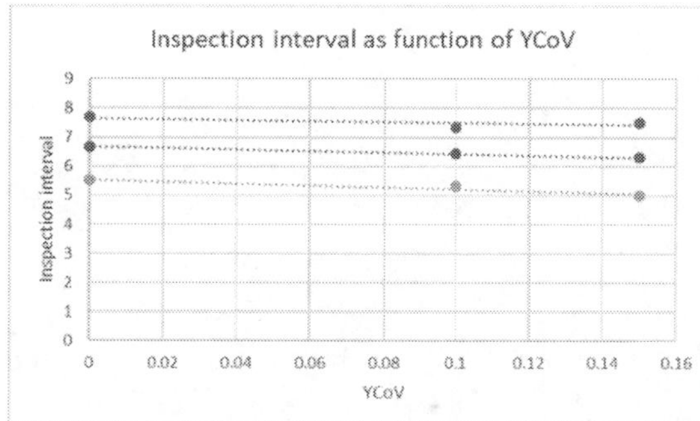

Figure 12. Inspection interval as a function of YCoV.

The effect of the CoV on the initial crack size is shown in Figure 13. Essentially, the CoV on the initial crack size have surprisingly very low influence on the calculated inspection intervals. However, this can be explained by the very narrow distribution of initial crack sizes used as basis in these simulations. The effect of the "acceptable probability of failure" Pacc on the inspection interval is shown in Figure 14. As expected, a lower Pacc result in more frequent inspection intervals.

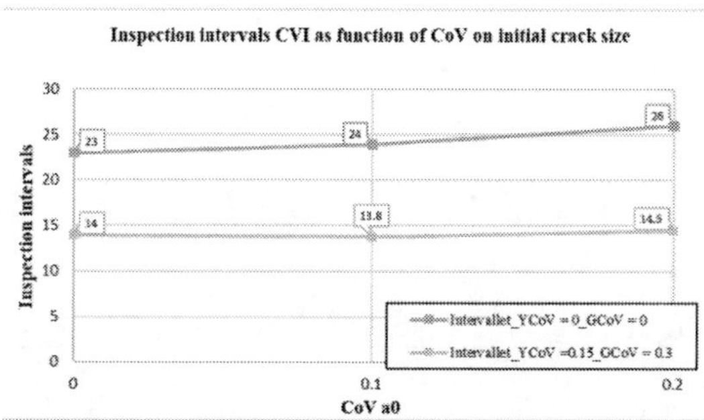

Figure 13. Inspection interval as a function of CoV on initial crack size.

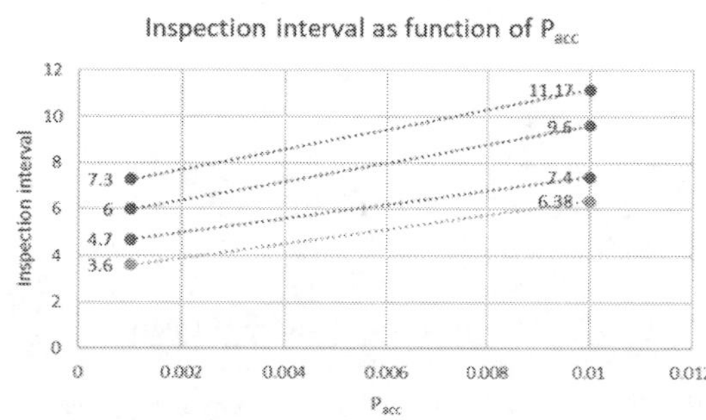

Figure 14. Inspection interval as a function of Pacc for MPI and CVI for CoV's equal to zero and CoV's equal to 0.15.

Figure 15 shows the inspection interval (as fraction of the calculated fatigue life) as function of the calculated fatigue life. A certain tendency for a longer inspection interval as fraction of the calculated fatigue life is seen. However, the increase is minor in the case with model uncertainty.

Table 1 gives an overview of a selected number of simulations performed in the master thesis of Neeraas [1]. As shown in Table 1 the inspection intervals for a detail with a calculated fatigue life of 20 years are as low as 3.6-5.2 years for CVI inspections and 6-8 years for MPI inspections. Such inspection intervals are not far from the present practice in the offshore industry. However, these suggested inspection intervals are a result of a master thesis without the quality assurance required in projects in the offshore industry. Hence, the results should be seen as indications of the capability of the method rather than actual suggestions for inspection intervals to be used in practice.

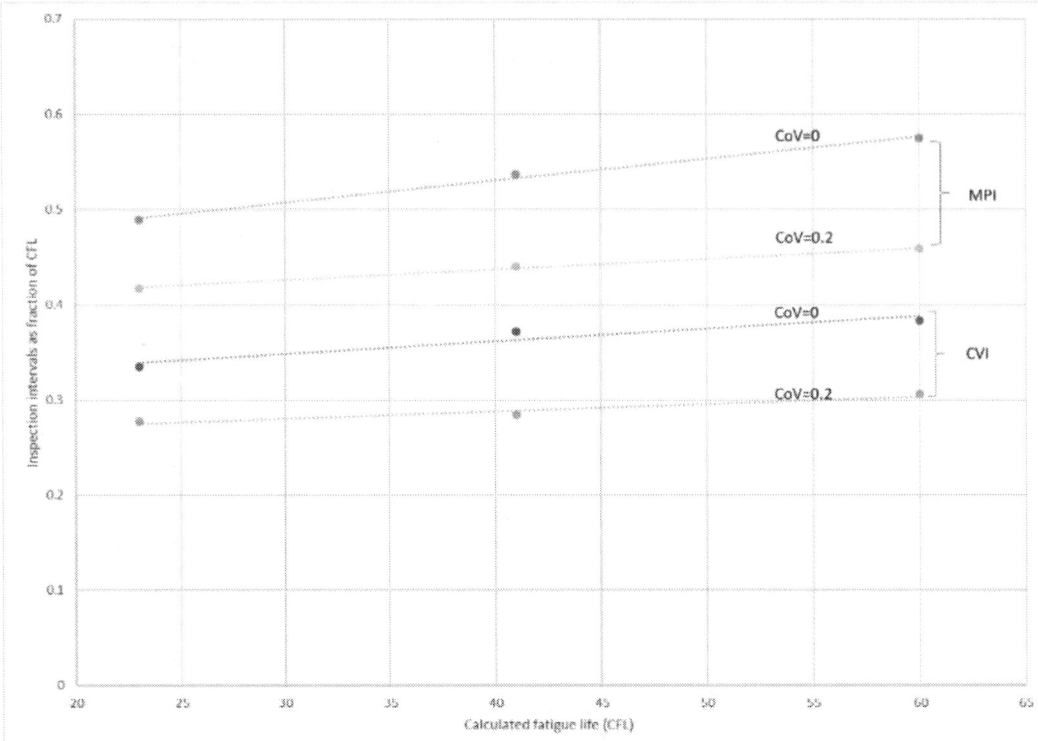

Figure 15. Inspection intervals as fraction of calculated fatigue life (CFL) as a function of the calculated fatigue life.

Table 1. Overview of results for inspection intervals for a jacket structure [1].

FL	P_{ACC}	Y_{COV}	G_{COV}	$A0_{COV}$	CVI INT	MPI INT	$A0_{COV}$	CVI INT	MPI INT
23	0.01	0	0	0	7.4	11.17	0.1	8	11.33
23	0.01	0.15	0	0	7.4	10.57	0.1	7.6	10.57
23	0.01	0	0.15	0	6.67	10.71	0.1	6.67	10.86
23	0.01	0.15	0.15	0	6.38	9.6	0.1	6.23	9.6
23	0.01	0	0.3	0	5.18	7.9	0.1	5.8	7.9
23	0.001	0	0	0	4.7	7.3			
23	0.001	0.15	0.15				0.1	3.6	6
41	0.01	0	0	0	14.5	22	0.1	16	22
41	0.01	0.15	0	0	14	19.7	0.1	13.5	20.7
41	0.01	0	0.15	0	13.2	17.75	0.1	12.8	18.3
41	0.01	0.15	0.15	0	11.33	18.3	0.1	12	17.8
41	0.01	0	0.3	0	9.4	14.6	0.1	9.3	14
41	0.001	0	0	0	8.75	13.4			
41	0.001	0.15	0.15				0.1	6.75	10.25
60	0.01	0	0	0	22	35	0.1	24	34
60	0.01	0,15	0	0	22.5	32	0.1	22	33
60	0.01	0	0.15	0	20	31	0.1	20.5	32
60	0.01	0.15	0.15	0	18.7	27.5	0.1	18	27.5
60	0.01	0	0.3	0	14	21	0.1	15	21.3
60	0.001	0	0	0	13.25	23			

Similarly, a study to establish guidance on inspection intervals were developed during the update of NORSOK N-005 [8] as shown in Table 2. The inspection intervals here is based on a generic reliability based analysis assuming no crack detection in all sub-sequent inspections. The inspection method assumed is above water Eddy Current inspection (EC) with probability of detection (PoD) as defined in DNV GL RP-C210 [6]. A wall thickness of 25 mm is used and the detail is assumed analysed with a DFF =1.0. SN curve for tubular sections in air and pure membrane stress are used. Based on DNV GL RP-C210 (with Weibull shape parameter 0.8, ln(A) = 2.3114 and coefficient of variation (CoV) 0.3, 0.2 and 0.15 respectively). The table provides generic structural reliability inspection intervals comparable with the results of the simulations in this paper with P_{acc} = 0.001. From the 4th inspection and onwards the requirement of NORSOK N-006 [9] of a maximum inspection interval of 5 years was used to overrule the probabilistic inspection intervals in this work.

Table 2. Generic reliability-based inspection intervals prepared during the development of NORSOK N-005.

Fatigue life (years)	Uncertainty of fatigue and hot spot stress calculation	1st (years)	2nd (years)	3rd (years)	4th and onward (years)
20	High (simplified)	4.7	4.3	5.0	5
20	Medium (parametric)	6.7	5.3	6.0	5
20	Low (Finite element calculation)	7.5	4.5	6.0	5

The inspection intervals presented in Table 2 is based on a long-term stress range distribution rather than the method used elsewhere in this paper. The case calculated in this paper with YCoV = 0.15, GCoV = 0.15 and Pacc = 0.001 giving MPI inspection interval of 6 years is comparable with the "Medium (Parametric)" case in Table 2 indicating an inspection interval of 6.7, 5.3 and 6. Although the results in ***Table 2*** are performed with SN-curves for air and with a long term distribution of stress ranges (and some other minor differences) the comparison might be described as a relatively good match.

5. Conclusions and discussion

The aim of the simulations presented here is to:

- Investigate the trend of the inspection intervals for older structures to see if lower inspection intervals were obtained with age by this method of updating. Rather than giving lower intervals with age, the method is predicting a constant inspection interval onwards from the theoretically first inspection. This is an interesting but also understandable conclusion and fully in line with the inspection intervals that have been used in shipping and offshore industry for several years.
- Use the Monte-Carlo simulation method to illustrate the probability-informed crack growth model and updating in a way that were easier to follow and understand. The method has shown that it is capable of producing results that are easy to follow. However, the way the method is programmed at the moment is rather slow and are not necessarily applicable for practical engineering purposes.
- Perform a parameter study to gain insight in how the inspection intervals change with variation in model uncertainty (modelled by the CoV of the model uncertainty). The uncertainty on the load model (YCoV and GCoV) behaved as expected and gave shorter intervals between inspection when the CoV values increased. Somewhat surprising the model uncertainty on the initial crack distribution had little influence on the resulting inspection intervals.
- Establish inspection intervals for calculated fatigue lives in the range between 20-60 years. These results are shown in Table 1.
- Evaluate if the proposed method gave reasonable inspection intervals. The analysis here is based on several simplifications (e.g. the model for equivalent stress and the stress intensity function used) and the analysis is performed as a part of a master thesis with no or little quality assurance. However, the inspection intervals produced by the method seems reasonable and comparable with

inspection intervals developed by the expert committee during the 2015 revision of NORSOK N-005 [8].

The calibration of fracture mechanics crack growth parameters to match SN-fatigue curves is a topic of debate in the industry and is not necessarily well standardized. In DNVGL RP-C210 some guidance on this subject is given, but this guidance is not fully followed in this paper. For example, DNVGL RP-C210 highlights the importance of achieving calibrated material parameters that produces reasonable crack growth curves. In addition, it may be more correct to calibrate the material parameters to the SN-curve at a 50% probability level rather than the 2.3% probability level as used herein (see ***Figure 3***).

6. Recommendation and further work

The updating process presented in this work is somewhat synthetic and do not follow a Bayesian parameter updating as described in e.g. Ang and Tang [10, 11]. A more correct Bayesian parameter updating is needed in further work and would require that a probability distribution of the crack size at the time of inspection is established. The updated probability distribution functions for the crack size after inspection can then be established by use of the likelihood function. As an example, if X is the random variable where updated information is available (e.g. the crack size at a location) with a density function$f(X)$, the updated density function for the variable after observing the experimental outcome ε is then given by:

$$f^U(X) = \frac{P(\varepsilon \mid X) \cdot f(X)}{\int_{-\infty}^{\infty} P(\varepsilon \mid X) \cdot f(X) \cdot dX} \tag{13}$$

where the function $P(\varepsilon|X)$ is commonly referred to as the likelihood function of X and denoted $L(X)$. $L(X)$ is the likelihood of observing the experimental outcome ε assuming a given X.

For this method to be used in practical engineering a version requiring less computation time is needed. For example, further work to establish a method for parameter updating using first or second order reliability method (FORM and SORM). Monte-Carlo simulation may still be an option if the code is streamlined to minimize simulation time.

The effect of the calibration of fracture mechanics crack growth parameters to match SN-fatigue curves is a topic of debate in the industry and need further investigation. The calibration of fracture mechanics crack growth parameters to match SN-fatigue curves should be done in accordance with DNVGL RP-C210. In addition, further work to study the effect on inspection intervals by using “in-air”, “free-corrosion” and “cathodic protection” SN-curves would be beneficial for practical engineering.

Acknowledgement

The results of the simulations presented in this paper is the product of the master thesis by Håkon Neeraas [1] and have limitations that imply that they cannot necessarily be used for inspection planning for jacket structures. The results presented herein cannot in any way be construed to be recommendation by the employers of the co-authors (Petroleum Safety Authority and Aker Solutions).

The authors would like to acknowledge Magnus G. Ersdal for the development of the simulation program used in this paper.

References

[1] Neeraas H 2019 Sprekkvekst og inspeksjonsplanlegging / crack growth and inspection planning University of Stavanger (Stavanger, Norway)
[2] Madsen H O, Krenk S and Lind N C 1986 Methods of Structural Safety: (Prentice Hall Inc. Englewood Cliffs, New Jersey, USA)
[3] DNV 1992 Classification notes 30.6 Structural reliability analysis of marine structures (DNV, Høvik, Norway)
[4] Ersdal G 2005 Assessment of existing offshore structures for life extension University of

Stavanger (Stavanger, Norway)
[5] Dalane J I 1993 System reliability in design and maintenance of fixed offshore structures Dept. of marine structures (Norwegian Institute of Technology, Trondheim, Norway)
[6] DNVGL 2015 DNVGL-RP-C210 Probabilistic methods for planning of inspection for fatigue cracks in offshore structures (DNVGL, Høvik, Norway)
[7] Bury K V 1975 Statistical Models in Applied Science (Wiley London)
[8] Standard Norge 2017 In-service integrity management of structures and marine systems. Edition 2 (Standard Norge, Lysaker, Norway)
[9] Standard Norge 2015 Assessment of structural integrity for existing offshore load-bearing structures. Edition 2 (Standard Norge, Lysaker, Norway)
[10] Ang A H S and Tang W H 1975 Probability Concepts in Engineering Planning and Design, Volume I – Basic Principles (John Wiley and sons New York, USA)
[11] Ang A H S and Tang W H 1984 Probability Concepts in Engineering Planning and Design, Volume II – Decision, Risk, and Reliability (John Wiley and sons New York, USA)
[12] Ersdal G and Oma N 2019 Investigation of updating methods for probability-informed inspection planning for offshore structures, COTech2019, University in Stavanger, Norway.

Acceptable fatigue crack occurrence rate

O T Vårdal[1]*, T Moan[2]

[1] AHPA as, Asset Health & Performance Analyses, Norway
[2] Norwegian University of Science and Technology, Trondheim, Norway

Corresponding E-mail: varole61@gmail.com

Abstract. In fatigue management, it is essential to know the fatigue crack growth potential. The lessons learned from use of refined fatigue analyses, fracture mechanics and probabilistic methods for platforms in-service are presented. For ageing offshore units of semi-submersible design, the inspection history of more than 20 000 NDT inspections and detection of close to 1000 fatigue cracks, are used in this study. These experience data are used to assess the fatigue failure potential. The fatigue crack growth capacities are influenced by the extent of gross errors in the design, construction and operation phase in addition to the normal uncertainties of design, construction and inspection. Corrosion in combination with not detected weakening of the capacity against fatigue crack growth is one of the main challenges. Assessments based on collected in-service observations are mandatory.

1. Introduction

Reliability measures may be expressed as component or system failure probabilities. The system failure is represented by sequences of component failure, which involve overload (collapse/fracture) and fatigue failure. In this paper we will focus on the probability for fatigue failure.

It is well known that fatigue analyses give a very uncertain estimate of the fatigue crack growth potential. Since 1989 we have applied a probabilistic model to give prognosis or forecast of the fatigue crack growth occurrence. A probabilistic fracture mechanics model makes it possible to get an updated estimate of the fatigue crack growth potential i.e. updating by use of information from in-service inspection, [12]. Such a method also allows for treatment of the interaction between design, inspection, maintenance and repair requirements, [4, 5, 9, 10].

A project to validate the probabilistic inspection analyses approach for jacket structures, was conducted during the period 1994 to 1999, [1 - 3]. The inspection and crack detection histories of around 40 jacket structures were used to identify the degree of correlation between theoretical predictions and observed fatigue cracks, by expressing predictions in probabilistic terms considering normal uncertainties. A calibrated model for the probabilistic forecast was made. However, validation of the calibrated model has not been done by use of a separate set of observation data. Attempts to update parameters in the probabilistic forecast model are presented in [8, 9]. The results for updated parameters for initial crack depth and probability of detection during an inspection were a significant contribution to the calibration, [8]. The validation of the weld improvement effect by grinding, focused in [9], is also a result included in the model for probabilistic forecast of fatigue cracks used during the last 15 years.

In [13] the conclusions from the validation study for the Jacket structures were discussed based on the collected experience data from 25 Semi-submersible of an average operation history of 20 years in the North Sea. It is significant deviations in the performance of the probabilistic forecast model for

Jacket Structures and Semi-submersibles, [13]. The systematic over prediction of fatigue crack growth in Jacket structures are not present for Semi-submersible. For Semi-submersible it is also a lower per cent of the fatigue cracks to occur on hot spot not scheduled for inspection than for the Jacket structures.

In this paper, we address the potential to obtain measures of the fatigue crack growth potential based on the in-service observations that include the effect of all uncertainties and gross errors not detected during the design, construction and first year in operation as well as corrosion.

The outline of this paper is as follows; Section 3, the next section, deals with the decision model for scheduling inspections, improvements and modifications. We focus the uncertainty elements associated with variation in workmanship and engineering. Section 3 give references to descriptions of probabilistic forecast models and assume familiarity with the concept of probabilistic fracture mechanics crack growth model as presented in [6, 12]. We also include considerations related to grouping of uncertainties and example of how these are included in our model.

Section 4 contains sensitivity studies of the uncertainty level in the fatigue load effect and the bias of calculated fatigue life. Section 4 does not include any updating based on inspection history. Section 5 discusses the selection of target reliability level focusing uncertainties in the fatigue load effect. Section 6 focuses the model including update based on the inspection history of No Find. The consequence of bias and variation in uncertainties for the fatigue load effect and PoD (Probability of Detection) values for the inspections are focused.

In section 7, we summarize the likely possibilities to achieve non-conservative results by use of a probabilistic forecast model. Measures and awareness to avoid non-conservative scope of improvement maintenance and inspection are given. To fully avoid non-conservative decisions we need to validate or calibrate the model forecasting the fatigue crack growth potential by use of in-service observations. Examples of such measures are discussed. Measures that include the added uncertainties and bias by the extent of gross errors in the design, construction and operation phase as well as corrosion.

2. The decision model for scheduling inspection, improvements and modifications

Figures 1 is a schematic illustration of how information from inspection and fatigue analyses results are combined for a specific hot spot location. The upper part of the figure illustrates the simulated crack size distribution and the lower part illustrates the development of the estimated probability of a fatigue failure to occur. This illustration is based on an inspection with result No Finding i.e. the event of crack smaller than the detectable crack size, is used for the probability update.

After updating, the estimate of the fatigue crack growth potential is improved. Updating based on inspection result of detecting a fatigue crack, will increase the probability of fatigue failure and not reduce the fatigue failure probability as shown in Figures 1.

The model and philosophy described in [4, 5] has since 1999 been applied for scheduling inspections, improvements and modifications for offshore floating structure. The used probabilistic fracture mechanics model gives results that are comparable with the probabilistic fracture mechanics model recommended used in [6]. Figures 2 contains comparison with curves in Figure 8-1 in [6].

The curves presented in Figures 2, are initially considered as notional probability levels. For these probabilistic fracture mechanics models to represent frequency of occurrence, we need to validate them by use of experience data. The validation is done by comparison of observed frequency of fatigue crack occurrence and forecasted probability of fatigue cracking. The presence of methodology presented in Figure 1 includes the normal uncertainties and not the influenced by the extent of gross errors in the design, construction and operation phase. The observed frequency of fatigue failure will however include the influence by the gross errors present but not detected.

IOP Conf. Series: Materials Science and Engineering **700** (2019) 012031 doi:10.1088/1757-899X/700/1/012031

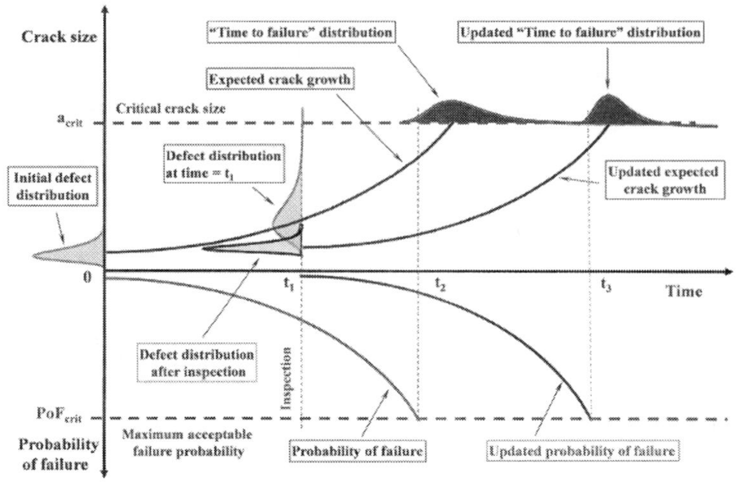

Figure 1. Schematic illustration of crack growth and probability of fatigue failure before and after an inspection, [6].

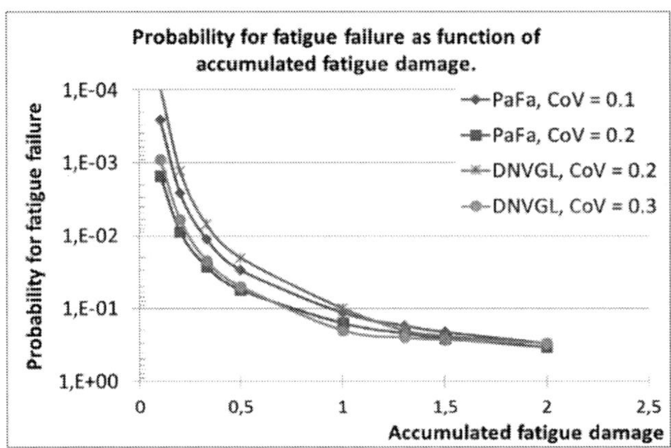

Figure 2. Comparison of probability of fatigue failure (TTC) between the PaFa model used since 2004 and the new model presented in [6] from 2015. The "PaFa" model does not include correlation between Weibull scale and shape parameter for the distribution of load effect. This effect is equivalent to a shift in the uncorrelated model of around 0.1 for the CoV-value of the load.

Modelling the uncertainties in loads and resistances for a hot spot in an offshore structure is a crucial task. The system approach in addition requires an estimate of the uncertainty of the system model as well as the correlation between variables in the different failure functions that represent the system. We will in our discussion only focusing the use of these probabilistic models for prognosis or forecast the fatigue crack growth potential. The system aspect and potential correlation effects are not included. The load and resistance uncertainties are of the same order of magnitude in fatigue crack potential in offshore structures. The uncertainties may in general be grouped as, [6]:

- Physical uncertainty, also known as inherent or intrinsic uncertainty which is a natural randomness of a quantity such as variability in current, uncertainty in yield stress etc.
- Statistical uncertainty is uncertainty due to limited amount of information such as a limited number of observations. Unlike physical variability, the statistical uncertainty arises solely because of lack of sample data. Hence, it will decrease and finally vanish as the amount of data increases.

- Measurement uncertainty is uncertainty caused by imperfect instruments and sample disturbance when observing a quantity such as a fatigue crack size.
- Model uncertainty is uncertainty due to imperfections and idealizations made in the applied physical and probabilistic models and reflects a general confidence in the model to describe "real life". It may further account for unknown effects of other variables and their interaction, which are not included in the analysis model.
- Bias is in general defined as difference or a ratio between expectation of an estimator ê and the quantity e being estimated. When assessing load effect and fatigue capacity it may be practical to use the ratio between these two parameters. In order to avoid misunderstanding it is recommended to define the nominator and the denominator in each case where bias is used.
- Gross errors are uncertainties related to human errors and normally not covered within the framework of structural reliability.

It is not straight forward to group the uncertainty elements present and implementing them in the probabilistic model. Traditionally the uncertainty elements grouped as gross errors are not included in the probabilistic forecast for fatigue crack growth. However, the effect of poor workmanship in design, structural analyses, construction, inspection and the in-service assessment are to be included in our probabilistic model to get reliable forecast on the fatigue crack growth. At least we need to understand how we potentially get non-conservative forecast on fatigue development by neglecting these effects. We will address the challenges of the following uncertainty elements:

a) Deviations between As-Is or As-Constructed and the model for calculation of fatigue life
b) Local design and fabrication quality
c) Inspection quality and QA (Quality Assurance) process for the construction work
d) Inspection quality for the In-service inspection
e) The estimate of accumulated fatigue damage at time of inspection

The fatigue crack growth is highly affected by the local design and the fatigue capacity is presented by different class of SN-curves. The probabilistic fracture mechanics crack growth models are calibrated towards a selected SN-curve [6]. To avoid model calibration towards a high number of SN-curves we use an adjustment on the model for the fatigue load effect.

Figure 3 illustrate the stress categories to be considered, [14]. In addition to the element of local geometry as shape of stiffener ending, weld profile, mismatch in the angular and linear alignment the uncertainty model for fatigue load effect to include potential presence of undercut, weld porosity, weld remains and scars from disk grinder that contribute to selection of SN-curve.

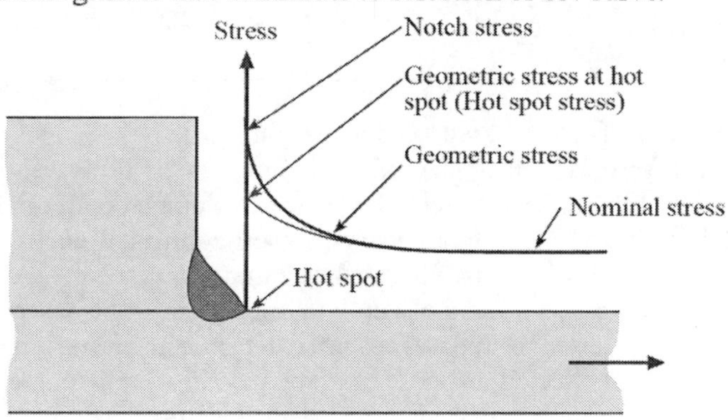

Figure 3. Definition of stress categories.

In offshore structures there are a high number of details that not fully are in accordance with the required fabrication quality. These mainly minor deviations according the requirement of fabrication are in the construction phase mainly assessed according the ULS (Ultimate Limit State) requirements. FLS (Fatigue Limit State) and fatigue crack growth potential is highly affected by measures in the operation phase. Hence, the fatigue challenges related to minor deviations according the requirement of fabrication, are often classified as a challenge to handle in the operation phase. The documentation of deviation reports and fitness for service assessments from the construction phase are traditionally not properly transferred from the construction project to asset integrity responsible in the operation phase. At hot spot level the analyses model frequently shows deviation with the As-Is condition. This is a model uncertainty we for practical reasons includes in the uncertainty level for the fatigue load effect. Without a detailed As-Is inspection and related adjustment of the calculated fatigue life, a higher level of uncertainty in the fatigue load effect is proposed, [13]. The As-Is mapping is also used to remove the deviations between As-Constructed and the construction drawings used as bases for the structural analyses models.

Table 1 and 2 contain results from the reported detection of crack indications. Hence, all flaws not visually detectable but reported by the in-service inspection of NDT (Non Destructive Testing). Table 1 contains data from inspection on Jacket structures and Table 2 contains data from inspection on Semi-submersible.

Table 1. Summary of the collected experience data used for the jacket validation study in [1], from 40 jacket structures in the North Sea. A total of 3366 NDE inspections have been reported.

	Non propagating cracks	Potentially propagating cracks	Most likely propagating cracks	Total number of cracks
Number of cracks	228	124	159	511
Percentage of inspections with cracks	6,77 %	3,68 %	4,72 %	15,2 %

Table 2. Summary of the collected experience data from 12 semi-submersible with more than 20 years of operation time in the North Sea. A total of 22282 NDE inspections have been reported.

	Non propagating cracks	Potentially propagating cracks	Most likely propagating cracks	Total number of cracks
Number of cracks	205	324	622	1164
Percentage of inspections with cracks	0,92 %	1,45 %	2,79 %	5,22 %

The study presented in [14] is based on the data collected from Jacket structures in Table 1. According [14] the average initial crack size of the fatigue cracks observed was close to 0.9mm. The average size of the fabrication defects detected had also an average depth of close to 0.9mm, [14]. The tubular joint in a Jacket structures are fabricated in a well-controlled environment that according [6] have an inspection quality of PoD (Probability of Detection) value of 0.4. An inspection with PoD of 0.4 has a 90 % confidence to detect a crack of depth 0.9mm. Hence, the observed fatigue crack growths in the Jacket structures have started from flaws assumed detected and removed in the construction phase.

Similar studies for the collected experience data for the Semi-submersible are yet not done. However, observations related initial crack size for Jacket structures are most likely also valid for Semi-submersible. As part of the data collected for Semi-submersible, closely 15 000 hot spot given a detailed As-Is mapping, have revealed a large number of deviations between actual fabrication quality and the bases for the structural analyses. Of the more than 1000 detected fatigue cracks in the Semi-submersible, closely to 500 are related to hot spot where the As-Is mapping shows deviations towards the assumptions in the calculation of fatigue life. Hence, the as-is mapping has resulted in change of selected SN-curve or SCF value estimated by the hot spot method.

According [8] the estimated PoD (Probability of Detection) value for the collected inspection data for Jacket structures was 1.9 i.e. an inspection quality of 90 per cent confidence to detect a fatigue crack of length 60mm. The recommendation in [6] is to use PoD values in the range 0.4 to 1.2 i.e. inspection quality of 90 per cent confidence to detect a fatigue crack of length 13 and 38mm.

From closely 25 yard stays and management of the inspection process for re-classification of mobile offshore drilling units of Semi-submersible design we experience a high scatter in the in-service inspection quality. The management and QA-process for the inspection process will affect the PoD values for the inspection. It is likely to assume a variation from 0.3 to 2.0 in the PoD value as function of the management and quality assurance in the inspection process i.e. the 90 per cent confidence limit to detect the fatigue crack will vary between crack length of 7 and 70mm.

Traditional design and construction focus to document sufficient long calculated fatigue life estimates. For non-critical areas these estimates may be very conservative. As highlighted in [6] the input to the probabilistic forecast models from the structural analysis are to be unbiased estimates of calculated fatigue life. For a Semi-submersible, it is typically more than 1000 hot spots assessed for a detailed follow-up scope. It is extremely time consuming to get accurate calculated fatigue life for all of these hot spots. Hence, it is impractical to secure unbiased and accurate calculated fatigue life for all hot spots. The hot spot with the highest fatigue failure consequence potential are focused.

For mobile offshore drilling units operating in the North Sea it is common to estimate the long-term distribution of the fatigue loading based on the worldwide sea scatter diagram i.e. the North Atlantic long-term sea scatter diagram. In our probabilistic models including updating based on years in service, we assume the experienced fatigue loading to be according the North Atlantic scatter diagram and not the sea scatter diagram of the location of operation i.e. a biased and increased model uncertainty for the North Sea operated units.

Based on the discussion above, we may have the following hypothesis: The probabilistic forecast model of fatigue cracks will normally have an unrealistic low uncertainty level and a conservative mean value in the model for the fatigue load effect.

In the following sections, we will use this assumption as a premise and investigate the potential to obtain non-conservative conclusions in our models for scheduling inspections and improvement maintenance.

3. No previous inspections and no probabilistic updating

Figure 4 shows the reliability index for fatigue failure as function of calculated fatigue life for 5 and 10 years additional service. The intersection with the target reliability level defines the limit of calculated fatigue life for inspection interval of 5 or 10 years. The curves denoted "PaFa-0,1 5 year", "PaFa-0,2 5 year" and "PaFa-0,3 5 year" give the limit for 5 year interval for inspection. The different curves represent st. dev. 0.1, 0.2 and 0.3 for the Ln(A), where A is the Weibull scale parameter of the long term distribution of the fatigue load effect. The curves in Figure 4 are used when we have a predefined follow up regime by inspection interval of 5 and 10 years.

For the uncertainty level in the fatigue load effect of st.dev. Ln(A) of 0.1, the limits for calculated fatigue life are 18.3 and 36.6 years, ref. the curves "PaFa-0,1 5year" and "PaFa-0,1 10year" Hence, all hot spots with calculated fatigue life shorter than 18.3 years are scheduled for modification to get an improved fatigue capacity. Hot spots with calculated fatigue life between 18.3 and 36.6 years get an inspection interval of 5 years. Hot spots with fatigue life longer than 36.6 years are recommended an inspection interval of 10 years. The limit values for calculated fatigue life are 30.8 and 61.7 years for st. dev. Ln(A) of 0.2. The limit values for calculated fatigue life are 76.3 and 153 years for st. dev. Ln(A) of 0.3. Table 3 summarizes the calculated fatigue life for hot spots in a typical mobile offshore drilling unit of Semi-

submersible design. Table 4 shows the scheduled scope for modification, inspection of interval 5 and 10 years by combining the scheduling criteria in Figure 4 and example data in Table 3.
If we underestimate the uncertainty level in the calculated fatigue load effect we will get non-conservative scope for improvement maintenance (modifications) and the extent of inspection. Our hypotheses in section 3 included a conservative mean value of calculated fatigue life. Table 4 shows scope also for a bias value of 2 on the calculated fatigue life. Combining a bias of 2 and increased level of uncertainty by st.dev. Ln(A) of 0.3 and not 0.1, will give a larger scope than bias of 1 and st.dev. Ln(A) of 0.1. The ratio between limit values for calculated fatigue life in Figure 4 give the bias value that balance the non-conservative effect by us of to low uncertainty value for the fatigue load effect. For the cases in Figure 4 these ratios are 1.7(=30.8/18.3=61.7/36.6) and 4.2(=76.3/18.3=153/36.6). Hence, even vary large bias values on the calculated fatigue life will not cancel the non-conservative effects by the underestimation of the uncertainty levels in the fatigue load effect.

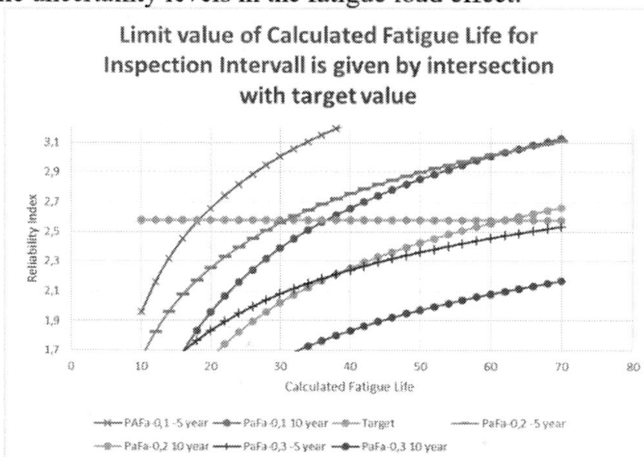

Figure 4. The reliability index as function of calculated fatigue life. For each of the three different level of uncertainty in the fatigue load effect the reliabilities against fatigue failure after 5 and 10 years additional time of operation are given by the curves "Inter 5 year", and "Int. 10 year". The uncertainty levels of fatigue load effect are given by st.dev. of 0.1, 0.2 or 0.3 for Ln(A), where A is the Weibull scale parameter of the long term distribution of the fatigue load effect. Hence, the curve denoted "PaFa-0,1 5 year" are the reliability against fatigue failure after 5 years additional service as function of calculated fatigue life when the uncertainty of the fatigue load effect is given by st.dev. lnA = 0.1.

Table 3. Typical example of number of hot spot grouped by range of calculated fatigue life.

# of Hot Spots within group	Range of Fatigue life in years	Cumulative list # of Hot Spots
0	1-5	0
5	5-10	5
5	10-15	10
4	15-20	14
19	20-30	33
50	30-40	83
128	40-60	211
132	60-100	343
205	100-300	548
507	300-∞	1055

Table 4. Scheduled scope for modifications, 5 years and 10 years inspection interval for different assumption of uncertainty level in the fatigue load effect. A bias value of 2.0 include the assumption

of systematic conservative calculated fatigue life estimate by a factor 2 i.e. reported calculated fatigue life of 20 year to represent a fatigue life of 40 year upon removal of the conservative assumptions.

st.dev. Ln(A)	Bias on calculated fatigue life	# for modifycation	#insp for 5 year interval	#insp for 10 year interval
0,1	1.0	10	73	64
0,2	1.0	33	178	99
0,3	1.0	244	201	31
0,1	2.0	5	8	15
0,2	2.0	10	23	90
0,3	2.0	74	191	93

4. The selection of target level

According to [6] and [7] the target reliability levels are deduced from the selected Design Fatigue Factor, DFF. The time for the first inspections are reached when the accumulated fatigue damage = 1/DFF. For fatigue failure with substantial failure consequence potential the DFF of 10 is used and for a reduced consequence potential we select DFF of 3. From Figure 4 we can identify the reliability level for reaching fatigue damage of 1/10 and 1/3 for the different classes of uncertainty level i.e. st.dev. Ln(A) of 0.1, 0.2 and 0.3. For the curve for 5 years inspection interval the target reliability level is given for the calculated fatigue life of 5*DFF i.e. 15 and 50 year for DFF=3 and DFF=10, respectively. Table 5 contains the target reliability level deduced based on DFF of 3 and 10 for the different assumption of uncertainty in the fatigue load effect.

Table 5: The deduced target reliability level from DFF for different level of uncertainty in the fatigue load effect as well as the recommendation in [6].

st.dev. Ln(A)	DFF = 3	DFF = 10
0.1	2.4	3.4
0.2	2.0	2.9
0.3	1.6	2.4
Ref. [6]	2.5	3.7

The concept of DFF was introduced before detailed FEM analyses were applied for the fatigue analyses. According [6] these fatigue analyses give uncertainty in the fatigue load effect as st.dev. Ln(A) of 0.3. The traditional class follow up regime of mobile offshore drilling units require inspection by 5-year interval and minimum 20 (15) years calculated fatigue life. From Table 5 the traditional class follow-up regime for mobile offshore drilling units results in reliability indexes as low 1.6. According [6] the target reliability level deduced from the DFF value of 3 and 10 are 2.5 and 3.7. The recommended target safety levels in [6] are based on calculations assuming a very low uncertainty level in the calculated fatigue load. To use the recommendation of target level in [6] as a universal target level is therefore more restrictive than traditional industry practice for operation of offshore structures. As an alternative to use universal target level are the target cases proposed in [4], [5] and [13]. The term target case is used when same probabilistic model is used for calculation of the target level as for the hot spot considered. For the concept of target case, the positive gap between the reliability curve for the hot spot and the target level, have low sensitivity w.r.t. potential variation in basic variables or error in the probabilistic fracture mechanics model, [13]. As long as the probabilistic forecast model for the fatigue crack growth potential not have been validated by use of in-service data the probability or reliability curves are notional values. It is challenging to establish universal target level for acceptable fatigue crack growth potential as we use notional probability estimates, [2]. For monitoring the actual safety level against fatigue crack growth, it is recommended to systematically compare the observed frequency of fatigue crack detection and the forecast from the probabilistic model. According [13] we may need to reconsider our follow up regime and model for estimation of the fatigue crack growth potential if the number of fatigue crack detections exceeds 3-15% of the scheduled inspections. The limit of 3-15% is deduced based on the requirements and methodology presented in [6] and [7].

5. Inspection history with no finding

After some years in operation, observations through inspections are obtained. A probabilistic fracture mechanics model makes it possible to get an updated estimate of the fatigue crack growth potential i.e. updating by use of information from in-service inspection, [6]. The calculated reliability against fatigue crack growth varies with the quality of the inspection method, procedure and the skills of the inspection Technicians. According [6] the PoD-value (median crack depth for probability of detection) are in the range 0.4 to 1.2mm for the inspection methods MPI (Magnetic Particle Inspection) and EC (Eddy Current). The study in [8] based on in-service observation indicate a PoD value of 1.9mm.

Figure 5 shows the reliability index for fatigue failure as function of calculated fatigue life for 5, 10 and 15 years additional service after inspection. The curves in Figure 5 are for hot spots that after 20 years' service have been inspected with the result No Find. The intersections with the target reliability level define the limit of calculated fatigue life for inspection interval of 5, 10 or 15 years.

Figure 5 as Figure 4 are used when we have a predefined follow up regime by inspection interval of 5 and 10 years. The intersections with the target reliability level define the limits for calculated fatigue life for inspection interval of 5, 10 or 15 years. The intersection between the curves and the target value are for the calculated fatigue life of 3, 10 and 40 years. Hence, all hot spots with a calculated fatigue life shorter than 3 years are scheduled for modification to get an improved fatigue capacity. Hot spots with calculated fatigue life between 3 and 10 years have an inspection interval of 5 years. Hot spots with fatigue life between 10 and 40 years have an inspection interval of 10 years.

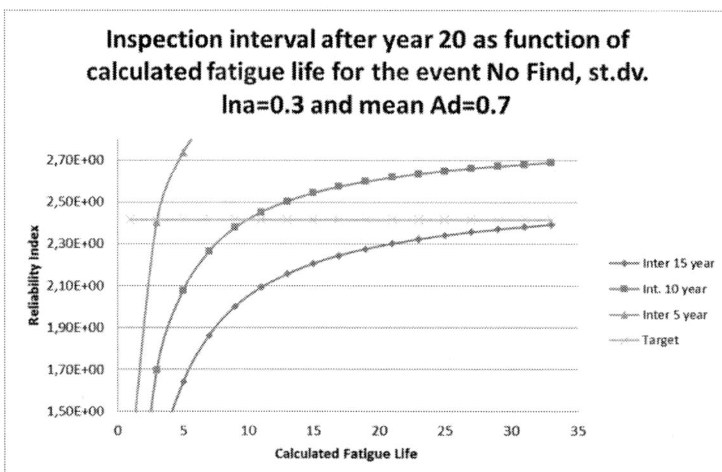

Figure 5. The reliability index as function of initial estimated fatigue life for a given inspection history. The three curves "Inter 5 year", "Int. 10 year" and "Inter 15 year" give the reliability against fatigue failure after 5, 10 and 15 years additional time of operation after inspection with result No Find. The uncertainty level of fatigue load effect is given by st.dev. of 0.3 for Ln(A), where A is the Weibull scale parameter of the long-term distribution of the fatigue load effect. The inspection is done according to procedure and skills securing mean detectable crack depth to be 0.7 mm i.e. 90% confidence of detection of cracks of length 11mm.

Table 6 contains the limit values for calculated fatigue life as presented in Figure 6 in addition to the cases of 25, 15 and 10 years' service at time of inspection with the result No Finding.

The second last column in Table 6 contains the accumulated fatigue damage at time of the event inspection with result No Finding. The last column contains the added accumulated fatigue damage at time of the scheduled inspections. The calculated reliabilities against fatigue failure as function of accumulated fatigue damage, are invariant with years in service to reach the level of accumulated fatigue damage. Hence, the results in the last two columns can be used to obtain a general representation of time to next inspection as function of accumulated fatigue damage at time of inspection with result No Find. The time to next inspection is the product of acceptable additional fatigue damage and calculated fatigue life of the hot spot. We extend the Table 6 with other combination of values for st.dev. Ln(A) and PoD as well as service years before inspection. Based on the last two columns in an extended Table 6 we establish a functional relationship between accumulated fatigue damage at time of inspection with result No Find and the acceptable amount of additional fatigue damage before new inspection is required. See Figure 6.

Each curve represents a combination of uncertainty level in fatigue load effect and PoD value for the inspection method. The PoD values of 0.4, 0.7, 1.2 and 1.6 represent 90% confides to detect fatigue cracks of length 13, 22, 37 and 51mm. The uncertainty levels in the fatigue load effect vary between st.dev. Ln(A) of 0.1 and 0.3.

According Figure 6 the future inspection intervals are highly affected by the inspection quality. The curves in Figure 6 for equal PoD values have different inclination and these curves have comparable limit for added fatigue damage when the accumulated fatigue damage is approximately 2. As shown in Figure 4 and no updating by inspection, high uncertainties in the fatigue load effect reduce the time to next inspection. However, high uncertainties level in the fatigue load effect will give a higher updating effect by the inspection result No Find. These 2 effects balance for the accumulated fatigue damage of approximately 2 in our model as shown in ***Figure 6*** and the intersection of curves with equal PoD.

Table 6 Limits of Calculated fatigue life for scheduling scope for improvement maintenance and inspection. The service period with result, No fatigue crack detection, vary from 25 to 10 years. Inspection quality and uncertainty in the fatigue load effect are given by PoD value 0.7 and st.dev. Ln(A) of 0.7. The last two columns give value of accumulated fatigue damage at service years with the No Find inspection results and the added accumulated fatigue damage before new inspection is scheduled.

Service years before a "No Find" inspection	Inspection interval (Year)	Sorting Criterion for serv act. for Calculated Fatigue Life	Fatigue damage at last inspection	Added fatigue damage in next interval
25	5	3	8,23	1,65
25	10	8	3,22	1,29
25	15	16	1,52	0,91
20	5	**3**	6,52	1,63
20	10	**10**	1,99	0,99
20	15	**40**	0,50	0,38
15	5	4	3,50	1,17
15	10	18	0,84	0,56
15	15	112	0,13	0,13
10	5	6	1,75	0,87
10	10	29	0,34	0,34
10	15	595	0,02	0,03

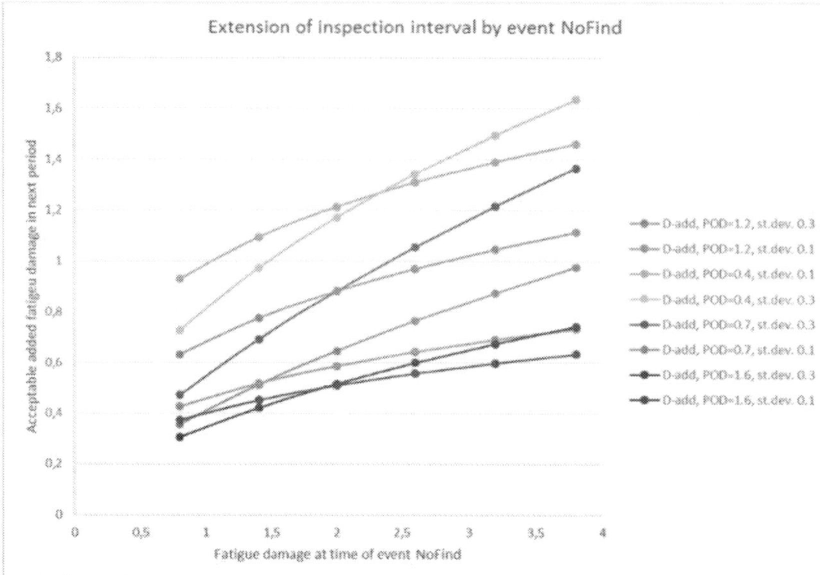

Figure 6. The acceptable level of additional fatigue damage for the scheduled inspection interval as function of the accumulated fatigue damage at time of last inspection with result No Finding.

For the case of no updating based on inspection history it was a non-conservative approach to select a low level of uncertainties in the fatigue load effect. For the cases of updating based on the event of no fatigue crack detection this will not necessarily be correct.

From [8] and [1] the review of in-service data indicates a much higher PoD value for offshore MPI / EC inspection than proposed in [6]. The recommendation in [6] highlights that the recommended uncertainty levels do not include gross errors. The deviation between PoD values from [8] and [6] indicate the effect of including variation in workmanship. The quality controls and management of the inspection campaign are important to focus if the lower PoD values are wanted to obtain. The potential for non-conservative conclusions by selection a too high uncertainty level for the fatigue load effect are also reduced by high PoD values as shown by the curve in Figure 6. If we have high PoD values it is low sensitivity with respect to uncertainties in the fatigue load effect.

Figure 6 may also be used to discuss the systematic bias of assuming fatigue loading of world-wide sea scatter diagram for units operating in the North Sea. The experienced annual fatigue damage of a mobile offshore drilling unit operating in the North Sea will typically be 2 to 4 times lower than the estimates based on the word-wide sea scatter diagram [7]. As long as the units continue to operate in the North Sea the overestimation of fatigue damage at time of inspection with result No Find will be as a systematic conservative bias on the calculated fatigue life. We get non-conservative inspection interval if the unit moves to harsher environment.

6. Example cases

The methodology presented above is used for a number of different semi-submersibles since 2004. The table below shows three cases of units operated more than 20 years in the North Sea. The age, operation history and procedure for fatigue analyses shows high degree of correlation between these units.

The following tables (Table 7 – 12) summarize the collected inspection history of crack detections. The first columns indicate time period. The following three columns group the findings in per cent belief of the cracks to be propagating cracks. The following columns contain the number of NDT inspections done i.e. EC or MPI, and the cracks grouped according crack depth. Minor indicate depth less than 40% of the wall thickness, than larger than 40% of wall thickness and finally number of cracks with leakage or leak potential. The columns "Minor", "> 0.4*wt" and "Leakage" exclude the cracks classified with 30% or lower belief to be propagating cracks.

Case 1 presented in Table 7 and 10 shows a limited number of larger fatigue cracks. For the last 5 years in operation it has only been detected smaller cracks. Case 2 presented in Table 8 and 11 had server cracking in the years 1998 to 2007. Efforts implemented in this period were not sufficient to secure per cent of cracking with leak potential to be less than 0.5% according the requirement deduced from the regulations and recommended practice for fatigue management. Additional efforts were required in the period year 2008 to 2010. For the year 2011 to 2014 we secured a sufficient low fatigue crack growth level for unit Case 2. Case 3 presented in Table 9 and 12 have server fatigue cracking in the year 2011 to 2014. The efforts to improve the fatigue capacity implemented in the period 1998 to 2010 were comparable with the investment done for unit Case 1 and 2.

Table 7. The collected inspection results for Case 1. The number of cracks detected.

Year	< 30%	Between 30-70%	>70%	# of NDT Inspect	Minor	> 0.4* wt	Leakage
1991-2007	23	16	14	988	21	9	0
2008-2012	27	5	0	955	5	0	0
2012-2017	0	0	19	535	19	0	0

Table 8. The collected inspection results for Case 2. The number of cracks detected.

Year	< 30%	Between 30-70%	>70%	# of NDT Inspect	Minor	> 0.4* wt	Leakage
1991-2007	0	2	19	125	0	21	19
2008-2012	1	4	16	479	10	11	7
2012-2017	13	20	24	516	46	11	0

Table 9. The collected inspection results for Case 3. The number of cracks detected.

Year	< 30%	Between 30-70%	>70%	# of NDT Inspect	Minor	> 0.4* wt	Leakage
1991-2007	26	34	18	451	56	22	5
2008-2012	37	60	27	1014	110	14	4
2012-2017	4	25	35	937	17	47	22

Table 10. The collected inspection results for Case 1. The per cent of inspection to detect crack.

Year	# of NDT Inspect	Minor	> 0.4* wt	Leakage
1991-2007	988	2,13 %	0,91 %	0,00 %
2008-2012	955	0,52 %	0,00 %	0,00 %
2012-2017	535	3,55 %	0,00 %	0,00 %

Table 11. The collected inspection results for Case 2. The per cent of inspection to detect crack.

Year	# of NDT Inspect	Minor	> 0.4* wt	Leakage
1991-2007	988	0,00 %	16,80 %	15,20 %
2008-2012	955	2,09 %	2,30 %	1,46 %
2012-2017	535	8,91 %	2,13 %	0,00 %

Table 12 The collected inspection results for Case 3. The per cent of inspection to detect crack.

Year	# of NDT Inspect	Minor	> 0.4* wt	Leakage

1991-2007	988	12,42 %	4,88 %	1,11 %
2008-2012	955	10,85 %	1,38 %	0,39 %
2012-2017	535	1,81 %	5,02 %	2,35 %

Case 1 and 2 have more or less the same number of fatigue cracks detected. However, the number of minor fatigue cracks and cracks reported that is classified with low per cent of belief to be propagating, is significant lower for the Case 1 unit. The Case 1 unit has been given a slightly different site follow-up regime during the inspection and implementation of the improvement maintenance scope. A difference with high impact on the operation cost.

7. Considerations

The fatigue crack growth potential is highly influenced by gross errors in the design, construction and operation phase in addition to the normal uncertainties of design, construction and inspection. Collected data and experience indicate that the theoretical predicted fatigue crack growth potentials are generally conservative without including the influence of gross errors. Hence, the safety of the regulatory requirements is in average adequate. However, by use of the probabilistic approach and Bayes updating non-conservative estimates may easily be obtained if the potential influence of gross errors isn't properly considered. The acceptable fatigue crack growth potential or fatigue failure rate deduced from present codes represents 3-15 per cent of the schedules inspections to detect fatigue cracks and 0.5 percent to detect fatigue failure. As long as the inspection history documents a lower per cent of detection for the scheduled inspections the failure rate is within the acceptable level. The scheduled inspections represent the inspections selected by the theoretical models as presented.
The observations are the key book. Independent of the calculated characteristic fatigue life a detection per cent higher than 3-15 per cent of fatigue crack detection and 0.5 per cent for fatigue failure indicate a failure rate for fatigue larger than the acceptable limit and actions is required.

References

[1] HSE OTO-0059, 1999 Validation of inspection planning methods.
[2] HSE OTO-0060, 1999 Target levels for reliability based assessment of offshore structures during design and operation
[3] HSE OTO-0061, 1999 Review of probabilistic inspection analyses methods
[4] Vårdal O T, Moan T and Bjørheim L G 2000 Application of probabilistic fracture mechanics analysis for reassessment of fatigue life of a floating production unit – Philosophy and target levels, OMAE2000/MAT-2078.
[5] Johannesen J M, Moan T and Vårdal O T 2000 Application of probabilistic fracture mechanics analysis for reassessment of fatigue life of a floating production unit – Theory and validation OMAE2000/MAT-2079
[6] DNVGL-RP-C210 Probabilistic methods for planning of inspection for fatigue cracks in offshore structures, November 2015
[7] DNVGL RP-C005 Fatigue strength analyses of offshore steel structures, June 2014.
[8] Moan T, Vårdal O T, Hellevig N-C, and Skjoldli K 1997 In-service observations of cracks in North Sea jackets. A study on initial crack depth and PoD values." Proc. OMAE-97, Paper1335, Yokohama Japan.
[9] Haagensen P J, Larsen J E and Vårdal O T 2014 Long term effectiveness of life extension methodologies applied to offshore structures, 20th European Conference on Fracture (ECF20), Elsevier Procedia Materials Science, **3** 2187-94.
[10] Haagensen P J, Larsen J E and Vårdal O T 2011 Fatigue damage repair and life extension of a floating production unit – The VFB platform revisited, OMAE Paper 2011 50319
[11] Moan T 1981 The Alexander L. Kielland Accident, Proceedings from The first R.B. Wallace Lecture, MIT 1981

IOP Conf. Series: Materials Science and Engineering **700** (2019) 012031 doi:10.1088/1757-899X/700/1/012031

[12] Moan T, Johannesen J M and Vårdal O T 1999 Probabilistic inspection planning, Paper No. 10848 Offshore Technology Conf., Houston, USA.
[13] Vårdal O T and Moan T 2016 Lessons learned from predicted versus observed fatigue of offshore steel structures in the North Sea, OSRC2016, Stavanger, September 2016
[14] DNV Cl.N. 30.7 Fatigue Assessment of Ship, April 2014

IOP Conf. Series: Materials Science and Engineering **700** (2019) 012032 doi:10.1088/1757-899X/700/1/012032

Synthesis and characterization of fly ash and slag based geopolymer concrete

B Kallesten*, **S Kakay and K Gebremariam**

University of Stavanger, Norway

* Corresponding author: kallesten@kral.no

Abstract. This paper presents the formulation and characterization of composite fly ash and slag based geopolymer concrete. Potassium hydroxide (KOH) and potassium silicate (K_2SiO_3) liquid alkaline activators were used in the preparation of the geopolymer concrete. The mechanical response of the best recipe was investigated through compressive, tensile and flexural strength tests after 3, 7 and 28 days of curing. One batch was tested for compressive strength after 16 and 32 weeks of curing. The microstructure of the concretes was characterized using scanning electron microscopy (SEM). The testing results are compared with a reference ordinary portland cement-based concrete. After 28 days of curing, test results showed that the geopolymer concretes can approach the compressive strength of the reference ordinary concrete. Moreover, the tensile strength and flexural strength attained reached up to 68% and 80% of the reference cement, respectively. The results indicated the potential application of geopolymer that can satisfy the standard strength requirement. After longer time of curing, 32 weeks, some of the geopolymer preparations exhibited 111% compressive strength, compared to the ordinary cement reference.

1. Introduction

Concrete is found nearly everywhere. Among others, concrete is used in many areas of civil engineering works such as for bridges, foundations, dams, commercial and residential buildings, tunnels, pipes, offshore structures etc. Mechanically, concrete can resist high compressive forces, also tensile forces when reinforced. It is resistant to rough weather conditions and moisture, it has a medium to high fire-resistance and can be cast in almost any shape or form. However, it has some drawbacks linked to its relative mechanical properties like resistance to deformation, efflorescence, extremely high energy intensive (clinker formation and grinding) and high CO_2-emitting production processes, etc. Especially to address the environmental issues that are getting more and more considerations, it is critical to find an alternative with solutions to the problems associated with this most important and the top commercial product in the world after water.

A geopolymer concrete is an emerging inorganic polymer produced by alkali activation/alkalination of silicate, aluminate and aluminosilicate materials [1, 2]. These groups of materials are one of the most abundant on earth, besides being components of the enormously produced by-products released in hundreds of millions of tons per year, like fly ash from thermal power stations and slag materials from metal productions. It is a promising endeavor that can pave way for utilization of solid waste and by-products with the ensuing multiple benefits.

The formation of geopolymers involves different stages: release of silicates and aluminates, gelation, polymerization and hardening [3]. They involve the formation of macromolecules after polymerization involving the aluminate, silicate and aluminosilicate groups. Geopolymers have similarities with zeolites with regards to their chemical compositions, but are amorphous in their microstructures [4-6].

Besides its low carbon footprint and minimized energy utilization, geopolymer concrete has a number of advantages. They have attractive mechanical, thermal and physicochemical properties such as low density, micro-porosity, high mechanical strength, good surface hardness, thermal stability, and resistance to chemical exposures, and weathering like freezing and thawing [7-9]. Geopolymerization is considered an easy, cheap way of producing concrete with attractive characteristics that can, thus, be utilized for the handling of diverse ground granulated or powdered inorganic waste materials for a cleaner and sustainable environment [10]. As a consequence, it has grabbed increasing interest in the past couple of decades academically and commercially, leading even to large scale productions and regulatory standardization [11]. Geopolymer concrete is, thus, an interesting direction to explore that can potentially yield cementitious products capable of solving some of the issues raised with Ordinary Portland Cement based concrete (OPC).

This paper is not intended as a contribution towards the general climate change debate, although it is difficult to mention geopolymer concrete without touching the subject of CO_2 and the impacts of replacing OPC with geopolymer binders could have. During the cement production process, a significant amount of CO_2 is released into the atmosphere. If these anthropogenic CO_2-emissions are a driving factor towards a long-term climate change on a global scale, or whether the anthropogenic greenhouse gases are negligible compared to nature's own cyclic activities can be discussed. However, it is estimated that OPC production alone is responsible for approximately 5% of the world's anthropogenic CO_2-emissions [12]. Thus, if reduction of CO_2-emissions is a proclaimed goal the world has decided to work towards, curbing the emissions from the cement industry seems like a good place to start.

In the recent years, geopolymer research gained increased interest by both industry and the academic world with many new discoveries and patents. The concept and usage of geopolymer concrete or synthesized rocks is, however, not recent discoveries. Alkali-activation have been utilized sporadically in the past hundred years, but gained more recognition recently due to the environmental pollution concerns [11]. Even much earlier, in the Roman time, hydraulic concrete materials were produced based on naturally occurring volcanic fly ash and limestone capitalizing on the pozzolanic characteristics [13]. Portland cement, the main binding ingredient in modern concrete (OPC), was invented in 1824 [14]. Geopolymer concrete has already been used for load bearing structures in some large-scale projects, and some of the precursors of alkali-activated concrete (blast furnace slag and fly-ash) are increasingly being commercially utilized as additives to regular Portland cement in concrete. However, as the building industry tends to be a conservative one, geopolymer is not yet widely used as a substitute for regular cement. Geopolymer concrete can be made using many naturally occurring materials. Some materials have natural pozzolanic properties, such as volcanic ash, while other materials such as kaolin clay can be calcined into metakaolin (at much lower temperatures than Portland Cement, thus requiring less energy for calcination process). Metakaolin is a well tried out precursor for geopolymer concrete, but the use of such material as a geopolymer precursor can be a source of conflict with other industries, as kaolin clay is also used to make for instance ceramics, and it is not an abundantly available resource.

Diverse materials were used for the preparation of geopolymer concretes at ambient and elevated temperatures along with sand and other different aggregates. Among the recent attempts in such preparations are the use of copper slags [7], fly ash [15], fly ash and furnace slag [16-17], waste olivine and silica sands [10], construction and demolition wastes [18], volcanic ash [19], geothermal nanosilica waste [20] and natural pozzolan feldspars [21] to mention few. There are also recent attempts to modify the properties of the geopolymer concretes through the use of additives like fibers, carbon nanotube and graphene [22]. Along with the experiment with diverse materials, optimization of the experimental parameters and elucidation of the mechanisms of the geopolymerization processes are current areas of research.

IOP Conf. Series: Materials Science and Engineering **700** (2019) 012032 doi:10.1088/1757-899X/700/1/012032

This paper presents the formulation and testing of alkali-activation of Ground Granulated Blast Furnace Slag (GGBFS) and Fly Ash (FA) as main binders in geopolymer concrete. Both ingredients are industrial waste products, and require no extra calcination, and very little extra processing (GGBFS is milled into fine powder, FA is used as-is). There is an estimated yearly production of 367 million tons of fly ash alone, and only about 55% are utilized in other ways than ending up in landfills [23].

Both FA and GGBFS are being commercially used by the construction industry today, but mostly mixed with regular Portland cement in various percentages. The recipes applied in this paper are however completely without Portland cement (except for the reference batches), thus indicating that the potential of these materials is not only in the use as additives or partial substitutes for Portland cement, but also as completely stand-alone binders in concrete.

Other advantages to geopolymer concrete are that the hydration process seems to continue much longer and at a steeper rate after a few weeks, than that of OPC. When the hydration process of OPC slows down, the geopolymer concrete can still hydrate and gets harder over time. It can also have a much higher fire resistance [24] and a higher sulfur and acid resistance than OPC [25].

In this work, several recipes of geopolymer concrete have been formulated and characterized. However, the paper presents only some of the formations that led to good results. Moreover, the formulated geopolymer will also be compared with the ordinary Portland cement. Destructive mechanical testing and SEM analysis were carried out to evaluate the strength and the internal structure of the samples.

2. Theory and Methods

The mechanical strength of the geopolymer and OPC-reference cubes were investigated through Compressive strength testing, tensile splitting test and flexural strength tests.

2.1 Tensile split test

Concrete in general is weak in tensile strength and its strength determined indirectly by a split test (Brazilian test). The specimen is prepared in 32mm diameter by 66mm length cylindrical cup. Tensile testing was carried out according to standard NS-EN 12390-6:2009 [26]. During testing, the specimen is loaded between two parallel plates and loaded along the axial direction. The load continuously increases until failure. The tensile strength is given as:

$$\sigma_t = \frac{2F}{\pi DL} \tag{1}$$

where F is the maximum load, D is the diameter and L is the length of the specimen. The unit of the tensile strength is in N/mm2 or MPa.

2.2 Compressive tests

The compressive strength of the specimen describes the measure of the maximum load carrying capacity for the applied compressive load using Zwick Z/020 test machine. Unlike the tensile testing, the specimen is loaded between parallel plate and the loading is on the surface of the cube. The size of the cubes was (LxWxH) 50x50x50mm, where the contact surface that the uniaxial loading applied is 2500mm^2. The testing procedures is according to NS-EN 196-1 standard [27]. The compressive strength is given by:

$$\sigma_c = \frac{F}{L*W} = \frac{F}{2500} \tag{2}$$

where F is maximum compressive load in N, LxW is the contact surface area of the cube.

2.3 Flexural strength

The flexural strength of the beam specimen is the measure of the beam's ability to withstand bending loading. The measurement is a three-point testing frame and conducted by inserting in the compression chamber. This three-point test frame directs the forces to a middle point of a specimen of 40x40x200mm, which is supported by to supports distanced 100mm apart. The reference standard is: NS-EN 12390-5:2009– single point load [28]. Flexural strength is calculated by:

$$F_{cf} = \frac{3FL}{2d_1d_2^2} \tag{3}$$

F_{cf} (N) is the flexural strength in N/mm^2 or MPa, L is the distance between the supports (100 mm in this case) d_1 is the cross-sectional width of the beam, and d_2 is the cross-sectional height of the beam.

2.4 Scanning Electron Microscopy

Scanning Electron Microscopy with Backscatter Electron Detector (SEM-BSED) was used to obtain high-resolution micrographs of compositional maps that reveal the internal structure of the cured geopolymer components. Fragments from batches 14 and 18 were coated with palladium and were analysed in a Zeiss Supra 35VP instrument. TEAM™ Software Suite was used for data processing.

3. **Experimental works**

The Eurocode Part 2, NS-EN 1992-1-1:2004 (EC2) [29], is the governing standard for concrete design work. The EC2 defines the standard strength classes and overview of their class parameter values and limitations. Eurocode Part 2 (EC2) defines all use of concrete, as from material, to design ULS/SLS and ULS, to execution.

3.1 Materials

A total of 21 different geopolymer mix designs have been formulated. In this paper, only a few selected recipes are presented. The slurry formulation was according to mix design, which is according to the procedure described in NS-EN 196-1:2016 – Methods of testing cement [27].

Table 1A presents batch 14 recipe formulations, while Table 1B presents some calculated properties of the same mix design. Table 2A and Table 2B present the same formulations and calculated properties for batch 18 while Table 3A and 3B show formulations and calculated properties for the reference batch made with OPC. The geopolymer concrete was prepared by mixing aggregate (quartz sand), binder (GGBFS and FA) and binder/hardener (KOH and K_2SiO_3). The ratio of aggregate and solid binder is constant; however, the ratios of the hardeners, water and other additives vary.

The composition proportion of geopolymer concrete is very different from OPC since the liquid hardener, also known as liquid binder/activators contains water – which needs to be deducted from the total amount of the required water. The solids that are blended in the liquid hardener are considered as parts of the binder and should be added to the binder calculations for an accurate proportioning. After mixing, the batches are cast in cubes of 50x50x50mm, and wrapped in plastic film to prevent evaporation of liquids.

The reference batch (OPC-based) was mixed with cement and aggregates proportionately the quantities of GGBFS+FA and quartz sand respectively. Furthermore, it was mixed to resemble the same workability/consistency as the geopolymer batches – as a mix with a w/b-ratio as low as batch 18 would have produced a non-workable mix. All batches were de-moulded next day and placed in a water bath keeping a consistent temperature of 40°C.

IOP Conf. Series: Materials Science and Engineering **700** (2019) 012032 doi:10.1088/1757-899X/700/1/012032

Table 1A. Batch 14 recipe formulation

Ingredients	Quantity (kg)	Solid (g)	Liquid (g)	Density (kg/m^3)	Recipe (kg/m^3)
FA	0.468			2300	337.31
GGBFS	0.468			2920	337.31
QS	1.8			2650	1297.36
KOH 4M	0.14528	28.28	117	1153	104.71
K2SiO3 3.4M	0.1655	66.1	99.5	1315.26	119.29
H20	0.0926			1000	66.74
Sx-N				1050	0.00
Sugar				849	0.00
Sum, ex addit.	3.14				2262.73

Table 1B. Batch 14 calculated mix properties

Total water content (kg/m^3)	Total solids (kg/m^3)	Total binder (kg/m^3)	Total liquids (kg/m^3)	Water/ binder ratio	Solid/ liquid ratio	Modular ratio (K2SiO3/ KOH)	Matrix volume (l/m^3)
222.74	2039.99	742.62	290.74	0.30	7.02	2.34	510.43

Table 2A. Batch 18 recipe formulation

Ingredients	Quantity (kg)	Solid (g)	Liquid (g)	Density (kg/m^3)	Recipe (kg/m^3)
FA	0.468			2300	343.69
GGBFS	0.468			2920	343.69
QS	1.8			2650	1321.89
KOH 4M	0.14528	28.28	117	1153	106.69
K2SiO3 3.4M	0.23002	91.8	138.37	1315.26	168.92
H20	0.0178			1000	13.07
Sx-N	0.0211			1050	15.50
Sugar	0.002			849	1.47
Sum, ex addit.	3.129				2297.96

Table 2B. Batch 18 calculated mix properties.

Total water content (kg/m^3)	Total solids (kg/m^3)	Total binder (kg/m^3)	Total liquids (kg/m^3)	Water/ binder ratio	Solid/ liquid ratio	Modular ratio [K2SiO3/ KOH]	Matrix volume (l/m^3)
200.55	2097.42	775.52	288.69	0.26	7.27	3.24	501.17

Table 3A. Reference OPC batch recipe formulation.

Ingredients	Quantity (kg)	Solid (g)	Liquid (g)	Density (kg/m^3)	Recipe (kg/m^3)
OPC (IND.)	0.936			3130	716.75
QS	1.8			2650	1378.37
H20	0.3276			1000	250.86
Sx-N	0.01866			1050	14.29
Sum, ex addit.	3.06				2345.99

Table 3B. Reference OPC batch calculated mix properties.

Total water content (kg/m^3)	Total solids (kg/m^3)	Total binder (kg/m^3)	Total liquids (kg/m^3)	Water/ binder ratio	Solid/ liquid ratio	Modular ratio [K2SiO3/ KOH]	Matrix volume (l/m^3)
250.86	2095.13	716.75	250.86	0.35	8.35	N/A	479.86

Aggregates: The aggregate used is a high purity, very well sorted medium quartz sand with grain sizes ranging between 200 μm and 500 μm. The sand used for these experiments is called "H 33" made by the company "Quartzwerke GmbH" [30].

Binders: Fly Ash (FA) is an industrial waste product that is precipitated from the flue gases from industrial furnaces burning any solid fuel. A waste product such as FA is an environmentally friendly alternative to regular Portland cement, which requires calcination of limestone at high temperatures and therefore, a high amount of energy is used. Additionally, the CO2 emissions from this process are formidable. As for FA, the calcination process has already taken place during the furnace burning, and therefore no extra time and energy is needed for neither calcination nor milling/grinding. ASTM C618 defines FA into class C (typically Ca content over 20%) and class F (Calcium content less than 7%) [31]. ASTM C618 Class F can roughly be translated into NS-EN 450-1 Class A. The fly ash used in this thesis is supplied by Norcem [32] and is classified as class A according to NS-EN 450-18 [33].

Ground Granulated Blast Furnace Slag (GGBFS): Ground Granulated Blast Furnace Slag (GGBFS) is an industrial waste product obtained from steel production. As opposed to FA, it needs to be ground into fine powder before being utilized as an alkali activated precursor. The slag used in this paper is a commercial product branded "Merit 5000" - supplied by SSAB Merox AB and is a by-product from the steel plant in Oxelösund. The slag is rapidly cooled in water, and thereafter dried and milled at a mill in Grängesberg [34].

Liquid activators: Two liquid activators were used. The first one is Potassium Hydroxide (KOH), colloquially called caustic potash. KOH is a strong base. The Potassium hydroxide-solutions were prepared, high purity KOH-pellets produced by "Riedel-de Hëin" [35] were dissolved in distilled water according to the desired Molar concentration. The second one is Potassium silicate (K2SiO3), which is an inorganic aqueous solution containing approximately 62% H2O (approx. 3.4 M-solution). The potassium silicate solution used is a commercial product from Univar AB in Malmö, Sweden [36].

Superplasticizer: Dynamon SX-N is a commercial superplasticizer made by the company Mapei [37] designed as a concrete additive to increase its workability. In other words, one can reduce the water/cement (water/binder)-ratio, and therefore increase the final strength of the concrete, while preserving the workability of the mix. It can be added in quantities of 0.2-2% (by weight of binder) without any significant effects on the concretes properties when set.

Retarder: There are several commercial retarders available, but regular table sugar is a known retarder in OPC-based concrete. Even very small amounts in the mix can retard the setting time significantly. Regular granulated table sugar was introduced as an attempted retarder in some of the geopolymer batches, to see if we can get any similar results as it would cause in OPC-based mix designs.

4. Results and Discussion

4.1 Mechanical testing

4.1.1 Compressive strength after 1, 7 and 28 days of curing: With reference to NS-EN 12390 testing of concrete, the geopolymers were tested after one day, seven days, and 28 days. Even though testing after 21 days is not a part of the testing standard, including the earlier mention days, the OPC were tested 21 days as well. Figure 1 shows the comparisons of the compressional strength of the geopolymer recipes to OPC reference batch. All datapoints represents the average value of three cubes.

In terms of strength development, the figure shows that the reference concrete curing rate is higher until seven days of curing and fully developed its strength gently after 28 days. On the other hand, the geopolymer batches strength development is relatively lower than the reference until seven days, but the strength increases with a higher rate and reaching to nearly the strength of the reference concrete after 28 days. This observation leads to the conclusion that it is possible to make FA- and GGBFS-based geopolymer concrete fulfilling most of the strength classes. The higher strength slope of the geopolymers after day seven suggests that at some point, the compressive strength of the geopolymer concrete could be equal, or even above, that of OPC. Based on these results the decision was made to leave more cubes of batch 18 and the OPC reference to cure in similar conditions over a long period of time to test the long-term strength development.

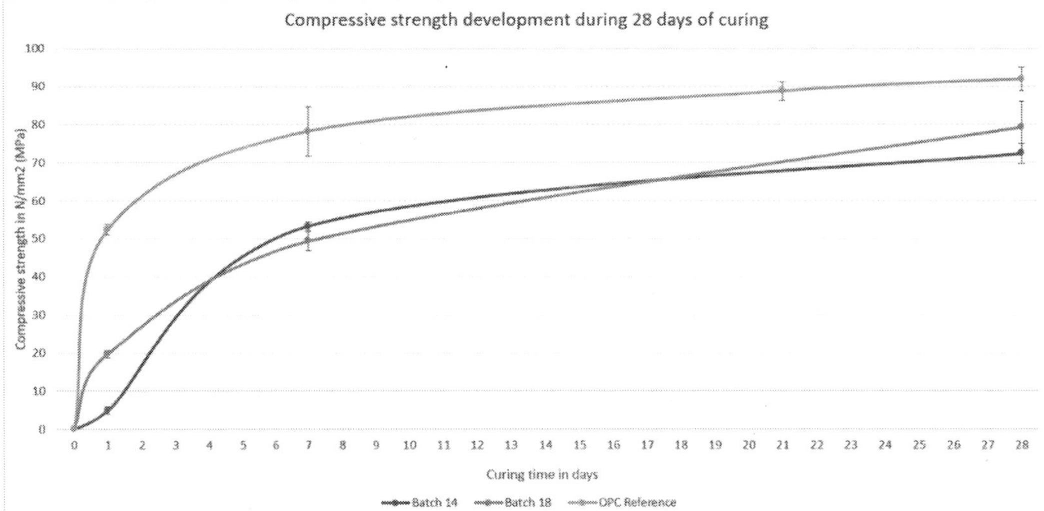

Figure 1. 28 days compressive strength development of batches 14 and 18, compared with the OPC-reference.

4.1.2 Flexural strength. Here, the best geopolymer recipe (Batch 18) was selected for testing of flexural strength. The result is compared with the reference OPC based concrete. A repeat test was carried out and the average value is reported. The test results in Table 4 shows that the geopolymer recipe obtained 80% of flexural strength relative to OPC after 28 days of curing.

Table 4. Comparison of flexural test results for OPC and Batch 18.

#	Concrete systems	Flexural strength, MPa
1	OPC Reference batch	7.8 ± 0.6
2	Geopolymer Batch #18	6.22 ± 0.06
	% Reduction	-20%

4.1.3 Tensile split strength. Table 5 shows the tensile strength of Batch 18 compared with the tensile strength of the reference ordinary Portland cement-Batch-R1. As shown in the table, the geopolymer recipe obtained 62% of tensile strength relative to OPC after 28 days of curing

Table 5. Comparison of tensile test results for OPC and Batch 18.

#	Concrete systems	Tensile strength, MPa
1	OPC Reference batch	4.0 ± 0.1
2	Geopolymer Batch #18	2.5 ± 0.3
	% Reduction	-38%

4.1.4 Compressive strength after long term curing (112 days/16weeks and 224 days/32 weeks). Exceeding the standards testing interval for what is considered final strength, we decided to see what kind of strength development we would get from long term curing. Cubes from both batch 18 and the reference batch was left to cure under similar conditions (40°C, immersed in water). As Figure 1 suggests, the curing rate of batch 18 has a steeper curve than the OPC-reference after seven days. Figure 2 starts with the same dataset as figure 1, but with a much longer timespan. It shows how the geopolymer mix design, batch 18, continues the hydration process long after the first 28 days. The three cube average values after 224 days were 106.6 MPa for batch 18 and 95.8 MPa for the OPC reference batch. The trend seems to be that the geopolymer concrete does indeed exceed the OPC reference but given the error margins further experiments should be conducted.

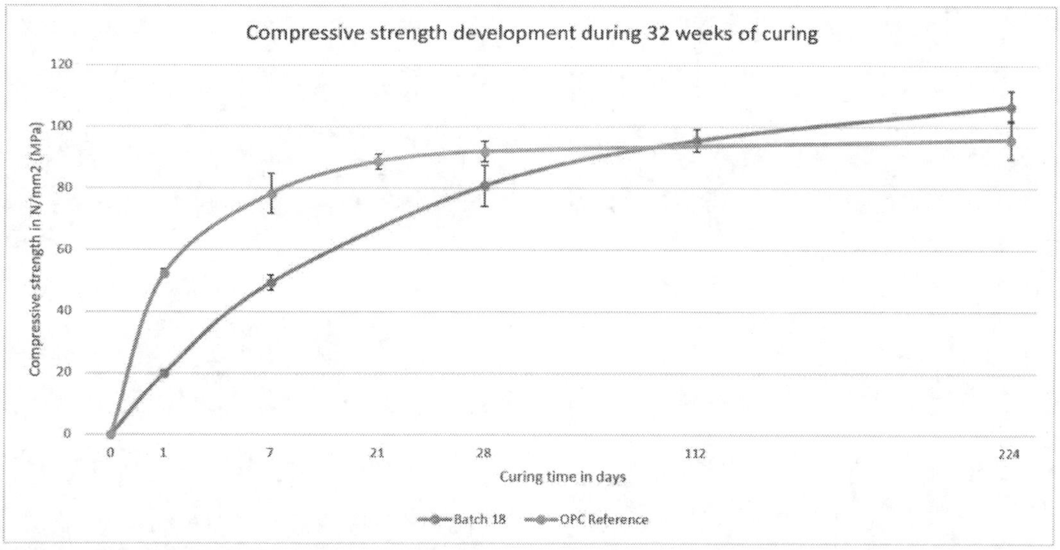

Figure 2. 224 days compressive strength development of batch 18, compared with the OPC-reference.

4.2 SEM analysis

Furthermore, the internal structure of the crushed specimen was analyzed through SEM. For this, it is wise to evaluate the best slurry (Batch 18) with respect to a relatively good one (Batch 14). Figure 2 shows the comparisons of the compressive strength. As shown after 28 days Batch #18's compressive strength is 10MPa higher than the Batch #14.

Figure 3 shows a SEM images on 1.5k x magnification of batches 14 (A) and batch #18 (B). The pictures clearly indicate the correlation between compressive strength and how the particles are bonded on a microscopic level. A higher degree of cohesion at microscopic level has led to a greater compressive strength. As shown in the figure 3A, batch 14 contains some micro cracks and fissures. However, one can observe a uniform and well-developed bonding between the particles. It is also interesting to see that the largest particles, of both FA and GGBFS have remained partially or completely undissolved/unreacted. Figure 4 displays the same trends and the same batches, but at higher magnifications (500x and 10,000x).

Batch 18 (B) shows relatively a better degree of polymerization, filling of the pore spaces and good boding among the diverse components as well. It can also be observed that Batch 18 has some partially unreacted particles (Figure 4C and 4D, at low and high magnifications, respectively), but the surrounding matrix shows a very good and uniform bond. Even the bond to the larger aggregates looks intact and of very good quality.

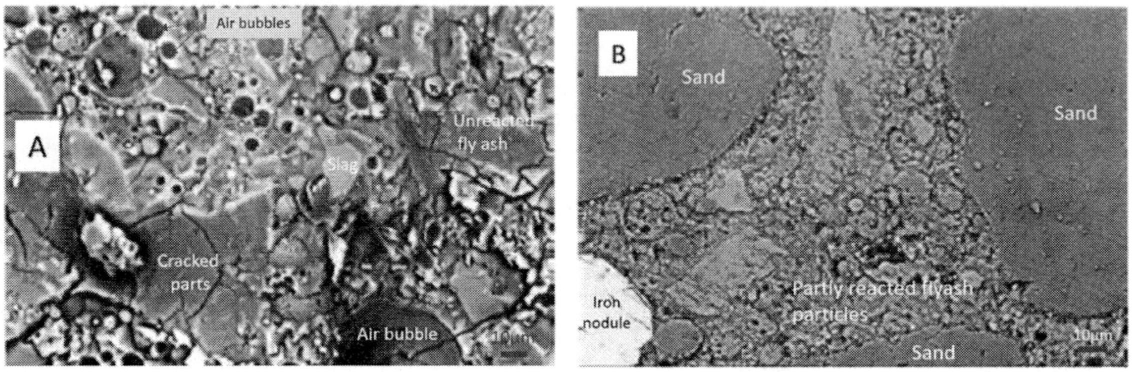

Figure 3. Back scatter electron (BSE) images showing internal structures of batch 14 (A) and Batch 18 (B) at 1.5k x magnification.

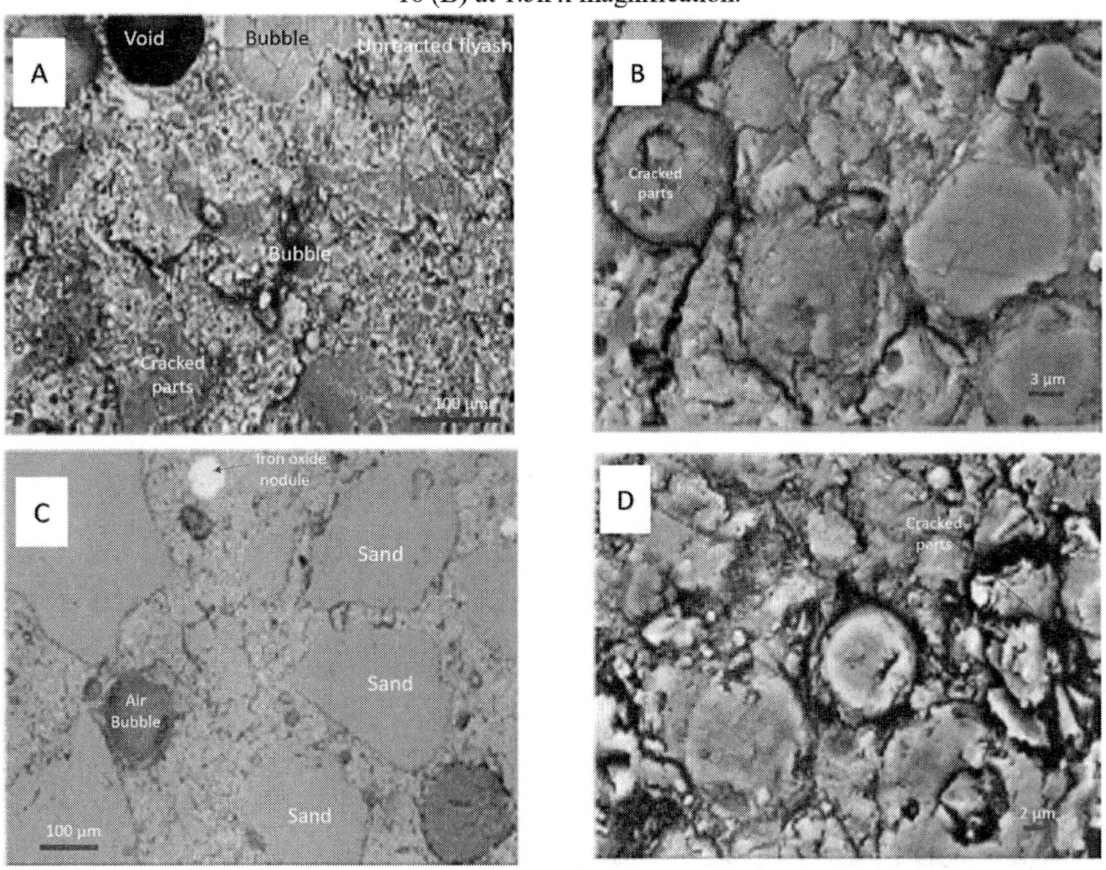

Figure 4. BSE images showing internal structures of batch 14 at 500x (A) and 10^4 x (B) magnification and Batch 18 seen at 500x (C) and 10^4 x (D) magnifications.

5. Conclusion

Several fly ash and slag- based geopolymer slurries were formulated and tested. The mechanical compressive strength is one of the primary properties of concrete. There are many other important properties of a successful mix design, but if a concrete mix does not perform within its designed strength class, all other properties are irrelevant. That does not mean that stronger is always better. In civil construction today, the most used strength classes are B20, B30 and B45, which would place characteristic cube compressive strength after 28 days test at 25, 37 and 55MPa respectively. The compositions and microstructures of the raw materials and geopolymer concretes produced were also characterized using light microscopy, electron microscopy and spectroscopic techniques.

Based on the prepared geopolymers and the reference OPC, the 28th day test results can be summarized as follows:

- The compressive strength of the geopolymer attained the strength of the OPC.
- The tensile and flexural strength of the geopolymer attained 68% and 80% of the OPC.
- Fly ash- GGBFS-based geopolymer has a potential to fulfill most of the strength classes, even the less used high-strength classes.
- The strength development of certain geopolymer mixes looks to be at a steeper rate than OPC-based concrete after day seven and up towards the standard defined final strength measured after 28 days.

Based on the long-term curing results, we can predict that the reactivity in this FA and GGBFS mix design (Batch 18) goes on unabated compared to the OPC reference, and ultimately producing a product that yields higher and higher compressive strengths than the reference. It has reached already 111% strength compared to the OPC reference after 32 weeks. This continuity in the strength gain is one of the interesting characteristics of the geopolymers.

The main drawback to the mix designs witnessed is the short setting time. Sugar was implemented as a retarder, but it resulted in little effect in improving the setting time. Other retarders should be investigated in the future. Through an intensive research, it is plausible to formulate and optimize a better and more economical production of geopolymer concretes that can fulfill standard requirements. With the materials selected for the test, it was possible to acquire a user and environmentally friendly geopolymer formulation. Further research should be conducted towards optimizing matrix volume, aggregate types and curing temperatures. Geopolymer concretes can potentially replace OPC in the future, with advancements in their formulations, characterizations and evaluations of their performances in practical settings outside the laboratory.

References

[1] Davidovits J 1989 Geopolymers and geopolymeric materials, *J. therm. anal.* ***35***(2), 429-441.
[2] Davidovits J 1999 In Chemistry of Geopolymeric Systems, Terminology, *Proc. of the 2nd Int. Conf. on Geopolymer*, Saint Qunentin, Saint Qunentin, p 39.
[3] Duxson P, Fernández-Jiménez A, Provis J L, Lukey G C, Palomo A, van Deventer J S J 2007 Geopolymer technology: the current state of the art, *J. Mater. Sci.* ***42*** (9), 2917-33.
[4] Xu H, Van Deventer J S J 2000 The geopolymerisation of alumino-silicate minerals, *International J. Miner. Process.* ***59*** (3), 247-266.
[5] Davidovits J 1991 Geopolymers, *J. Therm. Anal. Calorim.* *37* (8), 1633-1656.
[6] Mustafa Al Bakri A M, Muhammad Fahee M T, Sandhu A V, Alida A, Salleh M A A M, Ruzaidi C M 2013 Microstructure Studies on Different Types of Geopolymer Materials, *Appl. Mech. Mater.* ***421***, 384-389.
[7] Miltiadis S K, Giannopoulou I, Tahir M F M, Hashim M F A, Panias D 2019 Upgrading Copper Slags to Added Value Fire Resistant Geopolymers, *Waste Biomass Valorization*. 1-10
[8] Cheng T W, Chiu J P 2003 Fire-resistant geopolymer produced by granulated blast furnace slag, *Miner. Eng.***16** (3), 205-210.
[9] van Jaarsveld J G S, van Deventer J S J, Lukey G C 2002 The effect of composition and temperature on the properties of fly ash- and kaolinite-based geopolymers, *Chem. Eng. J.* **89** (1), 63-73.

[10] Furlani E 2019 Waste olivine and silica sands boost geopolymers' performances: experimental investigation, *The Int. J. environ. studies.* **76** (3), 491-506.
[11] Provis J L 2014 Geopolymers and other alkali activated materials: why, how, and what? *Mater. Struct.* **47** (1), 11-25.
[12] Andrew R M. 2018 Global CO_2 emissions from cement production, *Earth Syst. Sci. Data.* **10** (1), 195-217.
[13] Brandon CJ, Bottalico L, Oleson J P 2014 Building for eternity: The history and technology of Roman concrete engineering in the sea, (UK:Oxbow books) p 368.
[14] Cabrera J G, Woolley G R 1996 Life cycle benefits of calcium silicate replacements, *Waste Manage.* **16** (1), 215-220.
[15] Sharma P K, Alex T C, Kumar A, Bhasin M 2019 Physical Characteristics of Fly Ash Based Geopolymers, *Adv. Sci. Eng. Med.* **11** (1-2), 104-109.
[16] Mohajerani A, Suter D, Jeffrey-Bailey T, Song T, Arulrajah A, Horpibulsuk S, Law D 2019 Recycling waste materials in geopolymer concrete, *Clean Technol. Environ. Policy* **21** (3), 493-515.
[17] Pilehvar S, Cao V D, Szczotok A M, Carmona M, Valentini L, Lanzón M, Pamies R, Kjøniksen, A -L 2018 Physical and mechanical properties of fly ash and slag geopolymer concrete containing different types of micro-encapsulated phase change materials, *Construct. Build. Mater.* **173**, 28-39.
[18] Dadsetan S, Siad H, Lachemi M, Sahmaran M 2019 Construction and demolition waste in geopolymer concrete technology: a review, *Magazine Concr. Res.* 1-21.
[19] Luhar S, Rajamane N P, Corbu O, Luhar I 2019 Impact of incorporation of volcanic ash on geopolymerization of eco-friendly geopolymer composites: A review, *IOP Conf. Series: Mater. Sci. Eng.* **572**, 012001.
[20] Gomez-Zamorano L Y, Vega-Cordero E, Struble L 2016 Composite geopolymers of metakaolin and geothermal nanosilica waste, *Constr. Build. Mater.* **115**, 269-276.
[21] González-García D M, Téllez-Jurado L, Jiménez-Álvarez, F J, Balmori-Ramírez H 2017 Structural study of geopolymers obtained from alkali-activated natural pozzolan feldspars, *Ceram. Int.* **43** (2), 2606-2613.
[22] Huang Y 2019 A Review on the Modification of Geopolymer with Inorganic Mineral Materials, *Mater. Sci. Forum.* **953**, 223-228.
[23] Basu M, Pande M, Bhadoria, B P S, Mahapatra S C 2009 Potential fly-ash utilization in agriculture: A global review, *Prog. Nat. Sci.* **19**(10)1173-1186
[24] Konstantinos S, Alexandros S, Pavlos N, Dimitrios P 2015 Behaviour of Passive Fire Protection K-Geopolymer under Successive Severe Fire Incidents, *Mater.* **8** (9), 6096-6104.
[25] Mehta A, Siddique R 2017 Sulfuric acid resistance of fly ash based geopolymer concrete, *Constr. Build. Mater.* **146**, 136-143.
[26] Standard: NS-EN 12390-6:2009
[27] Standard: NS-EN 196-1:2016
[28] Standard: NS-EN 12390-5:2009
[29] Standard: NS-EN 1992-1-1:2004 (Eurocode 2)
[30] Data sheet, Quartzwerke GmbH: Leistungerklärung kennummer 13139-2013-1
[31] Standard: ASTM C618
[32] Data sheet, Norcem/Heidelberg Cement Group: Flygeaske 10/2015
[33] Standard: NS-EN 450-1
[34] Data sheet, SSAB Merox AB: Merit 5000 Product specification. (2010)
[35] Data sheet, Riedel-de Hëin: 000000020767 (2018)
[36] Data sheet, Univar AB: Potassium Silicate 3,4M Solution
[37] Data sheet, Mapei: Dynamon SX-N

Predictive maintenance (PdM) analysis matrix: A tool to determine technical specifications for PdM ready-equipment

I El-Thalji

Department of Mechanical and Structural Engineering and Materials Science, University of Stavanger, Norway

Contact E-mail: `idriss.el-thalji@uis.no`

Abstract. Predictive maintenance (PdM) and operations optimisation are expected to generate the highest industrial and societal impact within the oil and gas industry. Such an optimistic expectation requires several changes in asset design and maintenance management. Nowadays, design for maintenance and maintenance support needs are mainly guided by the IEC 60706-2 standard. However, Designing for PdM ready-equipment is not yet part of that standard. To design PdM ready-equipment a specific analysis method shall be performed to evaluate the technical requirements and specifications of designed equipment to be PdM ready. Therefore, the purpose of this paper is to propose and demonstrate a PdM analysis method that helps to specify the technical specifications to monitor and predict the health of a specific physical asset. The proposed matrix is an evolution of further development of failure Mode, effect and criticality analysis (FMECA) and failure mode symptoms analysis (FMSA) rather than a revolutionary analysis. The case study method is used to extract stakeholders needs of what they expect from PdM analysis (PdMA) and how practical such type of analysis shall be. The developed PdMA matrix shows a simple relation between failure (their modes/levels) and measured abnormal symptoms and tracking and prediction indicators. The electric generator is used to demonstrate the use of the matrix. The PdMA matrix can be developed further to be more quantitative by including the probability of detection and probability of prediction.

1. Introduction

Predictive maintenance (PdM) and operations optimisation are expected to generate the highest industrial and societal impact within the oil and gas industry [1]. To gain the potential benefits of predictive maintenance in industry 4.0 era, two aspects shall be considered at the design phase: (1) design a PdM ready-equipment, and (2) develop a robust maintenance management architecture. Designing for PdM ready-equipment is not only to allocate sensors, but all physical and cyber items shall also be allocated to support the planned detection and prediction purposes. Especially, prediction in industry 4.0 context is about system-level prediction where physical components: (1) interact with each other and with the cyber-infrastructure using the Internet of thing technology, and (2) share data coming from several data acquisition systems with clear variety (process parameters, health parameter, i.e. value/waveform, meta-data) that acquired at different velocities and of different levels of veracity (Quality of data). Thus, the technical requirements and specifications that shall be allocated while designing PdM ready-equipment are complicated and require critical evaluation.

Nowadays, design for maintenance and maintenance support needs are mainly guided by the IEC 60706-2 standard [2]. However, Designing for PdM ready-equipment is not yet part of that standard. To design PdM ready-equipment specific analysis shall be performed to evaluate the technical requirements and specifications of the designed equipment to be PdM ready. The general maintenance requirements and specifications are commonly determined by applying reliability centred maintenance (RCM) analysis. RCM determines the specific maintenance policy to prevent specific failures based on the Failure Mode, Effect and Criticality Analysis (FMECA). NORSOKZ-008 Standard [3] defined work-flow for establishing preventive maintenance (PM) program. The FMECA can also be used to determine the most critical components and failures that shall be designed-out, monitored or frequently inspected. For monitoring purpose, Failure Modes and Symptoms Analysis (FMSA) ISO-13379-1 [4], ISO-17359 [5] is used to determine the most effective symptoms to detect abnormalities in the asset health. Gonzlez Nava[6] considered FMSA as core step within the entire work process of PdM evaluation at the design phase. However, the FMSA currently looks like (as shown in figure 1) a matching matrix between faults and their general symptoms. It is not offering any kind of quantification for the symptoms to select the most critical one. Thus, Lemme and Furseth [7] developed a procedure to quantify the symptom effectiveness within FMSA matrix.

Symptoms / Faults	Current	Torque	Speed	Vibration	Temperature	Coast down time	Axial flux	Oil debris	Acoustic emission
Eccentric rotor	•			•			•		
Bearing damage	•	•		•	•	•		•	•
Unbalance				•					
Misalignment				•					

Figure 1. Part of FMSA of electric motor faults

The idea is that after preforming FMSA the design process continues with allocating a measurement or monitoring system for each symptom. At this stage, several questions may arise based on FMSA: Are we going to measure all symptoms, do we have measurement/monitoring systems that can measure all symptoms, shall we select the most critical symptoms? How effective the measurement/monitoring systems can detect abnormalities of the measured symptoms, how effective the measurement/monitoring systems can handle fluctuations of the measured symptoms, and how early the measurement/monitoring systems can detect the abnormalities. In fact, these questions point out the detection specifications that shall be determined. Moreover, the FMSA is mainly handling the detection part of predictive maintenance and not the prediction part. To determine the specifications of prediction (How current symptoms will be diagnosed and how future symptoms shall be predicted), a further analysis shall be performed. It is worth to highlight that symptoms might changes over time as the fault is developing. Failures according to NORSOKZ-008 and Offshore and Onshore Reliability Data Handbook (OREDA) [8] are classified into three categories in terms of their evolution (earlier to later): incipient, degraded and critical.

The current prediction requirements are of digital type, i.e. software, cyber-ware, and might not lead to any physical requirements that shall be considered during designing PdM ready equipment. However, some hybrid diagnostic and prognosis techniques might require additional measurements techniques (in physical terms sensors). Thus, selecting some critical symptoms might limit the diagnostic and prognosis process. This means we need to purposefully select the symptoms that will be measured and monitored to support the diagnostic and prognosis process rather than selecting and measuring all symptoms. Moreover, edge computing technology (fog computing) might change the cyber-physical configuration of Predictive maintenance systems. It aims to allow processing and storage to occur at or near edge devices rather than at a centralised cyber infrastructure, e.g. cloud. Svorobej et al., [9] highlighted that edge computing is expected to "reduce latency for time-sensitive applications, support IoT performance in low-bandwidth environments and ease overall network congestion". However, this means also that several physical requirements for prediction shall be designed and allocated at the design phase.

Therefore, the purpose of this paper is to propose and demonstrate a PdM analysis method that helps to specify the technical specifications to monitor and predict the health of a specific physical asset. The case study method is used to extract stakeholders needs of what they expect from PdM analysis and how practical such analysis shall be. The electric generator is used to demonstrate the use of the method. Therefore, in the following sections, the needs for PdM analysis is first extracted and described, followed by the description of the proposed PdM analysis matrix and an illustrative example. Finally, the proposed method is discussed, and some conclusions are drawn up.

2. PdM needs and options

The Predictive maintenance might have several scenarios based on the physical asset needs and expectations of industrial managers. Predictive maintenance scenario is mainly combined of two parts: detection techniques and prediction methods. The Detection part aims to detect fault symptoms before the failure occur so that the failure consequences will be minimal (avoid unplanned stoppage, high downtime and level of repair). Detection is very effective to plan preventive maintenance. However, the maintenance work will be scheduled "when an asset needs" and not "when industrial manager prefers". In other words, detection helps to avoid sudden and sever down-times, but it might not provide flexibility to schedule it during low-production intervals or any opportunistic maintenance interval. In many cases, where the production system has several planned production stoppages, redundancy systems (means maintenance stoppage does not influence production), and maintenance crew are on board; detection is sufficient. However, for production systems that are operated in the unmanned manner (Offshore Wind farms, Offshore Unmanned Oil and Gas platforms), maintenance visits are quite costly and required medium/long time intervals for planning these visits, the detection is not sufficient alone and prediction is required.

Ultimately, prediction aims to estimate whether the faulty asset can survive until the next opportunistic maintenance interval. If prediction achieves this goal, then the unintended maintenance visits can be avoided or at least reduced. The detection and prediction scenarios are illustrated in Figure 2. Figure 2 illustrates four items: (1) the health status (black solid lines) of one physical component over the entire lifetime, (2) Planned intervals for maintenance visit (Yellow columns) and time between planned maintenance visits, (3) several levels of health status (Incipient, degraded, critical and full damage), and (4) detection points (green and yellow circles). In fact, critical faults usually have strong and clear symptoms that most of the condition monitoring instruments and tools are able to detect them. However, the effectiveness and earliness of these instruments and tools may significantly vary in detecting symptoms of degraded and incipient faults. Thus, the detection capabilities differ for each fault level, and they are more complex and demanding for degraded and incipient faults. The component shows healthy status until point 1, after that there are three different fault growth scenarios: scenario 1 (points 1-2-3-D), scenario 2 (points 1-4-6-D), and scenario 3 (points 1-5-7-8-D). It is clear that scenario one was detected by several detection levels and luckily, all before the planned maintenance visit. So, in this case, the condition monitoring system that just capable of detecting critical faults was sufficient for this scenario, even though the degraded and incipient symptoms could provide longer planning interval. However, if the planned maintenance interval was done just before the critical fault was detected, at point 6 in scenario 2, then the critical fault detection will not be effective as in scenario 1. At point 6, the industrial manager had to plan and fix the fault as soon as possible. So, scenario 2 shows that critical fault detection is good to detect and prevent damage/unplanned downtime, however, it could not help to avoid unintended maintenance visit (still we have to visit and fix the fault after point 6). Scenario 3 shows the most unfortunate cases, where both critical and degraded fault detection systems have not detected any fault symptoms before the planned maintenance visit. Based on Figure 2, the yellow points between the two planned maintenance visits means that two unintended maintenance visits are required. The potential solution to avoid the unwanted maintenance visits is to predict them at some points (i.e. points 1, 4 or 5) before the planned maintenance interval. The prediction at point 1 or 5 requires long-term prediction compare to point 4.

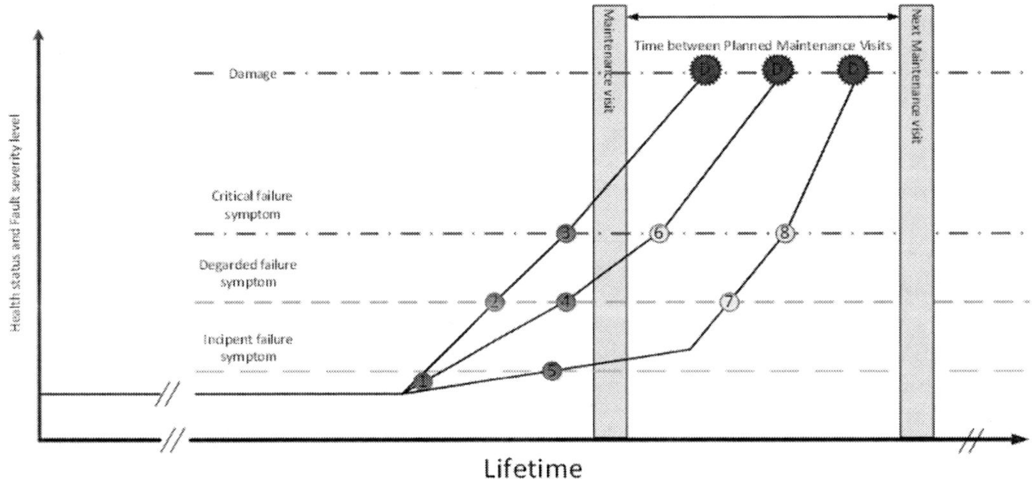

Figure 2. PdM analysis matrix

Prediction can be made in different ways. For example, many managers and engineers who have long experience, they prefer to perform heuristic prediction once the condition monitoring system detects a critical fault. Based on their experience of a specific system, fault type and severity, they might decide to fix right now or within the next few weeks. That might be sufficient for reaching an immediate goal, as in scenario 1 where the planned maintenance visit is quite soon. However, for cases like in scenario 2 and 3, the heuristic uncertainty is quite high. The second prediction way is statistical based, and the ISO 13381-1 (2015) listed two statistical-based prediction methods: extrapolation and projection. Predicting the remaining useful-time using extrapolation means you are going to make a regression fit curve based on the historical measured points (e.g. point 1 and 4). The intersection between the fit curve and what is called a "trip set-point (a little bit before damage/failure value)" defines the end of the remaining useful-time. The drawback for most of the statistical prediction is that you predict the future based on the past. So, in case, the future loading is changing, or the fault growth in the following stage is different than previous stages, such prediction will be misleading.

Projection method to predict the remaining useful-time is also listed in ISO 13381-1, and might be more reliable than extrapolation. It is based on using a ready-made curve of component health evolution (similar to the bathtub curve). The measured fault indicators will be projected on that curve, and the remaining useful-time will be estimated. The health evolution curves of specific physical component/system are usually generated using massive industrial data sets until a robust evolution pattern is extracted. For example, the SKF bearing rating life curve is quite reliable for bearings since it was created based on long-term and huge industrial cases and experimental tests. In fact, SKF extracted a mathematical equation based on these tests that allow us to estimate the bearing rating life with variable operating conditions and loads. However, the question is do other original equipment manufacturers have such models to be used for prediction.

There are other advance data-driven prediction methods, beside physics-based and hybrid methods [10], [11]. Instead of presenting each prediction method and discussing the advantages and drawback, it is better to discuss the main challenges of predicting remaining useful-time for physical components. In fact, these challenges might not be presented in each physical components. So it is more useful to know what are the main challenges of your component and then determine the appropriate prediction method. In Figure 3, four main challenging scenarios are illustrated. It is assumed that a degraded fault is detected at point D, and we are going to predict the health evolution curve. Scenario 1 considers potential overloading and reinforced failure mechanism due to multi-faults (either within the same component or other connected components). The effect of overloading and reinforced failure mechanism can both accelerate the fault evolution and make the remaining useful-time shorter. Scenario 2 represents the case of fault evolution when it influenced by either overloading or reinforced failure mechanism. In this scenario, the fault evolution is accelerated, but not as fast as scenario 1.

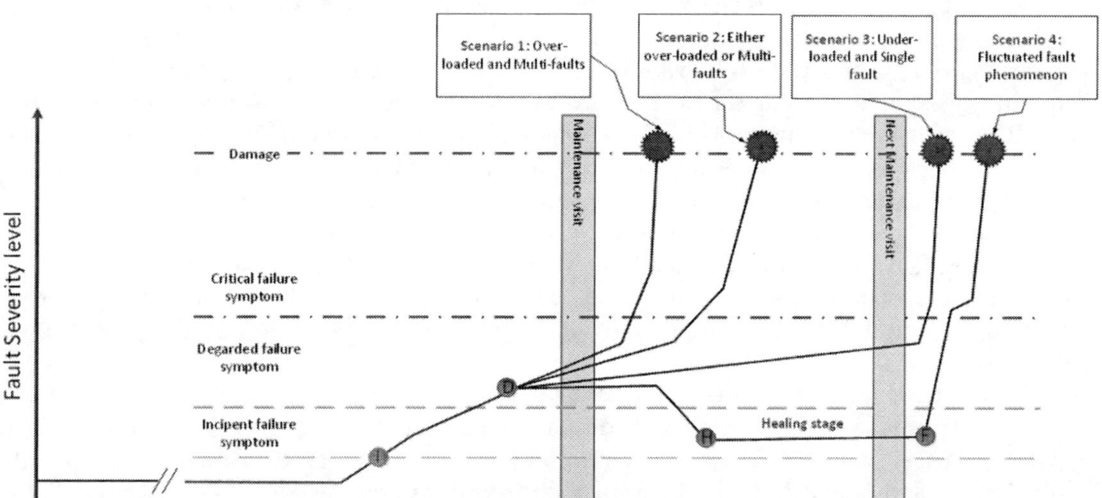

Figure 3. PdM analysis matrix

Scenario 3 represents the mild evolution of the detected fault where the component is going to be either normally loaded or even under-loaded. Moreover, there is only one fault and is not influenced by other faults of other components. Therefore, the fault evolution in this scenario is quite slow. These three scenarios illustrate two challenges of prediction: an estimate of future loading and failure mechanisms interactions among multi-fault situation and at system-level (several faulty components).

IOP Conf. Series: Materials Science and Engineering **700** (2019) 012033 doi:10.1088/1757-899X/700/1/012033

The third challenge of prediction is the fluctuation phenomenon, as illustrated in scenario 4 in figure 3. Scenario 4 starts at point "I" and getting higher until point "D" where it stabilises and goes down for a while (between points "H" and "F") before it goes rapidly at point "F" to the damage level. This phenomenon that is shown between points "H" and "F", is named as "Healing phenomenon" and was explained in a great deal of literature [12]. In fact, there is no physical healing, rather than the symptoms disappear due to an interaction between several failure mechanisms and they cancel the symptoms of each other. However, the fault growth is continuing and therefore it is rapidly accelerated at the end. To understand how symptom might be cancelled or disappeared, let us take the symptoms of localised defect inside a bearing. When the defect is initiated, the rolling element in the bearing will repetitively hit that defect (especially its asperity) and generate impact signals (as a symptom). Since this is a repetitive process, the defect after a while will be smoothened (asperity is getting lower and less sharp) due to abrasive wear. Therefore, the impact of the hits is getting smaller and hence, the symptom as well. However, the failure growth process is progressing where cracks are propagating and branching until the larger amount of material is detached and more prominent defect(s) with sharper asperities are created. Therefore, at that stage, the symptom is very evident and high. Thus, a symptom is an output of a specific physical phenomenon. However, it is hard to separate or isolated that specific phenomenon from its context (physically connected with other components) while we are taken measurements. Therefore, the measured symptom is a collective measurement at a specific location, not an individual and insulated measurement.

The main risk of healing phenomenon is to believe that the component is healing by itself and take no action toward that. Moreover, the development of failure (at the end of the healing stage, point "F") toward the damage state is very fast (almost sudden) and leave very short or even no time to act. In terms of prediction, this scenario is a nightmare. Several prediction models, especially statistical ones, when they try to predict at point "D", they end up with too early remaining useful time (almost like scenario 1 or 2). Getting earlier prediction or shorter remaining useful time than the real values are not acceptable as we might loss useful-time if we follow such prediction (and fix the component before it is required). Being early in fault detection is something to look for, but predicting and fixing the damage earlier than needed is something to be avoided, it shall be precise. Other group of prediction models estimate the precise useful-time at point "F", however, at that stage, such prediction has very little advantage for planning, since the situation needs very fast action. The best case is to predict at "D" or "H" a precise remaining useful-time. It is worth to mention that such a scenario is the most common scenario for several mechanical components as industrial and experimental cases have indicated [12],[13].

The fourth challenge of prediction is related to the diagnostic and tracking performance. As you might notice that all predictions are based on the diagnosed indicators of current health status as points "I" and "D". However, it is significantly vital to understand what is diagnosed and tracked. We don't track the symptom rather than its indicator "Symptom indicator". To illustrate that, let us have one example. Most of health monitoring symptoms are measured as waveform i.e. signals, for example, vibrations, acoustic emission, ultrasonic. Therefore, diagnostic engineers try to extract a clear and relevant indicator to track the symptom evolution, for example, Root mean square (RMS) value in time-domain or the peak-amplitude value at a specific frequency in the frequency domain. For each measurement data set, these value (indicators) will be estimated and tracked over time. So, the values of points "I" and "D" are in fact values of one selected fault indicator such as RMS, Kurtosis, Crest factor, Peak-to-Peak, or peak-amplitude value at the natural frequency, etc. The challenge is that the performance of almost all time-domain and frequency domain indicators varies over time due to the involvement of several physical failure phenomenon and limitation in their mathematical structure. For example, if you use Kurtosis for single fault case might show a clear indication of the fault. However, for a multi-faults case, it might indicate that everything is normal (it lose it's effectiveness) and mislead your detection and further prediction.

Predicting remaining useful time based on a detected incipient fault is further challenging. Incipient failure is defined in ISO 14224 [14] as "imperfection in the state or condition of an item so that a degraded or critical failure might (or might not) eventually be the expected result if corrective actions are not taken". Therefore, it is more complicated and uncertain process to predict the fault initiation and evolution over a long time interval. In fact, The prediction of critical and degraded faults was mainly based on modelling a specific failure mechanism(s). However, prediction of incipient faults deals with failure causes and vast potential combinations of failure mechanisms. Physics-based prediction is quite challenging as the physics of such a situation is unknown or highly uncertain. In this context, the data-driven prediction is commonly used to correlate and recognise a pattern in the fault causes (inputs) and the evolution value (measured outputs), where the fault mechanisms and their physics considered as a black box. It is worth to mention that most of the causes are related to the operating and loading parameters that are measured using process parameters instruments, e.g. SCADA, while the fault evolution is mainly based on health parameters measured by condition monitoring instruments. Therefore, it requires integrated data analytics.

3. PdM Matrix and Illustrative example

In the previous section, the author tried to describe three detection and four prediction challenges that were extracted from industrial practitioners at the Norwegian Continental Shelf and discussed them inline with the academic literature. These extracted challenges are going to formulate the basic needs for development PdM analysis method. In other words, the proposed PdM analysis method shall enable the PdM designer to identify these challenges for their physical system and determine the related requirements.

First, the detection requirements shall be determined to design PdM ready equipment. The detection requirements can be extracted from the previously illustrated scenarios (in figure 2), and summarized as follows: (1) the abnormal symptoms related to critical faults shall be detected before damage occur and spread/affect other systems, (2) The abnormal symptoms related to degraded and incipient faults should be early detected, i.e. detection earliness, (3) The detection technique (symptom measurement) shall be highly effective i.e. clear that show abnormality value, (4) The detection technique shall be highly dependable, that is not influenced by other sources of abnormalities within the monitored system (5) The detection technique shall be highly dependable to track the symptoms evolution (i.e. fluctuation in the appearance of the symptoms) and (6) The detection technique shall be compatible with the environmental conditions, loading patterns and equipment characteristics, e.g. high/low-speed equipment.

Second, the prediction requirements shall be determined in order to design PdM ready equipment. The prediction requirements can be extracted from the previously illustrated scenarios (in figure 3) and are summarised as follows: (1) The tracking indicators for critical, degraded and incipient faults shall be defined, (2) The diagnostic technique shall be effective to extract clear indicators for tracking purposes, (3) The diagnostic technique shall be effective to extract indicators as early as possible, (4) The diagnostic technique shall be reliable to extract effective and early indicators over the entire fault evolution process i.e. fluctuations in fault symptoms or loading patterns, (5) The prognostic (Prediction) indicators for specific faults shall be defined, (6) The prognostic technique shall be effective to predict future fault evolution and remaining useful-time related to the determined prognostic indicator (or a combination), (7) The prognostic technique shall be precise in predicting future fault evolution and remaining useful-time under fluctuated symptoms situations, (8) The prognostic technique shall be be precise in predicting future fault evolution and remaining useful-time under different future loading patterns, (9) The prognostic technique shall be be precise in predicting future fault

IOP Conf. Series: Materials Science and Engineering **700** (2019) 012033 doi:10.1088/1757-899X/700/1/012033

evolution and remaining useful-time under multi-fault situation or fault-interactions or both., and (10) the prediction scope (interval) and sensitivity shall be as high as possible.

The Matrix in Figure 4 is based on the six detection and ten prediction requirements. The sequence of PdM analysis matrix is based on traditional predictive maintenance procedures (described in ISO 13374-1). It links failure modes and mechanisms, with fault symptoms (detection), fault indicators (diagnosis) and prediction indicators (prognosis). The Matrix in Figure 4 helps to roughly map specific physical assets and determine what are required to detect, diagnosis, and predict their faults. Moreover, the PdM analysis as highlighted by industrial practitioners, shall provide a kind of ranking procedure to prioritise the critical faults, critical symptoms, and critical (essential) indicators for diagnoses and prognosis. Such ranking might help in decision making to determine the optimal level of PdM implementation. Therefore, the matrix has three additional columns to facilitate such ranking: detection criticality, tracking (diagnosis) criticality and prediction criticality. At this stage of development, it is not recommended to screen out some component or symptoms from design consideration rather than prioritise the implementation. So the criticality factor might not be a multiplication of several critical factors as it was done for Risk priority index (RPI) in FMECA. Moreover, it is preferred to use a qualitative scale for this matrix instead of quantitative at the moment, since it might be hard for the original equipment manufacturer to provide that. However, this is something good to push toward it. In this regards, a simple scale (High, medium and low) is used to evaluate the criteria in this matrix. The illustrative example is an electric generator, and for simplicity, only some components are selected to illustrate the PdM analysis.

4. Conclusion

The paper shows how the requirements for developing predictive maintenance analysis can be inductively extracted from industrial practitioners and their challenges. These challenges were illustrated and discussed in details to determine the scope, expectations, capabilities and limitations of predictive maintenance (detection and prediction). The PdM analysis matrix built with the aim to help the industrial practitioners to determine or map the requirements to monitor and manage the health of their physical assets. The matrix covers the entire predictive maintenance process, i.e. detection, diagnosis and prediction. However, it is up to the analysts to determine how detailed analysis they would like to perform based on customer needs, operating scenarios, requirements of the physical asset, design opportunities and limitations. In general, the specifications for PdM ready equipment depend on: (1) whether you are going to detect or also predict, (2) how early to detect, (3) how early to predict and (4) time between planned maintenance visits, e.g. short or long, wind farm, and um-manned.

It is worth to highlight that the purpose of this paper is to show how the proposed matrix is developed. Thus, it is hard to claim how effective it is, at this stage of development. It can be mentioned that the matrix is simple and considered as an extension to well-known methods like FMECA and FMSA. However, the scales for criticality determination might need further development. Moreover, the level of implementation for such matrix highly depends on the business model and how detailed the designer would like to go during the design process.

IOP Conf. Series: Materials Science and Engineering **700** (2019) 012033 doi:10.1088/1757-899X/700/1/012033

System ____
Subsystem ____
Component ____
Design Lead ____
Core Team ____

PdM Analysis (PdMA)

Key Date ____

FMEA Number
Prepared By
Date
Revision Date
Page

Scale
H High
M Medium
L Low

System	Maintainable Item	Potential Failure Mode(s)	Probability of Occurrence	Severity of Consequences	Criticality rank	Failure symptoms	Detection Earliness: Detectability of Critical failure (Completion)	Detection Earliness: Detectability of Degraded failure (Propagation)	Detection Earliness: Detectability of Incipient failure (Initiation)	Effectiviness (Clarity)	Dependability (Fluctuation)	Detection Criticality	Tracking Indicators	Tracking Clarity	Tracking Earliness	Tracking reliability	Tracking Criticality	Prediction Indicators	Precision (Future loading	Scope (how far)	Sensitivity (Fluctuation)	Prediction Criticality
Electric Generator	Rotor	Eccentric rotor				Current	H	M	L	M	L											
						Vibration	H	M	H	H	H	Required	RMS	H	M	L	sufficient	Extrapolated RMS	L	L	L	sufficient
		Unbalance				Vibration	H	H	H	H	H	Required	FFT, Peak at 1X	H	H	H	Sufficient	Extrapolated Peak-Value (FFT)	M	M	M	sufficient
		Misalignment				Vibration	H	H	H	H	H	Required	FFT, Peak at 1X, 2X,3X	H	H	H	sufficient	Extrapolated Peak-Value (FFT)	M	M	M	sufficient
	Bearing	Bearing fault				Current	H	M	L													
						Torque	H	M	L													
						Vibration	H	H	M	M	M	Required	FFT, Peak at Natural Frequency	H	H	M	sufficient	Projected Peak-Value (FFT)	M	M	M	sufficient
						Temperature	H	L	L	L	L											
						Oil-debris	H	M	M	M	M											
						Acoustic emission	M	H	H	M	M											
	Other components																					
Other systems																						

Figure 4. PdM analysis matrix

IOP Conf. Series: Materials Science and Engineering **700** (2019) 012033 doi:10.1088/1757-899X/700/1/012033

References

[1] World Economic Forum, 2017
[2] IEC60706-2, 2006
[3] NORSOKZ-008, 2017
[4] ISO13379-1, 2012
[5] ISO17359, 2018
[6] Gonzlez Nava, H A 2019. A work process model for PdM evaluation in study phase: A case study. Master Thesis, University of Stavanger.
[7] Lemme, A and H J Furseth, 2019. Expansion of the Condition Monitoring Strategy for ESD and PSD Valves on Johan Sverdrup Phase 1 A case study in Equinor ASA. Master Thesis, University of Stavanger.
[8] OREDA, 2015. Offshore and Onshore Reliability Data Handbook. Topside Equipment. SINTEF and NTNU Vol. 1(6th Edition).
[9] Svorobej, S, P Takako Endo, M Bendechache, C Filelis-Papadopoulos, K Giannoutakis, G Gravvanis, D Tzovaras, J Byrne and T Lynn 2019. *Future Internet* **11**(3).
[10] Jardine, A K S, D Lin and D Banjevic 2006 *Mechanical Systems and Signal Processing* **20**(7): 1483-1510
[11] El-Thalji, I and E Jantunen 2015 *Mechanical Systems and Signal Processing* **60-61**, 252-272
[12] El-Thalji, I and E Jantunen 2014 *Engineering Failure Analysis* **45** 204-224.
[13] Li, H, Y Li and H Yu 2019 *Sensors* **19**(9).
[14] ISO14224, 2016

IOP Conf. Series: Materials Science and Engineering **700** (2019) 012034 doi:10.1088/1757-899X/700/1/012034

Investigation of updating methods for probability-informed inspection planning for offshore structures

G Ersdal[1,2,*] and N Oma[2]

[1] University of Stavanger, Norway
[2] Petroleum Safety Authority, Stavanger, Norway

[*] Corresponding author: gerhard.ersdal@uis.no

Abstract. The paper presents a novel process for probability-informed inspection planning based on the following principles. 1) A description of the strength of knowledge about a problem shall be included and be accounted for in the decision making, 2) Probabilistic methods should be used to predict quantities expressing the physical reality of nature (observable quantities) and their uncertainty, and 3) Any update of the probabilistic model should be limited to the actually observed quantities. The process for updating the probability of failure after inspection is programmed in accordance with these principles based on Monte-Carlo simulations and Bayesian parameter updating. The application of these principles and the proposed process is illustrated by an example calculation resulting in an example of inspection intervals for a jacket structure.

1. Introduction

Planning the extent and frequency of inspections is a key topic in in-service integrity management for offshore structures. In many cases this is being performed in accordance with prescriptive standards typically requiring first a base-line inspection during the first year of operation and then periodic inspections at fixed time intervals for specified structural details. In addition, an operator is typically required by such standards to do special inspections to monitor the development of known defects and anomalies and to perform un-scheduled inspection after special events such as major storms or collisions. This prescriptive time-based scheme for inspection planning has been in use for many years in the offshore industry. In addition, it has been the fundament for the in-service inspection program (IIP) provided by class societies for mobile offshore units (MOUs) and in the shipping industry.

A prescriptive time-based inspection programme is claimed to require a quite extensive inspection scope. Operators of offshore jacket structures seldom find signs of deterioration and damages corresponding with such an extensive inspection scope in the primary structure. As a result, a condition-based inspection programme (CBI) has been implemented by many operators to reduce the extent of the inspection scope if earlier inspections have found that the condition is good.

The inspection-intervals may be determined by previous experience, but probabilistic methods rapidly proved to be a more rational approach for determining the inspection intervals, especially for fatigue cracks. The general idea for determining the inspection intervals by probabilistic methods is that all uncertain parameters are modelled by a representative probability distribution. These parameters may include:

IOP Conf. Series: Materials Science and Engineering **700** (2019) 012034 doi:10.1088/1757-899X/700/1/012034

- The cyclic stress distribution
- The initial crack size in the welds
- The crack growth parameters in the Paris crack growth law
- Detection accuracy of the inspection method

Model uncertainties are normally included into the probabilistic model to account for the uncertainty in the various models used for calculating the fatigue crack growth, such as determining the long-term cyclic stress distribution and determining the uncertainty in the Paris crack growth model.

The probability distribution of the crack growth of various details can then be predicted by various probabilistic methods such as Monte Carlo simulation, first-order and second-order reliability methods (FORM and SORM). In a Monte Carlo simulation this can be performed by randomly selecting values for each uncertain variable from their probability distribution and calculate the crack growth based on each set of variables. The probabilistic methods can then produce a probability distribution of the expected crack size with time as indicated in Figure 1. The probability of failure can be defined as the probability of the crack exceeding a certain value (e.g. the thickness of the specimen as shown in Figure 1).

The next step is to implement an updating of the distribution of the crack size after an inspection has been performed. If the inspection has not revealed any indications of fatigue cracks, the expected crack size can be assumed to be lower than the crack size that can be determined by this inspection method. However, the crack size that is detectable by an inspection method is also an uncertain parameter and this is normally illustrated by a probability-of-detection (POD) distribution rather than a fixed number. In a Monte Carlo simulation, the updating of the crack-size distribution can be performed by filtering the possible expected crack size after the inspection.

As an example: Assume a case where one simulation is performed resulting in a crack of 3 mm in depth at the time of inspection. Further, let's assume that the POD curve gives a 90 percent probability of detecting a crack of this size, then there will be a 90 percent chance that this will be discovered. If we assume that the actual inspection did not find any crack, this simulation must be rejected. In the simulation this is included by a so-called Bayesian updating. The Bayesian updating in this context is performed by requiring a new crack size after inspection that is taken from the distribution of the cracks that was not detected in the inspection. The updated distribution of cracks can be illustrated as shown in Figure 2.

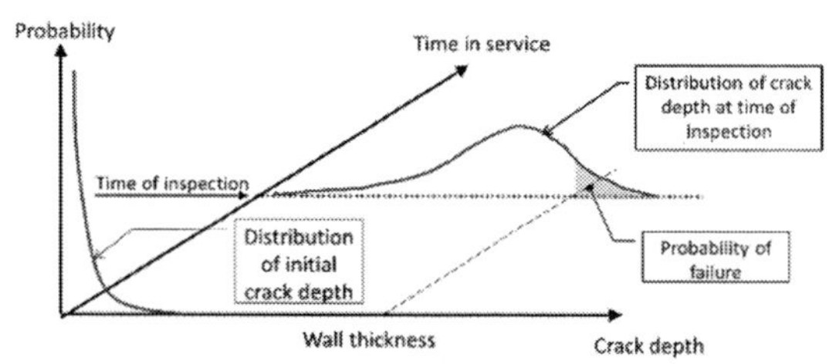

Figure 1. Illustration of development of crack depth distribution over time in service from an assumed initial crack depth distribution to the expected distribution of crack depths at time of inspection.

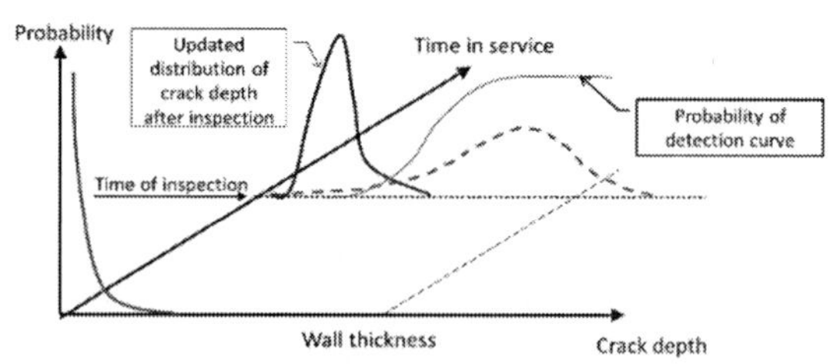

Figure 2. Updated distribution of crack depth after inspection taking into account the new information gained by the inspection result.

An example of the results of an analysis using such updating is shown in Figure 3. In this example inspection has been determined to be needed after approximately 18 years. After the inspection, assuming no cracks found, the probability is updated, and the next inspection is then needed after an additional 12 years. The inspection intervals indicated in this example, assuming no cracks found in any of these inspections, can then be found to be 18, 12, 20 and 25 years respectively (see Figure 3).

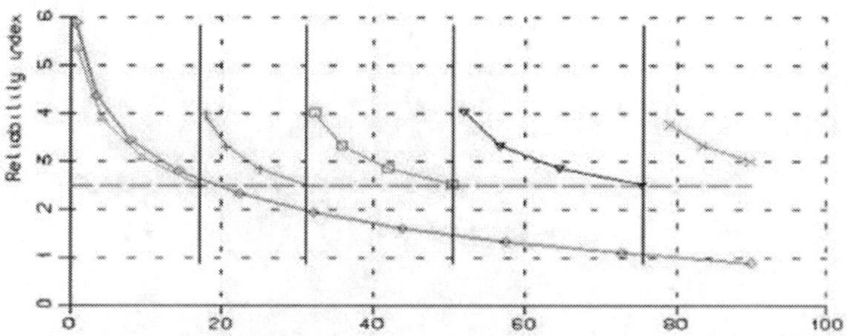

Figure 3. Updated probability of fatigue crack in structure (www.dnvgl.com).

The industry approach to structural design may introduce a bias on the load response. This is important in the design of new structures in order to provide sufficiently safe structures. However, a bias larger than unity may contribute to unrealistic strong effects in a general updating process.

Although the example in Figure 3 is purely an illustration it indicates the trend that is often the result of probability-based inspection planning, namely that the inspection intervals for ageing structures seems to increase with age rather than to decrease as one would intuitively expect. Hence, an alternative method of updating the failure probability is presented in this paper to investigate a different method for establishing inspection intervals. This method updates the probability distribution functions for the parameters where new information is available [1, 2] rather than performing a general updating [3].

In addition, the probability-informed inspection planning presented herein is a suggestion for a method that is believed to be in line with PSA's Risk and Risk management Memorandums [4, 5].

2. Proposed principles and method for probability and risk-informed analysis

Risk and probabilistic methods for decision making is a way to account for the inherent uncertainties in a problem. Such uncertainties can be in the physical parameters like the loads from waves and the material parameters that describes the fatigue crack growth. These uncertainties will also be in the model we use to determine the crack growth based on the Paris' equation. In addition, the presence of gross errors (design errors and fabrication defects) are uncertain. Some of these uncertainties can be included in the probabilistic analysis based on available knowledge. This knowledge is however often taken from other structures and from laboratory experiments not necessarily representing the structure being analysed.

The uncertainty can be illustrated as shown in Figure 4 where the inner green circle is the uncertainties that is included in the analysis based on good and relevant data (wave data for the specific field may fall into this category if measured properly for a significant long period). The next circle (in orange) illustrates other uncertainties that is included in the analysis but based on more uncertain data (the assumed uncertainty in the wave loading and crack growth model may fall into this category). The next circle (in blue) illustrates known uncertainties that are not included in the analysis (e.g. uncertainties related to gross errors). The area outside this circle illustrates the unknown uncertainties which for obvious reasons are not included in the analysis, but it may be relevant for the safety of the structure.

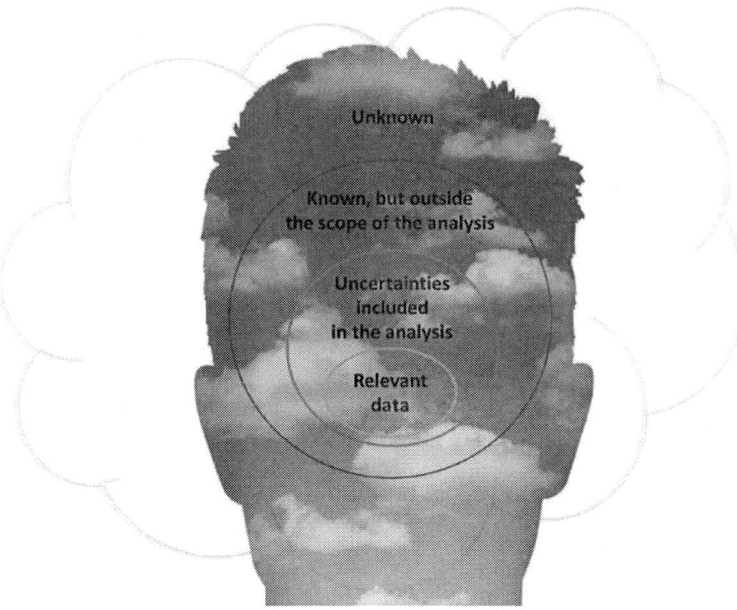

Figure 4. Uncertainty related to a decision-making problem [5].

If one accepts the illustration in Figure 4 and the fact that there are uncertainties that are not included in the probability- and risk-informed analysis it is rational to conclude that the probabilities and consequences that we describe will have limitations. Because of these limitations, the principles presented in this paper is suggested to be used for probability- and risk-informed analysis for inspection planning of structures.

The strength of knowledge behind a calculated probability and the consequences of an event are limited by the modelling of uncertainties mentioned above. Hence, assigned probabilities and consequences are not sufficient to make decisions alone, e.g. by comparing the calculated probability with an acceptance criterion and making the decision solely by this comparison without evaluations of sensitivity and strength of knowledge behind the calculated results. Uncertainties by means of the strength of knowledge behind the physical problem need to be investigated and accounted for. This leads to:

Principle 1: *A description of the strength of knowledge about a problem shall be included and be accounted for in the decision making.*

Taking principle 1 into account it is rational to argue that the probabilistic analysis should focus on determining observable quantities[1] (such as crack sizes and wave heights) rather than probabilistic values ($P_{failure}$). This is because the decision maker is better fit to evaluate such observable quantities, but also that data about future events can be used to validate these observable quantities (while experience indicate that a $P_{failure}$ seldom can be validated). This leads to:

Principle 2: *Probabilistic methods should be used to predict quantities expressing the physical reality of nature (observable quantities) and their uncertainty.*

This principle's focus on observable quantities rather than abstract probabilistic values for a crack growth and inspection planning analysis is an attempt to highlight uncertainty about e.g. the crack size, the inspection intervals and the load model rather than a discussion about the calculated probabilities. Modelling can then be used to get insight into factors influencing the safety of the structure, identify

[1] Observable quantities are values that are unknown at the time of the analysis but may become known at a later stage (Aven 2012). These observable quantities can be predicted by the probabilistic analysis and uncertainties related to these observable quantities can be described.

major contributors to the safety of the structure and see the effect of changes by use of sensitivity analysis.

The 3rd principle is an extension of principle 2 and is requiring that also Bayesian updating based on e.g. performed inspections is focusing on the parameter that is being observed (e.g. the crack size), while traditional updating methods (general updating) in probability analysis may be used to update several parameters (including parameter that is not actually observed). This leads to:

Principle 3: *Any update of the probabilistic model should be limited to the actually observed quantities.*

For structures, example of such possible observable parameters includes e.g. wave heights, measured strains, measured number of load cycles and crack sizes. In this paper the primary observable quantity used is the crack size found during inspections. Also, the suggested inspection intervals can be observed to detect if insufficient or excessive inspections is proposed.

Based on these three principles it is proposed to perform the probability-informed inspection planning analysis as follows:

- Inspection intervals should not be determined directly from the acceptance criteria. The uncertainty in the estimated inspection intervals should be evaluated and experience from integrity management of real structures should be taken into account in the evaluation.
- The strength of knowledge about the problem should be described and accounted for in determining the inspection intervals.
- The analysis should be performed to illustrate the uncertainty in the observable quantities (crack size and necessary inspection intervals).
- The updating procedure should exclusively update the observed quantities (e.g. crack size).

3. Proposed method for inspection planning

A process for probability-informed inspection planning following the principles presented in section 2 for inspection planning of an offshore jacket structure is illustrated in ***Figure 5***.

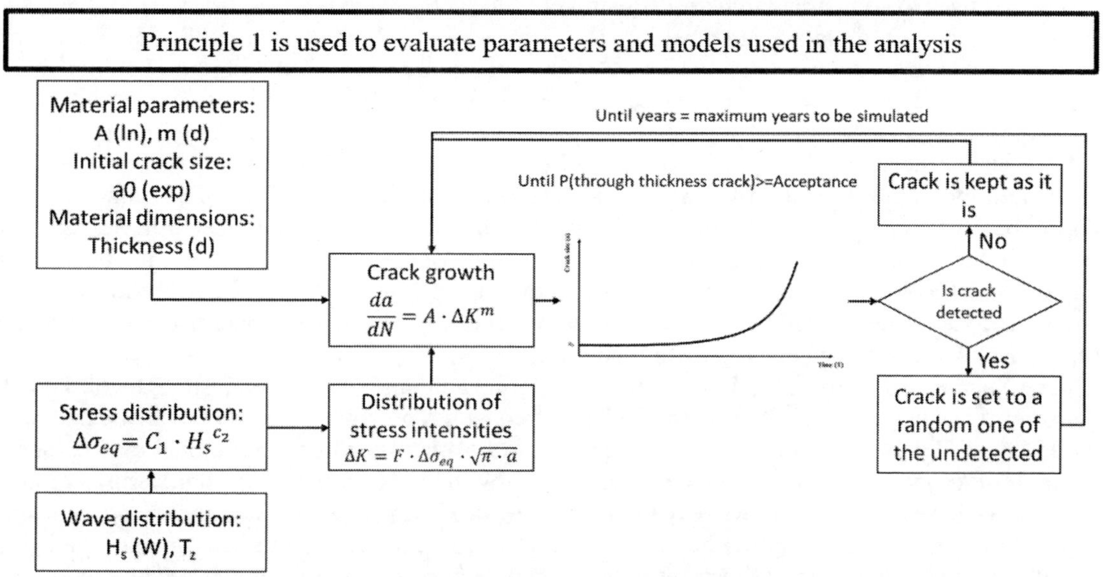

Figure 5. A proposed process for probability-based inspection planning based on the three principles proposed. Principle 2 and 3 are accounted for by the focus on the observable parameter crack size.

In this process the crack growth is simulated until inspection. The inspection is simulated by calculating the probability of detecting (P_D) the simulated crack size based on the PoD-curve. Then, by drawing a random number $\hat{r}$ between 0 and 1 we can simulate the event. If $\hat{r} > P_D$ the simulated

IOP Conf. Series: Materials Science and Engineering **700** (2019) 012034 doi:10.1088/1757-899X/700/1/012034

inspection is classified as a NO-FIND event and, if $\hat{r} \leq P_D$ the simulated inspection is classified as a FIND event.

If the simulations indicate that a crack is detected the crack size is updated by setting the crack size to the largest of 1) the crack size of a random one of the undetected and 2) a random value of the initial distribution of cracks. The undetected cracks are retained as they are. Both detected and undetected cracks are then simulated further until the maximum simulation time is achieved.

The proposed method for updating will differ from the general updating approach [3]. In a general updating also the stress distribution, stress concentration factors etc. will be updated as part of the process. This will in the traditional approach in the case of no-finding lead to increased inspection intervals and in the case of findings lead to decreased inspection intervals. In the proposed method in this paper this increase and decrease of inspection intervals will not occur as the stress distribution, stress concentration factors and all other not-observed parameters are left as is in the updating.

4. Proposed method for evaluating strength of knowledge

Several models for the description of strength of knowledge exists. In general, the following reasoning may be implemented [6]:

- If risk is found negligible according to probability, and the strength of knowledge is strong, the risk is judged as negligible.
- If risk is found in negligible region according to probability with large margins, the risk is judged as negligible unless the strength of knowledge is weak.
- If risk is found negligible according to probability with moderate or small margins, and the strength of knowledge is not strong, the risk is judged as not being in the negligible region.
- If risk is *not* found to be negligible according to probability, the risk is judged as not being in the negligible region.

In the context of this paper this may be understood as:

- If strength of knowledge behind the parameters and models are strong the results of the probabilistic analysis can be regarded as trustworthy and the calculated failure probability can be used to compare with reasonable target failure probabilities.
- If strength of knowledge behind the parameters and models are fair the results of the probabilistic analysis should be used with care and the calculated failure probability should be compared with reasonable target failure probabilities with large margins.
- If strength of knowledge behind the parameters and models are poor probabilistic analysis should be avoided.

In addition to the strength of knowledge, also the influence of the parameter and model on result of the probabilistic analysis and the final decision is of importance, as also indicated in Selvik et al [7]. Hence, in this paper it is proposed to include an evaluation of the influence of the parameter and models and classify these as Minor, Medium or Large. Further, a simplified approach is used to describe the strength of knowledge regarding parameters (e.g. initial crack size, material parameters etc.) and models (e.g. wave to stress model, crack growth model etc.).

The strength of knowledge is classified as Good, Fair and Poor. Good strength of knowledge indicates that a large database is available and that frequentist-probability can be established. Parameters based on good knowledge may be used directly into the analysis without additional evaluations. Poor strength of knowledge indicates mostly guestimates by the analyst or by the engineering community. Parameters based on poor knowledge should not be used in probabilistic analysis and decisions on inspection intervals should rather be based on trustworthy experience and deterministic models. Fair knowledge is used to illustrate situations in between poor and good. Parameters based on fair knowledge may require additional parameter sensitivity evaluations.

To combine the strength of knowledge and the influence of the parameter and models, action categories as indicated in Table 1 may be used.

In category 1 (green) the strength of knowledge is good in combination with minor or medium influence or, the strength of knowledge is fair in combination with minor influence. Parameters or models in this category could in most cases be used as they are, but the evaluation of the strength of knowledge and the assumptions herein should be clearly stated.

For categories 2-3 a further evaluation is needed. If possible, more data should be collected to improve the strength of knowledge. Limitations in the strength of knowledge should be clearly stated. The results of the probabilistic analysis should be used with care and the calculated failure probability should be compared with reasonable target failure probabilities with reasonable margins.

For category 4, probabilistic analysis should not be used unless significantly more data is available for probability- and risk-informed evaluations.

Table 1. Categories based on strength of knowledge and influence on result of evaluations.

	Influence on result of decision making		
Strength of knowledge	Minor	Medium	Large
Good	1	1	2
Fair	1	2	3
Poor	3	4	4

Model uncertainty are often used in probability-informed methods to account for formulation inexactness (e.g. in the Paris crack growth formula), measurement errors and insufficient data. Model uncertainty is typically included by a parameter with a mean value of 1.0 (or a bias value if relevant) and a coefficient of variation (CoV). Sensitivity evaluations of parameters are often similarly used to evaluate the effect of changes in a parameter on the result of the probabilistic analysis. The use of such model uncertainty and sensitivity analysis is a common method to include uncertainties.

After inspections it may be argued that the strength of knowledge has improved as more information about the actual crack size has been gathered. This increased knowledge is however included in further analysis by the updating process.

5. Illustration of the proposed model by an example

An example of probability-informed inspection planning for a node on an offshore jacket structure is used to illustrate the proposed method. The example is based on the master thesis by Neeraas [8]. A short overview is given in section 5.1.

5.1. Example of probabilistic crack growth analysis

Monte Carlo simulations of crack growth and results of inspections are programmed in MathLab and used to evaluate the need for inspections. The input parameters to the analysis is given in Table 2.

Based on the wave height and period the width of the stress intensity, ΔK, can be defined by the equivalent stress range $\Delta\sigma_{eq}$, e.g. as described in Ersdal [9] for a tubular node in a jacket structure.

$$\Delta\sigma_{eq} = \frac{C_1}{1.702 - 0.138 \cdot m_a} \cdot (H_s)^{C_2 - 0.03} \tag{1}$$

where C_1 and C_2 are coefficients, H_s is the significant wave height and m_a is the slope of the crack-growth curve in the Paris equation.

Crack-growth is modelled by the Paris equation:

$$\frac{da}{dN} = A \cdot \Delta K^{m_a} \tag{2}$$

where A and m_a is material parameters and:

$$\Delta K = \Delta\sigma \cdot F(a,t) \cdot \sqrt{\pi \cdot a} \tag{3}$$

where ΔK is the stress intensity range, $\Delta\sigma$ is the stress range of each cycle, $F(a,t)$ is the geometric stress function for the detail and a is the present crack depth. As an example, for a tubular member node the geometric stress function may be set to [10]:

$$F(a,t) = \left(1.08 - 0.7 \cdot \frac{a}{t}\right) \cdot \left(1.0 + 1.24 \cdot e^{-22.1\frac{a}{t}} + 3.17 \cdot e^{-357\frac{a}{t}}\right) \tag{4}$$

where a is the present crack depth and t is the thickness of the material. For each sea-state the crack growth can be calculated as [10]:

$$da_i = A \cdot \left[\Delta\sigma_{eq} \cdot F(a_{i-1}, t) \cdot \sqrt{\pi \cdot a_{i-1}}\right]^m \cdot N_w \tag{5}$$

where N_w is the number of cycles in the sea-state.

Table 2. Input parameters for the Monte Carlo simulation.

Input parameters	Comment	Distribution
Initial crack size	Exponential distribution with an average initial crack size of $\mu_{ao} =$ 0.11mm (DNV 1992)	$f(x) = \lambda \cdot e^{-\lambda \cdot x}$ $F(x) = 1 - e^{-\lambda \cdot x}$ where $\lambda = \frac{1}{\mu_{ao}}$
Material parameters in Paris equation	Normally determined by calibration to SN-curve	m is fixed to of 3.0, i.e. one-slope curve is used. Mean ln(A) = 29.1 with a standard deviation of 0.64.
Long-term distribution of significant wave heights	Given by the Weibull distribution where H_C, H_0 and γ are statistical values in the Weibull distribution and should be taken from proper metocean specifications.	$F_{H_s}(h) = 1 - exp\left[-\left(\frac{h-H_0}{H_C-H_0}\right)^\gamma\right]$
Mean zero up-crossing period of the sea-state	Given as a function of the significant wave height.	$T_Z = 3.3 \cdot \sqrt{H_s}$ with a minimum value of $T_Z = 5.72\ s$

As sea-states are simulated the crack growth is calculated and the crack increases. At the end of each year the number of cracks larger than the thickness (i.e. through-thickness cracks) is counted and if the number of through thickness cracks divided by the total number of simulations exceeds the acceptance requirements (0.01 and 0.001) an inspection is performed.

An inspection will always have a certain probability of detecting a crack, and the probability of detection (PoD) is normally expressed by curves e.g. given by DNVGL-RP-C210 [11].

$$PoD(a) = 1 - \frac{1}{1+\left(\frac{a}{x_0}\right)^b} \tag{6}$$

where a is the depth of the crack and X_0 and b is parameters depending on the individual inspection methods. For close visual inspection the PoD curve is given by the crack length. The ratio between crack depth and crack length $\frac{a}{c}$ may be assumed to be 0.15 [10].

Figure 6 shows the crack growth of 20 random simulations of a detail in an analysis. The updates after inspection can normally be seen as a sudden drop in the crack size. However, some examples of the opposite can be seen, as the updated crack size (selected as a random value of the undetected cracks) may be larger than the detected crack.

Figure 7 shows an example of the distribution of crack sizes before and after updating due to inspection at the second inspection [8]. The distribution of cracks prior to updating is shown in red, and the distribution of cracks after updating is shown in blue (note that blue over red becomes purple in these figures). It can easily be seen that the number of large cracks is reduced, and it is also possible to see that the number of small cracks is increased after the updating.

IOP Conf. Series: Materials Science and Engineering **700** (2019) 012034 doi:10.1088/1757-899X/700/1/012034

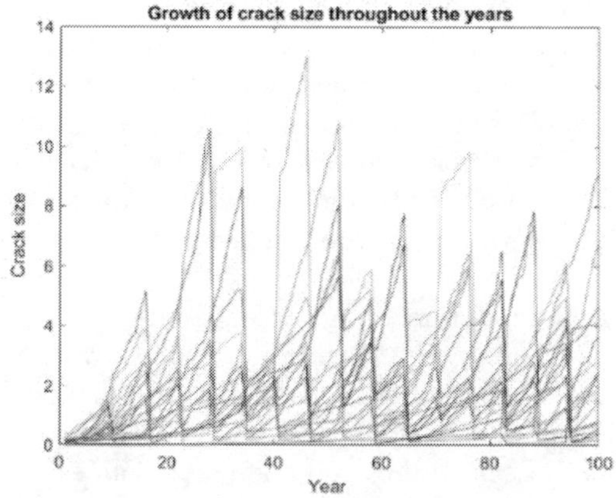

Figure 6. Crack growth using the proposed method for updating after inspections [8].

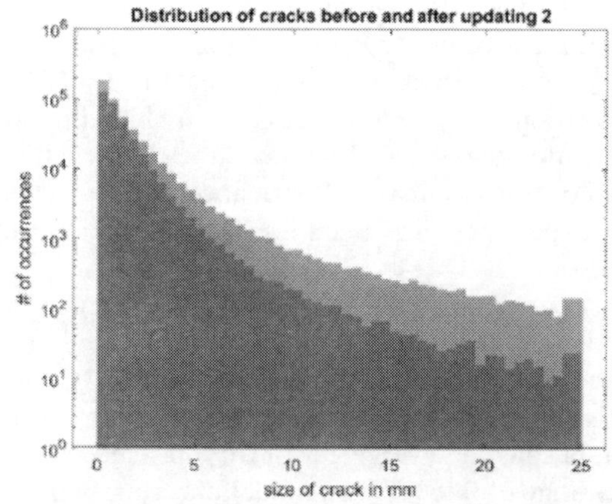

Figure 7. Examples of distribution of crack sizes before and after updating due to inspection for second inspection [8].

Essentially, the distributions of crack sizes before and after inspection turns out to be very similar at all inspections by using the proposed method for updating, as shown in Figure 7. This is the underlying reason for the time history of probability of failure (through thickness crack) as shown in Figure 8. As indicated in Figure 8 this method seems to always provide constant inspection intervals after the first inspection. Hence, this process will produce one "mean" inspection interval in addition to the time to the first inspection. The time to the first inspection is relevant from a methodical point of view but in practise, inspections prior to the calculated first inspection will be needed (e.g. base line inspection) to determine gross errors, fabrication defects and other outliers.

As an additional benefit the likelihood of a crack being detected during the inspection (in Figure 9 called repair required) can be calculated using this simulation method. The trend indicated in Figure 9 is quite similar in all cases simulated. For CVI the curve typically flattens out at a plateau of 0.4 and for MPI (also representative for Eddy Current inspections) at a plateau of 0.6-0.67 as shown in Figure 9.

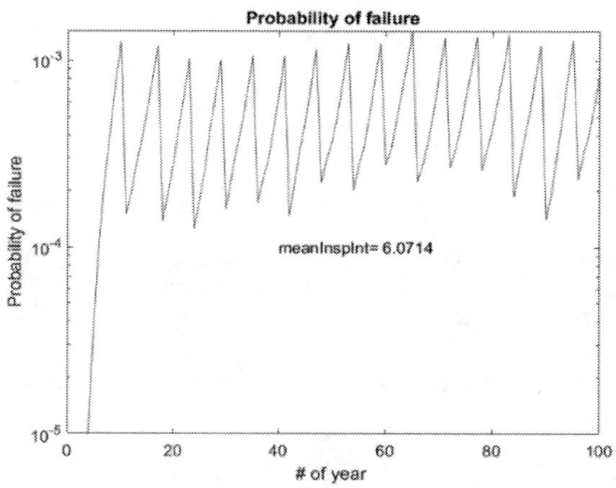

Figure 8. Probability of failure as a function of time with updates.

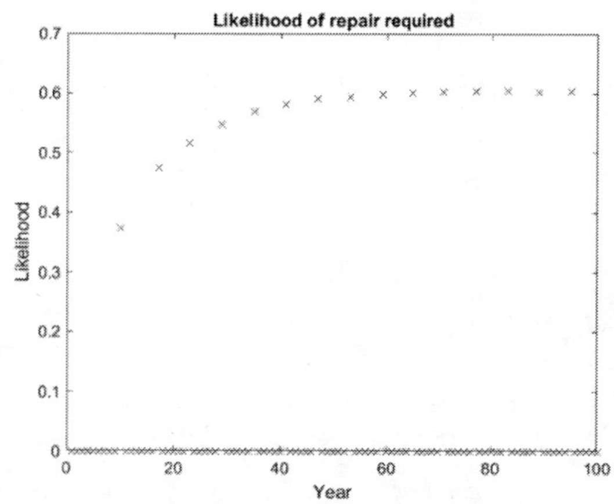

Figure 9. "Likelihood of repair required" using MPI is illustrating the probability that an indication of a crack is detected during the inspection.

The calculated likelihood of finding an indication of a crack (LoIC) as shown in Figure 9 can be used to verify the input parameters to the analysis and possibly the condition of the structure. For example, if the LoIC is calculated to be 0.5 for 20 "identical" details (identical with respect to geometry, stress level etc.) are inspected, we would expect to find 10 cracks. If more that 10 cracks are found it may be an indication of a problem that is not included in the analysis. If less that 10 cracks are found it may be an indication that the analysis is on the conservative side, the inspection quality is lower than predicted by the PoD curves, that the fabrication quality is better than assumed or that the weather that the structure has experienced has been less than predicted from the long-term distribution of sea-states used in the analysis.

A parameter sensitivity study is performed including the model uncertainty (load and crack growth model and the initial crack size). The model uncertainty is included by stochastic variables with mean value of 1.0 and a coefficient of variation (CoV). The model uncertainty in the load model is modelled by a stochastic variable that is kept for the full life of the simulation (GCoV) and a stochastic variable that is changed every year (YCoV) as shown in Equation 7. Each of these are modelled with a CoV in the range of 0.0-0.3. The model uncertainty in the initial crack size is modelled by a stochastic variable with a CoV in the range of 0.0-0.2.

$$\frac{da}{dN} = A \cdot \left[\alpha_G \cdot \alpha_Y \cdot \Delta\sigma_{eq} \cdot F \cdot \sqrt{\pi \cdot a}\right]^{m_a} \tag{7}$$

In addition, the effect of two probability levels for when inspections should be performed is studied ($P_{acceptance}$ equal to 0.01 or 0.001). The $P_{acceptance}$ values are based on normally used acceptance criteria for important and medium important structural nodes and members.

In Table 3 the base case is given with a yearly load model uncertainty (CoV) of 0.15, a general load model uncertainty (CoV) of 0.15 and, a model uncertainty in the initial crack size (CoV) of 0.1.

Table 3: Overview of results from Neeraas (2019) for inspection intervals for a jacket structure

FL	PACC	YCOV	GCOV	A0COV	CVI INT	MPI INT	A0COV	CVI INT	MPI INT
23	0.01	0	0	0	7.4	11.17	0.1	8	11.33
23	0.01	0.15	0	0	7.4	10.57	0.1	7.6	10.57
23	0.01	0	0.15	0	6.67	10.71	0.1	6.67	10.86
23	0.01	0.15	0.15	0	6.38	9.6	0.1	6.23	9.6
23	0.01	0	0.3	0	5.18	7.9	0.1	5.8	7.9
23	0.001	0	0	0	4.7	7.3			
23	0.001	0.15	0.15				0.1	3.6	6
41	0.01	0	0	0	14.5	22	0.1	16	22
41	0.01	0.15	0	0	14	19.7	0.1	13.5	20.7
41	0.01	0	0.15	0	13.2	17.75	0.1	12.8	18.3
41	0.01	0.15	0.15	0	11.33	18.3	0.1	12	17.8
41	0.01	0	0.3	0	9.4	14.6	0.1	9.3	14
41	0.001	0	0	0	8.75	13.4			
41	0.001	0.15	0.15				0.1	6.75	10.25
60	0.01	0	0	0	22	35	0.1	24	34
60	0.01	0,15	0	0	22.5	32	0.1	22	33
60	0.01	0	0.15	0	20	31	0.1	20.5	32
60	0.01	0.15	0.15	0	18.7	27.5	0.1	18	27.5
60	0.01	0	0.3	0	14	21	0.1	15	21.3
60	0.001	0	0	0	13.25	23			

5.2. Evaluations of the results of the probability analysis

The results presented in Table 3 gives the probability-based inspection intervals. In practise these are seldom used directly and, as indicated in the principles in section 2 the decision on the time between inspections should also include a discussion of the sensitivity to parameters and models used and the influence of the probability of acceptance. Such a discussion should include the CoV of the crack growth model (YCoV and GCoV), the calculated fatigue life, the CoV of the initial crack size and the influence of the acceptable probability of "failure" ($P_{acceptable}$). Further, the strength of knowledge should be evaluated and taken into account in the decision making as proposed in Table 6.

The load and crack growth model are in most probability-informed inspection planning analysis the primary uncertainty. Other important uncertainties might be:

- The presence of fabrication and design imperfections are not included in the analysis. The presence of gross imperfection is believed to have less impact on later inspection intervals as they most likely will materialize at an earlier stage. However, fabrication standards at different yards may have influence on later inspection intervals.
- The cyclic stress distribution is a part of the uncertainty in the fatigue analysis modelled by the mentioned CoV's.
- The initial crack sizes in the welds is studied by Neeraas [8] and is shown to have minor impact on late inspection intervals.
- The crack growth parameters in the Paris crack growth law is based on a calibration towards SN-fatigue curves. These SN-fatigue curves do also have significant uncertainty. However, it is believed that the SN-curves used are most likely to be conservative and it is believed that inspections results can be used to validate the crack growth and possibly the crack growth model parameters.
- Detection accuracy of the inspection method is believed to be modelled reasonable with the probability of detection curves in DNVGL RP-C210. However, there exist experience indicating that these curves may be non-conservative.

It is believed that the CoV's described in DNVGL RP-C210 [11] is reasonable for this purpose. However, if the strength of knowledge is stronger than assumed in DNVGL RP-C210 it can be argued to decrease the respective CoV's. The square root of sum of squares combination of GCoV and YCoV can be seen in the context of uncertainty as described by DNVGL [11] where the model uncertainties as shown in Table 4 are suggested depending on the quality of the analysis determining the long-term distribution of stress ranges.

Table 4. Suggested model uncertainties [11].

Uncertainty of fatigue and hot spot stress calculation	COV	Combinations of GCoV and YCoV
High (simplified)	0.3	YCoV = 0.3, GCoV = 0.0 YCoV = 0.0, GCoV = 0.3
Medium (parametric)	0.2	~YCoV = 0.15, GCoV = 0.15
Low (finite element calculation)	0.15	YCoV = 0.15, GCoV = 0.0 YCoV = 0.0, GCoV = 0.15

In Figure 10 and Figure 11 the calculated inspection intervals for CVI and MPI respectively is presented by increasing CoV in accordance with DNVGL. In addition to the proposed CoV's from DNVGL the base case with CoV = 0.0 is included primarily for comparison. As expected, the time between inspections decreases with increasing CoV and more frequent inspections are needed for lower values of P_{acc}.

There is a slight tendency for longer inspection intervals as ratio of the calculated fatigue life (CFL) when the CFL is increasing. The minimum values indicated in Figure 10 and Figure 11 are as a result of this representing the lowest calculated fatigue life. For longer calculated fatigue lives a slightly higher fraction of the calculated fatigue life is obtained (0.42, 0.44 and 0.46 times CFL for MPI and 0.28, 0.29 and 0.31 times CFL for CVI representing 20, 40 and 60 years CFL respectively).

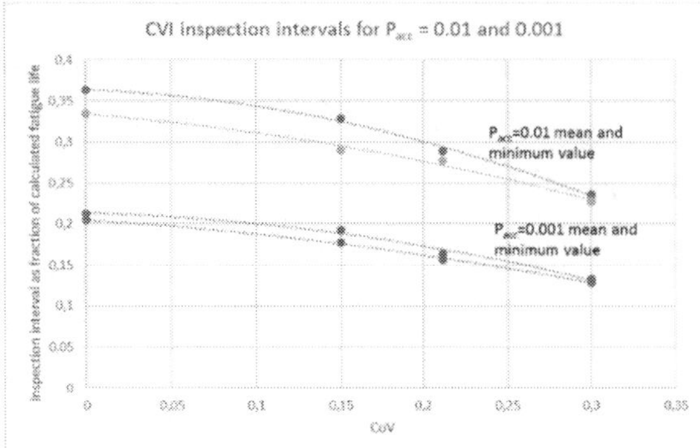

Figure 10. CVI inspection intervals for Pacc equal to 0.01 and 0.001.

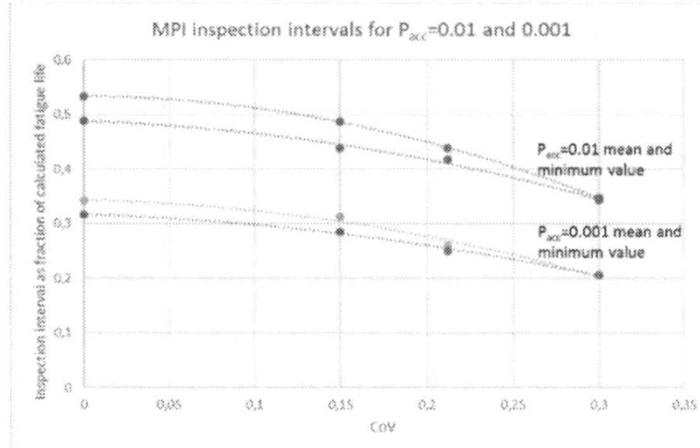

Figure 11. MPI inspection intervals for Pacc equal to 0.01 and 0.001.

The different $P_{acceptable}$ is normally linked to the importance of the structural member or node. In general, the importance is a measure of the consequence of the failure of the member or node. Traditionally a $P_{acceptable}$ equal to 0.001 is used for high consequence of failure members and nodes and $P_{acceptable}$ equal to 0.01 is used for medium consequence of failure members and nodes. For a jacket structure a suggested understanding of medium importance may be that failure of this member increases the utilization check (UC) less than 5% (in any member or node in the structure). Similarly, a failure of a high importance member and node increases the UC less than 25%.

In some cases, reserve strength ratios (RSR) are used to determine the importance of members and nodes. If RSR values are used medium and high important members can be defined as medium important member: reduces the RSR value less than 5% if it fails and high importance member: reduces the RSR value less than 25% if it fails respectively.

5.3. Decision making based on the probability analysis

The aim of this paper is to describe a method for the analysis of inspection intervals and to facilitate decisions-making taking into account the strength of knowledge, assumptions and the parameters importance on the result of the decision. The decision should therefore be based on evaluating observable quantities (inspection intervals), the probability-based analysis, the analysts experience and understanding of the problem, evaluation of strength of knowledge and assumptions. An example of such evaluation is given in Table 6 (based on the parameters and models described in Section 5.1).

The evaluation of strength of knowledge in parameters and models as shown in Table 6 indicates that the calculated probabilities of through-thickness cracks must be seen as indicative and not as absolute values. Hence, the inspection intervals need to be carefully adjusted to experience and continuously monitored in order to within a reasonable level of confidence determine the existence, extent and consequence of damage, degradation and deterioration. For example, information and experience on imperfections from fabrication and design flaws from the relevant yard will be of importance for the final decision-making.

In Table 3 it is shown that for a jacket structure with approximately 20 years fatigue life and design fatigue factor (DFF) of 1, an inspection interval of 3.6 years using CVI for high importance nodes and members is calculated. In practice, such inspection intervals will be rounded to match the inspection campaigns during the summer season. For high importance jacket structural members with a fatigue life of 20 years and DFF of 3 an inspection interval of 10 years for CVI and 16 years for MPI are suggested. Such intervals may be regarded as large; however, it must be admitted that few cracks have been detected in critical areas on jacket structures. Hence, the inspection intervals found in the analysis is regarded as reasonable based on the authors' experience. However, this evaluation does not take into account the fabrication performance of the specific yard.

IOP Conf. Series: Materials Science and Engineering **700** (2019) 012034 doi:10.1088/1757-899X/700/1/012034

Table 5: Evaluation of strength of knowledge in parameters and models, assumptions and the parameters and models influence on the result of the probability-informed inspection planning

Parameter	Description	Strength	Influence	Conclusion / comments
H_s	Long-term distribution of significant wave height.	Good	Large	Based on some 50 years of hindcast and measurements. No further action required.
T_Z	Zero-up crossing period in sea-state.	Fair	Minor	Low influence in the uncertain range
$\Delta\sigma_{eq}$	Load model to transfer wave loading to stress in detail	Fair	Large	Model uncertainty is included in the analysis.
a_0	Initial crack size	Fair	Minor	Based on laboratory tests of different welds but assumed to be relevant. However, model uncertainty is included.
log(a)	SN curve parameter	Good	Large	Based on reasonable datasets although not same steel. No further action required
m_{SN}	SN curve parameter	Good	Large	
ln(A)	FM curve parameter calibrated from SN-fatigue calculation	Fair	Large	No further action required. Uncertainty in FM curve parameters are included by modelling ln(A) as a normal distributed variable
m_a	FM material parameter	Fair	Large	
t	Thickness of material	Good	Large	No further action required
PoD	Probability of detection of defects for various NDT methods.	Fair	Large	Based on inspection trials performed in several industry projects included in the proposed PoD curves in DNVGL [11]. Possibly optimistic, and this optimism should be taken into account in the decisions.

It should also be noted that the present revision of NORSOK N-006 [12] requires that inspection intervals on high importance members and nodes in jacket structures should not exceed 5 years independent on inspection method used. Although there are limitations to the strength of knowledge in some parts of these analysis and evaluations, the resulting inspection intervals (the observable quantity) are reasonably in line with current practice and recommendations given in class rules and standards. The experience using such inspection intervals is also reasonably good. Hence, based on these evaluations taking into account the analysis, experience, strength of knowledge, and the understanding of the problem a set of inspection intervals can be developed as indicated in Table 6 for a detail in a jacket node with 20 years calculated fatigue life based on the illustrative and simplified simulations performed by Neeraas [8].

It should be noted that the inspection intervals given in Table 6 is based on sea-states following the given long-term distribution. If several years of worse sea-states are experienced the inspections should be more frequent and special analysis may be required.

Table 6. Proposed inspection intervals given the analysis in Neeraas [8], experience, strength of knowledge and understanding of the problem.

	Importance	
Uncertainty of fatigue analysis	Medium	High
High (simplified)	4 years interval (CVI)	2.5 years interval (CVI)
CoV = 0.3	7 years interval (MPI)	4 years interval (MPI)
Medium (parametric)	5 years interval (CVI)	3 years interval (CVI)
CoV = 0.2	8 years interval (MPI)	5 years interval (MPI)
Low (finite element calculation)	6 years interval (CVI)	4 years interval (CVI)
CoV = 0.15	9 years interval (MPI)	6 years interval[1] (MPI)

1) Note that NORSOK N-006 recommends a maximum of 5 years intervals

IOP Conf. Series: Materials Science and Engineering **700** (2019) 012034 doi:10.1088/1757-899X/700/1/012034

6. Conclusion and recommendation and further work

The probability-informed results used in this paper is based on a probabilistic analysis performed as a part of a master thesis by Neeraas [8]. These analyses have limitations and cannot be construed to provide general inspection intervals for jacket structures. The suggested inspection intervals in this paper should be seen as examples and not as recommendations. However, the indications of constant inspection intervals are not affected by this limitation. The process used in the master thesis is based on the method suggested in this paper.

The paper illustrates the importance of taking into account the strength of knowledge, assumptions and the parameters importance on the result of the decision when determining inspection intervals for structures. Further, the importance of basing the analysis and decision-making on evaluating observable quantities as an addition to traditional probability-based analysis is highlighted. Especially, it is highlighted that the analysts experience and understanding of the problem should be taken into account.

The calibration of fracture mechanics parameters from the SN-curves needs further investigation. The calibration is performed with no model uncertainty (CoV = 0.0) for both the fracture mechanics crack growth model and the SN-fatigue model. This choice may not be correct and may be un-conservative.

The calculations performed in the master thesis by Neeraas [8] is based on a model using long-term distribution of sea-states and equivalent stresses. In most cases, probability-informed inspection planning analysis is rather based on a simplified long-term distribution of stress ranges. However, both methods have been shown to provide similar results [9]. A future work to implement the standard long-term stress range distribution into this type of analysis is recommended.

Acknowledgement

The results that is presented in this paper should not be seen as a suggestion for inspection intervals for jacket structures in general. This paper describes a methodology for decision-making of inspection intervals based on a probability- and risk-informed approach.

The authors would like to acknowledge the valuable comments to this paper Nils-Christian Hellevig. In addition, the authors would like to acknowledge the analysis performed by Håkon Neeraas and the simulation program developed by Magnus G. Ersdal.

The opinions expressed herein are those of the authors, and they should not be construed as reflecting the views of the Petroleum Safety Authority Norway or any of those persons mentioned in this paper.

References

[1] Ang A H S and Tang W H 1975 *Probability Concepts in Engineering Planning and Design, Vol I – Basic Principles* (John Wiley and sons, New York, USA)

[2] Ang A H S and Tang W H 1984 *Probability Concepts in Engineering Planning and Design, Vol II – Decision, Risk, and Reliability* (John Wiley and sons, New York, USA.)

[3] Madsen H O, Krenk S and Lind N C 1986 *Methods of Structural Safety* (Prentice Hall Inc. Englewood Cliffs, New Jersey)

[4] PSA 2016 *Risikobegrepet i petroleumsvirksomheten* (Petroleum Safety Authority PSA, Stavanger, Norway) in Norwegian

[5] PSA 2018 *Integrert og helhetlig risikostyring i petroleumsindustrien* (Petroleum Safety Authority PSA, Stavanger, Norway) in Norwegian

[6] Aven T 2013 Practical implications of the new risk perspectives *Reliab. Eng. Syst. Saf.* **115** 136-45

[7] Selvik J T, Scarf P and Aven T 2011 An extended methodology for risk based inspection planning *Reliab. Theor. and Appl.* 115-26

[8] Neeraas H 2019 *Sprekkvekst og inspeksjonsplanlegging / crack growth and inspection planning* (University of Stavanger Stavanger, Norway)

[9] Ersdal G 2005 *Assessment of existing offshore structures for life extension* (University of Stavanger, Stavanger, Norway)

[10] Dalane J I 1993 *System reliability in design and maintenance of fixed offshore structures* Dept. of marine structures (Norwegian Institute of Technology, Trondheim, Norway)
[11] DNVGL 2015 DNVGL-RP-C210 *Probabilistic methods for planning of inspection for fatigue cracks in offshore structures* (DNVGL, Høvik, Norway)
[12] Standard Norge 2015 *Assessment of structural integrity for existing offshore load-bearing structures*. Edition 2, Standard Norge (Lysaker, Norway)

State-of-the-art of crack propagation modelling in tubular joints

M Atteya*, O Mikkelsen and H G Lemu

Dept. of Mechanical and Structural Engineering and Materials Science, University of Stavanger, Norway

* Corresponding author: mostafa.a.atteya@uis.no

Abstract. Offshore structures are conventionally designed for a service life between 20-30 years and mostly used beyond their service life. They are operated up to the point where costs of operations, maintenance and repair exceed the revenue generated by the offshore structure. The cyclic nature of environmental loading induces multi-axial stresses with high gradients in welded tubular joints that lead to fatigue cracks emanating from imperfections in the welded materials. Tubular joints may exhibit significant residual life after fatigue cracks formation due to crack propagating around the weld circumference. It is important to understand tubular joints behaviour and estimate the residual life beyond crack initiation as it promotes the safe operation of the offshore installation, inspection planning and life extension of the asset. This paper presents an overview of the residual life estimation of cracked tubular joints using numerical methods. The recent developments and possible enhancements to the modelling methods of cracked tubular joints are presented.

1. Introduction

In fracture mechanics theory, fatigue life is generally divided into three stages, namely, crack initiation, crack propagation and final failure, though the traditional S-N curves based fatigue design methods make no clear distinction between crack initiation and crack propagation. For tubular joints, fatigue life may be divided broadly into four stages from N_1 to N_4: N_1 is the point at which the crack is first noted by any inspection method, N_2 denotes the first visual crack, N_3 denotes through-thickness crack, and N_4 represents the actual failure of the joint. Failure may be defined as the appearance of through-thickness crack (N_3) as the available experimental data showed that it is difficult to conclude N_4 from S-N curves [1-4]. Tubular joints may exhibit significant residual life after through-thickness crack formation due to crack propagating around the weld circumference. Zhang and Stacey [2] reported that for a total of 281 test results the mean ratio between N_4/N_3 is 1.38. Residual life of damaged joints depends on joint type, chord thickness, loading mode, as well as environmental exposures. It is important to understand tubular joints behavior and estimate the residual life beyond crack initiation as it promotes the safe operation of the offshore installation, inspection planning and life extension of the asset [2].

The behavior of cracked tubular joints can be studied using Fracture mechanics [5]. The behavior of each tubular joint is unique in a way as many variables can affect the joint behavior, i.e. the crack location, geometric dimensions, joint type, loading modes, stress ratio, weld profile, and boundary conditions [6]. Full-scale laboratory testing of cracked tubular joints is expensive and time-consuming while numerical simulation of fracture process of tubular joints has been a major task in recent research. The advancement of numerical methods and computational power made it possible to assess cracked tubular joints considering the uniqueness of each joint. Numerical methods such as finite element methods (FEM), extended finite element method (XFEM) and boundary element method (BEM) are typically used for fracture mechanics analysis. FEM is widely used for stress analysis and fatigue life estimation of cracked tubular joints [7-16], while applications for cracked tubular joints with the use of XFEM has not been identified. BEM was encountered only once by Borges et al. [17] for comparison between SIFs for cracked tubular joints estimated from FEM and BEM.

This paper presents a review of the residual life estimation of cracked tubular joints using numerical methods. Based on the observations discussed above, this paper focuses on the numerical modelling of fatigue life using FEM. In general, the numerical models are based on the assumptions of constant-amplitude fatigue loading with crack propagating from a given initial crack within the domain of linear elastic fracture mechanics (LEFM).

2. Fatigue life evaluation of cracked tubular joints

The assessment of structures using fracture mechanics approach requires sizing of a crack-like defect. The size of the crack is determined mainly by the inspection method used. If the crack cannot be observed, the crack size can be assumed as the largest undetectable crack. Once the initial crack size is determined, fracture mechanics can be employed to evaluate the crack propagation until the onset of unstable crack extension, at which the structural element will lose its structural capacity. Fracture assessment of the component rely on the techniques of LEFM or elastic plastic fracture mechanics (EPFM) depending on material properties and plate thickness [1]. The fracture mechanics approach provides mathematical relationship between three critical variables, namely, flaw size, stresses in the vicinity of the crack and toughness.

Fracture mechanics analysis of tubular joints is demanding. The tubular joint geometry with boundary conditions induce mixed-mode loading of the crack [18]. In this respect, the detailed analysis of stresses requires three-dimensional models of the joint [1, 2]. In the vicinity of the crack, the size of the finite elements must be comparable to the size of the crack increments, thus the meshing procedure becomes challenging.

Conventional FE approach can be divided into four main steps: (1) development of numerical model with an initial crack-like defect; (2) evaluation of fracture mechanics parameters; (3) estimation of crack propagation rate, crack propagation direction and selecting crack increment size; (4) updating the numerical model with the new crack front. These steps are repeated for successive crack sizes until the structure loses its load carrying capacity as shown in Figure 1.

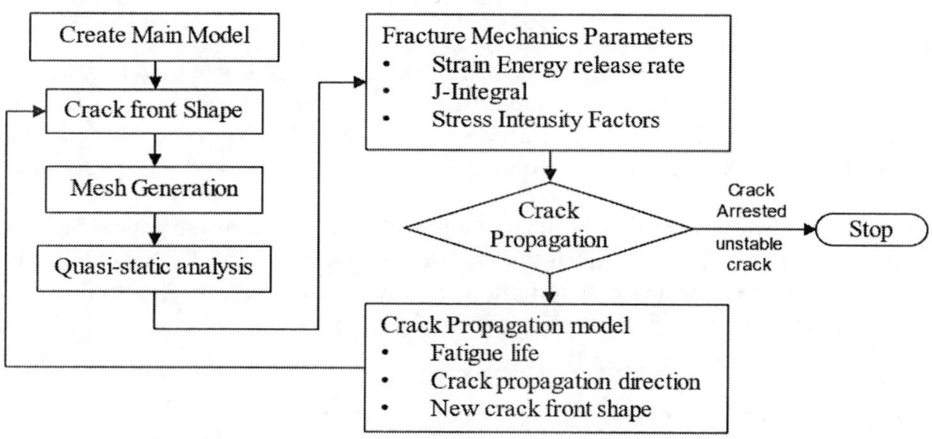

Figure 1. Conventional FE approach for fatigue life estimation of cracked tubular joints [19].

3. Numerical model generation

Numerical modelling of cracked tubular joints for fatigue life assessment is a computational expensive task as computational procedures are repeated for each successive crack increment. In this literature study, only simple tubular joints with focus on T/Y- joints are considered.

IOP Conf. Series: Materials Science and Engineering **700** (2019) 012035 doi:10.1088/1757-899X/700/1/012035

3.1. Type of finite elements
Stress concentration is generated at crack-tip and mesh refinement is needed in the vicinity of the crack-tip to capture the large stress and strain gradients. For sharp crack-tip, an inverse square root singularity exists. The singularity must be considered for reliable analysis of stress. First attempt in modelling tubular T-joint with a shallow weld crack was carried out using a wedge-shaped singular element [20]. The element was obtained by collapsing faces of 8-node brick element and choosing a displacement interpolation function. Elements with 6, 5 and 4 faces produced from the brick element with the displacement interpolation function updated accordingly. Later Rhee el al. [7] modelled K-joint for fatigue crack growth analysis using singular elements for crack-tip with quarter point.

The most common finite elements used in recent literature are isoparametric elements with quadratic shape functions. These elements can represent curved shapes with small number of elements. To capture crack-tip singularities, solid elements such as those shown in Figure 2(a) – (d) are typically used. While 20-node hexahedral elements (Figure 2(a)) are generally used for structured mesh, 10-node tetrahedral elements (Figure 2(b)) can be used for unstructured mesh [18, 21]. In the vicinity of a sharp crack-tip, square root stress singularity can be obtained by collapsing the 20-node hexahedral element (Figure 2(c)) to 15-node wedge elements and moving the mid-side nodes to quarter points.

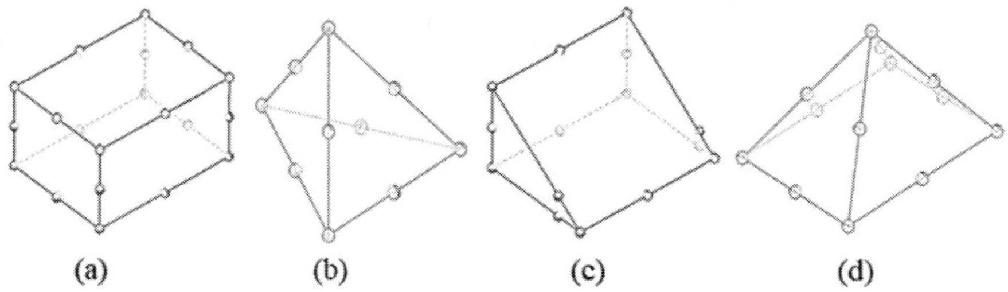

Figure 2. Typical solid mesh elements (a) 20-node hexahedral element (b) 10-node tetrahedral element (c) collapsed quarter-point hexahedral element and (d) 20-node pyramid – collapsed hexahedral.

The usage of tetrahedral elements allows decreasing the number of elements through thickness, away from sensitive areas, which reduces the computational time. On the other hand, this approach requires dense mesh around the crack-tip to produce convergence of the high stress-gradient expected and to accurately evaluate stress intensity factors (SIF).

3.2. Mesh Generation
Finite element simulation can be facilitated using an automated mesh generation system to handle the remeshing procedure during simulation. Many researches developed their meshing algorithms to carry out the remeshing procedure [9-12, 15, 16, 22]. Generally, the mesh generation of cracked tubular joints can be divided into two groups, structured mesh approach and unstructured mesh approach.

3.2.1. Structured mesh approach. One of the first automated structured mesh approaches was developed by Bowness and Lee [11, 23]. Their algorithm was used to assess cracked T-Joints considering a doubly curved semi-elliptical crack emanating from the weld toe. The algorithm is of 'building block' nature starting with a discretization of a plain plate containing a semi-elliptical crack as presented in Figure 3(a) below. A modelling sequence, as illustrated in Figure 3 (a) – (f) [23], is implemented using the following steps:

a) Adding an attachment to the plain plate,
b) Mapping to provide a weld profile at its base,
c) Making the mesh curved round 90° to become a brace,
d) Mapping the main plate into a square,
e) Making the main plate curved around the crack plane by 90° to form the chord,
f) Finally, adding the rest of the chord to form a quarter tubular T-joint.

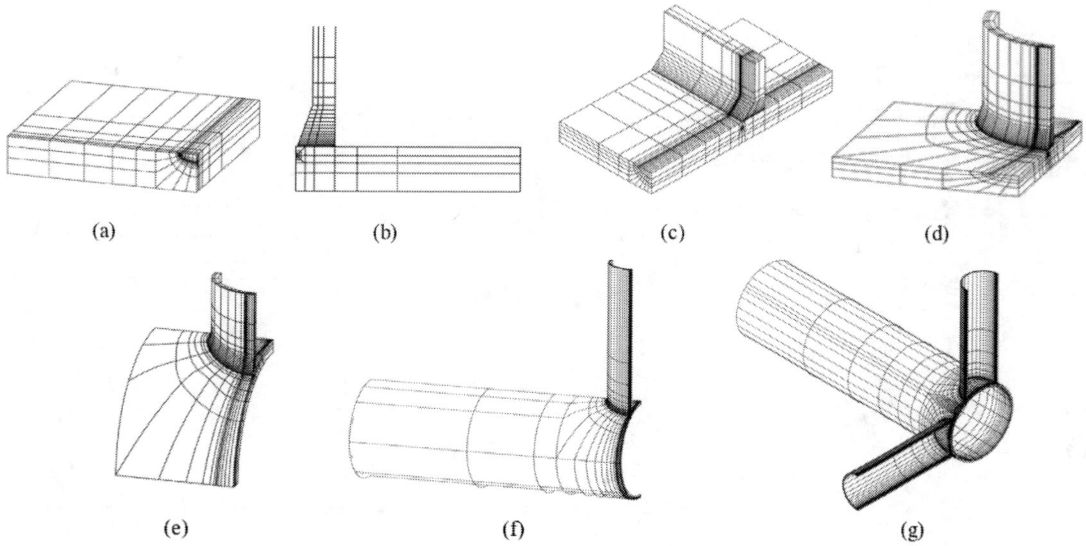

Figure 3. Bowness and Lee modelling sequence, (a) quarter plain plate with half surface crack, (b) mapped attachment and weld profile, (c) half T-butt joint, (d) curved T-butt plate to form the brace and weld, (e) chord-brace assembly from curved configuration d, (f) quarter T-joint, (g) Final mesh.

Lie et al. [15] developed a structured meshing algorithm to study cracked T/Y-shape joints. The algorithm creates a surface crack block which is divided into three parts: (1) crack tube part, where the crack front is located, (2) part which acts as transition from crack tube to the rest of the model and (3) part which contains the crack faces. Then it creates the chord-brace intersection zone by double mapping a flat vertical attachment and fillet weld.

3.2.2. Unstructured mesh approach. The advantage of unstructured mesh approach is that no focus mesh is required around the crack-tip, which significantly reduces the time and effort needed to develop the meshing algorithms. Yagi et al. [16] used tetrahedral elements to study surface-cracked T-shape joints. The approach was based on creating an intact three-dimensional CAD model for the joint then inserting an initial semi-elliptical surface crack at the weld toe. Solid elements meshing with quadratic tetrahedral elements were applied with double nodes on the crack surface to simulate a seam crack. The meshing procedure was repeated for each crack increment.

3.3. Weld profile

Typical tubular brace to chord joints are full penetration butt welded, characterised by changing weld profile from crown heel to crown toe due to the projection of the tubular brace on the tubular chord. It was found that weld toe radius and weld angle affect the SIF [24, 25]. Hence, it is important to model the weld profile if cracks emanate from weld toe or root. The influence of the weld leg size and weld angle to the SIF is prominent for smaller crack depth [26]. Typical weld details across the joint are shown in Figure 4.

Lie et al. [15] and Bowness et al. [11, 23] map the attachment to include fillet weld, then the attachment and the weld is being curved to simulate the brace. Mapping fillet weld does not conform with butt-weld profile at the saddle location see Figure 4, especially for joints with high brace diameter to chord diameter ratio. However, SIF at the crack deepest point was in a good agreement with experimental work. The effect of weld profile is not dominant at the crack deepest point. In fact, it will be more dominant towards the crack corners.

IOP Conf. Series: Materials Science and Engineering **700** (2019) 012035 doi:10.1088/1757-899X/700/1/012035

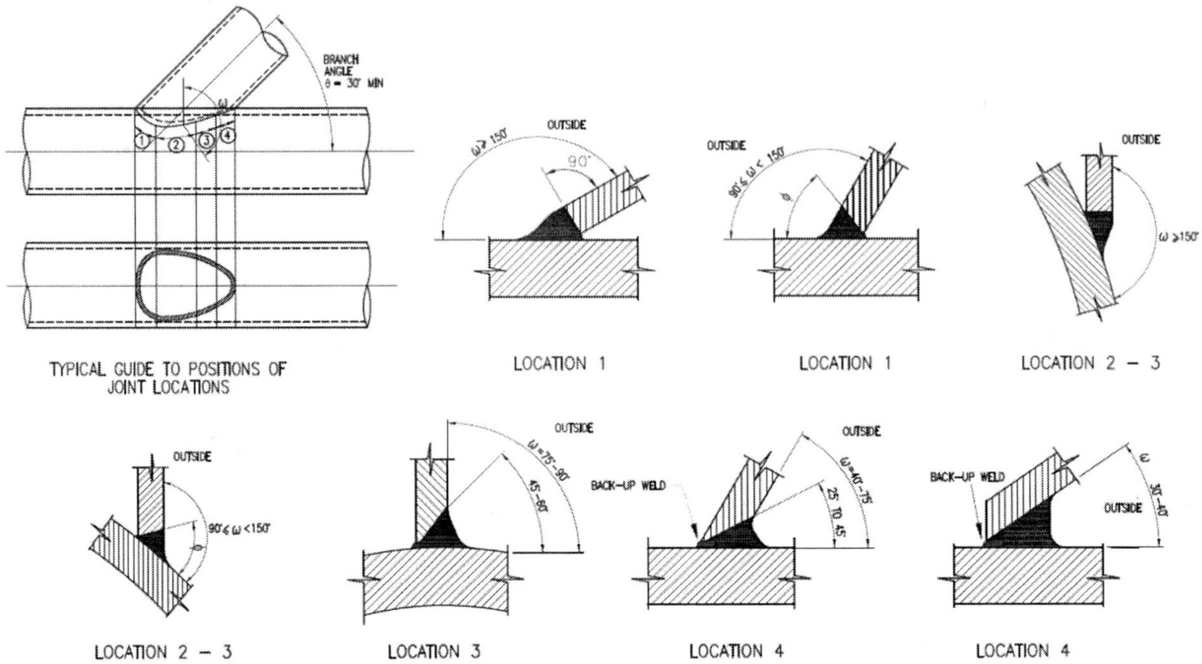

Figure 4. Weld profile of a typical offshore tubular joint.

3.4. *Crack front shape*

Crack front shape is generally double curved, and difficult to describe and predict [27]. Under fatigue loading, cracks emanate from the surface then evolve to through-thickness cracks. Through the literature, surface cracks are idealized as semi-elliptical cracks [14, 25, 28, 29]. In reality, cracks initiate at multiple locations and then coalescence to form a part-elliptical shape [30, 31]. Qian et al [3] showed variation up to 7% between semi-elliptical crack shape and the true-crack front profile. The variation translates to more than 20% in the crack growth rate.

Modelling surface cracks is more complex than modelling through-thickness cracks. For surface cracks, if the cracks exist on the chord side, the crack surface will be of inclined plane and the crack geometry will lead to mixed-mode condition [32]. The crack surface was assumed to be perpendicular to the free surface all the time by YongBo et al. [14]. While Bowness et al. [33] showed that the crack surface is not perpendicular to the free surface and that it curves away from the axis perpendicular to the free surface by a fourth degree polynomial function.

3.5. *Singularities in cracked bodies.*

For a surface crack in a homogeneous material not restrained by a weld, two sources of stress field singularity exist along the crack-front, near field stress singularity and vertex stress singularity. Near field stress singularity extend over most of the crack front and diminish to zero at the corner points. Vertex stress singularity exists at the corner points. The power of the vertex stress singularity is weaker than the near field stress singularity. The power changes monotonically from -0.5 at the interior to -0.37 at the free surface [34]. The weaker stress singularity at the corner point eventually leads to drop in the SIF from the interior of the crack front. Practical approach can be used by extrapolating the SIF from the interior part to the corner point and use it to evaluate the crack growth rate and direction [16]. Vertex singularity is a function of the material Poisson's ratio and the angle between the crack front and the free surface β. If the angle is small then SIFs tends to zero at the corner point, while if the angle is large SIFs tends to infinity. The crack front is observed to intersect the free surface at a critical angle β_c for which SIFs are finite. The critical angle is a function of material Poisson's ratio (equation 1) and type of loading. For steel materials with $\nu = 0.3$, under mode I loading, the crack front intersects the free surface at a critical angle of 100.4° and an angle of 67.0° for antisymmetric loading [27]. Hence, the assumption that surface cracks are of semi-elliptical form (Figure 5) may not be valid.

The critical angle β_c for mode I case is given by:

$$\beta_c = \tan^{-1}\left(\frac{\nu - 2}{\nu}\right) \quad (1)$$

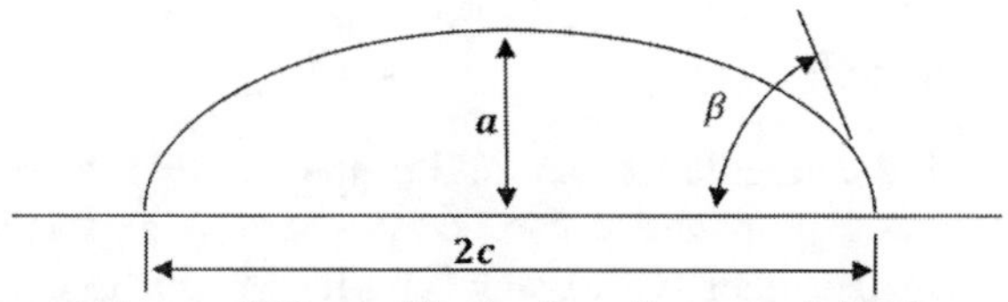

Figure 5. Part-elliptical surface crack [27].

4. Stress intensity factor evaluation methods

For crack-like defects in structural components under loading, there will be relative movement between the two crack faces, and stress field will develop at the crack tip. SIF describes the stress field in the vicinity of the crack and represents the tendency of the crack to advance under static loading conditions. As a minimum, to accurately determine the SIF for a crack in a tubular joint, the following parameters must be taken into account: crack position, crack front shape, coalescence of cracks, variation of stresses, and residual stresses.

Analytical solutions of SIF for a three-dimensional surface crack in tubular joints do not exist in the literature. Approaches that can be used to determine the SIF of cracked tubular joints are; classical solutions for idealized geometries modified to represent tubular joints, Semi-empirical methods based on full-scale testing or numerical methods. Numerical methods are the only approaches that can take into account all the parameters required to accurately determine the SIF [1].

The evaluation of SIF using conventional FEM requires that the mesh conform to the crack tip geometry. The accuracy of stress intensity factors depended on the quality of the FE mesh [3, 15-17, 26]. Several numerical methods exist for the evaluation of SIF in three-dimensional cracked components, the methods can be divided into the two main categories: (1) stress and displacement matching methods and (2) energy-based methods [19].

4.1. Stress and displacement matching methods.

These methods match the stress or displacement field calculated by the FEM with the analytical stress or displacement field which contain the SIF. For instance, using the *displacement extrapolation technique* [7, 8, 11, 15, 22, 23] the SIF is estimated from the elastic displacements behind the crack-tip. Then the displacements are extrapolated along a radial path to the weld toe as illustrated in Figure 6. For plane strain conditions, K_I can be estimated from the following extrapolation [35].

$$K_I = \lim_{r \to 0}\left[\frac{E u_y}{4(1-\nu^2)}\sqrt{\frac{2\pi}{r}}\right] \quad (\theta = \pi) \quad (2)$$

where K_I denotes SIF for mode I loading, r is the radial distance between the crack-tip and the point of displacement extrapolation, E is Young's modulus of elasticity, ν is Poisson's ratio, and θ is the angle between the crack plane and the point of displacement extrapolation.

For plane stress loading, K_I can be estimated from the following expression.

$$K_I = \lim_{r \to 0}\left[\frac{E u_y}{4}\sqrt{\frac{2\pi}{r}}\right] \quad (\theta = \pi) \quad (3)$$

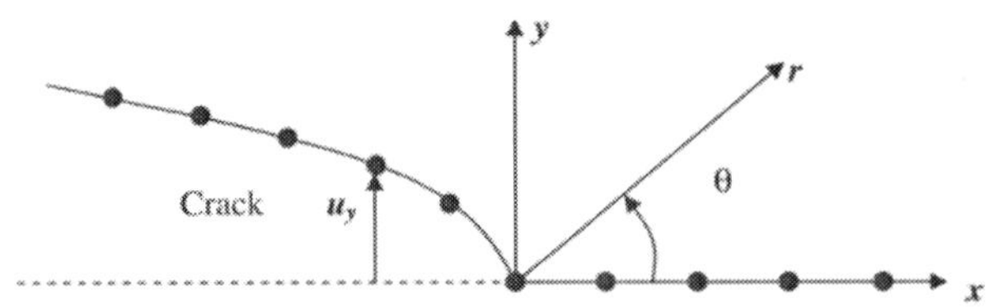

Figure 6. displacement field at crack tip [35].

4.2. Energy-based methods.
Energy-based methods relate the singular stress field to the energy release rate. The change in energy release rate associated with crack propagation can be evaluated by performing two numerical analyses, one with crack length a, and the other with crack length $a + \Delta a$. Energy-based methods are more-efficient than the stress and displacement matching methods, as energy is insensitive to mesh refinement.

As an energy-based method, the *virtual crack extension* (VCE) method was used [10, 15, 16, 23] based on the assumption that the energy release rate (G) for a growing crack from length a to $a + \Delta a$ is equivalent to the work needed to close the crack by Δa. It requires the calculation of stresses and displacements at the crack-tip. Numerical representation of VCE in a two-dimensional FE problem is shown in Figure 7.

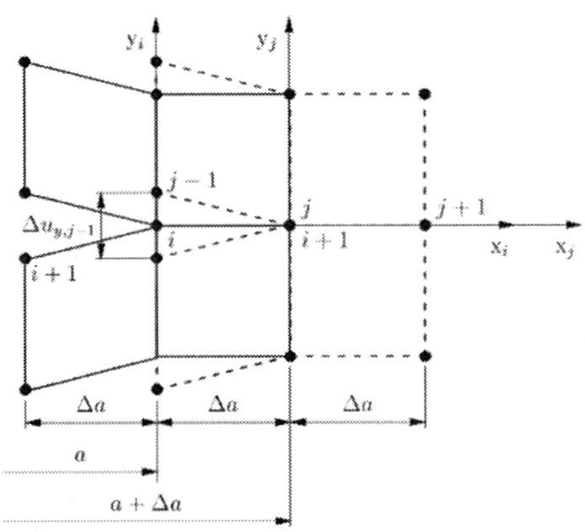

Figure 7. numerical representation of the virtual crack closure integral method [36].

Two stress or displacement calculations are required to evaluate the SIF using VCE for one crack configuration and the procedure can be numerically expensive. The analysis can be performed only once if the crack intervals Δa are small [35] and hereafter the procedure is called the modified virtual crack closure method [36].

Energy domain integral methods. It is the general framework for numerical analysis of J integral. For a rate-independent quasi-static fracture analysis, a single analysis is required to determine the energy release rate in the vicinity of a growing crack. J-integral is a function of the energy release rate associated with crack advance. The J-integral yields an accurate solution to the stress intensity factors within the LEFM domain, though it cannot easily used in the condition of mixed mode loading which is present in the tubular joints. In addition, J-integral lacks path independence at the crack corners because of the presence of two stress singularities, one caused by the crack-tip and the other caused by weld [14, 23].

5. Crack propagation models

Under static loading, the necessary condition for crack propagation is determined by material properties and the stress field in the vicinity of the crack expressed with SIF. Typical fatigue crack growth behaviour in metals can be described by a log-log curve of da/dN and ΔK as shown in Figure 8. da/dN is the crack growth rate and ΔK is the stress range expressed with SIF's $K_{max} - K_{min}$. The curve consists of three regions; the first region, includes the threshold value ΔK_{th}, below which the crack will not grow.

The second region describes the linear relation between log da/dN and log ΔK. In the third region the crack accelerates eventually until final failure of the joint.

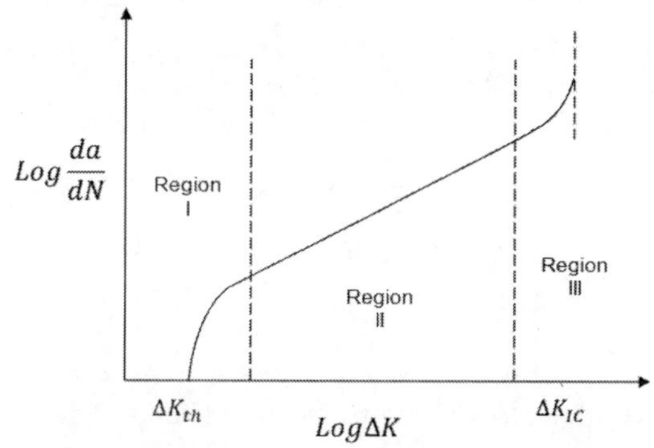

Figure 8. Typical fatigue crack growth curve

Paris-Erdogan proposed a power relation of the form;

$$da/_{dN} = C\Delta K_{eff}{}^{m} \tag{4}$$

where C and m are material constants, and $\Delta K_{eff} = K_{max} - K_{op}$, K_{max} is the maximum stress intensity, and K_{op} is the stress intensity at which the crack opens.

Paris-Erdogan law was being used extensively in cracked tubular joints literature [3, 14, 16, 26, 37, 38]. Paris-Erdogan law is insensitive to the stress ratio R defined by K_{min}/K_{max} and considers only region II (Figure 8). The most common expression to describe fatigue crack growth is called NASGRO [39].

$$\frac{da}{dN} = C\Delta K^{m} \frac{\left(1 - \frac{\Delta K_{th}}{\Delta K}\right)^{p}}{\left(1 - \frac{K_{max}}{K_{Ic}}\right)^{q}} \tag{5}$$

where $C, m, p, and\ q$ are material constants and K_{Ic} is fracture toughness

NASGRO expression considers the three regions for crack propagation rate and is sensitive to the stress ratio. The number of cycles (N) required for a given increase in crack size from an initial length a_o to a final length a_f is given by:

$$N = \int_{a_o}^{a_f} \frac{da}{f_1(\Delta K)} \tag{6}$$

5.1. Fracture criterion

Under mixed-mode loading condition, the use of equivalent stress intensity K_{eq} is necessary to consider the effect of K_I, K_{II} and K_{III}. Cracks can grow under the following conditions.

$$K_{th} < K_{eq} \leq K_{IC} \tag{7}$$

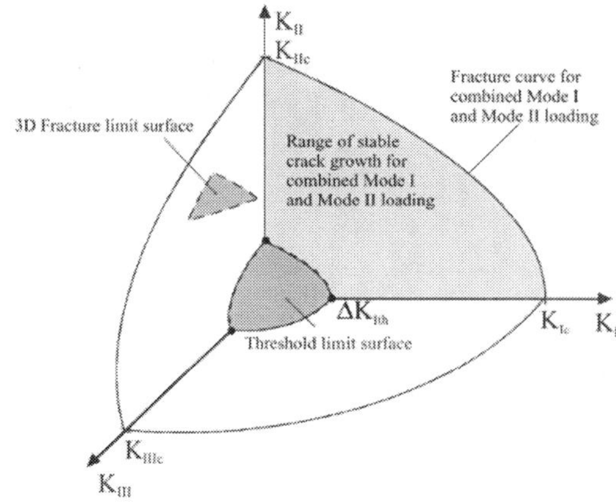

Figure 9. Mixed mode loading with fracture limit and threshold surface [40].

Rhee [18] considered the effect of mixed-mode on the SIF by converting the energy release rate relation which is given by

$$\frac{\Delta K_{eq}^2}{E} = G = \frac{1-\upsilon^2}{E}\left(K_I{}^2 + K_{II}{}^2 + \frac{K_{III}{}^2}{1-\upsilon}\right) \tag{8}$$

Rhee [41] proposed another parameter to describe the effect of mixed-mode on the SIF and it was used by Qian et al. [3], YongBo et al. [15] and Chiew et al. [3, 12, 14].

$$K_{eq}{}^2 = \frac{K_I{}^2 + K_{II}{}^2 + K_{III}{}^2}{1-\nu} \tag{9}$$

Another criterion was proposed by Richard [42], which is based on the MTS criterion,

$$K_{eq} = \frac{K_I}{2} + \frac{1}{2}\sqrt{K_I{}^2 + 4(1.155K_{II})^2} \tag{10}$$

5.2. Crack propagation direction

Crack propagation in solids under mixed-mode loading tends to eliminate mode II and mode III loading conditions. For crack propagating under mode I and mode II, the crack tends to eliminate mode II condition by adopting a crack path direction corresponding to the Mode I loading. When a crack propagates under mixed-mode loading, the crack front tends to both deflect and twist the orientation of the crack to comply with mode I loading [32, 33].

Crack propagation direction criteria can be divided into two groups, the first group describes the crack propagation direction at each point on the crack front by using the values of the SIFs and can be called *local type criteria* [32]. Example of these criteria is the MTS criterion, proposed by Erdogan and Sih [43] and extended to three dimensions by Schollmann et al. [44] and simplified by Richard et al. [45]. The second group describe the crack propagation by considering the mean value of SIFs along the whole crack front and can be called *global type criteria*. Examples of these criteria are the criteria by Pook [27] and Lazarus et al. [32, 46, 47]. Crack deflection and crack twisting angle by Richard et al. [45] can be described as;

$$\theta = \mp\left[A\frac{|K_{II}|}{K_I + |K_{II}| + |K_{III}|} + B\left(\frac{|K_{II}|}{K_I + |K_{II}| + |K_{III}|}\right)^2\right] \tag{11}$$

where $\theta < 0°$ for $K_{II} > 0$ and $\theta > 0°$ for $K_{II} < 0$ and $K_I \geq 0$.

$$\psi = \mp\left[C\frac{|K_{III}|}{K_I + |K_{II}| + |K_{III}|} + D\left(\frac{|K_{III}|}{K_I + |K_{II}| + |K_{III}|}\right)^2\right] \tag{12}$$

where $\psi < 0°$ for $K_{III} > 0$ and$\psi > 0°$ for $K_{III} < 0$ and $K_I \geq 0. A = 140°$, $B = -70°$, $C = 78°$ and $D = -33°$

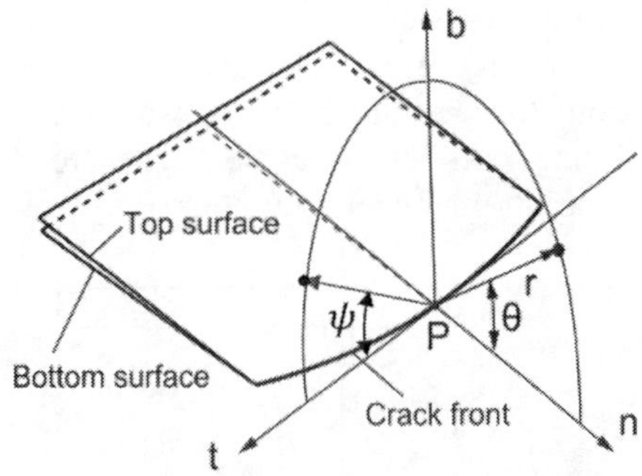

Figure 10. Polar coordinates at point P along the crack front with crack deflection angle $\boldsymbol{\theta}$ and crack twisting angle $\boldsymbol{\psi}$ [42]

Yagi et al. [16] used the three dimensional MTS criterion proposed by Gerstle [48], and formulated the crack deflection angle in the form of;

$$\theta = 2\tan^{-1}\left[\frac{-2\Delta K_{II}}{\Delta K_{Ieq} + \sqrt{(\Delta K_{Ieq})^2 + 8(\Delta K_{II})^2}}\right] \tag{13}$$

where ΔK_{Ieq} is an equivalent SIF range including the mode-III effect, and given by;

$$\Delta K_{Ieq} = \Delta K_I + B|\Delta K_{III}| \tag{14}$$

where B an empirically determined factor = 1.0.

Criterion by Pook [27] can be calculated in a step-by-step process.

$$K_I \sin\theta = -K_{II}(3\cos\theta - 1) \tag{15}$$

$$K_{vI,II} = \frac{0.83K_I + \sqrt{0.4489K_I^2 + 3K_{II}^2}}{1.5} \tag{16}$$

$$\tan 2\psi = \frac{2K_{III}}{K_{vI,II}(1 - 2\upsilon)} \tag{17}$$

In Pook's criterion the equivalent stress intensity K_{eq} considers the effect of K_I, K_{II} and K_{III}. K_{eq} can be described by the following relation.

$$K_{eq} = \frac{K_{vI,II}(1 + 2\upsilon) + \sqrt{K_{vI,II}^2(1 + 2\upsilon)^2 + 4K_{III}^2}}{2} \tag{18}$$

Another crack propagation criterion in use is the minimum strain energy criterion as presented by Sih [49, 50]. The crack is then assumed to propagate in the direction of minimum strain energy density, S, expressed as;

$$S = \frac{a_{11}K_I^2 + 2a_{12}K_I K_{II} + a_{22}K_{II}^2 + a_{33}K_{III}^2}{\pi} \tag{19}$$

where a_{ij} are coefficients relating crack deflecting angle θ, elastic modulus and Poisson's ratio.

5.3. Crack increment step

Typically, the crack growth increment and direction are calculated for each node at the crack front based on SIFs, and the crack is then extended for the next step. The number of nodes at the crack front is kept the same for all increments [16, 29].

Crack increment step is influential to fatigue life estimation. There are two approaches for crack increment step; the first is based on a constant crack length increment and the second is based on a constant number of cycles increment (Figure 11). Due to the geometry and loading of the tubular joints, it is expected that SIF will vary along the crack front. As a result, the crack will not advance equidistantly along the crack front for the same number of cycles. Therefore, for accurate crack propagation estimation, crack increment step based on constant number of cycles may be used.

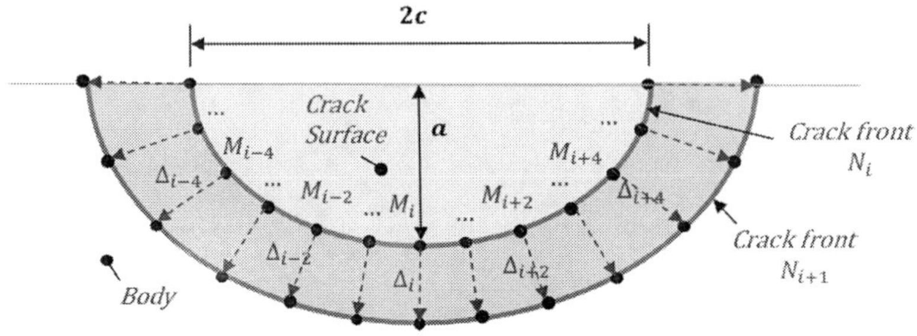

Figure. 11. Surface crack and schematic illustration of the crack propagation [29].

6. Conclusion

The objective of this paper is to provide an overview of the fatigue life estimation of cracked tubular joints using fracture mechanics and FEM. With respect to mesh generation, the methods proposed over the last two decades considered two different types, structured and unstructured mesh generation. Structured mesh generation is considered as systematic approach and provide reliable results. However, the method is limited to simple geometries. Unstructured mesh has the potential to be extended to more complex geometries due to the less effort needed for automation and model creation. To the knowledge of the authors the meshing algorithm which accommodate the automatic transition from surface cracks to through-thickness cracks has not yet been implemented and documented.

Through the literature, surface cracks are idealized as semi-elliptical cracks. In reality, cracks initiate at multiple locations and then coalescence to form a part-elliptical shape. That idealization may cause a discrepancy in crack growth rate up to 20%. The crack front in tubular joints is of double curved nature and subjected to mixed mode loading due to the joint geometry. The assumptions for calculation of crack propagation angle was limited to the use of MTS criterion. The MTS criterion provides crack deflection and crack twisting angles for each point on the crack front which fits well with meshing algorithms.

The methods for SIF evaluation of cracked tubular joints, found in the literature were displacement extrapolation, VCE and the domain integral methods. VCE method for a crack configuration requires calculations of two stresses or two unknown displacements to evaluate the SIF. The method is regarded numerically expensive. The domain integral method performs poorly under mixed-mode loading conditions and at crack corners it is not path independent. The displacement extrapolation method is thus recommended for SIF evaluation of cracked tubular joints

The advantage of the LEFM is that each parameter that may affect the fatigue life may be assessed individually and it is possible to extend the test data to other conditions than those under which the tests were performed [1]. However, none of the studies reviewed considered the influence of the following parameters on the residual fatigue life of cracked tubular joints: residual stresses, stress ratio effects, environment (air, seawater) exposures and variable amplitude loading.

The complexity of evaluating the residual fatigue life of cracked tubular joints is due to the inherent shortcomings of FEMs, including;

- Finite element discretization at crack tip area in 3D is complex.
- The assumption of crack propagation direction prior to calculation of SIF.
- Explicit definition of the crack front.

The literature study shows that the XFEM can overcome the limitations listed above. The method is independent of crack initiation or propagation and it can provide solution-dependent crack initiation and propagation path. The accuracy of XFEM is however not tested with applications for tubular joints.

Acknowledgements
This work is part of a PhD study supported by the Norwegian Ministry of Education and Research. The financial support is acknowledged.

References
[1] CIRIA Underwater Engineering Group 1985, Design of tubular joints for offshore structures, UEG, London
[2] Zhang Y H and Stacey A 2008 Review and assessment of fatigue data for offshore structural components containing through-thickness cracks *Int. Conf. Offshore Mech. Arc. Eng.*, Estoril, Portugal
[3] Qian X and Nguyen C T 2012 Prediction of the fatigue crack propagation in large-scale tubular joint specimens *22nd Int Offshore Polar Eng. Conf.*, Rhodes, Greece
[4] Sharp J V, Stacey A and Wignall C M 1998 Structural integrity management of offshore installations based on inspection for through-thickness cracking *17th Offshore Mech. Arct. Eng. Conf.*, Lisbon, Portugal
[5] Rhee H C and Kanninen M F 1988 Opportunities for application of fracture mechanics for offshore structures *Appl. Mech. Rev.* **41** 23-35
[6] Haswell J and Hopkins P 1991 A review of fracture mechanics models of tubular joints Fatigue *Fract. Eng. Mater. Struct.* **14** 15-29
[7] Rhee H C 1989 Fatigue crack growth analyses of offshore structural tubular joints *Eng. Fract. Mech.* **34** 1231-9
[8] Laham S and Fm B 1997 The ultimate strength of cracked tubular K-joints
[9] Cao J J, Yang G J, Packer J A and F.M B 1998 Crack modeling in FE analysis of circular tubular joints *Eng. Fract. Mech.* **61** 537-53
[10] Bowness D and Lee M M K 2002 Fracture mechanics assessment of fatigue cracks in offshore tubular structures: HSE Books
[11] Lee M M K and Bowness D 2002 Estimation of stress intensity factor solutions for weld toe cracks in offshore tubular joints *Int. J. Fatigue.* **24** 861–75
[12] Chiew S P, Lie S T, Lee C K and Huang Z W 2003 Parametric equations for stress intensity factors of cracked tubular T&Y-joints *13th Int. Offshore Polar Eng. Conf.*, Honolulu, Hawaii, USA
[13] Lie S T, Lee C K and Wong S M 2003 Model and mesh generation of cracked tubular Y-joints *Eng. Fract. Mech.* **70** 161–84
[14] Yongbo S and Tjhen L S 2005 Parametric equation of stress intensity factor for tubular K-joint under balanced axial loads *Int. J. Fatigue.* **27** 666-79
[15] Lie S T, Li T and Shao Y B 2016 Stress intensity factors of tubular T/Y-joints subjected to three basic loading *Adv. Steel Constr.* **12** 109-33
[16] Yagi K, Tanaka S, Kawahara T, Nihei K, Okada H and Osawa N 2017 Evaluation of crack propagation behaviors in a T-shaped tubular joint employing tetrahedral FE modeling *Int. J. Fatigue.* **96** 270-82
[17] Borges L, Chiew S P, Nussbaumer A and Lee C K 2012 Advanced numerical modeling of cracked tubular K joints: BEM and FEM comparison *J. Bridge Eng.* **17** 432-42
[18] Rhee H C 1986 The behavior of stress intensity factors of weld toe surface flaw of tubular X-joint *Offshore Technol. Conf.*, Housten, Texas
[19] Branco R, Antunes F V and Costa J D 2015 A review on 3D-FE adaptive remeshing techniques for crack growth modelling *Eng. Fract. Mech.* **141** 170-95

[20] G.S H, M A and Munaswamy K 1985 Analysis of tubular joint with weld toe crack by finite element methods Communications in *appl. numer. methods.* **1** 325-31
[21] Du Z Z and Hancock J W 1989 Stress intensity factors of semi-elliptical cracks in a tubular welded joint using line springs and 3D finite elements *J. Pressure Vessel Technol.* **111** 247-51
[22] Burdekin F M, Yang G J and Cao J J 1998 Assessment of fracture strength of cracked tubular joints, HSE Books, Manchester
[23] Bowness D and Lee M M K 1995 The development of an accurate model for the fatigue assessment of doubly curved cracks in tubular joints *Int. J. Fract.* **73** 129-47
[24] Niu X, Glinka G 1987 The weld profile effect on stress intensity factors in weldments *Int. J. Fract.* **35** 3-20
[25] Delft D R V V, Dijkstra and Snijder H H 1986 The calculation of fatigue crack growth in welded tubular joints using fracture mechanics *Offshore Technol. Conf.*, Houston, Texas, USA
[26] Lie S T, Chiew S P, Lee C K and Huang Z W 2004 Fatigue performance of cracked tubular T joints under combined loads. II: numerical *J. Struct. Eng.* **130** 572-81
[27] Pook L P 1995 On fatigue crack paths *Int. J. Fatigue.* **17** 5-13
[28] Skallerud B, Eide O I and Berge S 1990 Fatigue crack growth in complex tubular joints
[29] Tanaka S, Kawahara T and Okada H 2014 Study on crack propagation simulation of surface crack in welded joint structure *Mar. struct.* **39** 315-34
[30] Pang J H L, Tsang K S and Hoh H J 2016 3D stress intensity factors for weld toe semi-elliptical surface cracks using XFEM *Mar. struct.* **48** 1-14
[31] Pang J H L and Chew Y X 2014 Fatigue crack growth and coalescence algorithm starting from multiple surface cracks *Adv. Mater. Res.* **891-892** 1003-8
[32] Lazarus V, Buchholz F G, Fulland M and Wiebesiek J 2009 Comparison of predictions by mode II or mode III criteria on crack front twisting in three or four point bending experiments *Int. J. Fract.* **153** 141-51
[33] Bowness D and Lee M M K 1998 Fatigue crack curvature under the weld toe in an offshore tubular joint *Int. J. Fatigue.* **20** 481-90
[34] Shivakumar K N and Raju I S 1990 Treatment of singularities in cracked bodies *Int. J. Fract.* **45** 159-78
[35] Anderson T L 2017 Fracture mechanics: fundamentals and applications: CRC Press
[36] Schoollmann M, Fulland M and Richard H A 2003 Development of a new software for adaptive crack growth simulations in 3D structures *Eng. Fract. Mech.* **70** 249–68
[37] Shen W and Choo Y S 2012 Stress intensity factor for a tubular T-joint with grouted chord *Eng. Struct.* **35** 37-47
[38] Etube L S 2001 Fatigue and fracture mechanics of offshore structures vol 4 London: Professional Engineering Publ.
[39] Forman R G and Mettu S R 1990 Behavior of surface and corner cracks subjected to tensile and bending loads in Ti-6Al-4V alloy *NASA technical memorandum*: NASA
[40] Richard H A, Fulland M and Sander M 2005 Theoretical crack path prediction Fatigue Fract. *Eng. Mater. Struct.* **28** 3-12
[41] Rhee H C, Han S and Gibson G S 1991 Reliability of solution method and empirical formulas of stress intensity factors for weld toe cracks of tubular joints *Offshore Mech. Arct. Eng. Conf.* 441-52, New York
[42] Richard H A, Schramm B and Schirmeisen N H 2014 Cracks on mixed mode loading – theories, experiments, simulations *Int. J. Fatigue.* **62** 93-103
[43] Erdogan F and Sih G C 1963 On the crack extension in plates under plane loading and transverse Shear *J. Fluids Eng.* **85** 519-25
[44] Schöllmann M, Richard H A, Kullmer G and Fulland M 2002 A new criterion for the prediction of crack development in multiaxially loaded structures *Int. J. Fract.* **117** 129-41
[45] Richard H A, Schirmeisen N H, Fulland M and Sander M 2001 Experimental and numerical simulation of mixed mode crack growth Proc. *Int. Conf. biaxial/multiaxial fatigue fract.* 623-

30
[46] Lazarus V and Leblond J B 1998 Crack paths under mixed mode (I + III) or (I + II + III) loadings
[47] Lazarus V, Leblond J-B and Salah-Eddine M 2001 Crack front rotation and segmentation in mixed mode I+III or I+II+III. Part II: Comparison with experiments *J. Mech. Phys. Solids.* **49** 1421-43
[48] Gerstle W 1985 Finite and boundary element modelling of crack propagation in two and three dimensions using interactive computer graphics: Cornell University, PhD Thesis
[49] Sih G C 1974 Strain-energy-density factor applied to mixed mode crack problems *Int. J. Fract.* **10** 305-21
[50] Sih G C 1991 Mechanics of fracture initiation and propagation, Springer Netherlands.

IOP Conf. Series: Materials Science and Engineering **700** (2019) 012036 doi:10.1088/1757-899X/700/1/012036

Time-variant rule-based reliability of corroded structures by Monte Carlo simulation

K M Neumann[1,*], B Leira[2], O T Vårdal[3] and S Ehlers[4]

[1] Wood, Structural & Marine Technology department, Kokstadflaten 35, Bergen, Norway
[2] NTNU, Institute of Marine Technology, Otto Nielsens veg 10, Trondheim, Norway
[3] AHPA,Sildaberget 22a, 5171 Loddefjord, Norway
[4] TUHH, Institute for Ship Structural Design and Analysis, Am Schwarzenberg Campus 4 C, Hamburg, Germany

*Corresponding author: karoline.neumann@woodplc.com

Abstract. General widespread corrosion is of increasing concern for structural reliability of ageing marine structures, particularly for semi-submersible pontoons. Conditional failure rate, also termed hazard function is sought to describe the increasing maintenance required with age. Literature is reviewed connecting corrosion degradation to hazard rate, either through a time degrading limit state function, or statistical data for time to failure. Here, a rule based failure definition is applied based on corrosion tolerance levels, together with a linear corrosion degradation model to make a time variant limit state function. Monte Carlo simulation is applied based on statistical models for the basic variables including importance sampling, to obtain the time to failure distribution from which the hazard curve is deduced. A more realistic hazard function is also produced, requiring average of 3 simulations to exceed the tolerance in order to fail. It is seen that this approach delays the increase of the hazard curve, at the expense of higher failure rates.

1. Introduction

General widespread corrosion is a threat to structural reliability of ageing floating marine offshore structures. Particularly, the ballast tanks in pontoons of old semi-submersibles are of concern because they are not available for inspection or repair when in operation, and the environment is susceptible to corrosion. Additionally ballast tanks are closed spaces which have to be gas freed, and entering is associated with hazard. Sometimes scaffolding is necessary for close inspection, which is imposing increased coast and risk. A search on the ship-info.com database gives 42 results for drilling rigs, and 12 results for accommodation rigs built prior to 1999. These are 54 semi-submersibles which currently have exceeded their initial design period which is typically 20 years. Corrosion in ballast tanks may be kept at bay for long periods with coating and sacrificial anodes, despite this, general widespread corrosion is common in ageing pontoons. Considering that the coating system after several years gradually breaks down, and the corrosion from this time starts to progress at slightly various rates, more and more corrosion will exceed our tolerance limits. The objective is to describe the increasing failure rate using the hazard curve. Ageing can be described by the hazard function, often termed bath-tub curve, and is according to Ersdal poorly utilized or forgotten [1]. A report produced for the Norwegian Petroleum Safety Authority (PSA) states that additional information is needed on deterioration and its modeling to determine the form of the bath-tub curve, and the start of wear out phase [2]. It will be analysed if the shape of the hazard curve will follow the usual bath-tub shape with a probabilistic linear corrosion model. The hazard curve for a corroding structure will be obtained by simulating the parameters coating breakdown, corrosion rate and original thickness by Monte Carlo simulation and calculating notional probability of failure through the limit state function, in Section 3. During operation of a unit observations can be gathered and used to update the distributions of the input parameters to increase the relevance of the prediction. The resulting hazard curve is presented in Section 4. First, a review is performed relating hazard function to corrosion through the application of time-variant structural reliability methods.

2. Structural reliability methods applied on corrosion

Garbatov and Guedes Soares analysed a dataset of corrosion on tankers and bulk carriers subject to permissible tolerance levels to obtain time to failure, T. Reliability $L = P(T \leq t)$ was presented as a function of time, and used to decide between different maintenance strategies [3]. Barone and Frangopol considers a bridge described by a limit state including corrosion degradation, and performs bi-objective optimization between total maintenance cost and several key performance indicators (kpi), one of them being the hazard function, h(t) [4]. It is found that hazard function gives similar results as availability as kpi, but hazard function is computationally cheaper [4]. Akpan considers general corrosion degradation of the ultimate hull girder bending moment capacity as limit state, and presents failure probability considering varying corrosion distributions and parameters [5] [6]. The reliability or probability of survival, L, is calculated through the hazard function, h(t), according to Equation 1. The deduction of this formula is presented in Rausand and Høyland's book page 19 [7]. Ellingwood and Mori developed the model of ageing structural reliability of concrete components to assess probability of failure with time in nuclear plants [8]. The hazard function was presented for various degradation models and parameters. Moan and Ayala-Uraga incorporates the effect of corrosion on the fatigue limit state of components of a ship. The hazard rate and the cumulative probability of failure is presented, and it is concluded that the hazard rate is a better measure, and that not considering corrosion effects on fatigue is non-conservative [9].

$$L(t) = exp(-\int_0^t h(t)dt) \tag{1}$$

3. Method

The limit state is reached when crossing from the safe to the failure region, in other words when load effect S, exceeds resistance R. The difference between the resistance and the load effect is termed the safety margin, or just margin M, and describes the limit state function. An illustration of the time dependent limit state with constant limit S and time variant remaining thickness R(t) is seen in Figure 1. Cumulative probability of failure at time t, F(t) is defined by the probability that the limit state function or margin, M is less or equal to zero, or the probability that a failure will occur any time prior to t, according to Equation 2. The former is typically used when a theoretical model is available such that M can be computed, the latter is used when statistical data of time to failure, T of a number of components is available. The reliability is in any case L(t)=1-F(t). Here we will use a model to define M, according to Equation 3, and simulate time to failure using Monte Carlo simulation of the involved variables as will be explained in Section 3.2. Resistance, R is here remaining thickness, which is original thickness t_{orig} minus corrosion diminution, d(t), which will be elaborated in Section 3.1. Load effect S is a rule based discrete minimum allowable remaining thickness based on DNVGL-CG-0172 Thickness Diminution for Mobile Offshore Units [10]. For secondary category area for column-based units minimum 85% of original thickness remaining is required. This gives

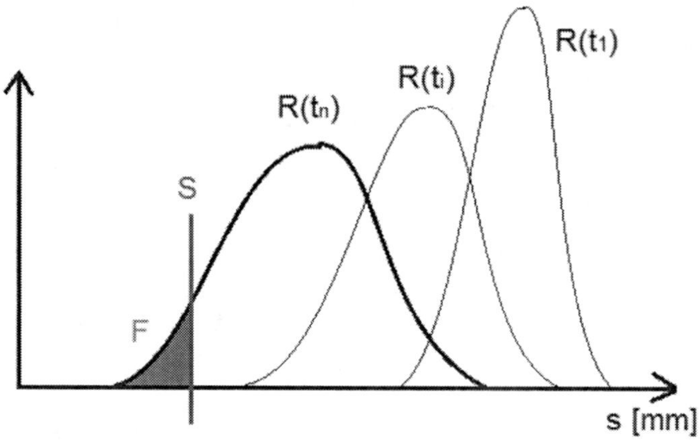

Figure 1. Illustration of limit state where S is load effect and R is resistance

a deterministic value of the acceptable limit of S =10.2mm for a 12mm thickness structure considered here. Different categories of structures such as primary or special would yield another percentage requirement, and other thicknesses would result in other discrete mm failure limits. For simplicity, only a 12 mm secondary structures is considered here. A secondary structure gives a higher allowance than primary or special members, so the results will give an optimistic scenario relative to other categories. The definition of failure here is corrosion diminution exceedance of this rule based limit, similar to Garbatov's failure definition based on tolerance levels, rather than the physical load effect [3]. Unlike ships where global failure modes of the ship hull beam can be defined, every failure mode of every component have to be separately addressed for the semi-submersible. To avoid this, the rules based limit is more practicable and more useful from a maintenance perspective. The notional cumulative failure probability, F for the given input is calculated by counting numbers of failures divided by total number of simulations in the Monte Carlo method as explained in Section 3.2. By derivation of F(t), the corresponding failure density f(t) can be calculated, termed the instantaneous failure rate in the interval $[-0.5\Delta t < t < 0.5\Delta t]$, which can be computed numerically for discrete years according to Equation 4. The conditional failure rate in the same interval, given that no failure has occurred prior to t is termed hazard function, and defined as in Equation 5, see eg. [7] [9].Another illustration of time-variant thickness and deterministic tolerance level is seen in Figure 2.

$$F(t) = P(M(t) \leq 0) = P(T \leq t) \tag{2}$$

$$M = R(t) - S = t_{orig} - d(t) - S \tag{3}$$

$$f(t) = F(t) - F(t-1) \tag{4}$$

$$h(t) = \frac{f(t)}{1 - F(t)} \tag{5}$$

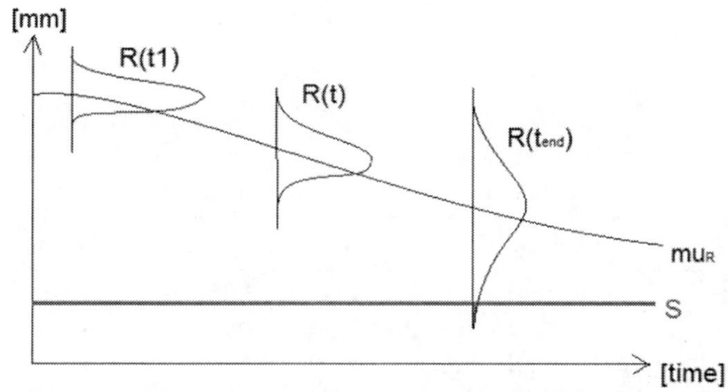

Figure 2. Illustration of time variant reliability in the case of corrosion degradation of resistance and constant rule based limit S.

3.1. Corrosion modeling and parameters

A simplification of Paik's model in Equation 6 is used to model the corrosion degradation with time t in years, with exponent C2 = 1 so that it presents a linear model, and transition time $T_t = 0$, so that corrosion is assumed to suddenly start progressing. This is in line with Melchers and Southwells suggestions and C1 equals corrosion rate [11] [12]. The applied model for thickness diminution d(t) in mm, is presented in Equation 7. Here, the original thickness t_{orig} is represented by the variable X1 in mm, the time of coating breakdown (Tc) by the variable X3 in years, and the corrosion rate c_{rate} by the variable X2 [mm/y]. Then Equation 3 can be written in the form of Equation 8. The applied target distributions for each variable is presented in Table 1. Original thickness is also modeled, since measurements of old corroded structures can show a substantial amount of 'increasing' thicknesses compared to nominal, such as the sunken MV Chester, or more than half of the readings in a campaign reported by Luque [13][14]. The mean is taken as the mid of upper and lower tolerances of class A steel of 12 mm according to EN10029, and the standard deviation of X1 is 1/3 of the tolerance span from the mean to each side at 1.7/2 mm (lower=-0.5, upper=+1.2). Then the tolerance is interpreted as 99.7 percentile which equals 3*standard deviation. Corrosion rate, X2, is also assumed to follow a truncated normal distribution.

The values are based on measurements of more than 1000 locations on uncoated ballast tanks of an old semi-submersible, measured at the same location with 5 years interval. The mean value may seem low at first glance, compared to literature on ships' ballast tanks, reporting from 0.043 mm/y to 0.1 mm/y [15] [13]. However, ballast tanks in ships are more often emptied and filled compared to semi-submersibles which maintain the same ballast for long periods on location to keep submerged. Hence smaller corrosion rates may be expected due to a less corrosive environment with less mixing of water and air. Also the applied standard deviation for X2 at 0.1 may seem high, but is taken the same as Yamamoto, although Wang reports values as low as 0.038 [16] [17]. Since the corrosion rate was deduced from measurements that were taken in an old structure, the high standard deviation is due to a highly uneven surface. Different distributions of coating breakdown time is suggested in the literature, such as a lognormal [18], [11] and normal [19], [20]. Here a truncated normal distribution will be applied for X3 with parameters as seen in Table 1 where mean and standard deviation are loosely inspired by [20]. Each parameter is simulated using Monte Carlo (MC) simulation, with importance sampling (IS) for variables X2, and X3.

$$d_{paik}(t) = C_1 * (t - T_c - T_t)^{C_2} \tag{6}$$

IOP Conf. Series: Materials Science and Engineering **700** (2019) 012036 doi:10.1088/1757-899X/700/1/012036

$$d(t) = (t - T_c) * c_{rate} = (t - X3) * X2 \tag{7}$$

$$M(t) = X1 - (t - X3) * X2 - S \tag{8}$$

Table 1. Target distributions and parameters of variables

Variable	Distribution	Mean	Standard deviation
X1 (original thickness)	Normal	12.35 mm	0.85/3
X2 (corrosion rate)	Normal truncated at 0	0.01 mm/y	0.1
X3 (coating breakdown)	Normal truncated at 0	9.5 y	3

3.2. Monte Carlo simulations

Brute Monte Carlo simulation involves choosing (pseudo) random values between 0 and 1 on a uniform distribution, and finding the corresponding value of each variable x by use of the inverse cumulative distribution function of the variable, as illustrated in Figure 3. The discrete combination of x values are used to calculate the margin. Each discrete calculation of M is checked for failure or no failure. After many simulations, probability of failure can be calculated by dividing number of failures by total number of simulations. This approach will be applied here, and is illustrated for 3 variables X1, X2 and X3 and constant S in Figure 1.

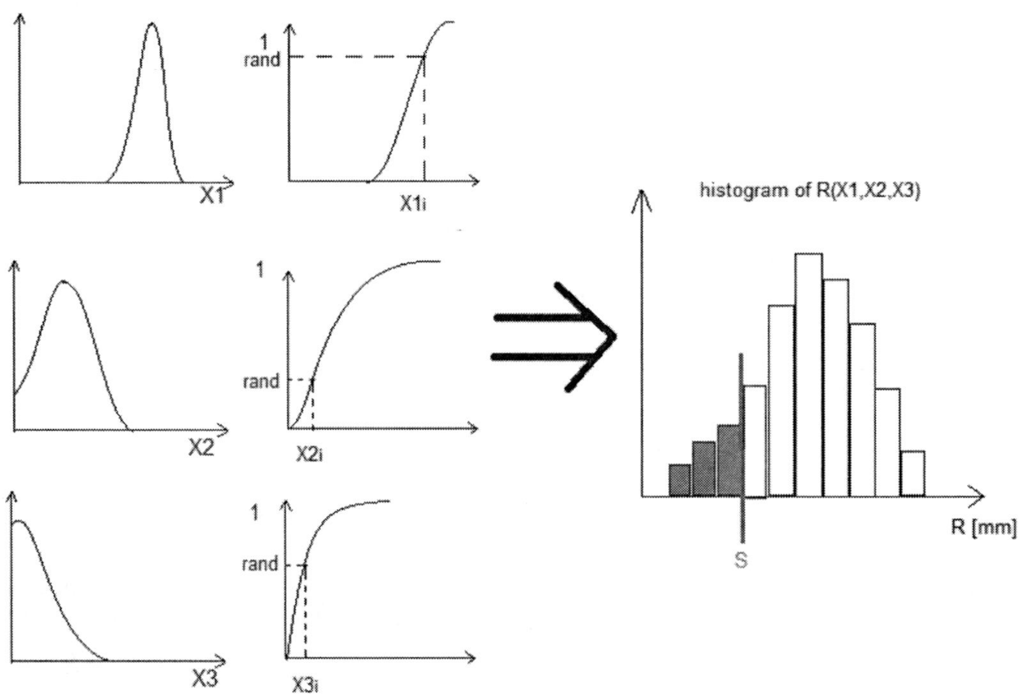

Figure 3. Illustration of Monte Carlo simulation of M with 3 variables and discrete S

3.2.1. Importance sampling Importance sampling is performed to get a better estimate of the small probabilities of failure. This is particular in the cases where X3 is small and X2 is large, or X3 is large and X2 small. Since these combinations are so unlikely, a usual Monte Carlo sampling from X2 and X3 target distributions would give us very little failed combinations in these cases. The solution is to sample from a different proposal distribution that will pick combinations both resulting in failure and safe combinations on a more equal basis, as will be seen later in Figure 8. When computing the probability of failure, F, the fact that the target distributions (which are the realistic ones) is not used has to be compensated for by a weighing factor p/q as in Equation 9, where p is the target distribution and q is the proposed distribution for weighting. Here, the proposed distributions for X2 and X3 are the same as for the target distributions, but with standard deviation 5 times as high. Figure 4 illustrated this for a few simulations for variable X2, showing a much more uniform distribution that will give more weight to higher corrosion rates than the target distribution.

$$F = \frac{1}{N}\sum_{n=1}^{N}(M \leq 0) \cdot \frac{p(x)}{q(x)} \tag{9}$$

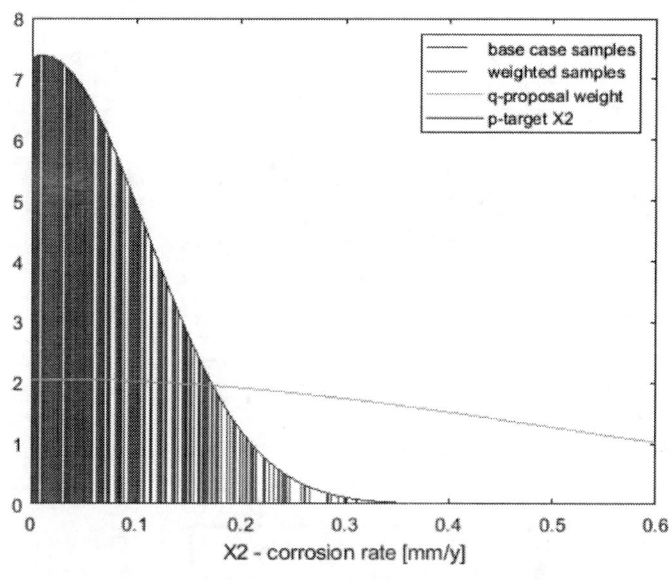

Figure 4. Importance sampling of variable X2, where q is the proposal distribution which IS is tanken from, and p is the target distribution of X2 given in Table 1.

3.3. Pseudo-code

A flow chart for the matlab script to calculate probability of failure is presented in Figure 5. First the the discrete values, S, original thickness, n iterations and y total years as well as counters i and t are initialized. Also the distribution for the variables X1, X2 and X3 is assigned, together with corresponding parameters according to Table 1. Using a pseudo-random algorithm, an array of n random realizations are generated for each variable, assuming they are all statistically independent. Then diminution d, remaining thickness R and margin M are calculated according to Equations 7 and 3 for the first set of simulated variables. If the margin

IOP Conf. Series: Materials Science and Engineering **700** (2019) 012036 doi:10.1088/1757-899X/700/1/012036

is less than 0, it is considered a failure and will be counted as $fail = fail + 1$, before it will increment to next simulated variable and perform the same check until M is calculated for all simulated combinations of variables. Then probability of failure for the given year is calculated as a relative sum of failures $F(t) = fail/n$. The probability of failure is cumulative with years and corresponds to the cumulative distribution function of the instantaneous failure density, calculated by Equation 4, hence the hazard function can be calculated according to Equation 5. Then the procedure is to repeat this for each year, since corrosion diminution d is time variant. The counter for failures, 'fail' will be set to zero as t iterates to the next year.When this is performed for all years, the loops end, and relevant figures are plotted.

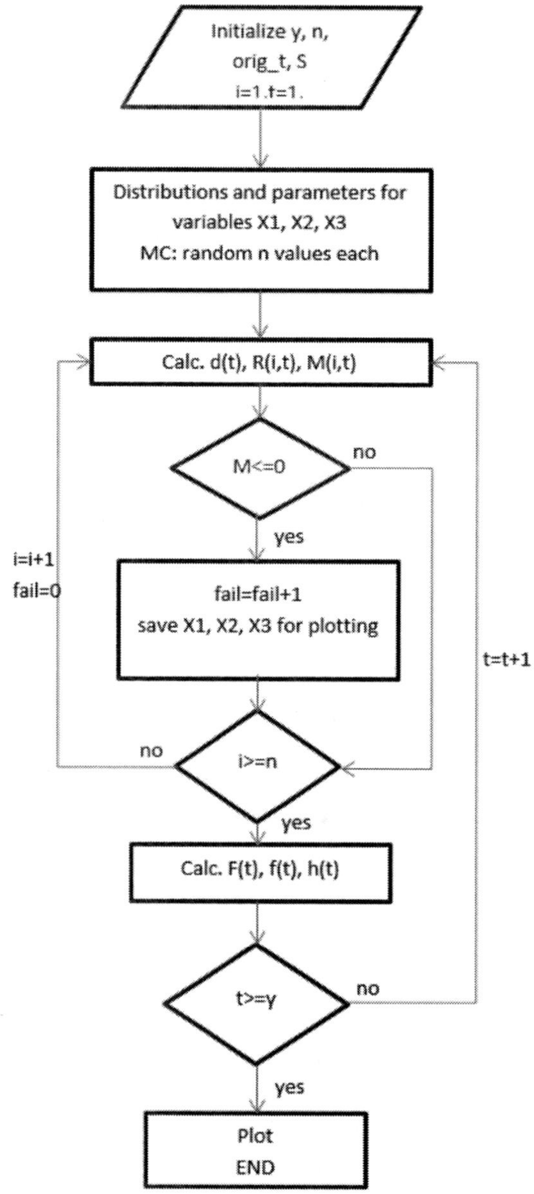

Figure 5. Flow chart for computation of F in matlab script

IOP Conf. Series: Materials Science and Engineering **700** (2019) 012036 doi:10.1088/1757-899X/700/1/012036

4. Results and discussion

4.1. Distribution of diminution and remaining thickness

An impression of the corrosion progression and distribution according to the simulation is presented in Figure 6, where the histogram is normalized to an area of 1. It can be seen that both the mean and the variance increases with time. The graph is truncated in the vertical direction to get a better look at the flatter distributions. The 5 years distribution is much higher near zero, explaining the seemingly small area. In Figure 7, simulated remaining thickness R is plotted for a selection of years. Again it is normalised, and cut in the vertical axis, as well as in the lower end of the horizontal axis. At year 40, the probability to measure thicknesses exceeding the original nominal thickness, 12 mm, $P(d = 0 \mid t = 40) = 0.09$. Similarly there is a 13% chance at 30 years and a 75% chance at 10 years of measuring values greater than the nominal thickness, with the values applied in this simulation.

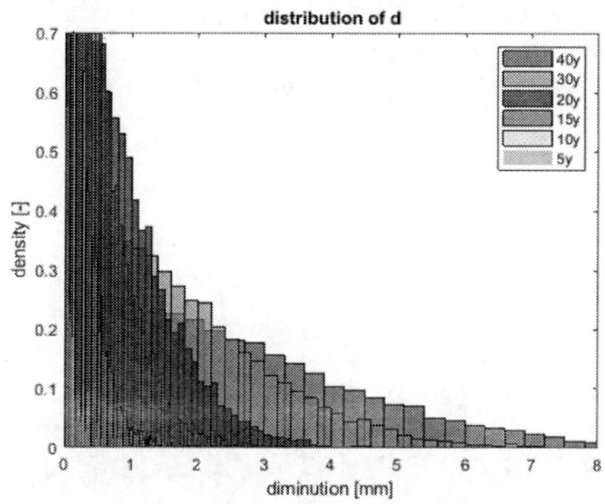

Figure 6. Simulated corrosion diminution d, for a selection of years.

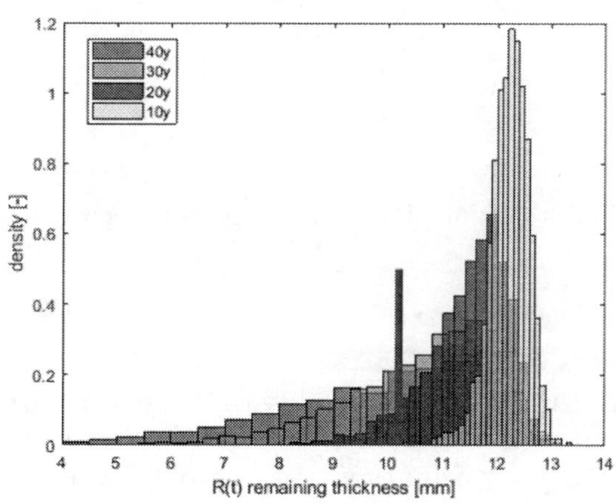

Figure 7. Simulated R, remaining thickness, and minimum allowable remaining thickness S is shown as a red line.

IOP Conf. Series: Materials Science and Engineering **700** (2019) 012036 doi:10.1088/1757-899X/700/1/012036

4.2. Failure regions
The failure planes in the physical plane for a diminution corresponding to 10,20,30 and 40 years is seen in Figure 8, together with 3000 samples simulated using importance sampling. IS samples from the proposed distributions are distributed more evenly between the failure regions and fall on both sides of the planes. Samples from the target distribution would have concentrated on the mean, giving few points close to the edges. Upper right corner represents safe side, and opposite side of the planes, towards left lower corner, represents failure region. As the years pass, the failure planes move closer and closer towards upper right corner, increasing the probability of failure. It can be seen that the planes lean slightly, in X1 axis, showing initially high original thicknesses survive slightly longer. In Figure 9 the failure planes are plotted in 2D, seen from above. Here safe region is towards lower right corner, and fail towards upper left. The leaning of the planes in X3 axis is then transformed to thickness of the lines representing variations in original thickness. It can be seen that for high corrosion rate, failure will occur almost immediately after coating breakdown Tc, while for low corrosion rates, failure is dominated more and more by time to coating breakdown as years pass. For the 10 year plane, et least the area spanned by 8 years coating breakdown or more, and less than 0.1 mm/y corrosion rate is safe. When 40 years have passed, at least all combinations with corrosion rate larger than 0.25 mm/y and coating breakdown less than 29 years have failed.

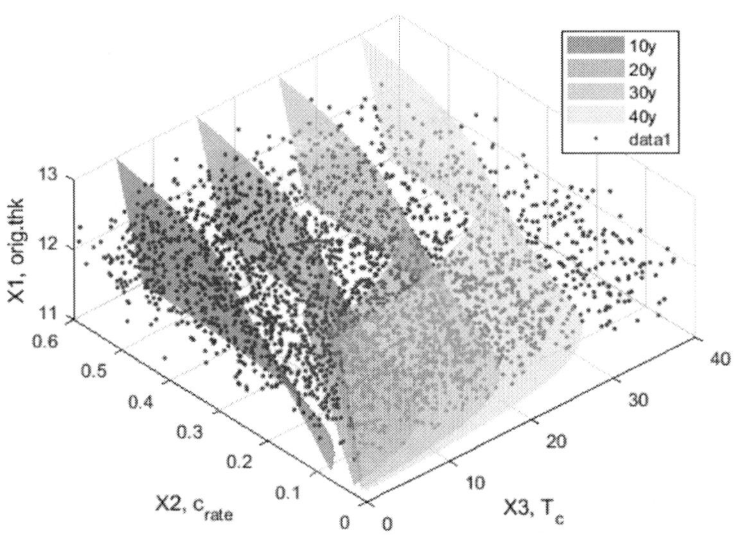

Figure 8. Failure planes in physical space together with IS samples

4.3. Time-variant probability of failure
The simulated time-variant cumulative probability of failure is shown in Figure 10, using 3 million simulations. This graphs shows that the failures are slow in the beginning, only starting after 10 years. Then the number of failures increases rapidly, reaching an inflection point before 25 years. This inflection point corresponds to the peak of the instantaneous unconditional failure rate, f(t), which is plotted in Figure 11 together with the instantaneous conditional failure rate, h(t), called hazard function. While f(t) δt is number of failures in interval δt divided by total simulations, h(t) δt is number of failures in the same interval, divided by number of simulations

IOP Conf. Series: Materials Science and Engineering **700** (2019) 012036 doi:10.1088/1757-899X/700/1/012036

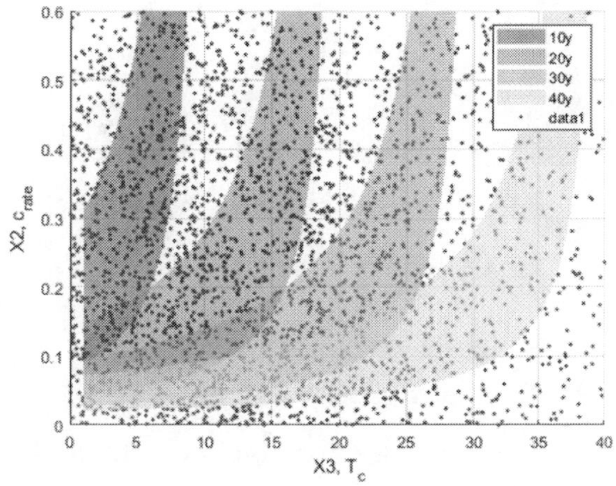

Figure 9. Failure planes in 2D in physical space together with IS samples

that have not failed yet. Failure as referred to until now represent too low individual thicknesses.

For a more useful case, the failure criterion can be defined when average of three thicknesses are below the rule-based limit, S, hereafter termed the average of three definition of failure. The three measurements are spaced equally in a component of size in the range of meters and are considered independent. According to Melchers, independent samples of corroded thickness are obtained already after 5 mm spacing, while a more recent study finds no correlation over 40 mm spacing [21] [22]. In practice the average of three approach may be the case in the yard. Measuring three thicknesses at a plate, stiffener or any other component reduces the uncertainty. A component failing according to this definition requires remediation or repair. This average of three definition is used when producing Figure 12. This Figure is shifted in relation to Figure 11, meaning that in practice, although individual measurements start to exceed the regulatory limit around 10 years, this is still allowed, because single measurements does not represent the component as a whole. Components regulatory failures are not starting until age of 15. Here only original components are accounted for, without regard of what happens to the failed components. It is seen that the average of three approach delays the onset of the hazard curve, at the expense of higher failure rates.

4.4. Errors and convergence

In Figure 13 the simulated probability of failure F is plotted as a function of number of simulations for selected years. For low amount of simulations, the graph is fluctuating. When reaching $3 * 10^6$ simulations the values have stabilised. Importance sampling contributes to this. The variance in the probability of failure is presented in Table 2 and is considered sufficiently low.

5. Discussion

Some may be surprised that the presented hazard curve do not have the shape of a bath-tub. This is due to the input, where all the corrosion is modeled according to the same distributions. In further work, multiple distributions for X1, X2 and X3 could be considered, representing different structures and corrosive areas. Then a few areas may be very prone to corrosion, which will all fail very early causing an early bump. The increasing wear-out part of the curve is

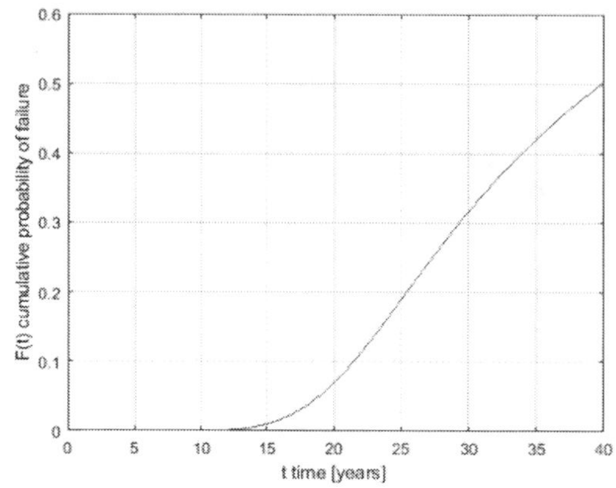

Figure 10. Time variant probability of failure, F, in [0 t].

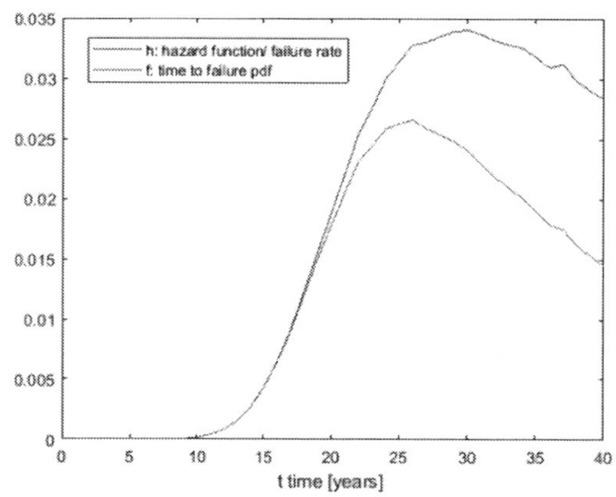

Figure 11. Instantanous failure density function f(t), and conditional failure rate h(t).

Table 2. Variance and probability of failure for selected years at 3 million simulations

Years	F	Variance
5	$1.4 * 10^-6$	$4.7 * 10^-13$
10	$9 * 10^-4$	$3.1 * 10^-10$
15	0.0221	$7.4 * 10^-9$
20	0.1184	$3.9 * 10^-8$
30	0.4006	$1.1 * 10^-7$
40	0.5749	$1.3 * 10^-7$

expected, but yet another surprise is the subsequent decreasing part. Does it mean that once the bump is passed, the maintenance can be decreased? Overall, most likely no. The presented hazard curve represents only the failure frequency of the original remaining parts that have not failed yet, not taking into account what happened to all those failures in the past. In reality,

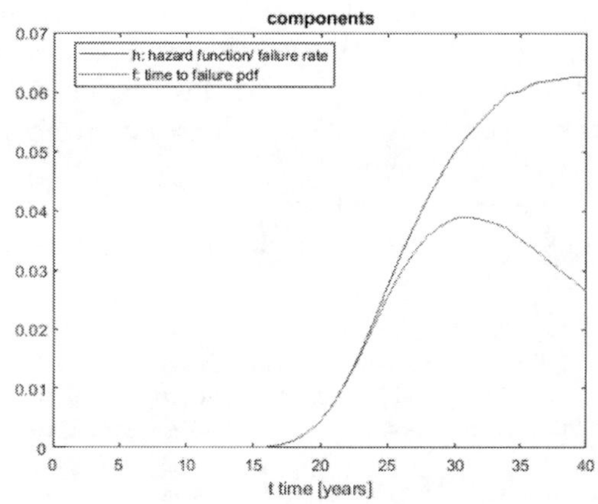

Figure 12. Instantanous failure density function f(t), and conditional failure rate h(t) for components failures, applying average of three failure definition.

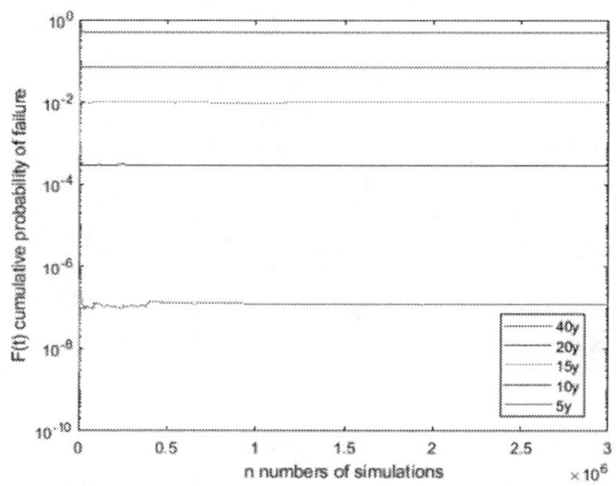

Figure 13. Probability of failure as a function of simulations for selected years

the failed parts were fixed, and by the time the presented hazard curve decreases, then the parts that were fixed early on probably starts failing again. These failures are not accounted for. However the required maintenance of the remaining non-failed parts past the bump may be decreased. Consider an example where, after 40 years, a plate in a protected void space may be shifted for the first time which is accounted for in the hazard rate, while in the same year several plates in the trim tank are replaced for the second time, which is not counted the second time in the hazard rate. Then again, if the previously failed components are fixed forever to never fail again, the hazard rate would decrease. In further studies the effect of combining various structural components, failure tolerances and corrosive environments could be investigated, in addition to how different corrosion models than the linear approach would affect the hazard curve.

6. Conclusion

The hazard rate is presented for a corroding semi-submersible pontoon applying Monte Carlo simulation to model the time variant probability of failure. It is seen that the average of three approach delays the onset of the increase of the hazard curve, at the expense of higher failure rates. The hazard curve may be a tool to help management decisions regarding future maintenance regimes or scrapping. It may also be used as a performance indicator for comparing assets, or optimizing cost. It can also be concluded that the expected number of measurements higher than nominal thickness is low for units exceeding their design life according to this model. It is proposed that further work may include the implication on hazard curve by using different corrosion models, including spatio-temporal descriptions. Also, various distributions could be applied for the variables X1, X2, X3 to represent different structures and corrosive environments.

References

[1] Ersdal G, Hörnlund E and Spilde H 2011 *Proc. of the ASME 2011 30th Int. Conf. on Ocean, Offshore and Arctic Engineering* January 2011 pp 517–522 ISBN 9780791844359
[2] Galbraith D and Sharp J 2007 Recommendations for Design Life Extension Regulations Tech. rep. PSA Stavanger
[3] Garbatov Y and Guedes Soares C 2010 *Maintenance planning for the decks of bulk carriers and tankers* (Taylor & Francis)
[4] Barone G and Frangopol D M 2014 *Struct. Saf.* **48** 40–50
[5] Akpan U O, Koko T S, Ayyub B and Dunbar T E 2003 *Nav. Eng. J.* **115** 37–48
[6] Akpan U O, Koko T S, Ayyub B and Dunbar T E 2002 *Mar. Struct* **15** 211–231 ISSN 09518339
[7] Rausand M and Høyland A *System reliability theory: models, statistical methods, and applications* 2nd ed (Hoboken, New Jersey: John Wiley & Sons) ISBN 3175723993
[8] Ellingwood B R and Mori Y 1993 *Nucl. Eng. Des* **142** 155–166 ISSN 00295493
[9] Moan T and Ayala-Uraga E 2008 *Reliab. Eng. Syst. Saf.* **93** 433–446 ISSN 09518320
[10] DNV-GL 2015 DNVGL-CG-0172 Thickness Diminution for Mobile Offshore Units
[11] Paik J K, Lee J M, Park Y I, Hwang J S and Kim C W 2003 *Mar. Struct* **16** 567–600 ISSN 09518339
[12] Melchers R E 2013 *Corros Sci* **68** 186–194 ISSN 0010938X
[13] Stambaugh K A and Knecht J C 1991 SSC-348 Corrosion experience data requirements Tech. rep. Ship Structural Committee Washington DC
[14] Luque J, Hamann R and Straub D 2014 *Proc. of the ASME 2014 33rd Int. Conf. on Ocean, Offshore and Arctic Engineering*
[15] Paik J K, Lee J M, Hwang J S and Park Y I 2003 *Mar. Technol.* **40** 201–217
[16] Yamamoto N and Ikegami K 1998 *J. Offshore Mech. Arct. Eng.* **120** 121–128
[17] Wang G, Spencer J and Sun H 2005 *J. Offshore Mech. Arct. Eng.* **127** 167–174 ISSN 08927219
[18] Guo J, Wang G, Ivanov L and Perakis A N 2008 *Mar. Struct* **21** 402–419 ISSN 09518339
[19] Qin S and Cui W 2003 *Mar. Struct.* **16** 15–34 ISSN 09518339
[20] Melchers R E and Jiang X 2006 *Ships Offshore Struc* **1** 61–70 ISSN 1744-5302
[21] Melchers R, Ahammed M, Jeffrey R and Simundic G 2010 *Mar. Struct.* **23** 274–287 ISSN 09518339
[22] Neumann K M and Ehlers S 2019 *Proc. of the ASME 2019 38th Int. Conf. on Ocean, Offshore and Arctic Engineering* (Glasgow: ASME)

A web based solution to track trawl vessel activities over pipelines in Norwegian Continental Shelf

M Farmanbar[1], A Palanisamy[1,*], A B Høydal[2], A Keprate[2] and G Haug[2]

[1] Faculty of Science and Technology, Department of Electrical Engineering and Computer Science, University of Stavanger, 4036, Stavanger, Norway
[2] DNV GL, Høvik, Norway

[*] Corresponding author: a.palanisamy@stud.uis.no

Abstract. Vessel activities such as trawling and anchoring potentially affect pipeline integrity. Therefore, the detailed information about the trawl activity in the area is essential for accurate assessment of pipeline, where to inspect and where to implement corrective intervention, based on up to date trawling intensity and equipment used. The main contribution of this paper is to present a detailed analysis of trawling vessels activities based on the automatic identification system (AIS) data combined with DNV GL ship register, data from HIS Maritime World Register of Ships and information gathered on the trawl equipment. In this aspect, an interactive web-based solution has been designed and developed to monitor and track trawl vessel activities in the Norwegian Continental Shelf which can be used for assessing integrity of pipelines. Data Analysis and visualization techniques have been utilized to transfer high-dimensional data by encoding it as visual objects contained in graphical presentations. The extracted knowledge result in the density mapping, vessel tracking, computing crossing points of trawl vessel paths, computing a buffered polygon around the pipeline for the chosen threshold and cross-over statistics of pipeline and vessels, etc. The use of such a framework is mostly advantageous for the port authorities, marine transportation, relevant fisheries management authority, etc.

1. Introduction

Trawling is an approach for fishing that includes pulling a fishing net through the water behind at least one vessel. Trawlers change in size from little open vessels with as meagre as 30 hp (22 kW) motors to enormous processing plant trawlers with more than 10,000 hp (7.5 MW). Trawling is dubious in view of its effects on the subsea condition including coral breaking, harm to environments and expulsion of ocean growth [1]. Trawling can be classified into mid-water trawling also called pelagic trawling and bottom trawling [2].Bottom trawling is of worry to subsea structures and pipelines as both offshore oil and fishing ventures are regularly working in similar zones. As a result, a lot of attention have been paid to the safety and security of marine transportation. Currently, vessels are equipped by Automatic Identification System (AIS), sonar, radars and other conventional aid-navigation equipment assisting the deck officers in making rational judgements to improve the navigational safety and security [3]. However, such equipments are not always reliable to ensure safety in maritime navigation.

Data visualization techniques have been utilized to communicate data or information by encoding it as visual objects (e.g., charts, lines or bars) contained in graphical presentations. Data visualization and information visualization are useful for referring to any visual representation of data. The objective of visualization systems is to reveal the structure within massive high-dimensional data clearly and efficiently to users acquire a qualitative information knowledge and better identifying patterns and detecting outliers. As for maritime transportation, several exciting research in the field of simulation of marine traffic have been presented [4-9]. A framework for maritime data analysis and simulation has been developed in [4]. The proposed framework analyzes historical data about ships and area conditions, of varying time and space granularity, measures critical parameters that could influence the levels of hazard in certain regions and clusters such data according to their similarity. Data visualization methods in the areas of spatiotemporal vessel trajectories and vessel traffic density were extensively investigated within the short review [5]. Kernel Density Estimation-based density visualization, massive vessel density visualization, visualization of fishing activities and visualization-guided ship safety domain have been fully investigated.

A probabilistic model is proposed in [6] to analyze the risk of two frequent types of marine accidents, such as collision and grounding. The probability of vessel colliding is evaluated in terms of a model based on minimal distance to collision. The model uses a mathematical vessel motion model to define the collision zone and considers traffic flow as a non-homogeneous process. Similar approach is followed for the probabilistic assessment of grounding. Furthermore, Montewka et al. [7] presented a simulation model that efforts to evaluate the potential impact of deepening the terminals and ports of Delaware River and Bay areas, by considering multiple parameters such as tide, navigation, terminal and anchorage rules as well as vessel profiles. A procedure for processing, analyzing, and visualization of AIS data with example outputs and their potential uses is demonstrated in the work of [8]. The knowledge is extracted in terms of density mapping, vessel tracking, interpolations of vessel dimensions, and ship type analysis. Willems et al. [9] demonstrated the visual presentation of the AIS data to assist operators of coastal surveillance systems and decision-making analysts to have a comprehensive understanding of the vessel movements, thus improving monitoring effectiveness. A noteworthy remark is that amongst the reviewed publications, a major common point was identified: that the information service depends heavily on the quality of AIS information and information processing, integration, and presentation applications. It has also been shown that AIS alone has never been sufficient to define a complete, extensive and a full, comprehensive maritime picture.

The present research focuses on the Norwegian Continental Shelf, the continental shelf that Norway exercises sovereign privileges as established by the United Nations Convention on the Law of the Sea. The shelf region is four times the metropolitan area of Norway and represents about one-third of the continental shelf of Europe. It is wealthy in oil and gas and is the basis of Norway's petroleum economy. Currently, the Norwegian Petroleum Directorate (NPD) regulations demand that all subsea establishments on the Norwegian division of the North Sea be planned with the goal that angling rigging won't be harmed [1]. During ongoing years, it has been recorded that the pattern for trawl apparatus plan and weight has expanded. Especially the utilization of cluster loads frequently expands the proficiency and is relied upon to be famous and regular later. New submarine pipelines should be structured by as of late utilized trawl apparatuses and recently introduced pipelines exposed to obstruction must be re-examined [1]. Towards this direction, DNV GL consistently utilizes AIS information and vessel explicit data on trawl gear as a contribution for breaking down potential hazard to marine structures which begins from ship action. The fundamental driver of the new technique is to secure the pipelines, outsiders, for example, trawl vessels and the earth by guaranteeing safe plan and task of pipelines [2].

The main objectives of this paper are summarized as follows: 1) Develop an interactive web-based solution to track and monitor trawl vessel activities in the Norwegian Continental Shelf which can be used for assessing integrity of pipelines for a chosen area of interest. 2) Implement web scrappers to automate the process of collecting vessel details. The remainder of this article is organized as follows: Section 2 briefly presents the datasets used in this work. Section 3 details the proposed web solution and data visualization system, and Section 4 describes our experiments based on available use cases and the last section concludes this study.

2. Datasets

The AIS data [10] used in this study is combined with DNV GL ship register, data from HIS Maritime World Register of Ships and information gathered on the trawl equipment. These datasets are used to track trawl vessel activities over pipelines in Norwegian Continental Shelf. As the ship registers, do not contain information about trawling equipment and size, the ship owners/operators are contacted and asked to provide information about the type of trawling performed (bottom or pelagic), if they use single or double trawl and type, size and weight of equipment used. Density maps are then made based on categorization related to the risk to the pipeline from each type of equipment. The study area for this research is the Norwegian Continental Shelf and the overview is given in Figure 1.

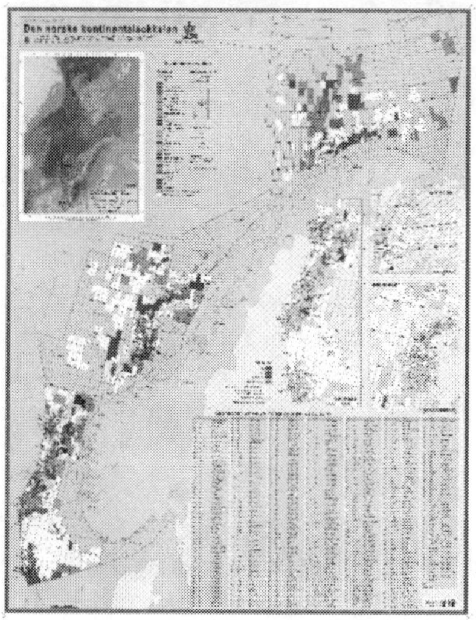

Figure 1. Norwegian Continental Shelf Area

Table 1. Pipeline Dataset Overview.

Field	Description	Type
Pipe Id	Unique Id of the pipe	int
Pipe Name	Name of the pipeline	text
Pipe Dimension	Dimension of the pipe	float
From Facility Id	Id of the starting pointing of pipeline facility	float
From Facility Name	Name of the starting pointing of pipeline facility	text
To Facility Id	Id of the ending pointing of pipeline facility	int
To Facility Name	Name of the ending pointing of pipeline facility	text
Pipe Operator Name	Name of the pipeline operator	text
Pipe Medium	Medium Information of the pipeline	text
Pipe Line Geometry	Geometry of location information of the pipeline	WKT

2.1 Pipeline dataset

This Dataset contains information of the pipelines constructed in the Norwegian Continental Shelf which are extracted from Norwegian Petroleum Directorate (NPD Facts) [11]. The data contains various information about pipes, such as the name, operator, pipe dimension, spatial information in WKT format. The original downloaded NPD dataset did not include any information regarding the type of pipelines, but it will be mostly trunk pipelines. In addition, DNV GL has discharged the DNVGL-RP-F114, which gives all-encompassing direction and proposals inside a geotechnical system for pipe-soil association assessments [1]. The overview of the data fields in pipeline dataset is shown in Table 1.

2.2 AIS dataset

This Dataset contains information of the vessel crossing in the Norwegian Continental Shelf for the period of 2013 to 2018 [10]. The data contains various information about the vessels such as the vessel name, mmsi number, imo number, utc time, latitude and longitude coordinates, etc.

2.3 Web scrapped vessel information

Web scraping technique is used to extract vessel information from various web sources. This module collects vessel specific information from the following five sites: 1) Norwegian Fisheries Directorate Website, 2) Marine Traffic Website, 3) Vessel Finder Website, 4) Proff.no, 5) 1881.no, 6) Google images.

3. Visualization of vessels trajectories and pipelines

This section presents a detailed insight regarding pre-processing, data transformation to track trawl activities as well as data visualization and visual mapping which can be used for assessing integrity of pipelines for a chosen area of interest. The overall workflow of the various components associated with the proposed web solution is demonstrated in Figure 2.

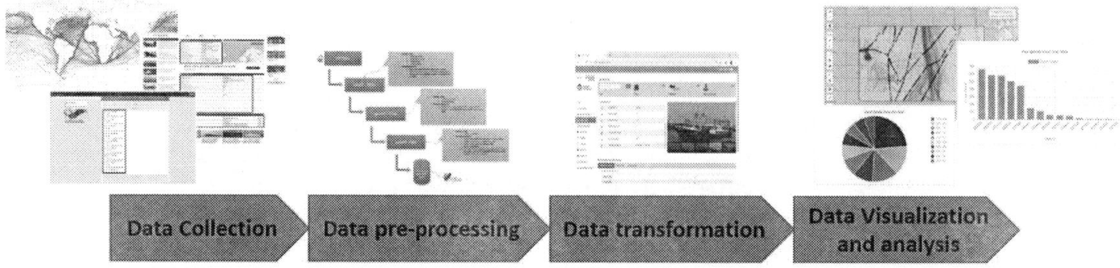

Figure 2. General processing pipeline for data visualization.

3.1 Data collection from different resources

We collected pipeline information and vessel information passing through Norwegian Continental Shelf area from various sources. Furthermore, additional information about crossing point information of trawler vessels over the pipelines, all trawl related statistics including clump door weight distribution, and trawl door weight distribution and KPI distribution are gathered. All this information is then stored in a Cross-Over Statistics Database in a SQLite3 database management system. Finally, we combine all gathered vessel details form different sources and categorize the vessels based on the availability of information according to the priority summarized as follows: 1) Equipment Source Info, 2) Fisheries Directorate Info, 3) Marine Traffic Info, 4) Vessel Finder Info.

3.2 Data transformation and pre-processing

In this step, we deal with data transmission errors and the problems of missing values to transform raw source data into desired standard format. It is well known that the raw maritime traffic data is often degraded with undesirable artifacts, i.e., transmission errors, missing values, duplicated records, outliers and conflicting values. For instance, there are some ambiguities in AIS data to identify unique vessels. First, automatically we drop all duplicates based on the Mmsi Id, IMO Number, Vessel Call Sign and Vessel Name columns. Thus, to group these duplicates, we create a primary key called DnvGl Id which can be used to identify the unique vessels. In addition, according to the specification from DNVGL-RP-F111 [1], trawling activity takes place at a speed of less than 6 knots. The raw AIS data is pre-processed to identify trawl vessel paths i.e. path traversed by the vessels. The vessel paths of the ships are split as: 1) Trawl Path - speed lesser than or equal to 6 knots. 2) Transit Path - speed greater than 6 knots.

The ship paths are then stored in WKT format for quicker access during the later period of computation. The last step of the data gathering approach is to update the internal database with information from the above process. It is worth noting that, data visualization and web-based dashboard often suffer from high computational cost due to the large volume of vessel trajectories. To reduce the computational cost, the size of the pre-processed files is reduced to a greater extent compared to the raw AIS data while preserving the crucial features.

3.3 Data visualization

The automation of data reporting plays an important role in achieving the goal of tracking trawl vessel activities over pipelines. In this step, data are converted into the appropriate visual structures for statistical graphs including line charts, bar charts and pie charts, etc. A sample screenshot of the vessel information page, containing the all vessel details is shown in the Figure 3.

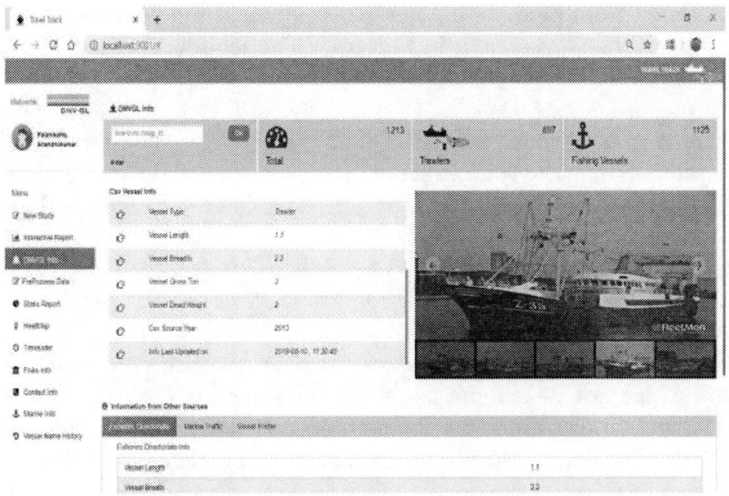

Figure 3. Web application - Vessel information.

3.4 Web-based visual mapping of vessel traffic density

We have developed a visualization framework for tracking trawl vessel activities over pipelines in Norwegian Continental Shelf which is capable of monitoring with high efficiency. By projecting the geographical position of trawl vessel activities into world map, the visual analytics of fishing activities have been analysed from global point of view. The interaction course is depicted in the two figures Figure 4. By presenting data visually, i.e., pipeline locations and vessels activities, it is of great convenience for relevant fisheries management authority to monitor and assess the activity of fishing vessels.

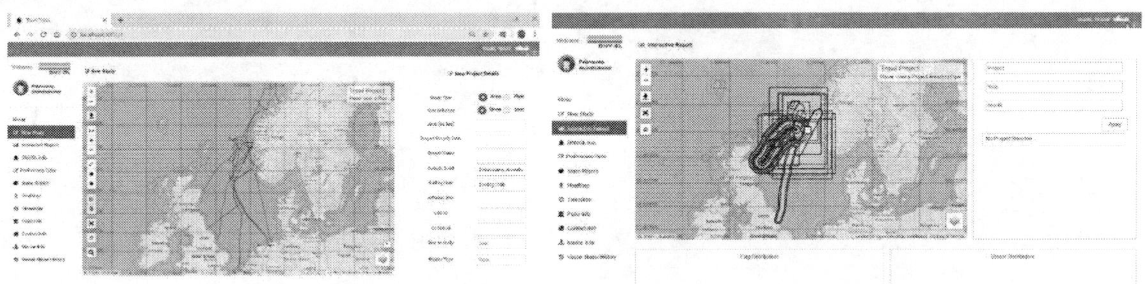

Figure 4. Web application – Dashboard.

4. Experimental use case

This section discusses the outcome of the experimental evaluation of the simulation web tool. The choice of creating new study may be either specific to any chosen area of interest or an area bounding a specific pipeline of interest. Therefore, we have conducted our experiments for one of the cases and present our analysis and discussion based upon the results.

Case study: Area Specific Study; the polygon area for the chosen area of interest can be specified in two ways by the web application: By using the drawing tool on the map or by manually entering the coordinates in the respective field.

Step 1: Selecting the Polygon Area. This is a user interactive tool and hence as soon as we have specified our polygon selection, the area covered by the polygon will be available in the respective field (or) will be shown on hovering over the polygon.

Step 2: Extracting pipelines passing through the polygon: The path of the pipeline lying inside the chosen polygon is then extracted from pipeline dataset. The extracted pipeline paths inside the polygon area will be highlighted in red lines as shown in Figure 5. The lengths of each pipeline path will then be available in the respective fields.

Step 3: Extracting trawl vessel paths passing through the polygon: These pipelines paths are extracted from the pre-processed AIS dataset. The extracted trawl paths inside the polygon area will be highlighted in green lines as shown in Figure 5. For this experiment we have selected a polygon with the following specifications:
Polygon Coordinates: [[60.08,1.48], [58.92,1.48], [58.92,4.76], [60.08,4.76]], Polygon Area: 22588.69 km2, Years Studied: 2015, Total Pipelines under this area: 15

Step 4: Density Map: Crossing points of trawl vessel paths over the pipelines are then computed and will be highlighted in blue dots. User can use the layer filters to view the chosen layer of interest as shown in Figure 6. User can also filter out the path of a particular vessel using the vessel path search functionality. These density maps also provide a detailed inference about the most affected and least affected zones and necessary measures can be taken to alert inspection on those areas. For example, in this experiment, a dense narrow strip of trawl vessel paths can be visualized, indicating a suggestion for inspection of pipelines along that area.

Step 5: Visualizing and statistical analysis of the vessels distribution: Overview of top 10 ships which are doing frequent trawl activity over the pipelines in the chosen area of interest can be obtained in the proposed framework as shown in Figure 7. This indicates a suggestion of focusing more in monitoring the activity of these vessels. In addition, an overview about the months which exhibit more trawling activity over the pipelines by the top 10 vessels and overview about the pipelines which exhibit more trawling activity by the trawler vessels can be plotted as shown in Figures 8 - 10.

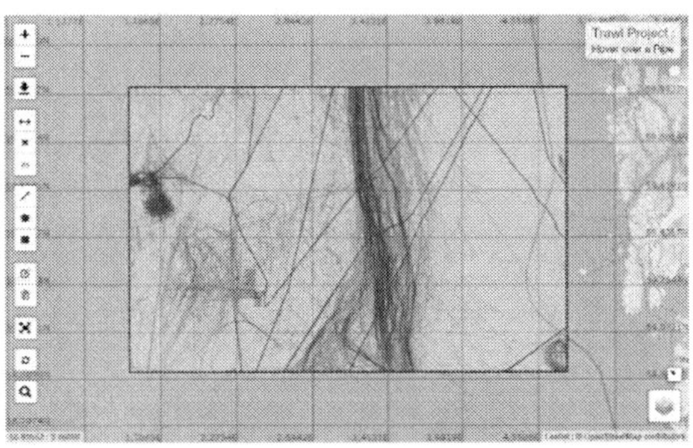

Figure 5. Pipelines and trawl vessel paths passing through the polygon.

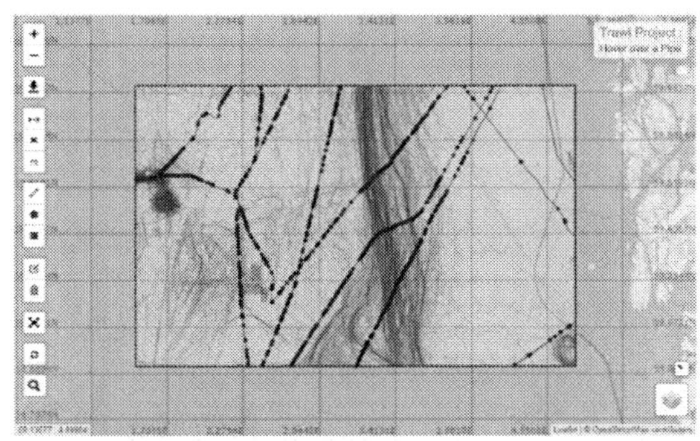

Figure 6. Crossing points of trawl vessel paths over the pipelines.

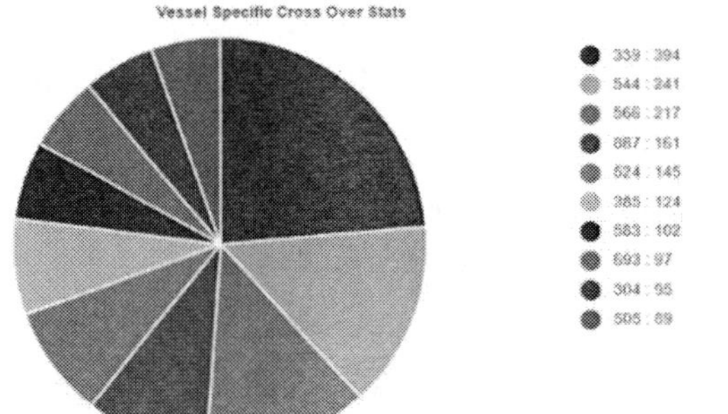

Figure 7. Top 10 ships which are doing frequent trawl activity.

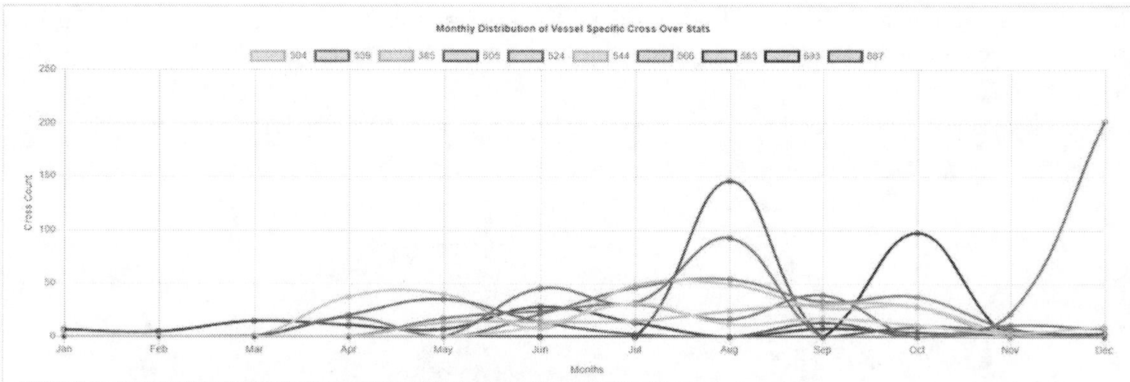

Figure 8. Monthly distribution of top 10 vessels.

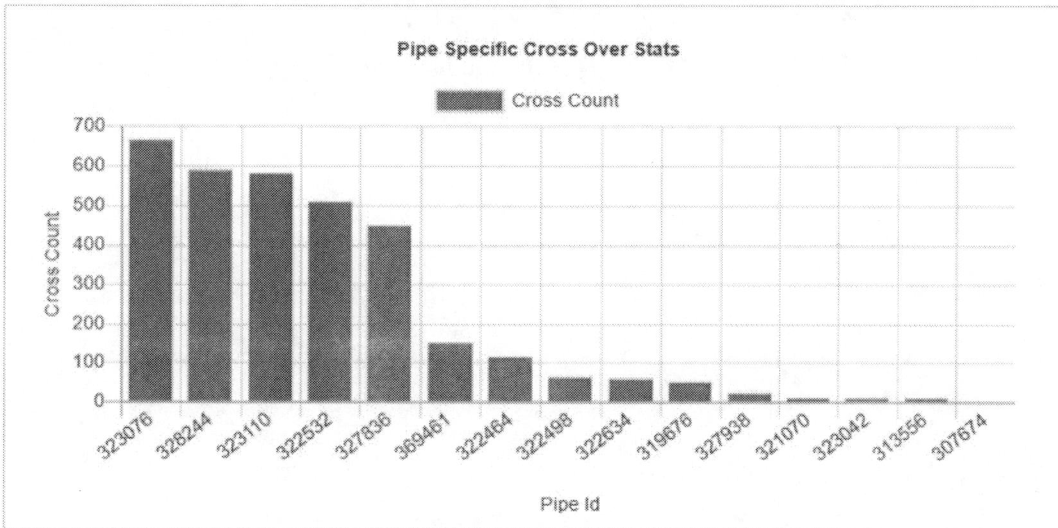

Figure 9. The cross over counts over pipes.

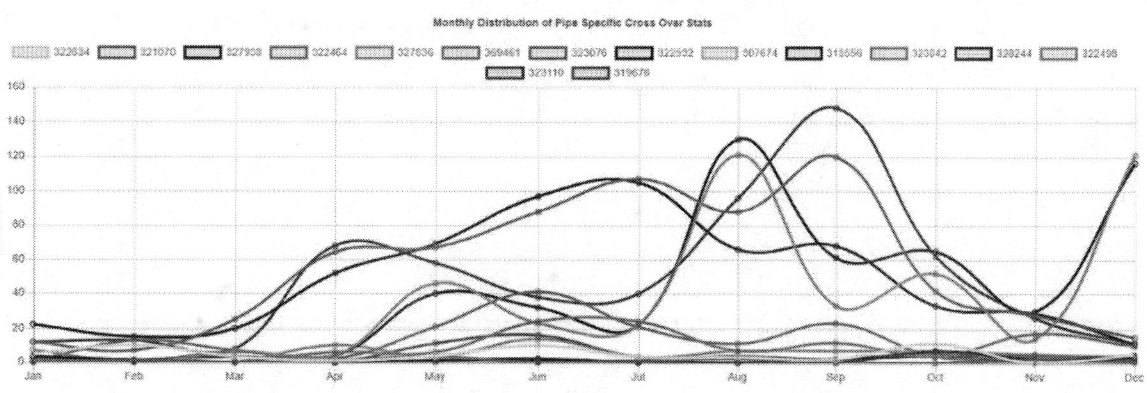

Figure 10. Monthly distribution of pipes based on the cross over counts.

Overview about the cross over statistics of trawler vessels for each km along the pipeline, in the chosen area of interest can also be available. This graph indicates a suggestion of kilometre intervals along the pipelines that need more focus in monitoring the integrity of the pipeline shown in Figure 11 and Figure 12. The key performance indicator (KPI) distribution of all pipelines inside the polygon is available

IOP Conf. Series: Materials Science and Engineering **700** (2019) 012037 doi:10.1088/1757-899X/700/1/012037

through a drop-down filter. The starting km of the KPI is measured by computing the difference between the starting coordinates of the whole pipeline with the starting coordinates of the pipeline inside the polygon. The KPI distribution covers the length of the pipeline that lies inside the chosen polygon area.

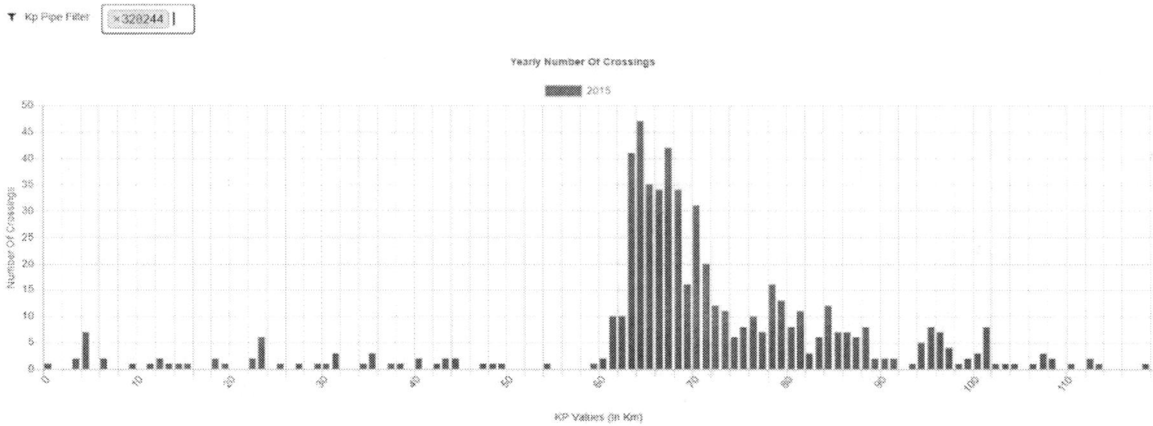

Figure 11. KPI Distribution for a sample pipe

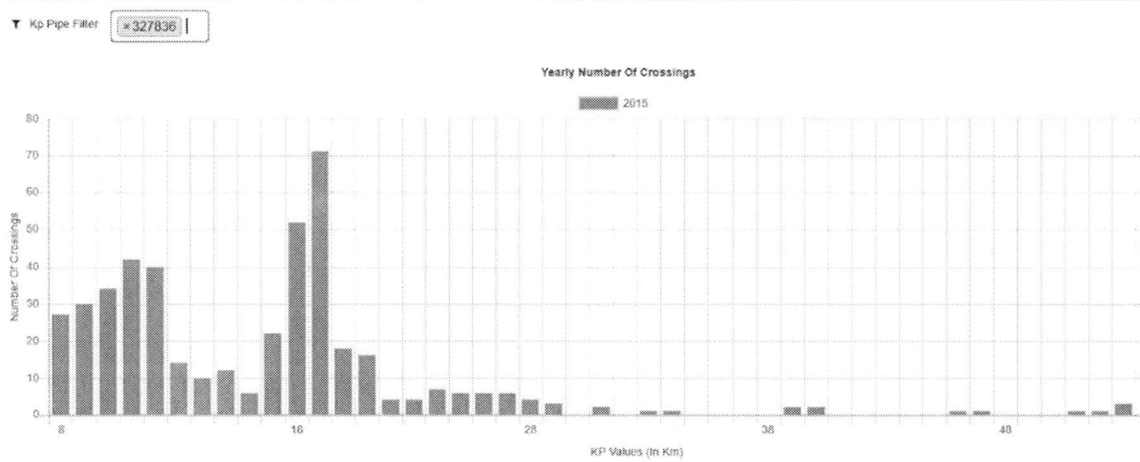

Figure 12. KPI Distribution for another sample pipe

Trawl Door Weight Distribution, Pelagic Door Weight Distribution, Door Weight Range Distribution and Clump Weight Range Distribution can be also seen in our web tool Figures 13 - 16.

IOP Conf. Series: Materials Science and Engineering **700** (2019) 012037 doi:10.1088/1757-899X/700/1/012037

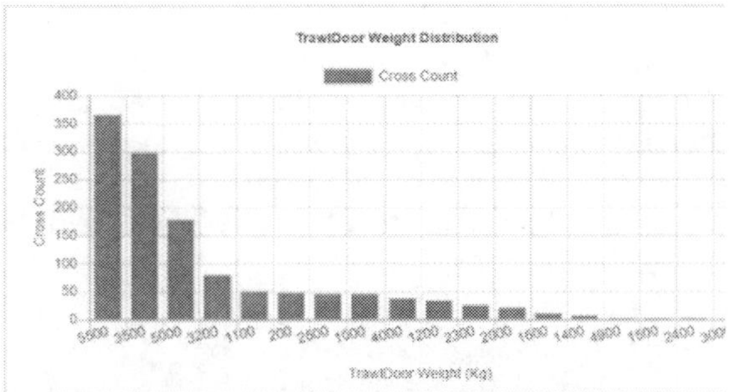

Figure 13. Trawl door weight distribution

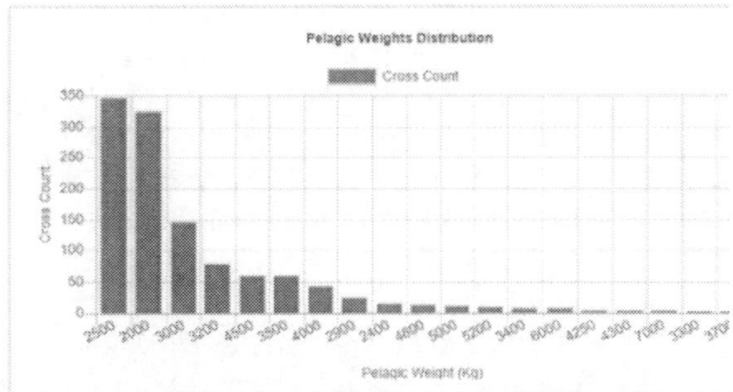

Figure 14. Pelagic door weight distribution

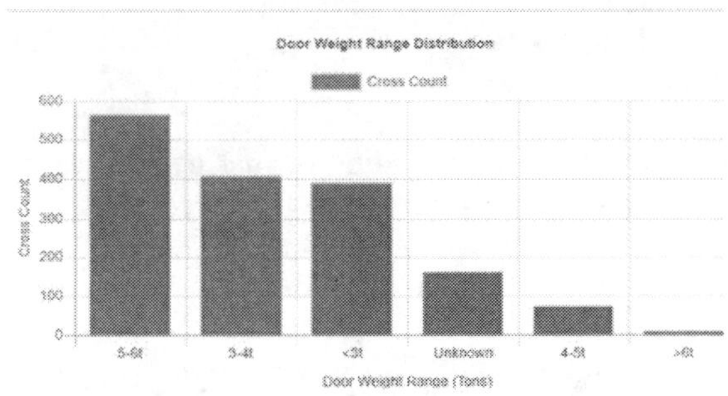

Figure 15. Door weight range distribution

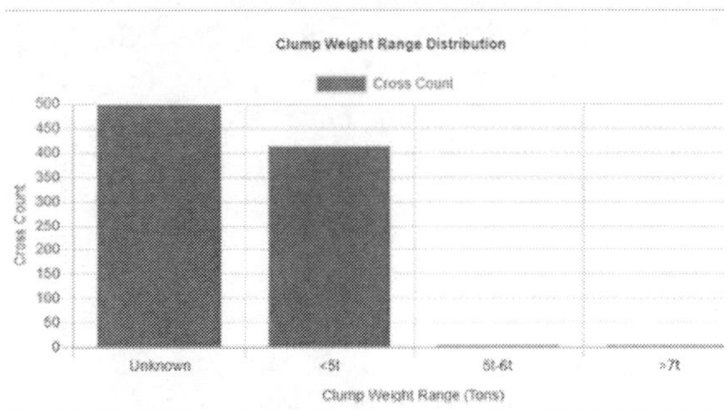

Figure 16. Clump weight range distribution

5. Conclusion

Monitoring and tracking the marine and transportation data are of major importance and the main goal of the current research. The advances of data analytics and visualization methods could be exploited for monitoring vessels activities over the pipelines to discover the knowledge from vast number of vessels trajectories. In this direction, we have developed a visualization framework for tracking trawl vessel activities over pipelines in Norwegian Continental Shelf which is able to achieve highly efficient monitoring. The AIS system, along with other sensor data such as pipeline dataset provided by DNVGL-RP-F114 can fully support the above actions. Existing and recently developed solutions have been investigated in order to create an efficient framework. Based on all the above arguments, it is mature to state that the proposed work, is utilizing data mining techniques, requiring less programming and deploying effort from the user's perspective, recent web technologies and Information and Communications Technology (ICT) services in order to ensure effective modeling, analysis and simulation of marine traffic data and provide experts with useful insight on the data.

6. Acknowledgments

This work was fully supported by DNV GL (Norway).

References

[1] DNVGL-RP-F111 2014, *Interference between trawl gear and pipelines*. 2014.

[2] DNV-OS-F201, *Offshore Standard Submarine Pipeline Systems*. 2000.

[3] Tu E, Zhang G, et al. 2018 Exploiting AIS Data for Intelligent Maritime Navigation: A Comprehensive Survey From Data to Methodology. *IEEE Trans. Intell. Transp. Syst.* **19**(5) 1559-82. *doi:10.1109/tits.2017.2724551*

[4] Maragoudakis M 2019 Data analysis, simulation and visualization for environmentally safe maritime data, *Algorithms,* **12**(1) 27. *doi:10.3390/a12010027*

[5] Wang K, Liang M, et al. 2019 Maritime traffic data visualization: A brief review. *IEEE 4th Int. Conf. Big Data Analytics (ICBDA)*. 2019. *doi:10.1109/icbda.2019.8713227*

[6] Montewka J, Krata P, et al. 2011 Marine traffic risk modelling – an innovative approach and a case study. *Proc. Inst. Mech. Eng., Part O: J. Risk Rel.* **225**(3) 307-22. *doi:10.1177/1748006X11399988*

[7] Almaz O A and Altiok T 2012 Simulation modeling of the vessel traffic in Delaware River: Impact of deepening on port performance, *Simul. Modell. Pract. Theory*, **22** 146-65. *doi:10.1016/j.simpat.2011.12.004*

[8] Shelmerdine R L 2015 Teasing out the detail: How our understanding of marine AIS data can better inform industries, developments, and planning. Mar. Policy, **56** 17-25. *doi:10.1016/j.marpol.2014.12.010*

[9] Wijk N W H V D W J J V 2009 Visualization of vessel movements. *Comput. Graphic Forum* **28**(3) 959-66. *doi:10.1111/j.1467-8659.2009.01440.x*

[10] Haug G, Høydal A B, Bjørgen H P, Rodrigues de Miranda D 2017, *Pipelines and trawling Risk reduction from detailed assessments of vessel activities*, in *Offshore Pipeline Technology (OPT)*: Amsterdam.

[11] Directorate N P 2005 *Norwegian Petroleum Directorate,*
Available at: *https://www.npd.no/en/facts/publications/.*

An efficient approach for ship collision design of reinforced concrete pontoon walls

Y Sha [1,*], J Amdahl [2] and K Yang [3]

[1] Department of Mechanical and Structural Engineering and Materials Science, University of Stavanger, Stavanger, Norway
[2] Department of Marine Technology, Norwegian University of Science and Technology, Trondheim, Norway
[3] Department of Energy and Process Engineering, Norwegian University of Science and Technology, Trondheim, Norway

[*]Corresponding author: yanyan.sha@uis.no

Abstract. This paper presents an efficient method for collision design of prestressed concrete pontoon walls. High fidelity finite element models of a container ship bow and a prestressed concrete pontoon wall are first established. Integrated numerical simulations are conducted to study the structural deformation of the pontoon wall during the collision. Parametric studies are also carried out to investigate the effect of the pontoon wall thickness. Based on the failure mode of the pontoon wall, a punching shear check procedure is proposed. This method yields a good accuracy while significantly reduces computational efforts. Hence, the proposed approach can be used for collision analysis in the preliminary design phase of floating pontoons.

1. Introduction

The Norwegian Public Roads Administration is planning to build a highway along the west coast of Norway. New installations must be constructed to cross the wide and deep fjords along the route. Convectional structures with fixed foundations cannot be constructed and novel floating bridges should be designed and constructed alternatively [1-4]. For floating bridges, one of the main concerns is the pontoon safety against accidental ship collisions. Ship collisions may result in structural damage in the pontoon wall and subsequent flooding in the pontoon compartments [5, 6]. Further, progressive collapse of the bridge may also occur due to the excessive flooding in the pontoon.

Floating support mudules, i.e. pontoons, pylons, platforms, are commonly constructed with reinforced concrete or steel material. Compared with steel floating foundations, reinforced concrete floaters are easier to construct and the cost is also generally lower. Therefore, reinforced concrete structures are widely used to construct large scale floating structures. However, normal strength concrete floating structures are vulnerable to accidental ship bow collisions. High strength ship bows may induce excessive punching shear damage in the concrete structures [6]. Consequently, progressive collapse may occur to the whole floating structure if the floating unit is flooded. Therefore, it is important to design the floating concrete units against accidental ship collisions.

Three approaches are generally used for ship collision analysis and design of concrete structures. Experimental tests are the most direct approach for obtaining the collision force and structural damage during a collision [7]. However, this method is seldom conducted due to site limitations and cost.

Another approach which is widely used for ship collision is numerical analysis. Reliable results can be obtained with validated numerical models [8-10]. Numerical analysis, however, requires significant modelling and simulation efforts. Hence, it is not suitable for practical use in the preliminary design phase. Validated analytical method can provide efficient and accurate results and thus can be used efficiently in practical engineering practice [11, 12].

This paper aims to investigate the ship collision response of RC pontoon walls and propose simple but reliable design check procedures for both structures. In this work, high fidelity finite element models of an RC pontoon wall and a container ship bow are first established. The impact demand and failure mode are identified for RC pontoon walls, respectively. Based on the failure mode of the pontoon walls, a simple design check procedure, which can be used in the preliminary design phase, is proposed for RC pontoon walls.

2. Finite element models

In this study, finite element models of a reinforced concrete pontoon wall and a container ship bow are established based on the structural drawings for numerical collision analyses.

2.1. Pontoon wall model

The pontoon model is established for a cable-stayed continuous floating bridge as shown in Figure 1 [13]. The cable-stayed navigation span in the middle is supported by two towers resting on two main pontoons. These two pontoons are particularly exposed to the passing ships. Hence, the two main pontoons have a higher possibility of subjecting to accidental ship collisions compared to other side pontoons. In addition, they are essential supporting structures regarding the overall stability of the whole bridge. Hence, the focus is placed on the collision analysis of the main pontoons in this study.

Figure 1. Rendering of the floating bridge concept.

The main pontoons are divided into sub-compartments so that the damage of one or two compartments shall not lead to overall flooding of the pontoon. A finite element model of the pontoon was developed based on this prototype. It should be noted that only the front wall of the first middle compartment was modelled in detail as shown in Figure 2. This is because developing the whole pontoon model requires extensive modelling efforts and the computational time would be excessive. It is not necessary to model the whole pontoon for a local ship collision analysis as most of the structure is far away from the impact region and is thus not expected to have any influence on the local collision response of the front wall [6]. The pontoon-supported bridge load was not included in the analysis as the collision duration is very short compared to the natural frequency of the bridge. Therefore, a ship collision accident can be considered as a transient process and thus the loads can be ignored for local collision analysis. The load effects from the construction and fabrication stage were also neglected.

To accurately model the RC pontoon wall, a detailed modelling was applied to the concrete, reinforcements and prestressing tendons in the pontoon as shown in Figure 2. The concrete part in the pontoon wall was modelled with 8-node solid elements while the reinforcements and tendons were all modelled by circular beam elements. The reinforcements are embedded inside the concrete cover of the pontoon wall. The prestress in the concrete was introduced by tendons made by strands of high strength steel wires. The height, width and thickness of the pontoon wall are 20 m, 13.3 m and 0.9 m, respectively. The diameter of the rebar and the stirrup is 15 mm and 8 mm respectively, while the vertical and transverse tendons have a diameter of 90 mm and 70 mm respectively. A detailed description of the structural modelling including prestressing modelling can be found in Sha and Amdahl [6].

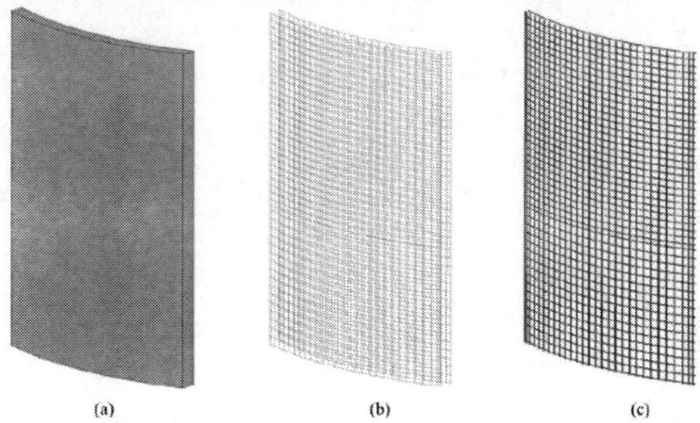

Figure 2. Finite element model of the pontoon wall portion, (a) concrete, (b) rebar and stirrups, and (c) prestressed tendons [6].

2.2. Ship bow model

According to the statistical and risk analysis of passing ships, the design ship is selected as a 20,000-ton container ship. The ship bow model as shown in Figure 3 is developed based on the actual structural drawing. This ship has an overall length of 166.62 m and a moulded breadth of 27.4 m. The depth and the scantling draught of the ship are 13.2 m and 9.6 m, respectively. In this work, the first 20 meters of the ship bow structures are modelled. The various decks, stringers and transverse frames are modelled in addition to the outer shell panels. The vertical stiffeners have a spacing of 0.6 m. The thickness of the steel components in the ship bow varies from 7.5 mm to 20.5 mm.

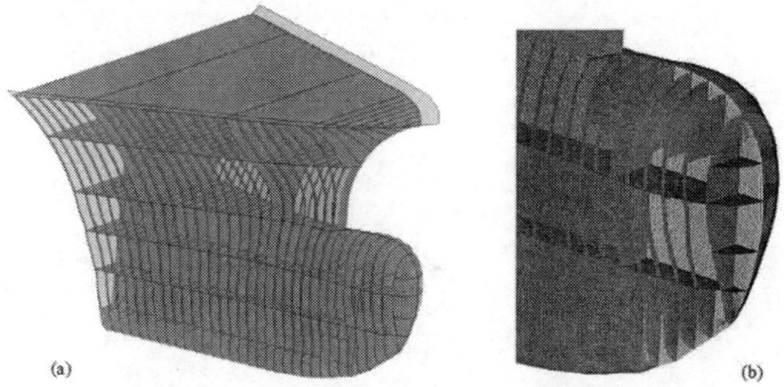

Figure 3. (a) FE model of the ship bow and (b) structural mesh.

2.3. Material modelling

The extended Karagozian & Case Rel3 (MAT_72Rel3) material model was used for the concrete in the pontoon [5]. This material model has been widely employed to model the dynamic behaviour of concrete including plasticity and damage softening after failure. This material model can automatically generate stress and strain parameter by only specifying the compressive stress. Therefore, no stress-strain curve is required for the concrete material. The model has been proven yielding reliable numerical simulations of concrete material damage to shock and impact loads. For the concrete in the pontoon wall, compressive strength of 60 MPa is used. Concrete failure is considered by utilizing the erosion algorithm MAT_ADD_EROSION with a failure strain of 0.1 [14].

The material model MAT_PIECEWISE_LINEAR_PLASTICITY was employed for the steel reinforcements and tendons. The reinforcements are made of normal mild steel while the tendons are made by high strength steel strands. A user-defined power-law hardening material model was for the steel in the ship bow which has a characteristic strength of 275 MPa. The stress-strain curve of the steel used for the ship bow is shown in Figure 4. Detailed parameters for all materials are listed in Table 1.

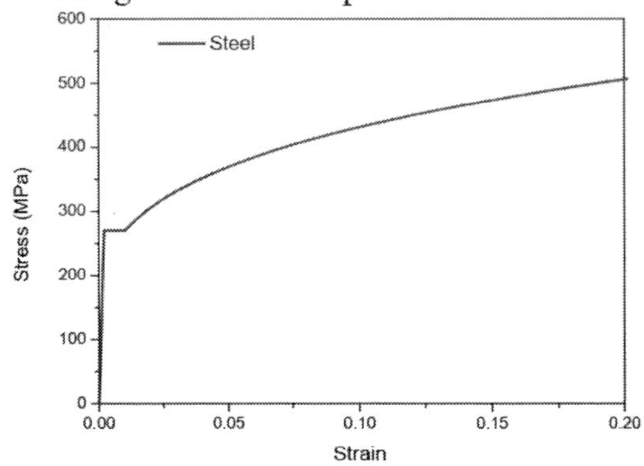

Figure 4. Stress-strain curve of the steel in the ship bow

Table 1: Parameters of concrete and steel materials.

Materials	Items	Values
Concrete	Density	2400 kg/m^3
	Poisson's ratio	0.2
	Compressive strength	60 MPa
	Failure strain	0.1
Steel (Reinforcements)	Density	7850 kg/m^3
	Poisson's ratio	0.3
	Young's modulus	2.1E11
	Yield stress	275 MPa
	Failure strain	0.35
Steel (Tendons)	Density	7850 kg/m^3
	Poisson's ratio	0.3
	Young's modulus	2.1E11
	Yield stress	1860 MPa
	Failure strain	0.35
Steel (Ship bow)	Density	7890 kg/m^3
	Young's modulus	210 GPa
	Poisson's ratio	0.3
	Yield stress	275 MPa
	Strength index	740 MPa
	Strain index	0.24

3. Numerical results

3.1. Impact force and structural damage

In the simulation, the ship bow collided with the pontoon walls at a constant speed of 10 m/s. All nodes at the four edges of the pontoon wall were fixed in all degrees of freedom.

As shown in Figure 5, the impact force increases from zero until 35 MN at 0.7 m ship displacement due to the crushing of the ship bow. Later, the general force level is between 30 MN and 40 MN due to the gradual engage and damage of internal bow frames and decks. The RC wall with a thickness of 0.9 m can resist the bulb impact with very limited spalling damages on the surface. Major damage occurs in the ship bulb, which dominates the energy dissipation as shown in Figure 6.

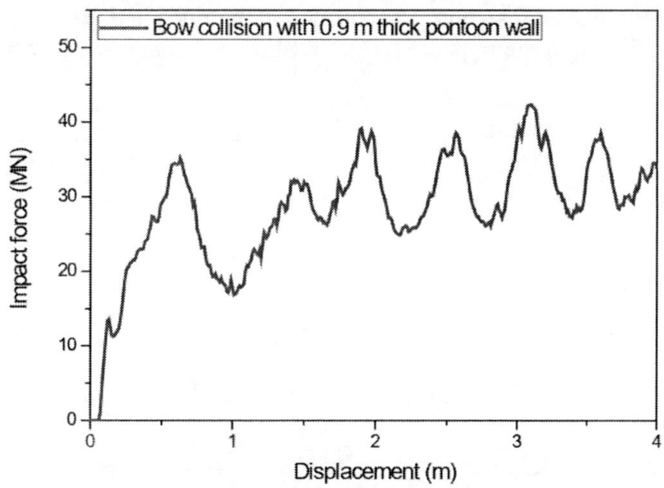

Figure 5. Force-displacement curves for ship bow collision with the 0.9 m thick pontoon wall.

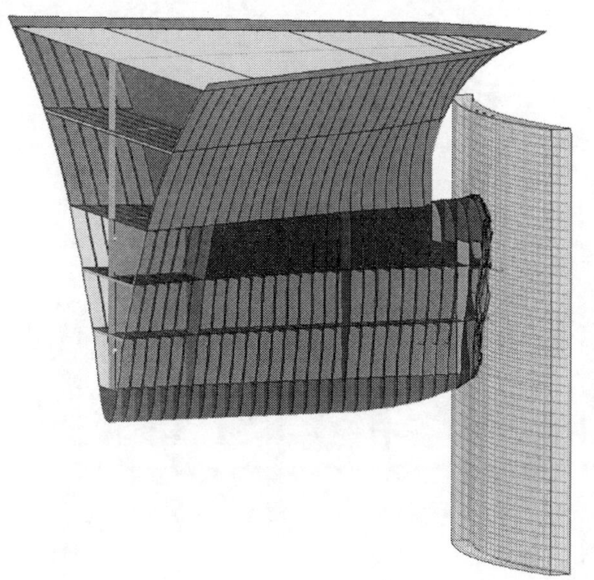

Figure 6. Structural damage for ship bow collision with the 0.9 m thick pontoon wall.

3.2. Effect of pontoon wall thickness

Further, it is interesting to analyze how the pontoon wall thickness affects the collision response of the pontoon wall and the ship bow. Numerical simulations are conducted for pontoon walls with smaller thicknesses of 0.7 m and 0.8 m.

Figure 7 shows that both the 0.7 m and 0.8 m thick pontoon walls are unable to resist the impact of the ship bow. The punching damage of the pontoon wall is associated with a sudden drop in the resistance. A thinner wall fails earlier and results in a lower energy absorption ability. The energy dissipated for 0.8 m thick wall is about 60 MJ while only 18 MJ collision energy is dissipated for the 0.7 m thick wall as shown in Figure 7 (b). This shows that the structural damage is very sensitive to the relative strength of the ship bulb and the pontoon wall. The major deformation switches from the ship to the pontoon when the wall thickness is reduced from 0.9 m to 0.8 m or less.

Figure 8 shows the side views of the 0.7 m thick pontoon at various ship displacements. It is obvious that the bulb collision-induced damage is concentrated at the impacted area. A localized punching shear failure can be observed around the impacted region. Similar damage mode is also observed for the 0.8 m thick pontoon wall.

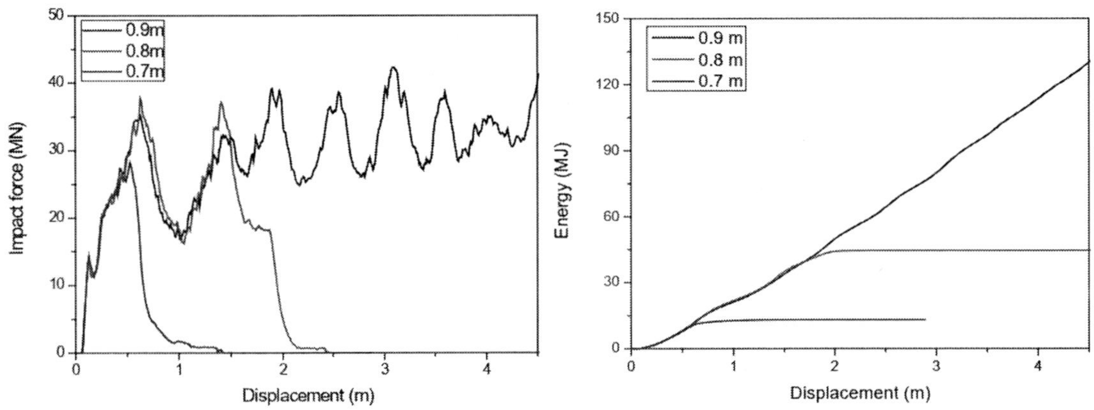

Figure 7. (a) Force-displacement curves and (b) energy dissipation curves for RC walls with different thickness.

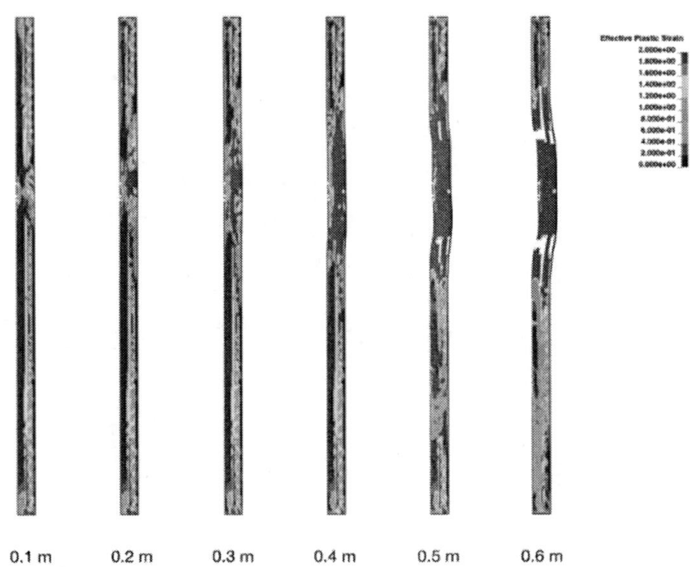

Figure 8. Side view of the strain profiles for the 0.7 m RC wall at various ship displacement.

4. Simplified method

In the preliminary bridge design phase, it is often difficult to conduct detailed numerical analyses with high-resolution FE models. It is more desirable that simplified design approaches can be used in the early design phase. In this paper, a design procedure, which can be used for punching shear failure prediction of concrete pontoon walls subjected to ship bulb collisions, is proposed.

For punching shear design of concrete slabs, it is reported [15] that the critical shear crack theory (CSCT) proposed by Muttoni [16] which includes the size effect, yields better results than the design codes ACI 318-14 [17] and Eurocode 2 [18]. Consequently, the CSCT method is used to predict the punching shear capacity of pontoon walls against ship bulb collisions in this study.

In the CSCT method, the punching shear capacity is determined as the intersection between the failure criterion and the load-rotation curve. The failure criterion is defined by a relationship between the punching shear capacity (V_C) of a slab and the rotation (ψ) of the slab. The shear demand is expressed by the impact load (V_d) - slab rotation curve. A semi-empirical failure criterion which relates the punching shear capacity to the rotation of the structure was proposed by Muttoni [9] as

$$V_C = \frac{3/4}{1+15\dfrac{\psi d}{d_{g0}+d_g}} \cdot u_0 d \sqrt{f_c} \tag{1}$$

where u_0 is the perimeter of the critical section for punching shear and d is the effective depth of the wall. f_c is the average cylinder compressive strength of concrete. $\frac{d}{d_{g0}+d_g}$ is a size coefficient that accounts for the wall thickness and aggregate size. d_g is the maximum aggregate size and d_{g0} is a reference size equal to 16 mm. For a pontoon wall structure, it is recommended to take the maximum aggregate size d_g no larger than one-third of the effective depth d [19].

The load-rotation relationship can be simplified by assuming a parabola with a 3/2 exponent for the rotation as a function of the load-flexural strength ratio $\frac{V_d}{V_{flex}}$. The flexural strength V_{flex} is equal to $8m_d$, where m_d is the yield moment of the slab. The load-rotation relationship can be obtained by

$$\psi = 0.33\frac{L}{d}\frac{f_s}{E_s}\left(\frac{V_d}{8m_{Rd}}\right)^{3/2} \tag{2}$$

where L is the main span of the slab. f_s and E_s are the yield strength and modulus of elasticity of the reinforcements.

The failure criteria of the RC walls with 0.7 m, 0.8 m, and 0.9 m calculated with Eq. (1) while the load-rotation relationships are obtained from Eq. (2). The punching shear capacity can be readily obtained as the intersections of the failure criteria and the load-rotation curves. Here, we obtain a punching shear capacity at the intersections of 33 MN, 44 MN, and 59 MN for the 0.7 m, 0.8 m, and 0.9 m thick pontoon wall, respectively.

The obtained punching shear capacity of the three RC pontoon walls are compared with the ship bulb strength as shown in Figure 9. It can be observed the punching shear capacities of 0.7 m and 0.8 m thick pontoon walls are smaller than the maximum ship bulb strength. The 0.9 m thick RC pontoon wall, however, has a sufficient capacity compared with the ship bulb strength. The predictions are in line with the numerical observations where punching shear failure occurs in the 0.8 m and 0.7 m thick pontoon walls while the 0.9 m thick pontoon wall remains intact.

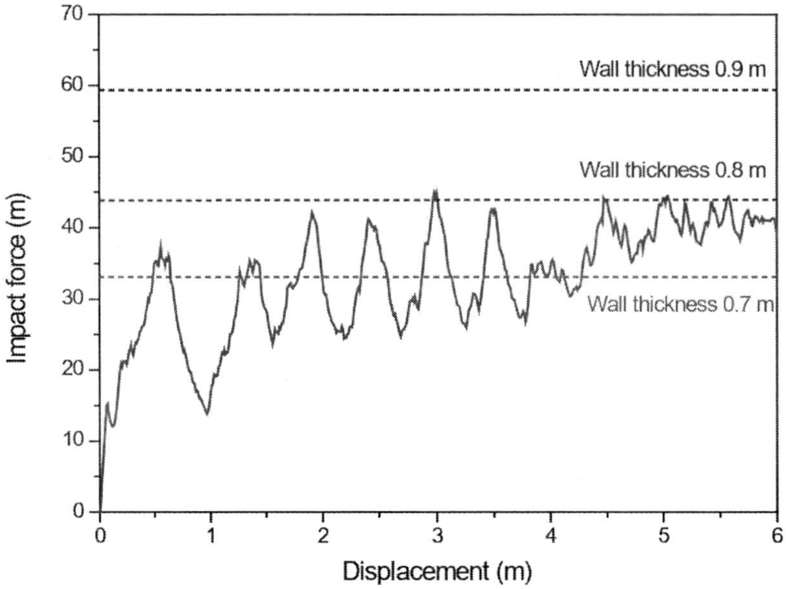

Figure 9. Comparison of impact demand and punching shear capacity for the RC pontoon walls.

5. Conclusions

The findings of the current study are summarized as follows:

1. The 0.9 m thick RC pontoon wall can resist the bow impact of the 20,000-ton container ship. Severe damage occurs in the ship bow which dissipates the majority of the collision energy. The pontoon wall, however, has very limited damage and can maintain its integrity after the collision.
2. Further reducing the pontoon wall thickness to 0.8 m or 0.7 m will result in a punching shear failure in the pontoon wall under the collision from the same container ship bow. The ship bow has a small local deformation upon contact.
3. A design check procedure that can be used for design RC pontoon walls against ship bulb collisions is proposed. The impact force and damage initiation obtained from the design check procedure agree well with the numerical results. This indicates that the proposed analytical punching shear failure model captures the essential physical effects of the collisions.

Acknowledgements

This work was supported by the Norwegian Public Roads Administration (project number 328002) and in parts by the Research Council of Norway through the Centres of Excellence funding scheme, project AMOS (project number 223254). The support is gratefully acknowledged by the authors.

References

[1] Sha Y, Amdahl J and Dørum C 2018 Local and global responses of a floating bridge under ship-girder collisions, *J. Offshore Mech. Arct. Eng.* **141**(3) 031601

[2] Sha Y, Amdahl J, Dørum C and Yu Z 2018 Numerical investigation of the collision damage and residual strength of a floating bridge girder. In: *36th Int. Conf. Ocean, Offshore Arct. Eng.* American Society of Mechanical Engineers), June 17 - 22, Madrid, Spain

[3] Kvåle K A, Øiseth O and Rønnquist A 2017 Operational modal analysis of an end-supported pontoon bridge, *Eng. Struct.* **148** 410-23

[4] Wang J, Cheynet E, Snæbjörnsson J Þ and Jakobsen J B 2018 Coupled aerodynamic and hydrodynamic response of a long span bridge suspended from floating towers, *J. Wind Eng. Ind. Aerodyn.* **177** 19-31

[5] Sha Y and Amdahl J 2017 Ship Collision Analysis of a Floating Bridge in Ferry-Free E39 Project. In: *36th Int. Conf. Ocean, Offshore Arct. Eng.* American Society of Mechanical Engineers) V009T12A21, June 25 - 30, Trondheim, Norway
[6] Sha Y and Amdahl J 2019 Numerical investigations of a prestressed pontoon wall subjected to ship collision loads, *Ocean Eng.* **172** 234-44
[7] Consolazio G R, Cook R A, McVay M C, Cowan D, Biggs A and Bui L 2006 *Barge impact testing of the St. George Island causeway bridge*: Department of Civil and Coastal Engineering, University of Florida)
[8] Fan W and Yuan W 2014 Numerical simulation and analytical modeling of pile-supported structures subjected to ship collisions including soil–structure interaction, *Ocean Eng.* **91** 11-27
[9] Sha Y and Hao H 2012 Nonlinear finite element analysis of barge collision with a single bridge pier, *Eng. Struct.* **41** 63-76
[10] Storheim M and Amdahl J 2014 Design of offshore structures against accidental ship collisions *Mar. Struct.* **37** 135-72
[11] Sha Y and Amdahl J 2018 A simplified analytical method for predictions of ship deckhouse collision loads on steel bridge girders, *Ships Offshore Struct.* **14** (1) 121-34
[12] Sha Y and Hao H 2014 A simplified approach for predicting bridge pier responses subjected to barge impact loading, *Adv. Struct. Eng.* **17** 11-23
[13] Sha Y, Amdahl J, Aalberg A and Yu Z 2018 Numerical investigations of the dynamic response of a floating bridge under environmental loadings, *Ships Offshore Struct.* **14** (1) 113-26
[14] Sha Y and Hao H 2013 Laboratory tests and numerical simulations of barge impact on circular reinforced concrete piers, *Eng. struct.* **46** 593-605
[15] Goswami A, Adhikary S D and Li B 2019 Predicting the punching shear failure of concrete slabs under low velocity impact loading, *Eng. Struct.* **184** 37-51
[16] Muttoni A 2008 Punching shear strength of reinforced concrete slabs without transverse reinforcement, *ACI Struct. J.* **105** 440-50
[17] Committee A 2014 Building code requirements for structural concrete (ACI 318-14) and commentary (ACI 318R-14). American Concrete Institute)
[18] Institution B S 2004 Eurocode 2: Design of Concrete Structures: Part 1-1: General Rules and Rules for Buildings. British Standards Institution)
[19] Bažant Z P, Pan J and Pijaudier-Cabot G 1987 Softening in reinforced concrete beams and frames, *J. Struct. Eng.* **113** 2333-47

IOP Conf. Series: Materials Science and Engineering **700** (2019) 012039 doi:10.1088/1757-899X/700/1/012039

Impact from ice floes and icebergs on ships and offshore structures in Polar Regions

J Amdahl

Dept. of Marine Technology, Norwegian University of Science and Technology, Centre for Autonomous Marine Operation and Systems (AMOS)

Corresponding e-mail: jorgen.amdahl@ntnu.no

Abstract. The principles for ULS design and ALS design for offshore structures are discussed. The use of pressure-area curve versus force-area curves for analysis for estimation of ALS impacts is deliberated. Aspects of local and global shape of the ice feature are discussed in view of external mechanics (demand for energy dissipation) and internal mechanics (local damage). Simplified methods for structural damage assessment are reviewed for ice loads that may move both transverse to and along the shell plating. Material modelling of ice for nonlinear finite element analysis (NLFEA) of ice-structure interaction is reviewed. The material models must be calibrated against design curves and the sensitivity of the material parameters with respect to mesh size is investigated. Results from simulation of impacts of ice with different shapes on the column of a floating platform are presented, and the critical shape for penetration of the front plating for the present structural configuration is identified.

1. Introduction

Oil activity, shipping and cruise traffic in Arctic regions increase, partly sparked by global warming. This instigates safety concerns with respect to environmental pollution, fatalities and economic loss. With large distances to infrastructure it may be challenging to assist in case of critical events. Structural damage due to impacts from ice floes and icebergs may become fatal if excessive flooding and loss of stability occur. Ships and oilrigs operating in permanent ice cover will need to be ice-strengthened. Lightly ice-strengthened or non-ice strengthened structure may operate close to the ice edge or may need to move into light ice-conditions, e.g. during search and rescue operations.

The design against ice may be carried out in all general limit states, SLS (acceptable deformations), FLS (cyclic ice actions), Ultimate Limit State (ULS) and Accidental Limit State (ALS) In most cases the assessment of the resistance to ice action is carried out in ULS and ALS (for offshore structures, ISO 19906 [1], ALS design is mostly indirect for ships. ULS design (annual probability of occurrence typically 10^{-2}) is important for structural safety. This control ensures that all foreseen loads can be resisted without damage with an adequate margin. In ALS design (annual probability of occurrence typically 10^{-4}), damage to the structure is allowed as long as there is sufficient residual strength to prevent progressive collapse, and the safety of the crew and environment can be maintained.

Three design principles may be applied to enable structures to resist accidental or abnormal loads including extreme ice actions from impacts with ice floes and growlers/bergy-bits : strength design, ductile design and shared-energy design (see Figure 1).

Strength design assumes that the ice will dissipate most of the collision energy and that the structure will undergo only limited yielding and small permanent deformations. This assumption is virtually identical to that used for ULS design.

Ductile design assumes that the entire collision energy is dissipated by the structure. If the collision energy is dissipated with moderate damage to the structure (e.g., no penetration of cargo tanks or buoyancy compartments essential for hydrostatic stability) and with satisfactory residual strength, the ALS acceptance criteria may be satisfied

Shared-energy design implies that both the ice and the structure undergo significant deformations and dissipate energy. At any instant, the weaker structure (or ice) will deform

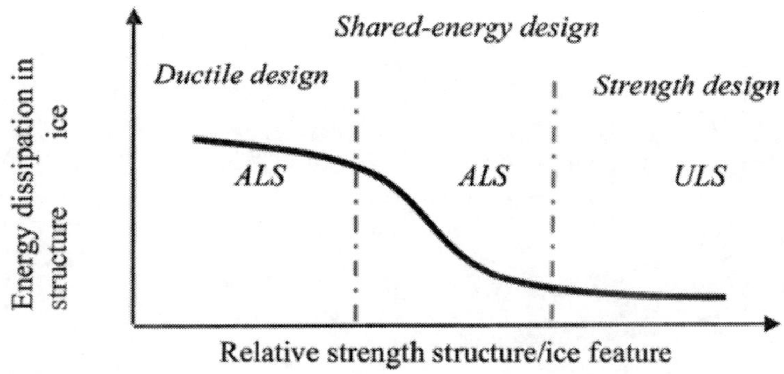

Figure 1 Characterization of design principles in terms of the share of energy dissipation versus relative strength of structure and ice

1.1. Pressure-area relationships

ISO 19906 specifies that design for ULS ice actions and ALS ice actions shall include both local and global actions. Global actions represent the total action on the structure and are applied over the nominal contact area, while local actions are applied to specific areas within the nominal contact area.

According to ISO19906 [1], the ice pressure distribution for massive ice features is highly non-uniform within the nominal contact area and is rapidly changing during ice-structure interaction. Ice at the boundaries may spall off reducing the actual contact area relative to the nominal contact area. Smaller areas within the actual contact areas will be subjected to pressures considerably larger than the average pressure. According to ISO 19906, this should be addressed by assessing local design separately using pressures representative for the high-pressure zones. The actions are often given as local pressure-area (*p-A*) relationships ($p = CA^{ex}$ where *C*, and *ex*<0 are empirical coefficients). This is relevant for structures undergoing small or very moderate plastic deformations, which is normally assumed in ULS design.

p-A relationships are also often specified for ALS design (e.g. ISO19906 [1]). Since the *p-A* relationships depend on the recurrence period and they will be higher in the ALS than in the ULS as illustrated in **Figure 2**. The application of *p-A* curves is, however, not straightforward in ALS. A principle sketch of ULS and ALS *p-A* relationships is shown in Figure2. The resistance domains for a structure (i.e. the structural capacity versus loaded area) are also illustrated in **Figure 2**. There is no clear transition between them so they are overlapping. In ULS the resistance must be in the strength domain as the behaviour shall be essentially elastic with only moderate plastic deformations. When the *p-A* demand increases, the structure will undergo finite deformations and thus move into the shared energy domain or, ultimately, into the ductile domain. In both cases the structural resistance will govern the interface pressure; i.e. plate and stiffeners may deform before the interface force reaches the local AL pressure. This implies that the ice maintains its shape in the local high-pressure zones, the local indentation into the structure increases slightly locally, but simultaneously the contact area expands and mobilizes more structural members. Consequently, in the ALS, we are not so much concerned about the resistance to local, rapidly varying pressures, but rather the resistance to the global ice actions.

It is, of course, possible to strengthen the (offshore) structure so that the strength domain becomes larger than the ALS pressures so that the same ULS strength check techniques can be utilized, but this is essentially not required if the structure can survive the ice load with the calculated ALS damage and also has sufficient residual strength.

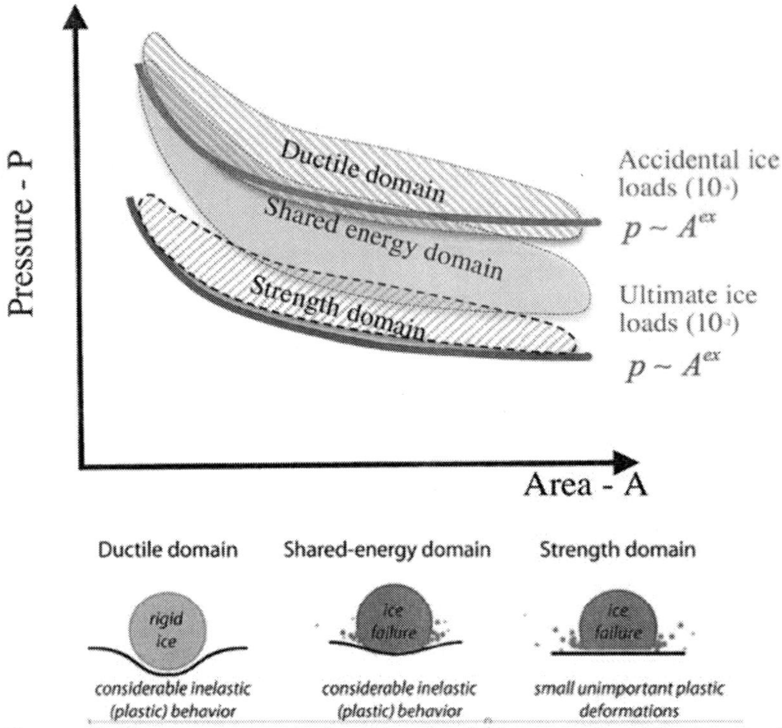

Figure 2 Pressure-area relationships for different probability levels. Strength, shared-energy and ductile domains for structural resistance are also indicated.

1.2. Damage and energy dissipation

The principle of estimating the damage and energy dissipation in the structure is illustrated in Figure 3, which is reproduced from ship collision with offshore installation according to DNVGL RP-C204 [2], but here adapted to ice impacts. The basic ice force-deformation relationship, F_{ice}, is the force that would be obtained when the ice is crushed against a rigid structure. F_{ice}, should essentially incorporate the ALS *p-A* curve and would correspond to that obtained when the resistance is in the strength domain. Conversely, assuming the ice to be rigid, the force-deformation curve for the structure can be calculated. The resistance, which is in the ductile domain, will depend on the assumed shape of the impacting ice feature. Since the force acting on the ice and on the structure must be equal, the resulting damage is obtained when the energy dissipation (equal to the area under force deformation curves) reaches the demand for energy dissipation, as required by the external mechanics analysis.

It is obvious that this approach is very simplified and does not account for interaction effects: when the structure undergoes finite deformations locally, it tends to "wrap around" the ice feature, which becomes more confined and the pressure needed to crush it is likely to increase. This is illustrated by the dotted force curve for the ice. Conversely, if the structure is strong enough to crush the ice, the contact area becomes blunter and larger; consequently, the resistance increases compared to that for rigid ice as shown by the doted resistance curve. This demonstrates the complexity of the damage process when the resistance is in the shared-energy domain; increasing structural deformation or ice crushing may switch forth and back depending on which body that is the instantaneously the stronger one. The ISO 19906 ice *p-A* relationships have been derived using force time histories and converting force values to pressure. The design curves are based on the peaks of the force/pressure data. In addition, the pressures are based on the mean plus three times the standard deviation for each contact area. For contact areas larger than approximately one square meter, the pressure values drop so fast ($A^{-0.7}$) that it is questionable whether they are can be used for ALS assessment. Kim et. al [3] argue that the physical limits to the energy absorption capacity of ice become more important than the underlying pressure-area relationship. It may therefore be useful to consider the amount of energy spent on crushing a unit mass of ice.

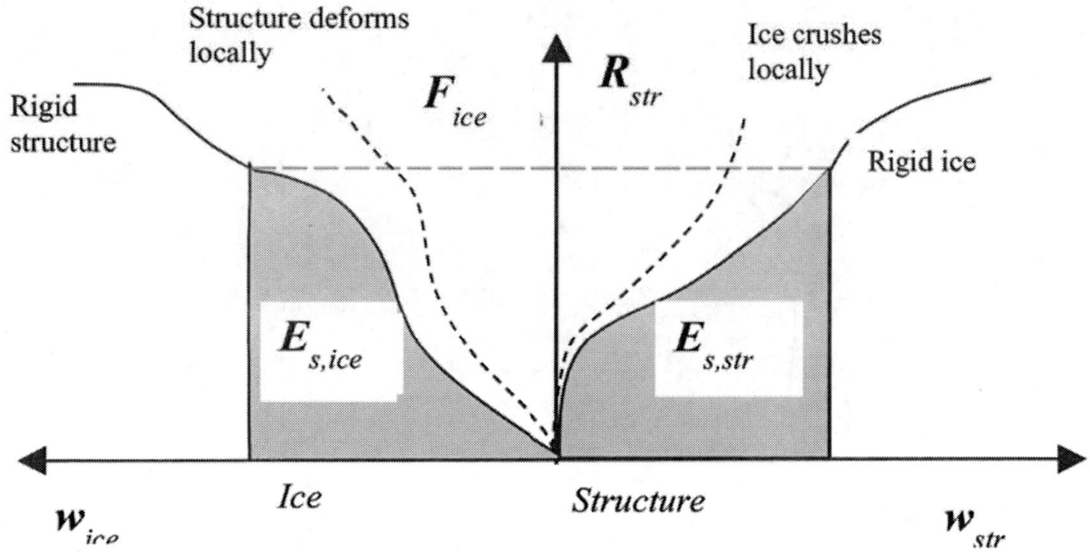

Figure 3 Estimation of force and damage (deformation) for ice-structure impacts adopting the approach for ship collision (DNMVGL RP-C204, 2019). Ice force curves to the left, resistance curves to the right. $E_{s,ice}$, $E_{s,str}$ – strain energy dissipation in ice and structure, w_{ice}, w_{str} – deformation of ice and structure, F_{ice} – ice force, R_{stru} – structure resistance

For fully confined iceberg ice at Pond Inlet (Masterson et al. [4]), when the effect of the sample size and boundaries are minimized, the specific crushing energy is maximum 6.09 kJ/kg (average value being 2.96 kJ/kg). This value was derived using force-time history records obtained in the Pond inlet tests with 900, 1280 and 2300 mm radius indenters.

While the specific crushing energy value can be used to calculate the energy spent on ice crushing, the process *p-A* relationship during an ALS event could be also established. The question is: can we use the ISO curve with *C=7.4* and *ex=-0.7*? The majority of the data for this curve comes from the Pond Inlet tests on iceberg ice with 2300 mm radius indenter. The pressures obtained in the Pond Inlet test set are a series of maxima during individual tests (four tests in total), the data were binned according to the fully confined contact area and $p = 7.4A^{-0.7}$ is the mean plus three standard deviations. Thus, it is questionable whether this curve can be used in the context of ALS. Instead, the process *p-A* relationship underlying the local ice-load calculations PC 3 vessels in IACS PC code [5] (i.e., $p = 3.2A^{-0.1}$) was found to better represent the physical value of the energy absorption of iceberg ice during crushing and spalling failure. The experimental data and force curves corresponding to crushing energy of 2.96 kJ/kg for the Pond Inlet tests (2300 mm radius indenter) are shown in **Figure 4**. Hence, if the force curve $F = 3.2A^{0.9}$ is used in ALS design, the local pressures and forces will be smaller than the ISO 19906 p-A curves for $A < 4m^2$, and larger for $A > 4m^2$

Thus, when ice material models are used in finite element analysis it is recommended that the ice models be calibrated/verified against force deformation curves (e.g. $F = 3.2A^{0.9}$) and not *p-A* relationships. Ideally, the ice material models should be calibrated by trying to replicate the tests conditions. In the Pond inlet tests a rigid, spherical indenter was pushed into an ice block, so the ice was subjected to considerable confinement. In the case ice impact with a structure the degree of confinement is more uncertain and is likely to be less, (unless the structure is subjected to severe deformation). It is likely that the force crated by crushing is smaller, but experimental evidence is lacking. A more careful approach is to calibrate to the force curve for ice crushing against a plane, rigid surface.

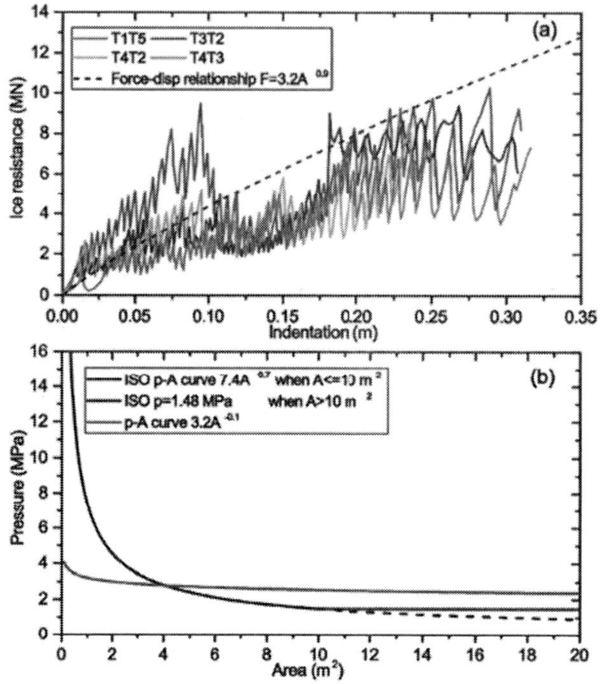

Figure 4. Force indentation curves in Pond inlet tests and "average" force curves

2. Impact geometry

It is very important to distinguish between local and global geometry of the ice feature, refer Lu et. al [6]. The **global** geometry is decisive with respect to the mass of the ice feature as well as the exposure to collision of the various parts of the structure, i.e., the column and the floater. In addition it is governing the demand for energy dissipation, which is expressed as a fraction of the total kinetic energy at the instant of collision. The local geometry is essential with respect to the resistance of the ice to deformation (crushing, extrusion etc.) at the contact point.

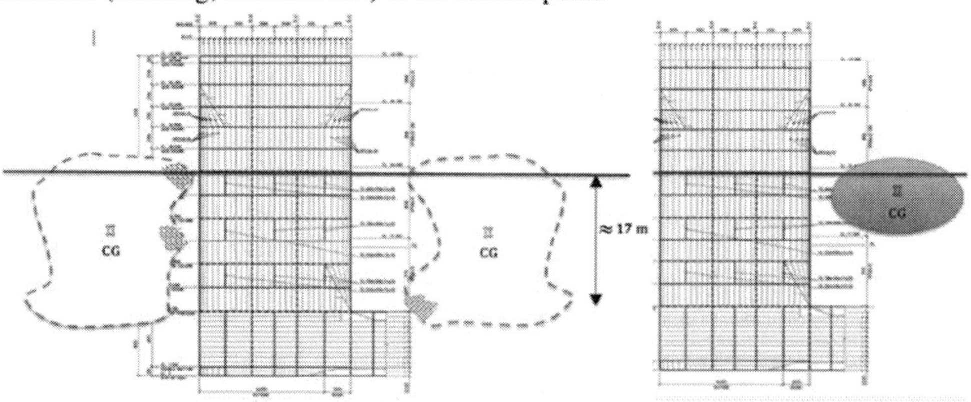

Figure 5. Left: Illustration of global geometry for collision between a large ice feature (5-10,000tons) and the column and pontoon of a semi-submersible platform in still water. The hatched areas indicate possible contacts where crushing will take place and where information of local geometry is needed. Right: Collision with a small ice feature.

Figure 5 illustrates the importance of global geometry for the exposure of a column and pontoon of floating platform at still water (or relatively small waves). The pontoon can only be hit by 'large' ice feature at still water. The collision will in most case be eccentric, i.e., the force vector has a large arm with respect to the centre of gravity (CG), and substantial energy may be transferred into pitch/roll motion of the ice feature. This reduces the demand for energy dissipation. The same holds true for collisions around water level.

Depending on the ice features shape, collision may take place at intermediate depths and may be close to the CG; and most of the kinetic energy must be dissipated by deformation of structure and/or the ice feature. For smaller ice features, like the spheroidal ice in Figure 5 right, only the column may be exposed to collision, which in this case is centric. It is true that the mass of the ice feature here is small; but it is also prone to larger wave induced motions leading to considerable collision speeds. No agreed global shape for bergybits/growlers exists at present and the shapes used in various studies

differ greatly. An interesting case is shown in Figure 6. The ice is washed out by the waves in the waterline and gets a " foot" that protrudes significantly from the visible part above waterline. This shape has been of particular concern to seal hunters according to Gudmestad and Alme [7]

Figure 6 Small iceberg with a "foot" Photo: Bjørn Kvermo

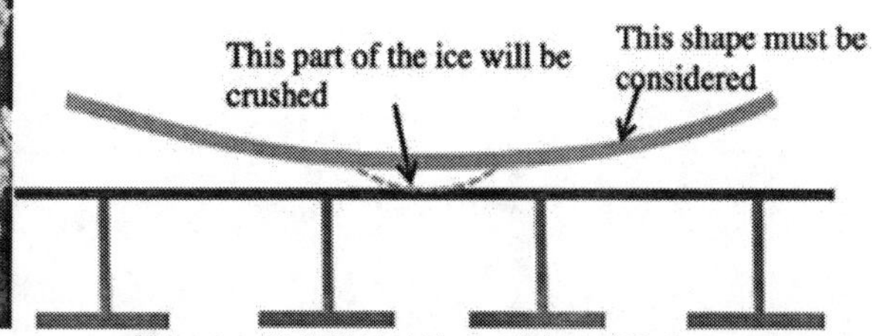

Figure 7 Illustration of local ice geometry

There is also no common agreement on the local shapes that should be used for ice features. The issue of local shapes is illustrated in Figure 7. The ice has a small protrusion. If this is not crushed it may be detrimental with respect to indentation and puncturing of the shell plating between adjacent frames of the structure. If the ice is crushed ice becomes more blunt and the worst shape that should be considered could be the one with large curvature; this could prevent penetration of the shell plating and allow contact to spread over several frames. Evidently, the shape the local ice shape should be considered in view of the structural layout in the contact area.

Other potential ice geometries could be shaped as discs representing ice floes, but only impacts from spheroidal ice is analysed in this paper.

2.1. The effect of relative motions

The presence of waves induces oscillatory motions in six Degrees of Freedom (6 DOFs) for both the ice feature and the structure. Depending on the motion characteristics, the vertical exposure of the structure will vary in different wave conditions. This is illustrated in Figure 8.

If only the horizontal motions of the ice are considered, the likelihood of hitting the pontoon in a wave trough will most likely be considerably smaller than the likelihood of hitting the column in a wave crest. This depends on the mean drift of the ice feature versus the relative wave induced surge/sway motion of the ice feature and the structure. On the other hand, vertical collision on the pontoon due to the relative heave motions can be significant.

Lu and Amdahl [8] investigated the drift, impact velocities and vertical collision exposure for a small, spheroidal iceberg hitting a semi-submersible platform. They adopted the method introduced by Fylling [9] for drifting ship impacts. Figure 9 shows the temporal variation of the track when the mean drift speed is small. Impact will only occur when the sphere advances towards the platform indicated by the red lines. Obviously it is more likely to hit the platform during a wave crest and at a high location.

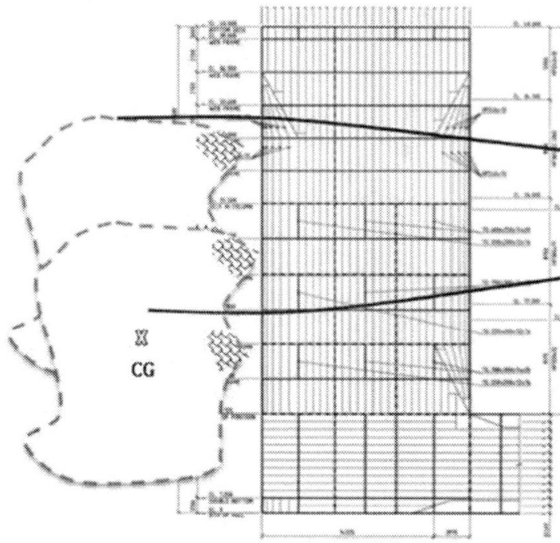

Figure 8. Illustration of potential global geometries for collision between 'large' ice feature (5,000-10,000 tons) and structure's column and pontoon in waves. Upper sketch for high position: collision close to wave crest, lower sketch for low position: collisions close to wave trough.

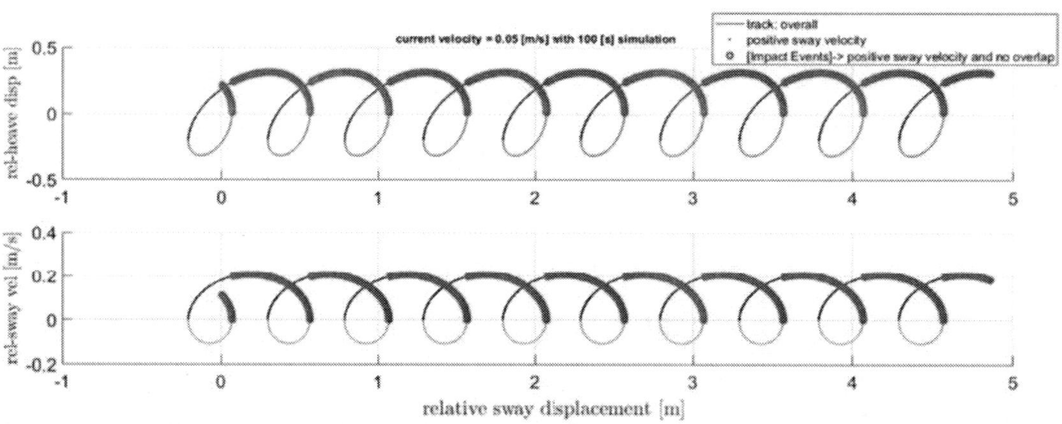

Figure 9 Simulation with coupled motion track and potential impact events for slow current velocity (0.05 m/s.) [8]. The red part of the track indicates when horizontal impact is possible.

Figure 10 shows an example of the probability distribution for the vertical location of impact with the column and associated values of the impact speed at various probability of exceedance levels. The most likely place to hit the column is below still water line. The largest impact speeds occur at a high level, however, the probability of occurrence is small.

The study by Lu and Amdahl [8] was based on linear wave theory. Wave drift forces were not calculated, the nonlinearity of the vertical restoring force was accounted for as well as change of the hydrodynamics when the ice feature is close to the column. Extensive hydrodynamic analysis of the ice impact was carried out by Ommami et. al. [10], where among others the nonlinearity of the Froude-Krylov force was taken into account and the repellent action when the ice approaches the column was accounted for.

In a follow-ip study [11] the method used by Lu and Amdahl [8] was improved in the sense that the restoring force, hydrodynamic added mass, the Froude-Krylov force and drift force all were based on the actual submersion of the glacial ice, which for simplicity was assumed to have a cubic shape in trhis case. The improved analyses remove some conservatism form the previous calculations, and the impact speeds are of similar order of magnitude.

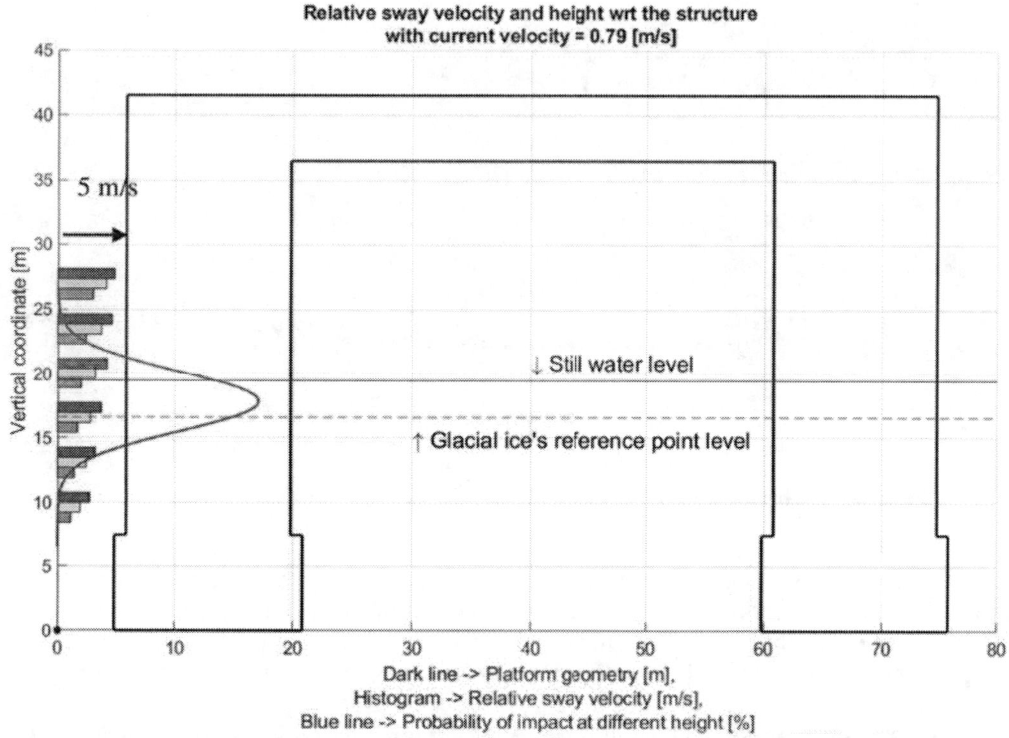

Figure 10 Probability distribution for the vertical location of impact. Conditional impact speed shown for different heights; colour indicates non-exceedance levels: red 99%, yellow 90%, green 50%

2.2. *External mechanics*

The *external mechanics* deals with the rigid body motions of the ice and the structure during impact. It is vary favourable when the external mechanics can be separated from the structural deformation estimation, because the latter, commonly denoted *internal mechanics,* can be determined by a single nonlinear finite element analysis of a given impact scenario with detailed local models of the structure and the ice. The result of the external mechanics is the demand for strain energy dissipation, which is determined by the internal mechanics.

If the impact is not centric part of the kinetic impact energy will be transferred to other forms of translational and rotational energy, reducing the demand for strain energy dissipation in the structure and the ice. Popov et al. [12] presented a 6DOF+3DOF model for ship and ice floe collisions. It was assumed that the impact impulse always acts in the normal direction at the impact point, and the friction between ship and ice was ignored. A "reduced mass" concept was introduced to simplify the 6DOF+3DOF problem to a 1DOF +1DOF problem. Popov's model has been extensively used to calculate the effect of ice loads for ship navigation in ice-covered waters; see e.g. IACS [5], Daley [13, 14] and the SSC report by Dolny [15]. The SSC report acknowledged that Popov's model presupposes that the collision process is quick, that there is no sliding along the hull, and that sliding should be considered for more severe limit states. Daley [13] noted also the limitation of Popov's model in handling sliding impacts]. Liu and Amdahl [16] removed this limitation and presented a new approach to multi-planar ship-ice impacts. It is based on the theory developed by Stronge [17] and can handle motions in 6DOF for both the ship and the ice. Closed-form solutions for the dissipated strain energy were obtained explicitly for the so-called "stick" and "slide" mechanism

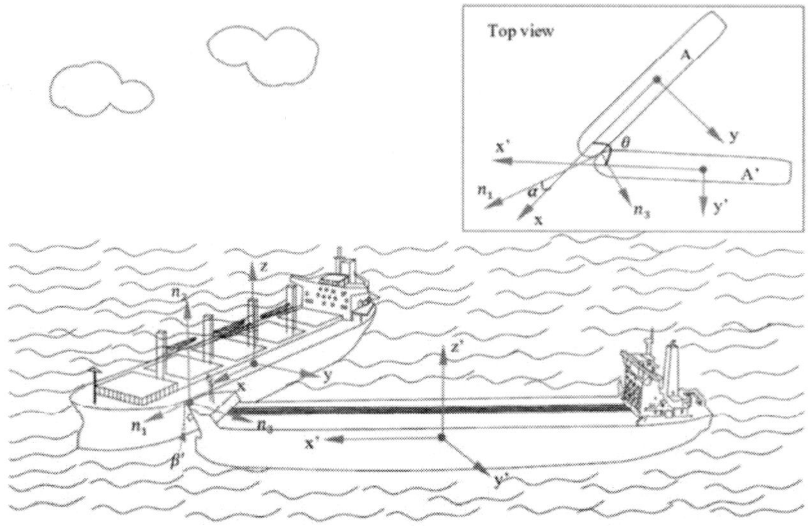

Figure 11 Illustration of external mechanics for ship collision [16].

The method is illustrated for ship –ship collisions as illustrated in Figure 11 but is also relevant for structure–ice collisions. The method was formulated by considering the equilibrium and conservation of momentum in the directions normal and parallel to the tangential collision plane (n_1,n_2,n_3) in Figure 11. Closed form solutions for the required strain energy dissipation during impact was determined for two different cases, depending on whether the ice sticks to the structure or whether it slides along the side for a given limiting friction factor. The friction factor concerns generally "normal" steel-to-ice friction but it could also be increased to take into the increased force to transverse deformation of the ship side. From numerical simulations of grounding Alsos [18] and, from ice impacts, Quinton [19] observed that the normal force during sliding is approximately half of the value required for lateral penetration of the structure to the same indentation level. They also found that the penetration force is virtually independent of the friction factor, while the sliding force is very sensitive to it.

3. Constitutive modelling of ice

Constitutive modelling of ice seems to follow two tracks; the crushable foam model and plasticity based model. The crushable foam model was developed was used in numerical simulations of a ship-ice collision and to estimate the damage to the vessel; see Gagnon [20] and Gagnon and Wang [21] and LSTC [22]. In this model, the material strength increases as it deforms, and this is viewed as an analogy so that the essential behaviour of the actual ice can be realized in the simulations. The analogue model ensures that a relatively small region in the centre of contact experiences high pressure (a hard zone) and the surrounding contact material exerts a somewhat lower pressure (soft zone) – an established fact from various experimental studies (Gagnon [23]; Riska et al. [24] The unknown material parameters were determined through trial and error by comparing the results from simulations of full-scale ship/bergy bit impacts with the actual field data (i.e., the actual peak load, impact duration and contact pressures). The predictive capabilities of the model were demonstrated by simulating growler impact tests at laboratory scale. The model successfully produced reasonable values for the total load and pressure when compared with the instrument measurements.

Plasticity based modelling has been adopted by several authors, e.g. Kierkegaard [25], Liu et. al [26], Shi et. al [27] and Xu et. al. [28].

3.1. *Plasticity based modelling*

The continuum mechanics model for ice developed by Liu et al. [26], was intended for practical design of structures against abnormal ice loading. It consists of an elastic-plastic isotropic material model, where the yield stress is dependent on the hydrostatic pressure so as to represent ice confinement. The model has the advantage of capturing several ice deformation characteristics. It was implemented in the explicit finite element software LS-DYNA. The behaviour of ice is modelled by means of of the 'Tsai-Wu' elliptic yield criterion [29] and a strain based failure criterion that is dependent on the hydrostatic pressure. The yield surface is a function of both the second invariant J_2 of the deviatoric stress and the hydrostatic pressure,:

$$f(p, J_2) = J_2 - \left(a_0 + a_1 p + a_2 p^2\right) = 0$$

where a_1, a_2 and a_3 are material coefficients to be specified by users, the values adopted in subsequent calculations are those suggested by Kierkegaard [25] a_1, = 2.588 a_2 = 8.63 and a_3 = 0.163. To avoid unreasonable tensile pressure a cut-off criterion equal to -2 MPa was adopted.
In order to simulate the mechanical effects of ice crushing, an empirical failure criterion is adopted based on the effective plastic strain and hydrostatic pressure (pressure positive in compression) is adopted:

$$\varepsilon_f = \varepsilon_0 + \left(\frac{p}{p_2} - \gamma\right)^2$$

where, p_2 is the larger root of the yield function, and ε_0 and γ are the parameters that will be calibrated to get the desired for-contact area relationship. In the present case Lu et. al. [11] carried ot the calibration with respect to the force curve $F = 3.2A^{0.9}$ discussed in Section 1.2. In numerical simulations, when the failure strain is exceeded, the element is eroded in subsequent analysis.

3.2. Calibration to design curves

There are two ways to calibrate to the fitted force-displacement relationships. The natural choice would be to mimic the Pond Inlet tests by pushing a rigid indenter into the ice block. When this was done the red curve in Figure 12 was obtained [11]. Next, the same ice failure parameters were used by crushing a spheroidal ice against a plane rigid wall. The corresponding force contact area relationship that was obtained is shown by the blue curve. It appears the force is reduced by approximately 50 % relative to the red curve. It is possible that the large deviation is physically correct; the ice is considerably more confined in the mimicking of the Pond Inlet tests. However, experimental evidence is missing. As the ice, would become quite soft compared to the structure in subsequent integrated analysis, it was decided to calibrate for the spherical ice crushed against a rigid, plane plate. This is also more in analogy with the method used in the IACS PC code [5]

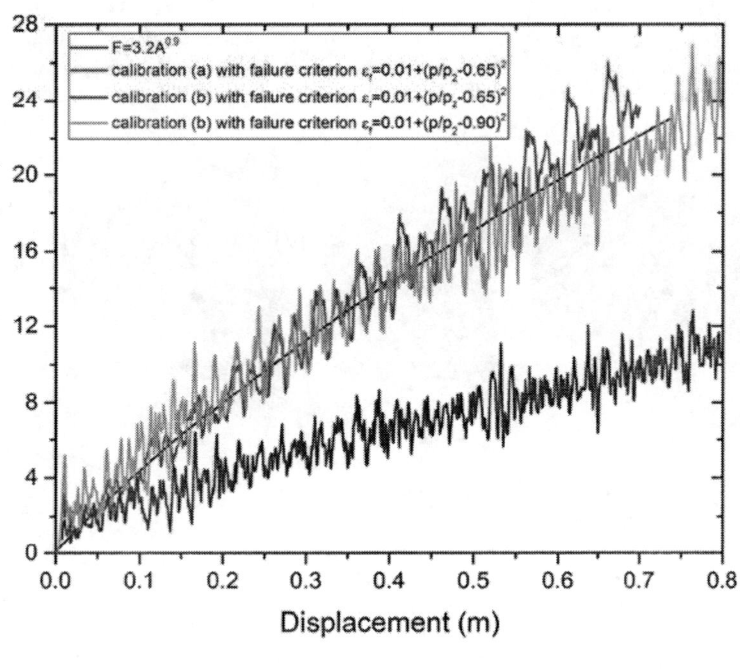

Figure 12 Calibration to force curve F = 3.2A$^{0.9}$. Red curve from calibration with rigid block pushed into ice, blue curve when same calibration is used for spheroidal ice crushed against a rigid, plane surface. Green curve from calibration of spheroidal ice crushed against a rigid, plane surface. Element size 50 mm.

By this calibration the green curve was obtained. Thus, in integrated analysis, where the structure may fold around the ice and create confinement, it is likely that the resistance will be larger than the green curve, and this method is therefore conservative. Conversely, for the first calibration method, it is likely that the force curve in integrated analysis will fall between the red - and the blue curves.

It is not surprising that the calibration is sensitive to size of the finite element mesh. Figure 13 shows force-contact area curves for five different mesh sizes: 100 mm, 75 mm, 50 mm, 35 mm and 25 mm and failure strain parameters calibrated with the first method [11]. The force curve reduces with decreasing mesh size. This is expected because a fine mesh is more likely to capture concentrated strains while a coarse mesh averages the strains over the larger distance. Therefore, element erosion takes place at an earlier stage with a fine mesh. This behaviour is, in fact, analogous to the well-known mesh size sensitivity of shell elements for fracture prediction of of steel plates.

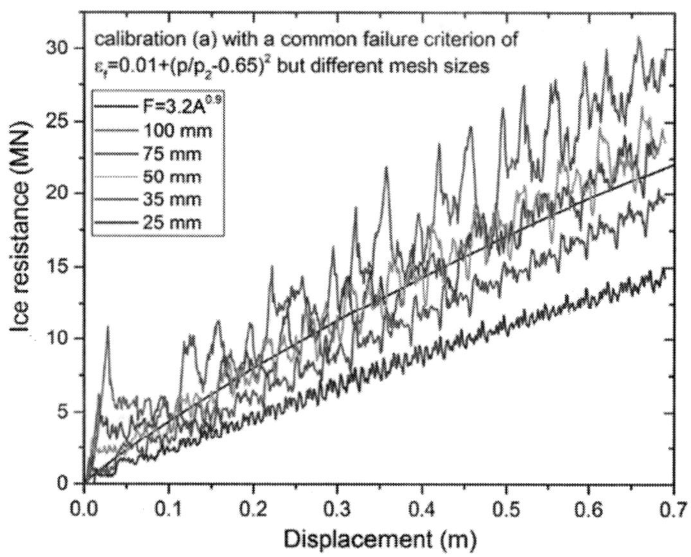

Figure 13. Mesh size sensitivity for calibration against force curve $F = 3.2A^{0.9}$. Parameters calibrated for element size 50 mm.

When different ice shapes shall be analysed it is essential to have the same mesh size. This may, however, be difficult to achieve in practise. If different mesh sizes are used in the ice modelling it is therefore necessary to calibrate for each shape. This was done for the local ice shapes shown in Figure 14: The curvature radius varies in the range of 0.68 m – 3.51 m. The calibrated parameters and the target force curves are shown in Figure 15. For the same ice deformation level, the "ice shapes with small curvature radius produce less force, but the force-area relationship are essentially identical.

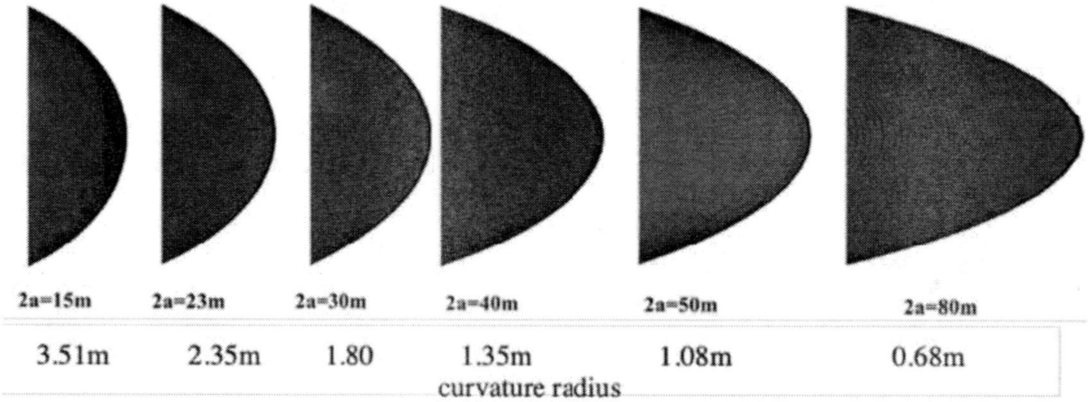

Figure 14. Assumed local shapes for iceberg colliding with platform column. The curvature radius in the front part is indicated [11]

IOP Conf. Series: Materials Science and Engineering **700** (2019) 012039 doi:10.1088/1757-899X/700/1/012039

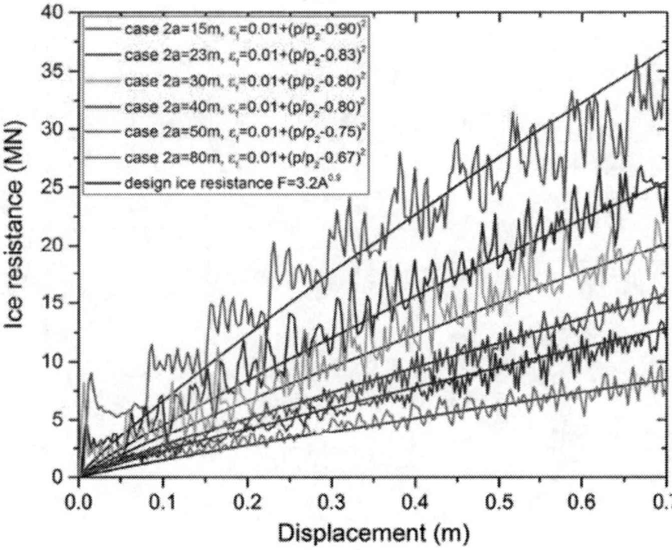

Figure 15. Target curves and force versus deformation from simulation during calibration of the failure criterion for each local ice shape. [11]

4. Local shapes

Using the calibrated material parameters for the various shapes the local resistance to ice impacts on the front plating of the column of a floating semi-submersible platform as analysed. The finite element model of the front panels is shown in Figure 16.

The resistance versus deformation of the ice and the structure is shown in Figure 17. It is observed that for the smallest curvature radius 0. 68 m (2a=80 m), the ice undergoes crushing while the structure deformation is very moderate. The ice resistance is virtually identical to that of a rigid structure. For increasing curvature radii (decreasing 2a), the ice resistance increases somewhat as the structure is subjected to increasing deformations, but no puncturing of the plating occurs, until the curvature radius is 1.8 m (2a=30 m). This represents a transition. When the curvature radius is 2.35 m (2a = 23 m) the mutual behaviour is reversed. The ice undergoes very little deformation and the structure must absorb most of the energy. The resistance is virtually identical to that of a rigid ice. This shape is particularly critical wrt. penetration of the front plating, which takes place at 0.8 m deformation. When the curvature becomes larger, the ice engages more of the front panel, so structure resistance increases. The deformations of the ice and the structure at two deformation levels are

illustrated in Figure 18. It appears that the ice is little crushed before it contacts directly with decks and transverse frames, but the contribution to energy dissipation is small.

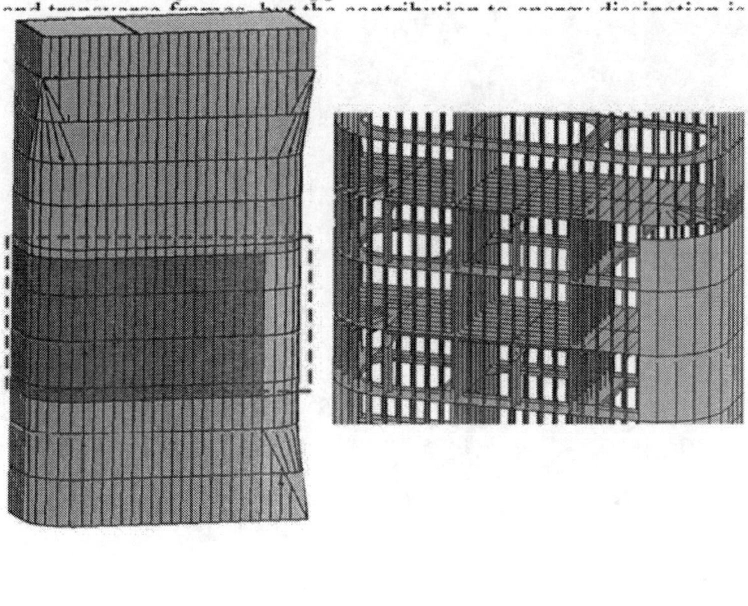

Figure 16. Finite element model of front panel of a column in semi-submersible platform. Typical deck spacing 2.9 m, frame/bulkhead spacing 2.5 m, plate thickness 17 mm, stiffeners L275•100•11.5/15, spacing 625 mm.

IOP Conf. Series: Materials Science and Engineering **700** (2019) 012039 doi:10.1088/1757-899X/700/1/012039

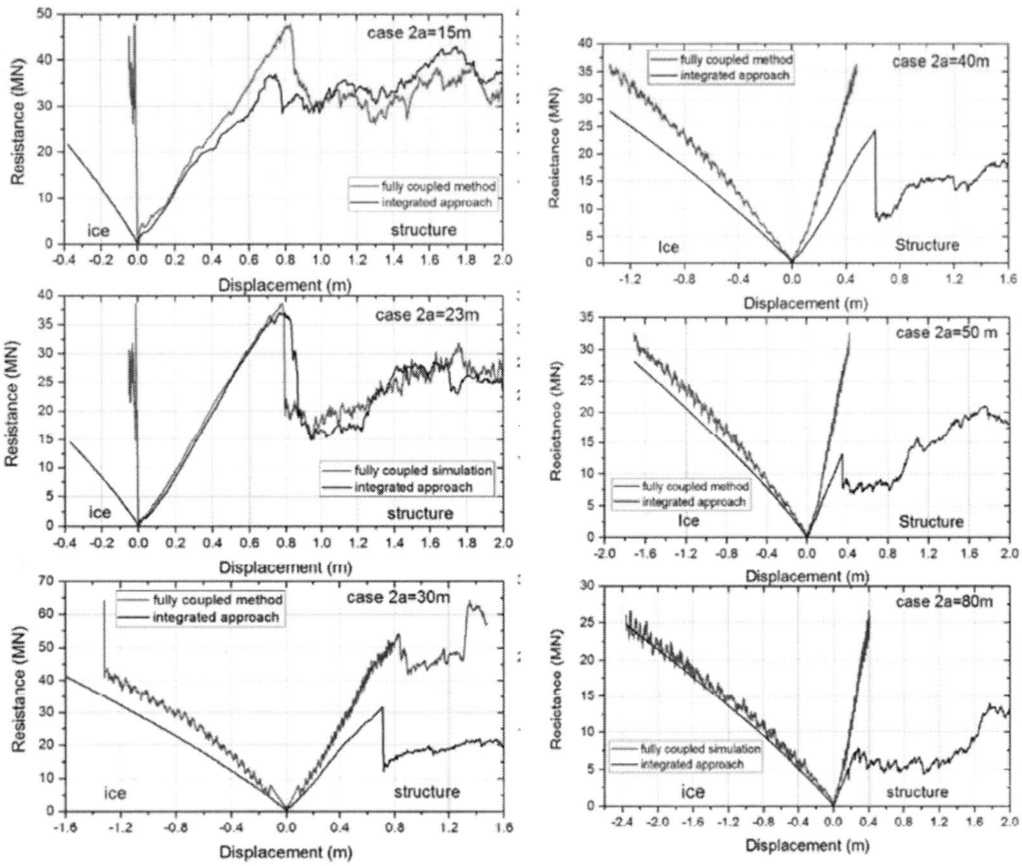

Figure 17. Force versus displacement of ice and structure from integrated analysis with deformable ice and structure (red curves) and from analysis where either the ice or the structure is assumed rigid (black curves). The curvature radius of iceberg varies [11].

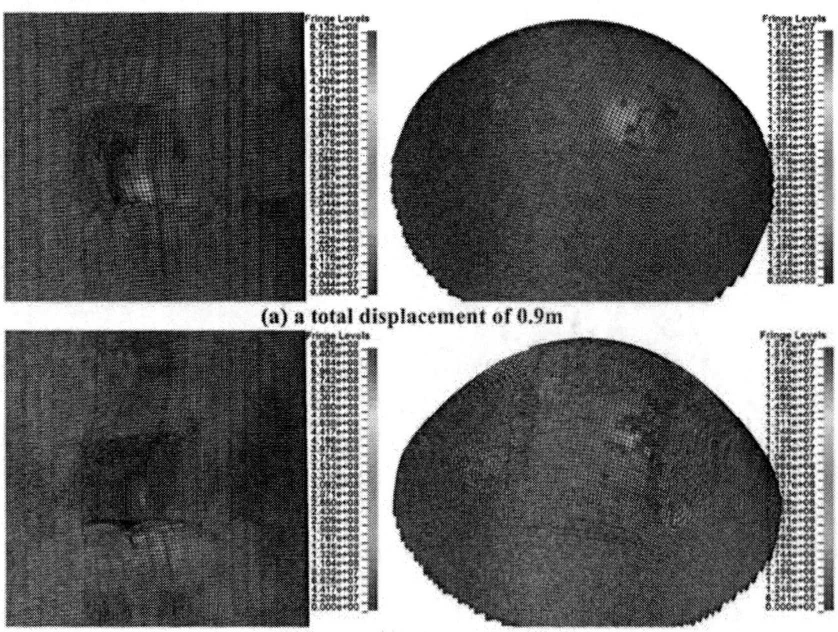

Figure 18. Structural damage and ice crushing for the critical ice shape [11].

IOP Conf. Series: Materials Science and Engineering **700** (2019) 012039 doi:10.1088/1757-899X/700/1/012039

5. Conclusions

Ships and offshore structures are typically designed against extreme ice loads (annual probability 10^{-2}) in the ultimate limit state (ULS) where only small yielding and plastic deformations are allowed. Explicit partial safety factors or in-built conservatism shall prevent the structures to enter the ULS. The design is carried on the basis of pressure-area (*p-A*) relationships. For plates between stiffener/frames or stiffener with associated plate flange the design pressures can be very high. For larger areas corresponding to stiffened panels between frames and deck/stringers, the pressures become moderate.

Offshore structures are also often designed for abnormal ice loads (annual probability 10^{-4}) in the accidental limit state (ALS). These pressures follow typically the same area dependency, but with higher intensities. In general, local plates and stiffeners do not need to be designed to resist these load levels because:

- Plates have generally considerable strength reserves when they undergo large deformations
- Stiffeners are generally too weak to support the plate in the large deformation range, but the shell plating can still carry significant loads in two directions
- When high local pressures exceed the resistance of the plate or stiffener, they yield and the ice contact is gradually transferred to adjacent structures components
- Locally deformed plates and stiffeners are accepted in the ALS

The ALS check should therefore normally be carried out for larger units, such as panels between decks/stringers and frames/bulkheads. The contact area will typically be larger than, say $4m^2$. The check should be based on average force–contact area relationships, where the pressure is only weakly decreasing with increasing contact area. Average pressure–area curves (force-area) curves implicit in the IACS code for Polar vessels may be adopted: The curve for PC 3 class vessel is e.g. fairly close to the average force curve in the Pond Inlet tests

Integrated analysis of structures subjected to ALS loads is becoming more common, where the ice is modelled as a material as well. Various constitutive models have been developed. At best they capture the essential features of ice crushing, but they need calibration, notably w.r.t. the strain failure criterion adopted. Plasticity-based models are highly mesh size sensitive and the failure criterion must be calibrated for each size.

It was considered to calibrate the ice parameters in tests mimicking the Pond Inlet tests. However, it was found that this would give unverified ice behaviour under less confined conditions. It is therefore recommended to use the more conservative approach; to calibrate the ice for crushing a rigid, planer surface..

Characterization of global and local ice shape is essential. The global shape governs the demand for strain energy dissipation as a fraction of the impact energy. The local shape is important w.r.t. analysis of structural damage and ice crushing. Unfortunately, no commonly accepted procedure for ice shape characterization exists. Very simplistic geometries are often assumed.

The effect of local ice shape, expressed as the curvature in the contact area, on the ice-structure interaction was investigated from a structural resistance point of view. It was found that ice with relatively large curvatures was crushed by the front panel in the column of a semi-submersible platform, and thus the structure resisted the ice loads well. When the curvature was reduced, the ice crushing resistance became larger due to the growing confinement caused by the increasing structural deformations. At a certain curvature, the ice was relatively little deformed and could penetrate the front panel virtually as a rigid object. This was denoted the critical shape for the given structural configuration, ice shape and ice strength. Ice with smaller curvatures became less critical in spite of less crushing, because the ice mobilized larger areas of the front panel.

The critical curvature radius was 2.35 m. This was in the same range as the distance between deck girders and vertical bulkheads/frames. More studies are needed to clarify if this finding can be generalized,

Acknowledgement

The work of the author was supported by the Research Council of Norway through the Centres of Excellence funding scheme, project number 223254 – NTNU AMOS.

IOP Conf. Series: Materials Science and Engineering **700** (2019) 012039 doi:10.1088/1757-899X/700/1/012039

References:

1. Organization, I.S., *ISO19906 Petroleum and natural gas industries – Arctic offshore structures*. . 2018.
2. DNVGL, *RP-C204 Design against accidental loads*. 2019, DNVGL: Høvik, Norway.
3. Kim, E., et al., *Uncertainty quantification in the ice-induced local damage assessment of a hull section*. 2019.
4. Masterson, D., et al. *The medium scale iceberg impact test program*. in *Proceedings of the International Association of Hydraulic Engineering and Research (IAHR) Symposium on Ice*. 1992.
5. IACS, *Requirements colncerning Polar Class*, in *I2: Structural requirements*. 2016, Int. Association of Classification Societies.
6. Lu, W., Yu, Z, Van den Berg, M, Lubbad, R, Amdahl, J, Løset, S and Kim E, *Assessment of Structural Damage due to Glacial Ice Impact*, in *PTIL - KONSTRUKSJONSSIKKERHET I NORDOMRÅDENE (NORD ST19)*. 2018, ArcIso: Trondheim, Norway.
7. Gudmestad, O.T. and J. Alme, *Implementation of experience from the Arctic seal hunter expeditions during the late 19th and the 20th century*. Ocean Engineering, 2016. **111**: p. 1-7.
8. Lu, W.A., nJ. *Glacial Ice and Offshore Structure Impacts under Wave and Current Excitations*. in *25th Int. Conf. on Port and Ocean engineering under Arctic Conditions*. 2019. Delft, The Netherlands.
9. Fylling, I. *On the Statistics of Impact Velocities and Hit Positions Related to Collisions and Mating Operations for Offshore Structures*. in *BOSS*. 1994.
10. Ommami, B., Berthelsen, P A and Firoozkoohi, R, *Loads, design and operation of floaters in the Arectic*. 2018, Sintef Ocean: Trondheim, Norway.
11. Lu, W., Yu, Z, Van den Berg, M, Lubbad, R, Amdahl, J, Løset, S and Kim E, *Assessment of Structural Damage due to Glacial Ice Impact*, in *PTIL - KONSTRUKSJONSSIKKERHET I NORDOMRÅDENE (NORD ST20)*. 2019, ArcIso: Trondheim, Norway.
12. Popov, Y.N., et al., *Strength of ships sailing in ice*. 1969, ARMY FOREIGN SCIENCE AND TECHNOLOGY CENTER CHARLOTTESVILLE VA.
13. Daley, C. *Energy based ice collision forces*. in *Proceedings of the 15th International Conference on Port and Ocean Engineering under Arctic Conditions (POAC)*. 1999.
14. Daley, C., *Oblique ice collision loads on ships based on energy methods*. Ocean Engineering International, 2001. **5**(2): p. 67-72.
15. Dolny, J., *Methodology for defining technical safe speeds for light ice-strengthened government vessels operating in ice. Technical report SSC 473*. 2017, Ship Structure Committee: Washington, DC.
16. Liu, Z. and J. Amdahl, *On multi-planar impact mechanics in ship collisions*. Marine Structures, 2019. **63**: p. 364-383.
17. Stronge, W.J., *Impact mechanics*. 2018: Cambridge university press.
18. Alsos, H.S., *Ship grounding - analysis of ductile fracture, bottom damage and hull girder response* in *Dept. of Marine Technology*. 2008, Norwegian University of Science and Technology: Trondheim, Norway.
19. Quinton, B.W.T., *Progressive damage to ship's structure due to ice loading* in *Ocean and Naval Architectural Engineering*. 2008, Memorial University.
20. Gagnon, R., *A numerical model of ice crushing using a foam analogue*. Cold Regions Science and Technology, 2011. **65**(3): p. 335-350.
21. Gagnon, R. and J. Wang, *Numerical simulations of a tanker collision with a bergy bit incorporating hydrodynamics, a validated ice model and damage to the vessel*. Cold Regions Science and Technology, 2012. **81**: p. 26-35.
22. LSTC, *LS-DYNA theory manual*. . 2015, Livermore Software Technology Corporation.
23. Gagnon, R., *Consistent observations of ice crushing in laboratory tests and field experiments covering three orders of magnitude in scale*. 1999.
24. Riska, K., Rantala, H and Joensuu, A., *Full Scale Observations of Ship-ice Contact: Results from Tests Series Onboard IB Sampo, Winter* 1989, Helsinki University of Technology, Laboratory of Naval Architecture and Marine Engineering: Helsinki, Finland.
25. Kierkegaard, H., *Ship collisions with icebergs*. 1993, Denmark Technology University: Lyngby, Denmark.
26. Liu, Z., J. Amdahl, and S. Løset, *Plasticity based material modelling of ice and its application to ship–iceberg impacts*. Cold regions science and technology, 2011. **65**(3): p. 326-334.

27. Shi, C., et al., *A nonlinear viscoelastic iceberg material model and its numerical validation*. Proceedings of the Institution of Mechanical Engineers, Part M: Journal of Engineering for the Maritime Environment, 2017. **231**(2): p. 675-689.
28. Xu, Y., et al., *Nonlinear viscoelastic-plastic material modelling for the behaviour of ice in ice-structure interactions*. Ocean Engineering, 2019. **173**: p. 284-297.
29. Tsai, S.W. and E.M. Wu, *A general theory of strength for anisotropic materials*. Journal of composite materials, 1971. **5**(1): p. 58-80.

IOP Conf. Series: Materials Science and Engineering **700** (2019) 012040 doi:10.1088/1757-899X/700/1/012040

Choice of fuel system for the offshore support fleet

M L Eliassen[1] and O T Gudmestad[1, 2]

[1] Western Norwegian University College, Haugesund Campus
[2] University of Stavanger, Norway

* Corresponding author: eliasenmartin@hotmail.com

Abstract. The stringent requirements of the IMO´s emission regulations call for alternative fuels and new ways of powering ships. However, the IMO predicts that technology which could reduce ship emission to zero, will either not be available, nor cost-efficient over the next 40 years. However, technological innovations in ship power & propulsion systems; such as duel-fuel LNG engines, and the utilization of energy storage technology in the form of lithium-ion batteries have been found to reduce harmful emissions; and are already utilized to quite some extent in the offshore support fleet. This paper elaborates on the reliability and sustainability of newer ship power technologies, where equal to existing levels of reliability are expected as a minimum.

1. Introduction

Forces calling for "greener" maritime- offshore operations have resulted in the development of and utilization of newer technologies related to ship´s propulsion systems. The vision of the International Maritime Organization is to reduce greenhouse gas (GHG) emission from international shipping, and to eventually eliminate GHG emission during the current century [1]. More specifically, the IMO pursues a 50% GHG emission reduction by the year of 2050, compared to 2008 emission data [2]. The IMO also aims at significantly reducing emission of air pollutants such as oxides of Sulphur (SO_x) and oxides of Nitrogen (NO_x) from ships [2], where ship emission is today limited by MARPOL Annex VI. Ship emission limitations will become more and more stringent. A tendency which is often referred to as decarbonization [3]. However, according to [3] and [4]; there is no single pathway to decarbonization of the shipping industry, for now, they point towards Liquefied Natural Gas as the most appropriate current and short-term mitigation of emission reduction demands.

Platform supply, anchor handling, and subsea vessels are often termed Offshore Support Vessels (OSV) [5]. These vessels typically service offshore units, floating or submerged, and other related equipment on the continental shelf. From a practical standpoint, these vessels can be regarded as a necessity in order to conduct any operations on the continental shelf. Here, the North Sea is classified as an Emission Control Area (ECA) which stipulate the most stringent emission limits, and thus vessels that operates in the North Sea ECA therefore need to be particularly technologically innovative, in order to be compliant. Starting at the origin, the choice of power & propulsion system heavily relies on a vessel´s operating profile [6]. As OSV´s have a rather varied operating profile, diesel-electric propulsion architectures are commonly used, as diesel-electric propulsion offer a high level of flexibility [6]. In diesel-electric propulsion, all consumers are principally connected to the same electrical network, which can lead to blackouts under fault conditions [6]. However, with a propulsion system with separated busbars, the risk of a full blackout is significantly reduced [6], [7]. This will be emphasized later on.

Further, Marine operations offshore can be quite complex [8], thus involving multiple interconnected aspects such as; operational safety, client´s requirements of power availability, preferences of high levels of manoeuvrability, and also the application of Dynamic Positioning (DP) [8]. In general terms, the design principle of an offshore vessel´s propulsion system revolves around designing a system which ensure safe operations at nearly any sea state [9]. As a consequence, highly reliable propulsion systems are pursued as certain failures such as power failures may compromise marine operations and can further cause high economical losses [10], starting at the low-consequence end.

There are in general two structural approaches towards improving vessels reliability. Either using components of higher quality, or designing-in redundancy in such as the power/propulsion system [10]. This either – or approach can be seen as an implicit compromise [10] as high-quality components are costly. However, when it comes to offshore support vessels, their operational requirement of a highly reliable propulsion system therefore involves a combination of both design approaches [10]. As a consequence of designing-in redundancy, power availability is increased, which is a typical attribute for these types of ships, [6, 7].

Reliability and safety in some ways walk hand in hand [11]: Reliability deals with the concept of failures, while safety concerns the consequences of these failures [11]. Continuing our focus on redundancy; there are different categories as can be seen in *Figure 1*. A common offshore support vessel design involves, among other things, including structural redundancy [12] for key systems such as generating sets. These gensets feed two or even three main switchboards (MSWB) with power, which is distributed to electrical consumers. The MSWB´s can be physically isolated from another by so-called bus-tie breakers, enabling the vessel to operate with two independent "sides" at the power & propulsion system. These conditions are normally operation dependant; and/or can also be a client requirement. This design philosophy is in connection to the principle of *spinning-reserve*, which is generally aimed at reducing the risk of blackouts [8]. More specifically, spinning-reserve means the running of more gensets than what is required in order to cope with unintentional shutdown of gensets, or other power failures [13].

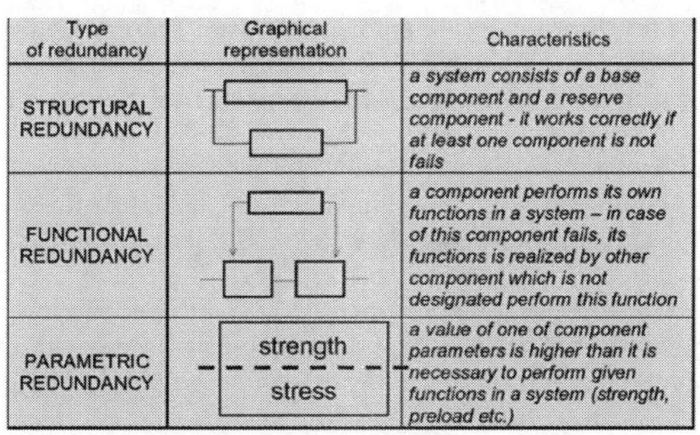

Figure 1. Types of redundancy [12].

However, designing-in redundancy and meeting operational requirements of *spinning-reserve* have a significant downside. They generally lead to the running of marine engines at low to medium loads [5, 7 - 9]; somewhat inefficient operational patterns. During these engine loading conditions, generating sets (gensets) generate each kWh at a higher fuel consumption and with less favourable emissions compared to when engines run at medium to high loads [9]. Further, these operating conditions generally lead to vessel operational patterns that generate a lot of emission [5].

Meanwhile, as offshore support vessels are subject to increasingly more stringent emission limits, this have led to the utilization of newer and alternative technology related to offshore vessels propulsion system. However, with respect to the above, and according to [14]; new technologies need to pertain the same level of reliability and safety as the more conventional propulsion system [14]. Hence; ship-owners and builders needs to account for multiple aspects, where several trade-offs can be identified. In this paper, we will focus on the choice of fuel system and alternative ways of powering offshore support vessels, considering reliability, operational safety, and sustainability in terms of emission. What options related to offshore support vessels propulsion system are there amongst existing technology today that can accommodate these aspects?

2. Methods

In development of this paper, a literature study was carried out, which was supplemented by raw data as found in the preparation of the first author´s master thesis [15]; which focused on similar topics. In addition, in order to provide a more pragmatic approach, the first author´s operational experience as a deck officer and Dynamic Positioning Operator (DPO) on board a variety of offshore vessels such as geotechnical drilling, construction, subsea, and platform supply is referred to, where found applicable.

2.1. Literature study

The majority of the used literature was found during the development of the first author´s master thesis. The strategy therein involved pursuing the application of published works as references, and to apply other sources such as regulatory publications, paper articles, manufacturers' statements, to supplement the above. However, as the initial literature study was done for the sole purpose of the master thesis, additional literature was needed in order to elaborate more specifically on the title of this paper. The literature study was primarily carried out in databases such as Web of Science, Research Gate, and Science Direct.

3. Results

In presentation of our results, the domain considering ECA and general approaches towards compliance are elaborated on. Then, applied technologies within the offshore support fleet is elaborated on, with regards to the focus areas of this paper.

3.1. Emission Control Area compliance

According to [16] there are three different approaches toward ECA compliance. The first involve using Heavy Fuel Oil (HFO) with emission reducing technology such as Selective Catalytic Reduction (SCR) and seawater scrubber technology. However, HFO, which often fuel large two-stroke marine engines, cannot in our case be considered to be relevant for OSV´s where diesel-electric propulsion is predominately applied; in which four-stroke diesel engines are dominant [17]. In addition, the emission regulations abolish this approach, i.e. sulphur limits of 0,1% in the North Sea ECA applies, which should render HFO inapplicable, Hence; is not considered here. The second approach involve using low sulphur fuels such as MGO (Marine Gas Oil) combined with SCR (Selective Catalytic Reduction). There are MGO´s today that meet the 0,1% sulphur content limit [18]. Further, if a low-sulphur MGO is combined with SCR technology which reduces NO_x emission by exhaust treatment in a urea plant [18], hence; MGO is still a viable fuel. The third approach towards ECA compliance, involve using LNG as a marine fuel [16].

Having elaborated on approaches towards ECA compliance, the significance of LNG as a marine fuel is here considered. Other fuels should, however, be mentioned; even though it can prove difficult to predict whether these fuels are suited for offshore support vessels or not. A scientific comparison of alternative fuels for was carried out by [19], with a strong emphasis on environmental and economic performance. The alternative fuels that was included in the study [19] were methanol, ethanol, LNG, and hydrogen. These fuels were evaluated on the basis of numerous criterions: Ship safety, global usage, bunker capability, durability, ease of application, rules & regulations, engine performance & emission, effect on engine components, as well as commercial issues. On the basis of their analysis, [19] found LNG to be the most applicable alternative fuel for ships.

Hydrogen came out second in their evaluation, however, it was found that more studies and improvements are required with regards to emission regulation compliance and with regards to concerns involving hydrogen´s effect on engine components. This need for further Research & Development (R&D) are also mentioned by others [20]. In continuance of our paper, LNG and possible combinations will therefore receive focus. In addition, the first author has served on board several LNG powered OSV´s, and therefore possess some insight.

3.2. Liquefied Natural Gas

For approximately 15 years ago, some offshore shipping companies decided to utilize LNG as a marine fuel. LNG medium speed engines can be regarded as a proven technology [21], and can be divided into three different categories when turbine engines are not considered (Balcombe et al., 2019). LNG powered offshore support vessels, utilizing low-pressure dual-fuel engines is one, while the other categories are; high-pressure dual-fuel engines, and lean burn spark-ignition (LBSI) engines [3], of which the latter is installed on board a couple of ferries [21]. LNG is a fossil fuel, which for the most part consist of Methane (CH_4) [22]. According to [23] LNG is heading into oversupply, which makes it cheaper than other fossil fuels. But, in order to utilize natural gas effectively, it is converted to its liquid form (LNG), and then takes up 1/600th of the volume, compared to its gaseous state [22]. Natural gas becomes liquid at approximately-162°C and is therefore considered to be a *cryogenic liquid* (-100°C and colder) [22]. Further, and according to GIIGN [22] it is highly important to bear in mind that it is the very properties of LNG as a good source of energy that also makes the fuel hazardous if not contained properly. Using LNG as a marine fuel involve additional safety measures compared to conventional diesel-electric propulsion; both structural/technical, and operational measures. Characteristically, the use of LNG as a marine fuel involve a higher sensitivity towards variable propulsive loads, as these types of engines typically have a narrower operating window (air-fuel ratio between 2.1 and 2.3) [21], than other fossil marine fuels.

The low-pressure dual-fuel LNG engine does not require any additional technology for compliance with IMO tier III emission requirements [9], the fuel is Sulphur-free [18], but, however, CO_2 reductions can only be said to be marginal [21]. The utilization of LNG also meets the Energy Efficiency Design Index (EEDI) baseline requirements [23]. However, literature [9], [3] identifies challenges such as *methane-slip*, and as a consequence, GHG reduction is limited to 8-20% [3]. Methane-slip is caused by misfire and incomplete fuel combustion as a consequence of a too lean gas/air mixture [21]. Methane-slip tends to occur in LNG combustion engines during low-load engine operation [18], which is a typical condition for these vessels when in ports [18]. Interestingly and according to the Intergovernmental Panel on Climate Change [24]; in a one-hundred years perspective, the emission of methane impacts the climate 28-34 times more than CO_2. Because of these conditions, [3] finds that the application of LNG as a fuel alone is not sufficient for meeting the future 50% GHG emission reduction as depicted by the [1]. However, it is as mentioned by [21] a proven technology, and, should the LNG supply feeding the DF engine be interrupted, the DF engines seamlessly switches from *gas mode* over to *Diesel mode*, without any effect on ongoing operations. Further, and according to [25]; marine engine technology can be regarded as technologically mature and reliable, where LNG can be utilized as a fuel in such applications as DF engines, without compromising safety at sea. However, capital investment is higher than for the more traditional marine fuels [21], [25].

Another mentionable challenge, is that LNG bunkering may; depending on the chosen method, be quite ineffective [26], i.e. bunkering from trucks where bunker facilities don't exist can be quite time-consuming. Apparently, a lack of infrastructure has been found to be one of the main reasons that the utilization of LNG as a fuel has not reached the heights as first predicted [27]. However, for offshore support vessels, arguably, this does not need to be the case. The first author knows from experience that many of the offshore bases in Norway today offer suitable LNG bunkering facilities. On the other hand, some does not, and LNG needs to be transported by trucks, which involve the bunker ineffectiveness as mentioned by [26]. When it comes to the choice of marine fuel, there are as depicted many aspects to consider. For now, focusing on commercialized fuel technology applicable for offshore support vessels; the options are somewhat confined to low-sulphur diesel oils, or LNG. However, commercialized technology such as energy storage can be combined with the use of these types of marine fuels.

IOP Conf. Series: Materials Science and Engineering **700** (2019) 012040 doi:10.1088/1757-899X/700/1/012040

3.3. Energy storage technology

It is reported in [6] that hybrid systems involving the application of energy storage technology supported by advanced control strategies, can reduce fuel consumption and emission in the range of 10-35%, while at the same time improve maintainability, manoeuvrability, comfort, as well as reducing noise. Further, energy storage technology has been found to be well-suited for vessels with a diverse operating profile [6], [28], [29], hence; a particular relevancy for offshore support vessels [8]. As operational requirements during DP operations usually call for the application of spinning-reserve, engines will as mentioned typically run at low loads, which result in fuel-inefficiency and a lot of emission [6]. Importantly, even if spinning-reserve is principally related to the design philosophy involving redundancy, it is not the same thing. Nor is it equal to such efforts as DP operations with an open bus-tie. Spinning-reserve, can, however, be regarded as the additional generating capacity that is easily available, and that is already connected to the switchboard.

The rise of energy storage technology for ships is said to originate from the FellowSHIP project [30]. In this project, which endured for 15 years, fuel cell technology and battery energy storage were installed and tested on board an LNG powered platform supply vessel.

At first, the project focused on fuel cell technology, but later shifted towards applying lithium batteries in a battery energy storage system, because fuel cells had a slower response time, where batteries could compensate [30]. Today, there are numerous OSV's that utilize battery energy storage in form of lithium batteries. To the knowledge of the first author, there are, however, no OSV's that currently utilize fuel cell technology. Fuel cells can be fuelled by natural gas, hydrogen, and other fuels [20] to generate electricity, however, and according to [31] fuel cells are technologically immature. The main problem according to [31] concerns the storage of hydrogen. As a testimony to these concerns, the explosion of a hydrogen tank at a Norwegian gas station is here mentioned [32].

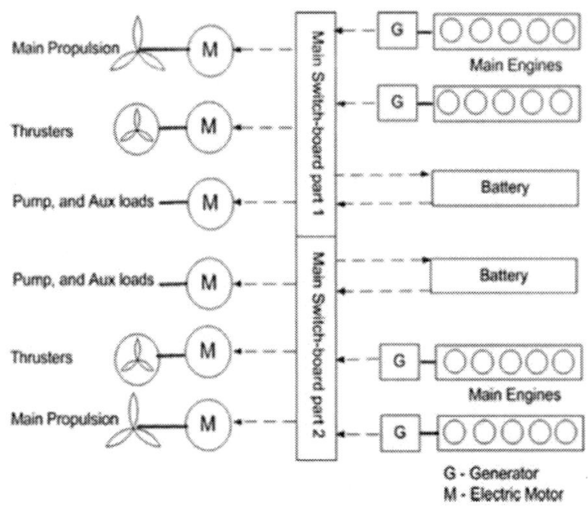

Figure 2. Hybrid propulsion layout [9].

There are many types of energy storage systems, and their applicability can partly be summarized by *Figure 3*. The various types of batteries are significantly more mature than fuel cell technology, hence; the trend today of installing battery energy storage.

Most offshore shipping companies today that have pursued energy storage technology utilizes some kind of lithium batteries in their systems [16], of which there are several [31, 33]. Battery energy storage systems are either installed as containerized solutions, or they are integrated in the ships structure. Usually, retrofits involve the containerized solution, and new builds involve integrating the battery energy storage system in the ships structure.

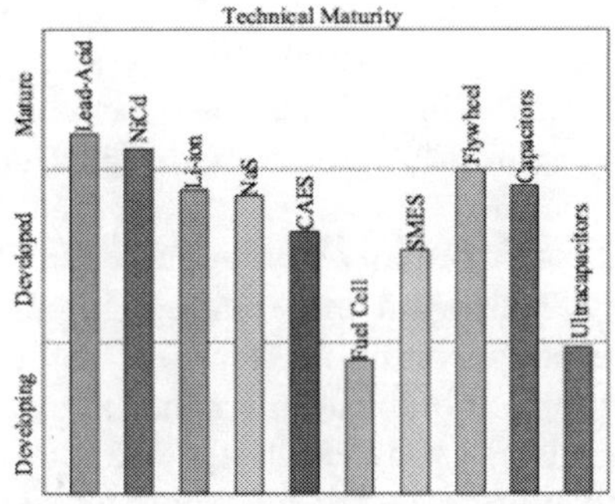

Figure 3. ESS technical maturity [31].

The main purpose of the battery energy storage system is shown in Figure 4, by applying so-called *peak-shaving* and *load-levelling* [31], which are controlled by rather advanced automated systems, such as the Energy Management System (EMS) [6]. Continuing our line on LNG, combined with battery energy storage, the following quote sums up an important benefit of this combination, which was discovered during the FellowShip project: *"... burning gas at low loads produces higher amounts of methane. Using batteries to cover low load requirements allowed us to reduce methane emissions".* Øystein Alnes, Principal Maritime Engineer DNV GL [34].

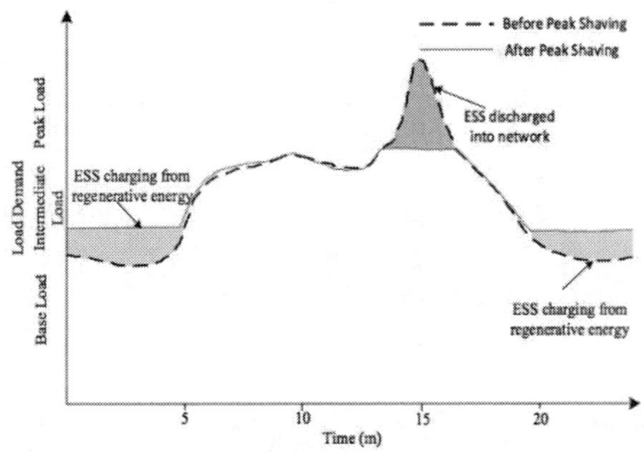

Figure 4. Peak-shaving and load-levelling [31].

In more general terms, which involve both LNG and MGO; ship owners' report of a reduction in fuel consumption in the range of 9-25%, as well as reduced maintenance costs due to the utilization of battery energy storage [16]. However, there are other benefits which are more relevant for this paper, which concerns the reliability of these systems. First, battery energy storage in the form of lithium batteries can be regarded as commercialized technology, as class notations do exists [35]. Here, we distinguish between the *battery power* and *battery safety* class notations. The main difference is that the battery power class notation allows for the battery energy storage system to serve as means of propulsive power, while the battery safety class notation does not.

However, the battery safety class notation have the potential of energizing the propulsion system for a shorter period of time and may therefore in this context be regarded as *functional redundancy* [12], 2018b). An example of this use, as mentioned by [7] may be to maneuver a platform supply vessel to a safe location, should failures at the propulsion system occur that could potentially lead to blackouts while the vessel is working alongside an installation. The other class notation, battery power, as mentioned, allows for the use of battery energy storage for purposes such as spinning-reserve. Hence, a battery energy storage system can replace the function of one or more gensets, during DP operations, and in other operational modes as required. This is well illustrated in *Figure 2*, but there are, however, other possible setups [6]. In this case, batteries act as additional redundancy, and can be considered as *structural redundancy* [12]. Considering *Figure 2*, where in the case of unintentional genset shutdown(s), or other power failures, the battery energy storage offer an increased level of power availability, further reducing the risk of blackouts. As an example, a captain of a platform supply vessel expressed that their everyday operations have been made safer, due to the application of battery energy storage [36].

Ultimately, the utilization of battery energy storage in vessels propulsion systems, as a rule of thumb support compliance with the newer stringent emission regulations for ECA´s as per MARPOL [12], and generally offer "healthier" operating conditions for gensets [16]. The utilization of battery energy storage here minimizes the running of combustion engines at low-loads [4], [5], [8], [9]

Further, it should be mentioned that even though it is clear that operating with an open bus-tie breaker (see *Figure 2*) reduces the probability of a full blackout significantly [7], this operating mode on the other hand becomes quite counter-productive with respect to the utilization of the battery energy storage system, depending on battery configuration [16]. Whether to operate with a closed or open bus-tie breaker depends on the specific maritime operation in question and/or client's requirements and appear to be a rather congested subject in the industry [16]. Aside from this, the additional redundancy that battery energy storage offers should not be underestimated, as marine operations offshore can be carried out even safer.

On the ship side, directly related to utilizing battery energy storage, a discussed aspect in literature revolve around the risk of *thermal runaway* [14], [37], [38], [39] and [40]. Thermal runaway is caused by such as erroneous operation, and internal faults as a consequence of faulty manufacturing [37], [40] Conditions such as overcharging, short-circuits, and submission to heat can lead to an increase in temperature at the battery cell level [37], leading to so-called *exothermic reactions* [40]; which causes several battery components to degrade; further leading to batteries emitting toxic and flammable gases [37], [40]. The stages of thermal runaway are informatively presented below by *Figure 5*.

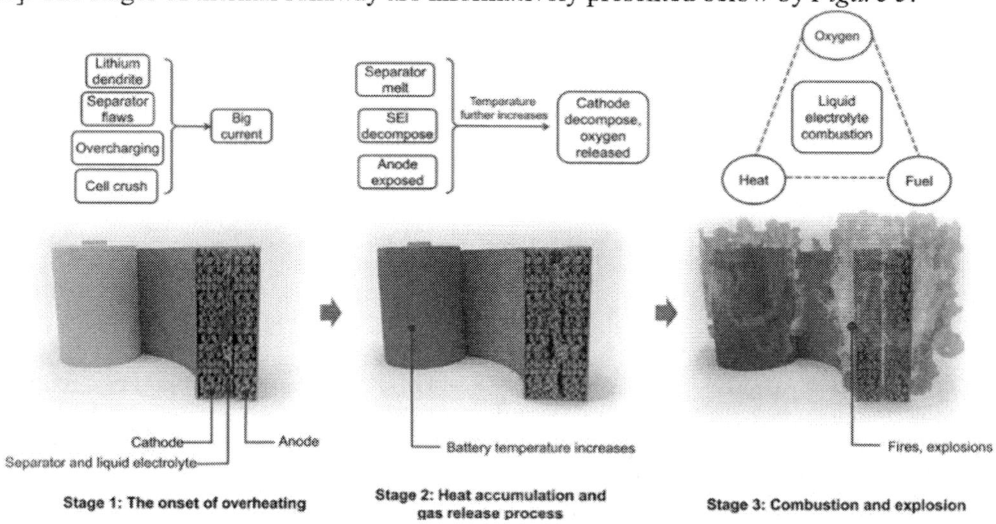

Figure 5. Stages of thermal runaway [37].

There exist several barriers, both battery internal and external, stipulated by class requirements [35] to prevent the occurrence and escalation of thermal runaway. However, even though this technology is considered to be safe [16], and even though battery energy storage is commercialized, the occurrence of thermal runaway is according to the [14] still possible; as failures of an internal kind within the battery cells might occur, in spite of the battery energy storage system´s internal safety systems. Suggestions to further increase battery energy storage safety and the prevention of thermal runaway revolve around the choice of materials for battery chemistry [37], which is regarded as the best prevention of the occurrence of thermal runaway. According to [41], the two most applied battery chemistries on board ships are

Lithium Manganese Cobalt Oxide (NMC) and Lithium Iron Phosphate (LiFePO4). NMC is the most applied chemistry per number of ships [41], considered to pertain a good stability and safety [42]. LiFEPO4 is the second most applied chemistry [41], and considered to be relatively safe during most conditions [37]. Other important aspects for the choice of battery chemistry are naturally energy density (kWh/kg) and power density (kW/kg) [31]. Therefore, and as mentioned by [42], when it comes to the choice of Li-ion battery, a trade-off between "*...efficiency, cost, and safety*" arises. Several other kinds of li-ion batteries exist [33]), with different characteristics; however, safety should always have the main priority.

On the more external side, other ways of improving the safety of the utilization of li-ion batteries in energy storage systems on board ships are suggested in literature. Reference [38] calls for battery water-cooling, and to abolish the use of air-cooling, as the water-cooling option is able to transport much more heat from the batteries, more effectively. Further, [31] highlights aspects such as preventing over-charging and over-discharging to prevent the rise of thermal runaway.

Erroneous use of battery energy storage might as mentioned lead to thermal runaway. In addition, erroneous operation involving deep discharges [40] can damage the battery cells, and reduce their lifetime, which is normally considered to be a period of minimum 10 years [16], and should therefore be avoided.

Interestingly, DNV GL (2018b) has reported of a planned initiation of a joint development project which will pursue the increase of knowledge concerning the use of li-ion batteries in the shipping industry; involving a lot of partners in different parts of the battery energy storage chain. Further, and importantly, the partners will pursue higher inherent safety levels for li-ion battery energy storage [43].

3.4. Other approaches

Other approaches today related to OSV's propulsion system involve whether to apply Alternating Current (AC) or Direct Current (DC) electrical distribution [6], [7], a rather complex subject, where the main difference can be said to be the use of fixed speed engines (AC) or variable speed engines (DC).

[31] Highlights the combination of several energy storage technologies as the future solution of the utilization of the technology, this is referred to as *hybrid energy storage systems*'.

Other approaches not directly associated with the propulsion system, but as efforts to reduce fuel consumption and emission involve strategies such as *slow steaming* [4], *cold-ironing* [44] the use of anti-fouling paint, various hull designs, and several others [3], [4]. There are as mentioned, and according to [3], [4] no single pathway to decarbonization, and a combination of several approaches are therefore suggested in literature.

However, a greener approach does not in all cases necessarily grant any market premium in the offshore segment [45]. Towards the ending of this paper, the following quote is here found to be befitting: "*The ultimate goal of any new technology is to be able to perform the same job more efficient, less polluting, with reduced cost and improved safety*" [13].

4. Discussion

In the preparation of this paper, we set out to elaborate on the choice of fuel system for offshore support vessels. Given the continuing and increasing focus towards greener maritime- offshore operations, and in the shipping industry in general, from alternative ways of powering OSV's an emphasis on LNG as a marine fuel and the electrification trend arose. There exist, according to the International Maritime Organization, no alternative fuel which could eliminate harmful emission from ships. They predict that it will either not be available, nor cost-efficient over the next 40 years. Therefore, a combination of approaches (policy, strategy, technology etc.) are called for. Whichever chosen technology, it shall pertain the same level of safety as the more conventional technology. A suitable approach for offshore support vessels today, based on the aspects as presented in this paper, may be to combine dual-fuelled LNG-engines with battery energy storage, utilize the *cold-ironing* (shore power) concept while in port, and to adapt/apply strategies such as slow steaming when reasonably practicable.

Using LNG as a marine fuel for these vessels involve additional safety measures and other challenges related to the efficiency of bunkering operations, as well as methane-slip. To use LNG as a marine fuel is considered to be safe, given the technical maturity of marine engines. However, should failures related to the LNG fuel system occur, the dual-fuelled engines seamlessly switch to diesel mode, not affecting ongoing maritime- offshore operations. Further, methane-slip can be significantly reduced by the application of battery energy storage; as low-load engine operation is minimized. This can also be the case when these vessels are in port, as cold-ironing can be applied, if the port offer sufficient shore-power facilities. Most importantly, the additional redundancy and back-up power offered by the application of battery energy storage should not be underestimated, as operational safety increases.

Failure conditions within the propulsion system which could normally cause a partial or full blackout can be mitigated either indirectly or directly by the battery energy storage system, as a function of the increased redundancy it offers. Therefore, the probability of blackouts can be significantly reduced.

However, even though safety barriers are installed as required by classification societies, the occurrence of thermal runaway is still a possibility, and has already received focus, where corrective measures are sought for, referring to the launch of the planned project orchestrated by DNV GL.

5. Conclusions

Regulators stipulate emission limits that will become more and more stringent over time. Technologies that have the potential of accommodating these limits must therefore be pursued. Meanwhile, there exist no clear pathway towards decarbonization, therefore, a combination of approaches is advised.

Using LNG as a marine fuel contributes to reducing harmful emissions such as oxides of sulphur and oxides of nitrogen. However, LNG only marginally reduce GHG emissions. However, combing LNG and battery energy storage further reduce emission, as low-load engine operation becomes significantly reduced, amongst other effects. Application of battery energy storage result in even more reliable offshore support vessels, due to the additional redundancy the system offers, which further enhances safety in marine operations offshore.

References

[1]. International Maritime Organization (2018a) Note by the International Maritime Organization to the UNFCCC Talanoa Dialogue. Adoption of the initial IMO strategy on reduction of GHG emissions from ships and existing IMO activity related to reducing GHG emissions in the shipping sector.
Retrieved from: http://www.imo.org/en/MediaCentre/HotTopics/GHG/Pages/default.aspx

[2]. International Maritime Organization (2018b) The International Convention of Prevention of Pollution from Ships (MARPOL). School access: https://vp.imo.org/

[3]. Balcombe P, Brierley J, Lewis C, Skatvedt L, Speirs J, Hawkes A and Staffel I 2019 How to decarbonize international shipping: Options for fuels, technologies and policies. Energy Convers. Manage. **182** 72-88.

[4]. Bouman E A, Lindstad E, Rialland A I and Strømman A H 2017 State-of-the-art technologies, measures, and potential for reducing GHG emissions from shipping – A review, *Transp. Res. Part D* **52** 408-21.

[5]. Lindstad H E, Eskeland G S and Rialland A 2016 Batteries in offshore support vessels – Pollution, climate impacts and economics, *Transp. Res. Part D.* **50** 409-17.

[6]. Geertsma R D, Negenborn R R, Visser K and Hopman J J 2017 Design and control of hybrid power and propulsion systems for smart ships: A review of developments, *Appl. Energy* **194** 30-54.

[7]. Skjong E, Volden R, Rødskar E, Molinas C, Maria M, Johansen T R and Cunningham J 2016 Past, present and future challenges of the marine vessels electrical power system, *IEEE trans. Transp. electr.* **2**(4) 522 – 37

[8]. Valeria-Garcia J J and Atutxa-Lekue 2019 On the optimal design of hybrid-electric power systems for offshore vessels, *IEEE trans. Transp. electr.* **5**(1) 324-34.

[9]. Lindstad H E and Sandaas I 2016 Emission and fuel reduction for offshore support vessels through hybrid technology, *J. Ship prod. Des.* **32**(4) 195-205.
[10]. Tarelko W 2018a Application of redundancy in ship power plants of offshore vessels, *New Trends Prod. Eng.* **1** 443-470.
[11]. Verma A K, Ajit S and Karanki D R 2016 *Reliability and Safety Engineering.* Second Edition. Springer Verlag. UK: London. DOI: 10.1007/978-1-4471-6269-8.
[12]. Tarelko W 2018b Redundancy as a way increasing reliability of ship power plants, *New Trends Prod. Eng.* **1** 515-22.
[13]. Odegard S 2018 *Energy Storage – Now relevant for any vessel type*. Marinelink. Marinenews 11/2018. https://www.marinelink.com/news/energy-storage-relevant-vessel-type-443389
[14]. Norwegian Maritime Authority (2016) Guidelines for chemical energy storage – maritime battery systems. https://www.sdir.no/en/shipping/legislation/directives/guidelines-for-chemical-energy-storage---maritime-battery-systems/
[15]. Eliassen M 2019 *Reliability & Sustainability of Battery Technology in Maritime Applications.* Master thesis, Western Norwegian University College, Haugesund Campus, Norway.
[16]. Brynolf S, Magnusson M, Fridell E and Andersson K 2014 Compliance for the future ECA regulations through the use of abatement technologies or change of fuels. *Transp. Res. Part D Transp. Environ.* **28** 6-18.
[17]. Wärtsilä (Undated) Combustion engine for power generation: Introduction. https://www.wartsila.com/energy/learn-more/technical-comparisons/combustion-engine-for-power-generation-introduction
[18]. Styre L and Winnes H 2019 Emission from ships in ports. Chap. 6 in: B*ergqvist, R and Monios*, J (2019) Green ports. Inland and seaside sustainable transportation strategies. Elsevier. ISBN: 978-0-12-814054-3.
[19]. Deniz C and Zincir B 2015 Environmental and economical assessment of alternative marine fuels. J. Cleaner Prod. **113** 438- 49.
[20]. Inal Ö B and Deniz C 2018 *Fuel cell availability for merchant ships*. INTNAM 2018 conference. Retrieved from: https://www.researchgate.net/publication/324910580_Fuel_Cell_Availability_for_Merchant_Ships
[21]. Æsøy V, Einang P M, Stenersen D, Hennie E and Valberg I 2011 LNG-fueled engines and fuel systems for medium-speed engines in maritime applications. NTNU/MARINTEK. Retrieved from: www.sae.org.
[22]. The International Group of Liquefied Natural Gas Importers (GIIGN, undated). LNG information paper no.1. Retrieved from: https://giignl.org
[23]. Attah E E and Bucknall R 2015 An analysis of the energy efficiency of LNG ships powering options using the EEDI. *Ocean Eng.* **110** 62-74.
[24]. IPPC, Intergovernmental Panel on Climate Change. Retrieved from: https://www.ipcc.ch/reports/
[25]. Abadie L M and Goicoechea N 2019 Powering newly constructed vessels to comply with ECA regulations under fuel market prices uncertainty: Diesel or dual fuel engine? *Transp. Res. Part D: Transp. Environ.* **67** 433-448
[26]. Rozmarynowska-Mrozek M 2015 *The Development of the LNG-Fueled Fleet and the LNG-Bunkering Infrastructure within the Baltic and North Sea Region*, Naval Engineering Conference (INEC).
[27]. DNV GL (2017) Uptake of LNG as a fuel for shipping. https://www.dnvgl.com/article/uptake-of-lng-as-a-fuel-for-shipping-104195
[28]. Späth N 2015 Update on "Shipping 2020": DNV GL sees hybrid propulsion and connectivity as emerging trends. https://www.dnvgl.com/news/update-on-shipping-2020-dnv-gl-sees-hybrid-propulsion-and-connectivity-as-emerging-trends-25931
[29]. Radan D, Southall M, Benatmane M, and Butcher M 2016 Integration, optimization and benefits of energy storage for marine applications. GE Power Conversion, *Int. Naval Eng. Conf.*

Bristol: IMarEST, pp.1-13
[30]. DNV GL 2018a FellowSHIP project concludes 15 years of maritime battery and fuel cell research. https://www.dnvgl.com/expert-story/maritime-impact/FellowSHIP-project-concludes-15-years- of- maritime-battery-and-fuel-cell-research.html
[31]. Mutarraf M U, Terriche Y, Niazi K A K, Vasquez J C and Guerrero J M 2018 Energy storage systems for shipboard microgrids – A review, *Energies* **11**(12).
[32]. Aftenposten 2019 Stenger ti hydrogenstasjoner midlertidig etter eksplosjon, (Norwegian) https://www.aftenposten.no/norge/i/P9OlAp/Stenger-ti-hydrogenstasjoner-midlertidig-etter-eksplosjon
[33]. Stan A-I, Stroe D-I, Teodorescu R and Andreasen S J 2014 Lithium ion battery chemistries from renewable energy storage to automotive and back-up power applicants – An overview, Int. Conf. Optim. Electr. Electronic Equipment (OPTIM). DOI: 10.1109/OPTIM.2014.6850936
[34]. Alnes Ø 2018 Principal Maritime Engineer DNV GL, private communication, 2018.
[35]. DNV GL 201) *Rules for classification. Ships*. Edition October 2015. Part 6 Additional class notations. Chapter 2 propulsion, power generation and auxiliary systems.
[36]. Equinor (undated) This battery-hybrid ship was designed to cut emissions. It exceeded their wildest expectations. https://www.equinor.com/en/magazine/battery-hybrid-supply-ship.html
[37]. Liu K, Liu Y, Lin D, Pei A and Cui Y 2018 Material for lithium-ion battery safety, *Sci. adv.* **4**(6)
[38]. Brown G 2018 Safety concerns for hybrid, electric ships. Sea technology. Buyers guide/Directory 2019. pp. 7-7. Retrieved from: https://pbes.com/safety-concerns-hybrid-electric-ships/
[39]. Gardner E 2018 Lithium-ion batteries: A new safety issue for ships? https://www.ship-technology.com/features/lithium-ion-batteries-new-safety-issue-ships/
[40]. Kurzweil P 2014 *Lithium battery energy storage: State of the art including lithium-air and lithium-sulfur systems*. In: Garche, J and Moseley, P T (2014) Electrochemical energy storage for renewable sources and grid balancing.
[41]. Henningsgård S 2018 Maritime battery forum (2018) Batteries in ships. Landstrømsforum 12 April 2018. Retrieved from: https://www.nek.no/landstromsforumsmote-april-2018/presentasjoner-fra-landstromsforumsmote-2018/
[42]. Bhatt A, Withers R and Wang G 2016 Lithium-ion batteries. Australian Academy of Science. https://www.science.org.au/curious/technology-future/lithium-ion-batteries
[43]. DNV GL (2018b) New JDP looks to drive marine battery safety and adoption. https://www.dnvgl.com/news/new-jdp-looks-to-drive-marine-battery-safety-and-adoption--113380
[44]. Bergqvist R and Monios J 2019 *Green ports in theory and practice*. In: Bergqvist, R and Monios, J (2019) Green ports. Inland and seaside sustainable transportation strategies.
[45]. Kyvik Ø and Gjøsæther S 2017 Environmentally sustainable innovations in offshore shipping: A comparative case study. *J. Innov. Manage. JIM* ,**5**(1) 105-31.

IOP Conf. Series: Materials Science and Engineering **700** (2019) 012041 doi:10.1088/1757-899X/700/1/012041

The influence of the composition of water on the results of accelerated aging resistance testing of low voltage subsea cables insulation

R Erfurth[1] and O T Gudmestad[*]

[1] Department of Mechanical and Structural Engineering and Materials Science, University of Stavanger, Norway

[*] Corresponding author: ove.t.gudmestad@gmail.com

Abstract. The composition of water changes with temperature, for instance, the saturation rate of water in polyurethane membranes is related to the temperature. Thus, it is necessary to evaluate the influence of the composition of water on absorption of water in subsea cable insulation. This paper is suggested to represent a supplement to existing standards regarding "Electrical insulation materials - properties with regard to thermal long-term behaviour."

1. Introduction

Rahmati et al. [1] reported that the increase in temperature does not have a significant effect on the percent of adsorption of water in polyurethane membranes. This means polyurethane will not adsorb more water at higher temperature, but faster. The Arrhenius approach is based on a constant diffusion according to Ficks' Law. However, the composition of water changes with temperature, for instance, the saturation rate is related to the temperature. Therefore, it is necessary to evaluate the influence of the composition of water on absorption of water in subsea cable insulation.

In a nutshell, when water is used as a conductor, chemical reactions are induced. These chemical reactions can involve the specimen, the electrodes or the water itself. In literature [2, 3] the dielectric constant (relative static permittivity) and the ion product of water are descriptive parameters of electrochemical behavior. A chemical reaction is hydrolysis. This paper is suggested to represent a supplement to existing standards regarding "Electrical insulation materials - properties with regard to thermal long-term behavior" [4].

To simulate the chemical degradation of the sheath material over time and quantify the insulation properties and application limits of low voltage cable they are artificially aged. The recommended practice to assess insulation properties advises the use of non-destructive measurement of insulation resistance. This approach is based on the Arrhenius behavior for chemical reaction of first order. Most common standards, which are referred to, are: ASTM D3032, VG 95218, NEK 606 and IEEE 1580. [5–8]. Standard test methods which include artificial aging separate aging, so-called pre-treatment and the actual measurement. If cables are artificially aged, then this is done almost exclusively in a heating cabinet in air or in other gas mixtures. Most standards do not define water in more detail, because the water is used exclusively for insulation measurement.

In the present work, the cables were aged at an accelerated rate in artificial seawater and the resistance insulation testing was in situ. The lack of testing guidance along with a wrong theoretical approach result in overly conservative and restrictive design parameters [4]. Hydrolysis is a chemical reaction of an order greater than one and accordingly, adjustments should be made

In this context, improving the validity of insulation resistance testing in combination with artificial aging is the very important. Measurements in water are predominantly of a unique nature and have not yet undergone sufficient assessment. The aim of the present work is to make the measurement results transferable. In previous tests, the electrical insulation resistance was determined according to the known method, Conventional short-term method for characterizing long-term thermal behavior [9, 10]. Great uncertainties have arisen in the process. The aim of the present study is to investigate to what extent the following parameters influence the measurements:

• Dwell time and the exact point in time of the initial/reference measurement,
• The water composition,
• The temperature coefficient and the convertibility of measured results.

Furthermore, the general set-up for the insulation resistance test is examined.

2. Measurements

The method employed in the present study is the point measurement testing, also known as short time measurements, which is the most frequently used method. A test voltage of 500 V is applied for a period of 120 seconds. This method returns a curve as a function of resistance over time.

By comparing periodically determined point measurement with the initial point measurement, it is possible to recognize trends of degradation. In addition to the statements regarding defined limit values that are considered "safe", this allows observations on changes in the condition of the insulation material over time.

Dielectric absorption behavior and polarization provides more information about the condition of the insulating material. However, if these were measured in situ, the mobility of the water molecules would have a misleading effect.

2.1 Initial point measurement

In point measurement the trend is established by comparing the change R_i (t = 120 s)/$R_{Reference}$. For the evaluation of long-term characteristic values, a limit value of 0.5, or 50 % relative to the initial value is often used [11] to define the end of service life.

The exact time at which the reference value is recorded must be identified. According to IEC 60092-350 [12] the first measurement should be made 1 hour after immersion. The pre-treatment taken into account according to DIN VDE 0472-502: "The measurement starts at the very latest 5 minutes after the end of the pre-treatment" [13]. The time of immersion in the water bath is defined as pre-treatment in DIN EN 60216-1: "For determining the initial value of the property [...], these specimens shall be pre-treated by storage for two days (i.e. for 48 hours +/- 6 ours) at the lowest stage of the storage temperature intended for the test [...], in order to achieve a stable value" [14]. Then the initial value is to be measured.

This means that further processing of the data is refined, for example in accordance with DIN VDE 0276-605: "Test report with diagrams showing capacity increases expressed as a percentage of the first day value" [15]. In Figure 1 the percent change compared to the previous measurement is shown, as one can see after 48 hours the variation is sinks significantly to less than 30% change. For more specifics regarding the sheets, see [19]. Note that details of sheet manufacture are confidential. Data points are given in the figure (Figure 1) and lines are shown to connect data points.

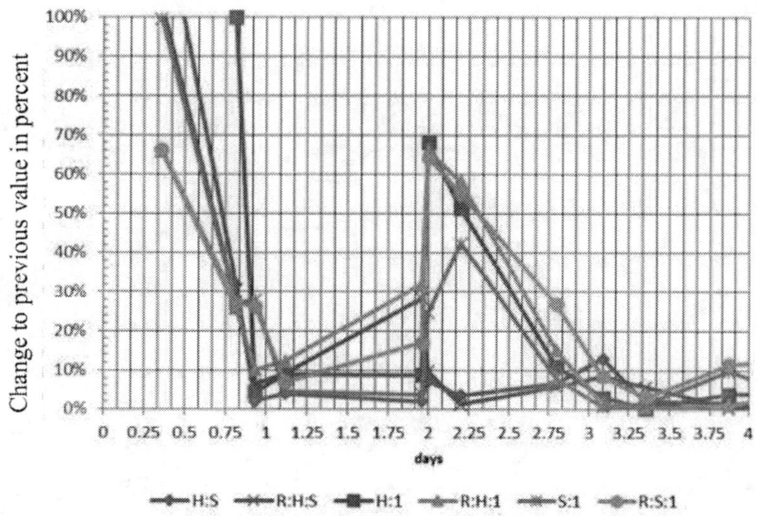

Figure 1. Resistance change in percent to previous readings of various types of cable sheaths.

2.2 Water composition

The electron neutrality condition generally applies for water. The sum of all cation equivalents, respectively concentrations, is equal to the sum of all anion equivalents, respectively concentrations. This means that water behaves electrically neutral. Thus, the pH value provides information about the activity of the hydrogen ion. Pure water (CO_2 free) has a pH value of 7, being neither an acid nor a base. The CO_2 present in the ambient air dissociates to carbonic acid and accordingly reduces the pH value.

$$H_2O+CO_2\rightarrow H_2CO_3$$
$$H_2CO_3\rightleftharpoons H^++HCO_3^-$$

Through buffer solutions, as present in real and artificial seawater, the H+-ions compound with the OH--ions of the calcium hydroxide to form undissociated H_2O. Therefore, pure water (pH 5.6 with dissolved CO_2) is chemically more active, Table 1.

Table 1. Properties of water.

	Ultra-pure water	Pure water	Purified water	Sea water
Resistivity [MΩ.cm]	10-18	1-10	1-0,02	~20
Conductivity [µS/cm]	0,1-0.055	1,0-0,1	1-50	~4.8 10^4

Gases diffuse from the surrounding air into the water with different saturations points which depend on the type of water, Table 1. Saturation is a factor regarding the transferability of the measurement results, since, for example, passive transport processes such as diffusion, take place until the concentration gradient is balanced, see also Figure 2.

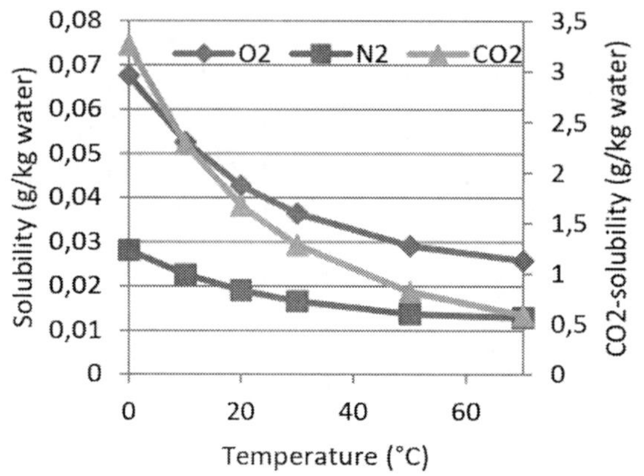

Figure 2. Solubility in water.

2.3 Temperature

Conductivity is temperature-dependent, and the change is described by temperature coefficients. In most insulation-resistance measurements, the range of temperature differences is in a relatively small interval and changes of material properties are neglected. At limits within which the material properties change little, a linear change in the temperature difference is assumed. However, the rate of change in a chemical system depends exponentially on temperature. Based on this relation, the resistance is derived:

$$R_T = R_{T\_0} \cdot e^{-b\left(\frac{1}{T_0}-\frac{1}{T}\right)}$$

Where

R_T	Resistance at temperature T
R_{T_0}	Resistance at reference temperature
b	Correction of insulation resistance

For a NaCl solution, in a small temperature interval, change in conductivity as a function of temperature can be assumed to be linear, Figure 3.

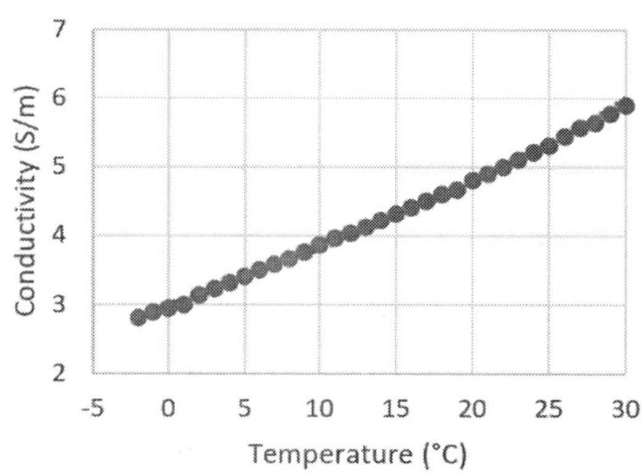

Figure 3. Conductivity of a NaCl solution (35‰) as a function of temperature.

However, if a trend-line is derived from a temperature interval of 30°, i.e. Figure 3, the following coefficient of determination is obtained:

Linear	$y = 0.0959x + 2.9204$	$R^2 = 0.998$
Exponential	$y = 3.0186e^{0.023x}$	$R^2 = 0.996$

In a paper by Ellison [15], the conductivity of seawater was measured. (Here we should note the differentiation between seawater and artificial seawater: According to investigations by Riley and Skirrow in 1965 [16], the density of the organic content of seawater is in the range of 1.50-3.00 g/m^3, which applies to most seawater areas. The particle sizes ranged from $1*10^{-6}$ m to $50*10^{-6}$ m. It was estimated that 14% of particulate suspensions were of organic origin. Organic particles included bacteria, fungi, phytoplankton, zooplankton and detritus. The inorganic suspensions consisted of different types of mineral particles. Trace elements of various complexes of metals, mercury, cadmium etc. were also detected. The dielectric constant measured in the study showed minimal deviations. However, these were in the range of the measuring error of 1 %.

The relation between salinity and conductivity is shown in Figure 4 at four different temperatures. Within a temperature interval of 30° C and a relatively small salinity-range a linear model of conductivity is reasonable. It is assumed that there is a direct proportional increase of the conductivity with the number of free ions dissolved in the water. Ellison summarizes that there is no significant difference between natural seawater and synthetic seawater, prepared according to the standard recipe [17]. However, he clarified that there is a significant difference in the conductivity and permittivity between natural seawater and an aqueous NaCl solution of the same salinity, [16].

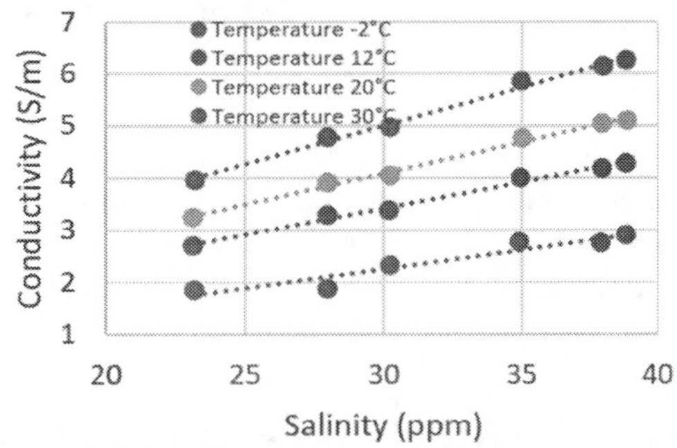

Figure 4. Conductivity as function of salinity at four temperatures with values taken from [16].

Measurement should be carried out in similar temperature conditions. If it is not possible, the results can be compared by conversion to a reference temperature. However, as shown in Figures 4 and 5, temperature and salinity are identified factors, which have a direct effect on the measurement result, a standardized measurement method is essential in order to achieve the best possible comparability.

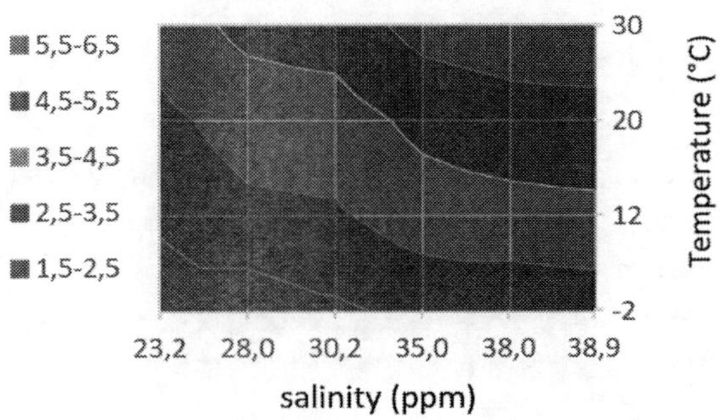

Figure 5. Conductivity as function of salinity and temperature.

2.4 Temperature coefficient

"The long-term thermal behavior of an electrical insulation material always applies only to a fixed property and a certain limit value," [14]. Since different properties change at different rates, a material may have different temperature indices. For materials with non-negligible changes in material properties

IOP Conf. Series: Materials Science and Engineering **700** (2019) 012041 doi:10.1088/1757-899X/700/1/012041

within the temperature interval, the resistance value varies quasi exponentially. In order to determine a dominating temperature coefficient for the specimen, a ramp test is carried out. The water temperature is successively increased until the defined apex is reached and then temperature is gradually reduced to the initial temperature in the same way. The insulation resistance is measured during this procedure in four hour intervals. In Figure 6, the averaged curve of insulation resistance versus time must be described by an exponential function (Too few data points are available to suggest formula at the moment).

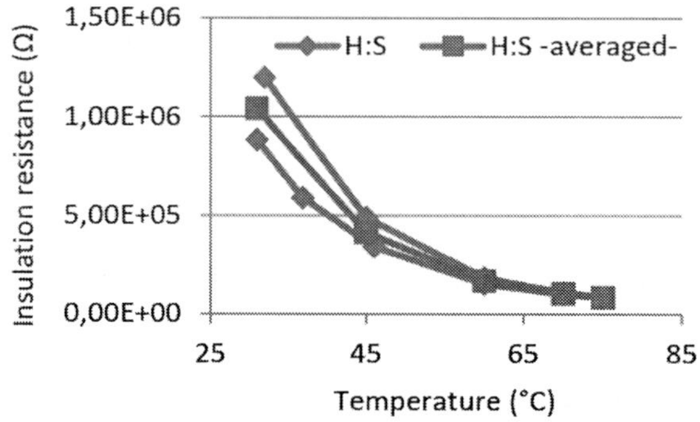

Figure 6. Ramp to determine the temperature function.

The arrangement of the cables is mentioned for the sake of completeness, as the standards do not provide any general recommendations. The geometry and position relative to the electrode in the water tank etc. have no significant influences. However, the attachment point of the guard terminal is noteworthy. To exclude the surface leakage current from the measurement a guard terminal must be attached to the cable.

3. Results and discussion

The total resistance is the sum of the internal resistance of the conductor, the insulation resistance in the cable and the resistance of the artificial seawater. The artificial seawater contributes to the total resistance due to the measurement of the specimen, taking place in situ. When a current is applied, the water turns into an electrolyte. Figure 7 shows the fluctuation of voltage right after an insulation resistance measurement with 500V over 2 minutes. The potential then decreases for about an hour and a half. The measured insulation resistance between two stainless steel electrodes, which were 15 cm apart, drops to a value of 4 mega ohm within an hour. The noise in the measured signal can be attributed to the formation of bubbles at the electrodes caused by water electrolysis.

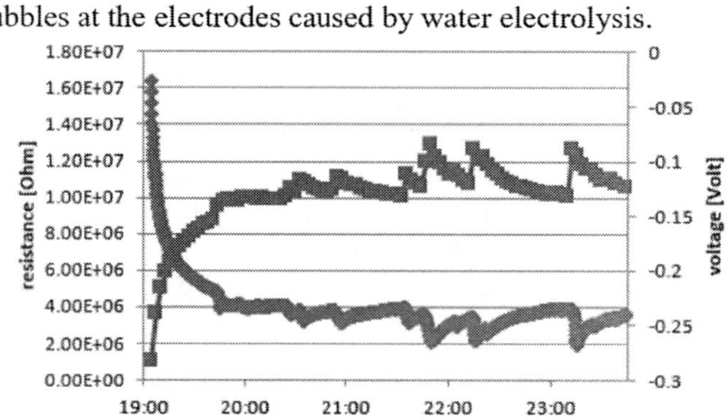

Figure 7. Resistance versus voltage fluctuation caused by electrolysis over of time.

For the electrolyte, one assumes an exponential temperature coefficient. Measured resistance from 19.5 °C to 80 °C result in a graph shown in Figure 8, approximated as a function:

$$R(T) = 227.05\Omega \cdot e^{-(T/71.42)}$$

Taking diffusion and hydrolysis into account, the composition of materials and water changes at elevated temperatures of 70 °C and more, resulting in a temperature coefficient that changes over time. This leads to an error, which is increasing over time. The ramp tests of the insulation resistance leads to a temperature-dependent function of the insulation resistance as shown in Figure 9.

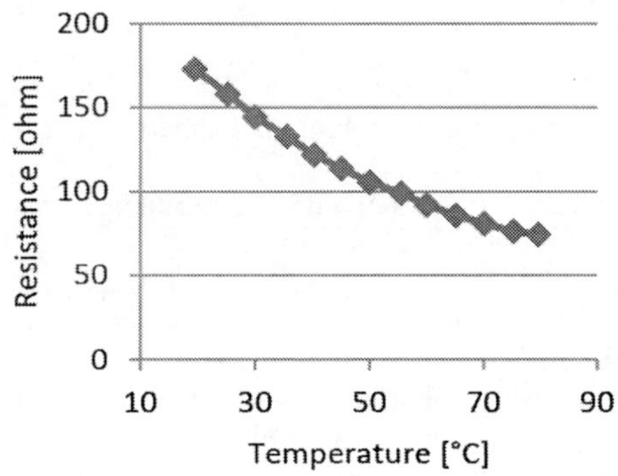

Figure 8. Resistance of water versus temperature (19.5 °C - 80 °C).

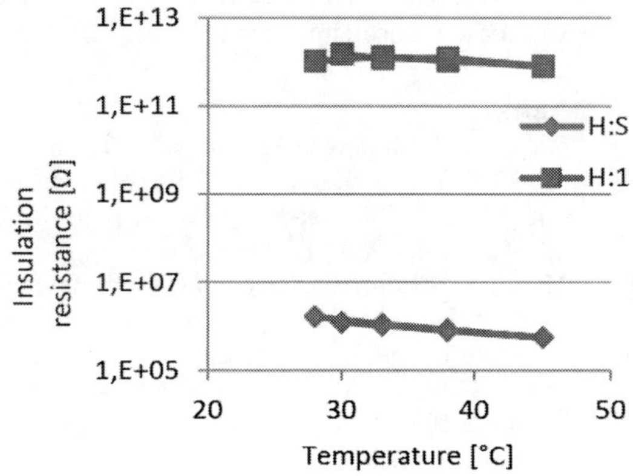

Figure 9. Dependence of insulation resistance as function of temperature.

$$R_{sheat}(T) = 8M\Omega \cdot e^{-(T/16.67)}$$
$$R_{core}(T) = 3T\Omega \cdot e^{-(T/41.67)}$$

In our experiments, the gradient of the exponential temperature coefficient was decreasing over 320 days by 13%. Rudorf [18] proposed a calculation method to compare the measured values from the short-term bearings with those of the base curve. The measured values of the time to failure to be regarded as provisional are to be corrected by comparing the calculated values with the values from the long-term storage - the basic characteristic:

$$t_N = t_R + R^k \cdot t_E$$

t_N Service life at normal temperature
t_R Period of time at normal temperature. Conditions
t_E Period of time at elevated temp conditions
R Extension factor
K Temp-coefficient with $k = \frac{T_{elevated} - T_{normal}}{10}$

The extension factor was originally assumed to be 2. By the changes of the temperature coefficient function over time presented before, the constant R should be adapted to the exponent dependent on time.

$$R = \frac{-0.13}{320} \cdot t_E + 2$$

As shown in Rudorf [18], this makes it possible to evaluate the measured values and to predict the estimated service life more precisely.

4. Conclusions

The resistance of water is several orders of magnitude lower than the resistance of the insulation material and can be neglected. However, the type of water dictates the extent of hydrolysis and chemical by-products. The complexity of the salinity/ temperature/conductivity ratio is enhanced by a further dimension caused by the ions dissolved during hydrolysis. This limits the comparability of test results despite materials with identical compositions and additives have been tested. Furthermore, the degree of chemical activity of the water will directly influence the process of artificial ageing.

The temperature at which the water is set up for testing differs among studies, or is unspecified. In conclusion, it can be said that the reference temperature of the insulation test is different among the reviewed standards. It is recommended to measure the reference value at an elevated temperature. Artificial seawater containing inorganic salts according to ASTM-D 1141 [17] is the recommended practice for use for in laboratory testing where reproducible simulation seawater is required.

A ramp test at the beginning and at the end of artificial aging can determine the temperature function of the material and the change of the gradient over time. The subsequent modification of the extension factor of Rudorf's adjustment function [18] increases the accuracy of the prognosis.

References

[1] Rahmati M, Modarress H, Gooya R 2012 Molecular simulation study of polyurethane membranes, *Polymer* **53**(9) 1939–50. doi: 10.1016/j.polymer.2012.02.051

[2] Grigull U 1983 Dielektrizitätskonstante und Ionenprodukt von Wasser und Wasserdampf, *Brennstoff-Wärme-Kraft* **35**(6), In German

[3] Hütter L A 1994 Wasser und Wasseruntersuchung: Methodik, Theorie und Praxis chemischer, chemisch-physikalischer, biologischer und bakteriologischer Untersuchungsverfahren, 6., erw. und aktualisierte Aufl. Laborbücher Chemie. Salle, Frankfurt am Main, In German

[4] DIN, VDE 2015 VDE 0304-7-1 Elektroisolierstoffe - Eigenschaften hinsichtlich des thermischen Langzeitverhaltens Teil 7-1: Beschleunigte Ermittlung der relativen thermischen Beständigkeit mit analytischen Prüfverfahren (RTE_A) - Anleitung für die Berechnung auf der Grundlage der Aktivierungsenergie(60216-7-1), In German

[5] ASTM International (2016) Standard Test Methods for Hookup Wire Insulation 29.060.10(D 3032). http://www.astm.org/cgi-bin/resolver.cgi?D3032-16

[6] Bundesamt für Ausrüstung Informationstechnik und Nutzung der Bundeswehr (2009) Kabel und Leitungen — Teil 2: Fachgrundnorm 29.060.20(95218-2), In German

[7] Incore cables NEK 606 standard. http://www.incore-cables.com/home/free-ebooks/ebook-nek-606-standard/

[8] IEEE 2011 Recommended Practice for Marine Cable for Use on Shipboard and Fixed or Floating Facilities (1580)

[9] DIN, VDE (2009) VDE 0304-25 Elektroisolierstoffe - Eigenschaften hinsichtlich des themischen Langzeitverhaltens Teil 5: Bestimmung des relativen thermischen Lebensdauer-Indexes (RTE) von Elektroisolierstoffen 29.035.01(60216-5), In German

[10] DIN (2002) Spezifikationen für Prüfverfahren: Teil 1-4: Elektrische Prüfverfahren - Isolationswiderstand 33.120.20(50289-1-4), In German

[11] DIN, VDE (2006) VDE 0304-22 Elektroisolierstoffe - Eigenschaften hinsichtlich des thermischen Langzeitverhaltens - Teil 2: Leitfaden zur Bestimmung thermischer Langzeiteigenschaften von Elektroisolierstoffen - Auswahl der Prüfmerkmale 29.035.01(60216-2), In German

[12] IEC Electric installations in ships: Part 350: General construction and test methods of power, control and instrumentation cables for shipboard and offshore applications 47.020.60; 29.060.20(60092-350)

[13] DIN, VDE (1986) VDE 0472-502 Prüfung an Kabeln und isolierten Leitungen - Isolationswiderstand und spezifischer Durchgangswiderstand 29.060.20(0472-502), In German

[14] DIN, VDE (2014) VDE 0304-21 Elektroisolierstoffe - Eigenschaften hinsichtlich des thermischen Langzeitverhaltens Teil 1: Warmlagerungsverfahren und Auswertung von Prüfergebnissen 29.035.01(60216-1), In German

[15] DIN, VDE (2009) VDE 0276-605 Starkstromkabel - Ergänzende Prüfverfahren 29.060.20(0276-605), In German

[16] Ellison W, Balana A, Delbos G et al. 1998 New permittivity measurements of seawater, *Radio Sci.* **33**(3): 639–48

[17] ASTM International (2003) Standard practice for the Preparation of Substitute Ocean Water 13.060.10(D 1141-98). http://www.astm.org/cgi-bin/resolver.cgi?D1141-98 (2013).

[18] Rudorf U 2016 Systemtheorie und Lebensdauerbewertung für Geräte der Meerestechnik: Teil 2 Nutzungsdauerberechnung für Kunststoffmaterialien in Unterwasserkabeln, 1. Auflage. Vorlesungen zur Meerestechnik. Shaker, Herzogenrath. In German.

[19] Erfurth R 2018 Time to failure of electric cables in seawater. *MSc Thesis*, University of Stavanger, Stavanger, Norway

Rationale for development of design basis for Barents Sea field developments

O T Gudmestad

Department of Mechanical and Structural Engineering and Materials Science, University of Stavanger, Norway

Corresponding author: otgudmestad@gmail.com

Abstract. Developments of hydrocarbon fields in the Barents Sea would normally include surface process units; thus, we must take into account the ice conditions in the area. Alternatively, smaller fields can be developed as satellites to existing production units. Full well stream to shore facilities may also be considered, in the case that one can document the flow conditions in the pipeline(s). Although glacial ice historically has been present all over the Barents Sea, the probability of meeting glacial ice in the southern part is very low, regarded by many as "negligible". At which latitude one could expect sea and glacial ice in the future is, however, uncertain, as past experience has seen glacial ice on the Coast of Finnmark County and there are ongoing processes related to global warming which might increase the probability of iceberg encounter at the location. We will discuss concerns related to the selection of the design basis for the Barents Sea, including the Russian part of the Barents Sea, and discuss the term "negligible". Of particular interest are criteria for the need for disconnection options for production units. The paper concludes with recommendations for thorough preparation of the design basis for the entire Barents Sea area.

1. Introduction

The importance of a design basis that adequately covers the relevant meteorological and oceanographic conditions for the design and operation of built facilities is of concern. The database of past events is normally used to statistically determine possible future events, with a certain annular probability of exceedance. The selected value of the annular probability of exceedance reflects the safety level chosen by the authorities. The exceedance probability for the "design event" is to be chosen with regard to the consequence of the event. Safety factors are, furthermore, included to cater for uncertainties related to the database and the possible consequences of the event. Standards distinguish between the importance (according to the importance class) of the facilities, by selecting a lower value of exceedance probability for the most important facilities. For example [1], hospitals, schools and the most important public facilities such as firefighting facilities are rated in a higher importance class than storage areas for nontoxic goods and houses for farm animals.

In the international oil and gas industry, the selection of a minimum safety level is set by the International Organization for Standardization [2]. Countries can decide to implement a higher safety level if desired.

During the process of selecting the safety level and safety factors and identifying design events, the database of past events is critical for ensuring that safe design is obtained. In this regard, the concepts of the known-knowns, the unknown-knowns, the known-unknowns and the unknown-unknowns play an important role [3]. In this paper the following terms and explanations will be used:

- Known knowns: Things we are aware of and understand (aleatory uncertainty like waves, wind...)
- Unknown knowns: Things we understand but are not aware of (hidden knowledge and hidden likelihood of occurrence like earthquakes)
- Known unknowns: Thing we are aware of but don't understand (epistemological uncertainty like elections)
- Unknown unknowns: Things we are neither aware of nor understand (ontological uncertainty like black swans)

We know about the waves in an area, and we can make statistical predictions about future wave conditions by using the present database. Data analysis shows that, using databases of past wave conditions, the extreme waves predicted for the south-western Barents Sea represent a stable estimate [4]. Using data assumed to represent future storms being caused by a presumed warming climate, the predicted extreme waves are higher [5]. So, we have a situation of known-unknowns: Should we use models which will give rise to larger design forces or should we base ourselves on data representing the known-known situation?

In this special case, it is difficult to give advice; however, the designer could choose to prepare a redundant design, which could be somewhat damaged, however, not to a level causing progressive collapse, should a worse prediction occur.

Alternatively, to ensure that all possibilities are covered, the designer could choose to implement a *Maximum Possible Event*, this would represent the maximum possible event restricted by physics. This event would be different from the *Maximum Credible Event* that is linked to a certain probability of annual exceedance, for example 10^{-4}.

For example for the wave condition, as based on physics, i.e. the geography and the extreme meteorology of the Barents Sea: Assume a low pressure from the west and a high pressure to the east. How large a gradient can be set up, and how large could the waves be in the area? A similar exercise was done for the northern North Sea [6], with a conclusion that the maximum possible wave condition is close to the wave found by using extreme statistics when selecting an annual exceedance probability of 10^{-4}. In this situation, the unknown-unknown situation represents an event we have no knowledge about, or we have no meteorological or mathematical model to predict it. In risk analysis, such an event is often called a "black swan" [7].

There is no way we can identify a "black swan" or an unknown-unknown; however, we can enquire amongst the local population and those using the area for fishing or other activities whether there have been events not reported in scientific or historical literature. Such events can be found described in historical archives, during interviews with "the elders" (see for example [8], where the author interviews sealer captains regarding the ice conditions in the Arctic Seas).

2. Examples of insufficient design basis

There are numerous examples of insufficient design basis. One example is the earthquake code of Kobe, Japan. Prior to the Kobe earthquake of 17 January 1995, the design basis for infrastructure was based on a comparison with US earthquakes [9], attempting to represent a known-known situation; however, insufficient attention was given to local historical information.

The Fukushima earthquake of 2011, which caused damage to the cooling system of nuclear reactors, might be considered in the class, unknown-knowns, as the combination of the earthquake and the huge tsunami (13 to 15 m high) following the earthquake was not considered in the design basis. The tsunami caused a large number of fatalities and damaged the cooling system of the nuclear plant, eventually leading to the escape of radioactive gasses and a large number of fatalities, caused by the radioactive exposure.

In intraplate areas around the world, large earthquakes are not expected; however, there may be local dormant faults, where historical records do not indicate any earthquakes. A devastating earthquake in any such an intraplate area might be considered an unknown-unknown event that it is absolutely impossible to warn against. However, large disruptions of geological formations, like water flooding or water injection into rocks, could trigger large earthquakes.

In Basel, Switzerland, a geothermal powerplant project was abandoned, as there were suddenly earthquake records with magnitudes up to Richter scale 3.4. There was grave concern that the project could trigger larger earthquakes, like the 1356 earthquake of magnitude 6.7, severely damaging the city [10]. It should be noted that one may in general not get a warning, as in the case of Basel, such that a known-unknown effect can be transferred into a known-known effect.

3. The safe design basis

In the oil and gas industry, the design conditions are given by the Ultimate Limit State (ULS) requirement [2]: "The ULS requirement ensures that no significant structural damage occurs for actions with an acceptably low probability of being exceeded during the design service life of the structure. Significant damage is associated with impairment of structure function". An annual probability of exceedance equal to 10^{-2}, with a factor of safety equal to 1.35 on the loading and a factor of safety of 1.15 on the material (steel) is selected in [2]. Furthermore, a condition for check of accidental collapse limit state (ALS) shall be investigated with an annual probability of exceedance equal to 10^{-4}, with safety factors set to 1.0. In the case of such an event, the structure shall still be able to withstand the 10^{-2} condition (with safety factors 1.0).

For fixed structures, the combination of these criteria should ensure that a complete collapse would have an annual probability lower than 10^{-5} [11]. The complete collapse of an offshore production unit would represent a very severe situation with respect to loss of life and environmental pollution. In the case of offshore structures exposed to waves, wind and currents, it is assumed that the extreme wave conditions are predicted by the use of statistical analysis of data from previous storms. It should be noted that the wave conditions give rise to loading on steel space frame structures that is proportional to wave height squared. There is no large jump in wave loading when the waves are increasing (as we will discuss later, the situation for interaction with ice floes is different, as the ice floes, and in particular glacial ice, vary greatly in size). Uncertainties related to possible future sea level rise and the effects of very large storm surges are implemented in the Norwegian NORSOK-N003 standard [12], by prescribing a certain air gap (higher than in previous versions of the standard), as wave-in-deck loading leads to excessive forces and moments.

Based on the above discussions, taking precautions for wave-in-deck loads, there should be no reason for selecting a wave design basis based on uncertain modelling of future wave situations [6] or on maximum possible events [7].

For physical load situations where the load, given a selected annual probability of exceedance, increases greatly when selecting a lower annual probability of exceedance, the situation is different. In this case, the ALS loading is much higher than the ULS loading. For the design of nuclear plants, where a very large number of people are at risk, the Maximum Possible Event [13] is selected as the design basis for structural design. This event is estimated with a basis in the potential size of maximum earthquakes on known faults. The local effects due to soil amplifications must be included. Although the estimates may be uncertain, the estimated earthquake accelerations may represent the best possible estimate to be used as design basis. It should be noted that the term Maximum Credible Event (MCE) is linked to a certain annual probability of exceedance [14] and is different from the Maximum Possible Event discussed above.

The question then arises as to whether an event larger than the MCE could be termed "*negligible*". The term "negligible" is defined as "so small, trifling, or unimportant that it may safely be neglected or disregarded" [15]. Aven [7] focuses on the probability of an event; however, the risk is often identified as the product of the probability of an unwanted event multiplied with the consequences. Thus, when using the term "negligible", one should consider both the probability and the consequences. Bearing in mind the term "black swan" or the term "unknown-unknown", we might not be in a position to predict the maximum possible event; however, we might be in a better position to imagine the maximum possible consequences. And when the maximum possible consequences are unbearable, the event should *not* be termed "negligible". Therefore, in such situations, measures should be implemented to reduce the consequences from "As low as reasonably practical" (ALARP) to "As low as possible", by introducing a sufficient number of barriers.

It should be noted that risk is at present e.g. in the PSA regulation [16] defined as "consequences of the activity with related uncertainty". PSA further indicate that if the uncertainty is large (unknown unknowns) the precautionary principle should be implemented. This statement is to a large extent a different way to express the same conclusion.

4. Design basis for Barents Sea oil and gas developments

We have discussed above the design basis for waves, wind and current relevant for all seas, including the Barents Sea. We have also discussed the handling of global warming effects, with respect to these meteorological and oceanographic conditions. The discussion reflects considerations related to unknown knowns and known unknowns.

We have also discussed the design basis for earthquake load effects and shared the concern that the ALS load effects are much higher than the ULS load effects for wave loading. A factor for ALS/ULS wave height of 1.3 (or 1.35) is normally quoted for harsh environmental seas, while a factor of 5 can be the case for earthquake load effects [14].

In the case of the Arctic Seas, and also for the Barents Sea, the question is whether sea ice is expected and in particular whether glacial ice is expected to occur. For a period of 30 years, the author has followed discussions related to a database for field development studies of the Stockman gas/condensate field at 73°N 44°E / 73°N 44°E in the Russian Barents Sea (where an estimated 3.8 x 10^9 Sm^3 of gas is in place). The design basis has been a matter of controversy. In the early 1990s, it was claimed that there was no need to design against icebergs, as drifting icebergs had not been sighted for many years. A fixed Tension Leg Platform (TLP) was suggested for processing the gas volumes. In 2003, however, a swarm of icebergs drifted into the area, due to winds and currents from the north east [16]. It should be noted that previous studies [18] had identified that drifting icebergs could be expected in the area. Short-term observations were considered sufficient to neglect the historical information. All assessments of a platform to support the processing equipment had to be re-evaluated after the 2003 events, and a dis-connectable production ship was selected for further design studies. In the case of a dis-connectable vessel, the vessel can leave the location should threatening ice conditions appear. This could be the case if the ice management implemented [18] is not capable of moving drifting icebergs away from a path where collisions may occur, or in the case of drifting sea ice threatening to cause the vessel to drift off location, causing damage to the anchor system. In this way (by ice management and the dis-connection option), there are several barriers to prevent the consequences of an unwanted ice event from escalating into a catastrophe.

Based on the experience with the Stockman development studies, the question remains regarding the parts of the Barents Sea in which one should consider the probability of drifting icebergs. According to [20], icebergs have even been located off the coast of Norway in historical times: in 1881, 1929 and 1939. Are these observations of relevance today? Could we expect icebergs all over the Barents Sea? The Atlas of Abramov [18] gives a positive answer to the question; however, it could be argued that the climate is warmer, so icebergs will melt more quickly as they are transported southwards. On the other hand, it could be argued that even an average temperature rise may not eliminate outbreaks of very cold weather. Furthermore, with an average warming, one could expect that more glaciers will melt and cause the calving of more icebergs in the future. For information, see also [12]. As a consequence of the uncertainty, the author recommends that the potential for drifting icebergs is evaluated for all developments in the Barents Sea, in accordance with recommendations in [12]: "For structures with annual probability of iceberg impacts in the range 10^{-4} to 10^{-5}, operational mitigations such as physical iceberg management and/or dis-connection should still be considered in accordance with the ALARP principle and actions from possible iceberg impact scenarios should be evaluated".

Based on the above discussion and recommendation, it is unclear what is meant by a statement in the Plan for Development and Operations (PDO) for the Johan Castberg oilfield in the south-western Barents Sea [21], where the term "negligible" is used for the unwanted event "drifting sea ice and icebergs". It is, furthermore, stated that "An Ice Risk Management System (IRMS) shall be implemented according to the PDO. The ice conditions will be continuously supervised. In the case of sea ice drifting south of 73°N and the ice being forecasted to drift further south, the production shall be stopped temporarily and shall not be started up until all ice west of 30°E on the Norwegian Shelf is retracted to the north of 73°N" (author's translation of the document). In the case of any oil spill, an Emergency Response Analysis is prepared [22]. For this field, the requirements of the standard [12] are thus fulfilled; however, there is no possibility to move the production facilities, should there be problems associated with the ice management. On the other hand, the ice drift will be forecasted many days ahead; thus, it will be possible to mobilize the necessary vessels for ice management.

For future developments even further north or east, where the likelihood of drifting ice and icebergs is higher compared to the Johan Castberg field, it is very important that dis-connection of the facilities be considered. There are two possible design options:

- To consider a vessel that can be dis-connected and moved away whenever there is a possibility of drifting ice or icebergs. Such facilities exist on the Grand Banks on the Newfoundland Shelf (the Terra Nova and White Rose floating production and offloading vessels, FPSOs). A key consideration is the re-connection and start up of the facilities after a dis-connection. Ice management will be vital (as on the Grand Banks), to ensure that the vessel is moved only in the case of non-successful ice management. It should be noted that Equinor (formerly Statoil) has considerable experience with dis-connectable production vessels, through operations of the Lufeng FPSO in the South China Sea (1997 to 2009), where several dis-connections were carried out to avoid typhoon weather situations.
- To consider a vessel that can be disconnected the "hard way", by investing a minimum amount in a turret that would be damaged in case of dis-connection. The investments will be small; however, the repair costs and the value of delayed production can be large. Net present value analysis would be necessary to choose between the two design options.

In the case of future developments of oil and gas resources in the Barents Sea, there is a need to discuss safety measures to reduce the risk of a loss of production facilities, with the subsequent possibility of pollution of large parts of the Barents Sea area, possibly with oil pollution emerging under the Arctic ice. It should be noted that the direction of currents is towards the north east; thus, an Arctic disaster is the ultimate consequence. The ice conditions are regarded as being of the type, known-unknown, as global warming causes large uncertainties. The risk can in no way be considered "negligible"; however, for the ongoing development of the Johan Castberg field, plans are in place to ensure that the probability of an unwanted event is low and there is an attempt to mitigate the consequences to a large degree.

5. Conclusions

We have reviewed the design basis for oil and gas developments in the Barents Sea. The design basis for wave loading on offshore facilities is considered adequate, using present-day statistical analysis of wave data, provided we ensure that wave in deck of any fixed offshore structure is avoided. We have, thereafter, discussed the need to reduce the consequences of an unwanted interaction between any surface facilities (for example, an FPSO vessel) and surface ice or glacial ice (icebergs). It is recommended that dis-connection possibilities are considered as the last barrier, should ice management fail. The need to consider such an event is very relevant, as the future of drifting ice is unknown and as the historical database documents drifting ice in most of the Barents Sea. It will give us a false feeling of safety to conclude, before a thorough analysis, that the risk is "negligible".

For development of future design basis for the Barents Sea, we suggest that:

- To deal with unknown unknowns (high uncertainty) a Maximum Possible Event should be used as design basis

- If we are not able to predict the Maximum Possible Event the Maximum Possible Consequence should be predicted.
- If the Maximum Possible Consequences are unbearable the event should not be termed "negligible" and the Maximum Possible Event should be used as design basis.

As to whether any unknown-unknown situation could arise with respect to the physical environmental conditions, and in particular with respect to drifting ice, this is by the selected terminology (unknown-unknown) impossible to foresee. The characteristics of "black swans" are that they are impossible to predict. However, we must realize that known-unknown situations may appear in the Barents Sea.

6. Acknowledgements

The author acknowledges discussions with colleagues during the preparation of the paper. In the oil and gas industry, there is a concern that a "too high level of safety" might be disadvantageous for the competitiveness of the industry. There is, however, among both industry partners and authorities, a joint interest in reducing the probabilities and consequences of unwanted events to an absolute minimum, and the term "negligible" is being used, although not accurately defined.

References

[1] CEN Standard 1998-1 2004 Eurocode 8: *Design of structures for earthquake resistance Part 1: General rules, seismic actions and rules for buildings*, European Committee for Standardization, Brussels, Belgium, http://web.ist.utl.pt/guilherme.f.silva/EC/EC8%20-%20Design%20of%20structures%20for%20earthquake%20resistance/64-1998-1_e_stf.pdf

[2] ISO (International Organization for Standardization) 2013 *Petroleum and natural gas industries, General requirements for offshore structures*, Geneva, Switzerland.

[3] Aasen R, Undated, Known Knowns, Unknown Knowns, and Unknown Unknowns, *Shipweight*, https://shipweight.com/blog/known_knowns_unknown_knowns/.

[4] Orimolade P A, Haver S and Gudmestad O T 2016 Estimation of extreme significant wave heights and the associated uncertainties: A case study using NORA10 hindcast data for the Barents Sea, *Mar. Struct.* **49** 1-17.

[5] Natskår A, Moan T and Alvær P Ø 2015 Uncertainty in forecasted environmental conditions for reliability analyses of marine operations, *Ocean Eng.* **108** 636-47.

[6] Reistad M, Magnussen A K, Haver S, Gudmestad O T and Kvamme D 2005 How severe wave conditions are possible on the Norwegian Continental Shelf?, *Mar. Struct.* **18**(5-6) 428-50.

[7] Aven T 2015 Implications of black swans to the foundations and practice of risk assessment and management, *Reliab. Eng. Syst. Saf.* **134** 83-91.

[8] Alme J 2014 *Ishavsfolk si erfaring*, Ishavsmuseet Aarvak, ISBN 9788291003078; an extensive summary in English is icluded as Appendix.

[9] Katayama T 2004 Earthquake disaster risk mitigation before and after the 1995 Kobe earthquake, *13th World Conference on Earthquake Engineering*, Vancouver, B.C., Canada August 1-6, 2004, Paper No. 5005, https://www.iitk.ac.in/nicee/wcee/article/13_5005.pdf

[10] Edwards B, Kraft T, Cauzzi C, Kästli P and Wiemer S 2015 Seismic monitoring and analysis of deep geothermal projects in St Gallen and Basel, Switzerland, *Geophys. J. Int.* **201**(2) 1022-39,

[11] Moan T 1995 *Safety Level across Different Types of Structural Forms and Material – Implicit in Codes for Offshore Structures*. SINTEF Report STF70 A95210, Trondheim, Norway.

[12] Standards Norway 2018 Norsok N-003 Actions and action effects, Standards Norway, Lysaker, Norway.

[13] IAEA 2012 *IAEA Report on protection against extreme earthquakes and tsunamis in the light of the accident at the Fukushima Daiichi nuclear plant*, Vienna, Austria, https://www.iaea.org/sites/default/files/protection040912.pdf

[14] Harries K A and Davies J B 2004 Seismic design and the international codes - the South Carolina

experience, *13th World Conference on Earthquake Engineering* Vancouver, B.C., Canada August 1-6, 2004, Paper No. 3235, https://www.iitk.ac.in/nicee/wcee/article/13_3235.pdf

[15] Dictionary.com, https://www.dictionary.com/browse/negligible

[16] Petroleum Safety Authority of Norway 2019 *Regulations relating to design and outfitting of facilities, etc. in the petroleum activities, (The Facilities Regulations)*, PSA, Stavanger, Norway.

[17] Zubakin G Z, Naumov A K and Buzin I V 2004 Estimates of ice and iceberg spreading in the Barents Sea, presented at *14th International Offshore and Polar Engineering Conference*, Toulon, France.

[18] Abramov V 1996, *Atlas of Arctic Icebergs*, Backbone Publishing Company, ISBN-13: 978-0964431140, ISBN-10: 0964431149.

[19] ISO (International Organization for Standardization) 2018 *Petroleum and natural gas industries -- Arctic operations* -- Ice management, ISO standard Number 35104, ISO, Geneva Switzerland.

[20] Kvitrud, A., 2002, *Environmental conditions in the Southern Barents Sea*, Web based information page, http://kvitrud.no/Environmental-condtions-Barents-Sea.htm

[21] Statoil (now Equinor) 2017 *PL532 Johan Castberg PUD del II* – Konsekvensutredning (in Norwegian).

[22] Statoil (now: Equinor) 2017 *Beredskapsanalyse for oljevern* Johan Castberg, https://docplayer.me/68526738-Beredskapsanalyse-for-oljevern-johan-castberg.html

IOP Conf. Series: Materials Science and Engineering **700** (2019) 012043 doi:10.1088/1757-899X/700/1/012043

Development of a method for evaluation of measures used for winterization of offshore facilities and units

S R Jacobsen[1,*], S Viddal[2], K G Dørum[1], O T Gudmestad[3] and H S Wiencke[4]
[1] Principle engineer, Petroleum Safety Authority, Stavanger, Norway
[2] Senior Manager, Sopra Steria, Stavanger, Norway
[3]Professor Emeritus of Marine Technology, University of Stavanger, Norway
[4]Partner, Proactima, Stavanger, Norway

* Corresponding author: sigurd.jacobsen@ptil.no,

Abstract. The environmental conditions in the Barents Sea include low temperatures, wind, snow and precipitation influencing safety, working environment, operations and equipment functionality. The conditions are not extreme compared to other Arctic areas, however they do necessitate protection and winterization of the facility. A study performed for the Petroleum Safety Authority (PSA) in 2016 identified known measures for winterization of facilities in the Barents Sea. The study found uncertainties related to the effect the measures have on winterization when considering environmental conditions. Further work was recommended to evaluate the effect of measures when considering performance influencing factors. A new study was initiated by PSA to develop a method to evaluate the efficiency of measures and a means to document the results. The method considers how well a measure achieves a specific safety and winterization goal, effect of the measure on energy consumption, effect of the measure on investment cost and operating cost, and vulnerability of the measure, including operational and organizational factors. In order to refine and test the method, it was applied to case studies considering a year-round production facility, a light well intervention vessel and a drilling facility used in the winter season in the Barents Sea. The paper is not intended to be a research paper but is a presentation of a method that has been developed to perform an initial evaluation of the effectiveness of winterization methods.

1. Introduction

Based on previous work including a study [1] and a paper [2] related to commonly used winterization measures, the Petroleum Safety Authority (PSA). initiated further work in 2017 and 2018 to follow up identified recommendations. These included developing a method to evaluate the efficiency of measures used for winterization and testing of the method by applying it to selected operational cases. There is currently no common approach to evaluating needs and effects of winterization measures. The work is documented in a report issued in December 2018 [3] and is the basis for this paper.

The southern part of the Barents Sea is of particular interest regarding winterization requirements for petroleum activity. The meteorological and oceanographic conditions of the area have provided the basis for selection of measures, climatic, meteorological and oceanographic conditions considered in the work covered by this paper.

1.1. Definitions and abbreviations

For clarity and to avoid misunderstanding, we have chosen to use the following definitions and abbreviations in this paper:

Issue: used generically to describe "something that poses a challenge" to operation in a cold climate

Measure: used generically to describe an action, solution or measure put in place to mitigate the effects of an issue. The purpose of a measure in this context is to enable safe and prudent operation under winter conditions including low temperature, wind and precipitation.

Metocean: abbreviation for meteorological and oceanographic conditions

MTO: man, technology, organization, the effects and consequences of human (man), technology and organizational influence on the performance of a given measure

One-pager: collation of the various elements of the assessment documenting and visualizing the results

Performance influencing factor: any factor that may affect the performance of a measure, normally in a negative way.

Wind chill: the combined effect of low temperature and wind leading to a lower effective temperature and greater cooling effect

Winterization: the combined mitigating measures that are put in place on facilities to enable safe and prudent operation in winter conditions.

1.2. Context

We regard it as necessary to describe the metocean conditions relevant for the area considered. The Barents Sea covers a large area and presents considerable variations in temperature and ice coverage. The southern part of the Barents Sea (south of 74,5°N) opened by Norway for oil and gas activity covers approximately 313 000km^2 [4]. The conditions in this area are not considered extreme. However, low temperatures (Figure 1), wind, precipitation, including accumulation of snow and icing on structures and equipment, do affect operations and equipment functionality unless countermeasures are implemented.

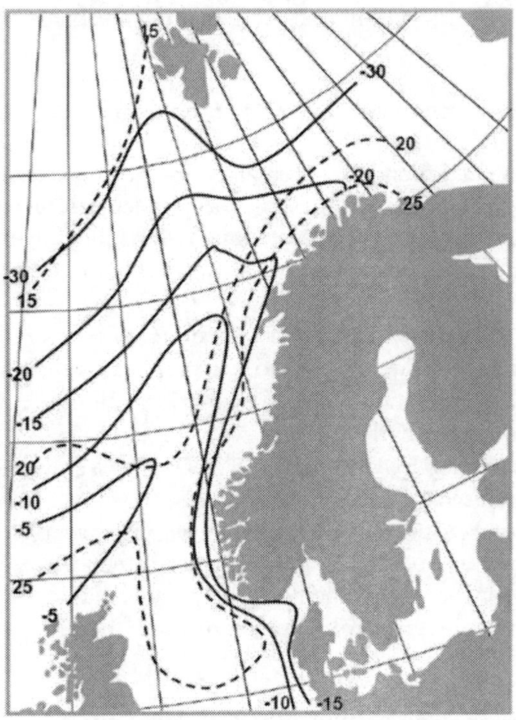

Figure 1. Highest and lowest air temperature with an annual probability of exceedance of 10-2 (the temperatures are given in °C) [5].

IOP Conf. Series: Materials Science and Engineering **700** (2019) 012043 doi:10.1088/1757-899X/700/1/012043

1.3. Metocean conditions

The metocean conditions for the Norwegian Continental Shelf are documented in NORSOK N-003 [5]. This standard contains a division of the opened area into five sub-areas having relatively uniform conditions and distinguishable from neighbouring areas. Recent studies performed by a collaboration of oil and gas operators in Norway have developed a comprehensive metocean report specifically addressing the relevant areas of the Barents Sea [6].

Ocean currents have a major influence on temperature and sea ice coverage in the Barents Sea. Warm Atlantic water from the south influences the climate in the Barents Sea. The Atlantic water keeps the southern part of the Barents Sea primarily ice-free during winter [6].

Intervals for temperature, wind and precipitation used in this work are based on metocean documents [5 & 6] and hindcast data from the Norwegian Meteorological Institute's NORA10 database [7]. The selection of the actual intervals for temperature, wind and precipitation are representative and do not include extremes.

Polar lows leading to relatively sudden increases in wind speed, changes in wind direction and heavy snowfall [8], represent a complicating issue that needs to be considered in winterization design.

Understanding the metocean conditions at a specific location is essential to assess the need for, and criticality of, different winterization measures. In order to perform a study for a given location and facility, the metocean data intervals should be adjusted to fit the specific conditions.

Air and seawater temperatures in the area affect the wind chill temperature and the potential for icing, both atmospheric and marine. Wind chill and icing are operational issues that are important to consider when operating in the Barents Sea.

1.4. Facilities and vessels in the Barents Sea

Currently there is only one permanent floating production facility, Goliat, in the Norwegian Area of the Barents Sea. There is a subsea facility for the Snøhvit field, which requires rig or vessel intervention. Exploration is performed using semi-submersible mobile offshore drilling units due to water depth. In addition, there are various types of vessels supporting the activities including emergency preparedness, transport of goods and equipment, well intervention, maintenance, construction and seismic surveys.

2. Methods and materials

The objective of the work has been to develop an assessment method and tool, "one pager", to evaluate the efficiency of a winterization measure. An experience-based approach is utilized in developing and testing the method. This has been found an acceptable approach based on previous work [1] and access to a broad experience base. Total involvement in the project has included persons from authorities, operating companies, drilling entrepreneurs, academia, risk management consultants, facility designers and suppliers of winterisation equipment. The work was divided into two stages.

Stage 1 - Development of method and tool:

- Develop a method to evaluate and analyse the effect of a winterization solution and measure,
- Include the performance influencing factors, which may strengthen or weaken the measure,
- Mature and refine the evaluation tool by performing a limited number of trials.

Stage 2 - Case tests utilizing method and tool:

- Test and refine the method by applying it to defined cases involving various types of facilities, locations and relevant measures.
- Evaluate optimal combinations of measures for varying weather conditions and locations.

The test of the method and the tool, one pager, was performed in a series of workshops. The participants in the workshops were supplemented with persons having operational experience with winterization of production facilities and exploration drilling, mainly from the Barents Sea.

2.1. Assumptions and limitations

Winterization measures are assessed individually. Combinations of measures are not included in the method but are discussed. Safety issues when designing winterization, are always of prime importance.

The winter season is not defined as any given months as this varies depending upon location. The geographical area is limited to the southern part of the Norwegian Continental Shelf in the Barents Sea opened for petroleum activities.

The effects of sea ice, icebergs, bergy bits, other forms of ice as well as fog have not been included as these are outside the project definition of winterization and are covered in other projects executed on behalf of the PSA. Polar lows have not been dealt with as a specific issue as they are a combination of low temperature, wind and precipitation and are thus inherently covered in the evaluation tool.

Climate data for temperature, wind and precipitation including snow was extracted from the NORA10 Hindcast database [7]. When using hindcast data for operational purposes, due conservatism should be applied as well as correcting for known biases.

The case studies have considered a selection of measures identified in the SINTEF report [1]. The selection has been made on the following criteria:

- All sub-issues must be represented by measures
- Similar measures are considered once rather than being duplicated.
- A balance between technical, operational and organizational measures is sought. Similarly, a balance is sought between measures that are suitable for modifications of existing facilities and measures that are only relevant to new builds.

2.2. Uncertainties and sources of error

As with all methods based on assessments made in a workgroup, one depends on the participants having both theoretical knowledge and practical experience. There is a degree of subjectivity when assessing the effect of a measure under different circumstances. There will be a considerable degree of subjectivity when distinguishing between "good effect" and "very good effect". Thus, the assessment may differ from person to person and between groups considering the same measures. This is a property, which the method shares with qualitative methods used for hazard and risk assessment, e.g. ISO 17776:2016 [9].

The estimates for investment costs (CAPEX) and energy consumption, operation and maintenance costs (OPEX), are generalized when considering what has a high or low impact. These assessments should be used with due consideration and be made specifically for a given facility when developing a winterization strategy.

2.3. Case studies – refinement and test of the method

The method was applied to three cases studies to retest, refine and confirm the viability and appropriateness of the method.

2.3.1. Case 1, Year-round production using a new build permanent production facility located in the Barents Sea area around the 73rd latitude towards the west. This case was selected to explore the flexibility that is available when taking the needs for winterization into account already in the design phase. The case allows for optimization of layout and design maximizing the effect of the selected measures. It also allows optimization of the electric generation systems required to power the winterization measures e.g. heating systems. A year-round operation case requires comprehensive measures to eliminate or minimise disruptions to activities.

2.3.2. Case 2, Winter drilling operation using an existing mobile offshore drilling unit located north of the 73rd latitude towards the east. This case was selected to explore the constraints on the choice of measures available when an existing unit is winterized by modification. This case illustrates the balance between what is desired and what possible, selection of measures, efficiency of measures, available electric power, CAPEX and OPEX. The case also provides an insight of what can be achieved when modifying an existing facility and the tailoring of measures employed on a new build facility, case 1.

2.3.3. Case 3, Well intervention using an existing light well intervention vessel in the same location as case 2. This case was selected to explore the constraints of modification of an existing unit (case 2) and the flexibility available when an operation can be planned with regard to weather and the possibility of suspending the activity if the environmental conditions exceed the design criteria. The expectation is that less stringent measures can be put in place compared to case 1 and case 2. This could reduce electric power requirements and costs, CAPEX and OPEX, while achieving an acceptable efficiency of the winterization measures. The operational window and up-time for the vessel will be lower than the facilities considered in the previous cases.

3. Results

The result of the work has been the development of a method and the "one-pager" providing a systematic approach to the assessment, documentation and visualization of the results [3]. An illustration of the process 'Figure 2' and the one-pager 'Figure 11' have been included to provide clarity.

The method was tested by application to three case studies considering a year-round production facility, a light well intervention vessel and a drilling facility utilized in the winter season in the Barents Sea. A workgroup including personnel with relevant competence and experience participated in the case studies.

The aim of the case studies has been to refine, test and confirm that it is possible and proper to use the method that has been developed. The actual results of the case studies as such are of less importance than confirmation of the viability and appropriateness of the method.

3.1. Evaluation method

The method is designed to evaluate an individual solution or measure. It is also possible to perform a case study assessing several measures for an installation of facility. The actual steps in the process are described below.

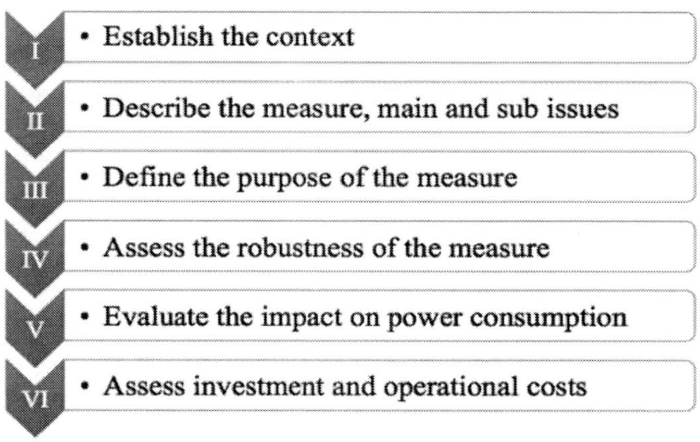

Figure 2. Method process chart.

3.1.1. Establish the context (I): The process is similar to that of a risk management process [10 & 11] and the purpose is to ensure that basic parameters and assumptions are defined, documented and agreed for the assessment.

The context has been established by identifying climatic conditions, such as minimum temperatures and their return periods, wind, precipitation, atmospheric icing etc. for the case studies used to test the method. The design assumptions and intended operations are established based on climate data for the specific locations with regard to operation, regularity requirements, barrier strategy including need for winterization of safety critical functions and design limits for the equipment. If the methodology is to be used to develop a concept for a specific project, it is natural that the level of detail needs to be increased.

3.1.2. Describe the measure, main and sub issues (II): Step II provides a short description of the measure and identifies the main and sub issues to be mitigated 'Figure 3'. The robustness of a measure is evaluated relative to the issue it is to mitigate and affect. For example, the cooling effect on a helicopter deck is greater than for a walkway, which is partially protected against the weather and wind by other structures.

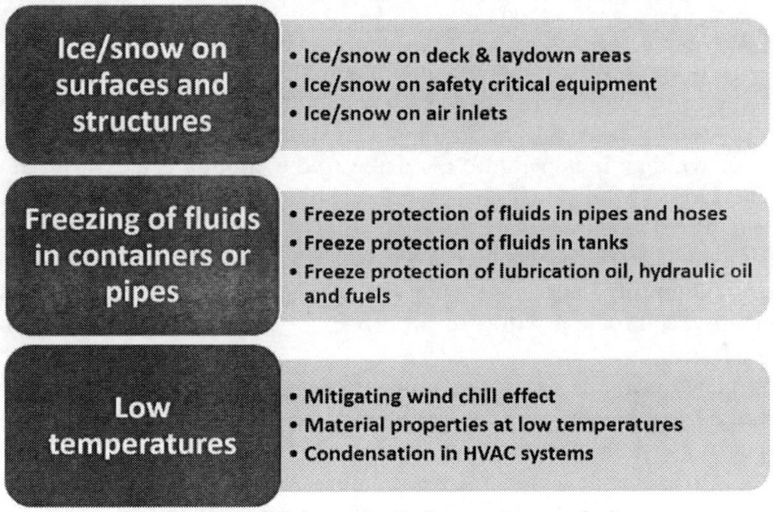

Figure 3. Main and sub-issues (examples).

3.1.3. Define the purpose of the measure (III): The purpose of a measure is to enable safe and prudent operations under winter conditions. Step III identifies the purpose of the measure, i.e. maintain normal operations or provide a safety critical function, i.e. prevent and limit the impact of unwanted events. Individual measures may take care of several issues. The most critical issue, in a barrier context, will define the performance requirement for robustness, capacity, reliability, efficiency, integrity etc. [12].

3.1.4. Assess the robustness of the measure (IV): Step IV is the core of the assessment taking into consideration performance influencing factors. The method uses the concept of robustness to describe how a measure is affected by conditions such as temperature, wind, precipitation and MTO factors. The opposite of robustness is vulnerability, and the terms are interchanged where appropriate. A robust measure will have the same or almost equivalent effect at a broad range of temperatures, winds and precipitation, as well as combinations of these.

Rather than assessing impact and robustness separately, we have chosen to assess robustness by looking at how well the measure performs under increasing severity of influence of temperature, wind, precipitation, and MTO factors. The effect of the measure describes how well it mitigates an issue. There is a significant element of subjectivity, especially in distinguishing between the three highest categories. The effect of the measure is shown in 'Figure 4'.

1	None or very little effect
2	Some effect, but insufficient
3	Useful effect, possibly sufficient
4	Good effect, sufficient
5	Very good effect

Figure 4. Effect of a measure.

The performance influencing factors, temperature, wind and precipitation are considered most relevant in a general context. When assessing measures for a given facility it will be necessary to make specific analyses for all performance influencing factors. This method does not limit the number of influencing factors that can be incorporated if appropriate and necessary. The following intervals have been chosen for temperature, wind and precipitation 'Figure 5'.

Interval	Temperature (°C)	Wind (m/s)	Precipitation (mm/day)
1	5 to 0	0	0
2	0 to -5	0 to 5	0 to 5
3	-5 to -10	5 to10	5 to 10
4	-10 to -20	10 to 20	10 to 20
5	< - 20	> 20	> 20

Figure 5. Intervals for temperature, wind and precipitation.

The intervals can occur in different combinations of wind, temperature and precipitation. Under real circumstances, some combinations are more likely than others e.g. the lowest temperatures usually occur in the case of clear weather and light wind.

The influence of wind and temperature is related to the cooling effect or wind chill. At any given temperature, an increase in wind velocity will increase both heat loss and energy consumption to achieve the same effect e.g. use of electric heating.

It is important to consider the robustness of a measure taking into account combinations of wind and temperature. Precipitation may also contribute to the cooling effect. For active and energy-intensive measures, the energy requirement will also be driven by the amount of energy required to melt or evaporate precipitation.

The dominant factor in terms of winterization is temperature. Most often, there will be a need to consider temperature in relation to wind to assess actual heat loss. Nevertheless, there may be issues and corresponding measures where it is the combination of temperature and rainfall that is of interest. In those cases, it is easy to exchange wind with precipitation on the y-axis using precipitation as the variable factor.

The method assesses the effect of the measure as a function of temperature (x-axis) and wind (y-axis) in a two-dimensional matrix as illustrated in 'Figure 6'. The example in the matrix illustrates that the measure is vulnerable to decreasing temperature and increasing wind. The coloured fields in the matrix categorize the vulnerability and robustness of the measure when combining temperature and wind.

Wind m/s						
> 20 m/s	5	Useful	Useful	Some	None/little	None/little
10 - 20	4	Good	Useful	Some	Some	None/little
5 - 10	3	Good	Good	Useful	Some	Some
0 - 5	2	Very good	Good	Good	Useful	Some
0	1	Very good	Very good	Good	Useful	Useful
		0 - 5	0 - (-5)	(-5) - (-10)	(-10) - (-20)	< (-20)
		1	2	3	4	5
		Temperature °C				

Figure 6. Robustness as a function of wind and temperature.

Precipitation is assessed independently of temperature and wind as shown in 'Figure 7'. This was done to avoid the complex presentation of a three-dimensional matrix.

None/little	>20 mm/day	5
Some	10 - 20	4
Useful	5 - 10	3
Good	0 - 5	2
Very good	0	1
	Precipitation	

Figure 7. Robustness as a function of precipitation.

An assessment of the effect of human (M), technology (T) and organizational (O) factors has been included. MTO factors influence the performance and the ability of personnel to act as required. It is used as a coarse assessment of vulnerability for issues like:

- Fatigue, inattention, habit, e.g. oversight of signs and warnings
- Resource requirements (persons) for the measure, e.g. manual removal of ice and snow
- Physical working environment factors, e.g. keeping personnel warm, adequate lighting, etc.

IOP Conf. Series: Materials Science and Engineering **700** (2019) 012043 doi:10.1088/1757-899X/700/1/012043

Human performance is affected by cold, including wind chill and rainfall [13]. This is assessed in the same way as vulnerability to temperature, wind and precipitation. An example is manual removal of snow and ice, which requires a lot of personnel resulting in a poor MTO score. Manual removal of snow and ice is also vulnerable to temperature, wind and precipitation requiring special protective clothing and limited exposure time for each person. In total one would not consider manual removal of ice and snow as robust.

The scale of the MTO factors 'Figure 8' is not divided into intervals and differs from the assessment of other physical factors, which affect a measure.

1	Little or no sensitivity to MTO factors
2	Some sensitivity to MTO factors
3	Moderate sensitivity to MTO factors
4	Sensitive to MTO factors
5	Highly sensitive to MTO factors

Figure 8. Categorisation of MTO vulnerability.

For categories 3, 4 and 5, it is necessary to identify compensating measures to reduce the vulnerability of the measure by MTO factors. The grading is coarse and includes many types of assessments. However, it is considered sufficient to highlight vulnerability of a measure to factors that affect human performance.

3.1.5. Evaluate the impact on power requirements (V): It is relevant to study the impact of the measures on energy consumption e.g. electric heating cables, may require significant amounts of power. This may be challenging for existing facilities and for new installations if the required energy consumption is not given sufficient consideration in the design phase.

A five-category scale without intervals was used for assessment of energy consumption 'Figure 9'. Although using a generic approach, it is easy to adapt the categories to suit a specific project.

1	No effect on energy consumption
2	Some effect on energy consumption
3	Moderate effect on energy consumption
4	High effect on energy consumption
5	Very high effect on energy consumption

Figure 9. Categorization of energy consumption.

In almost all cases, energy consumption will be linked to compensating for the effect of cooling caused by the combined effect of wind and temperature in 'Figure 6' and robustness in 'Figure 7'.

3.1.6. Assessing investment and operational costs (VI): The cost of implementing a measure has been included in the method. The assessment of cost impact is coarse but sufficient for the purpose of developing and testing a method. In operational use, it would be necessary to refine the basis and intervals for costs.

Costs may be a limiting factor and it is assumed that the winterization measures would be subject to a cost/benefit analysis. One would aim to achieve adequate and robust solutions with efficient investment and operational costs. Investment cost (CAPEX) and operating cost (OPEX) have been included in the method as shown in 'Figure 10'.

	CAPEX	OPEX
Very low	< 1 MNOK	0 – 0,5 MNOK
Low	1 – 5 MNOK	0,5 – 1 MNOK
Moderate	5 – 50 MNOK	1 – 5 MNOK
High	50 – 100 MNOK	5 – 10 MNOK
Very high	>100 MNOK	>10 MNOK

Figure 10. Categorisation of costs CAPEX and OPEX.

3.1.7. Evaluation tool, one-pager: Based on the method described above, a "one-pager" was developed to guide one through the process, facilitate documentation and visualization of the assessment results illustrated in Figure 11.

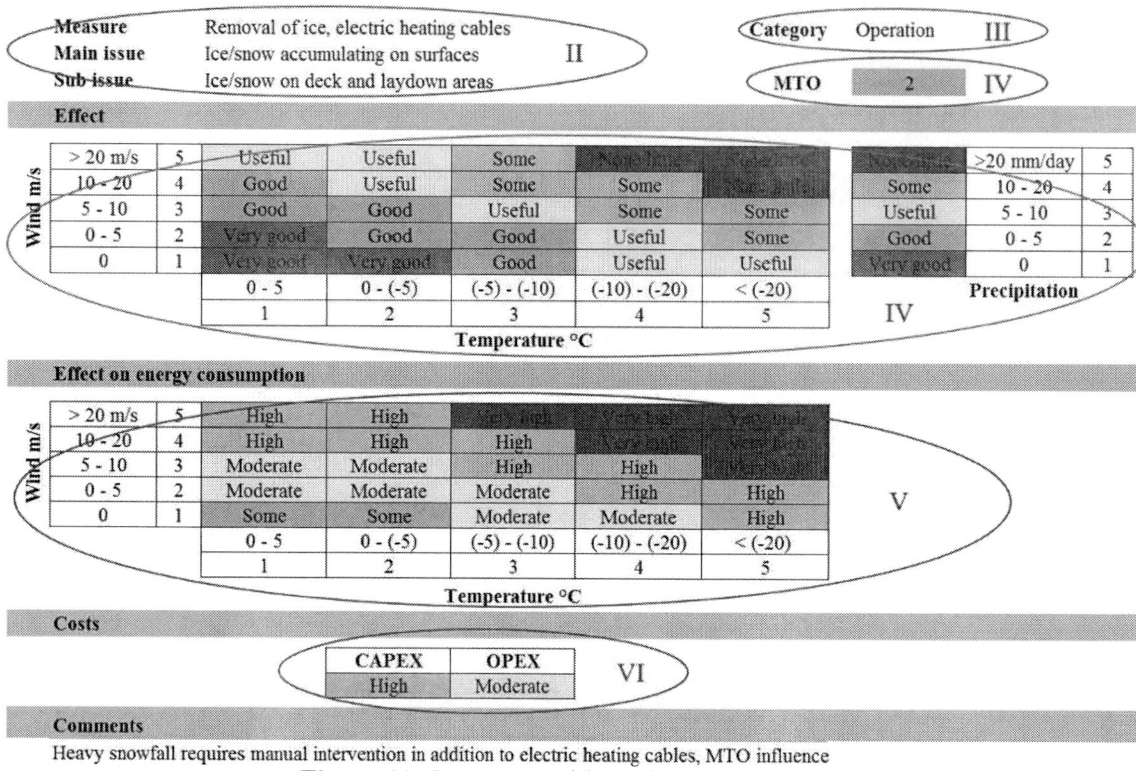

Figure 11. One-pager with method process steps.

3.2. Observations and discussion

The following observations are the main results and findings of the study:

- The method is practical, effective and replicable for evaluating individual measures.
- The use of a one-pager is suitable for documenting, visualizing and communicating results.
- The method has no built-in approach to handle combinations of measures.

As identified previously, the method is based on assessing measures individually. In practice, several measures will overlap and work together while others may have a negative interaction with each other. The method was found to be a useful tool for a coarse selection of measures and assessing their effect given a set of climatic conditions. In a real case, there is a need to assess the individual and combined effect of the measures comprising a winterization package. At present, a qualitative assessment for each specific unit or installation and the operating location needs to be performed in addition to the method.

Using case studies, albeit generally defined, demonstrated that the method could identify potentially useful measures and eliminate those with insufficient effect. Further selection of measures will then have to be done qualitatively based on the unit, location specific metocean conditions and combined effect of other measures.

The selected temperature and wind intervals worked and there has been no need to adjust these in the test. One can argue that the method has not been tested at temperatures below -20°C. Extremely low temperatures provide additional challenges, especially in terms of material quality and keeping fluids at the correct viscosity. On the other hand, the most demanding conditions for winterization tend to be associated with moderate cold in combination with strong winds and heavy precipitation e.g. during a polar low. It is demanding to have a defined method that captures absolutely every conceivable relationship within the same framework.

An important observation is the need to fully describe a measure. The better a measure is described; the better and more accurate assessments can be made. The commentary field proved more important than was envisaged when the method was developed. It was necessary to describe the context, important assumptions and consequences of the measure in addition to a schematic assessment of the effect of the measure under different circumstances. It is emphasized that an assessment of a measure should not be done purely according to a template. However, the method is well suited for a coarse or first pass selection of measures for further assessment as part of developing a winterisation strategy.

The method has primarily assessed the effect of the measure for a combination of temperature and wind. This is justified, as most often the cooling effect of temperature and wind is the main issue. However, it may be natural to evaluate the combination of temperature and precipitation and assess wind separately for certain measures and conditions. No such case was identified during the test of the method. Temperature appears to always be a key factor in evaluating winterization.

The use of measures requiring manual intervention seems widespread. According to the experience of the workgroup, it has been found that the consequences may not be sufficiently addressed in conjunction with other operational and emergency tasks. It is necessary to evaluate this more comprehensively than this MTO assessment is designed for.

Review of winterization manuals reveals that measures relying on manual intervention are used in response to issues associated with operation under winter conditions. These measures involving the use of personnel can include:

- Removal of ice with sledgehammers and bats
- Use of steam lances for ice removal
- Removal of snow by shovelling and sweeping
- Use of tarpaulins on helicopter decks and other structures or equipment.

The wind chill affects mobility and performance of personnel thereby reducing the efficiency of manual operational measures. Restrictions on exposure of personnel to cold also reduce available time for the persons to work in these conditions [13]. It is uncertain whether there is any correlation between use of these measures and the number of personnel in the organization on a facility.

4. Recommendations

The method is designed to assess individual measures and, based on experience, select optimal combinations of measures that also consider:

- health, safety and environmental impact
- measures that may counteract each other
- overlapping or similar measures
- issues that are insufficiently covered by existing and available measures

The method could and should be developed to include a systematic approach to assessing the totality and optimal combinations of measures regarding sufficient robustness to remedy the issues encountered in winter operations. Further development of the method should consider performance requirements related to winterization measures as part of a comprehensive barrier strategy [12]. The winterization strategy should not be detached from the barrier strategy and overall design strategy.

It is recommended to pay attention to equipment that is defined as safety-critical and/or has a barrier function e.g. emergency shutdown, ballast, riser or mooring systems. It is necessary to ensure that these functions are maintained under the environmental conditions with the selected winterization measure. This is required in order to reveal important issues that may be inadequately handled or not identified by an overall assessment.

5. Conclusions

The method has shown encouraging results for assessing individual winterization measures when used under prevailing environmental conditions in the Barents Sea. However, the method does not provide the optimal combinations of winterization measures. It is possible that several combinations of measures could provide a robust winterization package. It is anticipated that robust combinations with low energy requirements and negligible effects on health, safety and the environment would be preferred.

It is imperative to collect location specific climate data e.g. hindcast data and evaluate these against the design of the facility and planned operations. Several measures may provide the required effect within the climate conditions. After applying the method to specific facilities and operations, one sees that passive measures e.g. custom designs and enclosures, are in general, more robust than active measures.

There is sufficient flexibility in the method allowing it to be used for any type of facility and location. The method provides a good basis for evaluating, selecting and combining winterization measures required for an operational envelope in a cold climate environment. The method also lends itself to use in areas other than the Barents Sea. To do this, relevant metocean data is required for the area or location where a facilities winterization needs will be analysed.

6. References

[1] Bodsberg L, Guttormsen S, Bruun P K and Wærø I 2016 Vinteriseringstiltak for inn-retninger i Barentshavet (Winterization measures for installations in the Barents Sea), SINTEF report A27929, Trondheim, Norway

[2] Guttormsen S, Bruun P K, Jacobsen S R, Dørum K-G and Bodsberg L 2017 Winterization issues and measures related to low temperatures, snow and icing for installations operating in the Barents Sea, OMAE2017-62403, Published at OMAE 2017, Trondheim, Norway

[3] Viddal S, Gudmestad O T, Spande T, Nielsen L, Vartdal T, Wiencke H S and Holmen H K 2018 Vurdering av vinteriseringstiltak, Metodeutvikling og effektvurdering av vinteriseringstiltak (Assessment of winterization measures, Methodology development and impact assessment of winterization measures), Proactima report 1072862-RE-02, Stavanger, Norway

[4] The Ministry of Petroleum and Energy 2014 Facts 2014 The Norwegian Petroleum Sector, Oslo, Norway

[5] NORSOK N-003:2017 2017 Actions and action effects, Edition 3, Standard Norge, Oslo, Norway

[6] BaSEC 2016, BaSEC Rapport 1 - Fysisk miljø i Barentshavet sørøst (BaSEC Report 1 – Physical environment in the Barents Sea Southeast), Norwegian Oil and Gas, Stavanger, Norway

[7] Reistad M, Breivik Ø, Haakenstad H, Aarnes O J, Furuvik B R and Bidlot J-R 2011 A high-resolution hindcast of wind and waves for the North Sea, Norwegian Sea, and the Barents Sea, *Journal of Geophysical Research* **116**, C05019, doi: 10.1029/2010JC006402.

[8] Orimolade A P, Furevik B R, Noer G, Gudmestad O T and Samuelson R M 2016 Waves in polar lows *Journal of Geophysical Research - Oceans* 2016; Volume **121** (8) p 6470-6481

[9] ISO 17776:2016 2016 Petroleum and natural gas industries - Offshore production installations - Major Accident hazard management during the design of new installations, International Organization for Standardization

[10] NORSOK Z-013:2010 2010 Risk and emergency preparedness assessment Edition 3, Standard Norge, Oslo, Norway

[11] ISO 31000:2018 2018 Risk management – Guidelines, International Organization for Standardization

[12] Petroleum Safety Authority 2017 Principles for barrier management in the petroleum industry, Barrier memorandum, Stavanger, Norway

[13] NORSOK S-002:2018 2018 Working Environment, Edition 5, Standard Norge, Oslo, Norway

Evaluation of ballast failures during operations of semi-submersible rigs

N Unegbu, and O T Gudmestad

Department of Mechanical and Structural Engineering and Materials Sciemce, University of Stavanger, Norway

Contact address of authors: Nneomaaguta@yahoo.com, ove.t.gudmestad@uis.no

Abstract. Risk analysis/assessment is one of the challenges encountered during operations of offshore units on the Norwegian Continental Shelf (NCS). In recent years, the Petroleum Safety Authority (PSA) has focused on hazards relating to floating installations and thus requested that more attention should be made by the industry on hazards relating to buoyancy loss and stability. Ballast systems play a very vital role to ensure vessel stability. Various failure modes of semi-submersible ballast systems are identified and possible barriers and consequences due to the ballast system failure during drilling operation are considered. The failure mode effect and criticality analysis (FMECA) of the main components of the semi-submersible's ballast system is adopted to determine the failure causes and failure modes that could influence each components performance, and thus identifying the most critical component(s). The Structured What-If Technique (SWIFT) is used to compensate for hazard identification for the unidentified hazards (i.e., human errors), in the FMECA. By studying the most critical system components, a qualitative risk analysis is conducted to model accidental sequences by using the fault tree method to establish the chain of failure events. The result of the Structured What If Analysis (SWIFT) shows that maloperation of the ballast system is the main contributing failure cause. This involves, failure to properly describe ballast procedure, failure to follow ballast plan, wrong sequence of closing/opening valve, maloperation of valve, time pressure complacency, communication gap or general lack of knowledge of the system. The FMECA findings indicate that failure of valves to "close on demand" with a Risk Priority Number (RPN) of 60, is the most critical.

1. Introduction

The Norwegian Petroleum Safety Authority, PSA has focused on hazards relating to floating installations in the past couple of years and requested that more attention should be made by the industry on hazards relating to buoyancy loss and stability [1, 2]. Ballast systems play a very vital role to ensure vessel stability. The main function of the ballast system is to maintain stability and sufficient draft, and also to retain the shear forces and bending moments within required limits. The ballast system comprises of; ballast tanks, different networks of pipes, pumps and valves, hydraulic power system, electric power system and ballast control system.

A failure can be disastrous in nature. It also has tendencies to lead to other unwanted consequences even if it is not catastrophic. For instance, it could cause production loss in the event of downtime and prolonging of delivery deadlines. This therefore, affects projects in the sense of additional costs and wastage of resources hence, leading to the possibility of losing customer goodwill [3]. Failure to properly ballast may lead to accidents, which could potentially lead to loss of vessel, death of personnel and environmental disasters [4]. According to a research carried out by [5] on risk assessment of buoyancy loss (RABL), after vessel collision, the second main contributor to risk in terms of buoyancy loss and stability for offshore mobile drilling units is ballast system failure [6].

This paper aims to evaluate the failure of ballast systems' components during drilling operations. These evaluations include: consequence classification of the critical components of the ballast system; Identification of ways the systems, components, or processes fail to realise their design purpose; Identification and analysis and factors and conditions that cause to the occurrence of an undesirable event; Identification of safety barriers that aims to prevent; Control and mitigate effects of a hazardous event

2. Hazard identification

Hazard identification (HAZID) involves a thorough and comprehensive identification and documentation of hazards. It is very important to thoroughly carry out a comprehensive identification and recording of hazards because failure to identify any hazard at this stage might be detrimental, as it would not be considered in further assessment [1]. Therefore, a comprehensive and well-planned hazard identification is a critical basis for other elements in risk assessment.

2.1. Structured What-If technique

A structured What-If Technique (SWIFT) is a risk analysis method where a lead question "What if" is used to systematically identify potential deviations from normal conditions [7]. This technique is team-oriented and uses experienced personnel as team members. The hazard identification is based on brainstorming by utilising a generic checklist of elements to be reviewed. It is flexible for the use of any type of operations at any given lifecycle stage [8]. Although the SWIFT analysis is rarely used to identify hazards in the offshore oil and gas, it is used here to compensate for unidentified hazards in the FMECA (i.e., Human related errors). Table 1 presents the generic checklist and hazard brainstorming process hazard identification for the ballast system while Table 2 presents the Hazard identification based on SWIFT.

Table 1. Generic checklist and hazard brainstorming process of the ballast system.

Generic check lists	Hazard brainstorming
➢ Human factors (including operating errors)	1. Faulty ballast system design
➢ Maintenance	2. Vessel monitoring system failure
➢ Malfunction of equipment	3. Pump Failure
➢ Utility failure	4. Valve failure
➢ Measurement errors	5. Pipe failure
➢ Emergency operation	6. Remote System operation failure
➢ Integrity failure	7. Tank overpressure or under pressure
➢ External factors	8. Power failure
	9. Valve control system failure
	10. Maloperation of valves
	11. Poor maintenance
	12. Inadequate training
	13. Inadequate personnel selection process
	14. Tank over filling or under-filling
	15. Miscalculations

Table 2. Hazard identification based on SWIFT [8].

Ref.	Hazard Definition	Faulty ballast system design
1	Causes	Lack of regulation. Lack of experienced designer. Poor quality checking process. Financial constraints
2	Consequences	Failure to ballast efficiently. Low pump system capacity.
3	Safeguards	Approval process plan. Ballast tank capacity (Class/rules)
4	Recommendations	Design criteria must be considered.

Ref.	Hazard Definition	Ballast system failure
1	Causes	Failure or/and damage to pumps, pipes, valves etc. Suction Blockage. Insufficient/inefficient backup system.
2	Consequences	Inability to ballast. Inability regulate heeling. Unfavourable mass distribution.
3	Safeguards	Design. Maintenance. Limited redundancy
4	Recommendations	Adequate predictive maintenance strategy. Inspection. Performance testing and monitoring of ballast system

Ref.	Hazard Definition	Maloperation of ballast system
1	Causes	Failure to properly describe ballast procedure. Failure to follow ballast plan. Wrong sequence of closing/opening valve. Maloperation of valve. Time pressure. Complacency Communication gap. Lack of knowledge of the system
2	Consequences	Ballast system failure, Unfavourable heel or draft, unfavourable distribution of mass, Insufficient stability
3	Safeguards	Operating procedures. Monitoring. Training. Planning
4	Recommendations	Inclusion of performance monitoring in the ballast system procedures.

Ref	Hazard Definition	Inadequate planning of ballast operation
1	Causes	Lack of knowledge about the system. Missing description and training. Insufficient availability of personnel. Complacency. Failure to read accurate weather forecast.
2	Consequences	Ballast system failure, List, Structural damage, Loss of buoyance and stability
3	Safeguards	Training procedures, operational practice
4	Recommendations	Emphasis should be made on hazards regarding ballasting during training. Planning on competence availability of personnel

Ref	Hazard Definition	Loss of buoyancy/ insufficient stability
1	Causes	Flooding, structural failure, power failure, ballast system failure, large heel angle, loss of weather/water tight integrity. VCG movement and mass, free surface effects
2	Consequences	Failure of ballast system. Loss of platform. Power failure. Failure of ballast system to operate. Inability to launch live saving system.
3	Safeguards	Recognition of margins and regulations for stability
4	Recommendations	Emphasis should be made on hazards regarding ballasting during training. Planning on competence availability of personnel

Ref	Hazard Definition	Excessive heel during ballasting/ de-ballasting
1	Causes	Unfavourable mass distribution, Insufficient stability
2	Consequences	Failure of ballast system. Loss of platform. Power failure. Failure of ballast system to operate. Inability to launch live saving system.
3	Safeguards	Adequate design of ballast system
4	Recommendation	Design of ballast system should ensure adequate buoyance. Design miscalculations. Active response time with regards to effective intervention of the system

Ref	Hazard Definition	Loss of watertight integrity
1	Causes	Flooding through uncovered manhole, device with open and close functions
2	Consequences	Unwanted mass distribution. Ballast system failure. Insufficient stability. Total loss. Personnel injury/fatality
3	Safeguards	Operational procedures. Inspection
4	Recommendation	Procedures watertight integrity loss must be implemented

2.2. Failure Mode Effects and Criticality Analysis

A FMECA is carried out to reveal and analyse failure modes, failure causes and failure effects on the main components of the ballast system. This method systematically analyses all possible failure modes and its direct reflection on the system's performance [9]. The FMECA also enables predictions to be made on the failure effects on the system and how the failures could be avoided. This can be achieved by ranking the criticality of the failures. By knowing the critical components, improvements are made for reliability and safety purposes. A detailed description of the FMECA can be found in [9]. Table 3 presents a breakdown of the ballast system.

Table 3. Analysed components by FMECA technique.

Ballast tank configuration	
	Ballast tanks
Ballast control system	
	Ballast valves and pump room valves Sea chest valves and Discharge valves Ballast pumps and Ballast control logic unit
Pipes	
	Pipes
Electric power system	
	Main electric power generator Emergency backup generator UPS
Hydraulic power system	
	Main hydraulic power generator Hydraulic accumulator

It should, however, be noted that FMECAs are based on a single failure principle. Compliance with a single failure requirement is not sufficient to avoid incidents. This is in line with [10] concluding that FMECA analysis are not sufficient to identify and remove all relevant single failure modes. FMECA also considers only hazards arising from single-point failures and will normally fail to identify hazard caused by combinations of failures. Furthermore, the interactions between subsystems are not assessed in the FMECA when failure modes are reviewed separately in each subsystem. We will also refer to

[11] who indicates that FMECA, sea trials, and hardware-in-the loop testing, are insufficient and that the view on safety using these methods is too narrow. The safety constraints can be violated in other manners than component failures, as with human errors. It also assumes that failures are found and corrected before new errors occur. This is not in compliance with practice where errors are found, and put in a queue to be fixed together with other failures. After incidents, also other errors are typically found which were unknown before the incident. Still we suggest that FMECA is a method well suited initially to identify failures.

Functions of the elements in the ballast systems are considered together with their operational modes. For each of the functions and operational mode, possible failure modes are identified and listed. The failure modes are ranked according to their frequency of occurrence (O), severity (S), and the likelihood that the failure is detected on time (D). It is important to note that the failure modes were assigned subjectively based on sources including; RABL datasheet, OREDA reports, "Risikonivå i petroleumsvirksomheten" (RNNP) reports, reports on past incidents etc. The ranks are given ranging from 1 (lowest) to 5 (highest). The risk priority number (RPN) is therefore determined by multiplying the occurrence, severity and detectability. During the FMECA some assumptions were made. They include:

- It is assumed that one component fails at a time
- Human error contributions are neglected
- Failure modes analyzed are the more frequent failure modes but not the modes analyzed comprehensive
- The identified failure causes are not a full assessment of all the failure modes of the components

Based on the analysis, the components with the highest ranking are the valves, hence the most critical.

3. Risk reducing measure (barrier analysis)

As pointed out earlier, the oil and gas industry is faced with the risk of major accidents. Major accidents here mean accidents which has major consequences, capable of causing fatalities or/and environmental hazards. Fortunately, these unwanted accidents have low probability of occurrence due to presence of multiple layer of protection, otherwise known as barriers [12]. Although there may be possibility of a single failure to occur, it should not be allowed to lead to catastrophic events. Thus, the reason why multiple barriers are in place and need to be strategically managed all through the rigs' lifecycle [12].

Safety barriers are established and implemented with the aim of preventing, controlling and mitigating the effects of a hazardous event [13]. Depending on the scenario the ballast system is used as a safety barrier in order to prevent, control and mitigate unwanted lists of the vessel by means of ballasting. Barrier management is carried out, with the purpose of establishing and maintaining barriers to prevent unwanted event or in situations where unwanted events occurs, it can be properly handled [14]. Barrier management includes systems, processes solutions and measures that must be readily available to reduce risk by the implementation and follow–up of barriers [14].

3.1 Barrier analysis of past incidents/accidents

The event sequence is the basis of a barrier diagram. It is represented as rectangular text boxes that are linked. Adopted from [15], Figure 1 shows a sample and description of the barrier diagram used in this paper to provide an event sequence overview leading to the accident.

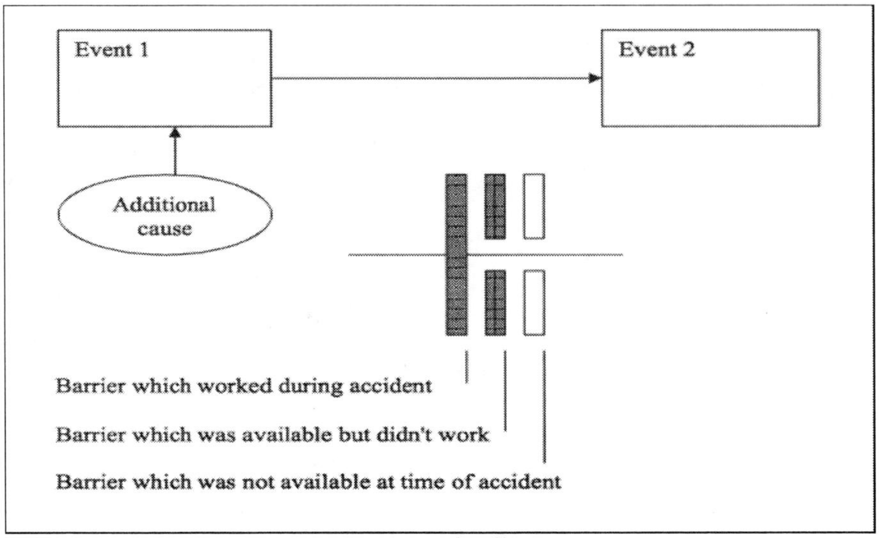

Figure 1: Barrier diagram

Additional causes are presented in the circular box. The shaded or white vertical bars represents the barriers. The shaded barriers mean the availability of barrier at the time of incident. White broken barriers represent barriers that were not available in the duration of time the incident occurred but were implemented in regulations. A full shaded barrier means that the barrier worked.

Reference [16] presents a detailed barrier analysis of accidents/incidents of nine selected rigs. An overview of the analyzed accidents for five of the rigs is shown in Table 4 and the barrier performance summary in Table 5.

Table 4. Overview of the analysed accidents.

Unit	Location	Year	Main cause
Ocean Ranger	Canada	1974	Ballast System, Portlight
Ocean Developer	West Africa	1995	Ballast system operation
Petrobras-36,	Brazil	2001	Operation of drainage, Hydrocarbon explosion, flooding
Thunder horse	Gulf of Mexico, US	2005	Hydraulic system operation
Scarabeo 8	Barents Sea, Norway	2012	Ballast system operation

Table 5. Barrier performance summary.

Barrier Function performance	Maintain Structural Integrity and Marine Control	Prevent escalation of initiating failure	Prevent loss	Prevent fatalities/ injuries
Ocean Ranger accident	Failure	Failure	Failure	Failure
Ocean Developer	Failure	Lack of Information	Lack of Information	Success
Petrobras-36 Accident	Failure	Failure	Failure	Partial Success
Thunder Horse	Failure	Success	Partial Success	N/A
Scarabeo 8 Incident	**Failure**	**Success**	**N/A**	**Partial Success**

The analysis shows the direct implications of barrier failures in terms of technical operational and organizational elements. In one way or other, human errors are significant as these are the main contributor to this failure. The initiating cause of failure for three of the incidents are directly linked to human involvement (i.e., Ocean Developer, Thunder Horse and Scarabeo 8). In the case of Ocean Ranger, the series of events that caused the accident was solely caused by poor design. The Petrobras P-36 series of events occurred as a combination of fires and explosions and a design that did not allow for operating the ballast system following the damages in the column.

4. Discussion

A considerably amount of information about past events is needed to prevent near misses and accidents in the future. However, some loss of stability incidents and accidents are not reported, or we lack full information about the events leading to the accidents. The downside to this problem is that detailed studies are not carried out to know how and why the event happened, especially for peculiar cases. This may be the reason why damage frequency on vessels have not improved over the years. Figure 2 presents a distribution of causes of nine selected past incidents/accidents that led to loss of buoyancy or loss of stability of semi- submersibles. Five of these are listed in Tables 4 and 5 while four additional cases were studied in [16]: Henrik Ibsen, Abel Peal, Flotel Superior and Island Innovator.

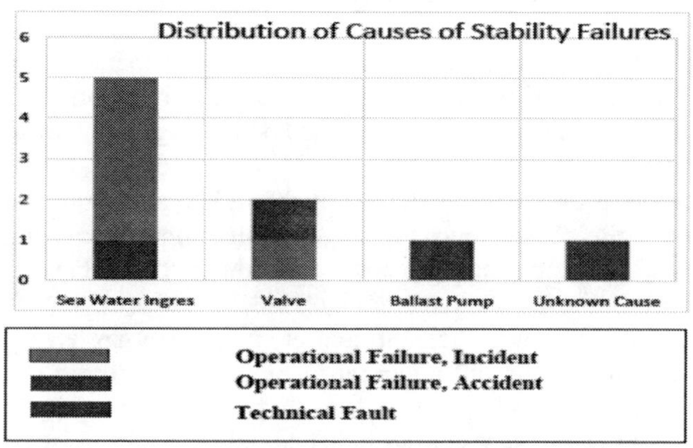

Figure 2. Distribution of causes of nine selected past incidents/accidents that led to loss of buoyancy or loss of stability of semi- submersibles.

A more detailed information about some semi-submersibles and their dimensions are given in the following. Five out of the nine incidents/accidents, were caused by uncontrolled water ingress. This is in line with the conclusion by [17], who noted that uncontrolled water ingress is the main common cause of accidents and incidents. On the other hand, a similar study carried out by [18] concluded that valve failures are the main cause category for incidents and accidents. This discrepancy may be due to the fact that in most cases where ballast valve failure is not the initiating cause of an event, however it seen to be among the casual factors on the incident chain. It is further observed that seven out of nine incidents/accidents occurred due to human errors. These accidents could have been prevented if the human interface (i.e., designers, operators and organization) had followed the guidelines in [19], PSA (2011) regulations. Therefore, when carrying out hazard analysis on systems such as this, it is important to incorporate human errors.

Risk assessment on ballast systems can be done by adopting either the qualitative approach, quantitative approach or a combination of both. However, this paper is limited to qualitative risk assessment of ballast failures during operations of semi-submersibles. The first step of this assessment method is aimed at identifying potential hazards that could be detrimental to operations. The techniques adopted are the SWIFT and FMECA.

The FMECA was adopted to systematically analyze all possible failure modes and their direct reflection on the performance of the ballast system. The SWIFT method was used to compensate for unidentified hazards in the FMECA (i.e., human related errors). The hazard identification is based on brainstorming by utilizing a generic checklist of elements in the ballast system. Table 2 shows a comprehensive hazard identification based on SWIFT. Hazards here are defined are follows;

- Faulty ballast system design
- Failure of ballast system
- Maloperation of ballast system
- Inadequate planning of ballast operation
- Loss of buoyancy /inefficient stability
- Excessive heel during ballasting /de-ballasting
- Loss of watertight integrity or weather-tight integrity

The causes and consequences for each of the defined hazard are identified. For instance, maloperation of the ballast system can occur due to failure to properly describe ballast procedure, failure to follow ballast plan, wrong sequence of closing/opening valve, maloperation of valve, time pressure complacency, communication gap or general lack of knowledge of the system. Controls otherwise known as safeguards are also identified as a risk-reducing measure. Finally, recommendations are made on how to achieve the safeguard (also see Table 2 for recommendations).

A FMECA was carried out to reveal and analyze failure modes, failure causes and failure effects on the main components of the ballast system. Information about the failure rates were acquired from the RABL data sheet, OREDA report and the RNNP report. The risk relating to the failure modes are presented by an alternative to the risk matrix, (i.e., Risk priority number (RPN). The RPN is determined by multiplying together the severity (S), occurrence (O) and detectability (D) of the failure modes. Numbers are subjectively assigned to the S, O, D based on the authors' degree of knowledge of the components. A detailed FMECA is presented by [16]. The findings show that failure of valves to "close on demand" is most critical.

It is established that the risk related to ballast failure can lead to fatalities or/and loss of platform. In order to prevent or reduce the consequences in the event the incident occurs, a risk reducing measure must be in place. The risk reducing measure adopted in this paper is barrier management. A detailed barrier analysis of five selected rigs accidents/incidents showed the direct implications of barrier failures in terms of technical operational and organizational elements. Human error was established to be the main contributor to this failure. The initiating cause of failure for three of the incidents are directly linked to human involvement (i.e., Ocean Developer, Thunder Horse and Scarabeo 8). In the case of Ocean Ranger, the series of events that caused the accident was caused by poor design. The Petrobras P-36 series of events occurred as a combination of fires and explosions and a design that did not allow for operating the ballast system following the damages in the column. In order to ensure that barriers are functioning, robust and available, it is important to have a defined barrier management strategy.

Figure 3 establishes an approach of barrier risk reduction. This approach starts with hazard identification of critical paths of the ballast system that may lead to a major accident. The second step aims to apply solutions that involve technical, operational or organizational aspects. This could be in the form of design modifications, improvement or changes in procedures and personnel selection process (i.e., to increase competence in ballast operations). A detection (e.g., sensors) and ballast control safety barriers must be available in order to detect events with critical deviations. In addition, mitigation barriers (i.e., reserve buoyancy in the form of buoyancy deck, air injection etc.) to prevent total loss should be established. Finally, performance monitoring must be an ongoing process. This will aim to continuously monitor the performance of components in the ballast system with the human interface.

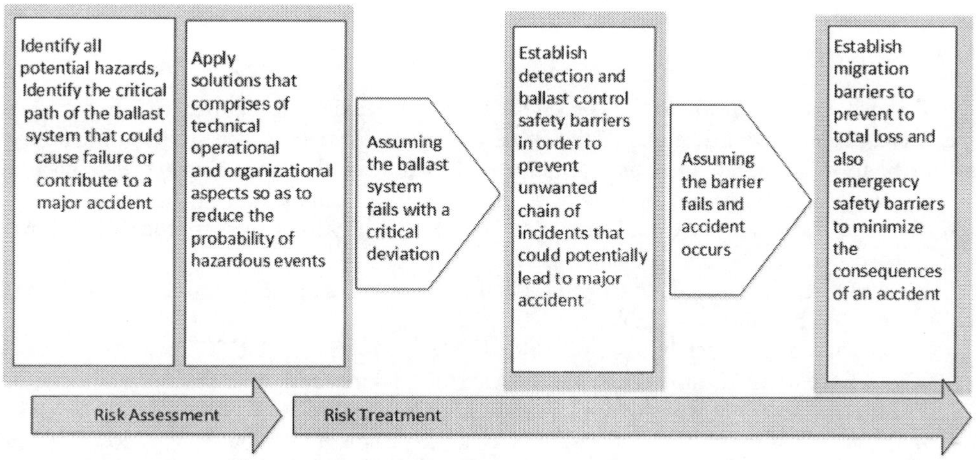

Figure 3. Barrier risk reduction.

5. Conclusions and future works

The consequences of instability of semi-submersible rigs during operations are considered to be severe. Hence, this paper is focused on integrating operational stability calculations of a semi-submersible rigs with risk analysis. The purpose of is to evaluate the failure modes of ballast system's components during drilling operation and suggest mitigation measures. To achieve this objective a qualitative risk assessment approach is adopted.

Some past incidents and accidents were reviewed in order to identify and understand causes and chains of accidental events. The reviewed literature included investigation reports on, Ocean Ranger, Ocean Developer, Petrobras P-36, Thunder Horse and Scarabeo 8. Also reports on accidental situations with the semisubmersibles Henrik Ibsen, Abel Peal, Flotel Superior and Island Innovator were considered in [16], however, not discussed in this paper.

Critical events were identified in the FMECA relating to changes in the amount of ballast water. Also, components in the ballast system were analyzed, and based on findings, the failures of valves to *"close on demand"* with a Risk Priority Number (RPN) of 60 was established to be the most critical. It is important to note that the valves regarded here are the valves in the ballast tank configuration. The SWIFT analysis identified human operational hazards that was not identified in the FMECA. A fault tree was then used to represent the relationship between events and component failures that may combine to cause an un-desirable event. Finally, it was established that, in order to ensure that barriers are functioning, robust and available, it is important to have a defined barrier management strategy.

As a future work, the study efforts and other past studies related to risk analysis of the ballast system of a semi-submersible during operations have identified fundamental information about reliability and risk analysis of the ballast system of a semi-submersible. However, further studies are required to improve the accuracy of the results of the study efforts and to reveal more efficient methodology for reliability and risk analysis. Therefore, future studies that might be considered are not limited to the following:

- Detailed quantitative risk and reliability analysis of potential ballast failures during operations of a semi-submersible
- Investigations that include integration of operational stability calculations of a semi-submersible rig with risk analysis. This thesis can serve as a foundation to such investigations
- Although this paper is limited to barrier management for risk reduction, it is recommended to integrate the risk acceptance criteria and ALARP principle so as to balance cost and safety of a selected risk reducing measure or strategy.

For more detailed analysis, quantitative risk analysis should be implemented in order to obtain a more accurate assessment of risks, the QRA would allow to implement assessment of single human errors, competence requirements, requirements to investigations and schemes for certification of the ballast systems etc.

References

[1] Vinnem J E 2013 *Offshoe Risk Assessment.* Springer Series in Reliability Engineering, vol. 1.
[2] Kvitrud A 2013 Modifications of the PSA regulations based on case studies of stability accidents. ASME 2013 32nd Int. Conf. on Ocean, Offshore and Arctic Engineering, OMAE. American Society of Mechanical Engineers Digital Collection, Nantes, France, June 9 - 14.
[3] Kumar S, Chattopadhyay G and Kumar U 2007 Reliability improvement through alternative designs: a case study, *Reliab. Eng. Syst. Saf.* **92**(7), 983-991.
[4] Sobena SOBENA - Sociedade Brasileira de Engenharia Naval (2007). [Online]: Marine Systems and Ocean Technology. Available: http://www.sobena.org.br/msot/downloads/MS&OT_v3_n1_jun2007.pdf [Accessed: 23.02.2017].
[5] Østby E, Berg M and Festøy B 1987 Ballast system failures and other faulty weight conditions. RABL Risk Assessment of Buoyancy Loss and Other Faulty Weight Conditions. PPS3: Ballast System.
[6] HSE 2003 [Online]: Review of the risk assessment of buoyancy loss (RABL) project. Available: http://www.hse.gov.uk/research/rrpdf/rr143.pdf [Accessed: 31.07.2019].
[7] Aven T 2008 *Risk analysis: assessing uncertainties beyond expected values and probabilities*. John Wiley & Sons Inc., Chichester, England.
[8] HSE 2001 [Online]: Marine risk assessment. Available: http://www.hse.gov.uk/research/otopdf/2001/oto01063.pdf [Accessed: 31.07.2019]
[9] Rausand M 2011 *Risk Assessment: Theory, Methods, and Applications.* Wiley & Sons Inc. Hoboken, New Jersey, 1st edition.
[10] Dong Y, Rokseth B, Vinnem J E, Utne I B 2016 *Analysis of Dynamic Positioning System Accidents and Incidents with Emphasis on Root Causes and Barrier Failures*. European Safety and Reliability Conference ESREL, NTNU.
[11] Rokseth, B., Utne, I. B. and Vinnem, J.E. (2017). A systems approach to risk analysis of maritime operations. Proc. of the Institution of Mechanical Engineers. Part O, *J. risk reliabil.* **231**(1).
[12] SINTEF 2015 [Online]: Towards a holistic approach for barrier management in the petroleum industry. https://www.sintef.no/globalassets/project/pds/reports/sintef-a26845-towards-a-holistic-approach-for-barrier-management-in-the-petroleum-industry-final.pdf [Accessed: 31.07.2019].
[13] Sklet S 2006 Safety Barriers: Definition, classification, and performance. Journal of Loss Prevention in the Process Industries, *J. Loss Prev. Process Ind.* **19**(5),494 - 506.
[14] PSA 2013 [Online]: "Causal relationships and measures associated with structural and maritime incidents on the Norwegian Continental Shelf". Available at: https://www.ptil.no/contentassets/0e200a7cce1a403eb08f49d16b7aeed4/causal-relationships-and-measures-associated-with-structural-and-maritime-incidents-on-the-ncs.pdf [Accessed: 31.07.2019].
[15] COWI 2003 Ageing rigs: Review of major accidents, Causes and barriers, Report no. P-055481-B32. https://www.ptil.no/contentassets/249745 71fd8442bea4c21d3679d93a8e/rapport-vedlikehold-for-aldrende-innretninger---en-utredning.pdf [Accessed: 31.07.2019].
[16] Unegbu N 2017 *Evaluation of Ballast Failures During Operations Of Semi-Submersible Rigs*, Master thesis, University of Stavanger, Norway.
[17] Tinmannsvik R, Albrechtsen E, et al. 2011 [On- line]: Deepwater Horizon-ulykken: Årsaker, lærepunkter og forbedringstiltak for norsk sokkel. SINTEF Report. https://www.sintef.no/globalassets/upload/konsern/media/deepwater-horizon---sintef---vedleggsrapport_mai-2011.pdf [Accessed: 31.07.2019].
[18] Vinnem J E, Kvitrud A and Nilsen L R 2006 [Online]: Stabilitetssvikt av innretninger på norsk sokkel - Metodikk for risikoanalyse Available: https://docplayer.me/11940906-Stabilitetssvikt-av-innretninger-pa-norsk-sokkel-metodikk-for-risikoanalyse.html [Accessed: 31.07.2019].
[19] PSA (2011). Guidelines for application for Acknowledgement of Compliance (AOC) for mobile facilities intended for use in the petroleum activities on the Norwegian Continental Shelf.

Survival in cold waters - learnings from participation in cold water exercises - a regulatory perspective related to the Norwegian offshore industry

J E Jensen[1], K E Solberg[2] and O T Gudmestad[3]

[1] Petroleum Safety Authority*, Norway
[2] University of Stavanger and DNV-GL, Norway
[3] University of Stavanger, Norway

* Corresponding author: *jan.erik.jensen@ptil.no*

Abstract: Participation in exercises lead by the University of Stavanger, the SARex 1 in 2016 and SARex 2 in 2017, provided valuable lessons related to survival in a cold maritime environment. The scenario of the exercises was a mass evacuation from a stranded cruise ship in Arctic waters but provided experiences relevant beyond the cruise industry. This article focuses on the relevance for the oil and gas industry in the Norwegian sector. SARex 1 and 2 were rescue exercises carried out north of Spitzbergen by industry, regulators and academia during the spring of 2016 and 2017. The Norwegian Coast Guard, the University of Stavanger and GMC, Stavanger, organized the exercises. The Petroleum Safety Authority found it important to participate in SARex 1 and 2, to better understand the factors enabling success in a cold climate survival situation. The survival tests in lifeboats and life rafts during SARex 1 and 2 showed the necessity of having trained and competent leadership on board. This issue has perhaps been neglected, compared to the development of equipment to ensure successful evacuation and avoid hypothermia. Some key factors for survival success were (1) Functioning of emergency equipment (2) Organising and teaching the function of various items of safety equipment (3) Operating evacuation means at sea, e.g. watch routines, operating hatches/venting routines and (4) Activities and maintaining mental awareness and motivation. The overall learning from SARex 1 and 2 is that survival is dependent on active participation from the survivors. There is a need for sufficient competence amongst personnel to micro-manage all the details required for survival. The mental factors following fatigue and seasickness, in addition to hypothermia, make leadership and competence in a cold maritime climate essential for survival.

1. Introduction

There has been a renewed interest in exploration in the Barents Sea, since the opening of licensing blocks further north and east more than a decade ago. These areas are further from the coast than existing discoveries in this region, thus creating new challenges compared to the more mature areas developed for oil and gas in the Norwegian sector. Today, resources for handling emergency situations are considerably scarcer in the Barents region.

The basis for SARex 1 [1] and 2 [2] was the International Maritime Organization's (IMO) new Polar Code [7] with a five-day survival requirement in the Arctic. There is no similar regulatory requirement in the oil and gas industry, as most of the requirements are functional. However, the industry has set its requirements through analyses of the different major accident scenarios. These requirements are

IOP Conf. Series: Materials Science and Engineering **700** (2019) 012045 doi:10.1088/1757-899X/700/1/012045

intended for areas with emergency response cooperation but have been regarded as industry standard throughout.

The survival requirement is established by IMO, based on the time of rescue in remote polar areas. These are applicable to all maritime activities under the flag regime, but not oil and gas activities. The criticality, difficulties working with arrangements to be used on demand, and state of the art regarding equipment is different from marine activities regulated by IMO-Polar code due to the urgency of acting quickly to escape in case of an incident onboard an oil and gas platform. The focus of the regulations and the oil and gas industry has been on technical inventions to ensure success in handling escape and evacuation during emergency scenario and not on the survival phase.

In an emergency, where evacuation is necessary, one is exposed to the following phases:

Escape – Evacuation – **Survival** – Rescue

Specifically, experiences from Norwegian Safety Authority Norway also shows that education of emergency personnel on board ships and oil installations focus mostly on the escape and evacuation but much less on the survival and rescue.

SARex 1 and 2 have shown the importance of validating technical requirements versus towards functional requirements. This article will show some of the challenges experienced in these two expeditions, in the frigid waters around Svalbard, held in spring 2016 and 2017.

One of the worst-case scenarios to put the search and rescue resources to the test is assumed to be the sinking of a cruise liner in polar waters. The *TS Maxim Gorkiy* incident, in which a cruise ship with nearly 1000 passengers hit drift ice on the evening of 19 June 1989 outside Svalbard, was such a scenario.

There are many heroic accounts of sealers and fishermen being stranded in their rescue craft for prolonged periods. Knowledge and equipment have been key factors in their survival. This article looks into the regulations, competence and equipment that are in place today, relevant for a survival situation in the northernmost areas of the Norwegian sector. It also includes experiences made during SARex 1 and 2, and recommendations based on these experiences.

2. Background

Most of the requirements in today's offshore regulations, guidelines and the progress in technology have been done in relation to the evacuation processes, for example with the introduction of free-fall lifeboats and marine escape chutes. With the prolonged survival phase being most likely to occur in northern waters compared to the more southern regions of the Norwegian continental shelf, it would be most relevant to look further into the evacuation means' ability to sustain survival, and to verify whether regulatory requirements are sufficient. Survival suits, and particularly helicopter transport suits, have also improved significantly over the last decade.

The Polar Code was implemented in January 2017 and is applicable to all vessels covered by the IMO in Arctic and Antarctic waters. These requirements are not applicable for offshore installations in the oil and gas industry, but they may be in the future (pers. com. Å. Waage, Norwegian Maritime Authority (resigned June 2019)). There have been exploration activities to date in the Polar Code area (Korpfjell- prospect), but only outside the harshest winter season.

3. Methods and materials

This article is based on review of regulations applicable as of August 2019 related to the competence in survival in the oil and gas industry in the Norwegian sector. The course content in safety courses required for work in the offshore industry in Norway will be compared to regulatory requirements and lessons learned from SARex 1 and 2. The author of this article was an active participant in tests during these exercises.

4. Limitations and assumptions

This article highlights the most relevant regulations for a survival scenario relevant for the offshore petroleum activities on the Norwegian continental shelf and compares how these requirements will support survival and rescue in a cold climate survival situation. Not all requirements related to emergency response are covered in this report.

The exercises were conducted with the leadership of naval officers with special training in leadership in emergency situations, in both the lifeboat and -raft. This competence would not be available in the case of civil vessels or oil and gas installations. The exercise participants also had personnel with special competence in the equipment used, with vendors' representatives on board. The privates onboard K/V Svalbard, civil servants and academics also had relevant knowledge to some extent.

There was also the continual presence of medical personnel on board during SARex 1 and partial presence during SARex 2. This can be assumed to produce less stress for the participants, compared to a real case, where it is unlikely that there would be medical personnel on board. The exercises were held in sheltered waters. The evacuations where dry evacuations.

5. Relevant regulations

The key regulations governing offshore petroleum activities are provided in the health, safety and environment regulations [3] that consist of the the Framework -, Management -, Facilities – and Activity regulations.

The regulations covering requirements for escape -, evacuation - and rescue means, and competence and training can be divided into different groups of barriers, which are all covered by the Norwegian offshore industry HSE-regulations; cf. the Management Regulations Section 5 about the different types of barrier elements:

- Technical
- Operational
- Organizational

"WHO is doing WHAT with what EQUIPMENT" illustrates the organizational, operational and technical barrier elements, respectively. All these types of barrier elements need to be in place, to ensure safe operations and a robust emergency preparedness in the offshore industry's operations. The Norwegian Petroleum Authority has written guidelines on how to interpret these requirements [4].

The offshore petroleum regulations in the Norwegian sector consists of mainly functional requirements. This is not the case for the maritime regulations. However, there is an exception with the recently introduced IMO Polar Code. Most of these requirements are of a functional nature, e.g. the five-day survival requirement mentioned in the introduction. It should, however, be noted that actors operating within the regulations are responsible for risk analyses and risk management in which design scenarios need to be developed; therefore, it is expected that the five-day requirement will be challenged by operators documenting safe survival when closer to rescue means.

5.1. Maritime regulations

On the Norwegian continental shelf, offshore mobile units can follow the maritime technical requirements in the design and equipment of maritime systems on board; cf. the Framework regulations Section 3. This includes the design of escape ways and all emergency response equipment on board. The maritime regulations related to petroleum activities offshore are stipulated through the Norwegian Maritime Authority (NMA) regulations or other flag state regulations to a similar or higher safety level. Where the other flag state requirements are below the NMA's regulations, then the NMA's regulations are to be followed. The NMA normally refers to the IMO requirements but, in some cases, has additional requirements. So, in practical terms, the NMA regulations form the basis for the design of mobile offshore units, including the evacuation equipment. As of today, most mobile units include semi-submersible and jack-up installations used for drilling or purely accommodation services (often referred to as "floatels").

IOP Conf. Series: Materials Science and Engineering **700** (2019) 012045 doi:10.1088/1757-899X/700/1/012045

The relevant regulations for the mobile offshore units are in the "red book", issued by the NMA [5]. In this collection of different regulations are the NMA regulations of 2 February 2016 No. 90 on evacuation and life-saving appliances on mobile offshore units (Rescue regulation), which gives the requirements for the design, placement and numbers of necessary life-saving equipment on board. In particular, the NMA Rescue regulations Chapter 4 (numbers and placement of evacuation and life-saving equipment) and Chapter 5 (design requirements) state the requirements for mobile units in terms of evacuation equipment. These regulations refer, to a large extent, to the IMO Life Saving Appliances Code (LSA-Code) [6].

Regarding life rafts, the LSA Code has some requirements that are seemingly useful features. However, the experience from SARex 1 and 2 has shown that the requirements have not been sufficient. For instance, there is not an absolute requirement of a double bottom in life rafts:

LSA Code, chapter 4.2.2.2 - *The floor of the life raft shall be waterproof and shall be capable of being sufficiently insulated against cold either:*

1. *By means of one or more compartments that the occupants can inflate, or which inflate automatically and can be deflated and re-inflated by the occupants; or*
2. *By other equally efficient means not dependent on inflation.*

The latter has in practice become a thin neoprene layer with silver foil/ heat reflective material on the top. The first requirement could be a double bottom, which in SARex 2 proved a lot better than the solution chosen by following item 2 above. However, the most common life raft type used on offshore installations today is the single bottom with a thin insulating layer.

IMO Polar Code: The IMO Polar Code [7] is a set of functional requirements that are adapted to the specific challenges of operating and handling emergency situations in polar regions. It covers the complete aspect of operating in these regions, but the most relevant sections related to emergency response are:

- Chapter 1 Definitions:
  - 1.2.4 *Habitable environment* means a ventilated environment that will protect against hypothermia.
  - 1.2.7 *Maximum time of rescue* means the time adopted for the design of equipment and systems that provide survival support. It shall never be less than 5 days.
- Chapter 8 Life-saving appliances and manning: That evacuation equipment shall function in the adverse conditions during maximum time of rescue. There is also Chapter 8.2.3 covering *survival*. Amongst several requirements, then the means of escape shall provide a habitable environment.
- Chapter 12 Manning and training: This chapter has the goal of ensuring that ships operating in polar waters are appropriately manned by adequately qualified, trained and experienced personnel.

5.2. Offshore Petroleum Industry

Within the petroleum industry, the most relevant regulations governing the design and operation of evacuation means are stipulated in Table 1 below. None of these regulations (Table 1) are specific about requirements for the survival phase of an emergency situation. However, as stated in the Activity regulations Section 77, item c): "*Measures shall be taken so that personnel can be rescued during accident situations*". "Rescued" in this context means that survivors shall be brought to a safe area either on board an installation, vessel or on land where necessary medical attention can be given.

IOP Conf. Series: Materials Science and Engineering **700** (2019) 012045 doi:10.1088/1757-899X/700/1/012045

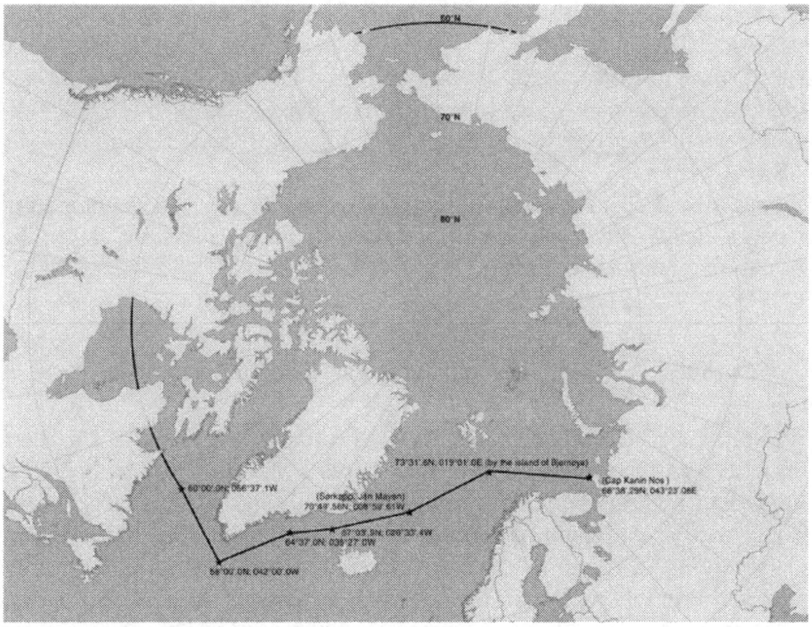

Figure 1. The maximum extent of the IMO Polar Code in the northern hemisphere (source: www.imo.org).

Table 1 Regulations governing emergency preparedness in Norway offshore oil and gas industry.

Regulation	*Description*
Framework reg. Section 20	Coordination of emergency preparedness resources, which state the operator's responsibility to establish and coordinate the emergency preparedness.
Management reg. Section 5	Barrier regulation.
Management reg. Section 14	Manning and competence, which, for instance, are relevant for ensuring that offshore personnel are competent to work in an Arctic environment (including emergency situations).
Management reg. Section 17	Risk analysis and emergency preparedness assessments.
Facility reg. Section 44	Means of evacuation, which refer to the requirements in NORSOK S-001 [8] Chapter 21 and for the design of free-fall lifeboats DNVGL-ST-E406.
Facility reg. Section 45	Survival suits and life jackets, etc.
Activity reg. Section 17	Transport to and from the offshore installations.
Activity reg. Section 21	Competence of offshore personnel, including the handling of situations of hazard and accident
Activity reg. Section 23	Training and drills, which require that personnel are trained and equipped to handle situations of hazard and accident in an efficient manner.
Activity reg. Section 73	Establishment of emergency preparedness, which state that there shall be a strategy for emergency preparedness in hazardous and accident situations based on analyses.
Activity reg. Section 75	Emergency preparedness organization, which state that this organization shall be robust, so that it is able to handle hazard and accident situations in an efficient manner.
Activity reg. Section 76	Emergency preparedness plans.
Activity reg. Section 77	Handling hazard and accident situations. Sub section c) here states that measures shall be taken so that personnel can be rescued during accident situations.

Standards and guidelines: The industry has developed standards and guidelines for the design of evacuation and safety equipment. NORSOK S-001 [8] states requirements related to lifeboats and life rafts in Chapter 21.4.3. However, there is no specific mention of requirements related to a prolonged cold-water survival phase. Survival suits are also covered here in NORSOK S-001. When it comes to the design of lifeboats, reference is made to the standard, DNVGL-OS-E406 [9]. For life rafts, SOLAS and flag state requirements are referred to. Chapter 21.4.4 covers requirements for survival suits. It also refers to SOLAS requirements.

The Norwegian Oil and Gas Producers Association (NOG) has also developed guidelines for both helicopter transport suits (A) and survival suits available at lifeboat stations (B) in their NOG guideline GL094. This guideline refers to the standard:

- NS-EN ISO 15027-1:2002 Immersion suits; Part 1, Constant wear suits, requirements including safety.
- NS-EN ISO 15027-2:2002 Immersion suits; Part 2 Abandonment suits, requirements including safety.

According to NOG GL094, both types of suits shall fulfill the requirements in the LSA Code (except item 2.3.2.1, where special Norwegian requirements apply). The suits shall be able to sustain a survivor's body heat and prevent drowning and near drowning for a period of at least 120 minutes. It is given in NOG GL064 [9] about area emergency preparedness. In Chapter 3.9, the guideline pays special attention to activities in the Barents Sea region. Other than recommending that special consideration be given to the performance criteria in this region, it does not find any reason to establish individual requirements for this region on a scientific basis. This statement can be understood based on the capacities of today's helicopter transport suits and specifications for the alternative IMO suits, but it does not mention anything about a most likely longer survival time during winter operations in the most distant parts of this region. The lack of daylight, the polar lows and significantly colder waters in the north-eastern part are not considered. *The industry has not yet established a common performance criterion for the survival time!* However, BaSEC group (BaSEC = Barents Sea Exploration Collaboration - established by Statoil, Eni Norge, Engie (GDF Suez), Lundin and OMV, in April 2015) has suggested a criterion for rescue in the Barents Sea of 24 hours [13] (pers. com. S. R. Jakobsen, Petroleum Safety Authority Norway).

6. Competence requirements for emergency response in the offshore petroleum industry

6.1. Onshore competence training

The onshore training centers in Norway provide a Basic Safety Training (GSK = nor. "Grunnleggende Sikkerhetskurs", English: "Basic Safety Training course") course for all personnel traveling to offshore petroleum installations in the Norwegian sector. On this course, there are only some basics on survival and rescue from lifeboats and life rafts, which are repeated in the refresher courses held at regular intervals (every fourth year).

The participants on these courses have:

- An equipment demonstration at an inflated life raft, exercises in the sea involving the turning of a capsized life raft.
- Demonstration of launching mechanisms, either by davit-operated rafts or by using the marine escape chute system.
- Evacuation exercise through the escape chute.
- Lifeboats, performing entering, lowering and evacuation
- Helicopter underwater evacuation training (HUET) in swimming pools (non-mandatory).

The lifeboat coxswain courses are mostly concerned with the functionality of the lifeboats, from mustering at the lifeboat stations to preparing for launching, launching techniques and moving away from the offshore installations, avoiding collision and re-entry. A minor part of the course focuses on operational/ organizational factors after entering the sea in a lifeboat:

- Get an overview on board, take care of passengers

- Keep the command and initiative, get an overview – provide information
- Delegate tasks to passengers (equipment, first aid, emergency systems)

However, these subjects are not elaborated to a large extent and, as from the lessons from SARex 1 and SARex 2, they are an important contributor towards increasing survival chances. These lessons are presented in a later chapter in this report. At the end of the lifeboat coxswain courses, this is the expected learning (source: A.-E. Sævareid at Falck-NUTEC, Norway):

- Recognize 10 items from the lifeboat equipment package
- Highlight position using available equipment
- Know the lifeboat's six emergency systems and be able to use these
- Manoeuvre, also by means of emergency tiller
    - Using compass heading
    - Towards objects in the sea
- Launching of marine escape chute
- Prepare davit-operated life rafts for launching
- Recognize the most common stress reactions in an evacuation situation

As a conclusion, there is very little or no focus, in either the Basic Safety Training courses or the Lifeboat Coxswain Training courses, on the knowledge needed for survival of a prolonged survival time in cold waters in lifeboats or life rafts.

6.2. Offshore competence training

The requirements related to training on site at each individual installation are stated in the Activity regulations Section 23 about training and drills. They require that each individual team member in an emergency team on board trains a minimum of once per offshore trip for each role he or she has in an emergency response team. The Norwegian Oil and Gas Producers organization (NOG) has focused on making the onshore training more efficient and reducing the time spent on these courses, while emphasizing the importance of local training on each installation. NOG has developed a set of modules (https://www.norskoljeoggass.no/drift/opplaring/treningsmateriell-for-beredskapslag/) to ensure that important aspects of handling different hazard and accident situations are covered in the training sessions offshore. These modules for coxswains do not focus much on survival phase.

7. Lessons learned in relation to survival in SARex 1 (2016) and SARex 2 (2017)

This section is based, to a large extent, on the author's contribution to the SARex reports [1] and [2]. Note that this paper does not provide a full list of experiences from SARex 1 and 2 but, rather, highlights those that are identified as particularly relevant for Norwegian offshore petroleum activities, as well as personal experiences.

7.1. SARex 1 and 2 in brief

SARex 1 and 2 considered evacuation and survival for maritime emergencies in Arctic waters. The scenario on both expeditions was an emergency on a cruise ship, involving evacuation. This section will look into lessons learned from SARex 1 and 2, which could be relevant for the oil and gas industry on the Norwegian continental shelf.

The author of this article was an active participant in the tests. Participation in the survival tests in both SARex 1 and 2 took place in the life raft. Of course, the best learning came from direct participation, which were the raft survival tests, the evacuation of the lifeboat by KV Svalbard personnel, helicopter evacuation from the lifeboat with the SAR-rescue helicopter and the use and testing of emergency equipment for maritime casualties in the Arctic.

Figure 2 Circles indicate areas for SARex 1 and 2 activities.

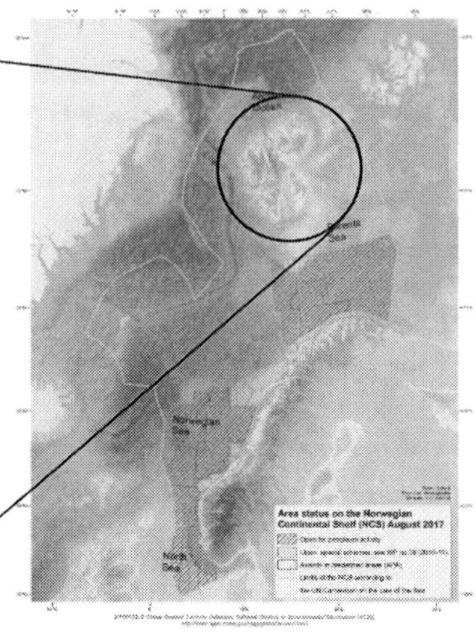

Figure 3 Illustration of the areas open for petroleum activity. Source: www.npd.no.

7.2. Climatic relevance
The sea temperatures were about 4-5 °C lower at SARex venues compared to normal winter temperatures at the northern part of the Norwegian continental shelf. However, in the far north and north east of the licensed area, there are occasionally sea temperatures close to 0 °C. The air temperatures where comparable. The wind conditions where for most of the time calm, with an exception towards the end of the SARex 2. The fetches were very limited at the venue of the trials.

7.3. Relevance of technical, operational and organizational barriers

7.3.1. Technical barriers
The equipment put to test for both SARex 1 and 2 was, for the most part, relevant for the offshore oil and gas industry, i.e. lifeboats, life rafts, survival equipment, survival suits and life vests, personal Protective Equipment (PPE), clothing underneath suits and life vests, hoods and gloves, as well as EPIRB and SART – detection equipment.

7.3.2. Operational and organisational barriers
Several operational and organisational barriers were tested during both SARex 1 and 2 relating to:
- Leadership in survival exercise on board life raft and lifeboat.
- Rescue of people from lifeboat to KV Svalbard with the use of a MOB boat (SARex 1).
- Rescue of people from lifeboat to shore with the use of a helicopter (SARex 2).

7.4. Differences and similarities in the exercise scenario of SARex 1 and 2, compared to potential oil and gas-related accident scenarios
There are some key **differences** between the scenario of SARex 1 and 2 (a large-scale mass evacuation scenario in the northern part of Svalbard) and the accident scenarios possible in the petroleum industry:
- Time to rescue: Oil and gas industry in Norway has suggested a criterion of maximum 24 hrs from launch of lifeboats to rescue [13] vs. 5 days IMO Polar Code requirement [7].
- Distance from marine resources: Offshore oil and gas activities offshore have a dedicated stand-by vessel with hospital with trained first aiders and man-over-board systems. Their rescue capability is of course influenced by the weather.

- Distance from shore-based resources: Today's distances from shore-based rescue resources to the offshore facilities, even in the furthest exploration areas, are considerably less than that of a potential cruise ship evacuation north of Svalbard (except exploration well Korpfjell in block 7435, where there was exploration drilling in 2017).
- Number of persons to be rescued: Less on oil rigs vs. cruise ships. Typical manning of mobile semi-submersible drilling rig is about 120 persons.
- Threat of solid ice: No areas on the Norwegian continental shelf with exploration drilling (or production) suffer an imminent threat of either sea- or glacial ice.

Other environmental differences: The areas chosen for SARex 1 and 2 were in an area with:

- Limited fetch, i.e. well protected, as opposed to the offshore oil and gas industry.
- Low average precipitation compared to areas of the offshore oil and gas industry.
- Lower sea water temperature compared to offshore industry (except far north east licenses).
- Wildlife threats (polar bears and walruses).
- Presence of daylight: Cruises have 24 hours light vs. offshore industry in dark winter months.
- Availability of emergency preparedness resources: Dedicated resources for offshore industry.
- Emergency preparedness training: The offshore industry provides more extensive emergency response training compared to cruise passengers.
- Average health of cruise passengers vs. offshore workers: Stricter health requirements offshore vs. cruise industry.

There are also some key **similarities** between the scenarios and equipment used in SARex 1 and 2 and what is relevant for the offshore oil and gas industry on the Norwegian continental shelf:

- Means of evacuation: Use of enclosed lifeboats and life rafts although free fall lifeboats are more common on semi-submersible rigs.
- Emergency preparedness organization: Trained dedicated emergency response personnel
- Survival suits and life vests: Similar insulation capacities although latest generation of helicopter transport suits for offshore industry are better.
- Air and sea temperatures: In the case of sea temperatures it is on relevant during winter in north eastern part of Norwegian part of continental shelf.

7.5. Learning and recommendations from SARex 1 and 2

7.5.1. Technical barriers

Lifeboats: See also SARex Spitzbergen Report April 2016 [1], Chapter 2.2.4 "The lifeboat's capabilities and capacity during the Phase 1 exercise" and similar chapter from SARex 2 [2].

1. Ice formation in vents in SARex 1. The low-pressure release valve that activates when the engine is running (using air inside) clogged up with ice. DNVGL-ST-E406 Chapter 3.6.1.2 states that ice formation shall be taken into account when designing a lifeboat. It also refers to DNVGL-OS-A101.
2. Ventilation – an issue related to condensation and also low O_2 plus high CO_2 levels. Part of DNVGL-ST-E406 Chapter 8.3.3. There was a CO_2 meter that indicated the need for ventilation. This, in turn, caused the lowering of temperatures inside and condensation.
3. Diesel heater provided temperatures too warm for wearing survival suits. Use of diesel heater needs to be such that survival suits could be worn in the case of need for quick evacuations. There is not enough room to store survival suits in a secure and proper manner. Part of DNVGL-ST-E406 Chapter 8.3.3.

IOP Conf. Series: Materials Science and Engineering **700** (2019) 012045 doi:10.1088/1757-899X/700/1/012045

4. Ability for lifeboat captain to communicate with passengers at the opposite/ far end of the vessel. The captain is often located at the helm, which was a raised seat in the aft of the lifeboat. There was additional noise from engine, wind, wave motion, etc., which made it difficult to make direct commands. A speaker system inside the lifeboat would be of great help. This subject is not covered by DNVGL-ST-E406.
5. Evacuation to helicopter from lifeboat rooftop not realistic in open seas, due to wave motion and risk of hooking into deluge system. Some sort of removable railing, together with anti-skid cover, would be a great asset to aid standing on lifeboat roof.
6. Toilet facility useful. Covered by DNVGL-ST-E406 Chapter 8.3.5.

SARex 1 and 2 utilized the same conventional lifeboat. The lifeboat used in 2017 was an improved version of the one used in the previous exercise, e.g. insulation on seats (styrofoam), toilet (type "Porta Potti") and diesel heater. These two improvements alone provided a huge impact on comfort and estimated survival rate on board.

Life rafts: See also SARex Spitzbergen Report April 2016 [1], Chapter 2.2.6 "The life raft's capabilities and capacity during the Phase 1 exercise" and similar chapter from SARex 2 [2].

1. The SARex 2 life raft also had near standing height and an air-filled beam to keep the tent in position, which provided support to hold on to during times of wave motion and when it was necessary to move around the raft.
2. The SARex 2 life raft, with features such as a double bottom and double tent, provides a major improvement, in terms of insulation from the single bottom and tent version used in SARex 1.
3. The SARex 2 life raft had a welded double bottom; the welds themselves created cold bridges.
4. The SARex 2 life raft also had some leaks in the seams that created a small but constant flow of sea water into the raft. This was removed by using sponges throughout the exercise.
5. Although the raft was not filled to maximum certified capacity, there was very little room to move. It was a similar experience in both SARex 1 and 2. However, personnel pulled out of the experiment at a far faster rate in SARex 1, leaving much more room to move. This room was used effectively last year to perform exercises, in order to produce heat. This was not possible to the same extent in SARex 2, where almost all participants lasted the whole duration of the experiment.
6. The zipper on the outside of the tent froze when waves hit the raft's tent. Over a prolonged stay in a raft, this could be an issue with oncoming hypothermia.
7. The radar reflector was not easily mounted on the raft ceiling outside. The reflector was also lost when the wind reached gale force.
8. There were too few waterproof torches on board for a winter situation with complete darkness. Then, it would be essential to have sufficient lighting to ensure all necessary activities and operations.

Survival suits & life vests: See also SARex Spitzbergen Report April 2016 [1], Chapter 2.2.7 "The capabilities of personal protection equipment the Phase 1 exercise" and similar chapter from SARex 2 [2].

1. Compared to SARex 1, all participants in SARex 2 wore full suits. Three participants in the survival test wore uninsulated suits, the rest fully insulated suits. Some of the insulated suits also had extra insulation/ padding in areas that had contact with the raft floor, when in sitting or lying positions. The insulated suits were comparable to the commonly used helicopter transport suits used on the Norwegian continental shelf. The expected time of rescue in the case of an oil and gas-related incident on the Norwegian continental shelf is about 24 hrs, although this is not an established dimensioning criterion. Together with wool clothing, these suits would most

IOP Conf. Series: Materials Science and Engineering **700** (2019) 012045 doi:10.1088/1757-899X/700/1/012045

likely provide enough insulation, in a dry environment inside a double-bottomed/ double tent type raft, to sustain body temperature in an Arctic environment.

2. The "one size fits all" suits that were used in SARex 1 and 2 do not have pockets, which could be useful for storing safety equipment and food rations.
3. The personnel on board used suits differently. It was obvious that some basic competence related to routines for opening the zipper to let moisture out would have been useful. The suits are water- and airtight. There is a build-up of moist inside the suits. How long time the suit was unzipped and how far down the zipper was seemed essential for the ability to maintain heat and stay dry. In addition, the use of the hood influenced the cold experienced. Those persons without an integrated hood noticed cooling at an earlier stage than those with integrated hoods.
4. The Velcro around the ankles and the waist strap was an essential feature for smaller persons using the "one size fits all suits".
5. Those participants with uninsulated survival suits without integrated hood, but separate hood, noticed a distinct cooling of the neck area.
6. All suits had separate gloves or mittens. The types provided were either three-fingered type crab-claw neoprene mittens or five-fingered thin (2 mm) neoprene gloves. The crab-claw mittens are far warmer but impossible to wear while performing the simplest tasks.
7. It is a requirement of the NMA to have life vests with a light, which automatically starts in contact with water. Lights on life vests might not be sufficient to detect personnel in the sea in snow or fog. See section below about detection equipment. There were no lights either on the "one size fits all" type survival suits used in this experiment.
8. The pontoons around the raft in SARex 2 provided a good bench, removing the need to sit on the cold – and at times wet – raft bottom.
9. The ballast tanks underneath the raft bottom provided a lot of stability, together with the very strong winds and short period wind-waves (chop) towards the end of the experiment. Nevertheless, there was sincere concern, when the winds reached more than 20 m/s, that the wind could have toppled the raft. At this time, the personnel on board the raft were taken on board KV Svalbard. The situation at least showed the importance of stability measures in place, particular for an open ocean situation.
10. When performing the helicopter pick-up, the inflatable life vest popped, as it should. However, it was not possible for the rescuer to thread the sling either from above or below. The blown-up floatation device on the life vest had to be cut open by the rescuer, using a knife.

Personnel Protection Equipment (PPE) according to trials done by Norwegian Maritime Authority and Memorial University of Newfoundland, St. John's.

1. There are no protective gloves with insulation for Arctic conditions that enable the operation of valves or small items in either a normal work or emergency situation: operating VHF-radio, zippers, fastening and mounting radar reflector, emergency flare, etc. The personnel on KV Svalbard used conventional Arctic clothing; however, this equipment did not provide protection against crush injuries.
2. The personnel on KV Svalbard used balaclavas combined with googles to provide adequate protection in e.g. operating the MOB boat in a considerable wind chill. They also used Gore-Tex waterproof working suits with neck seal.

7.5.2. Organizational and operational barriers

Leadership: See also SARex Spitzbergen Report April 2016 [1], Chapter 2.2.8 "Leadership on board life raft during Phase 1 exercise", Chapter 2.2.9 "Leadership on board lifeboats during Phase 1 exercise" for lifeboats and similar chapter from SARex 2 [2].

IOP Conf. Series: Materials Science and Engineering **700** (2019) 012045 doi:10.1088/1757-899X/700/1/012045

The participation in the life raft survival tests showed the necessity of having good leadership on board: for both physical and mental aspects. The life raft leader on both SARex 1 and 2 was one of the Navy's own officers, with the ability to organize the evacuated personnel to ensure:

1. Activities to sustain heat/warmth.
2. Operating the life raft at sea, watch routines.
3. Organizing water and food rations.
4. Organizing various safety equipment on board the raft.
5. Activities to maintain awareness, encouraging people to take responsibility, to come forward with good ideas and provide entertainment.

The life raft leader ensured that:

1. All survivors on board had responsibility for at least one or two pieces of equipment, making all persons on board explain the function and usage of the designated equipment.
2. Other tasks were distributed amongst the survivors, e.g. distribution of food rations, control of timing, taking turns on watch.
3. Bonding and engaging activities took place on board: introduction of each individual survivor onboard, games, etc.

Competence and fitness: Based on the operation of the life raft and the function of various pieces of equipment, it proved important to have either a competent life raft leader or competence amongst the evacuated passengers to operate the life raft and associated equipment. Knowledge on cold climate survival would also be an important asset in terms of increasing survival chances: activities for keeping warm, how to minimize exposure to cold surfaces (e.g. raft floor).

7.6. Other lessons and recommendations for the offshore oil and gas industry in cold and remote waters

7.6.1. Technical measures

- Wearing thin neoprene gloves in emergency situations during helicopter flights and before embarking, and underneath the mittens permanently attached on most helicopter transportation suits. Stricter guidance on wearing correct size suits to ensure too large suits are not given to passengers.
- Wearing hood zipped up during flights in cold waters, not only during take-off and landing as seen practiced in the Norwegian sector. It is common practice to not wear the hood during take-off and landing, only zipping up the neck seal. However, it is usual to see passengers not zipping up, even when instructed to do so by the pilot before take-off and landing.
- The industry should develop performance criteria and efficient patents to ensure better ventilation in lifeboats and life rafts, without affecting the watertight integrity. This is also to meet the functional requirement in the IMO Polar Code Section 8.2.3.3, sub-item .1: "*…resources shall be provided to support survival following abandoning ship, whether to water, to ice or to land, for the maximum expected time of rescue. These resources shall provide a habitable environment*". "*Habitable environment*" is defined in Section 1.2.4 "*…a ventilated environment that will protect you against hypothermia*".

7.6.2. Operational/ organizational measures

- Consider removing the N-notation on the GSK certificates in the Norwegian sector for offshore travel, which is given when personnel are exempted from HUET (see chapter 6.1).

- Ensure cold water acclimation before offshore helicopter transport in cold waters, e.g. introduce real scenario temperatures in HUET training or add an additional module for acclimation.
- Establish performance criteria for the Norwegian offshore oil and gas industry, in relation to maximum time until rescue by either helicopter or vessel, where necessary first aid treatment can be given; cf. BaSEC recommendations [13].
- Introduce specialized courses for oil and gas operations within the Polar Code area that educate participants in all aspects of emergency preparedness, especially in the survival and rescue phase, which is not part of the other safety courses available. This should include the management of a lifeboat/ life raft, trauma handling, survival techniques, etc.

8. Conclusion

The offshore regulations through Activity regulations Section 77 about handling hazard and accident situations, sub section c) states that measures shall be taken so that personnel can be rescued during accident situations. "Rescued" in this context means that survivors shall be brought to a safe area either on board an installation, vessel or on land where necessary medical attention can be given. The lack of regulatory requirements and focus from the industry to date, in relation to cold water survival, can mainly be attributed to the relatively short length of rescue times expected for the more established and developed areas of the Norwegian continental shelf. At least in theory. However, considering the exploration activities and possible new field developments in the more remote northern parts of the Norwegian sector and the lessons learned from SARex 1 and SARex 2, one must consider the extraordinary challenges that these marine environments impose on a marine survival situation.

Given the complexity of extended survival in cold waters, the petroleum activities taking place in the areas of the IMO Polar Code needs to have compensatory measures in place *to obtain a similar safety level to that of petroleum activities in the more southerly parts of the Norwegian continental shelf.* For example, if not considering the five-day survival requirement for the polar regions, then additional rescue resources should be in place to considerably reduce rescue time. To achieve a similar rescue time to that in the more southern parts of the Norwegian sector will be a difficult task, considering the limited infrastructure; therefore, effort needs to be put into developing individual requirements and equipment for operating in these areas. Due consideration needs to be given to technical, operational and organizational measures in this aspect, to maximize survival chances during petroleum activities within the area applicable for the IMO Polar Code.

References

[1] Solberg K E, Gudmestad O T and Kvamme B O 2016 *Search and rescue exercise conducted off North Spitzbergen (SARex1)*, exercise report published by University of Stavanger.

[2] Solberg K E, Gudmestad O T and Skjærseth E 2017 *Surviving a maritime incident in cold climate conditions (SARex2)*, published by University of Stavanger, report no. 69, p. 269.

[3] Regulations for the Norwegian Petroleum Industry, available at web page for the Norwegian Petroleum Safety Authority http://www.ptil.no/forskrifter/category215.htm/

[4] The Norwegian Safety Authority 2013 *Principles for barrier management in the petroleum industry* p. 34 available at web site: https://www.ptil.no/contentassets/ 43fc402b97e64a7cbabdf91c64b349cb/barriers-memorandum-2017-eng.pdf

[5] Norwegian Maritime Authority 2016 *Regulations for mobile offshore units*, issued by the Norwegian Maritime Authority.

[6] International Maritime Organization (IMO) 2017 *International Life-Saving Appliance (LSA) Code* 2017 Edition, IMO publication, ISBN 978-92-801-165-0, p. 288.

[7] International Maritime Organization 2016 *Polar Code - International code for ships operating in polar waters*, 2016 edition, IMO, ISBN 978-92-801-1628-1, p. 76.

[8] Standard Norge 2008 NORSOK Standard S-001 on technical safety, available at www.standard.no/petroleum

[9] DNVGL 2014 Standard for the design of lifeboats DNVGL-OS-E406, DNVGL, Høvik, Norway.

IOP Conf. Series: Materials Science and Engineering **700** (2019) 012045 doi:10.1088/1757-899X/700/1/012045

[10] Norwegian Oil and Gas Producers Organization GL064 2015 Recommended practices for the establishment of area emergency preparedness, rev. 3 dated 12.8.15, p. 29.
[11] Loeng H and Ingvaldsen R 2011 *Oseanografi og klima i Barentshavet*, University of Tromsø publication, p. 5.
[12] International Organization for Standardization 2008 Ergonomics of the thermal environment – Cold workplaces – Risk assessment and management ISO 15743, p. 31.
[13] Barents Sea Exploration Collaboration 2016 "BASEC SSEPA Report Barents Sea South West"– DNV-GL report no. 2015-1056, Rev. 1, doc. No 1RI9SV4-8, dated 16.3.2016, 18 p

Pipeline shore crossing approaches in Arctic conditions

L Kurbonshoeva[1*], O T Gudmestad[2], A Zolotukhin[1,2,3]

[1] Gubkin Russian State University of Oil and Gas, Russia
[2] University of Stavanger, Norway
[3] Northern Arctic Federal University, Severnaya Dvina Emb. 17, Arkhangelsk, 163002, Russia

[*] Corresponding author: kurbonshoeva.l@yandex.ru

Abstract. The development of the Arctic region will entail the construction of new infrastructure, in particular, subsea pipelines. Considering the lack of actual field practice, ecosystem vulnerability and lack of common Arctic international standards, their development will be a big challenge, in particular, in the shore crossing zone. The design and construction of pipelines in the shore crossing area require a special approach that takes into account environmental and technological aspects of development. This work is aimed at analysing and determining environmental and technological factors influencing the design of offshore pipelines in the Arctic coastline. The paper presents theoretical and analytical work and the research is applied to a specific case study (pipelines from the Leningraskoye field to shore), through engineering calculations. Currently, there are five Arctic projects with shore transition areas for which trenching has been implemented. In order to determine the best shore crossing approach, it is important to consider the following environmental conditions: ice encroachment; ice ridges; shoreline erosion; permafrost thawing. Environmental characteristics should predetermine the choice of approach. Among three existing methods: trenching, tunnelling and horizontal directional drilling (HDD), the micro-tunnelling method is recommended for the Leningradskoye field in combination with a cofferdam corridor to protect the buried pipe from waves and ice in the nearshore area. In order to protect the surrounded permafrost from melting seasonal cooling-device is recommended to be used. The burial depth is determined to be more than 3.52 m in accordance with Force model calculations of ice ridges scouring depth. On the basis of research, the general choice-making diagram was proposed for Arctic shore crossing areas.

1. Introduction

Based on the latest report from the International Energy Agency, the energy demand will increase by 25% by 2040. In these forecasts, the demand for natural gas and oil will grow by 42% and 10%, respectively [1]. To meet the global demand for hydrocarbons, the development of unconventional resources and the exploration of new regions are becoming critical, including the Arctic offshore. Despite all the current environmental and geopolitical challenges, the oil and gas upstream companies are therefore considering the Arctic region as a strategic play for current and future hydrocarbon exploration and production.

The development of the Arctic region will entail the construction of new infrastructure, in particular, subsea pipelines. Considering the lack of actual field practice, ecosystem vulnerability and lack of common Arctic international standards, their development will represent technological and environmental challenges including the shore crossing zone. The aim of this work is to analyze the influence of physical environmental factors and climate change processes on the pipeline shore crossing methods.

2. Existing Practices in Arctic Pipeline Shore Crossing Areas

2.1. Existing Arctic projects

Several Arctic projects with shore crossing transitions exist nowadays. One of the best- known is the Northstar project in Alaska (Beaufort Sea), the first subsea pipeline constructed in the Arctic. The shore crossing part consists of a vertical 90-degree transition between the below-ground subsea pipelines and the above-ground onshore pipelines. Such transition results in deep excavation between the shoreline and the pipeline's daylight location.

The offshore production facility Ooguruk shore crossing design is similar to the one implemented in the Northstar project. Opposite to Northstar pipeline, a long-radius vertical transition was designed [3].

Another Alaskan project is Nikatichuq pipeline located at a depth of 3m offshore in the Beaufort Sea. The uniqueness of this project is that the shore crossing location was placed at a man-made offshore gravel pad extended from the shore. The same techniques were also implemented for shore transition – with a vertical sweeping curve transition.

Another project located in the Arctic-like conditions is Sakhalin-I. In the shore crossing area, a cofferdam corridor with perpendicular wing walls was constructed in order to protect the trench and backfill soil from waves. One more Arctic project is Bovanenkovo – Ukhta pipeline, 67 km of which was laid offshore on the bottom of the Baydaratskaya Bay. The pipeline coastline transition area was constructed by trenching and backfilling. In the areas of the coast, a cofferdam corridor was made.

Thus, five subsea pipeline projects with shore crossing in the Arctic region (sub-Arctic for the Sakhalin I) were considered and analysed. The general information on these projects is presented in Figures 1-3 and Table 1.

Table 1. Pipeline shore crossing projects existing in the Arctic

Project	Area	Water depth	Ice gouge protection	Trench depth	Average gouge depth	Max gouge depth	Shore crossing method	Bluff height
Northstar	Beaufort Sea	11.3 m	Trenching	2.1 m	0.6 m	1.0 m	Trenching. Vertical 90-degree transition	0.6 (max 2.4)
Ooguruk	Beaufort Sea	1.4 m	Trenching	-	1 m	2.09 m	Trenching. Long-radius vertical transition	1.8 -3 m
Nikaitchuq	Beaufort Sea	3 m	Trenching	-	-	-	Trenching. Artificial offshore gravel pad	-
Sakhalin I	Okhotsk Sea	15 m	Trenching	5 m	-	-	Trenching. Cofferdam corridor with wing walls	-
Bovanenkovo – Ukhta pipeline	Kara Sea	22-23 m	Trenching	-	12-13 m	20 m	Trenching. Cofferdam corridor	-

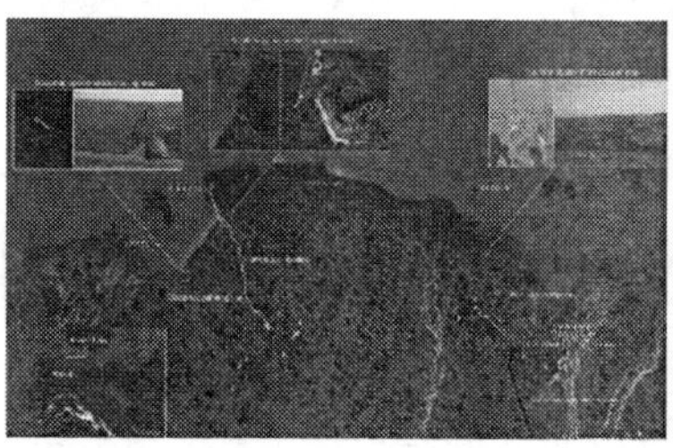
Figure 1. Beaufort Sea projects

Figure 2. Bovanenkovo-Ukhta pipeline

Figure 3. Sakhalin I

2.2. Methods of pipeline shore crossing

There exist three methods to design and construct pipelines in the shore crossing area: trenching, tunnelling and horizontal directional drilling (HDD). The method used for the above-mentioned projects is trenching which is based on the burial of pipelines in a trench and further, concrete or gravel covering can be used as well as pipeline strengthening or anchoring.

Tunnelling is based on building a tunnel initiated onshore and terminated at the seabed. The tunnel end should be at a depth where a pipeline is completely safe from scouring. For the onshore part, a vertical tunnel is required. Then pipes are placed inside the submerged tunnel lined with concrete. The tunnel gives protection role from water and permafrost thawing.

Horizontal directional drilling's (HDD) main advantage is the ability to use the method for long distances and quite large pipe diameters (up to 56'') [4]. By choosing the direction, areas with permafrost thawing can be avoided. Advantages and disadvantages of all methods are presented in Table 2.

Table 2. Advantages and disadvantages of different shore crossing methods

	Trenching	Tunneling	HDD
Advantages	less expensive easy to construct	less environmental impact ability to choose direction very big diameters pipes works in all soil types	less environmental impact long distances and big diameter ability to choose the direction
Disadvantages	affect the environment limited distances and diameters limited by coast type	expensive technologically difficult	expensive limited by coast type

3. Arctic challenges for subsea pipeline installation and design

Arctic harsh conditions such as severe cold temperatures, permafrost, presence of ice, presence of ice ridges, and erosion of the coastline should be taken into account when designing a subsea pipeline.

3.1. Ice ride-ups and pile-ups

During the beginning of the winter and during the spring the shore crossing area is characterized by ice ride-up and pile-up, which are known as ice encroachment. According to [5], the process when sheet ice remains intact or nearly intact as it is driven ashore is "ice ride-up". If the advance of the ice is halted by the slope and the ice fails in buckling or bending, it breaks up into individual blocks that form an "ice pile-up" either at the shoreline or somewhere on the above-water slope. The main factors that influence ice encroachment distance are wind stress, ice thickness and storm intensity. In [5] the encroachment distance (E) is calculated according to equation (1):

$$E = l + \frac{(h - H)}{tan\beta} \tag{1}$$

where h is a predicted pile-up elevation of the ice; l is the location of the peak of the ice encroachment, measured from the waterline; L is horizontal distance from the waterline to the elevation (at height H) of the land profile behind the shoreline; β is slope of the landward side of the ice rubble pile. For sheltered sites, it is appropriate to use l=0.5L, while for exposed areas the largest value can be obtained as l=0.67L.

The encroachment of the land behind the sloping shoreline is calculated according to equation (2):

$$E_{sp} = E - L \tag{2}$$

Since ice encroachment can damage the onshore pipeline and shore infrastructure, it is important to keep the pipeline below the ground at this distance when designing and installing pipelines in the shore crossing area. The total *set-back distance of the above ground pipeline* components from the shoreline is the additive sum of the estimated shoreline erosion and the estimated ice encroachment distance [2].

IOP Conf. Series: Materials Science and Engineering **700** (2019) 012046 doi:10.1088/1757-899X/700/1/012046

3.2. Ice Ridges and Ice Scouring

Ice ridges are ice features which were formed due to stress appearing within the ice when ice planes were colliding with each other under pressure. Ice ridges consist of two parts: above water part – «sail» made of small ice rubble accumulations; underwater part – «keel» which is formed as a chaotic conglomeration of broken ice. Typically, the height of the keel is four times bigger than the sail. The largest ridge to be recorded had a sail about 12 m and a keel of 45 m. Main drivers for ice ridges formation are winds and currents. In order to protect the pipeline, the correct burial depth should be determined. To calculate the gouge depth the Force model presented in [5], based on the assumption that friction forces depend on the gouging depth, is recommended (equation (3)):

$$F_{da} + F_{dw} + F_i - \mu F_{cy} - F_{cx} = 0 \tag{3}$$

where F_{da} is the wind drag force; F_{dw} – current drag force; F_i is ice force; F_{cy} and F_{cx} are friction forces; μ is friction coefficient. This equation can be solved using computer programs, for example Maple.

The more top sediments on the front surface, the greater the friction. At the maximum gouging depth, the steady forces are in balance with the resistance force.

It is suggested in reference [5] that the force model discussed above is conservative. Further research may conclude on the needs for lesser trench depth values. Until documented values are given in state of art literature, however, conservative values must be used. For the case studied in Charter 4, a lesser value of gauge depth may not change the conclusion, as the erosion of the coast line is a mayor driver to define the trenching depth.

3.3. Coastline characteristics. Coastline erosion.

The sea coastline is one of the most dynamic parts of the Earth because in this area natural phenomenon are actively interacting with each other: hydraulic, lithological, atmospheric and biological. The Arctic region is characterized by a large scale of coastline erosion due to climatic conditions as well as the structural-geological profile (stretches of soft sediments). Low temperatures and lack of sun radiation cause prolonged interaction of the coast by sea ice and also contributes to the development of permafrost.

Erosion-forming factors: wind, currents, storms largely depend on the ice regime of the sea. Therefore, it is important to take into account periods of open water, when the ice edge is located at a considerable distance from the coast, which entails an increase in wave and storm processes, and consequently, the level of coastline destruction. The average speed of the Russian seacoast erosion is about 1.2 m / year according to [6].

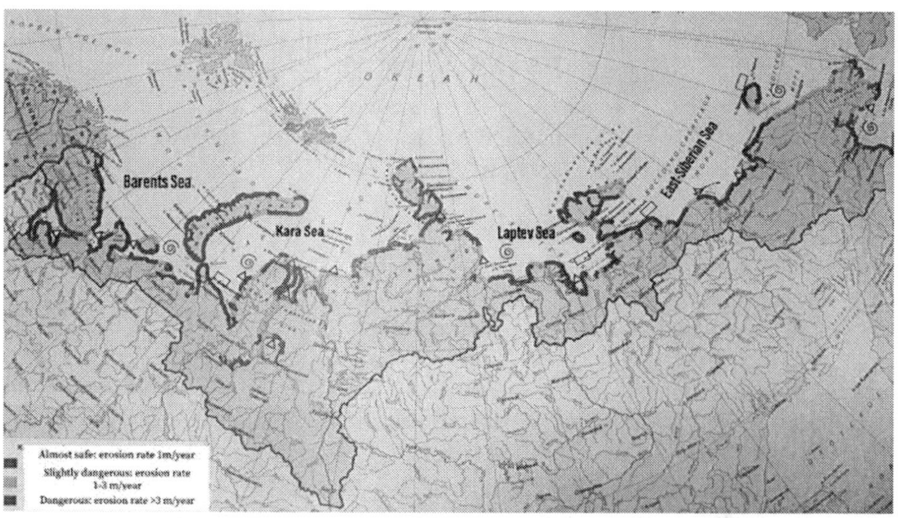

Figure 4. Morpho-dynamic map of the Russian Arctic seas [6].

The practically safe category with respect to coastal erosion (58% of the total length, green colour in Figure 4) is typical for the Arctic coast of Russia, 28% is of the second category - slightly dangerous and about 14% of the coastline is of the dangerous category (red colour in Figure 4).

Heat transfer from pipelines can also affect the shoreline erosion. When the pipeline is operating, its temperature is higher than the surrounding temperature, which leads to soil temperature increase. This process changes the bearing properties of the permafrost. As the temperature rises, the permafrost begins to melt, and the entire load carrying capacity is transferred to deeper soils. To calculate the soil thermal resistance for a single buried uninsulated pipeline the following equations (4-5) [7] are considered:

$$R_s = \frac{\ln\left\{\left(\frac{d}{r_{op}}\right) + \left[\left(\frac{d}{r_{op}}\right)^2 - 1\right]^{\frac{1}{2}}\right\}}{2\pi k_s}; \left(\frac{d}{r_{op}} > 2\right) \tag{4}$$

$$R_s = \frac{\ln\left(\frac{d}{r_{op}}\right)}{2\pi k_s}; \left(\frac{d}{r_{op}} > 4\right) \tag{5}$$

Where:

- R_S is thermal resistance of soil, $(\mathrm{m} \cdot \mathrm{K})/\mathrm{W}$;
- d – burial depth to centreline of the pipe, m;
- r_{op} – outer radius of pipe, m;
- k_S – Thermal conductivity of soil, $\mathrm{W}/(\mathrm{m} \cdot \mathrm{K})$.

3.4. The Effect of Climate Change on the Arctic Environment.

Due to effects of permafrost melting and soil settlement, as caused by the current world climate change, coastline erosion is accelerating. According to [8] mathematical modelling results showed that by the middle of the 21st century near-surface permafrost may shrink by 15-30%. That will consequently cause the complete frozen ground thawing in the upper «active layer». With the development of climate warming in the coming century, we should expect an increase in the degree of coastal processes. With raising water temperature, the rate of thermal abrasion and the intensity of coastline destruction will increase. In addition, climate warming will lead to a decrease in the ice cover of the Arctic seas, an increase in the duration of the ice-free period, increased winds and wind surges and increased storm activity, which will intensify the physical effects of waves on the shores and intensify the overall process of their erosion and destruction.

4. Choice of shore crossing approach method with example on Leningradskoye field

The Leningradskoye field located in the Kara Sea, close to the Yamal Peninsula, was chosen as a practical case for the analysis of a subsea pipeline shore crossing case. The initial gas reserves are estimated at 3.0 trillion m^3 [9]. One of the field development concepts introduced by Gazprom was the use of subsea production systems with the laying of subsea pipelines to a shore gas processing plant. [9]

4.1. Geotechnical characteristics of the territory

One of the main parameters for the selection of the shoreline crossing method is the geotechnical characteristics of the offshore and coastal zones. Abrasive – accumulative sediments compose the potential area. The height of cliff is 5 – 8 m and the maximum abrasion rate is 3 m/year - danger category red. The seabed of the studied area is also subjected to abrasion processes. The northern part of the Yamal Peninsula is composed of the Lower-Upper Paleocene, Tibasalinsk formation dominated by micaceous, silty clay. Sands with aleurolite and clay interlayers are observed to a depth up to 110 m [6].

The Yamal Peninsula is covered with permafrost with the thickness of 200 – 600 m and the annual average temperature is from -5 to -9 C°. According to reports from Yamal governor the issue of permafrost melting is a critical for the region nowadays. The governor reported that the permafrost on the territory of the Yamal Peninsula in the YNAO thawed by 40 cm in 2017 [10].

The studied area is characterized by high cliff height that makes the design and installation of shore crossing area more complicated. A good example in the case of a high cliff is the Langeledd pipeline laid between Norway and UK being the world's longest submarine pipeline. The height of the cliff in the UK shore approach area reached up to 5 m and the tunnelling method was used the transition area.

The Kara Sea is characterized by harsh ice conditions. The open water period lasts only 3 - 4 months and the sea is covered by solid first-year ice reaching a thickness up to 2 m from November to June. The most challenging aspect for the pipeline design in the Arctic conditions is the presence of ice ridges. The geometric parameters of the ice ridges observed in the Kara Sea are present in Table 3 [9].

Table 3. Geometric parameters of ice ridges in the Kara Sea area

Parameter	Ridge length	Sail width	Sail height	Keel depth	Keel width	Ridge thickness	Keel/ sail ratio
Min, m	24	7	1,5	6,0	21	7,7	3,0
Average, m	61	19	3,2	11,5	50	13,3	3,8
Max, m	95	34	4,5	15,7	72	19,8	6,7

According to the above discussion, it can be concluded that the shore crossing area is located in an area of harsh environmental and geotechnical conditions. During the design and installation of pipelines it is important to take into account the type of shore condition which is susceptible to coastline erosion; the presence of soft soils; the temperature of the permafrost which tends to melt with high rate; the high height of cliff and the presence of ice ridges in the nearshore area.

4.2. Choice of shore crossing approach

The coast of the Yamal Peninsula is not rocky, however, characterized by clay and sandy sediments. The absence of rocky sediments allows the use of all three methods: trench, tunnelling and horizontal directional drilling (HDD). However, the clays that make up the coastal territory are micaceous and silty, which characterizes them as structurally unstable, therefore, during the construction of the shore crossings it is necessary to create additional supports. Moreover, the clay conditions will obstruct achieving the necessary trenching depth required due to the harsh ice conditions in the region. The short Kara Sea open water period also restricts the open operations of trenching. In this case, the tunnelling method seems to be the most stable compared to the trench and HDD, since it will be possible to build a protective stable structure. The use of microtunneling will also help to protect the pipeline from scouring by ice ridges in the coastal zone. Thus, in the case of an unstable high cliff, the tunnelling method will be the most optimal.

In order to protect the stability of pipeline in the nearshore zone, subsea rock installation over the pipeline is recommended to be applied in order to guarantee safe operations regardless of the dynamics of coastal seabed.

The tunnelling method, however, is a complicated and expensive method. This technology was applied at the intersection of the coastline by the Langeled pipeline in Isington (UK). The microtunneling method, which is widespread in the Russian Federation, is described in TSN 40-303-2003 "Trenchless laying of communications using microtunnel penetration complexes and the reconstruction of pipelines using special equipment".

In this paper, the pipe-in pipe method has not been considered due to very large costs of the pipe-in pipe concept. The method could, however, be considered as a protection method for pipelines in Arctic shorelines at least for the most exposed sections of the shore approaches due to the large strength capacity of the outer pipe against mechanical action. Note also that the concept prevents heat transfer from product into the permafrost, if proper insulation is selected.

4.3. Comparative analysis for the further design of the pipeline access to the shore

To determine further the method for the optimal design of the pipeline access to the shore, a comparative analysis was carried out with existing projects for pipeline existing in the Arctic, as well as the Langeled pipeline. The method used is the integrated assessment method based on the calculation of generalized estimates, taking into account estimates for all weighted criteria. The results of the comparative analysis are presented in Table 4.

Table 4. Comparative analysis (high values are favourable)

Project	5 –Cliff Height	4 –Coast erosion rate	3- Permafrost	2 – Ice Conditions	Total
The studied project	0,07	0,1	0,07	0,05	0,29
Langeled	0,07	0,16	0,29	0,24	0,76
Northstar	0,23	0,32	0,09	0,07	0,71
Ooguruk	0,14	0,32	0,09	0,07	0,62
Nikaitchuq	0,14	0,32	0,09	0,07	0,62
Sakhalin 1	0,14	0,16	0,29	0,07	0,66
Bovanenkovo-Ukhta	0,11	0,32	0,07	0,06	0,56

Integrated assessment ranking	0 – 0,2 Severe conditions	0,2 – 0,4 Difficult conditions	0,4 – 0,6 Normal conditions	0,6 – 0,8 Moderate conditions	0,8 – 1 Favorable condition

Thus, 4 comparison criteria were determined and presented in accordance with the importance for the selection and design of the pipeline access to the shore. For the project under consideration, the main difficulties are the high values of coastal erosion and ice characteristics, which significantly distinguish the project from existing pipelines.

As can be seen, Alaskan projects are located in more moderate conditions due to smaller cliff height and relatively small erosion rate. The conditions of the Langeled project also vary from the studied one due to the absence of permafrost and ice. However, the cliff height and erosion rate are similar to the studied projects. The conditions of the Bovanenkovo-Ukhta project are considered as normal for the Arctic Seas which is caused by the severe ice conditions and presence of permafrost.

Therefore, since all existing projects with pipeline shore crossing area are located in more favourable conditions it is important to integrate the different technologies in order to meet the severe conditions of the studied area.

4.4. Construction phases

To lay a pipeline in a tunnel, it is necessary to build two shafts / mines: the starting one and the receiving one. The dimensions of the shafts should be set in accordance with the size of the working body. Before construction, the construction site of the starting shaft must be secured, and the soil in the places where the crane and other heavy equipment are placed should be compacted.

Next, in the shaft, it is needed to install a press frame with powerful jacks and install a tunnel shield. The jack moves the shield in the ground by an amount equal to its length. For the construction of the tunnel, it is necessary to use water that is supplied to the working body area by a feed pump located on the surface. After treatment, the water is fed back to the pump, where the soil must be removed [12]. The accuracy of penetration is carried out by a computer control complex using a laser shield system. By changing the size of the tunnel shield, it is possible to lay underground micro tunnels of different diameters from 250 mm to 3000 mm.

This paper presents the calculation of the pulling force according to SR 42-101-2003, "General provisions for the design and construction of gas distribution systems of metal and polyethylene pipes" [13]. For the calculation were taken some characteristics of the pipeline and the parameters of the tunnel (Table 5).

Table 5. Initial pipeline characteristics

Symbol	Description	Value	Unit
Do	Pipeline Outer Diameter	1,02	m
l	Tunnel length	400	m
Δ	Pipeline Wall thickness	0,018	m
ρm	Pipeline Material Density	7850	kg/m^3
ρi	Insulation density	975	kg/m^3
Δi	Insulation material wall thickness	0,003	m
F	The coefficient of friction of the pipeline on the finishing of the tunnel	0,45	
E	Elastic modulus for steel	2,06 *1011	N
μ	The coefficient of friction of the pipeline on the ground	0,25	

The pull force P is defined as the sum of all types of resistance to pipeline movement in a tunnel:

$$P = \sum_{i=1}^{4} P_i = P_1 + P_2 + P_3 + P_4, \tag{6}$$

Where P_1 – friction force from the weight of the pipeline (in the tunnel); P_2 - additional friction forces from the support reactions; P_3 - increased resistance to movement in the transition from straight to curved motion; P_4 - friction force from the weight of the pipeline outside the tunnel.

The friction force of the weight of the pipeline is calculated as in equation (7):

$$P_1 = q_w \times R \times (e^{f(l-l_i)/R} \times \cos\frac{l-l_i}{2\times R} - \cos\frac{2l-l_i}{2\times R}), \tag{7}$$

Where: q_w – linear weight of pipeline and insulation material, N / m^2; R – the estimated radius of curvature of the tunnel, m; l_i – current tunnel length, m;

The friction force P_1, due to the weight of the pipeline, according to equation (7) was P_1=204479 N.

Additional friction forces from P_2 support reactions are determined by the formula:

$$P_2 = 0{,}5 \times P_u \times (1 + e^{f(l-l_i)/R}), \tag{8}$$

Where: P_u – friction forces from the support reactions that determine the bending of the pipeline, which is calculated by the formula:

$$Pu = \frac{f\times\pi\times E}{16\times R\times B}\times(D_o^4 - D_i^4), \tag{9}$$

Additional friction forces from P_2 support reactions according to equation (8) is equal to be 43494N.

The increased resistance in the transition from straight to curved motion before leaving the pipeline from the tunnel P_3 is calculated by the equation:

$$P_3 = \frac{\pi\times E}{128\times R^2}\times(D_o^4 - D_i^4) \tag{10}$$

$$P_3 = \frac{\pi\times 2{,}06\times 10^{11}}{128\times 1224^2}\times(1{,}020^4 - 0{,}984^4) = 488{,}8N.$$

The frictional force of the weight of the pipeline on the soil outside the tunnel P_4 is determined by the equation (11):

$$P_4 = \mu\times q_w \times l_i = 108919{,}5N \tag{11}$$

The pull force P is determined by the equation (6):

$$P = 204479 + 43493{,}8 + 488{,}8 + 108919{,}5 = 357381{,}1 \text{ N}$$

Stresses that arise in the pipe during pull-in must satisfy the conditions of strength and deformability. The calculations showed that both of these conditions are satisfied. In accordance with the formulas, calculations were made of the thermal resistance of the soil for a single buried uninsulated pipeline and tunnel. The calculation results are shown in Figures 5 and 6.

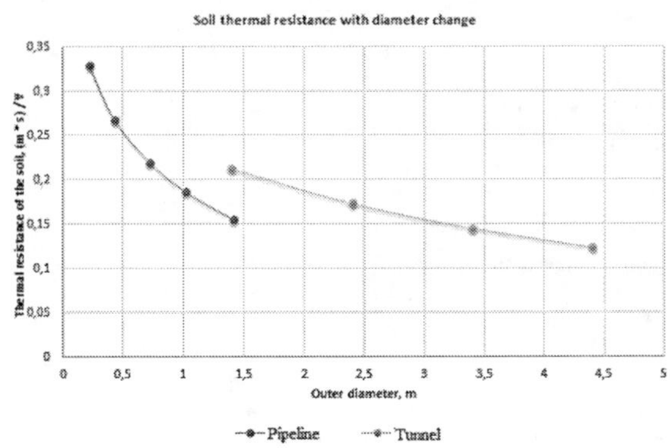

Figure 5. Soil thermal resistance vs diameter change

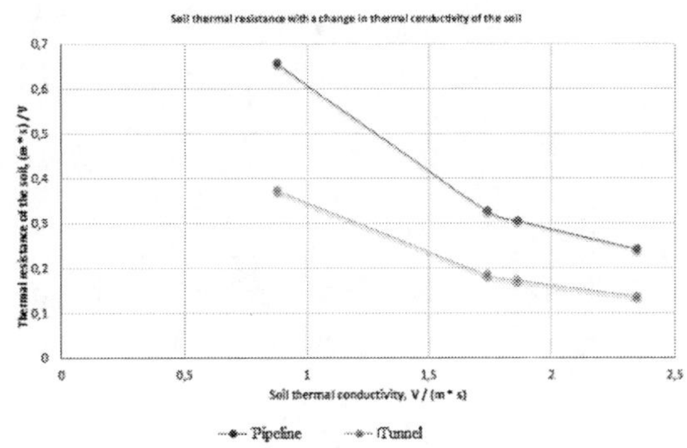

Figure 6. Soil thermal resistance vs soil thermal conductivity

Thus, from the calculations and the graphs it can be concluded that the thermal resistance is influenced by the composition of the soil, as well as the radius of the pipeline. With an increase in the thermal conductivity of the soil and its moisture content, the thermal resistance of the soil decreases. The project is located in the zone of wet soils and permafrost, which tends to melt in the summer. Also with an increase in the radius of the pipeline, the thermal resistance of the soil decreases. When designing a tunnel, it is necessary to introduce technologies to prevent the melting of soils under structures. Today there are several systems: horizontal tubular systems of freezing and temperature stabilization of soils, vertical tubular systems of freezing and tubular seasonal cooling devices (SCD) to maintain the bearing capacity of the soil. For the project under consideration, it is proposed to choose an SCD system to maintain the temperature of the soil. These devices may be installed along the entire length of the tunnel (NPO Fundamentstroyarkos LLC).

Another dangerous characteristic of the studied region is the presence of ice ridges and their scouring of the seabed; therefore, the sea part of the pipeline has been proposed to be buried in a trench. It is also proposed to build cofferdam for protection against waves, by analogy with the already existing projects. Equation (3) linked with Maple software and data characteristics of the Kara Sea were used to calculate the scouring depth. The calculation results are presented in Table 6.

Table 6. Results of scouring depth calculation

	Values	Units
Wind drag force	402.58	kN
Current drag force	24501.9	kN
Weight	518340.9	kN
Buoyancy	532219.4	kN
Ice Force	0.024	kN
Ice ridge scouring depth	**3.52**	m

Therefore, the depth of the trench pipeline in the near shore area and further offshore territory should be not less than 3.52 m.

An analysis of the territory of the prospective project has shown that the tunnelling method will be the most optimal method for getting the pipeline to shore. The main aspect of the choice of this method was the high value of the cliff height, as well as high rates of annual coastal erosion. For the coastal zone, it is proposed to lay the pipeline in a trench to avoid damage of the pipeline due to ice ridges. Also, to protect the coastal zone from erosion, it was proposed to use a cofferdam corridor. The approximate scheme of the studied project is presented in Figure 7.

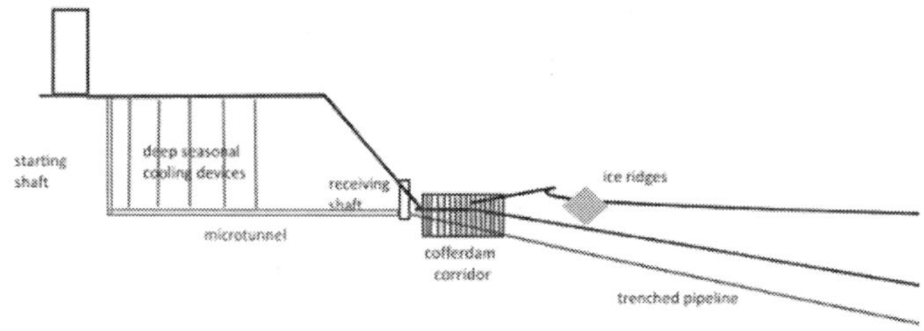

Figure 7. The approximate scheme of the project

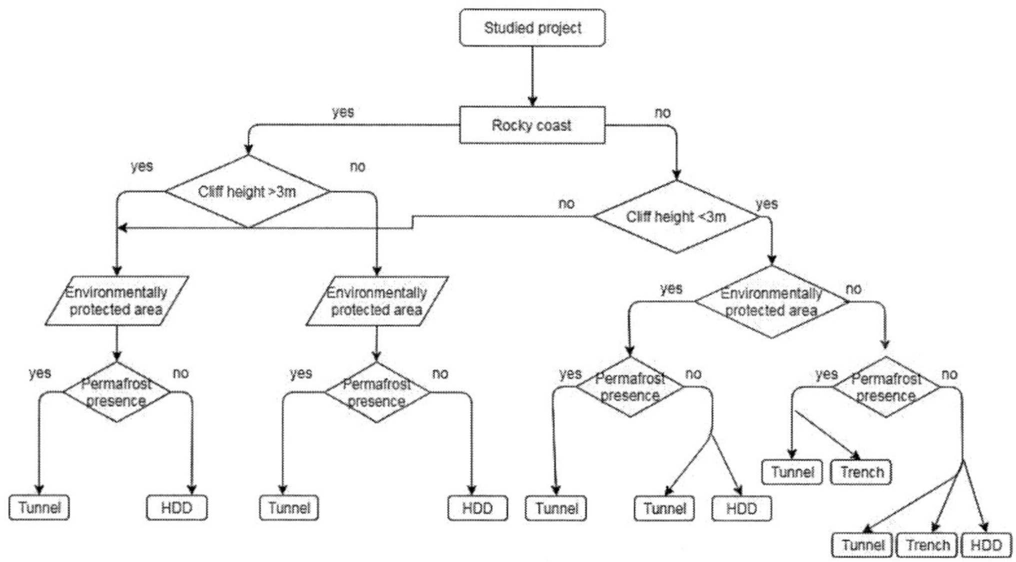

Figure 8. Decision making diagram

From a practical point of view and analysis of the literature we can conclude that the choice of the method of access to the pipeline depends on many natural and climatic factors. So, among the climatic factors, the factors most influencing the choice of method will be the following: (1) geological structure of the coastline; (2) composition of the coastline soil; (3) height of the cliff; (4) location of relatively protected natural areas; (5) coastal erosion rates; (6) the presence of permafrost; (7) ice conditions of the territory. In accordance with the identified criteria, a shore crossing method decision making diagram was drawn (Figure 8).

5. Conclusions

1. Currently there are five Arctic projects with shore crossing areas, for all of which the trenching method was implemented.
2. In order to determine the best shore crossing approach, it is important to take into account following environmental conditions: ice encroachment; ice ridges; shoreline erosion; permafrost thawing. Environmental characteristics should predetermine the choice of shore transition method.
3. Micro-tunneling method was recommended to be used for the Leningradskoye field in

combination with a cofferdam corridor to protect from waves and buried pipe in the nearshore area.

4. Calculations of soil resistance showed the decrease in case of tunnel diameter increase, therefore, in order to protect the surrounded permafrost from melting seasonal cooling devise is recommended to be used.
5. The burial depth is determined to be more than 3.52 in accordance with Force model calculations of ice ridges scouring depth.

References

[1] International Energy Agency World Energy Outlook 2018 (WEO 2018). [https://www.iea.org/weo2018/ accessed on 30.03.2019]

[2] Eisler B 2016 Shore Crossing Design Considerations & Solutions for Arctic Subsea Pipelines. OTC-27453-MS. In *Proc. of Arctic Technol. Conf.*, St. John's, Newfoundland and Labrador, 24-26 October 2016.

[3] Heuer C E, Caldwell J B and Zamsky B 2012 Design of Buried Seafloor Pipelines for Permafrost Thaw Settlement, Proc. of the Fourth Int. Conf. on Permafrost, Fairbanks, California, Final Report Ice encroachment in the Alaskan Beaufort Sea, April 2012.

[4] Duplenskiy S 2012 Protection of subsea pipelines against ice ridge gouging in conditions of substantial surface Ice, Master's Thesis, University of Stavanger/ Gubkin University 2012

[5] AO "Roscartography" National Atlas of the Arctic, Chief Editor - Nikolay Kasimov - Academician of the Russian Academy of Sciences, First Vice-President of the Russian Geographical Society, 2017

[6] Jianguang Y 2018 Methods of heat transfer analysis of buried pipes in district heating and cooling systems, *Applied Eng.* **2**(2), 33-38

[7] Anisimov O, Reneva S 2006 Permafrost and changing climate: The Russian Perspective. AMBIO, *J Hum. Environ.* **35**(4):169-175.

[8] Morev Y.A., Mirzoyev F D, Arkhipova O L, Ibragimov I E, Trudov S V 2017 Technical and Technological Solutions for the Development of Arctic Gas Condensate Fields using the Underwater Complex, GazpromVNIIgaz [https://www.znak.com/2017201/vlasti_yanao_obespokoeny_tayaniem_vechnoy_merzloty_kotoraya_negativno_vliyaet_na_stroyki]

[9] Zubakin G K, Egorov A G, Ivanov V V, Lebedev A A, Buzin I V, Eide L I, 2008 Formation of the severe ice conditions in the southwest of the Kara Sea. *18th Int. Offshore and Polar Eng. Conf.*, Vancouver, Canada, July 6 – 11.

[10] Prokopenko I A 2008, Determination of the optimal construction technology for offshore pipeline crossings at technical and economic criteria, Dissertation for the degree, Gubkin University, Moscow 2008.

[11] SR (set of Rules) Kodeks 42-101-2003 General provisions for the design and construction of gas distribution systems of metal and polyethylene pipes.

[12] NPO Fundamentstroyarkos LLC [https://www.nposa.ru/sites/default/files/sistemy_temperaturnoy_stabilizacii_vechnomerzlyh.pdf -Accessed on 26.05.2019]

[13] Gudmestad O, Loset S, Alhimenko A, Shkhinek K, Torum, A, Jensen, A 2007 *Engineering aspects related to Arctic Offshore developments*, St. Petersburg, Publisher "LAN" Design, 2007

IOP Conf. Series: Materials Science and Engineering **700** (2019) 012047 doi:10.1088/1757-899X/700/1/012047

Features of geotechnical surveys and leg penetration analysis for drilling platforms in the Arctic seas

I Marchenko[1,2]

[1] Gubkin Russian State University of Oil and Gas (National Research University), Leninsky prospect 65, Moscow, 119991, Russia

[2] JSC «MAGE», Business Centre «Krylatsky 2», 11, Osennyaya, St, Moscow, 121609, Russia

Corresponding author: ivanmarch94@gmail.com

Abstract. In recent years, one of the most important direction in the development of offshore hydrocarbon exploration is the expansion of the oil, gas and gas condensate fields of the Russian Federation to the Arctic shelf areas. Extensive and complex engineering surveys are actively performed in the Barents Sea and the Kara Sea. In the last year, a wide range of engineering investigations has been implemented on prospecting and exploratory wells.

The geotechnical surveys are mandatory before the drilling of exploratory and production wells to obtain the data for the upper part of the geological profile to provide necessary information for safe positioning of drilling rigs. This is extremely important as the upper part of the soil interacts with an engineering structure whether it is a drilling rig, pipeline, subsea production system or any other structure. The data of the stated surveys is used for optimal location selection for drilling rig positioning and the necessary stability calculations. The use of geophysical and geotechnical methods in engineering surveys provide an effective solution for the geological challenges and ensures the availability of sufficient information regarding soil composition, properties and strength conditions.

1. Introduction

The exploratory and prospecting wells are necessary components within the scope of work for hydrocarbon fields development in the Arctic seas. Drilling of these wells is performed with the use of jack-up and semi-submersible platforms. Prior to positioning of the platforms the complex investigation of geophysical, geotechnical, hydrometeorological conditions is to be performed, and necessary engineering calculations are to be completed.

An important part of the investigations is the identification and mapping of geological hazards on the site of the planned well. Some geohazards are typical for sea shelf generally. However a number of particular hazards have been discovered in the Russian Arctic shelf. A complex of methods has been developed for investigation of different geohazards which includes both traditional well-established marine survey methods and some special methods for Arctic seas.

Analyses and calculations for drilling platforms positioning are regulated by specific standards. In the current moment, many international standards have been implemented and used by most of the oilfield service companies. However, in the Arctic conditions, some additional factors, not described in the standards should be taken into consideration.

IOP Conf. Series: Materials Science and Engineering **700** (2019) 012047 doi:10.1088/1757-899X/700/1/012047

2. Geohazards on oil and gas fields of the Arctic shelf and methods for their investigations

Investigation of geological hazards, their presence and their possible propagation, is necessary for determination of the location for positioning of rigs for exploratory and production drilling. In areas of the Russian Arctic shelf, the following geohazards are distinguished:

- Gravity processes:
  - Slides/slumps;
  - Sediment creep;
  - Effects of turbidity currents;
  - Rockfalls.
- Sediments with increased gas content («gas pockets»), with channels («gas pipes») and with the presence of gas hydrates.
- Areas of focused fluid release (including mud volcanos);
- Areas with abnormal formation pressure;
- Presence of permafrost; sporadic and insular permafrost;
- Ice gouging (bottom exaration) by icebergs and stamukhas;
- Seismic – earthquakes;
- Areas with soft sediments like soft clay or silt

Because of the geological hazards, the typical challenges for engineering surveys in the Arctic seas are the following:

- Determination of the presence and expansion of dangerous natural processes and events (geohazards);
- Investigation of geotechnical profile features, including composition, texture, properties and conditions of the soils.

Several methods in complex engineering surveys are addressing these challenges. For example,

- Engineering-geodetic surveys (hydrographic works):
  - Multibeam bathymetric surveys;
  - Side screen sonar surveys (SSS) – definition of ground forms and objects on the sea bottom);
  - Hydromagnetic surveys – identification of metal-containing objects;
  - Navigation support of all types of works.
- Engineering-geological surveys (including geophysical investigations):
  - Continuous seismo acoustic profiling – high resolution and low resolution;
  - High resolution seismic;
  - Geotechnical works: soil sampling, drilling, laboratory testing (offshore and in an onshore laboratory).
- Engineering-Hydrometeorological surveys, including:
  - Hydrometeorological condition investigations: measurements of temperature, wind speed, pressure determination due to extreme values of the wind speed, identification of wind rose etc. (meteorological station);
  - Hydrological conditions investigations: measurements of high tides, current speed and salinity, determination of extreme values of currents, current rose, swell parameters, ice regime (autonomous bottom station).
- Environmental surveys, including:
  - Investigations of air and water pollution, radiation, ecosystem assessment, marine mammal observations, weather forecasts and recommendation issues.

Apart from standard geophysical (high resolution seismic, low/high frequency seismic-acoustic profiling, hydromagnetic surveys) and hydrographic (Multibeam, SSS) surveys, for the Arctic seas in regions with potential for permafrost, the following tasks are performed: transcendent electromagnetic soundings in near field, geotechnical drilling with in-hole thermometric investigations, and soil sampling with laboratory testing in an offshore laboratory (Figure 1).

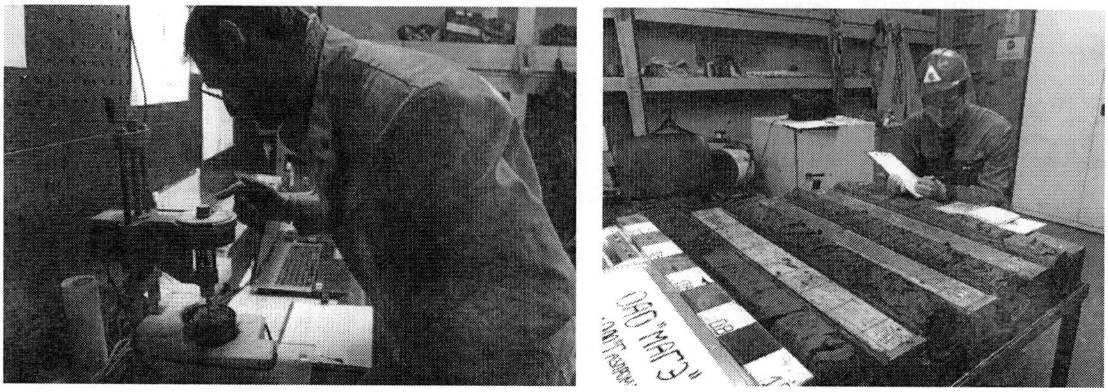

Figure 1. Soil testing in the offshore laboratory.

Geotechnical borehole drilling is performed in order to clarify the geotechnical data and geological conditions on-site for drilling rig positioning. The data needed include the determination of engineering geological profile; obtaining data regarding soil properties, texture and conditions as well as confirmation of electrical data for identification of frozen soils. To confirm the frozen soil presence and conditions, thermometric investigations are performed after the completion of the borehole.

During geotechnical drilling, the main goal is to obtain the maximum undisturbed core samples, appropriate for laboratory testing. The method for geotechnical borehole drilling in regions with permafrost expansion is characterized by many technological features special for the Arctic seas.

Engineering survey performance has several additional features in the Arctic, comparatively with similar works in the Southern regions. During the planning of fieldwork, including these methods it is necessary to take into account the specifics of the Arctic region and pay attention to the following aspects:

- Short navigation period. The navigation period is rather limited and depends on the particular conditions. In the Kara Sea - from the second half of July/ beginning of August – to the end of October. The suitable weather window is wider in the Barents Sea. However, the working period can shrink due to frequent storms. Thus, the preparation of the research vessel and equipment should start early, and considerable waiting time due to weather may be expected.
- Remoteness from the closest supply port. Murmansk and Arkhangelsk are the most developed ports in the Russian Arctic. The distance to the port from the oil and gas field could reach 600-700 km in the Barents Sea and more than 1500 km in the Kara Sea. The engineers have to identify the necessary reserve of equipment, spare parts, packing materials and other necessary details. Rather often in case of any difficulties and equipment malfunction or necessity to change the method of works, the engineers have to be ready to solve the problem only using the materials on the board of the vessel.

Soil samples treatment and testing also have some features. In the Arctic seas, the chilled soils are common. Soil testing has to be started immediately after the core has been pulled on board of the vessel to obtain the accurate properties. Measurements of temperature and investigations of thermal-physical soil properties also should be done fast.

3. Calculations for drilling rigs

Based on the data obtained after soil analysis in the offshore and stationary laboratory, the necessary calculations should be performed. The specific calculations depend on the type of platform.

The stability calculations and predictions of footing penetrations should be implemented in the course of the planning of drilling work with jack-up platforms. Prior to projects with semi-submersible rigs, the calculations for the anchors penetration and dragging distance as well as anchor holding force should be completed.

The drilling rig positioning process involves several companies, clients, contractors and operators, large human and financial resources are involved in the procedure. Due to these and many other tasks, a working day of the whole team costs a colossal sum of money. For this reason the calculations, including all possible factors influencing rig positioning, take an extremely important role.

For the current theme, a number of standards have been developed. In international practice, the most popular detailed standards are the following:

- SNAME Technical & Research Bulletin 5-5A. Guidelines for Site-Specific Assessment of Mobile Jack-Up Units, 2008 [1].
- ISO 19905-1 – Site-specific assessment of mobile offshore units – Part 1 – Jack-up platforms [2].

Some engineering standards, like SP 22.13330.2016 [3], SP 58.13330.2012 [4] and some others are used for calculations in Russian projects, although the named standards are not designed especially for jack-ups.

Generally, jack-up platform footing penetration analysis both in international and Russian standards can be summarized by following main stages:

- Design of the spudcan model;
- Computing of the vertical bearing capacity of the footing at various depths;
- Plotting of the vertical bearing capacity versus depth curve and comparison with the specified maximum preload and determination of the predicted footing penetration.

When comparing the calculation methods in Russian and international standards certain differences can be noticed. Generally the formulas for vertical bearing capacity calculations are nearly identical. However, in Russian SP 22.13330.2016 (1) a single formula is used for all types of dispersive soils, while the international standards, for example, SNAME 5-5A use two different formulas for clays (2) and sands (3):

$$N_u = b' \cdot l'(N_\gamma \xi_\gamma b' \gamma_I + N_q \xi_q \gamma'_I d + N_c \xi_c c_I) \tag{1}$$

where b' and l' – the breadth and the length of the foundation (spudcan); N_γ, N_q, N_c – bearing capacity non-dimentional coefficients, dependent on soil friction angle, ξ_γ, ξ_q, ξ_c – foundation (spudcan) form coefficients, γ_I and γ'_I - submerged unit weight of soil values, kN/m^3, c_I – specific cohesion of soil, kPa, d – foundation (spudcan) depth, m.

$$F_v = A(c_u N_c s_c d_c + p'_0) \tag{2}$$

$$F_v = A(0.5\gamma' N_\gamma s_\gamma d_\gamma + p'_0 N_q s_q d_q) \tag{3}$$

where A – spudcan effective bearing area, s_c, s_γ, s_q – bearing capacity shape factors, d_c, d_γ, d_q – bearing capacity depth factors, p'_0 - effective overburden pressure of soils, c_u – undrained cohesive shear strength, kPa.

In the Russian Standards, the specific cohesion of soil is used. The current parameter is determined in consolidated drained triaxial tests. In SNAME and ISO the values of undrained shear strength are applied. However the undrained shear strength value can be obtained by different methods including the following: undrained unconsolidated triaxial, laboratory vane, CPT and other. The method which is to be used for undrained shear strength determination is not specified in the standards and the values obtained by the named methods can be rather different. Thus, this is a point for discussion.

In both calculation methods the identical bearing capacity factors (N_γ, N_q, N_c) are used. However, in each standard the calculation methods and the obtained values for the current factors are quite different. The difference can be mostly distinguished in values for silica sands with friction angle 30-40 degrees and in sandy silts with similar properties [5].

It worth noting that in particular oil and gas service companies, for instance, in Fugro, some intradepartmental standards are developed. This kind of documents may be appropriate for specific regions where the company proceed with projects. However, these documents are used only inside the companies and are available for common use.

Proceeding the leg penetration depth determination, an engineer should pay attention to some factors specific for marine conditions and Jack-up rig type structures, including the following:

- Differences of forms and sizes of Jack-up leg spudcans. A precise model of the spudcan should be built prior to calculations.
- Specifics of Jack-up rig positioning procedure.
- Possibility of the appearance of alternative soil failure mechanism:
    - Squeezing - when a stronger soil layer is underlayered by soft soils.
    - Punch-through – a failure processes when a strong layer overlies a weak layer and hence a rapid additional penetration of the spudcan can occur, associated with a significant reduction in bearing capacity.
- Possibility of soil back-flow into a cavity over a spudcan.

The back-flow of the soils during spudcan preloading can occur by the following mechanisms: soil flows over the top of the foundation, cavity wall failure and sediment transport from the sea bottom into the cavity hole [6]. These three mechanisms are illustrated in Figure 2.

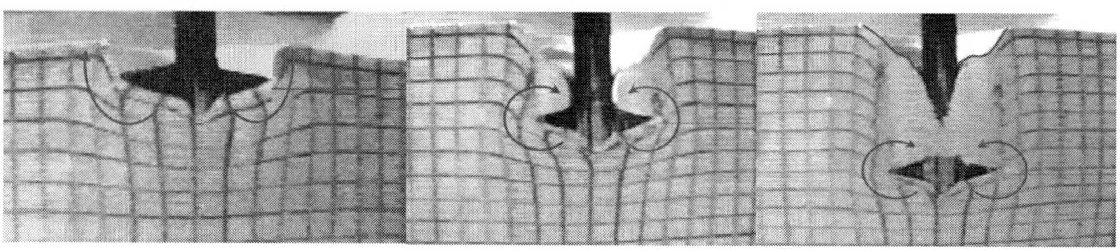

Figure 2. Illustration of possible back-flow mechanisms during spudcan penetration in centrifuge testing [6].

The above-named alternative soil failure mechanisms (squeezing and punch-through) are presented in Figure 3, where the designations in the figure stand for

Q_v – applied factored vertical load;

p_0 – effective overburden pressure at depth D.

A, V, B – Spudcan bearing area, volume and diameter

T – thickness of a weak clay lay.

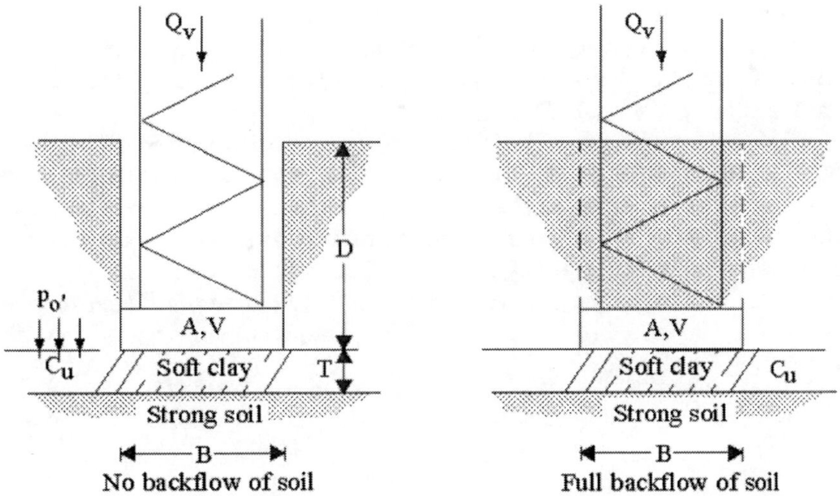

Figure 3. A typical scenario for the appearance of squeezing (left) and punch-through (right) soil failure mechanisms [1]

4. Conclusions

Geophysical and geotechnical methods are performed in order to obtain the fullest data for soil composition, texture, properties and conditions. The use of geological data obtained in a high quality manner and application of different calculation methods and standards makes possible performance of necessary assessment of foundation soils and comparison of the results taking into account different factors. All these applied methods make possible the completion of the main goal – determination of appropriate, safety and economically effective location for positioning of a drilling platform.

Jack-up rigs leg penetration analysis – is on of the challenges, which has to be performed prior to drilling works. A number of technological modifications of spudcans, legs and other mechanisms of a platform may be dependent on the calculation results. In certain cases it may be necessary to select an alternative platform instead of one preplanned earlier.

The appearance of the factors as back-flow, squeezing and punch-thorough is possible jack-up rig positioning procedure in any regions – all these factors should be taken into account. However, when planning any procedures and calculations for Arctic seas, several additional features should be taken into account. Engineers performing the calculations should pay attention that the bottom soils in the Arctic seas have the following features: possibility of permafrost presence and presence of chilled soils. In certain regions strong lateral variety of soil properties has been detected.

Due to the named above factors it can be concluded that a leg penetration calculation method should be updated for each specific region either by specific coefficient inputs or by application of a single method considering features of the region. In should be also noted that in order to avoid the uncertainty of geotechnical properties, especially in areas with strong lateral variety of soils, geotechnical investigations should be performed in the quality and precise manner. Scope of work for site survey should include geotechnical boreholes designated for location of each jack-up platform leg apart.

Acknowledgements
I am grateful to prof. O. T. Gudmestad and Prof. A. Zolotukhin for valuable advice given in the preparation of this work.

IOP Conf. Series: Materials Science and Engineering **700** (2019) 012047 doi:10.1088/1757-899X/700/1/012047

References

[1] SNAME T&R Bulletin 5-5A 2008 *Recommended Practice for Site Specific Assessment of Mobile Jack-up Units* (Jersey City, Society of Naval Architects and Marine Engineers, Revision 3).
[2] ISO 19905 2012 *Petroleum and natural gas industries – Site-specific assessment of mobile offshore units – Part 1: Jack-ups.*
[3] SP 22.13330.2016 *Soil bases of buildings and structures* Russian Standard.
[4] SP 58.13339.2012 *Hydraulic structures. Basic statements* Russian Standard.
[5] Kulikov S N 2017 *Bearing capacity factors impact on jack-up legs penetration analysis* (Materials of XIII All-Russian research-practice conference and exhibition «Development potential for engineering survey in Russian Federation) p 631.
[6] Hossain M S, Hu Y and Randolph M F 2003 *Spudcan Foundation Penetration into Uniform Clay* (Honolulu, Hawaii, Proceedings of The Thirteenth International Offshore and Polar Engineering Conference) p 647.

IOP Conf. Series: Materials Science and Engineering **700** (2019) 012048 doi:10.1088/1757-899X/700/1/012048

The Barents Sea of the Norwegian Continental Shelf: drilling through carbonates, modelling, risk management and well planning

M A Mosesyan[1,*], A Hartwig[2], K M Edin[2], A S Rabey[1] and L M Surguchev[1],

[1] LUKOIL Overseas North Shelf AS, Norway
[2] AkerBP ASA, Norway

* Corresponding author: michael.mosesyan@gmail.com

Abstract. There are a number of geological and hydrogeochemical modelling techniques used in order to ascertain formation of the hydrogen sulphide (H_2S) in the formation fluid within a geological setting in several blocks of the Norwegian Barents Sea. Careful consideration of the regional and chemical factors causing H_2S generation can also be used in the drilling context. Thermochemical sulphate reduction (TSR) is the most likely process of H_2S formation in this setting. Risk picture becomes different when presence of H_2S is closely coupled with occurrence of Permian and Carboniferous carbonate plays prone to causing severe drilling fluid losses and increasing exposure time of the downhole equipment to the H_2S. Using appropriate analogues and adequate well offset analysis tend to demonstrate that most of the risk factors can be well mitigated.

1. Introduction

The Norwegian Continental Shelf (NCS) is generally known for producing sweet, low-sulphur, petroleum. With first wells drilled mainly in the North Sea, exploration gradually moved further north in the Norwegian Sea with deeper target depths and high reservoir temperatures. Areas of the Barents Sea which have been opened for petroleum operations since the 1980's are frequently designated Barents Sea South, delineated by latitude 74°30'N.

As of 2019, a total of 185 wells have been drilled in Norway's Barents Sea sector, including 147 for exploration purpose only. Forty-nine discoveries were made in these waters from 1980 to 2016. With more discoveries potentially being evaluated for development and further production, e.g. Johan Castberg, Alta/Gohta and Wisting. Snøhvit and Goliat fields are currently the only two on stream.

2. Permian carbonates in the Norwegian Barents

Permian and Carboniferous carbonate plays started to gather a strong interest in the Norwegian part of the Barents Sea after the Gohta and Alta discoveries in 2013 and 2014. Drilling through carbonates can be challenging due to potential severe mud losses and hole instability. In case Hydrogen Sulphide (H_2S) is present in high concentrations in these geological conditions, this could lead to hazards that are difficult to cope with.

H_2S is a toxic gas and during drilling can be found and transported to surface in drilling fluids, recovered core samples or core barrels, and on downhole equipment. It is necessary to remove H_2S and soluble sulphites from drilling fluids as quickly as possible because of the destructive and corrosive effects of H_2S on personnel and materials.

Early well tests in the Norwegian Sea showed concentrations of hydrogen sulphide as high as 35-40 ppm, and the recent studies in the well planning phase in the Fedynsky High area of the Barents Sea East revealed a risk of possible H_2S presence in the Permian formation fluid.

Several wells with Palaeozoic carbonate reservoir targets either had minor shows (7128/6-1, 7130/4-1) or were dry (ex. 7229/11-1). The potential for high amounts of H_2S gas was recognized by the operators of well 7229/11-1, but this was not further quantified. A post drilling report on carbonate diagenesis in this well describes the initial plugging of primary porosity by anhydrite which is turned into calcite during Thermochemical Sulphate Reduction (TSR). A fluid inclusion stratigraphy report mentions the presence of pyrobitumen, sulfur species of thermal origin and fluid inclusion homogenization temperatures of ~ 140° C, all of which can form during the TSR process and may lead to substantial concentrations of hydrogen sulphide. The 3D regional basing modelling study allowed to estimate the reservoir paleo temperatures in the Barents Sea. Figure 1 shows the distribution of the maximum temperatures in the central region's geological history.

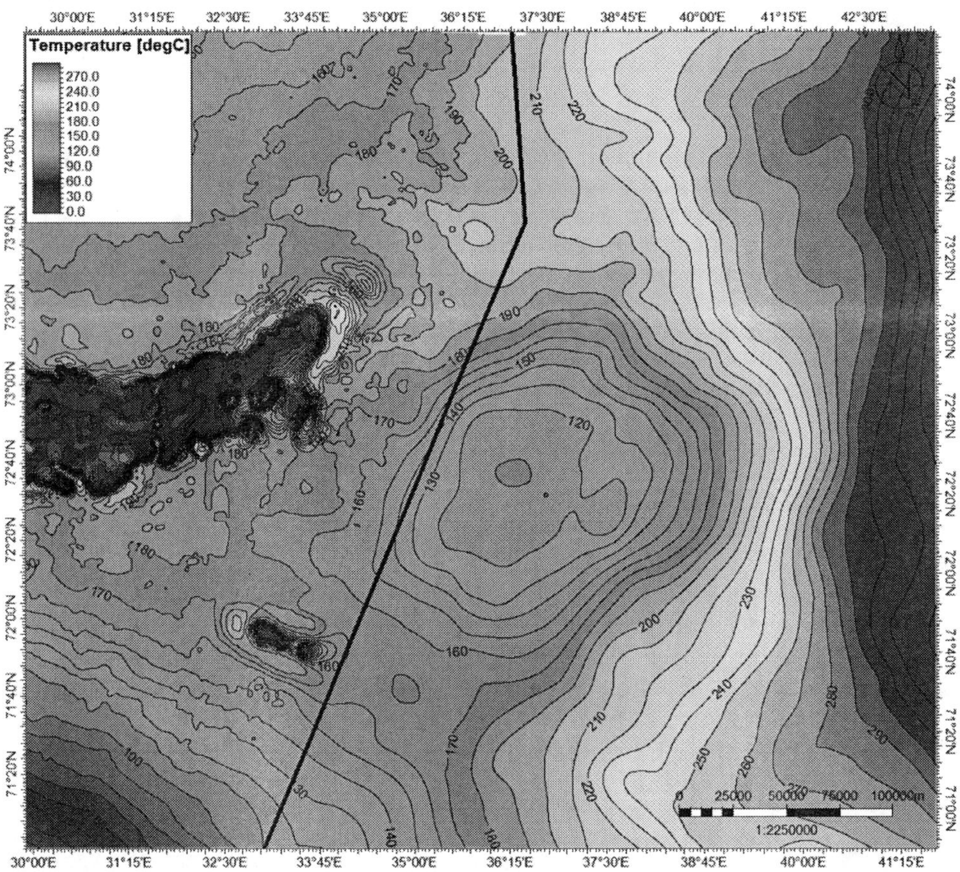

Figure 1. Distribution of paleo temperature in the central area of the Barents Sea.

The Palaeozoic sedimentary succession on the Finnmark Platform and in the Fedynsky High area consists of Caledonian basement overlain by Devonian and non-marine Early Carboniferous siliciclastics. A regional late Carboniferous transgression led to the development of an extensive carbonate platform that persisted throughout the Permian [1]. The Carboniferous–Permian ice age resulted in high-frequency, high-amplitude eustatic sea-level changes and in turn in intermittent flooding and subaerial exposure of the region. Siliciclastic shoreline progradation alternated with carbonate

IOP Conf. Series: Materials Science and Engineering **700** (2019) 012048 doi:10.1088/1757-899X/700/1/012048

deposition with phylloid algal thickets. The transition to the overlying warm shallow-marine calcareous to evaporitic facies of the Lower Permian carbonates was diachronous in response to palaeo-relief. Carboniferous-Permian carbonate build-up/reef growth has been reported from across the entire Arctic [1, 2]. Evaporites were deposited in sabkhas and salinas during sea-level lowstands in areas surrounding the then emerged build-ups. During the late Permian the carbonate depositional environment changed to a deep-water setting and the formation of cold-water carbonate build-ups. Based on analogue outcrop studies on Svalbard, well reports and data published in [3], the Carboniferous and Permian successions can contain source rocks of varying quality with terrigenous and/or marine organic matter.

During the middle to late Triassic the Fedynsky High area underwent rapid burial which initiated petroleum generation and migration from the Palaeozoic source rocks into the Permian aged carbonate reservoirs. Maximum burial likely occurred during the late Cretaceous, followed by tectonic uplift and several episodes of Cenozoic erosion and glacial tilting. The early charge, long residence time at temperatures above 120° C and the presence of anhydrite in or near the carbonate reservoir create a significant petroleum preservation and hydrogen sulphide risk for this Palaeozoic carbonate play type.

3. Bacterial sulphate reduction and thermochemical sulphate reduction

Sulphate can be reduced by hydrocarbons with concomitant oxidation of the organic compounds, either bacterially (BSR) or inorganically (TSR). BSR has been known for more than 50 years as acommon and widespread process in near-surface and shallow burial diagenetic. Petrographic and geochemical data on BSR are widely scattered in the geologic literature and are derived from dozens of geologic settings that include lakes, marine sediments, salt domes, and shallow burial diagenetic environments with base metal sulphide deposit. BSR is known from a multitude of geologic settings that range in temperature from 0 to about 80 deg. C [4].

TSR has been recognized much more recently as a process of geologic significance, and most literature on TSR has been published only during the last decade. A number of case studies and theoretical reviews have shown that TSR is common and widespread in deep burial diagenetic settings (also called high-temperature environments/settings). Based on empirical evidence, BSR and TSR occur in two mutually exclusive thermal regimes, i.e. low-temperature and high-temperature diagenetic environments [4].

Temperature is a critical but not the only element in the regional and chemical factor analysis. This paper addresses a number of other regional and chemical factors to further create a de-risking workflow.

4. TSR in the Permian play in the Norwegian Barents Sea

TSR is the most likely process of H_2S formation in the setting presented in this paper, i.e. Permian play in the Norwegian Barents. TSR comprises a complex web of hydro-geochemical reactions between the rock minerals and the aqueous and hydrocarbon species in the reservoir [5]. Data from well tests and from the literature indicate a correlation between H_2S concentration in the reservoir fluids and reservoir temperature.

Numerical modelling provides a first-pass estimate for assessing H_2S formation risk by TSR during the exploration phase in the absence of any calibration wells for the Fedynsky High area.

5. Modelling techniques as a proof of TSR

A combination of 3D petroleum system and 1D hydrogeochemical modelling allow to investigate the sensitivity of the TSR process regarding reaction kinetics, migration dynamics, as well as the role of reservoir mineralogy and formation water as an H_2S sink.

The 3D migration modelling lacks spatial and chemical reaction resolution, but handles 3-phase fluid flow and a simplified TSR reaction scheme [6]. The 1D hydrogeochemical modelling following the [5] approach, on the other hand, handles the very complex chemical interactions but does not simulate 3-phase fluid flow. This approach, however, still results in a large uncertainty range. One important modelling outcome for this geologic setting is that iron-bearing minerals that act as H_2S sinks exert a strong control on the H_2S gas content. Thus, the quantification of iron-bearing minerals in the Permian carbonates of analogue wells can provide a valuable tool for improving H_2S risk assessment.

6. Geological de-risking in the frontier region

The geologic de-risking went through 3 main phases going from regional to prospect specific scale.

The regional model first highlighted the H_2S risk due to the combination of the aforementioned factors. It constrained the thermal model, the petroleum migration dynamics and the overall timing of relevant events. The study was conducted together by LUKOIL Engineering and AkerBP. Unfortunately, the simplified TSR reaction scheme implemented in the modeling software could not help us de-risk the prospect, partially due to the absence of calibration data. One significant outcome was, that almost all of the formed H_2S can be dissolved in the formation water and thus be removed from the accumulation due to the long residence time.

But how much do we know about water flow in an undrilled carbonate reservoir in a frontier setting?

After the H_2S risk was flagged, an offset well analysis was conducted focusing on the products of TSR. For example, the 7229/11-1 well drilled in 1993, targeted the Paleozoic carbonates but was dry and the carbonates were not a reservoir facies in this location. The H_2S risk for this well was recognized at the time and drilling was prepared.

- Elevated H_2S was only encountered in the immature Jurassic Hekkingen Fm source rock. (probably related to bacterial sulfate reduction).
- The thin sections and diagenesis studies of the warm-water carbonate facies in the Ørn Fm mention pyrobitumen, dead oil staining and blocky pore filling calcite.
- Fluid inclusion studies mention the occurrence of several sulfur species that could be related to TSR and entrapment temperatures (aqueous inclusion homogenization temperatures) of 140° C or higher.
- There is an evidence that TSR has occurred in the area, and that H_2S might be a play risk.

In order to get a better understanding of the chemical process and main parameters that impact the present-day H_2S content, a 1D hydrogeochemical modelling study was conducted following the approach of [5]. It takes into account the complex web of (40+) chemical reactions between minerals, petroleum and various gaseous and aqueous species that make up TSR in a carbonate reservoir. Based on the modeling and evaluation techniques, a matrix for probability estimation on the H_2S formation was developed (Figure 2).

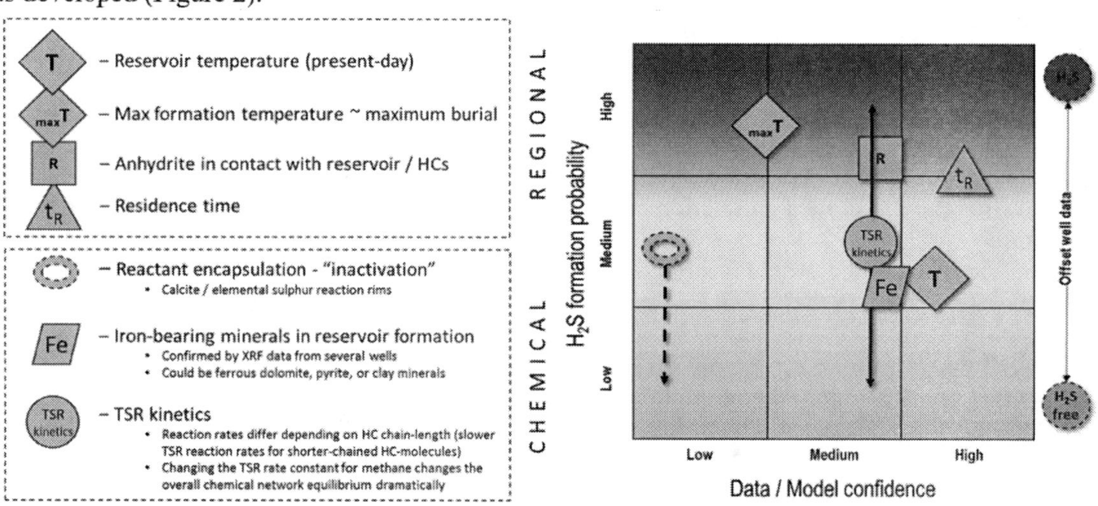

Figure 2. H_2S formation probability risk matrix. Regional and chemical risk factors.

7. Barents Sea regional overview

In order to study the frontier region in this area, experiences not only from the NCS were gathered, but also relevant analogues on the United Kingdom Continental Shelf (UKCS), on Karachaganak and Tengiz fields, Caspian region as well as in Timan-Pechora oil and gas province. The intention was to apply critical analysis on assumption around H_2S concentration vs. reservoir temperature. Real field examples were reviewed in order to assess the possibility for H_2S prediction, quantification and qualification prior to drilling a well. Among other critical elements in a well delivery process, factual drilling fluid design with H_2S scavenger, design of drilling tubulars and production casings with emphasis on prevention of sulfide cracking and brittleness have been analyzed in order to perform prudent risk assessment and develop a safe, robust and a cost-effective solution.

8. Experience from the Russian Barents and the Caspian regions

The Russian Barents Sea is yet to be explored however significant discoveries have been made in the Permian carbonates in the recent past with a profound effect on the entire region. These are Prirazlomnoye and Dolginskoe fields. Timan-Pechora oil province has been a source of hydrocarbons produced from the Permian and Devonian carbonate reservoirs for several decades. Exploration drilling in the Permian play on the Russian Continental Shelf in the Barents Sea that took place in the mid-80s, encountered at least 4 wells (Admiralteyskaya, Severo-Zapadnaya 202, Pomorskaya, Severo-Gulyaevskaya) with a very small amount of acid gas (H_2S) in a range of 1-3 ppm (Figure 3).

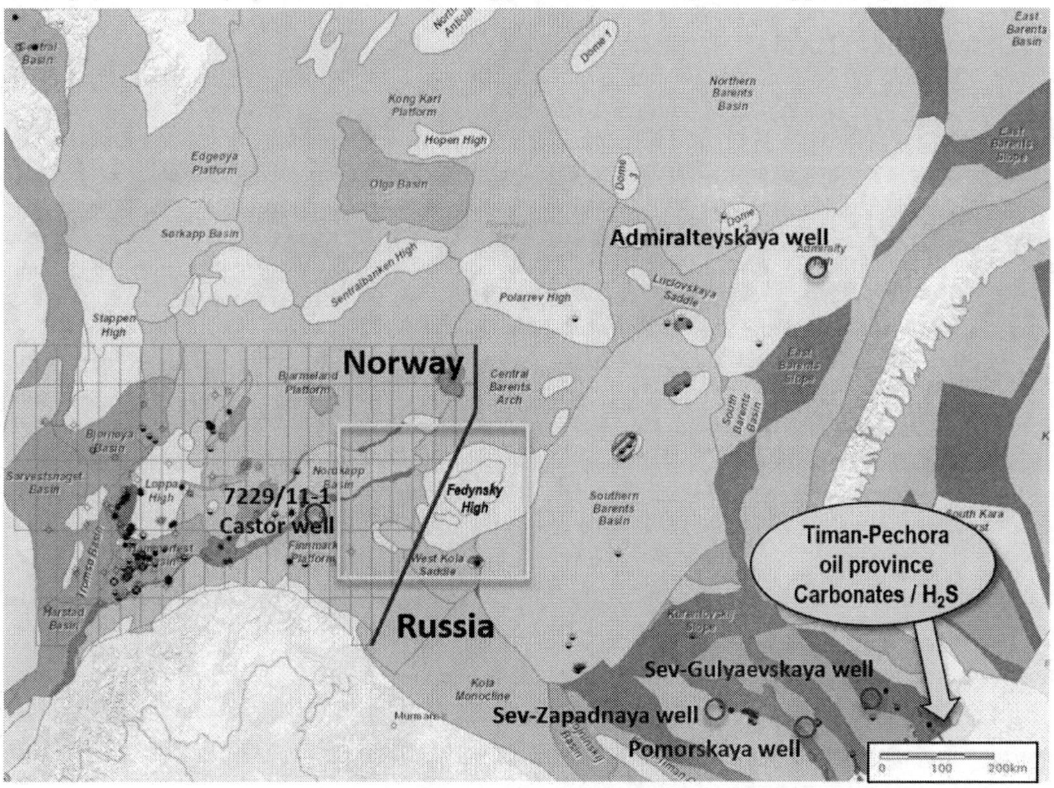

Figure 3. Barents Sea regional overview with exploration wells drilled through carbonates and H_2S traces encountered.

Recent exploration and production success in the Northern Caspian Sea revealed significant commercial potential for the hydrocarbon production from the terrigenous and carbonated reservoirs of Jurassic-Cretaceous age, e.g. Khazri offshore project. Exploration drilling within this area is taking place in the very challenging geological conditions in a fairly frontier region, high pressure (up to 1070 bar) and high temperature (up to 165 deg C) reservoir, carbon dioxide (CO_2) up to 4%, hydrogen sulphide up to 22% and a tight drilling window in the deeper sections of the well. There are also a number of world renowned production fields in this region in a similar geological setting, e.g., Karachaganak, Tengiz, Astrakhanskoe. Amount of H_2S on these fields normally reach 25% and sometimes higher whereas CO_2 level can be as high 22% [7].

9. Conclusions

The main objective of this paper is to demonstrate that modern modelling techniques allow comprehensive analysis of H_2S formation process in the reservoir fluid in the Norwegian Barents Sea.

Various modelling approaches have their strengths and weakness and are used to complement each other. The value of success in frontier exploration projects might determine the future of the entire region. Presence of H_2S in the reservoir fluid cannot be considered as a risk on its own but rather as a combination of risk factors caused by presence of anhydrite and carbonates in the geological cross-section, high CO_2, chemically balanced low pH, high level of inorganic and organic chloride's content, extremely high level of total dissolved solids, to name a few. Exploration and production drilling examples in the Caspian and Timan-Pechora regions demonstrate that these risk factors can be mitigated, when planned and managed appropriately.

References

[1] Ehrenberg S N, Nielsen E, Svånå T A and Stemmerik L 1998 Depositional evolution of the Finnmark carbonate platform, Barents Sea: results from wells 7128/6-1 and 7128/4-1, *Norsk Geologisk Tidsskrift.* **78**, 185-224.

[2] Stemmerik L, Larson P A, Larssen G B, Mørk A and Simonsen B T 1994 Depositional evolution of lower permian Palaeoaplysina build-ups, Kapp duner formation, Bjørnøya, Arctic Norway, Sedimentary Geology. 92(3-4), 161-174.

[3] Ohm S E, Karlsen D A and Austin T J F 2008 Geochemically driven exploration models in uplifted areas: examples from the Norwegian Barents Sea, *AAPG bulletin.* **92**(9), 1191-1223.

[4] Machel H G 1998 Bacterial and thermochemical sulfate reduction in diageneticsettings – old and new insights, *Sediment. Geol.* **140**, 144-148

[5] Fu Y, van Berk W and Schulz H M 2016 Hydrogen sulfide formation, fate, and behavior in anhydrite-sealed carbonate gas reservoirs: a three-dimensional reactive mass transport modeling approach, *AAPG Bulletin.* **100**(5), 843-865.

[6] Peters K E, Hantschel T, Kauerauf A I, Tang Y and Wygrala B 2013 Recent advances in petroleum system modeling of geochemical processes: TSR, SARA, and biodegradation, *J. Earth Sci.* **97**, 1115-1129.

[7] Mosesyan M A, Hartwig A, Bykousky V, Loparev D, Edin K M, Rabey A S, Surguchev L M 2018 The Barents Sea of the Norwegian Continental Shelf. Drilling through carbonates if hydrogen sulphide is present: modelling. Risk managemen, EAGE/ IFPEN.

IOP Conf. Series: Materials Science and Engineering **700** (2019) 012049 doi:10.1088/1757-899X/700/1/012049

Application of multicriteria fuzzy clusterization approach to assess the arctic seas oil and gas field development prospects

K N Pivovarov[1,*] A B Zolotukhin[1,2,3] and V V Streletskaya[1]

[1] Gubkin Russian State University of Oil and Gas (National Research University), Leninsky prospect 65, Moscow, 119991, Russia
[2] Northern Arctic Federal University, Severnaya Dvina Emb. 17, Arkhangelsk, 163002, Russia
[3] The University of Stavanger, P.O. Box 8600 Forus N-4036 Stavanger, Norway

* Corresponding author: pivovarov_k@mail.ru

Abstract. The article describes a unified approach to the assessment of resources and business planning. As an example, the oil and gas resources of the Pechora Sea are considered in the paper. Based on the specifics of the problem (the quantity and quality of the available data and their nature), the methods of multicriteria fuzzy clusterization are used. The assessment of the prospects for the development of oil and gas fields was carried out using 3 criteria, namely: availability of resources (natural and climatic conditions of the region); hydrocarbon resources and their degree of exploration; economic assessment of the development. The results of such an analysis are presented in the form of maps of the Pechora Sea area, illustrating the most important and accessible areas and the sequence of their development. Such a stepwise development based on minimizing the risk and maximizing the benefits can be the basis for the successful and trouble-free development of the resources of the entire region.

1. Introduction

Currently, the oil and gas industry of Russia is actively entering a new phase of the development of offshore Arctic fields. The reserves of natural hydrocarbons in the zone of the Russian Arctic are truly enormous. The degree of involvement of these reserves in the development will affect the economic growth of Russia, as well as the stability of the development of many other countries of the world. The Arctic region has an extremely challenging climate and as its development requires the highest technical and technological level of development, it becomes clear that the relevance of systematization of data and forecasting methods for the development of the Arctic is estimated to be extremely high.

Human activities in the Arctic include shipping, fishing, extraction of minerals, fast growing tourism and offshore oil and gas field development. Large oil and gas fields are already reasonably explored and ready for development.

Evaluation of the arctic offshore fields' accessibility is illustrated for the Pechora Sea region and is based on multi-criteria approach represented by 3 main axes:

- Availability of resources (natural and climatic conditions of the region);
- Hydrocarbon resources and their degree of exploration;
- Economic assessment of the development.

The methodology of the forecast is based on a fuzzy sets approach.

IOP Conf. Series: Materials Science and Engineering **700** (2019) 012049 doi:10.1088/1757-899X/700/1/012049

2. Accessibility of resources (natural and climatic conditions of the region)

The availability of resources, in terms of climatic, geographical and ecological conditions, is one of the fundamental characteristics for the analysis of the offshore regions. This represents a multi-criteria parameter challenge, consisting of the following criteria:

- Weather conditions: This parameter is used for understanding the climatic severity of regions, including the complexity of the working conditions. This parameter considers the influence of the weather environment, both on offshore structures and on the conduct of many marine operations. In determining the values of this factor, a comparative descriptive characteristic of the sea areas is given, based on the analysis of long-term observations of scientists (Figure 1(a)).
- Ice conditions: This parameter includes the ice type, the average width of the ice and so on. Its main parameter that has an influence on the concept of field development (type of platform, well completion, storage and transportation), investment costs (platform type) and operation costs (ice conditions control). This parameter points to the technological solution for the development, including the level of protecting offshore structures and the need for competent and timely ice management (Figure 1(b)).
- Icebergs and their occurrence probability in the region: Icebergs, as one of the most dangerous ice formations, threaten not only shipping and operations of offshore oil and gas facilities, but also significantly increase the risks associated with the safe conduct of the production operation processes. This factor is based on the analysis of historical data of long-term observations of ice conditions. But it should be borne in mind that this approach cannot fully describe the probability of an iceberg colliding with the platform, since the total movement of many icebergs is sufficiently random and chaotic, depending on currents, winds and the Coriolis force (Figure 1(c)).
- Gas hydrates accumulations in areas can influence exploration and development: larger volumes of gas hydrates deposited under the sea bed bring additional challenges to the production of HC. Gas hydrate areas are dangerous because when thawing, 1 cubic meter of gas hydrates in this zone can release more than 160 m^3 of gas [1]. This mainly occurs when drilling and cementing a well, as too hot drilling fluid and cement can significantly increase the temperature in the near-wellbore zone and change the pressure established in the reservoir (Figure 1(d)).
- Duration of the ice-free period: this parameter has a strong impact on exploration drilling which is usually performed in the ice-free period. In the areas with short ice-free navigation period the cost of exploration drilling is significantly higher than in ice-free areas. Accordingly, there is an escalation of risks of non-fulfilment of drilling obligations on time, because of storms, strong winds and high seas as often occur during ice-free periods. All this complicates the work and hinders its implementation (Figure 1(e)).
- Sea depth (Figure 1(f)): this is the second most important criterion for the analysis. The type of marine facilities is heavily reliant on the sea depth. In theory, the problems associated with depth increase exponentially with its increase, but practice shows that at shallow depths and in coastal areas other difficulties arise: high environmental sensitivity, ploughing the bottom with the keel of ice ridges (representing problems for subsea pipelines and sub-sea equipment) and coastal erosion [2].
- Proximity to shore (Figure 1(g)): this is used as criterion to choose the transport type (as pipeline from field to shore). In addition, it is important for health and industrial safety issues (search and rescue of personnel). This parameter is estimated as the shortest distance from the field location to the coastline. Offshore areas are characterized by the lack of infrastructure, satellite coverage and the possibility of rapid emergency response. It is important to note that coastal

areas are more environmentally sensitive than offshore areas, in particular is the coastal erosion of concern.

- Remoteness from the supply base is discussed for preliminary estimation of efforts for providing necessary equipment, materials and man power (transportation, supports, evacuation, etc.). In the Arctic, the logistics component of the project is also among the most important and complex tasks: the distance between the field and the supply base plays a significant role in the preliminary assessment of work to secure continuous supply of equipment, materials and labour (Figure 1(h)).
- Flora & Fauna: this criterion is related to the number and variety of animals and plants. This parameter means ecological balance in a region. Diversity of flora and fauna of the Arctic seas is quite large, and several animals are even included in the Red Book of endangered species (narwhals, polar bears, white whales, Atlantic walrus, Greenland whale, etc.) (Figure 1(i)).
- Potential environmental impact due to oil spill: this criterion shows the damage of the ecological balance due to an oil spill disaster and the possibility to return this balance to normal (Figure 1(j)). In addition to considering quantitative and qualitative indicators of flora and fauna, this factor also depends on other criteria: currents in the region and their vectors, meteorological conditions, remoteness of the area from the coastline and from operation bases, ice conditions, etc. Inclusion of two environmental factors in a row in the overall analysis is due to the fact that the care for the extremely sensitive and fragile northern ecosystem and understanding of possible consequences of disturbing the natural balance are fundamental criteria for the diligent and responsible development of oil and gas resources in the Arctic seas.

A technology accessibility map is, furthermore, shown on Figure 2:

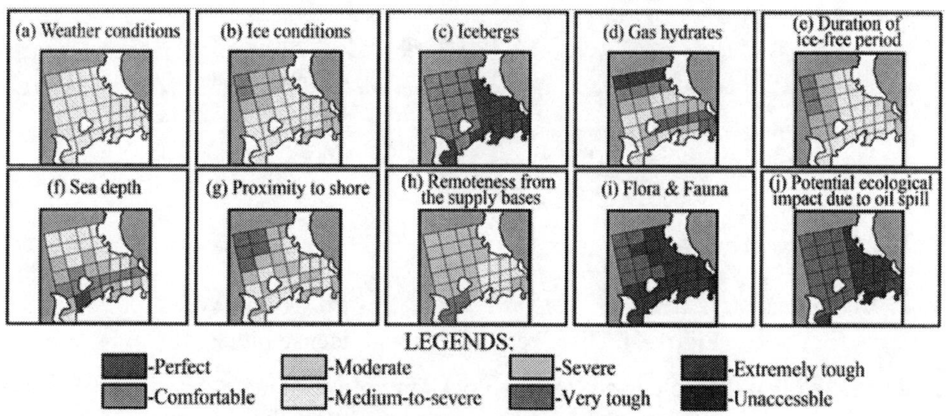

Figure 1. Features of the Pechora Sea

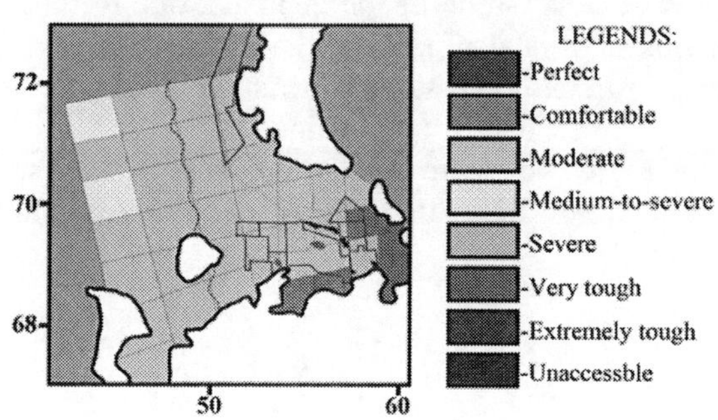

Figure 2. Map of technological accessibility

IOP Conf. Series: Materials Science and Engineering **700** (2019) 012049 doi:10.1088/1757-899X/700/1/012049

3. Hydrocarbon resources and their degree of exploration

The "Geological axis" shows the degree of exploration of the region in question and the region's reasonably assured oil and gas resources:

- The degree of exploration of the Pechora Sea region could be thought of as the density of the seismic exploration (lines per square kilometer) and the volume of the seismic survey (Figure 3)
- Resource assessment – oil, gas and condensate, expressed in tons of oil equivalent (TOE). It is based on the classification of structures (deposits) by the size of their reserves. Figure 4 shows the licensed areas of Rosneft and Gazprom, as well as the discovered deposits of the Pechora Sea. It depicts 6 fields, of which only the Prirazlomnoye field is currently in production.

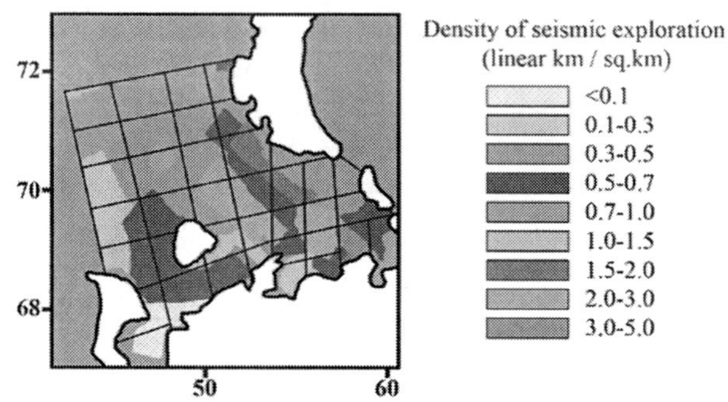

Figure 3. Density of seismic exploration, built on [3].

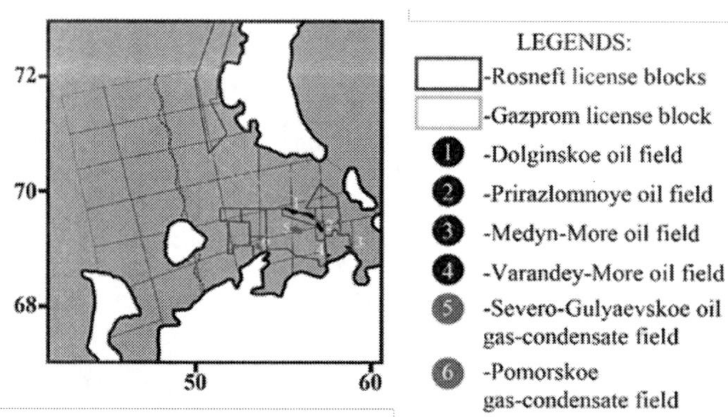

Figure 4. Discovered fields and license blocks.

It should be noted that the fields and license blocks fall into different zones, therefore their possible recoverable resources are spread over these zones proportional to the percentage ratio falling into one or another block. Such an approach enables us to present the zones as clusters of several sub-class deposits. It is possible, however, that some of the deposits, due to their size, are not profitable to be developed on a stand-alone basis. It is also possible that licensed blocks attributable to a specific zone will not contain commercial hydrocarbon reserves. For such blocks we assume that the degree of their exploration corresponds to a 50% probability (P_{50} value).

It is worth noting that the more sites there are and the smaller their sizes are, the more detailed will be the picture of the development of these sites and the need to use data averaging procedures will be reduced.

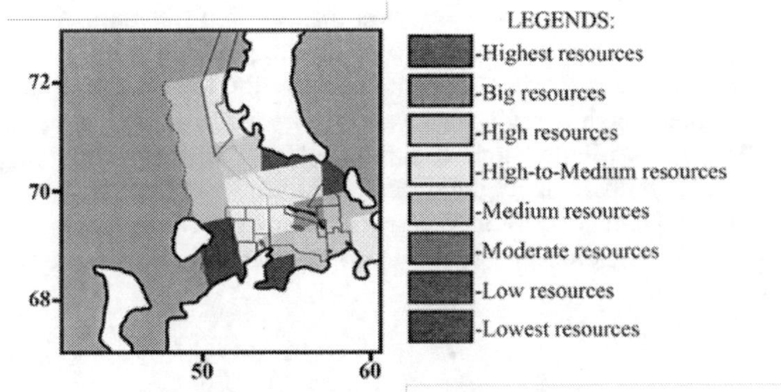

Figure 5. Map showing reserves

Because data on reserves or structures is not for the whole region, for the purpose of comparison, "empty" areas are excluded from further analysis (Figure 5). Combining the previous two parameters and using the fuzzy sets approach (see Section 5), we get a map of resources.

4. Economic assessment of the development

The "Economic assessment" is used as value development indicator:

- Possible revenue from a project is calculated as a product of the current price of the oil barrel and the preliminary estimated recoverable oil and gas reserves of a given territory (Figure 6). This economic indicator is not comprehensive and reflects only the very idea that an assessment of the economics of a block will allow a closer look at different territories in terms of their economic attractiveness.

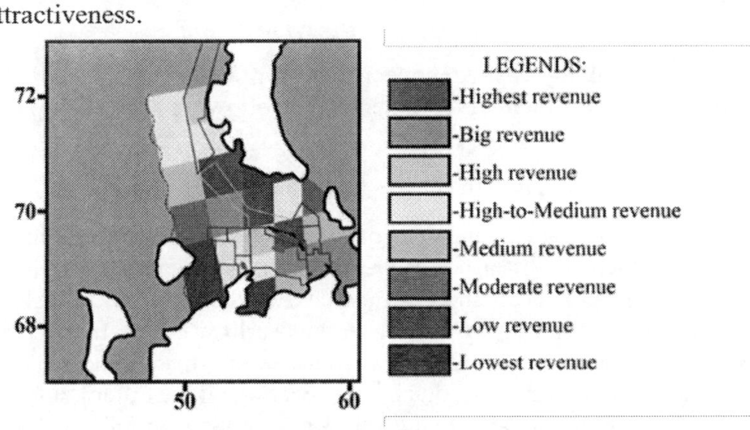

Figure 6. Economic assessment map.

5. Methodology

The methodology of the approach to assess the arctic seas oil and gas field development prospects is based on the fuzzy logic theory. The entire area of the Pechora Sea was divided into several zones. Those zones were studied and assessed by using the multi-criteria approach. The reason to use the fuzzy logic approach was that it enables us to operate with many parameters with limited knowledge of the subject and large uncertainty regarding the available data without any difficulty [4, 5].

In contrast to the Boolean logic, this approach differs from the standard binary logistic answers "true" or "false". It allows us to reflect on the degree of incompleteness and uncertainty of our knowledge [6] and provides effective tools for building models that most adequately reflect the quantity and quality of the available data [7].

The multiple criteria approach in assessing the complexity of the conditions means simultaneous analysis of many factors. This approach is discussed in detail in [8], where the task of mapping the technical accessibility has been analyzed. Consequently, the approach is only briefly described herein. As explained above, three main axes were selected for building the map. A whole range of values for each criterion was mapped on the axis of the corresponding membership function, which was further divided into eight different classes, where the 1st class had the lowest and the 8th – the highest value.

Subsequently, each membership grade was described by well-defined numerical values selected from the interval (0, 1), which is divided into 8 subintervals of 0.125 widths each. To aggregate these characteristics in order to build an aggregated map, the authors with respect to the existing approaches to fuzzy systematization of data, chose the Weighted Geometric Mean method (WGM). This method is based on the method of aggregation of a multiple criteria objective function into a single criterion one by using the following rule:

$$\mu_{A_0} = \mu_1^{\omega_1} \cdot \mu_2^{\omega_2} \cdots \mu_n^{\omega_n} \quad (1)$$

Here μ_i is an estimate by the i-th criterion, ω_i is the coefficient of significance of the *i*-th criterion, *n* – the total number of criteria, participating in the assessment [9].

In addition to the clustering of considered zones, the fuzzy approach can be used for ranking of the zones, i.e., their grading according to the degree of attractiveness or complexity of the conditions for their development [10]. The fuzzy ranking consists of 3 consecutive stages of calculations.

The first step is to calculate a non-strict preference when comparing fields or production zones (one is preferable to no less than the other) according to the equation [10]:

$$\mu_R(i,j) = 1 - \sum_{k=1}^{n} \omega_k \cdot [\mu_k(j) \ominus \mu_k(i)] \quad (2)$$

Here n – number of parameters in the evaluation, i, j – indexes of compared fields / production zones, symbol $\ominus$ – limited subtraction, $\mu_k(j) \ominus \mu_k(i) = \max\{0, \mu_k(j) - \mu_k(i)\}$ [10].
The second step is based on the principle of strict preference on compared criteria [10]

$$\mu_P(i,j) = \mu_R(i,j) \ominus \mu_R(j,i) \quad (3)$$

The last step of fuzzy ranking is a link between the two previous steps which rank the fields according to the degree of complexity of their development. The following relationship is used for this purpose [10]:

$$\mu_\Phi(i) = \frac{1}{(n-1)}\{\sum_{j\neq i} \mu_P(i,j) \ominus \sum_{j\neq i} \mu_P(j,i)\} \quad (4)$$

Here the first sum shows the fuzzy measure of the preference of the element i to all other elements j. The second sum shows the fuzzy measure of preference of elements $j \neq i$ to the element i. The map of approach to assess the Arctic Seas oil and gas field development prospects is shown in Figure 7.

Analysis of each zone according to the criteria described above made it possible to compile a summary table of the characteristics of all zones, which in turn made it possible to aggregate the multi-criteria assessment into a single-criterion one by aggregating all the criteria into the weighted geometric mean and construct an accessibility map that takes into account economic, technological and geological components (Figure 7). In addition to the division of production zones into clusters according to the levels of technical accessibility, degree of exploration and economic attractiveness, Figure 7 shows the ranking of territories.

It is easy to see that the Prirazlomnoye field is in the 1st ranking zone, which is confirmed by its industrial development. Organization of a systematic approach for the business planning enables creation of comprehensive and phased commissioning of oil and gas fields.

According to Figure 7, the next stage of the development of the region may well be the development of the Dolginskoye and Severo-Gulyaevskoye fields, then moving to the Varandey-More and Medyn-More fields. At the same time, it is necessary to continue exploration of license areas to identify promising structures and discover new deposits.

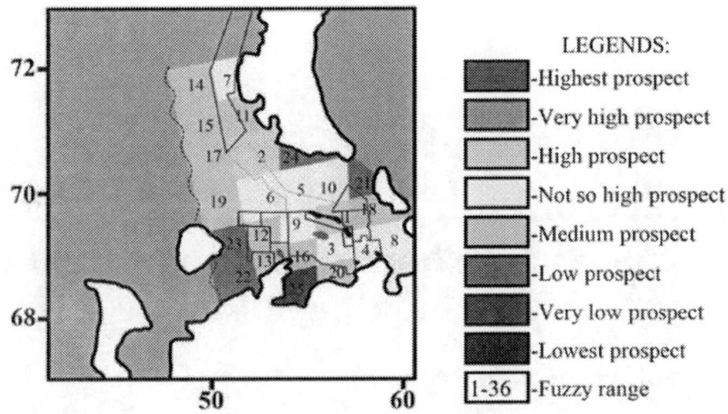

Figure 7. Fuzzy clustering and ranking of fields / production zones on 3 axes.

6. Conclusion

Pechora Sea is a region known for its environmental sensitivity and it is important to minimize negative anthropogenic impacts. Sequential field development planning allows a comprehensive approach to production optimization. The high commercial potential of the region and the uncertainty caused by the global climate change requires intensified research and development activities to achieve a stable and long-term development strategy. Due to the nature and intrinsic properties of fuzzy logic its application enables forecasting of development stages, based on changeable, incomplete and inaccurate information. Those specific mathematical properties allow us to assess potential development of oil and gas fields located in the arctic waters from different points of view, focusing either on a single parameter or conducting a comprehensive multi-criteria analysis.

It is worth noting that the example considered in the paper clearly demonstrates that the development of this area was started in the same manner as suggested by the system developed by the authors.

References

[1] Judzis A, Schofield TR and Yousif M 1997 Stabilization of In-Situ Hydrates Enhances Drilling Performance and Rig Safety *Proc. SPE Annual Technical Conf. and Exhibition (San Antonio)* Paper SPE 38568.

[2] Mørk K 2007 The challenges facing Arctic pipelines, design principles for extreme conditions, *J. Offshore Oil and Gas Magazine* **67** 9.

[3] Tkachenko M 2014 Geological structure and oil and gas potential of the Jurassic complex of the central part of the East Barents megaprogib (Moscow: FGBU VNIGNI).

[4] Zade L 1965 Fuzzy sets, *J. Inform. Control.* **8**(3), 338 – 353.

[5] Hurgin Ya 1995 Fuzzy methods in the oil and gas industry (Moscow: I. M. Gubkin State Academy of Oil and Gas), 131.

[6] Leonenkov A 2005 Fuzzy simulation in the MATLAB and fuzzyTECH (Saint-Petersburg: BHV Petersburg), 736.

[7] Golunov R, Dli M and Kruglov V 2001 Fuzzy logic and artificial neural networks (Moscow: Fizmatlit), 201.

[8] Zolotukhin A, Nesic S, Pivovarov K and Streletskaya V 2017 Mapping the main risks for offshore operations in the Pechora Sea *Int. Conf. on Safety and Security Engineering (Rome) SAFE*, **174**, 69–80.

[9] Zolotukhin A and Pivovarov K 2018 Use of multicriteria approach and methods of fuzzy mathematics for estimating development conditions of arctic seas in terms of the south-eastern part of the Barents Sea, *J. Arctic: ecology and economy,* **3**(31), 100–111.

[10] Zolotukhin A 2007 Engineering methods in petroleum science (Stavanger: Stavanger University) *Preprint* Lecture notes on the master's course MPE 140.

IOP Conf. Series: Materials Science and Engineering **700** (2019) 012050 doi:10.1088/1757-899X/700/1/012050

Problem and models of multicriteria decision making and risk assessment of the arctic offshore oil and gas field development

A B Zolotuhin[1,2,3] and Yu P Stepin[1]

[1] Gubkin Russian State University of Oil and Gas (National Research University), Leninsky prospect 65, Moscow, 119991, Russia
[2] Northern Arctic Federal University, Severnaya Dvina Emb. 17, Arkhangelsk, 163002, Russia
[3] The University of Stavanger, P.O. Box 8600 Forus N-4036 Stavanger, Norway

Corresponding author: anatoly.zolotukhin@gmail.com

Abstract. The article is devoted to the systematic solution of problems of multi-criteria assessment and the possibility of implementing projects for the development of hydrocarbon deposits on the Arctic shelf under conditions of uncertainty and risk. Risks are considered in a narrow and broad sense. It is shown that the solution of tasks, in this case, comes down: 1) to a multi-criteria assessment of the quality level of a field based on a hierarchy analysis method that takes into account uncertainties and risks of operating objects through the relevant necessary indicators (criteria); 2) determining, with the help of game theory adopted with the nature of the criteria for selecting the best Bayesian strategies, of development options under both partial and full uncertainty conditions; 3) selection using the rules of the Borda, Pareto and multi-criteria dispersive risk model for choosing the best development strategy.

1. Introduction

Oil and gas production in the fields of the Arctic shelf is now becoming one of the most pressing problems in the development of the world's hydrocarbon resources. The projected oil and gas resources of the Arctic shelf of Russia are quite large and are estimated at 700-1020 billion barrels of oil equivalent (BBOE) [1, 2]. However, they are concentrated mainly in hard-to-reach territories, far from industrialized regions covered by an integrated system for supplying the country with hydrocarbon resources.

These oil and gas basins are also distinguished by the fact that they have different and insufficient degree of knowledge, the presence of difficult climatic conditions, an incomplete and heterogeneous database, high economic, environmental and production risks, which ultimately determine large investments and low economic efficiency of the arctic shelf development. By taking this into account the following types of risks in the hydrocarbon production can be considered: geological, technological, organizational, economic and social. A detailed informative description of them can be found in [1, 2].

In the general case, risk is a measure of quantitative measurement of a hazard, which is a vector quantity that includes the following main indicators: the amount of damage from the impact of a hazard; probability of occurrence of the considered hazard; uncertainty in terms of both damage and the likelihood of a hazard.

Based on IEC standards, as well as foreign standards (see, for example, [3,4]), risk in a broad sense means the average damage to a system, process, etc. (usually in monetary terms) from the manifestations of a specific hazardous impact factor for a set period of time [0, t] (usually $t = 1$ year). The risk of the i – th hazardous effect (risk factor) is estimated using the expression:

$$R_i(t) = S_i \cdot \Phi_i \cdot t \tag{1}$$

If $t = 1$ year then $R_i = S_i \cdot \Phi_i$,

where S_i where S_i – is the average damage (loss) associated with the complete elimination of the i-th hazardous consequence (impact); Φ_i is the frequency (probability) of occurrence of the i-th hazardous effect (impact) or the average number of occurrences of this effect per year, the Φ_i dimension is 1/year.

In the classification adopted by the IEC standard, the following degrees of hazardous effects are distinguished: i = I - minor effects, II - boundary, III - critical, IV - catastrophic, which in this case correspond to the above states of nature Pj (see below).

Often in the literature on industrial safety, the risk in the strict sense of the i-th hazardous consequence means the quantity Φ_i – the frequency (probability) of the occurrence of a hazardous effect, while Φ_i is referred to as the risk or probability of the occurrence of a hazardous event per year. Typically, the value of Φ_i is used as a risk in a situation where there is no information about the magnitude of the damage S_i.

However, there are other models for risk assessment and analysis, among them – models based on the hierarchy analysis method [5 -7], on the theory of games with nature [8 - 10], the dispersive model of risk [11], as well as wider approaches with an in-depth philosophy of misconception of risk management [12-14]), models based on ISO standards and the achievements of the Council of the Society for Risk Analysis (SRA) [15].

Based on the methods of system analysis, we only consider problems and the implementation of models based on the IEC standard, the hierarchy analysis method, game theory with nature and the variance risk model.

2. Geological risks evaluation with probabilistic model

We will treat the geological risk as a non-confirmation of resources. To assess the risk of not confirming the resources of the arctic shelf deposit, a probabilistic model of geological risk assessment is proposed, the essence of which is explained below.

The probability of non-confirmation of resources (industrial reserves) will be called the risk. Then, the geological risk can be defined as the probability of geological failure – the risk of non-confirmation of the availability of resources, that is, the probability of a complex event, namely: *a)* the studied deposit is not productive; *b)* not found as a result of prospecting and *c)* its reserves are less than the minimum required value V, i.e.:

$$Pr = (1 - P_o) \cdot (1 - P_p) \cdot (1 - P_V), \tag{2}$$

where

P_o – probability 0f the exploration target productivity;

P_p – probability of the presence of reservoirs with net pay acceptable for commercial development;

P_V – probability that identified reserves are not less than the minimum required value V.

It is obvious that P_o, P_p, P_V in the lack of real data are subjective probabilities, i.e. probabilities assigned by the expert.

Thus, the problem is to expertly determine these probabilities with the least errors. As a rule, direct determination (assignment) of the considered subjective probabilities leads to erroneous results. However, today there is a method that allows to level out this shortcoming. This method includes the

IOP Conf. Series: Materials Science and Engineering **700** (2019) 012050 doi:10.1088/1757-899X/700/1/012050

principle of pairwise comparison of objects and the eigenvalue method, which is widely used in the hierarchy analysis method (AHP) [5 - 7].

The application of this approach is especially simple in case of subjective probabilities of two single unique and mutually complementary events (for example, events A and B, which determine the probabilities P_o and $(1-P_o)$ [5]. Obviously, the considered model can be applied to the definition of P_p and P_V. Note that there is also the likelihood that the probabilistic estimates made for different sections of the analyzed areas are different then it becomes possible to map the distribution of risks over the area.

3. Risk assessment (technological, organizational, economic and social) of projects for the development of the arctic offshore oil and gas field based on the hierarchy analysis process (AHP)

Let us consider a project for the arctic offshore hydrocarbon field development. There are three development strategies (scenarios), namely, $\boldsymbol{A_1}$, $\boldsymbol{A_2}$, $\boldsymbol{A_3}$. Please, note that each of the options considered may have its own uncertainties in evaluating performance indicators that can be associated with the degree (probability) of project implementation. In petroleum industry the levels of 90%, 50% and 10% probability ($\boldsymbol{P_{90}}$, $\boldsymbol{P_{50}}$ and $\boldsymbol{P_{10}}$ are used [2]. The decision maker, according to the results of calculations and estimates, has to select the best scenario and identify the associated risk – probability (degree) of the project failure (the risk of investment in the project).

A. Probabilistic model for comparing feasibility of the arctic offshore field development

Here we use the same methodical approach as in the geological risk assessment. We assume that all the strategies A_i, according to the risk factors presented above and the conditions in which they are implemented, generate their risk of project failure. Assuming (at first approximation) that these probabilities do not depend on each other, the overall risk of project failure to implement A_i can be defined as:

$$P_r^{Ai} = \left(1 - P_t^{Ai}\right) \cdot \left(1 - P_o^{Ai}\right) \cdot \left(1 - P_e^{Ai}\right), \tag{3}$$

where

P_t^{Ai}, P_o^{Ai}, P_e^{Ai} – probability of successful implementation of technical, organizational and economic solutions (respectively) of project scenario.

Obviously, in the conditions of uncertainty of occurrence of the considered risks – the lack of appropriate precedents and statistics, these probabilities should be considered and identified as subjective. Just as for geological risk probabilities, we will determine (evaluate) them on the basis of the hierarchy analysis method, which is based on the expert's compilation of matrices of paired comparisons of options (expert judgment matrices), asking the expert when comparing them to avoid errors [5] question: "How much more likely is the realization of event C (one scenario) than the realization of event D (another scenario) if these specific conditions (criteria, indicators, etc., describing this task, such as when assessing geological risk) are met. To quantify the fact that it is more likely – the implementation of C or D – it is necessary to use the fundamental point scale from 1 to 9 [5 - 7]. Using the matrix of pairwise comparison of options (strategies C and D) elements of the priority vector are calculated and interpreted as probabilities P_t^{Ai}. Then, based on the meaning of the risk of non-realization of the project scenario r_i, we get:

$$P_t^{Ai} = 1 - r_i. \tag{4}$$

In order to rely on the results provided by the matrix of pairwise comparisons, it is necessary to calculate its consistency index (*CI*) [5,6,7]:

$$CI = (\lambda_{\max} - n)/(n-1), \tag{5}$$

IOP Conf. Series: Materials Science and Engineering **700** (2019) 012050 doi:10.1088/1757-899X/700/1/012050

where n is the dimension of the matrix of pairwise comparisons; $\lambda_{\max}$ – the maximum value of the normalized eigenvector of the matrix.

In practice, the judgment matrix is considered consistent if $CI = 0 \div 0.15$ (it is known from matrix theory that the inverse symmetric matrix is absolutely consistent when $\lambda_{\max} = n$ [5]. Inconsistency occurs when an expert compares objects. For example, when comparing three objects C, D and E an expert writes: C = 5D, C = 6E, D = 4E, that is, C = 20E, but it writes down C = 6E – which breaks transitivity of estimates.

Further, it is obvious that in order to obtain $P_r^{A_i}$, it is necessary to compose two more matrices of pairwise comparisons with respect to obtaining $P_o^{A_i}$ и $P_e^{A_i}$. Multiplying the probabilities $P_t^{A_i}$, $P_o^{A_i}$ и $P_e^{A_i}$, we obtain the probability of successful implementation of the project scenario A_i.

However, such an approach requires a very high qualification of the decision maker and does not explicitly take into account the multi-criteria assessment of the viability of development scenarios. Therefore it can be considered as preliminary. To solve this problem, a set of criteria should be selected, and the most significant ones should be highlighted.

B. Model of ranking and criteria selection for the prospectivity assessment of the arctic offshore field

For ranking and selection of criteria, the technique proposed in [6,7] can be applied. It is based on the calculation of the values and significance (weights) of the criteria according to the following expression

$$K_i^* = \gamma_i F_i(K_i^D, K_i^S), \tag{6}$$

where K_i^D – the value of the i-th criterion, assessed by the expert (or calculated) at the time of its assessment – state D; K_i^S is is the value of the i-th criterion, estimated by the expert in points expressing his desire to improve the problem solution – the state S; γ_i is the weight (importance) of the i-th criterion, determined on the basis of the expert's experience and intuition, with $\sum_i \gamma_i$=1 (100%).

The connection of the i-th criterion value in points with the physical parameters for S and D states can be expressed using basic scales, allowing the dimensional indicator (criterion) to be converted into points by mapping the dimensional criterion on the basic scale [6,7]. The ratio of the Base Scale and the Scale of Indicators usually makes sense "the more – the better". For indicators "the less – the better" the scale is inverse.

The specific form of the function F_j can be the difference K_i^S and K_i^D, showing how much the situation needs to be improved, or quotient of their division showing how many times it is necessary to improve the situation. The sequence K_i^* gives a series of decreasing importance of criteria from the point of view of the leader and shows where to focus attention. The initial sequence K_i is significantly different from the sequence K_i^*.

The ranked sequence K_i^* can be used to reduce uncertainty while reducing the set of criteria in a decision-making process or, rather, when assigning objectives. It is performed on the basis of calculating the separation level $\alpha(n)$, which determines the total relative weight of the group of the most important criteria:

$$\alpha(n) = \sum_{i=1}^{n} K_i^* \Big/ \sum_{i=1}^{N} K_i^* \tag{7}$$

where N – the number of criteria considered, and n – the highest criteria number in a reordered sequence that will be taken into account by the decision maker.

Let us apply this methodology to obtain the ranking and selection of criteria for assessing the feasibility of hydrocarbon production from the arctic offshore field (Table 1). These criteria are based on the analysis of information presented in [1, 2, 6, 7] and their (new) estimates that has been performed by translating the real (dimensional) values of the criteria into grades by the base scale method.

As follows from Table 1 the level of separation α_{14} covers 14 of 26 criteria considered. These are criteria 1, 2, 3, 5, 6, 7, 8, 11, 14, 15, 18, 19, 21, 25 and they contribute with 76% of the "weight" of all criteria. If you add four more criteria that have the value $K_i^* \geq 6$, their total relative weight will grow to 92%.

It is obvious that the decision maker may decide to choose any group of criteria for evaluating the effectiveness of field development. Usually a decision maker operates with 70% – 77% of the total weight. However, his decision may be guided by other considerations, for example, the Pareto principle 20/80, then (in our case) the number of criteria could be reduced to three (α_3=26%). In this case it will be criteria with ranks 1-3, namely, (1) Net Present Value, (2) Environmental Safety and (3) Recoverable resources. In any case, the choice of the level of separation always depends on the decision maker, his experience, intuition, and information content.

Table 1. Ranking and selection of criteria for the prospectivity assessment of production assets.

No.	Criteria	K_i^D	K_i^S	$K_i^S - K_i^D$	γ_i %	$K_i^* = (K_i^S - K_i^D)\gamma_i$	Rank K_i^*
1	Recoverable oil (gas) resources	3	6	3	6	18	III
2	Reservoir porosity	6	8	2	6	12	IV-VIII
3	Social factors of development	1	3	2	6	12	IV-VIII
4	Total number of production wells	4	5	1	5	5	XIX-XXI
5	Net present value (NPV)	5	9	4	6	24	I
6	Internal rate of return (IRR)	4	6	2	5	10	IX-XI
7	Profitability index (PI)	4	7	3	4	12	IV-VIII
8	Pay-off period	5	7	2	5	10	IX-XI
9	Capital expenditures	3	4	1	5	5	XIX-XXI
10	Operating costs	2	4	2	3	6	XV-XVIII
11	Government take	3	5	2	5	10	IX-XI
12	Justification of selection of technical solutions	4	6	2	2	4	XXII-XXV
13	Reliability of field development control	5	6	1	2	2	XXVI
14	Options for the sales products	4	7	3	3	9	XII-XIV
15	Environmental safety	3	7	4	5	20	II
16	Preservation of flora and fauna	4	6	2	2	4	XVIII-XIX
17	Noise	2	4	2	2	4	XVIII-XIX
18	Reservoir permeability	4	7	3	4	12	IV-VIII
19	Reservoir depth	4	7	3	4	12	IV-VIII
20	Initial oil / gas saturation	5	8	3	2	6	XV-XVIII
21	Reservoir temperature	6	9	3	3	9	XII-XIV
22	Reservoir thickness	4	5	1	5	5	XIX-XXI
23	Lithology of reservoir rock	6	8	2	3	6	XV-XVIII
24	Rock metamorphism degree	4	6	2	2	4	XXII-XXV
25	Reservoir area	3	6	3	3	9	XII-XIV
26	Tectonics of the area / field	4	7	3	2	6	XV-XVIII

C. Method of analyzing hierarchies for multi-criteria risk assessment for the arctic offshore field development project

For the further analysis 14 criteria selected in a previous section will be grouped as follows:

B_1. Economic key performance indicators
C_1. Net present value
C_2. Internal rate of return
C_3. Pay-off period
C_4. Profitability index
C_5. Government take
B_2. Risks
B_3. Environmental protection
C_8. Environmental Safety
B_4. Geologic and technical conditions
C_9. Recoverable oil resources
C_{10}. Reservoir porosity
C_{11}. Reservoir permeability
C_{12}. Reservoir depth

C_6. Social factors of field development
C_7. Options for the sales products
C_{13}. Reservoir temperature
C_{14}. Reservoir area

Based on this set of criteria, the decision maker needs to select the best development scenario from A_1, A_2 and A_3 considered [1,2]. To assist in solving this task, the following graph can be built (Figure 1).

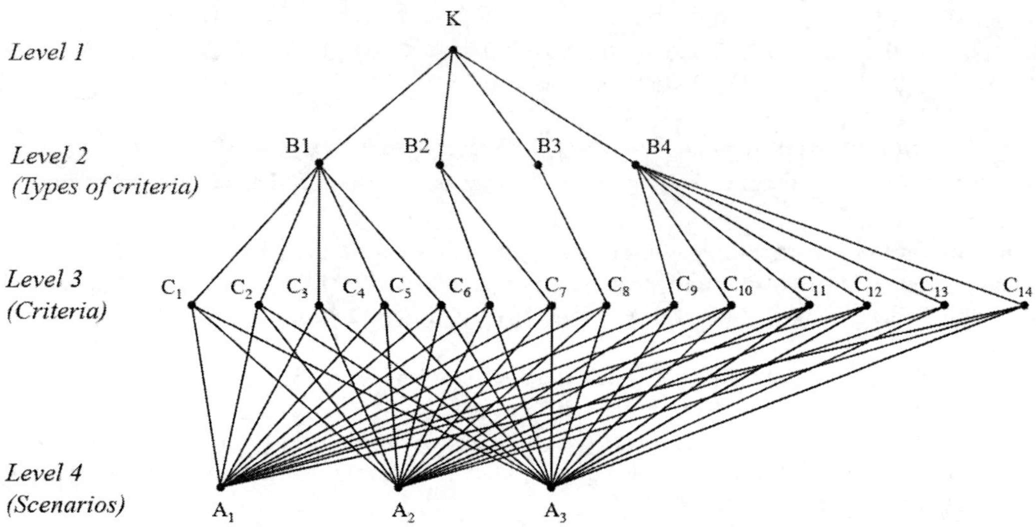

Figure 1. Hierarchical structure of selecting scenarios of the prospectivity assessment of the arctic offshore field development.
Legend: K – global criterion for the feasibility of field development strategies; B – indicators of field development efficiency; C_i – specific criteria from the above list; A_1, A_2, A_3 – scenarios (strategies).

For the hierarchy on Figure 1 the expert compiles a matrix of judgments (pairwise comparisons)

Level 2.

Priority matrix of criteria in relation to the main criterion K (Figure 1). The vector of priorities has the elements Π_j: 0.21; 0.26; 0.29; 0.24. $\lambda_{\max} = 4.2$; $CI = 0.07$.

Level 3.

This level considers three matrices of pairwise comparisons:

- Comparisons of economic criteria in relation to B_1 criterion. The priority vector has the elements: Π_i: 0.290; 0.195; 0.189; 0.162; 0.146. $\lambda_{\max} = 4.7$; $CI = 0.0$.
- Comparisons of risk criteria in relation to B_2 criterion. The priority vector has the elements: Π_i: 0.33; 0.66. $\lambda_{\max} = 1.88$; $CI = 0.0$.
- Comparisons of environmental criteria in relation to B_3 criterion. The priority vector has the elements: Π_i: 1.0. $\lambda_{\max} = 1.0$; $CI = 0.0$.
- Comparisons of geo-technical criteria in relation to B_4 criterion. The vector of priorities has the elements: Π_i: 0.248; 0.192; 0.165; 0.203 0.109; 0.083. $\lambda_{\max} = 6.786$; $CI = 0.157$.

Further, in accordance with the hierarchy of Figure 1, a compiled priority matrix of criteria priority column-vectors in relation to criteria B_1, B_2, B_3, B_4 should be multiplied from the right by a vector of priorities in relation to the main criterion K. As a result, the following final vector of criteria weights is obtained: α_{C_1}=0.061, α_{C_2}=0.041, α_{C_3}=0.040, α_{C_4}=0.034, α_{C_5}=0.031, α_{C_6}=0.086, α_{C_7}=0.172, α_{C_8}=0.290, α_{C_9}=0.060, $\alpha_{C_{10}}$=0.046, $\alpha_{C_{11}}$=0.040, $\alpha_{C_{12}}$=0.049, $\alpha_{C_{13}}$=0.026, $\alpha_{C_{14}}$=0.020

IOP Conf. Series: Materials Science and Engineering **700** (2019) 012050 doi:10.1088/1757-899X/700/1/012050

Level 4.

At this level of the hierarchy, 14 matrices of pairwise comparison of strategies A_i are compiled with respect to each criterion C_i. After that, the corresponding priority matrix is compiled, which is then multiplied from the right by the vector generated in a previous step, the priority column of α_{C1} – α_{C14} criteria, and the priorities of the development scenarios are calculated. This priority matrix will eventually have the following form:

	C_1	C_2	C_3	C_4	C_5	C_6	C_7	C_8	C_9	C_{10}	C_{11}	C_{12}	C_{13}	C_{14}
A_1	0.4	0.33	0.2	0.5	0.5	0.4	0.3	0.2	0.47	0.47	0.33	0.3	0.3	0.53
A_2	0.4	0.4	0.4	0.2	0.3	0.4	0.3	0.33	0.3	0.2	0.33	0.5	0.5	0.29
A_3	0.2	0.27	0.4	0.3	0.2	0.2	0.4	0.47	0.23	0.33	0.34	0.2	0.2	0.18,

and the priority vector of strategies A_i: α_{A_1}=0.325; α_{A_2}=0.351; α_{A_3}=0.324.

Thus, the most preferable development strategy is A_2. Accordingly, the risk of failure of this development scenario is $P_r^{A_i} = 1 - 0.351 = 0.649$.

In addition to the risk assessment models discussed above in the narrow sense, knowledge of risks in a broad sense is often required, that is, a possible loss of efficiency assessment. The theory that allows this is the theory of statistical solutions (theory of games with nature) [8,10].

4. Evaluation of the risk associated with the arctic offshore production asset based on the theory of statistical solutions (theory of games with nature)

It is known that in the theory of matrix games (games with nature) [8, 10] one player is active and "malicious", the other player (the "nature") is passive. Nature acts randomly, or rather, it is believed that its states are realized randomly.

Nature, for example, can be presented by oil and gas assets, natural conditions (land, sea, climate, etc.), geo-technical conditions (fluids and reservoir properties, etc.), and the uncertainty lies in the likelihood of possible outcomes of the states of nature. The task of the active player (decision maker) is to find a pure strategy that would secure the maximum benefit (or minimum loss), taking into account any state of nature. Note that the active player can't change the states of nature.

In this paper an approach is proposed for determining the states of nature, which, based on the data in the Table 1, allows a multi-criteria assessment of the characteristics of the field quality by applying one of the compromise schemes, for example, convolution of criteria [6, 7].

Further, dividing the entire range of possible values of a complex field quality criterion into a certain number of intervals, for example, four. This enables to determine the following levels Π_i of quality of the field: Π_1 – very high; Π_2 – high; Π_3 – medium; Π_4 – low quality level.

It is required to determine the best field development scenario (in our notation, the strategy A_i), using all possible criteria for evaluating the effectiveness and risks used in the theory of games with nature.

Table 2 contains the payment matrix of the games with nature. Parameters staying at the intersection of the columns and rows are complex indicators (criteria) of the field development options efficiency, generated by using the same methodical approach as the above-discussed states of nature. Note that these complex indicators are expressed in the base scale by a 20-point (French) grading system.

We will also assume that the greater the value of a complex indicator (regardless of whether it is a gain or a loss) – the better the solution. This follows from the properties of the inverse scale, which greatly simplifies calculations and their analysis.

Calculations of the field development effectiveness and risks can be carried out in two cases: (1) partial uncertainty situations, when the probabilities of the realization of the states of nature are given or can be calculated: (2) situations of complete uncertainty when probabilities are unknown or cannot be determined due to different conditions or when they are not important to the decision makers.

IOP Conf. Series: Materials Science and Engineering **700** (2019) 012050 doi:10.1088/1757-899X/700/1/012050

Table 2. Example of the payment matrix of the games with nature for various strategies for the arctic offshore field development.

Development options (strategies)	Field quality level			
	П1	П2	П3	П4
A1	14	10	13	15
A2	10	12	14	16
A3	9	12	14	16

In the first approach several criteria described below are used.

Bayes-Laplace criterion. According to his approach, there is such a strategy A_i, which gives the maximum value to the expectation of gain for all possible outcomes of the states of nature $P(\Pi_j)$.

$$\overline{a_i} = \sum_{j=1}^{n} a_{ij}P_j \rightarrow \text{max}, \tag{8}$$

where $\overline{a}_i$ – weighed arithmetic mean.

Minimum average risk criterion $\overline{r_i}$.

$$\overline{r_i} = \sum_{i=1}^{n} r_{ij}P_j \rightarrow \text{min} \tag{9}$$

Here the probabilities P_j are set in the same way as in a previous case.

The risk of decision-making when using the strategy A_i in the conditions of Π_j is the difference:

$$r_{ij} = \beta_i - \alpha_{ij} = \max_i \alpha_{ij} - \alpha_{ij} \tag{10}$$

where $\beta_i = \max_i \alpha_{ij}$, $r_{ij} \geq 0$.

If a decision maker considers that none of the states of nature is a priority, then the probabilities $P(\Pi_j)$ are considered equal, ie:

$P(\Pi_j) = P_j = {}^1/_n$ – so called principle of insufficient justification (Laplace criterion);

In this case the best strategy securing the highest gain is:

$$\max(1/n)\sum_{j=1}^{n} a_{ij} \tag{11}$$

For minimizing the risk this strategy gives:

$$\min\,(1/n)\sum_{j=1}^{n} r_{ij}\,, \tag{12}$$

where r_{ij} – risk of choosing the strategy A_i in the state of nature Π_j (see above).

Another approach to assign probabilities $P(\Pi_1)$ is to arrange states of nature in the order of their plausibility. Then the probabilities P_j are assigned to the proportional members of a decreasing arithmetic progression:

$$P(\Pi_1): P(\Pi_2): \cdots\ P(\Pi_n) = n: (n-1): \cdots : 1$$

Taking into account that $P(\Pi_1) + P(\Pi_2) + \cdots + P(\Pi_n)$=1, probabilities are calculated as follows: $P_j = \frac{2(n+j-1)}{n\cdot(n+1)}$ (Fishburne point estimate). Results are substituted into the Bayes-Laplace criterion.

In any of the cases considered, the optimal decision corresponds to strategy A_i^*, which gives the average maximum gain (or minimum total risk). Hodges-Lehmann criterion relies simultaneously on the maximin criterion and the Bayes-Laplace criterion. According to this criterion, the choice of the best strategy is determined by the expression:

IOP Conf. Series: Materials Science and Engineering **700** (2019) 012050 doi:10.1088/1757-899X/700/1/012050

$$Z = \max_i \left[\alpha \sum_{j=1}^{n} a_{ij} \cdot P_j + (1-\alpha) \min_j a_{ij}\right], \quad (13)$$

The choice rule for this criterion is described below:

1. The payment matrix with elements a_{ij} is complemented by a column composed of the sum of weighted average (with a weight α = const) of mathematical expectations and the smallest result of each row.
2. Solution with the largest value in the rows of this column are selected.

When α=1, the Hodges-Lehmann criterion converts into the Bayes-Laplace criterion, and with α = 0 – into the maximin Wald criterion. The parameter α expresses confidence in the probability distribution used.

Germeier's criterion is focused on the magnitude of the losses, that is, the negative values of all a_{ij} and is determined by the expression:

$$Z = \max_i \min_j (a_{ij} P_j) \quad (14)$$

Selection rule according to the Germeyer's criterion is formed as follows: the decision matrix with elements a_{ij} is supplemented with one more column containing in each row the smallest product of the result available in it by the probability of the corresponding state of nature Π_j, and then among the obtained values of this column a strategy with the largest value is selected.

Germeyer's criterion generalizes the Wald maximin criterion (see below). In the case of a uniform distribution of $P_j = 1/n$, they become identical.

In the second approach, in a situation of complete uncertainty, that is, when $P(\Pi_j)$ are unknown and they are not supposed to be calculated according to the above rules, the following criteria are used to make decisions: product criterion, Wald criterion, Savage test and Hurwitz criterion.

Product criterion. Here the selection rule is formed as follows: the decision matrix with elements a_{ij} is supplemented with a new column containing the products of all the results of each row, and then the option with the highest value is selected. The criterion makes sense when all the elements a_{ij} are positive. The product criterion is calculated as follows:

$$Z = \max_i \prod_{j=1}^{n} a_{ij} \quad (15)$$

Wald maximin criterion (pessimism criterion – always count on the worst) – the worst result is declared the minimum gain, i.e.:

$$\max_i \min_j (a_{ij}) \quad (16)$$

Savage's test (to avoid a big risk by all means) – the worst loss is declared not the minimum gain, but the maximum loss of gain compared to what could be achieved under given conditions:

$$\min_i \max_j r_{ij} \quad (17)$$

Hurwitz criterion (pessimism-optimism criterion) – the degree of pessimism is estimated by experts with criterion α:

$$\max_i \left\{\alpha \min_j a_{ij} + (1+\alpha) \max_j a_{ij}\right\}, \quad (18)$$

where $0 \leq \alpha \leq 1$ is the pessimism-optimism criterion. When α = 1, it turns into the Wald criterion.

The obtained calculation results are summarized in Table 3. In this table, the best strategies by criteria are marked with asteriks (*). However, speculatively, it is very difficult to choose the best strategy.

In order to do this, let us apply a multi-criteria ranking according to the Borda rule. The essence of this rule is that for each criterion the facts of the advantages of this strategy over the other ones are recorded and summed up by all the criteria. The best strategy is the one that has the greatest total number of advantages. As a result of this ranking in the case of both partial uncertainty and complete uncertainty the best strategy was to develop scenario A_2^* [6, 7].

The choice of the optimal strategy A_i^* will be found in a similar way, by ranking them over several criteria according to Pareto [6, 7] and solving the problem in the conditions of partial and complete uncertainty. As a result, the optimal strategy is that strategy A_i^*, which is the best in the aggregate of all criteria. Here, in order to rank the strategies, the Pareto matrices of pairwise comparisons are compiled according to the rule:

$$a_{ij} = \begin{cases} 1, & \text{if the } i\text{-th criterion has a strict preference over the } j\text{-th criterion} \\ 0, & \text{otherwise} \end{cases}$$

Table 3. Evaluation of performance and risk criteria for strategy A_i.

№	Criterion	Value of criterion		
		Strategy A_1	Strategy A_2	Strategy A_3
	a) Partial (incomplete) uncertainty			
1	Bayes-Laplace criterion, P_j are set	12.5	12.8*	12.6
2	Criterion of minimum mean risk, P_j are set	1.0	0.8*	1.0
3	Laplace uncertainty principle, P_j are equal	13.0*	13.0*	12.75
4	Laplace uncertainty principle, P_j are equal, minimum mean risk	1.0*	1.0*	1.25
5	Germeier's criterion, P_j are set	1.5	1.6*	1.6*
6	Hodges-Lehmann estimate, P_j are set	11.25	11.4*	10.8
7	Fishburne point estimate, P_j are set	12.7*	12.0	11.6
8	Fishburne point estimate P_j, minimum mean risk	0.6*	1.6	2.0
	b) Complete uncertainty			
1	Wald criterion (maximin)	10*	10*	9
2	Savage test (minimax risk)	2*	4	5
3	Hurwitz criterion (pessimism-optimism)	12.5	13*	12.5
4	Criterion of product	27300	28880*	24192

As a result, two groups of strategies are obtained: the first group of strategies: A_1, A_3, and the second group – where the strategy A_2^* is the best. However, it is easy to see that the set of criteria by which multicriteria ranking was carried out is meaningfully divided into two groups: criteria that have a sense of average gain, and criteria for average risk. We note here that it is not necessary to divide them into separate groups for multi-criteria ranking A_i, since it has been proven [8, 10] that, from the point of view of evaluating strategies, they give identical results. At the same time, to obtain a more systematic assessment of the effectiveness and risks of developing the deposits, it is necessary to consider another approach – the dispersion risk model.

5. Dispersion model for evaluating the risk of hydrocarbon field development located at the arctic offshore

There is another understanding of risk [11] – dispersion risk model. Let there be any operation, the income from which is a random variable, for example, the gain a_{ij} in the payment matrix, defined in Table 2, depending on the adopted strategy A_i of the field development and the state of nature Π_j. As mentioned above, according to the Bayes-Laplace and Hodges-Lehmann criterion, the average expected income is the expectation $M_i(a_{ij})$ of this random variable with development strategy A_i.

The standard deviation $\sigma(a_{ij}) = [D(a_{ij})]^{1/2}$ is a measure of the spread of possible values around the average expected income, where for a discrete random variable its variance that serves as a measure of risk is:

$$D(a_{ij}) = M\left(a_{ij} - M_i(a_{ij})\right)^2 = \sum_{j=1}^{n}\left(a_{ij} - a_{i\,mean}\right)^2 \cdot P_j \qquad (19)$$

In order to choose best option, it is necessary to use weighting formulas in the approach to determining risk. For example, the weighting formula may be as follows [8]:

$$\varphi(q,r) = q - r, \qquad (20)$$

where q – income, r – risk.

The results of calculations by the ratio (20) are presented in Table 4. To choose the best strategy A_i^*, in the rows of Table 4 we find the maximum value $\varphi(q,r)$, which will determine the Pareto-optimal solution – the minimum possible risk and the maximum possible income.

Table 4. Estimation of strategies A_i of the arctic offshore field development by two criteria: income (gain) and risk (standard deviation of income)

№	Criterion	The values of benefits, risks and their differences					
		Strategy A_1		Strategy A_2		Strategy A_3	
		Benefit; Standard deviation	$\varphi(q,r) = q - r$	Benefit; Standard deviation	$\varphi(q,r) = q - r$	Benefit; Standard deviation	$\varphi(q,r) = q - r$
1	Bayes-Laplace criterion, P_j set by expert	12.5; 1.746	10.754	12.8, 1.833	10.967*	12.6; 2.154	10.446
2	Bayes-Laplace criterion , Laplace uncertainty principle	13.0; 1.871	11.129*	13.0; 2.236	10.764	12.75; 2.8	9.88
3	Bayes-Laplace criterion, Fishburne point estimate	12.7; 1.83	10.87*	12.0; 2.0	10.0	11.6; 2.491	9.109
4	Hodges-Lehnmann criterion, P_j set by expert	11.25; 2.15	9.1*	11.4; 2.31	9.09*	10.8; 2.8	7.2
5	Hodges-Lehnmann criterion, Laplace uncertainty principle	11.5; 2.40	9.1	11.5; 2.25	9.25*	10.86; 3.41	7.45
6	Hodges-Lehnmann criterion, Fishburne point estimate	11.35; 2.30	9.05*	11.0; 2.24	8.76	10.3; 2.65	7.65

Comparing further the best strategies for specific criteria and conditions for implementing states of nature, marked with asterisk (*) in the table, the best among them can be chosen – this is the A_1^* strategy, having $\varphi(q,r) = q - r = 13.0 - 1.871 = 11.129$ by Bayes-Laplace criterion and the Laplace uncertainty principle. However, the final decision is for decision makers.

6. Conclusions

The article proposed a systems approach for assessing risks and the effectiveness of possible strategies for the development and ranking of potential arctic offshore production assets, based on their multi-criteria assessment.

Several groups of criteria are considered, such as geological, economic, technological, climatic, social and environmental, which determine the possibility and significance of projects for the development of deposits of the arctic offshore. The selected criteria provide a framework for building up their ranked list and the model for evaluating the expected risks based on the hierarchy analysis method. Depending on the tasks to be solved, multi-criteria adequate interpretations and risk assessment models are proposed and substantiated.

The risks and effectiveness of development options for the arctic shelf deposit are assessed for partial uncertainty (the probabilities of different states of nature are known or can be estimated) and the total uncertainties (the probabilities of the states of nature are unknown).

To improve the quality of the choice of the best development option, it is proposed to apply the rule of multi-criteria ranking developed by Borda and Pareto, as well as the dispersion risk model.

The proposed models combine rigorous mathematical methods with the experience and intuition of experts who are an integral part of the relevant computer decision support system and machine learning, the essence of which in this case is the computer's selection of the most significant criteria, the best compromise schemes and the corresponding optimization algorithms.

References

[1] Zolotukhin A and Gavrilov V 2011 *Oil & Gas Science and Technology* Rev. IFP Energies nouvelles (English edition).

[2] Gudmestad O T, Zolotukhin A B et al. 1999 *Basics of Offshore Petroleum Engineering and Development of Marine Facilities with emphasis on the Arctic Offshore* (Moscow: «Oil and Gas» Publ. Co) (English edition) p 344.

[3] Recommended practice DNV-RP-F116 *Integrity Management of Submarine Pipeline Systems* (Det Norske Veritas: Oct, 2009) (English edition) p 62.

[4] GOST R IEC 61511 2012 *Functional safety. Security systems instrument for industrial processes* (Moscow: Standard inform) (Russian edition) p 37.

[5] Madera A 2014 *Risks and odds: uncertainty prediction and assessment* (Moscow: KRASAND) (Russian edition) p 448.

[6] Stepin Yu and Trakhtengerts E 2008 *Computer support management of oil and gas processes and production* Book 1 2007 (Moscow: Vector TiS) p 384, Book 2 2008 (Moscow: MAX Press) (Russian edition) p 528.

[7] Stepin Yu 2016 *Computer support, the formation of multi-criteria ranking and optimization of management decisions in the oil and gas industry* (Moscow: Publishing house bowels) (Russian edition) p 421.

[8] Wentzel E 1972 *Operations Research* (Moscow: Soviet. Radio) (Russian edition) p 473.

[9] Stepin Yu 2016 *Selection of maintenance strategies for process equipment at oil and gas facilities under conditions of uncertainty and risk* (Moscow: Gubkin Russian State University of Oil and Gas) (Russian edition) **1/282** pp 106-21.

[10] Nevezhin V 2012 *Game Theory. Examples and Tasks* (Moscow: FORUM) (Russian edition) p 128.

[11] Kolemaeva V A 2008 *Mathematical methods and models of operations research* (Moscow: UNITY-DANA) (Russian edition) p 592.

[12] Aven T 2008 *Risk Analysis Assessing Uncertainties beyond Expected Values and Probabilities* (Norway: University of Stavanger).

[13] Aven T and Vinnen J 2007 *Risk Management With Applications from the Offshore Petroleum Industry* (Norway: University of Stavanger).

[14] Aven T 2010 *Misconceptions of Risk* (Norway: University of Stavanger).

[15] Aven T et al. 2018 *Society for Risk Analysis Glossary*

Remote inspection by the magnetic tomography method (MTM) to prevent the risks imposed by exploitation of Arctic offshore pipelines

S S Kamaeva[1, 2], I S Kolesnikov[1], N A Eremin[2, 3], L A Khusnutdinov[1, 2,*]

[1] Research and Development Center "Transkor-K", Moscow, Russia
[2] Oil and Gas Research Institute of Russian Academy of Sciences, Moscow, Russia
[3] Gubkin Russian State University of Oil and Gas, Moscow, Russia

* Corresponding author: lak@transkorworld.com

Abstract. Cold climate areas that provide opportunities for the remote inspection of pipelines include the Barents Sea, the Russian Arctic, the Alaskan Chukchi Sea, the Beaufort Sea and the Canadian Arctic offshore. First, an analysis of several actual projects of contactless diagnostics using the magnetic tomography method of pipelines in Arctic conditions is done. Second, the Risk-Based Inspection methodology for Arctic offshore pipelines is discussed. It involves ensuring pipeline reliability on the basis of data on the technical condition of the metal in actual operating conditions. The magnetic tomography method allows not only to remotely identify areas of anomalies with metal defects, but also to register mechanical stress levels taking into account actual loads. This reduces the risk for the structure to come to the critical state in terms of exceeding local loads. Finally, magnetic tomography technology allows managing risks in cases of local corrosion, stress cracking or loss of stability of underwater pipelines in areas with free spanning. The qualitative indicators of the inspection include the probabilities of identifying, interpreting the degree of danger, missing a dangerous defect. The pipeline diagnostics report provides the parameters of reliability forecasting: the period of incident-free operation, safe working pressure, and pressure coefficient.

1. Introduction

The Arctic territories are primarily cold-water seas: the Barents Sea, the Russian Arctic, offshore Sakhalin Island, the Alaskan Chukchi Sea, the Beaufort Sea, the Canadian Arctic islands and offshore northeastern Canada currently represent the most important area of development of energy for the world. On the other hand, these are territories with very sensitive ecosystems, where any inaccurate man-made intervention can lead to disastrous consequences for the environment.

The operation of the oil and gas production facilities without proper control of the technical condition, are dangerous in any environment. The extremities of operating conditions, and the enormous length of the pipeline infrastructure facilities operating under pressure, introduce additional risks. The risks introduced by breakdowns, as well as their consequences not only for the environment, but also for personnel, are also much more severe also because less time is available for rescue operations.

Considering individual characteristics of a human body, at a water temperature of 0 to + 5 ° C, typical for the Arctic latitudes and areas with a cold climate, only 20 to 40 minutes are available for the effective rescue. In view of all said, reliable methods of risk assessment based on diagnostics of the actual technical condition and on forecasting its dynamics (Risk-Based Inspection - RBI) are of particular relevance [1].

2. Remote inspection of pipelines in arctic conditions by the magnetic tomography method (MTM)

This report discusses the diagnostics of subsea pipelines in arctic conditions. To date, there is a number of methods for checking the technical condition of offshore fuel and energy facilities available at the market [2-6]. The main methods of inspecting such structures are visual inspection with the help of guided underwater robots (Remotely Operated underwater Vehicle – ROV), together with monitoring the state of electrochemical protection against corrosion (examining the state of sacrificial anodes), and in some areas – the LRUT (Long Range Ultrasonic Technology).

These methods are used for the so-called unpiggable pipelines. For some pipelines, it is possible to apply the in-line inspection that involves several magnetic or ultrasonic flaw detector shells, which allow highly accurate detection of different types of metal defects, their location and orientation relative to the axis of the pipeline [7]. On the basis of these conventional methods, the hazard degree of these defects is determined numerically, taking into account the information on the nominal loads impacting the object. This system of technical diagnostics is not without a number of drawbacks. Below are the main ones.

1. The primary (exploratory) detection of suspected cracks located in an unknown direction is difficult (they can occur in the longitudinal, transverse directions, and at an angle to the axis of the object, but the sensors are not omni-directional);
2. There is yet no generally accepted algorithm for assessing crack resistance and predicting the rate of development of crack-like defects based on the life-cycle mechanism in the form of a "bath";
3. There are no uniform quantitative criteria for assessing the risk of the in-line inspection (ILI) of pipeline defects of various types that would take into account the specific operating conditions of pipelines.
4. The conventional in-line inspection does not at all provide methods for assessing the degree of degradation of metal properties in aggressive embrittlement (hydrogen) conditions, such as microbial induced corrosion (MIC) or stress-corrosion (SCC) and sulfide-stress cracking.
5. The latter applies to conditions of increased static or cyclic loads with stress-deformed state (SDS) anomalies. These are pipeline sections with free spanning, corrugation, stress-strain-torsion levels above acceptable ones, with loss of stability, for example, due to washout (removal) of soil during heavy rains, in landslide areas, on precipices, in ravines, in mountainous terrain, in zones of seismic activity, permafrost and mountain soils, mine workings, as well as underwater (offshore) pipelines.
6. The main problem of the in-line inspection is that the degree of stress concentration in a particular section of the pipeline is not taken into account, i.e. local stress concentrators are not considered. The engineers of the pipeline integrity service of the operating organization assess the stress on their responsibility by expert assessment [8].

Pipeline operation possesses additional risks if there are stress-corrosion cracking defects (SCC) on unpiggable pipelines, where the use of the in-line inspection is impossible or inappropriate for any reasons. Such facilities include pipelines within gas compressor stations and oil pumping stations, underground gas storage (UGS) facilities, chemical and petrochemical plants, refineries, and a number of field pipelines. These unpiggable pipelines can comprise up to 70% of the total length of networks. The technology patented in the USA and Canada using the magnetic tomography method (MTM), regulated by the Russian guidance document RD 102-008-2002, is aimed at solving this problem [9].

The magnetic tomography method for such objects has certain advantages over the conventional diagnostic technologies that ensure its success [10-15], in particular:

1. Does not require special preparation of the pipeline, stopping it or changing its mode of operation;

2. Allows diagnosing areas inaccessible for in-pipe and contact methods of flaw detection;
3. Guarantees high accuracy of detection of areas with crack-like defects and stress-deformed state (SDS) anomalies (above 75%);
4. Demonstrates high performance (up to 7 km per day for 1 team of specialists with portable equipment, in some cases this can be even faster);
5. Wide range of detectable defects of the main and welded metal and high accuracy in assessing their danger degree, taking into account the existing loads;
6. Automatic route tracing followed by plotting the route on a topographic map of the area in absolute geographic coordinates;
7. Easy to organize automatic monitoring of the development of crack-like defects in areas of any length and within any time interval;
8. Low cost of services compared with the programs of pipe digging survey, taking into consideration the preparation of the metal surface for control and NDT for unpiggable pipelines.

As mentioned earlier, the pipeline facilities of the field infrastructure that transport gas, oily media, or water, are generally not fit for the in-line inspection. Sampling methods, conventional for such pipelines, be it the control of metal in pits, or with the help of witness samples (indicator plates inside the pipes), allow us to evaluate external and/or internal corrosion at certain points and are not able to reveal the state of the metal throughout, let alone the detection of the stress-corrosion cracking or sulfide-stress cracking (SCC) (figure 1).

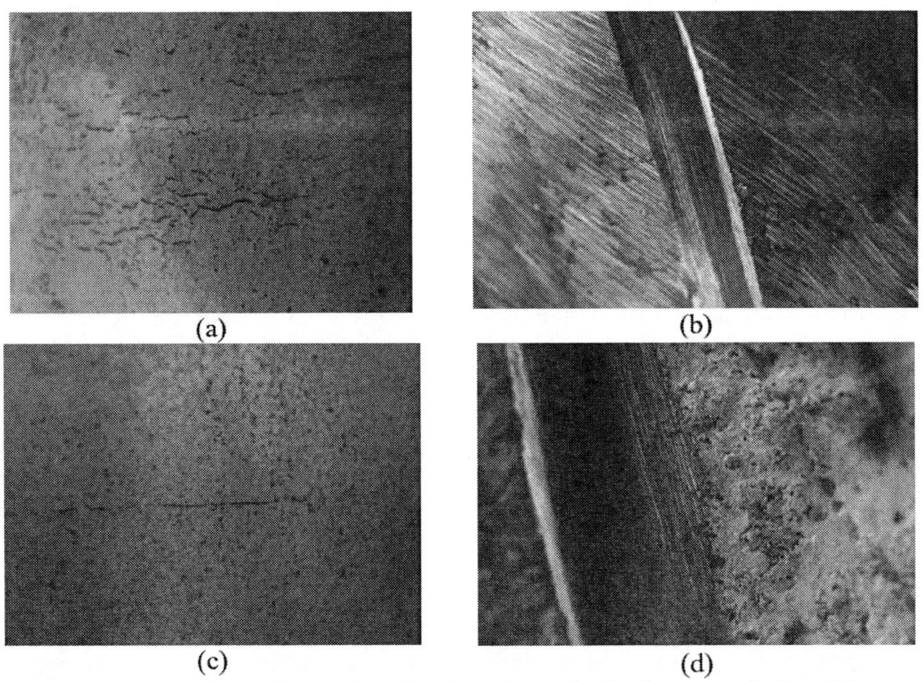
(a) (b) (c) (d)

Figure 1. Non-destructive evaluation (NDE) results in the control pits of the "Northern areas of Tyumen region - URAL-2" main gas pipeline with two stress-corrosion cracking defects (zones of transverse cracks) and their transverse cuts by a grinder to determine the depth. Where 1-st stress-corrosion cracking case: (a) - transverse crack zone and (b) - its transverse cut; 2-nd stress-corrosion cracking case: (c) - transverse crack zone and (d) - its transverse cut.

IOP Conf. Series: Materials Science and Engineering **700** (2019) 012051 doi:10.1088/1757-899X/700/1/012051

As an example of the detection of the stress-corrosion cracking defects, we present the results of the magnetic tomography method survey of a section of the main gas pipeline in Western Siberia (table 1). The total probability of detecting anomalies in areas with defects of the stress-corrosion cracking - zones of longitudinal cracks (ZLC) – the probability of detection (POD) reached 92%. All the stress-corrosion cracking defects indicated in the magnetic tomography method report were examined in control pits with the subsequent destructive control by means of cross-cutting with a tool (circular grinder) to determine the depth of cracks.

In addition to cracks in the longitudinal direction, the magnetic tomography method successfully detects both circumferential and angled defects, for example, micro-cracks of welded mounting connections and in spiral welded pipes.

Table 1. Verification of the magnetic tomography method data (detection of the stress-corrosion cracking defects) of the "Northern areas of Tyumen region - URAL-2" gas pipeline.

№ anomaly	Detected defects	Defects dimension, mm
5	ZLC	20×5×1
5, 6	ZLC	20×8×2,8
	ZLC	-
7, 8	ZLC	130×160×0,5
7, 8 and 9	ZLC	1100×150×0,3; 110×40×1.5
	ZLC	200×460×2; 130×5×0,3 Corrosion up to 1,0 mm
10	ZLC	60×200×0,8; 50×50×0,8
11	ZLC	150×40×1
11, 12	ZLC t	60×40×1; Corrosion up to 1,0 mm
	ZLC	100×200×0,5; Corrosion up to 0,6 mm
13	ZLC	390×200×0,3; Corrosion up to 1,0 mm

Note: ZLC - zones of longitudinal cracks.

Frequent causes of accidents are destruction due to micro cracks, and this is particularly relevant for newly built pipelines in unfriendly terrain (mountains, swamps). The reason is that such cracks appear in welded assembly joints in the zones of stress concentrators. That is why it is important not only to identify "defective" assembly joints, but also to identify those where, due to increased loads, there is a risk of an accident (figure 2). There are many of them, and it is impossible to repair them all at once, but it is important to point out the joints that create stress concentration, a possible origin of a breakdown. In such cases, it does not matter what caused the accident — deformation or imperfection of the metal, since the result is a destruction of the pipe.

Specialists on reliability conclude that such accidents at normal working pressure are caused by forces that imposed additional loads on the deformed bending areas during thermal cycles in the pipeline. As a result, the pipeline deforms again and again, which, in turn, leads to the growth of cracks until an accident occurs in the deformed area of the pipe. In world practice, pipeline accidents are frequent due to additional local loads that altogether exceed admissible load applied to the structure. In particular, the "zero inspection" of the Sabah–Sarawak Gas Pipeline (SSGP), a newly built main gas pipeline on Borneo Island (Malaysia), revealed a number of similar crack defects that were further confirmed by ultrasonic testing (UST) in verification pits.

According to the results of verification of the magnetic tomography method data, the probability of detection (POD) of defective joints requiring repair at this facility was 93%. Figure 3 shows the non-

IOP Conf. Series: Materials Science and Engineering **700** (2019) 012051 doi:10.1088/1757-899X/700/1/012051

destructive testing (NDT) pattern for the circular micro-crack in the control pit detected by the magnetic tomography method.

Figure 2. Destruction of the welded joint due to the "circular crack" of the welded assembly joint in mountain terrain.

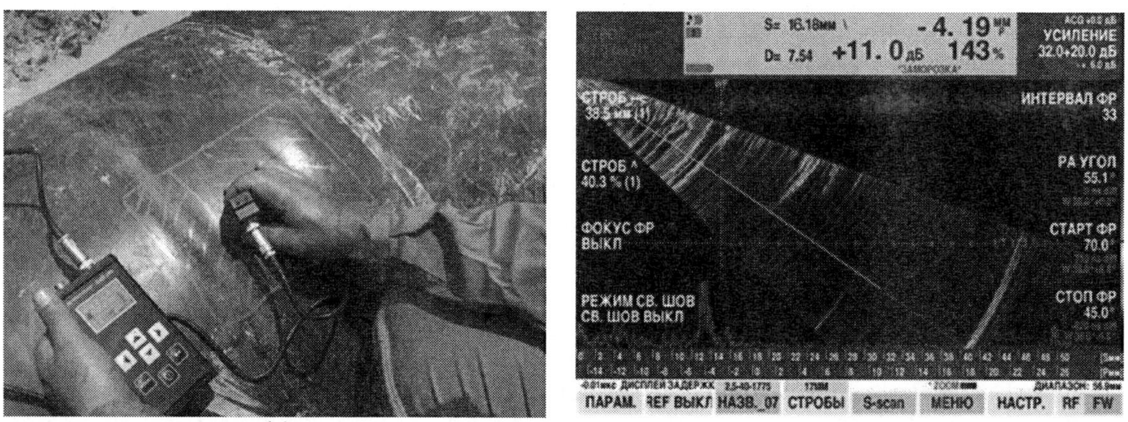

(a) (b)

Figure 3. The non-destructive testing (NDT) in the control pits of the Sabah–Sarawak Gas Pipeline (SSGP) in the areas of circular micro-cracks identified by the magnetic tomography method. Where (a) – eddy current testing in searching cracks and (b) – Ultrasonic testing in controlling whole wall thickness.

The undoubted advantage of the magnetic tomography method is the ability to monitor in real time the development of stress-corrosion cracking defects by installing special contactless sensors. The sensors monitor the growth of local mechanical stresses in the vicinity of the defects selected from the magnetic tomography method results that show the maximum stress concentration. For the pipes of the supply line of the gas-compressor station, by applying such sensors, it was possible not only to identify the stress-corrosion cracking defects (ZLC - zones of longitudinal cracks), but also correctly point out the particular defect among the seventeen other identified defects, the exact one which would lead to the destruction of the pipe during the resource tests (figure 4).

(a) (b)

Figure 4. Resource tests of identified gas pipeline defects using magnetic tomography technology. Where (a) – removing the pipeline from the test site after the destruction and (b) - general view of a brittle crack, indicated by MTM as the most dangerous defect.

The use of in-line inspection may be not possible in all sections of offshore pipelines. There might be a jam of a flaw detector on pipelines with complex seabed geometry, and this may lead to a complete stop in the operation of the pipeline until the flaw detector is removed. In order to remove it, first a place where it is stuck must be found, and then complex diving works underwater must be carried out. This always implies huge financial costs and environmental threats. Often, when preparing for the flaw detector run and pipeline cavity cleaning, the internal anticorrosion coating of the pipeline is damaged, and there is a risk of ignition in the event of electrostatic charges.

Not all pipelines are subject to the in-line diagnosis, for example, unpiggable pipelines include discharge pipelines from oil or gas wells, gas compressor and oil pumping stations, underground gas storage facilities, refinery plants, gas-liquefying plants. The inability to control the technical condition of the pipeline increases the risks of operation breakdown, incidents, and emergencies.

For such objects, where the in-line inspection is difficult to do, or for pipelines that are not fit to the in-line inspection at all, the inspection by magnetic tomography method with the help of contactless scanning magnetometers, and its underwater implementation adapted for subsea objects (AQUA MTM) is perfectly applicable.

The essence of the magnetic tomography method lies in the interrelation of the parameters of the magnetic field of a pipeline with mechanical stress levels in the metal of the pipe, and the method allows the detection of local concentrations of mechanical stress caused by the metal defects. This is a fundamental advantage over other technologies, especially relevant for offshore pipelines.

One of the "headaches" on offshore pipelines is the free spanning. In a free-span zone of a pipeline, the stress in the pipe metal increases significantly. Therefore, even a minor defect can be critical, since the overall stress level in the pipe can be in just a step away from the breakdown stress (often denoted as the specified minimum yield strength, or SMYS).

In case of free spanning, even on the pipelines equipped for the in-line inspection, it is not always possible to make a correct assessment of the danger degree of a defect, since it is impossible to take into account all the existing forces acting on the pipeline. And here the magnetic tomography method provides a direct answer to the question posed: yes, the magnetic tomography method is capable of estimating the stresses that arise in the pipeline during its operation in actual conditions.

This is what can be called the Risk-Based Inspection methodology, i.e. obtaining information about the pipeline based on all forces acting on it during operation. The stress in the pipe metal, calculated as a result of the magnetic tomography method, arising from defects, external impacts and/or stress-deformed state (SDS), reflects the actual technical condition of the pipeline, and the accuracy of this stress calculation was confirmed at the test site provided by Petronas, the largest oil and gas company in Malaysia, during the mock-up test while developing the AQUA-MTM technology.

While MTM technology shows locations on ground surface (in longitudinal, along the pipeline axis, and absolute geographical coordinates) and danger degrees of pipeline sections that have defects of metal, the Aqua Magnetic Tomography Method, the underwater MTM implementation, is a technology for assessing underwater pipelines technical condition.

Additional advantages of the magnetic tomography method and AQUA MTM that are especially beneficial for offshore pipelines are the following aspects:

- There is no need to connect any special equipment to the pipeline; magnetic field recording is performed remotely. The maximum distance from the axis of the pipeline to the magnetometer moving above the pipe should not exceed 15 diameters of the pipeline being examined;

- AQUA MTM inspection using ROV can be carried out in the free fly mode, it is only important to maintain the position of the ROV over the pipeline axis and maintain a vertical distance;

- As a result of the inspection, the pipeline operator receives a report reflecting the actual technical condition of the tested object based on the actual stresses in the pipe metal.

In the report, a pipeline operator receives, among all, the data on the effective stresses in each identified anomaly, and also the period of the incident-free operation - T_{safe}, the safe working pressure - P_{safe} and the estimated repair factor - ERF.

Based on the inspection data, at the request of the customer, a digital twin of the inspected pipeline can be created, or a risk map for the pipeline, indicating the most dangerous sections and forecasting the development of the situation over time, i.e. predictive and proactive functions. For subsea pipelines, the main criterion for the risk of transition to the critical state is the excess of local stresses with the possibility of local loss of stability due to an abrupt unpredicted increase in loads. This situation is reliably controlled by the magnetic tomography method.

Pipeline Integrity Services and Risk management of offshore pipelines at depths up to 200 m include:

- Control of metal pipelines with any type of insulation coating throughout the facility - remotely, without preparation for inspection;
- Registration of local mechanical stresses;
- Compliance with NACE ICDA – the standard of internal corrosion direct assessment (ICDA) of National Association of Corrosion Engineers and monitoring of metal defects of any nature (corrosion, stress-corrosion cracking, fatigue cracks);
- Identification of free spanning, general and local stability loss areas in the zone of inspection, or in seismically active zones;
- Report includes the data on safety parameters, including for the difficult & non-pigable pipelines: P_{safe}, T_{safe}, S_i, SCF, ERF;
- Removes restrictions on the minimum possible operation of the ROV - up to the shoreline.

Figure 5. AQUA-MTM operation system.

The AQUA-MTM system operates as shown in Figure 5. Vessel (2) locates its position on the GPS-navigation system (1). By «ROV Navigation and tracking equipment» (4) the system locates the relative position of ROV (with «Aqua SKIF Underwater Unit»), (3). Control Container (6) with the Vessel survey system calculates absolute coordinates of ROV in real-time "on-line" or "off-line" on the base of data of items 1,2. In 'on-line mode" Control Container transmits absolute coordinates of ROV in real-time to Aqua SKIF Control Unit or - in "off-line mode" - Aqua SKIF Control Unit shall receive the absolute coordinates of ROV by Flash Drive. Matching of the navigation data and SKIF magnetometers data in "off-line mode" shall be realized with the time stamps Aqua SKIF Control Unit receives from GPS Time Receiver. "Aqua SKIF Control Unit" (7) transmits the signal "Turn on" to "Aqua SKIF Underwater Unit" through interfaces: "Interface Control Container - Aqua SKIF Control Unit" (8), "Interface ROV-Control Container" (5). "Aqua SKIF Control Unit" (7) receives the data from "Aqua SKIF Underwater Unit" (magnetometer data – "magnetometer log") through "Interface ROV - Control Container" (5), and receives the absolute coordinates of ROV from Control Container.

The values of internal strain obtained using MTM technology were compared to stress values calculated according to other technologies of stress evaluation. Average relative accuracy is given in table 2.

Table 2. Results of comparing stress values obtained by MTM to other techniques of evaluating the stress-deformed state of the metal.

MTM-Strain Gauges	MTM - ASME B31.G	MTM - DNV RP F101 Part A	MTM - API RP 579	MTM-FEM
74%±10%*	97%±1%*	88%±2%*	91%±2%*	84%±3%*
MTM-Strain Gauges	MTM - ASME B31.G	MTM - DNV RP F101 Part A	MTM - API RP 579	MTM-FEM

Note: *Confidence intervals are given for confidence level 95%.

The probability of detection (POD) of defects was proved by 86±11%. Probability of Interpretation (POIn) of defects was calculated from MTM and FEM (finite elements model) comparison. The comparison results are as follows:

1. POIn for Corrosion defects: POI=83.3%±10.5%;
2. POIn for Dents: POIn=100%-10.5%;
3. POIn for Scratches: POIn=50%±RAs;
4. Gouges defects were identified as metal loss defects.

During correlation of MTM data on stresses with results of strength calculations by International codes, the following parameters of assessment were determined: all corrosion defects (100%) with ERF>0.98 was found; correlations of anomalies danger on stresses in accordance with ERF is ±14.52% with confidence level 95%.

The primary result of the huge Petronas-inspired test was the confirmation that AQUA-MTM technology is capable of high accuracy calculation of safety parameters of sub-sea pipelines, and it may serve as a basis for repair schedule. Consequently, Petronas became one of the largest end clients for MTM (Aqua MTM) inspection in the region.

To date, a number of projects that involved inspection of unpiggable pipelines by means of MTM technology in Arctic conditions have been carried out: 2018 year – Rosneft, Sakhalin, offshore oil pipeline, Nevelsky Strait - 18 km; 2017 year - water passages across rivers of the main gas pipeline of OJSC ALROSA-Gas, Republic of Sakha-Yakutia [16] (figure 6); 2017, 2016 and 2014, Statoil pipelines, Norway.

Figure 6. Non-contact FLY MTM diagnostics using a UAV of a water passage across the Bolshaya Batuobiya river of the main gas pipeline of OJSC ALROSA-Gas, Republic of Sakha-Yakutia, October 3, 2017 [16].

Other non-Arctic inspection projects involving offshore unpiggable pipelines (shelf depth of up to 200 m) applying the AQUA MTM technology include works for Petronas (Malaysia), offshore Oil pipelines from Platform oilfield (SKO, SBO, PMO); Chevron Pacific (Indonesia) offshore oil pipelines from Balikpapan, ADMA OPCO (UAE); offshore oil pipelines, and others. The total length of the inspected offshore pipelines by AQUA MTM totals to about 1,500 km.

3. Conclusion

To summarize, a convenient, reliable tool has been created to ensure the integrity of the underwater pipelines in the Arctic regions. This tool is particularly useful in view of the risks imposed by harsh climatic conditions.

The AQUA MTM inspection advantages are:

- Remote measurements up to 15 diameters away from the pipeline axis;
- The pipeline operation mode change or stop is not required;
- The pipeline is not affected in any way;
- No prior preparation of the pipeline for inspection is necessary;
- Detection of pipe sags, strains, and twists is possible;

As a result of inspection, the following information is provided:

- Location (longitudinal, GPS coordinates) of all defective (anomaly) zones;
- An assessment of danger degree (deviation of a pipe material stress-deformed state for each defective area is reported, the period of incident-free operation is estimated, safe working pressure for each anomaly area in the pipeline is determined);
- The absolute value of local stress in pipe material.

The MTM (Magnetic Tomography Method) Technology developed by R&D Center “Transkor-K” (Russia) and AQUA MTM, developed in a joint effort with Petronas (Malaysia) are both intended for metal stress-deformed state assessment of offshore steel pipelines of any purpose. The core method is based on remote registration of surrounding magnetic field of a pipeline using a subsea Remote Operated Vehicle (ROV). The technology allows inspecting the entire length of a pipeline with high probability of detection of stress-deformed state anomalies: POD >75%.

Acknowledgments

The authors would like to thank the management of the Oil and Gas Research Institute of Russian Academy of Sciences for the permission to present and publish this paper based on the results of the project "The Fundamental Basis of Innovative Technologies in the Oil and Gas Industry", No. AAA0139-2018-0006.

References

[1] Mahov V, Husnutdinov L, Goroshevskij V and Kamaeva S 2017 Risk-oriented approach and the criteria for assessment of quality of technical diagnostics of pipelines by magnetic tomography method, *Occup. Saf. Ind.* **5** 42–48.

[2] Henriksen T-E, Aarset K, Myrvang H, Haslum H and Nes R 2019 Aasta Hansteen Offshore Marine Operations – Deep, remote and harsh *Offshore Technol. Conf.* https://doi.org/10.4043/29350-ms

[3] Lindseth S, Røsby E, Vist B and Aarnes K 2019 Aasta Hansteen Subsea Production System for Deep Water and Harsh Environment *Offshore Technol. Conf.* https://doi.org/10.4043/29350-ms

[4] Paulin M and Caines J 2016 The Evolution of design tools for Arctic subsea pipelines. *Arctic Technol. Conf.* https://doi.org/10.4043/29350-ms

[5] Bybee K 2001 Are We Ready To Construct Submarine Pipelines in the Arctic, *Soc. Petrol. Eng.* https://doi.org/10.2118/0401-0055-jpt

[6] Palmer A 2000 Are we ready to construct submarine pipelines in the Arctic *Offshore Technol. Conf.* https://doi.org/10.4043/12183-ms

[7] Gorban N, Kamaeva S and Eremin N 2018 Sensors and equipment for the inspection of the offshore part of oil pipelines, *Sens. Syst.* **228**(8) 62-68.

[8] Gorban N, Kamaeva S and Eremin N 2018 The non-contact magnetometry of pipeline integrity: state and trends, *Sens. Syst.* **226**(6) 36-42.

[9] Instructions for diagnosing the technical condition of pipelines by the non-contact magnetometric method 2002 *The Russian guidance document 102-008-2002* (All-Russian Research Institute for the Construction and Operation of Pipelines, Fuel and Energy Complex Facilities, Moscow, Russia).

[10] Kamaeva S 2003 Biogenic sulfat-reduction as a bio-corrosion factor of underground pipelines *An abstract of a dissertation for the degree of candidate of technical sciences: 05.17.03* (All-Russian Research Institute for the Construction and Operation of Pipelines, Fuel and Energy Complex Facilities, Moscow, Russia) 28.

[11] Kamaeva S and Eremin N 2017 The risk-based approach to ensuring the safety of gas pipelines using the contactless technology for the technical diagnosis *Oil. Gas. Innovations.* **9** 75-82.

[12] Li Z, Jarvis R, Nagy P, Dixon S and Cawley P 2017 Experimental and simulation methods to study the Magnetic Tomography Method (MTM) for pipe defect detection *NDT and E Int.* **92** 59–66.

[13] Kolesnikov I 2014 Magnetic Tomography Method (MTM) & A Remote Non-destructive Inspection Technology for Buried and Sub Sea Pipelines, *OTC Arctic Technol. Conf.* https://doi.org/10.4043/24569-ms

[14] Liao K, Yao Q and Zhang C 2011 Principle and technical characteristics of non-Contact magnetic tomography method inspection for oil and gas pipeline *ICPTT 2011*.https://doi.org/10.1061/41202(423)111

[15] Eremin N, Kamaeva S, Chernikov A and Eremin Al 2018 Through digitization and quantization *Oil of Russia* **3-4** 62-65.

[16] Dmitrievsky A, Kamaeva S, Chernikov A and Eremin N 2018 Contactless diagnostics of oil and gas pipelines: status and prospects of development, *Actual Problems of Oil and Gas* **20**(1).

Digital transformation of gas production

A N Dmitrievskiy[1, 2], N A Eremin[1, 2*], V E Stolyarov[1]

[1] Oil and Gas Research Institute of Russian Academy of Sciences, Moscow, Russia
[2] Gubkin Russian State University of Oil and Gas, Moscow, Russia

* Corresponding author, E-mail: ermn@mail.ru

Abstract. The development of the digital transformation of the gas industry in the Russian Federation is mainly aimed at unique and giant fields. These fields provide the leading position of Russia in the world gas market and about 85 % of the total volume of gas production. The regulatory and technical base, the composition of the existing standards allows us to move to the practice of introducing digital gas technologies. The creation of a unified integrated model is being completed at the Bovanenkovo field. The creation of a digital twin of a unique gas field represents the transfer of a real field into virtual space through the use of Big data analytics and supercomputer technologies.

1. Introduction

In 2023, the size of global gas demand will exceed 4.1 trillion cubic meters compared to 3.74 trillion cubic meters in 2017 according to the International Energy Agency. About 40% of the world's gas production will fall in equal proportions to Russia and the United States in the next five to six years.

Until the mid-60s, automatic devices in gas fields were practically not used. The start of digitalization of oil and gas fields is associated with the joint project of «SmartFields» by FMC Kongsberg Subsea and Statoil, which started in January 1997 and in January 1999 A / S Norske Shell joined it [1]. The introduction of digital technologies in gas production allows reducing production costs, increasing technological and environmental safety and ensuring highly efficient gas recovery [2-11].

In recent decades, the process of digitalization has accelerated due to the large number of fields at the final stage of production. Distributed control technologies have become widely used in the gas industry according to the Program of phased deployment of works on the creation of an industry system of operational control of the Unified gas supply system of Russia, approved by Gazprom. At the late stage of field development, the role of operational information increases. The creation of technologies for managing residual reservoir pressure is required [3].

The introduction of simple pneumatic and mechanical devices, telemetry and automation systems provided an improvement in the technical and economic characteristics of the fields by at least 10-30%, saved from 5-15% of capital investments and operating costs of at least 10-30% of the design indicators [4, 5]. The activities carried out to equip production wells with digital monitoring and control systems can quickly optimize well operation in the well, as well as prevent the destruction of the bottom hole zone and increase daily production by 10-25% [2, 4, 7, 8 and 10].

Figure 1. The third gas production facility of the Bovanenkovo field. Source: Gazprom.

The practice of gas field operation shows that for the effective use of assets, priority is given to:

- new technologies for gas recovery and production management;
- Big data technology, for predictive analysis of the situation and the adoption of proactive impacts;
- reduction of technological and environmental risks of operation;
- widespread use of industrial Internet in the fields;
- application of distributed registry technologies (blockchain);
- providing virtual reality and augmented reality at technological facilities [12].

The existing regulatory and technical base, the composition of existing standards and regulations allows us to switch to the practice of the widespread use of digital technologies in gas production [12].

2. The digital transformation of gas fields

Today, PJSC Gazprom implements and successfully operates a unified control structure and the main production facilities at the following levels: the Central production dispatching service of PJSC Gazprom (the Company) - the production dispatching service Gazprom production region (the base enterprise) - the gas production administration (support base of the base enterprise) - gas field (field) - a cluster of gas wells (gas wells or remote technological facilities of the field). The architecture of the digital field includes technological facilities, a real-time monitoring system for gas operations; integrated gas production model; center for integrated production management; fiber-optic system for collecting and transmitting Big geological and geophysical data; geo-production data processing center [3, 4, 12].

Digital gas production technology allows for dynamic optimization and improving the quality of field operation management and includes:

- Digital adjustment and ensuring the adequacy of the integrated geological and technological model;
- Integrated calculation of material balance for wells, fields and the field as a whole;
- Calculation of reserves and execution of the corresponding reporting forms for writing off and paying the tax on natural resources production or severance tax for the field;
- Optimization of load distribution among wells and planning of capital repairs and stimulation;
- Adaptation of the field management system.

In recent years, 352 digital and automation systems have been created and modernized at 53 units of integrated and preliminary gas treatment at unique and large Arctic gas fields.

Figure 2. Well pad (cluster) in the Bovanenkovo field. Source: Gazprom.

At the meeting of the Council on the priority direction of the scientific and technological development of the Russian Federation on November 29, 2018, Dmitrievsky and Eremin made a short message. After the discussion, the council members approved the proposal for the formation of the "Scientific and technical program "Digital and technological modernization of the world's largest West Siberian oil and gas production center". As part of the implementation of this program, at a meeting at the Ministry of Energy of the Russian Federation on August 8, 2019, joint proposals of the OGRI RAS with partners for the creation of "digital twins" of a gas field and gas pipeline on land and at sea were presented. The creation of a digital twin of a unique gas field represents the transfer of a real field into virtual space through the use of Big data analytics and supercomputer technologies.

To fulfill the state task of creating the main gas production center in the Arctic, the Bovanenkovo field was commissioned, which is the base field in the center of Yamal gas production (Figure 1). The Bovanenkovo field has been developed by three gas production facilities, which were consecutively brought into operation in 2012, 2014, and 2018. Effective and safe management of complex gas fields is possible with the use of modern highly reliable digital control and management systems. As of 01.01.2019, the operating well stock was 437 out of 777 planned wells. Daily gas production will reach 317 in 2019 compared to 264 million cubic meters in 2018. The annual production of 115 billion cubic meters of gas is planned for 2022. Gas reserves at the Kharasavey field exceed 2 trillion m3. Full-scale development of the field with elements of digitalization began in March 2019. The gas production is going to start in 2023. The planned annual production level will be 32 billion m3 of gas per year from its Cenomanian-Aptian deposits. The program for the development of the giant Chayanda field assumes the widespread use of integrated control elements and a unified information space at all stages of the field's life cycle (collection of data, modeling, design, creation, operation, diagnosis and management of field objects, analysis of service life, risks of threats and safe operation).

The creation of an integrated system of this field is being completed. Every gas production facility with well pads (clusters) uses integrated digital systems and lightly manned technologies (Figure 2). The implementation of the integrated approach in production will provide modeling of the entire field, visualization of geological and geophysical information, dynamic optimization and improvement of the gas recovery management in real time (Figure 3).

Effective and safe management of the complex gas field is carried out on the basis of the use of automated process control systems "Promysel-1" and "Neman-R". In 2019, the production well stock amounted to 437 out of 777 planned wells. The daily gas production in 2019 will be 317, compared with 264 million cubic meters in 2018. The maximum level of project production from the Cenomanian (520–700 meters deep) and Apt-Albian (1,200–2,000 meters deep) deposits in the amount of 115 billion cubic meters of gas will be achieved in 2022. This volume is equal to the total supply of Russian gas to Germany, Italy, the UK, France, and the Czech Republic in 2018. In the medium term, it is expected to enter into the development of Neocom - Jurassic deposits, which will increase annual gas production to 140 billion cubic meters [12].

In 2019, the partly digital development of the Kharasavey field with gas reserves of 2 trillion m^3 was started; the beginning of gas production – 2023; the projected level of annual production from Cenomanian deposits is 32 billion m^3 of gas. In the future, it is planned to develop more deeply located Neocom-Jurassic deposits. The field is mainly located on land, partly in the Kara Sea. Wells for the offshore part of the field will be drilled from the shore.

Figure 3. Operator console Bovanenkovo field. Source: Gazprom.

In the medium term, it is planned to begin development of the Kruzenshtern field and the Tambey group of fields with its recoverable reserves of 7.7 trillion cubic meters.

The Chayanda field is the base field for the Yakutsk gas production center and a resource base for the "Power of Siberia" gas pipeline. At the field, 143 production gas wells were drilled as of 01/01/2019. The development program of the Botuobin and Hamakin horizons of the giant Chayanda oil and gas condensate field involves the widespread use of integrated control elements and a single information space at all stages of the field's life cycle (collection of field data, modeling, design, creation, operation, diagnostics and field management, threat risks and safe operation). Gazprom will use lightly manned technologies that provide for control equipment and integrated facility management. In the framework of providing design solutions for the project, the "Concept for managing field facilities (an integrated process control system of the Chayanda oil and gas condensate field)" was approved. Algorithms for integrated management of all remote objects of the field are being tested. The field provides for the use of a single information space along the entire technological chain "gas well pad or cluster - gas collection network - gas pre-treatment unit – field pipeline - integrated gas treatment unit - central booster compressor station - the receiving node of the main gas pipeline". Industrial gas production and gas supply from the Chayanda oil and gas condensate field to the «Power of Siberia» gas pipeline with a total length of 2150 km was provided for the planned dates for filling and commissioning of communication systems and telemechanic of linear facilities on the pipeline. The digital control and management system will ensure the technological and environmental safety of the export gas pipeline in real time. The control center for the digital technological complex of the Chayanda field was created on the basis of domestic software and hardware and high-speed communication channels that will provide modeling and monitoring of the state of technological objects, as well as information interaction and diagnostics of all elements of the production chain.

Taking into account the operating experience of gas production facilities implemented on the principles of remote management, it is necessary to develop the "Concept of creating digital gas fields" in accordance with the recommendations of the standard of organization Gazprom 2-2.1-1043-2016.

A global urgent task is to create an interdisciplinary design and research environment, i.e. integration and interaction of fundamental and applied science specialists, students and teachers, industrialists and researchers from different countries to solve specific problems of the gas production.

3. Conclusions

The use of digital technologies provides the ability to remotely, lightly manned and uninhabited to manage the unique and large gas fields in the harsh Arctic conditions. Among the innovative technologies in the Arctic can be identified quantization of the collection and transmission of Big volumes of geo-field data via secure fiber optic channels, a digital oil and gas university, artificial intelligence methods, petrorobotics, predictive analytics, oil and gas wireless and satellite petroleum Internet of things (PIoT) and domestic software PJSC «Gazprom avtomatizatsiya». The modern practice of developing gas fields demonstrates a gradual shift from the blind gas technocracy of the last century towards the digitalization and harmonization of gas business. The digital gas production technologies allow extending the terms of cost-effective exploitation of the unique gas fields at the stage of decreasing production. Current gas production appears to be hybrid, in which digital and traditional exploration, drilling and production technologies coexist equally, which contributes to increased labor productivity.

Acknowledgments

The authors would like to thank management of the Oil and Gas Research Institute of Russian Academy of Sciences for the permission to present and publish this paper based on the results of the project "The Fundamental Basis of Innovative Technologies in the Oil and Gas Industry", No. AAA0139-2018-0006.

References

[1] Halmoey S 2001 Smartcontrols For Smart Subsea Fields *Offshore Mediterranean Conference.* https://www.onepetro.org/search?q=OMC-2001-071&peer_reviewed=&published_between =&from_year=&to_year=&rows=25.

[2] Tubel P and Hopmann M 1996 Intelligent Completion for Oil and Gas Production Control in Subsea Multi-lateral Well Applications, *Soc. Petrol. Eng.* https://doi.org/10.2118/36582-ms.

[3] Abukova L, Dmitrievsky A and Eremin N 2017 Digital modernization of Russian oil and gas complex *Neftyanoe Khozyaystvo - Oil Industry,* **10** 54–58.

[4] Carvajal G, Maucec M and Cullick S 2018 *Intelligent Digital Oil and Gas Fields* (Gulf Professional Pub., Elsevier,Netherlands) 374.http://dx.doi.org/10.1016/c2015-0-02216-x.

[5] Ali Z, Al-Jasmi A and Qiu F 2013 Digital Oil Field Experience: An Overview and a Case Study *Soc. Petrol. Eng.* https://doi.org/10.2118/163718-ms.

[6] Cobb C 2004 The Oil Field of the Future, Now, *Soc. Petrol. Eng.* https://doi.org/10.2118/0504-0016-jpt.

[7] Erlandsen S 2000 Production Experience From Smart Wells in the Oseberg Field, *Soc. Petrol. Eng.* https://doi.org/10.2118/62953-ms.

[8] Denney D 2000 Integrating Seismic Monitoring and Intelligent Wells, *Soc. Petrol. Eng.* https://doi.org/10.2118/1000-0049-jpt.

[9] Al-Dhubaib T 2011 Intelligent Fields: Industry's Frontier & Opportunities, *Soc. Petrol. Eng.* https://doi.org/10.2118/141874-ms.

[10] Mjaaland S, Wulff A-M, Causse E and Nyhavn F 2000 Integrating Seismic Monitoring and Intelligent Wells, *Soc. Petrol. Eng.* https://doi.org/10.2118/62878-ms.
[11] Burda B, Crompton J, Sardoff H and Falconer J 2007 Information Architecture Strategy for the Digital Oil Field, *Soc. Petrol. Eng.* https://doi.org/10.2118/106687-ms.
[12] Eremin N, Melnikov I, Bobrikov N, Stolyarov V, Kogai A and Schegolev D The application of complex gas production control algorithms as the necessary elements to create a digital twin of the Bovanenkovo oil and gas condensate field technological complex, *Gas Industry Magazine* **785** (6) 42-49.

IOP Conf. Series: Materials Science and Engineering **700** (2019) 012053 doi:10.1088/1757-899X/700/1/012053

Proposals on 3D parallel edge-preserving filtration for x-ray tomographic digital images of porous medium core plugs

S S Arsenyev-Obraztsov, E A Volkov and G O Plusch

National University of Oil and Gas, 65 Leninsky Prospekt, Moscow, Russia

Corresponding author: arseniev@gubkin.ru

Abstract. Precise estimation of the effective petrophysical characteristics for the oil or gas-bearing reservoir plays a vital role in production control problems. It is preferable to use nondestructive measurement methods for porosity, conductivity, geomechanical modulus, and some other parameter fields. For the estimation of anisotropic permeability tensor flows in different directions need to be simulated in the core plug digital image void space, which is very difficult (if possible) to conduct with the real rock sample. The creation of the digital core image includes three stages: construction of its internal structure based on the computed tomography (CT) sinogram, filtration, and segmentation. Routine practice is the filtration of 2D slices pack of the core plug CT image because of a lack of computational power and memory limitations. Unfortunately, this can generate a directional orientation error orthogonal to the pack of slices. The total 3D edge-preserving filtration with the usage of two approaches: modified implicit anisotropic diffusion and discrete orthogonal transforms can reduce this error. Universal code based on the MPI+OpenAcc programming paradigm was tested on different high-performance computing systems incorporating various accelerators like GPGPU and heterogeneous processors such as Sunway 26010.

1. Introduction

Digital analysis of the oilfield core plugs can, to a great extent, replace traditional methods of laboratory examinations and significantly reduce the time required for them. Researchers have the opportunity to understand complex structure of the reservoir better since such an analysis allows correctly take into account the complex geometry of the void space, the morphology of the rock particles, the mineralogical makeup, and based on this information further simulate the flow of fluids at the scale of connected pores.

One of several tools for digital core analysis is CT, which is a method for constructing the internal structure of almost any transparent for some physical waves substance under the assumption that it comprises materials with different radiation absorption coefficients. Tomography is a method of visualization of an object by sectioning. A tomographic survey creates a stack of 2D images of slices of a certain thickness, from which the 3D digital model can be created. Micro-computed tomography (microCT) is a variant of the CT in which the spatial resolution is about a micron. If x-ray radiation is used to build a digital image, then it is called microcomputer x-ray tomography. Recently appeared microCT scanners with a nanoscale resolution which can be of interest for the examination of flow in the low permeability rocks.

For highly fragmented images, it is difficult to construct the boundaries of void space in the porous media. Impulse noises and fine image granularity makes the further flow simulation on the segmented pack of slices numerically very hard because of the time required for the calculations. That is why it is necessary to filter the images of slices before the segmentation. The noise in tomographic images has a diverse nature. It may occur because of the imperfectness of the CT scanner's matrix, the mechanics of x-ray exposition table motion, the computational round-off errors, and the ill-posedness of the inverse numerical Radon transform.

Most of the smoothing filters are not suitable for processing core plug tomograms, as they displace and flatten out boundaries. They change the morphology of pores, which makes almost meaningless determining effective petrophysical characteristics by numerical simulation of different physical processes on a digital 3D core plug model.

Since CT scan technology origins from medical applications, the common practice is to filter 2D slices in the pack independently [1]. It can be acceptable for large objects in the image, but for a porous medium, such filtering gives directional orientation error (Figure 1) when compounding slices into a 3D digital model. In Fig 1b the directional orientation error, generated by numerical 2D slices pack filtration procedure can be easily seen. Generated disturbances can severely influence the numerical estimation of petrophysical properties such as permeability, geotechnical modules, and some other.

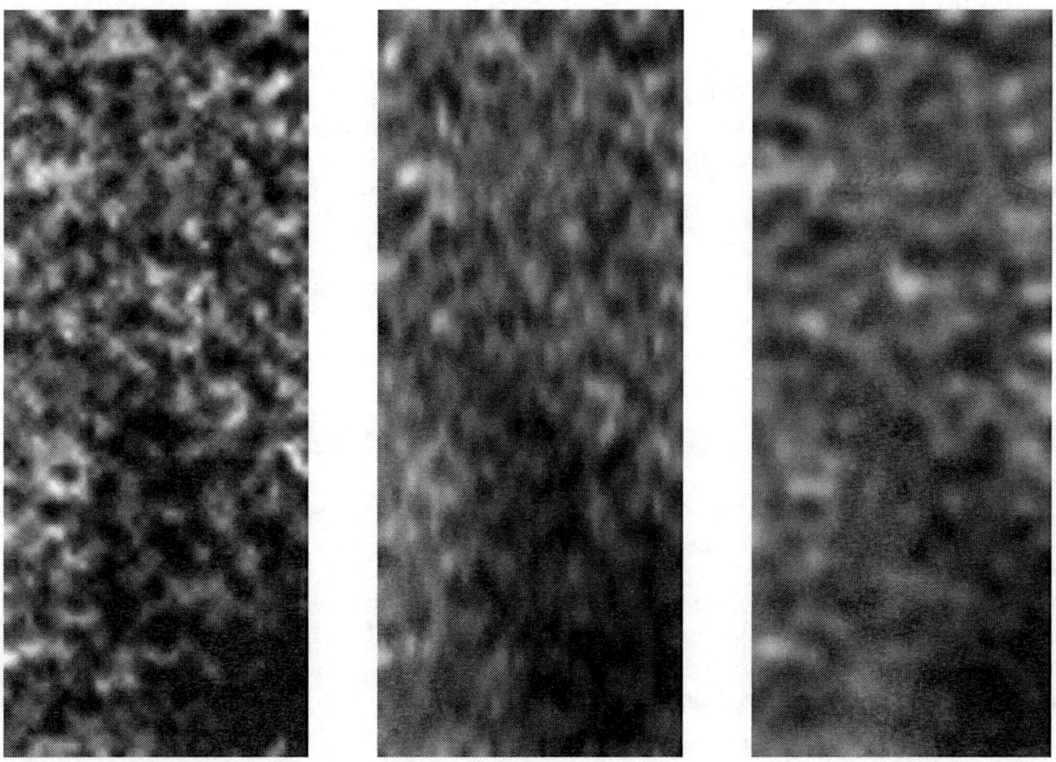

(*a*) (*b*) (*c*)

Figure 1. (*a*) original image, (*b*) 2D slice filtering with vertical orientation error, (*c*) 3D filtering

To increase the quality of the digital image, filtering, segmentation, and numerical simulation can be included in the cycle of inverse problem solution - adapting digital core plug simulation results to the real laboratory tests. RAM capacity required for constructing 3D images can reach

IOP Conf. Series: Materials Science and Engineering **700** (2019) 012053 doi:10.1088/1757-899X/700/1/012053

several dozens (hundreds) gigabytes even for an ordinary grayscale image coding. Such problems can be solved only by High-Performance Computing (HPC) clusters. For this purpose, modern heterogeneous HPC systems based on (CPU+GPGPU) or (CPU+accelerators) architecture or even on homogeneous clusters based on heterogeneous CPUs like the Sunway TaihuLight can be used. Because of many different architectures and approaches, the Fortran + MPI + OpenACC programming paradigm was chosen because of its portability and performance. Fortran was selected because of its internal parallelism achieved by including Coarray Fortran (usually based on MPI) in its new standards, which make parallel programming more straightforward.

Two approaches for a 3D edge-preserving filtration: modified implicit anisotropic diffusion and a family of discrete orthogonal transforms were tested.

2. Modified implicit anisotropic diffusion filter

One of the image smoothing algorithms that preserve edges is Perona and Malik anisotropic diffusion filter [2]. It models the diffusion of the voxels (pixels) intensity in a digital image with a specially selected diffusion tensor $c(r,t)$, which is defined so that the smoothing process does not affect the existing edges. The diffusion equation for our case is:

$$\begin{aligned} I(r,t)_t &= div(c(r,t)\nabla I(r,t)), \\ I(r,0) &= I_0(r), \quad r \in \Omega, \end{aligned} \tag{2.1}$$

where r is a point in the spatial domain Ω, $I_0(r)$ is the original (initial) image intensity field, and $I(r,t)_t$ is an image intensity time derivative.

The diffusion equation is defined in a finite domain Ω; therefore, boundary conditions on $\partial\Omega$ are needed. A natural approach is to use boundary conditions conserving image intensity, or exclude boundaries and solve the initial value problem on the n-dimensional tours as an initial value problem.

Perona and Malik proposed to use the norm of the image intensity gradient $||\nabla I||$ as an estimator for the edge presence at the interior points of Ω. The higher the gradient, the more possibility of edge presence in the point. So manual construction a diffusion coefficient (tensor) in the form that will not flatten gradient norm is needed:

$$c(r,t) = g(||\nabla I(r,t)||), \tag{2.2}$$

where $g(x)$ is a decreasing function mapping to the interval $(0,1]$. Perona and Malik proposed two functions:

$$g(x) = \exp\Big(-(x/\gamma)^2\Big), \quad g(x) = \Big(1+(x/\gamma)^2\Big)^{-1}, \tag{2.3}$$

where γ is a filter parameter determined either experimentally or by some heuristic "noise" estimator (for example, the Canny method [3]). Parameter γ acts as a "barrier" for the noise importance level. From (2.3) it can be seen that with a monotonous decreasing γ, the measure of the point presence in the edge increases. Therefore, its intensity will not change too much. There are some other forms of this function [4].

2.1. Explicit method

Perona and Malik proposed to solve the quasilinear parabolic equation (2.1) by the explicit form of the finite-difference (FD) method. The scheme was constructed by analogy to the spatial FD discretization for equations with variable coefficients. For example, a part of the FD operator

IOP Conf. Series: Materials Science and Engineering **700** (2019) 012053 doi:10.1088/1757-899X/700/1/012053

in the spatial domain for the x coordinate is:

$$\frac{\partial}{\partial x}\left(c(x,y,z,t)\frac{\partial I(x,y,z,t)}{\partial x}\right) \approx \frac{1}{(\Delta x)^2}\left(c(x+\frac{\Delta x}{2},y,z,t)\cdot(I(x+\Delta x,y,z,t)-I(x,y,z,t))- \right.$$
$$\left. - c(x-\frac{\Delta x}{2},y,z,t)\cdot(I(x,y,z,t)-I(x-\Delta x,y,z,t))\right). \quad (2.4)$$

The spatial discretization step size along each direction is set equal to the distance between pixels so $\Delta x = \Delta y = \Delta z = h$. For the simplicity let $h = 1$. In [2, 5] it was shown that scheme (2.4) will be numerically stable if $0 \le \Delta t \le 0.25$.

To diminish the number of calculations, Perona and Malik proposed to use the length of gradient projection on the corresponding axis of the FD grid instead of gradient norm [2]. In matrix form, the explicit 3D finite-difference scheme will be:

$$I^{p+1}_{i,j,k} = I^{p}_{i,j,k} + \Delta t\left(g(|N^{p}_{i,j,k}|)\ N^{p}_{i,j,k} + g(|S^{p}_{i,j,k}|)\ S^{p}_{i,j,k} + g(|E^{p}_{i,j,k}|)\ E^{p}_{i,j,k} + \right.$$
$$\left. + g(|W^{p}_{i,j,k}|)\ W^{p}_{i,j,k} + g(|D^{p}_{i,j,k}|)\ D^{p}_{i,j,k} + g(|U^{p}_{i,j,k}|)\ U^{p}_{i,j,k}\right), \quad (2.5)$$

where (i, j, k, p) are indices of the FD grid for (x, y, z, t), respectively. In this scheme the FD operators $N^{p}_{i,j,k}, S^{p}_{i,j,k}, E^{p}_{i,j,k}, W^{p}_{i,j,k}, D^{p}_{i,j,k}, U^{p}_{i,j,k}$ are:

$$\begin{array}{lll} N^{p}_{i,j,k} = I^{p}_{i-1,j,k} - I^{p}_{i,j,k}, & S^{p}_{i,j,k} = I^{p}_{i+1,j,k} - I^{p}_{i,j,k}, & E^{p}_{i,j,k} = I^{p}_{i,j+1,k} - I^{p}_{i,j,k}, \\ W^{p}_{i,j,k} = I^{p}_{i,j-1,k} - I^{p}_{i,j,k}, & D^{p}_{i,j,k} = I^{p}_{,j,k-1} - I^{p}_{i,j,k}, & U^{p}_{i,j,k} = I^{p}_{i,j,k+1} - I^{p}_{i,j,k}. \end{array} \quad (2.6)$$

The mnemonic notation N, S, E, W, D, U mean – North, South, East, West, Down, and Up.

This FD scheme is stable only for rather small time steps. It is unacceptable because CT of one core plug generates a bunch of slices, producing about a thousand of 2D images. The required computer memory depends on the used bit-coding.

2.2. Implicit method

Application of the implicit FD scheme can increase numerical stability:

$$I^{p}_{i,j,k} = I^{p-1}_{i,j,k} + \Delta t\left(g(|N^{p}_{i,j,k}|)\ N^{p}_{i,j,k} + g(|S^{p}_{i,j,k}|)\ S^{p}_{i,j,k} + g(|E^{p}_{i,j,k}|)\ E^{p}_{i,j,k} + \right.$$
$$\left. + g(|W^{p}_{i,j,k}|)\ W^{p}_{i,j,k} + g(|D^{p}_{i,j,k}|)\ D^{p}_{i,j,k} + g(|U^{p}_{i,j,k}|)\ U^{p}_{i,j,k}\right). \quad (2.7)$$

This discretization leads to a system of nonlinear equations, which can be solved, for example, by the Newton method, and this will increase the execution time. Such an approach will ensure the high accuracy of the PDE numerical solution, but our task is different: it is to prepare the image for the subsequent segmentation. Besides, filtration use double-precision arithmetic, but the result will be converted into the desired image bit-coding with a loss of accuracy because of rounding. Linearization of the system (2.7) is conducted by calculating the diffusion coefficient - c on the previous time layer. It is natural to suppose for the edge-preserving filter that the diffusion coefficient should not change too much in time. Therefore the resulting scheme (2.8) will be unconditionally stable [5, 6]:

$$I^{p+1}_{i,j,k} = I^{p}_{i,j,k} + \Delta t \Big(g(|N^{p}_{i,j,k}|)\ N^{p+1}_{i,j,k} + g(|S^{p}_{i,j,k}|)\ S^{p+1}_{i,j,k} + g(|E^{p}_{i,j,k}|)\ E^{p+1}_{i,j,k} + \\ + g(|W^{p}_{i,j,k}|)\ W^{p+1}_{i,j,k} + g(|D^{p}_{i,j,k}|)\ D^{p+1}_{i,j,k} + g(|U^{p}_{i,j,k}|)\ U^{p+1}_{i,j,k} \Big) \tag{2.8}$$

Instead of solving the boundary value problem, if the 3D digital image ($N \times M \times L$) is thought as a 4D torus surface (2.9) the problem for our PDE will be converted into the initial value one.

$$I^{\text{torus}}_{i,j,k} = I_{i \pmod N, j \pmod M, k \pmod L}, \tag{2.9}$$

In [7] it was shown that the linear operator of the system (2.8) with torus initial condition does not change the total image intensity it is symmetric and positive definite. Therefore, the preconditioned conjugate gradient (PCG) method can solve the system of linear equations (2.8). This approach gives the flexibility to choose the required accuracy in the solution process. It allows to control the termination of iterations and provides an additional acceleration of the filtering procedure. According to the routine method, the image obtained at the previous time step is used as the initial solution guess for the PCG method.

2.3. Filter stopping criteria and parameter estimation process

Usually, the relative residual is used as a stopping criterion for the PCG method. Let r^l_p be the residual of the iteration l at the time step p: $r^l_p = I^l_p - I^{l-1}_p$, where I^l_p is an image vector (1D representation of 2D/3D image). Then the stopping condition with the threshold value τ is: $||r^l_p||/||I^l_p|| < \tau$.

The non-parametric stopping criterion can be presented in the following form. Iterations will continue until the new value of the residual differs from the zero residual in the selected image bit-encoding unit.

$$\max_{1 \le i \le n} |(r^l_p)_i| = ||r^l_p||_\infty < 1. \tag{2.10}$$

Numerical experiments show that the application of this criterion significantly reduces the number of iterations required to solve the problem.

This approach is obvious. Pixel brightness (color) is an integer value, so if the infinity norm of the residual vector is less than one bit-encoding digit, then further refinement is not required since the fractional parts will be round-off when converting the result to the image integer presentation. The proposed filter scheme allows us to make arbitrary time steps, but impulse noise may remain in the image due to the large step sizes [7]. There exist some modifications of the Perona and Malik anisotropy functions, which give better results on impulse noise filtering [4]. However, such alterations are much more complicated and resource-consuming for the numerical calculation.

2.4. Filter modification

Two parameters are needed for the creation of an effective filter. The time step Δt can be interpreted as the desired one-step image details blurring or flattering level. The larger the time step, the less processed image minuteness. This means that small specific areas in the image will merge with the large ones containing them or adjacent to them. Parameter γ (scalar or vector) in functions (2.3) can be viewed as a noisiness level. For better results, it is proposed to conduct several filtering iterations with different parameters of the anisotropy function with the constant time step Δt.

Let us define the local maximum and minimum differences of intensities for an arbitrary point of the image at the time step p:

$$\eta_p(r) = \min_{v \in Q(r)} \{|I_v^p - I_r^p|\}, \quad \theta_p(r) = \max_{v \in Q(r)} \{|I_v^p - I_r^p|\}, \tag{2.11}$$

where $Q(r)$ is an intensity value for the set of N, S, E, W, D, U adjacent to this point neighbors.

The criteria for the point to be on edge Γ or belong to the flat region in Ω is:

$$\Gamma_p(r,\gamma) = \frac{g_\gamma(\theta_p(r))}{g_\gamma(\eta_p(r))}, \quad \Omega_p(r,\gamma) = g_\gamma(\eta_p(r)) - g_\gamma(\theta_p(r)). \tag{2.12}$$

Combination of these criteria into one can be an estimator of the classification ambiguity degree:

$$\phi_p(r,\gamma) = 2\min\big(\Gamma_p(r,\gamma), \Omega_p(r,\gamma)\big). \tag{2.13}$$

It can be shown that $0 \le \phi_p(r,\gamma) \le 1$. From inequality $0 \le g_\gamma(x) < 1$ can be easily deduced that $\phi_p(r,\gamma) \ge 0$. Let r be an arbitrary image point, and p is a number of filter iterations. Two cases are needed to be considered. First - let $\Omega_p(r,\gamma) \le \Gamma_p(r,\gamma)$, then $\phi_p(r,\gamma) = 2\Omega_p(r,\gamma)$ and $\Omega_p(r,\gamma) = g_\gamma(\eta_p(r))\Big(1 - \Gamma_p(r,\gamma)\Big) \le 1 - \Gamma_p(r,\gamma)$. The result of summing these two inequalities is $\Omega_p(r,\gamma) \le 1/2 \Longrightarrow \phi_p(r,\gamma) \le 1$. Second - let $\Gamma_p(r,\gamma) \le \Omega_p(r,\gamma)$, then $\phi_p(r,\gamma) = 2\Gamma_p(r,\gamma)$ and $\Gamma_p(r,\gamma) = 1 - \frac{\Omega_p(r,\gamma)}{g_\gamma(\eta_p(r))} \le 1 - \Omega_p(r,\gamma)$. Again summing these two inequalities results in $\Gamma_p(r,\gamma) \le 1/2 \Longrightarrow \phi_p(r,\gamma) \le 1$. As the point r was arbitrary therefore $(\forall r)(\phi_p(r,\gamma) \le 1)$.

If image filtration is needed with the desired flattering Δt and noisiness γ_0 levels. Then the ambiguity degree of an arbitrary image point will be $\phi_p(r,\gamma_0)$. One of the possible definitions of the ambiguity degree of the total image after p iterations with fixed γ_0 is:

$$\Phi_p(\gamma_0) = \underset{r}{\mathrm{median}}(\phi_p(r,\gamma_0)). \tag{2.14}$$

Filtration must continue until $\Phi_{p+1}(\gamma_0) \approx \Phi_p(\gamma_0)$ or $\Phi_{p+1}(\gamma_0)$ reaches the desired level.

Obviously, in the process of filtration, the noise level in the image will decrease. Therefore, it is necessary to change the parameter γ after each or some number of iterations. The computational experiment showed that the local value of the parameter $\tilde{\gamma}$ for an arbitrary image point r for the next time step is better to choose as a root of the equation:

$$g_{\tilde{\gamma}_{p+1}}(\theta_p(r)) = \phi_p(r,\gamma_0). \tag{2.15}$$

This choice expresses a natural desire to have a small diffusion on the domain internal boundaries. For the next filtration step, choose γ_{p+1}, which corresponds to the necessary blurring of all points in the image: $\gamma_{p+1} = \underset{r}{\mathrm{median}}(\tilde{\gamma}_{p+1}(r))$. The median estimator selection is justified by its stability to the existence of some points with a high noise level (outliers).

2.5. Numerical result and parallel implementation

It can be seen in Figure 2 that after applying filtration to the core plug CT image, pores edges become smoother because the filter removed impulse noise from the image. At the same time, the edges became sharper as the filter has flattened pixels intensity values both for pores and for solid-phase areas. Such flattening was achieved by selecting the appropriate filter parameters. The filter simulates the diffusion process in non-edge image areas. The nature of the blur can be considered to be the same type as for the blurring generated by the Gaussian filter [2].

During filtration, the global ambiguity of pixels decreases, which can diminish the computational complexity of the next step in the void space of the core plug generation

IOP Conf. Series: Materials Science and Engineering **700** (2019) 012053 doi:10.1088/1757-899X/700/1/012053

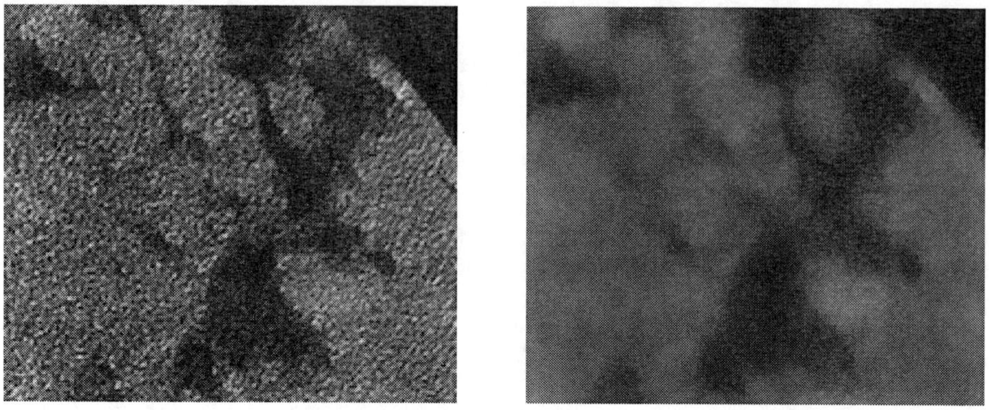

(*a*) (*b*)

Figure 2. (*a*) part of the original 2D slice image; (*b*) filtered image $\Delta t = 5$, $\gamma = 25$.

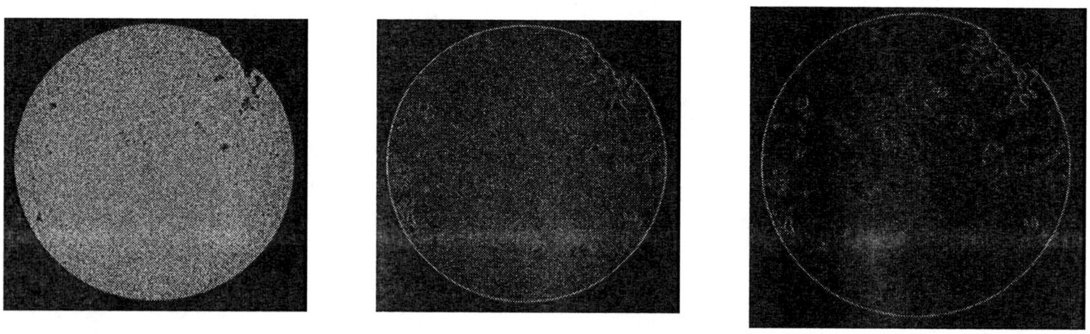

Figure 3. Change of the 2D slice image local ambiguity in the process of filter iterations (from left to right); black color means that the point is binary-determined (can be in plateau or in the edge), white color – vague points.

for hydrodynamic simulation – binary segmentation. Figure 3 visualizes the change of local ambiguity for every image pixel. For the plateau pixels (void space and porous medium i.e. solid-state skeleton) ambiguity approaches zero making their segmentation simpler and possibly much cheaper in terms of computational resources while the high ambiguity remains only for the edge pixels. This behavior of the ambiguity degree should be examined for possible usage in edge detection and also for the fast and effective segmentation.

Specialists in the experimental measurement of petrophysical rock parameters estimate filtration results as quite reasonable. However, any such assessment is subjective. For the objective evaluation, it is necessary to compare fluid flow simulation conducted on the digital core plug to the results of permeability and porosity tests executed in a laboratory.

The digital filter code was specially written for HPC usage under the paradigm "MPI+OpenACC" with implementation in Fortran. For codding, the most common OpenACC pragmas were used for better code portability across different CPU, heterogenic CPU and GPGPU architectures and various OpenACC compatible compilers. The proposed algorithm can be divided into two main sequential blocks: matrix generation for the linear algebraic system and filter parameters estimation, and then the calculation of the approximate solution for this system with the PCG method. Each of these two blocks can be easily paralleled. Independent

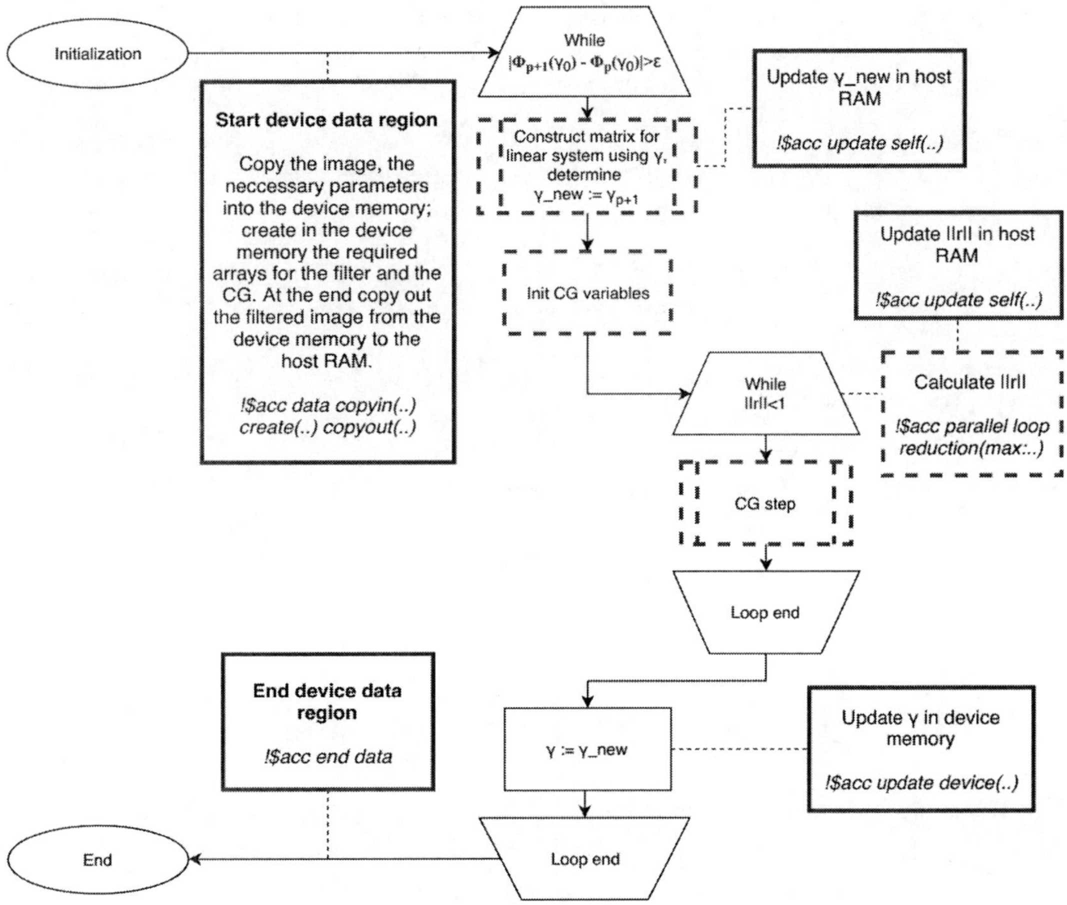

Figure 4. Dashed border is used for blocks that paralleled with directives *!$acc parallel* or *!$acc kernels*. The solid border is applied for blocks containing comments on the OpenACC usage.

threads associated with every image pixel can parallel the matrix generation process that gives a tremendous performance potential for GPGPU type processors. Due to the sparseness and symmetry of the matrix wide range of different storage schemes can be used. The flowchart of the parallelization algorithm for the proposed method is represented in Figure 4.

3. Filters based on an integral transformation

The quality and computational efficiency of the segmentation process depends on the level of noise in the digital image and the level of its high granularity. Therefore, it is necessary to pre-filter the CT sinogram – the result of the inverse Radon transform. This process will not only allow to remove the noise of different physical and computational nature but also perform a "preliminary segmentation" or diminish the high granularity of the digital core plug.

The positive feature of filters based on orthogonal transforms is the speed of numerical calculations. It is especially crucial if the raw data exceeds several hundred gigabytes.

This work aims to develop fast parallel algorithms for 3D digital images total filtration rather than filter a sequence (pack) of 2D sections further combined into the 3D image. This approach is more demanding for computer memory. However, it allows to get rid of the vertical orientation effect that occurs if filtering packs of planar slices. It is essential since this effect can lead to a severe anisotropy of petrophysical parameters fields obtained by computer simulation of various physical processes on 3D digital models of porous medium core plugs.

For initial testing, the direct and inverse discrete Fourier (DFT), and cosine (DCT) transforms, defined on the 3D mesh ($N \times M \times L$) were used. Direct transforms are:

$$\hat{I}(\hat{x},\hat{y},\hat{z}) = \sum_{x=0}^{N-1}\sum_{y=0}^{M-1}\sum_{z=0}^{L-1} I(x,y,z)\exp\left(-2\pi i\left(\frac{x\hat{x}}{N}+\frac{y\hat{y}}{M}+\frac{z\hat{z}}{L}\right)\right),$$

$$\hat{I}(\hat{x},\hat{y},\hat{z}) = \phi(\hat{x},N)\phi(\hat{y},M)\phi(\hat{z},L)\sum_{x=0}^{N-1}\sum_{y=0}^{M-1}\sum_{z=0}^{L-1} I(x,y,z)cos\left[A(x,\hat{x})\right]cos\left[B(x,\hat{x})\right]cos\left[C(x,\hat{x})\right], \tag{3.1}$$

where $A(x,\hat{x}) = \hat{x}(2x+1)\frac{\pi}{2N}$, $B(x,\hat{x}) = \hat{y}(2y+1)\frac{\pi}{2M}$, $C(x,\hat{x}) = \hat{z}(2z+1)\frac{\pi}{2L}$.

Inverse transformations for them are:

$$I(x,y,z) = \frac{1}{MNL}\sum_{\hat{x}=0}^{N-1}\sum_{\hat{y}=0}^{M-1}\sum_{\hat{z}=0}^{L-1} \hat{I}(\hat{x},\hat{y},\hat{z})\exp\left(2\pi i\left(\frac{x\hat{x}}{N}+\frac{y\hat{y}}{M}+\frac{z\hat{z}}{L}\right)\right),$$

$$I(x,y,z) = \phi(x,N)\phi(y,M)\phi(z,L)\sum_{\hat{x}=0}^{N-1}\sum_{\hat{y}=0}^{M-1}\sum_{\hat{z}=0}^{L-1} \hat{I}(\hat{x},\hat{y},\hat{z})cos\left[A(\hat{x},x)\right]cos\left[B(\hat{x},x)\right]cos\left[C(\hat{x},x)\right], \tag{3.2}$$

where

$$\phi(\xi,K) = \begin{cases} \frac{1}{\sqrt{K}}, & \xi = 0 \\ \sqrt{\frac{2}{K}}, & 1 \le \xi \le K-1 \end{cases}. \tag{3.3}$$

DCT can be expressed in terms of DFT [8] as:

$$\hat{u_k} = Re\{\phi(k,N)\cdot\exp\left(-j\frac{\pi k}{2N}\right)\cdot\mathcal{F}(y_k)\}, \tag{3.4}$$

where

$$\begin{cases} y_n = u_{2n}, & 0 \le n \le \frac{N}{2}-1 \\ y_{N-n-1} = u_{2n+1}, & 0 \le n \le \frac{N}{2}-1 \end{cases}. \tag{3.5}$$

A distinctive feature of these transformations is the ability to represent a multidimensional Fourier transform as a finite sequence of one-dimensional (here, $\mathcal{F}_i$ – DFT):

$$\hat{u}_{k_0,k_1,k_2} = \mathcal{F}_0(\mathcal{F}_1(\mathcal{F}_2(u_{j_0,j_1,j_2}))). \tag{3.6}$$

Illustration of the cosine transform-based filter is presented on a simple model with regular additive noise (Figure 5).

3.1. 3D orientation effect

Example of 3D orientation effect, generated by the conventional 2D slice by slice pack filtering is shown in (Figure 6 (*a*)). Synthetic input data was generated by superimposing regular additive noise on the 3D image created from a pack of identical 2D slices from the previous example. Then conventional 2D slice filtering was performed on a rectangular cuboid. From the resulting 3D array, the vertical section was taken. The presence of regular noise is visible on it.

For suppression of the "ort-directional" residual noise, it is necessary to apply 3D parallel filters created with the usage of some discrete transform. The selection of an integral transform depends on several properties: edge keeping, computational effectiveness, and easiness for

Figure 5. (*a*) original image, (*b*) noise disturbed image, (*c*) image, filtered by DCT.

Figure 6. vertical sample section: (*a*) after 2D slice filtering, and (*b*) after 3D total filtration.

parallelization. Figure 6 (*b*) presents the result of filtering based on the three-dimensional discrete cosine transform. This filtration severely diminishes regular noise.

For the filters mentioned above, there exist fast transformation algorithms: Fast Fourier Transform (FFT), Fast Discrete Cosine Transform (FDCT). The usage of fast transforms can significantly increase computational efficiency. Besides, they allow rather simple parallelizing. Furthermore, procedures can be called from precompiled libraries adapted for specific supercomputer architectures like xMath library for the SW26010 processor.

3.2. Parallel implementation

The amount of memory storing total 3D image raw data for further processing has significantly increased and needs the usage of high-performance computational systems. Therefore some "universal" approach for conducting calculations on different supercomputer platforms is needed.

One of the most advantageous programming paradigms is "MPI + OpenACC". The latest versions of the OpenACC allows to use it on both multicore and many-core processors. Also, it was adopted for several different computational accelerators. This approach gives the possibility to use the full computational power of modern GPGPUs (Nvidia, AMD), numerical accelerators such as Intel Xeon Phi, and heterogenic processors as Sunway SW26010. This paradigm provides painless portability from one high-performance computational system to another.

The basic idea of parallelizing fast multidimensional cosine transform for the shared/distributed memory systems is based on the following facts: DCT can be calculated by using FFT [8]; multidimensional transform can be executed by superposition of one-dimensional FFT [9]; existance of architecture-dependent parallelized 1D FFT precompiled routines [8].

Let us consider the implementation of this approach on the SW26010 processor, which can serve as a guinea pig for testing different algorithms. First, divide the input data into four parts for each MPE core interconnected by MPI. Second, on MPE cores with CPE matrix calculate FFT along X and Y axes for all subblocks of the 3D image. In switching from X to Y direction, it is necessary to transpose the array of pixels to speed up the calculation cycles. Third, conduct transform in the Z direction. For this, a new 3D image subdivision is needed, which induce massive amounts of data exchange between individual nodes (MPE cores). This data transfer can be executed by MPI_Alltoall function.

Sunway SW26010 processor memory is hierarchical; each MPE core has access to 8GB of main memory, 32KB L1 data cache, and 256KB L2 data/instruction cache. Each CPE core has 16 KB L1 instruction cache and 64KB Local Device Memory (LDM). Contrary to the cache memory, LDM can be managed explicitly by a programmer. The data movement between the LDM and the main memory is performed through Direct Memory Access (DMA) to guarantee high transfer efficiency.

In the Sunway version swOpenACC, they included several principal extensions that are new clauses, which can be of interest for parallel 3D FDCT calculation. The most important are:

- *local* – clause, to allocate space on the fast LDM memory of CPE thread;
- *swap/swapin/swapout* – clauses, to transpose array along specified index.

The *swap* directive which allows loading arrays transposed relative to a given index directly to the fast LDM memory significantly speeds up 2D blocks FFT computation.

All the above examples were calculated on the SW26010 processor using the previously mentioned approach.

Conclusion

Executed computer experiments show that conventional slice filtering is not suitable for processing tomographic 3D core plug images because of generated directional orientation error. Moreover, this does not depend on the type of edge-preserving filter. Two 3D filtering approaches have been proposed to improve the quality and speed up subsequent 3D image binary segmentation. For PDE type filtration, the modified implicit anisotropic diffusion filter, filtration stopping criteria, and an algorithm of estimating filter parameters were proposed. Criteria of local and global ambiguity can be used to estimate filtration quality. For the integral transform-based filters a template for further parallelizing them on heterogeneous computer architectures was created. All discussed filters were coded with the paradigm: Fortran + OpenACC + MPI. This approach increases program portability. Prototype programs were tested on cluster nodes equipped with Intel+Nvidia processors and on SW26010 heterogeneous processors.

Acknowledgments

We want to appreciate the National Supercomputing Center, in Wuxi, China, for giving us the possibility to conduct our calculation on the Sunway TaihuLight supercomputer.

References

[1] Pak T, Archilha N, Mantovani I and Butler I 2019 *Scientific data* **6** ISSN 2052-4463
[2] Perona P and Malik J 1990 *IEEE Trans. Pattern Anal. Mach. Intell.* **12** 629–639
[3] Canny J 1990 *IEEE Trans. on Pattern Analysis and Machine Intelligence* **8**(6) 679–698
[4] Michel-González E, Cho M H and Lee S Y 2011 vol 10 (BioMed Central) p 47
[5] Tauberab C *et al.* 1990 *Applied Numerical Mathematics* **60**(11) 1115–1130
[6] Rozhdestvenskii B and Yanenko N 1968 *Systems of quasilinear equations [Sistemy kvazilineinykh uravnenii]* (Moscow: Nauka)
[7] Arseniev-Obraztsov S S and Plusch G O 2019 *Automation, telemechanization and communication in oil industry* **1**(546) 30–40
[8] Plonka G, Potts D, Steidl G and Tasche M 2018 *Numerical Fourier Analysis* (Springer)
[9] Pratt W K 1991 *Digital Image Processing, New-York* (NY: John Wiley and Sons)

Field development optimization of waterflooding process using data assimilation methods

R Guliev[1*] and A Zolotukhin[1, 2, 3]

[1] Northern (Arctic) Federal University named after M.V. Lomonosov, Severnaya Dvina Emb. 17, 163002, Arkhangelsk, Russia
[2] Gubkin Russian State University of Oil and Gas (National Research University), Leninsky prospect 65, 119991, Moscow, Russia
[3] The University of Stavanger, P.O. Box 8600 Forus N-4036 Stavanger, Norway

* Corresponding author: ramilguliev1@gmail.com

Abstract. The application of data assimilation methods for field development optimization has been the subject of intense investigation during the past 10 years. Lately has seen remarkable progress in the ability of data assimilation approach in reservoir characterization and based on this, improvements of field development optimization. In this review paper, we have summarized key achievements in field development optimization of waterflooding process with data assimilation approach and review many of the achievements of the past time, including developments in the field of search for modifications of Ensemble Kalman Filter (EnKF) and Ensemble Smoothers (ES). An attempt has been made to discuss different data assimilation methods and to identify possible limitations of each. Current challenges and future research opportunities for improved data assimilation methods for field development optimization of waterflooding process are also discussed.

1. Introduction

Different studies have shown that model-based dynamic optimization of the waterflooding process improves the economic life-cycle performance of oil fields, see e.g., [1, 2]. One of the main challenges in this optimization is the high levels of uncertainty increasing from the modeling process of waterflooding. As a result, the potential advantages of dynamic optimization are not fully realized and the optimized objective value is not obtained.

Different approaches of waterflooding optimization under uncertainty can be broadly divided into two types. In the first type, also known as open-loop schemes, a decision maker selects a strategy without knowing the exact values taken by the uncertain parameters, and the exact values are assumed to belong to an uncertainty space. In the second type, also known as closed-loop (Closed-Loop Reservoir Management (CLRM)), the strategy is allowed to update/adjust to information that is revealed over time. This article discusses only CLRM type of waterflooding optimization.

Uncertainty quantification of the uncertainty space is one of the essential steps in CLRM. A general practice of quantifying uncertainty in waterflooding optimization in CLRM is implementation of data assimilation methods such as Ensemble Kalman Filtering (EnKF), Ensemble Kalman Smoothing (EnKS), Particle Filtering, variational approaches etc. The purpose of this work is to overview different types of data assimilation methods in task of closed-loop reservoir management and identify limitations of them.

The article is structured as follows. Section 2 discusses the basics of closed-loop reservoir management optimization. In Section 3, is discussed different data assimilation methods followed by conclusions in section 4 with discussion of current challenges and future research opportunities for improved data assimilation methods for field development optimization of waterflooding process.

2. Closed-Loop Reservoir Management

Closed-loop reservoir management (CLRM) is a cycle of production optimization and data assimilation (see Fig. 1). Aim of cycle of optimization at finding maximum of value. e.g. financial measure (Net Present Value (NPV)), over the producing period of the reservoir by optimizing the production parameters. For more information about CLRM see, e.g., [3], [4], and [5].

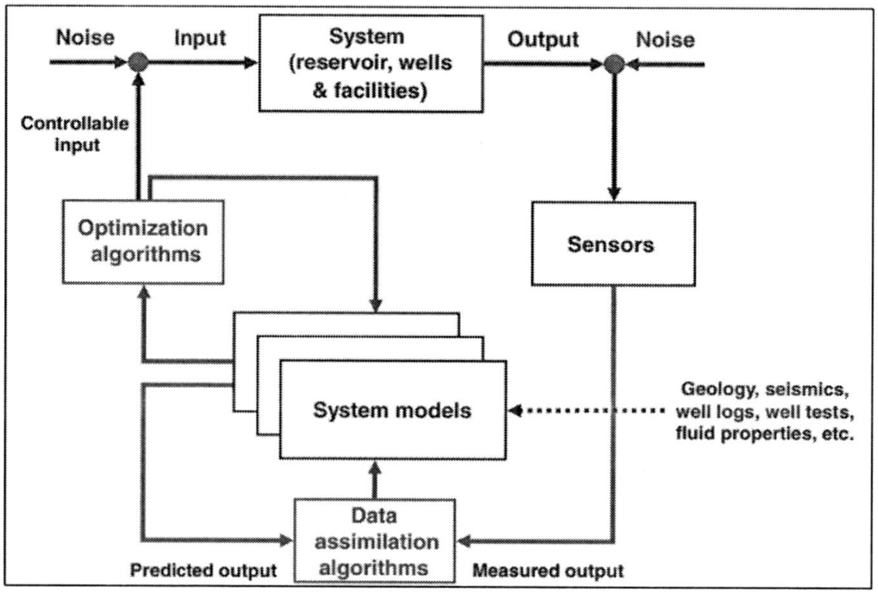

Figure 1. Closed-loop reservoir management (Source: [3])

NPV can be represented as:

$$J = \sum_{k=1}^{K} \left[\frac{r_o \cdot q_{o,k} - r_w \cdot q_{w,k} - r_{inj,k} \cdot q_{inj,k}}{(1+b)^{\frac{t_k}{\tau_t}}} \cdot \Delta t_k \right], \tag{1}$$

where r_o , r_w and r_{inj} are the oil price, the water production cost and the water injection cost, respectively. K represents the production period and Δt_k is the time interval of time step k . b is the discount rate for a certain time τ_t . $q_{o,k}$, $q_{w,k}$ and $q_{inj,k}$ is the cumulative production of oil, water and injected water at time step k .

Uncertainty is one of the main aspects of the model-based optimization of the waterflooding process. To quantify the uncertainty space Θ in waterflooding optimization, usually use an ensemble of uncertain model realizations. This set of ensemble-based uncertainty can be used with various schemes. One of the robust approaches is to maximize the average of the NPV objective over the model uncertainty ensemble [6].

IOP Conf. Series: Materials Science and Engineering **700** (2019) 012054 doi:10.1088/1757-899X/700/1/012054

Robust optimization can be formulated as:

$$J_O = \frac{1}{N_e}\sum_{k=1}^{N} J_i(\mathbf{u},\boldsymbol{\theta}_i), \qquad (2)$$

where J_i is the NPV objective and $\mathbf{u}$ is the input decision variable, $\boldsymbol{\theta}_i$ is a vector of uncertain model parameters.

Now we can formulate the optimization problem as finding a vector $\mathbf{u}$ including the set of the control parameters over the producing period of the reservoir. It should be noted, that although the optimization is based on N_e models, only a single strategy $\mathbf{u}$ is obtained. Typical elements of $\mathbf{u}$ are settings of well head pressures, water injection rates, valve openings, etc. Sometimes the problem is very nonlinear and nonconvex, i.e. it has multiple local maxima.

3. Data Assimilation

As we mention before, uncertainty quantification is one of the essential steps in closed-loop reservoir management. One of the types of uncertainty is ambiguity of geological properties. A prior distribution of the geological properties can be generated using various geostatistical techniques. Also, various measurements (y) suitable to unknown model parameters (θ) are available from surface and sensors. From a probabilistic point of view, Bayesian inference provides a good platform to compute the posterior probability density function (PDF) [7]:

$$p(\theta \mid y) = \frac{p(y \mid \theta)p_{pr}(\theta)}{p(y)}, \qquad (3)$$

where $p_{pr}(\theta)$ is the prior distribution, $p(y)$ is the density function of the measurements and $p(y \mid \theta)$ is the likelihood function for obtaining y given the model parameters θ. In our CLMR case, prior and posterior model parameters θ represented prior and posterior ensembles, respectively.
The main assumption in data assimilation is reducing uncertainty of the model parameters leads to an improved forecast capacity of the models, which, in turn, leads to improved decisions. In our CLRM case, decisions take the form of control vectors $\mathbf{u}$, aimed at finding maximum of the objective function J_O.

3.1 Ensemble Kalman Filter

Recently, the Ensemble Kalman Filter has gained popularity as a data assimilation method. EnKF is based on the Kalman filter [8], but simple Kalman filter can't be implemented for data assimilation of large models, especially with non-linear components.

Below shown the algorithm of EnKF implementation:

1. Generate N_e samples $\theta_i^{a,j}$ and $J_i^{a,j}$, $j = 1..N_e$ from the prior distribution of $p_{pr}(\theta)$ and the initial distribution of J.

2. For $i = 1..N_e$,

2.1 Generate N_e independent samples of J_i by integrating each sample from t_{i-1} to t_i using the forward model. In J_i, the most updated value for $p_{pr}(\theta)$, is applied. That is, for $j = 1..N_e$:

$$J_i^{f,j} = \begin{Bmatrix} J_i^{f,j} \\ \theta_i^{f,j} \\ g_i^{f,j} \end{Bmatrix} = \begin{Bmatrix} J_{i-1}^{f,j} \\ F(J_{i-1}^{f,j},\theta_{i-1}^{f,j}) \\ g(J_{i-1}^{f,j},\theta_{i-1}^{f,j}) \end{Bmatrix}, \qquad (4)$$

IOP Conf. Series: Materials Science and Engineering **700** (2019) 012054 doi:10.1088/1757-899X/700/1/012054

where F_i denotes the forward integration of equations of dynamic model from step of time $i-1$ to i and g_i denotes set of measurements. $\theta_i^{f,j}$ represents a sample from prior distribution). Superscript a denotes posterior solution from previous step.
2.2 Use Monte-Carlo approach for estimation of the first- and second-order moments from prior, using ensemble mean of matrix $\mathbf{Y}_i$ with each ensemble members

$$\overline{\mathbf{Y}}_i = \{\overline{J}_i^1..\overline{J}_i^{N_e}\} \tag{5}$$

and using covariance matrix:

$$\mathbf{C}_{J_i} = \frac{\Delta\mathbf{Y}_i\Delta\mathbf{Y}_i^T}{N_e - 1} \tag{6}$$

2.3 Generate N_e independent samples from the measurements distribution:

$$y_i^j = y_{obs,i} + \varepsilon_i^{d,j}, j = 1..N_e \tag{7}$$

2.4 Update each ensemble member using the traditional Kalman filter update:

$$J_i^a = J_i^f + \mathbf{K}_i(y_{obs,i} - \mathbf{H}J_i^f) \tag{8}$$

$$\mathbf{K}_i = \mathbf{C}_{J_i^f}\mathbf{H}^T(\mathbf{H}\mathbf{C}_{J_i^f}\mathbf{H}^T + \mathbf{C}_{y_i})^{-1}, \tag{9}$$

$$J_i^{a,j} = J_i^{f,j} + \mathbf{C}_{J_i^f}\mathbf{H}^T(\mathbf{H}_i\mathbf{C}_{J_i^f}\mathbf{H}^T + \mathbf{C}_{y_i})^{-1}, j = 1..N_e \tag{10}$$

where ε_i denotes measurements error, $\mathbf{K}$ denotes Kalman gain matrix, $\mathbf{H}$ denotes relationship between measurements and states.

Examples of implementation of EnKF: Consider a several examples of implementation EnKF: assimilating data of production history, assimilating seismic data and implementation of EnKF in CLRM.

Haugen et al. [9] implemented EnKF to improve history matching from production data from North Sea field. They used large nonlinear reservoir model. This work has been showed that we can improve model parameters (permeability and porosity) by assimilating production data such as bottom hole pressure and production rates of oil, gas, and water. It has been demonstrated that by using the EnKF, a better history-matched model could be obtained. In work [10] used to assimilation porosity fields used to model an oil reservoir with almost three years of production. Work was confirmed the possibility to use EnKF for history matching of real reservoir models.

Second group of works with implementation of EnKF considers works related with implementation of EnKF for data assimilation of seismic data. Skjervheim et al. [11] studied EnKF inversion scheme to assimilate interpreted seismic data into reservoir simulation models for both a 2D model and a real field and got ambiguous results.

Third and most interesting for us case is implementation of EnKF in CLMR. Chen et al. [12] implemented EnKF in closed-loop reservoir management for improvement waterflooding process. EnKF was in couple with optimization algorithm – EnOpt. The aim of this work was to maximize NPV through management of control devices.

The applicability of the EnOpt algorithm and the ensemble-based CLRM is assessed through the use of two examples. In the first example, the control devices are optimized by the EnOpt based on a known reservoir model. The results showed the ability of the EnOpt to improve reservoir management in the presence of complex features. The ensemble-based CLRM is demonstrated in the second example and the results are compared with other alternative control strategies. A good estimate of the permeability field (Fig. 2), which captured the features of the reference field. The net present value of the field is significantly increased by the CLRM.

IOP Conf. Series: Materials Science and Engineering **700** (2019) 012054 doi:10.1088/1757-899X/700/1/012054

3.2 Ensemble Smoothers

The Kalman filter is a recursive filter with the Markov property – it's estimate at step k is based only on the estimate from step $k-1$ and the measurement at step k . But this means that the estimate from step $k-1$ is based on step $k-2$, and so on back to the first epoch. Hence, the estimate at step k depends on all of the previous measurements, though to varying degrees. $k-1$ has the most influence, $k-2$ has the next most, and so on.

Smoothing filters incorporate future measurements into the estimate for step k The measurement from $k+1$ will have the most effect, $k+2$ will have less effect, $k+3$ less yet, and so on.
There are different types of ensemble smoothers: Fixed-Interval Smoothing, Fixed-Lag Smoothing, Fixed-Point Smoothing and more algorithms of smoothing and this is a topic for discussion of future research of ensemble smoothers.

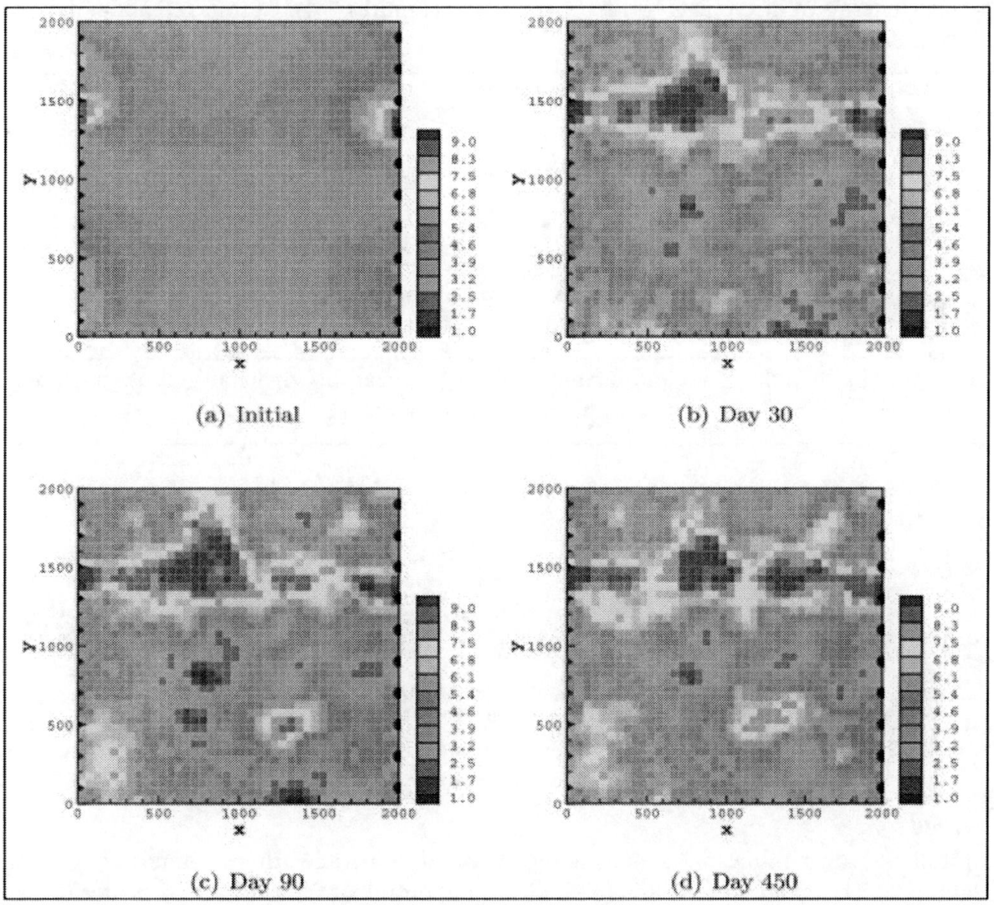

Figure 2. Mean of the initial $\ln k$ ensemble and mean of $\ln k$ ensemble updated at three different data times (Source: [12])

In our work we discuss results of implementation of the regularized Levenberg-Marquardt algorithm derived from minimum average cost (RLM-MAC [13]).

In study [13], the iterative Ensemble Smoother was applied to condition permeability field to well test data. 1D and 2D synthetic reservoir models were used to investigate the method performance with respect to measurement error and localization which we consider important from practical point of view.

At first, authors analyzed the influence of measurement error. Despite of high accuracy of the modern pressure gauges, the pressure data are often quite noisy. In practice, various filtering, denoising and smoothing techniques are employed in order to clarify the reservoir response and reduce data uncertainty. They evaluated several cases with different variance of measurement error. The comparison revealed that the pressure data noise has strong impact on the parameter estimation and the method convergence. In many cases, the noise caused the ensemble drifting away from the true solution.

Another important practical aspect is localization of model updates. During well test, pressure transients reflect pressure propagation away from the well. The propagation dynamic is governed by the formation properties within the disturbed reservoir domain. Therefore, the pressure measurements at a given time may be used for updating formation properties in the model only within the disturbed domain around the well. This would lead to the conclusion that a localization technique may be employed to relate model updates to relevant observations representing response from different reservoir areas. A time/distance dependent localization technique was tested to address this problem. The testing results showed that the proposed localization technique allowed for better estimation of permeability distribution (in terms of discrepancy with the true case) (Figures 3-4). Results tell about potentially good implementation of this approach in CLRM.

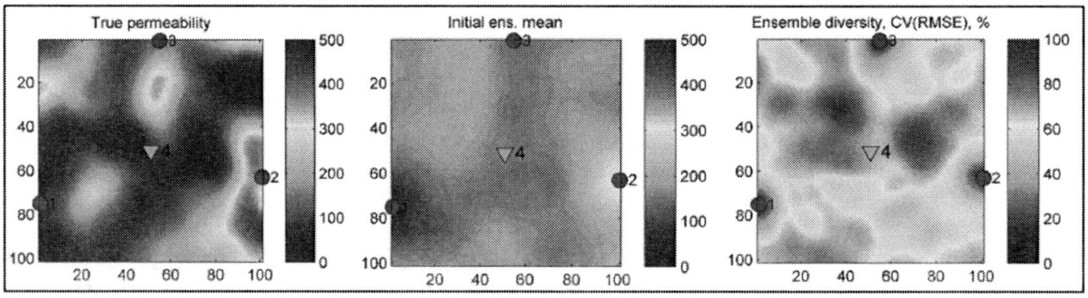

Figure 3. Initial ensemble (from left to right): a) true permeability distribution; b) ensemble mean; c) ensemble diversity (Source: [13])

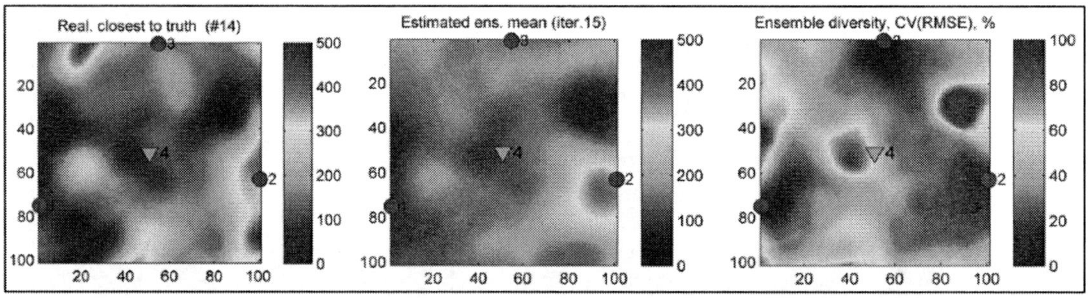

Figure 4. Conditioning with temporary localization (from left to right): a) realization closest to the true case; b) ensemble mean; c) ensemble diversity. (Source: [13])

3.3. Particle Filters

In the problem with nonlinear system, another method – particle filters offers the solution [14] for assimilate data. The main idea of this method is to approximate the prior and posterior probabilities using mixture of delta functions. The posterior is approximated by:

$$p(\theta \mid y) \approx \sum_{i=1}^{n} w_i \delta(\theta - \theta_i^a), \tag{11}$$

where θ_i^a are the particles at the update step that are needed to represent the whole state space, w_i are the corresponding weights, and n is the particle size. The particle filter method consists of the following steps:

IOP Conf. Series: Materials Science and Engineering **700** (2019) 012054 doi:10.1088/1757-899X/700/1/012054

Prediction step: The particles θ_i^a are evolved forward in time using the dynamical model to obtain the forecast estimates θ at the next time step. The corresponding weights w_i remain the same.

Filtering step: The particles remain the same. The weights are updated when receiving a new observation using Bayes rule.

Re-sampling step: In practice, the particle filter may have problems related to the weights known as weight collapse [14], especially for high-dimensional problems. You may have seen results of implementation of particle filtering in work [15] (Fig. 5-6).

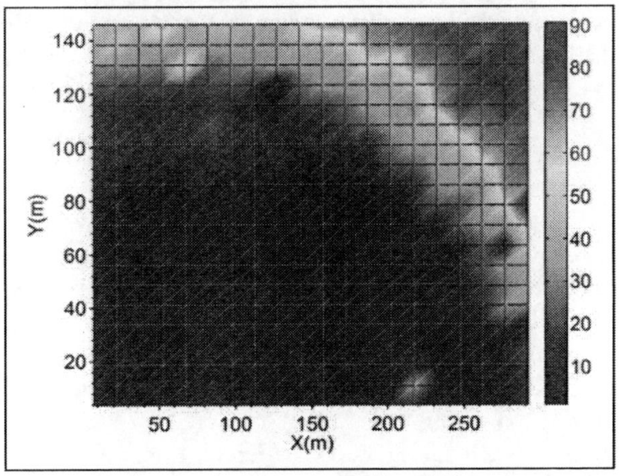

Figure 5. An initial ensemble of the absolute permeability field (Source: [15])

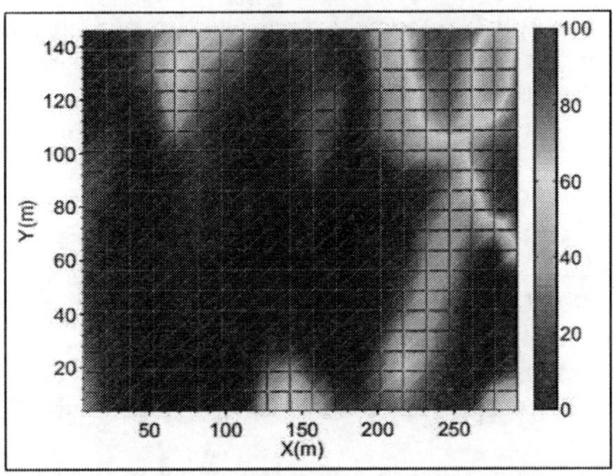

Figure 6. Estimates of absolute permeability fields at final time step using PF (Source: [15])

As can be seen, the permeability estimate is quite good, but it is necessary to conduct more numerical experiments to analyze the effectiveness of particle filters, but however this approach is applicable for CLRM.

4. Conclusions

Different methods, algorithms and approaches can be applied to assimilate different types of data (production data, well test data, seismic data).

- The most studied approach – Ensemble Kalman Filtering (EnKF) has very good results in data assimilation, but there is no sufficient research for evaluation of different schemes of EnKF such as: stochastic EnKF, singular evolutive interpolated KF, etc.

- Regarding the ensemble smoothing method the following should be noted: more research needed for evaluating the spatial/ temporal localization of the gain matrix for preventing ensemble collapse and retained ensemble diversity; studies about temporal localization based on the maximal extent of the disturbed domain over the entire ensemble; update of the localization matrix after each iteration.
- Particle filters contain the promise of fully nonlinear data assimilation. Method have been applied in numerous science areas, including geoscience, however, application of PF to high-dimensional geoscience systems has been limited due to its efficiency in high-dimensional systems in standard settings. However, huge progress has been made, and this limitation is disappearing fast due to recent developments in proposal densities, the use of ideas from (optimal) transportation, the use of localization and intelligent adaptive resampling strategies. However, more studies with implementation of PF to CLRM are required in the future.

References

[1] Brouwer D R and Jansen J D 2004 Dynamic Optimization of Waterflooding With Smart Wells Using Optimal Control Theory *SPE J.* **9** 391–402.
[2] Peters E, Arts R, Brouwer G, Geel C, Cullick S, Lorentzen R, Dunlop K, Vossepoel F, International S, Xu R, Sarma P, University S, Alhutali A and Reynolds A 2010 Results of the Brugge Benchmark Study for Flooding Optimization and History Matching.
[3] Jansen J D, Siep P and Douma S D 2009 Closed-Loop Reservoir Management System, SPE Reservoir Simulation Symposium, 2-4 February, The Woodlands, Texas, SPE 119098.
[4] Nævdal G, Brouwer D R and Jansen J D 2006 Waterflooding using closed-loop control *Comput. Geosci.* **10** 37–60.
[5] Chen Y, Oliver D S and Zhang D 2009 Efficient Ensemble-Based Closed-Loop Production Optimization *SPE J.* **14** 634–45.
[6] van Essen G, Zandvliet M, Van den Hof P, Bosgra O and Jansen J-D 2009 Robust Waterflooding Optimization of Multiple Geological Scenarios *SPE J.* **14** 202–10.
[7] G. D. Barros E, Van den Hof P M J and Jansen J D 2016 Value of information in closed-loop reservoir management *Comput. Geosci.* **20** 737–49.
[8] Kalman R E 1960 A New Approach to Linear Filtering and Prediction Problems *J. Basic Eng.* **82** 35–45.
[9] Haugen V E J, Naevdal G, Natvik L-J, Evensen G, Berg A M and Flornes K M 2008 History Matching Using the Ensemble Kalman Filter on a North Sea Field Case *SPE J.* **13** 382–91
[10] Bianco A, Cominelli A, Dovera L, Naevdal G and Valles B 2007 History Matching and Production Forecast Uncertainty by Means of the Ensemble Kalman Filter: A Real Field Application *Eur. Conf. Exhib.* 10.
[11] Skjervheim J-A, Evensen G, Aanonsen S I, Ruud B O and Johansen T-A 2007 Incorporating 4D Seismic Data in Reservoir Simulation Models Using Ensemble Kalman Filter *SPE J.* **12** 282–92.
[12] Chen Y, Oliver D S, Zhang D and California S 2008 SPE 112873 Efficient Ensemble-Based Closed-Loop Production Optimization 19–23.
[13] Khrulenko A, Shchipanov A and Berenblyum R 2018 Conditioning Reservoir Models To Pressure Transients By Iterative Ensemble Smoother.
[14] Gordon N J, Salmond D J and Smith A F M 1993 Novel approach to nonlinear/non-Gaussian Bayesian state estimation *IEE Proc. F - Radar Signal Process.* **140** 107–13.
[15] Aman B M 2012 *Reservoir History Matching Using Ensemble Kalman Filters with Anamorphosis Transforms.*

Coupling well hydrodynamic tests data with results of digital simulation for better identification of reservoir properties

M Khaydina* **and A Latysheva**

Russian State Gubkin University of Oil and Gas, Russia

* Corresponding author: mmp2003@inbox.ru

Abstract. In the study the existing approaches for stabilization time estimation are given and compared with the results of hydrodynamic simulation. A deviation from an assumed ellipsoid shape of drainage area has been observed for a horizontal well based on the simulation results. Then this geometrical definite form was taken into account in the gas influx equation. On its basis the time period of pseudo-steady state was estimated. As a result, stabilized horizontal gas well tests are considered to be not effective as the time period is too long when pseudo-steady state is reached and it's impossible to carry out the test in reality. However, it is suggested to continue a well test for a reduced period by adapting well tests on the model and matching simulation output with gauge measurements.

1. Introduction

A goal of the study is to develop an approach for gas well tests interpretation using geologic-hydrodynamic simulation of field development. In this work stabilized well tests of horizontal gas wells are observed. Multipoint well tests are for stabilized ones.

These gas wells tests are conducted in order to find recovery rates' dependence of pressure empirically to find coefficients for flow equation which allow estimating flow properties of the reservoir. A condition to conduct practical measurements is the achievement of steady-state pressure distribution in a reservoir from a well to an external boundary where constant pressure is maintained. As for well tests interpretation a reservoir is assumed to be homogeneous, a steady state implies that pressure at each point is constant [1]. As there is no maintenance of reservoir pressure, during gas production reservoir pressure decreases, therefore a concept of pseudo-steady state is used [2]. In a homogeneous reservoir during pseudo-steady state any change of pressure at each point depends on time linearly [1]. It is difficult to achieve such measurement conduction conditions for a horizontal well: in low permeable reservoirs the period to reach stabilization state can take much more time than is considerable for well testing, when a well test goes on at the rate being not equal to a usual operational one for this well.

As the result for well test interpretation it is necessary to identify the time period to reach pseudo-steady state to compare it with one spent on real well tests carried out on a field.

There are several methods to estimate stabilization time.

Authors of the study [3] suggest to calculate it using the following equation:

$$t_s = \frac{1200 \cdot \phi \cdot S_g \cdot \mu_g \cdot r_e^2}{k \cdot \overline{p_r}}, \tag{1}$$

where t_s – time period to reach stabilization, hours; ϕ – porosity; μ_g – gas viscosity, cP; S_g –gas saturation; k – effective permeability for gas, mD; $\overline{p_r}$ – average reservoir pressure, pounds per square inch; r_e - distance to an external boundary of a reservoir, ft. This time means the moment when pressure disturbance has reached boundaries of a reservoir or boundaries of zones of drainage created by surrounding wells.

In the publications of Russian scientists [4, 5] it is assumed that a state can be considered as stabilized one when pressure disturbance has reached one half of a distance between the well and an external boundary of the reservoir. In this case an analytical approach for stabilization time is:

$$t_{st} = \frac{C \cdot R_k^2 \cdot \phi \cdot \mu}{k \cdot P_{res}}, \tag{2}$$

where C – a coefficient varying in a range of 0,122-0,35; R_k – a radius of a reservoir, m; t_{st} – time required for pressure stabilization, seconds; ϕ – porosity; μ – viscosity, mPa·s; k – permeability, m²; P_{res} – reservoir pressure, Pa.

A problem of an analytical estimation of stabilization time is that values of permeability and a distance from constant pressure boundary are unknown. In reality pressure gauge readings are usually taken when the change of registered parameters becomes minimal, when a change of pressure is less than a measurement error.

2. The 3D hydrodynamic simulation to identify the conditions of pseudo-steady state

It was decided to use hydrodynamic simulator Eclipse 100 (Schlumberger) to identify a moment of achievement of pseudo-steady state according to the results of the numerical experiment.

A thin (relative to a reservoir extension), homogeneous, isotopic reservoir with impermeable top and bottom and an edge aquifer was simulated. A reservoir is penetrated by a horizontal well with a downward well full formation exposing. Relation of a gross pay to length of a simulated fragment equals to 0,027. A well path is shown in Figure 1. Table 1 and Table 2 demonstrate the parameters of the simulated reservoir, parameters of the aquifer, respectively.

Work of the horizontal well with a constant rates at 500 thousand cubic meters per day and at 1 million cubic meters per day was initiated on the model; three variants of the simulated fragment have been observed: variant 1 – with permeability of 1 mD, variant 2 – with 5 mD, variant 3 – with 10 mD.

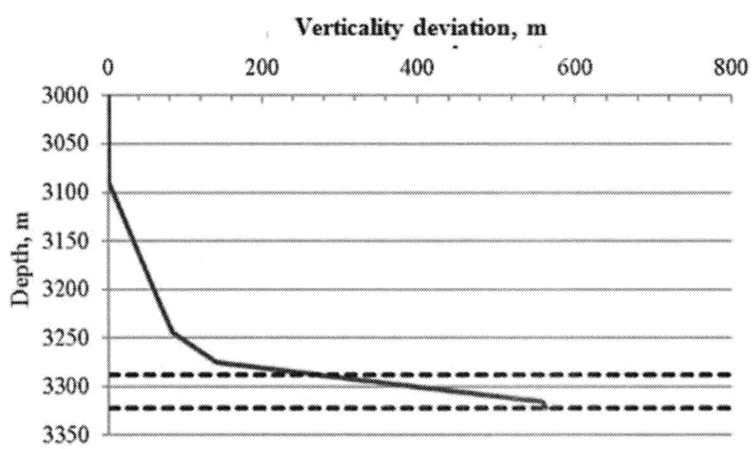

Figure 1. Well path of the tested well.

IOP Conf. Series: Materials Science and Engineering **700** (2019) 012055 doi:10.1088/1757-899X/700/1/012055

Table 1. Parameters of the simulated reservoir.

Parameter	Value
Number of cells	135x135x6
Size of cells, m	10x10x6
Net pay, m	29,2
Length of a horizontal part of the well, m	311
Depth of entrance of the well into reservoir, m	3287,4

Table 2. Parameters of aquifer.

Parameter	Value
Porosity. ϕ	0,125
Permeability, 1/Pa	14,504
External radius, m	152,4
Thickness, m	500
Angle of influence, degrees	360
Permeability, mD	equal to permeability of the fragment in every case

3. Analysis of simulation results

According to the simulation results it was identified that since the moment of well operation initiation a shape of drainage area was constantly changing. Its distinctive geometry samples are shown in Figure 2. Firstly, the boundary created inclined elliptical cylinder shape. Then it became straight elliptical cylinder. Truncated elliptical cone resulted the simulation output. These changes are logical and just prove the effect of mutual work of different physical forces. It means we can use the created shape sizes for special purposes. As well this data was used for well tests results interpretation.

It is worth mentioning that depending on reservoir properties of the simulated fragment a period when a drainage area is of a certain shape differs. For example, for 1 mD there was no transition of external drainage area boundary into a shape of a truncated cone. Different shapes of drainage areas correspond flow regimes towards a horizontal well after it started production (Figure 3) [6].

The first regime of flow towards a horizontal well is a pseudoradial flow in a vertical plane (Figure 3a). Pseudoradial flow has its name because of anisotropy: flow in a vertical direction moves slower than in a lateral one, therefore a shape of a perfect circle is absent.

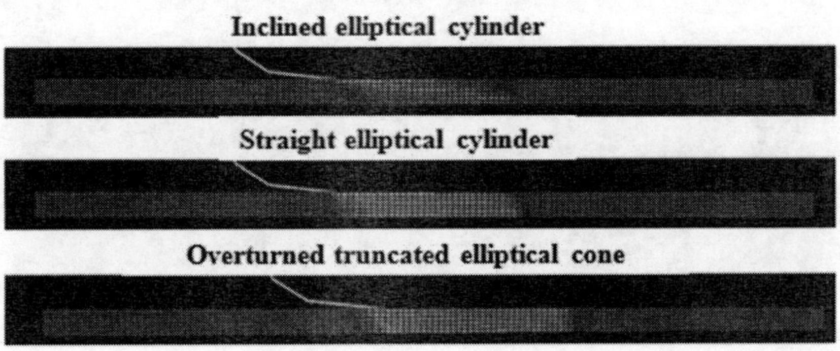

Figure 2. Shapes of drainage area during simulation

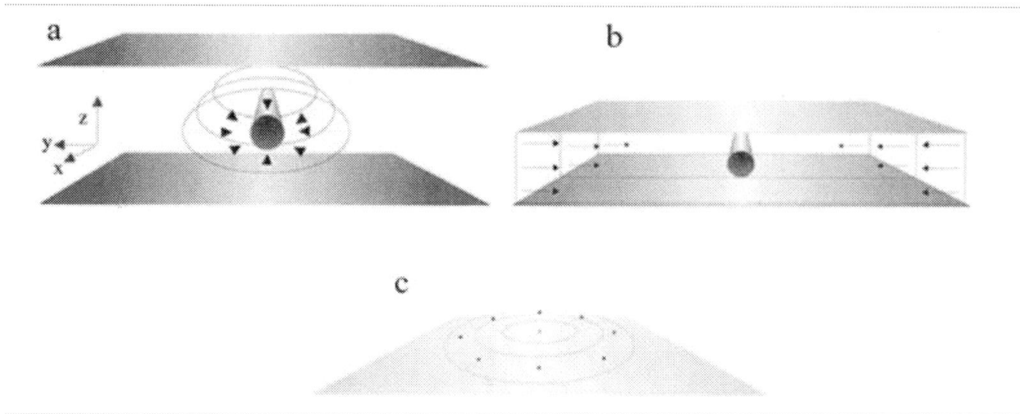

Figure 3. Flow regimes to a horizontal well [6]

As a boundary of a drainage area has reached top and bottom of a reservoir, flow becomes linear (Figure 3b), but it cannot stay linear forever because of a finite length of the horizontal part of the well [7]. When a shape of drainage area stops repeating the geometry of the well path, flow regime changes to radial one (Figure 3c). When boundaries of a reservoir start influencing gas flow, pseudo-steady state is supposed to be achieved.

In Figure 4 movement of drainage zone is shown. Colors characterize pressure distribution. According to the approach described in the study [6], pseudo-steady state is achieved when a drainage area reaches reservoir boundaries. Visualization of an external drainage area boundary movement via 3D hydrodynamic software post processing allows to see this moment realized though the time period is too long for real tests. After it meets the model boundaries, according to the approach from the work [6] pseudo-steady state is supposed to be reached.

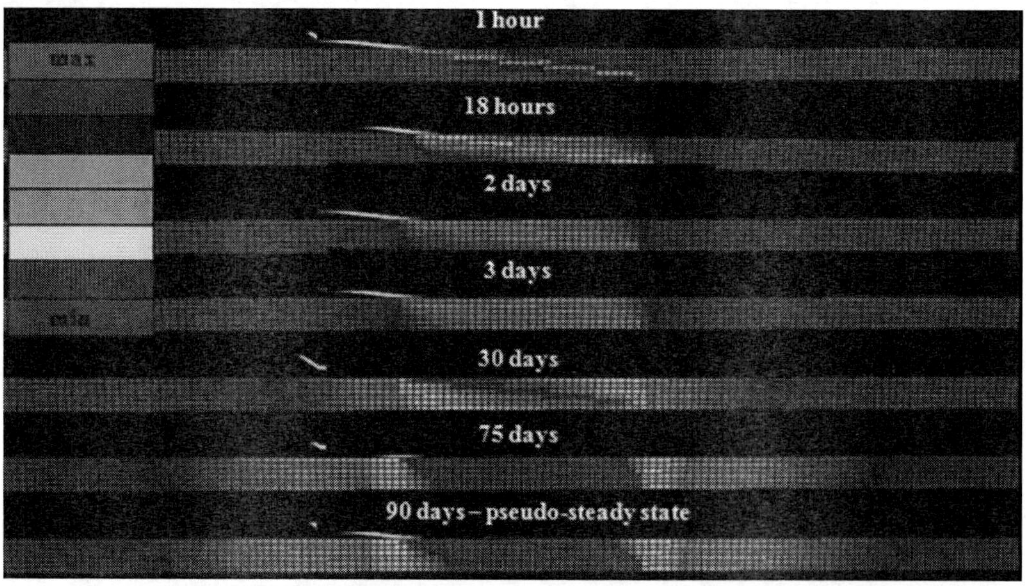

Figure 4. Evolution of shape geometry for a disturbance zone in a reservoir as time passes

4. The use of simulation results to estimate reservoir properties

To estimate a reservoir properties it is necessary to describe analytically a shape of drainage area formed in a pseudo-steady state regime and use it in a flow equation for gas flowing to a horizontal well which assumes action of a binomial flow law in the totally drained area, homogeneity and isotropism of rock porous media.

Mass flow rate [8]:

$$\rho\omega = Q\phi/F,$$

where $Q\phi$ – mass flow rate of gas, kg/s, F – area of a cross section of a reservoir orthogonal to gas flow, m^2. F is a side surface of a straight elliptic cylinder or a truncated overturned elliptic cone (Figures 5 and 6). In the case of a cone, area of a side surface can be calculated using an equation for a cylinder, lengths of semi axes are equal to lengths of semi axes of a medium plane of a cone (Figure 6).

$$F = P \cdot h,$$

where h – reservoir thickness, m; P – perimeter of an ellipse in a cylinder base, m.

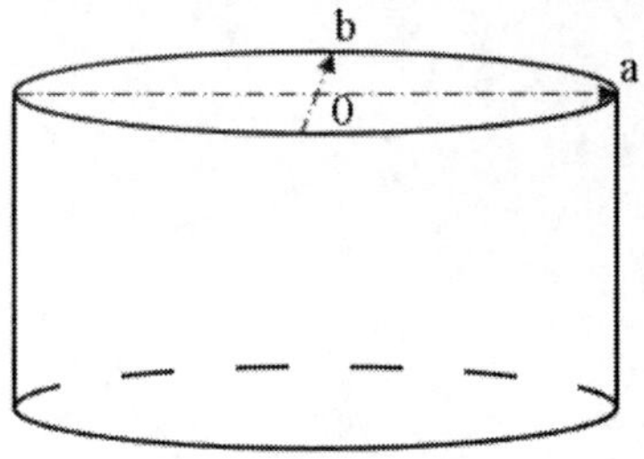

Figure 5. Shape of drainage area – straight elliptic cylinder

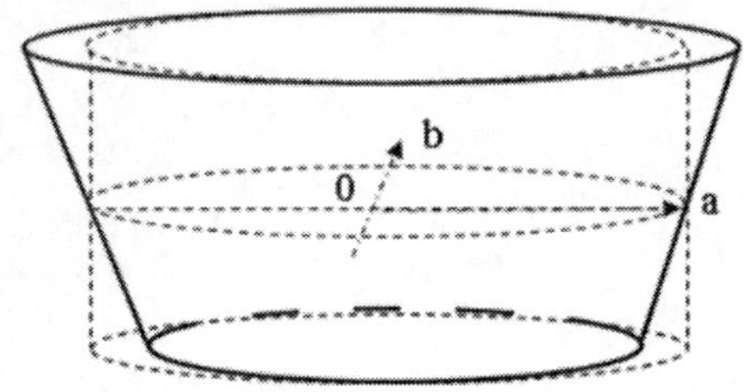

Figure 6. Shape of drainage area – overturned truncated elliptic cone

A perimeter of an ellipse can be found using one of approximation formulas:

$$P \approx 4 \cdot \frac{\pi ab + (a-b)^2}{a+b}, \tag{3}$$

where a and b – lengths of a big and a small semi axes of an ellipse in a base of cylinder, m.

After a mass flow rate is put into Forchheimer Law, we get the following flow equation for gas flowing towards a horizontal well:

$$P_k^2 - P_c^2 = \frac{\mu \cdot z \cdot P_0 \cdot \mathrm{T}_{res}}{2 \cdot k \cdot h \cdot \mathrm{T}_0} \cdot Q_0 \int_{a_c}^{a_k} \frac{a + b(a)}{\pi ab(a) + (a - b(a))^2} da + $$
$$+ \frac{\rho_0 \cdot z \cdot P_0 \cdot \mathrm{T}_{res}}{8 \cdot \varepsilon \cdot h^2 \cdot \mathrm{T}_0} \cdot {Q_0}^2 \int_{a_c}^{a_k} \left(\frac{a + b(a)}{\pi ab(a) + (a - b(a))^2}\right)^2 da, \tag{4}$$

where P_k and P_c – weight-averaged reservoir and bottom hole pressures, Pa; μ – gas viscosity, Pa·s; z – gas deviation factor; k – permeability, m^2; h – net pay, m; Q_{aT} – gas flow rate at standard conditions (P_0 = 0,1 MPa, T_0 = 293 K), m^3/s; a and b – major and minor semi axes of constant pressure surfaces (Figures 5,6), m. To estimate the length of a minor semi axis b is represented in dependence on a – the length of a major semi axis a has been concluded, therefore, $b = f(a)$. Index k corresponds a surface of equal pressure of weight-average reservoir pressure, index c – of a surface of bottom hole pressure.

Integration of the equation is carried out by numerical methods. Distinctive sizes of isobaric surfaces of bottom hole and weight-average reservoir pressures can be identified using pressure maps from simulation via Petrel software (Figure 7).

Following numerical experiment was conducted for two permeability values of the simulated fragment, 1.33 mD and 6.91 mD. The horizontal well production rate was laid down at 350000 m^3/days for both variants of permeability. Further, according to the equation (6) such value of a major semi axis a_k were estimated for different dates, so that permeability of a reservoir every time was equal to the prescribed one (1.33 mD or 6.91 mD depending on the variant). The experiment was carried out to identify pseudo-steady state of the system. The results of the numerical experiment are shown in Figures 8 and 9.

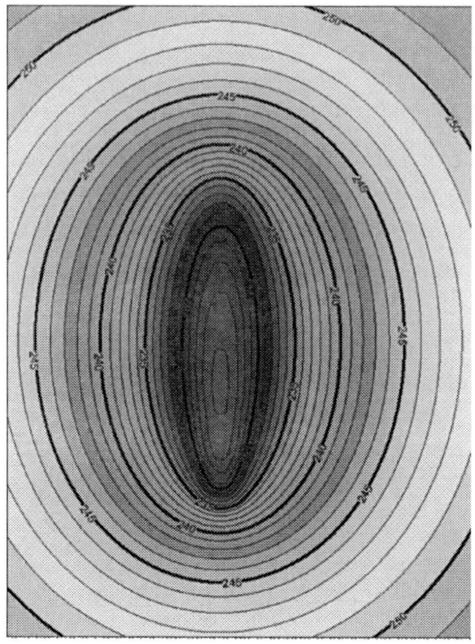

Figure 7. Map pressure from geologic-hydrodynamic simulation

Time of stabilization can be taken equal to the moment after which the length of a major semi axis a_k doesn't change practically. So, for 1.33 mD variant stabilization begins approximately after 119 days of operation, for 6.91 mD – in 58 days. Length of a major semi axis a_k relating to permeability of 1.33 mD in each case is less than a half of the length of the fragment, 675 m, which means that there is an effective drainage area existing.

Consequently, it was identified that a criteria for offence of pseudo-steady state of a gas-porous medium system was the formation of effective reduced pressure area, where gas comes from. It implies that after reaching this state a pattern of pressure distribution in a reservoir and a shape of drainage area do not change. In the Figure 10 shapes of drainage areas for variants with permeability of 1 mD and 7 mD at recovery rate of 500 thousand cubic meters per day are shown. In this picture a color scale is adjusted, maximum and minimum values of pressure at each date were changed so that a shape of drainage area could be identified visually.

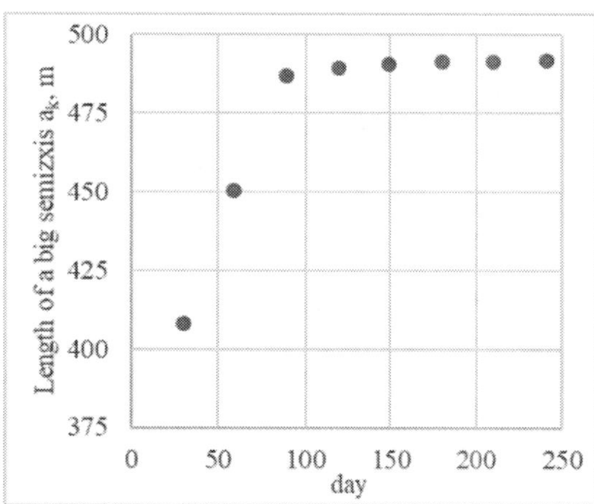

Figure 8. Change of major semi axis $a_к$ length in time for 1.33 mD variant

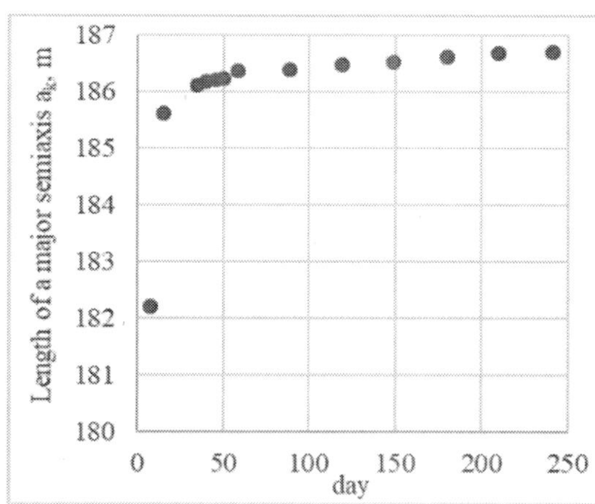

Figure 9. Change of major semi axis $a_к$ length in time for 6.91 mD variant

IOP Conf. Series: Materials Science and Engineering **700** (2019) 012055 doi:10.1088/1757-899X/700/1/012055

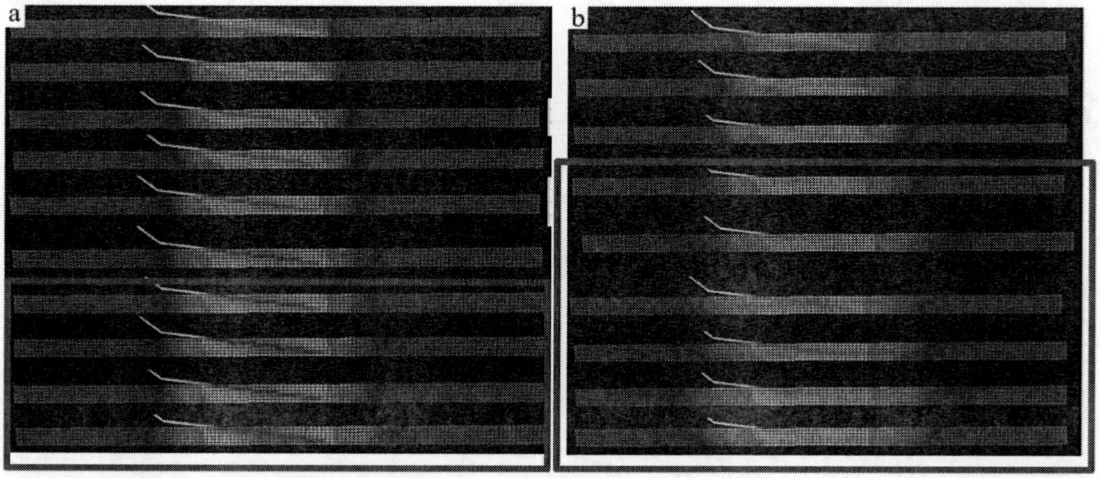

Figure 10. Shape of drainage area (a – 1 mD variant, b – 7 mD variant); red color means pseudo-steady state

Comparison of stabilization time values calculated via formulas (1), (2) and the one from simulation has been made. The results are shown in the Table 3.

Table 3. Comparison of approaches to estimation of time stabilization

Permeability, mD	Stabilization time (1), days	Minimal stabilization time (C=0,122), equation (2), days	Maximum stabilization time (C=0,35), equation (2), days	Stabilization time from modeling, days
1,33	108	57	164	119
6,91	21	11	32	58

As a result, for the variant with the smaller permeability, 1.33 mD, values of stabilization time according to equation (1) and from modeling, differ by 10.2 %. As for the variant with permeability 6.91 mD, the difference of stabilization time estimated from simulation and from equation (1) differ dramatically, in 2.76 times.

Usage of expression (2) becomes complicated because of the coefficient C which varies from 0.122 till 0.35. It gives variance of start time for pseudo-steady state. It is worth mentioning that in any case duration of well tests becomes equal to stabilization time from the Table 3 what is impossible in reality. There is possibility of well tests results adaptation using a hydrodynamic simulator, making a forecast of these results change as time passes, that allows to reduce duration of real well tests.

For this goal it is suggested to simulate a procedure of well tests on an actual model of reservoir, therefore, initially the well is operated at the same rate as during the real tests. Bottom hole pressure is adapted on the model to the value measured by a manometer during well tests.

While simulating time can align with real duration of a well test or be longer in case when achievement of stabilization requires it. Later stabilization time is estimated as a moment when bottom hole pressure starts to depend on time linearly. Also a change of the shape of drainage area should be mentioned. A generalized view of a plot is demonstrated in Figure 11, where values of bottomhole pressurs measured during well tests on the field are shown as a solid line, and a dashed line shows continuation of a case in simulation, which allows to actualize the well test with a numerical experiment and to find stabilization time.

IOP Conf. Series: Materials Science and Engineering **700** (2019) 012055 doi:10.1088/1757-899X/700/1/012055

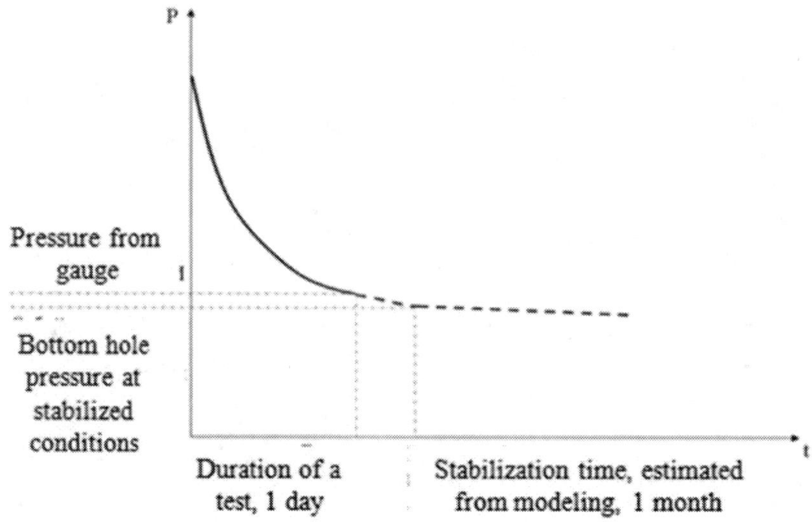

Figure 11. Chart of adaptation of a test on the model

5. Conclusions

Existing methods of stabilization time estimation have been observed after initiation of a well operation. The study showed that it's possible to use hydrodynamic modeling for estimation of time to reach pseudo-steady state. The flow equation for well tests of a horizontal well fully penetrating a reservoir has been modified taking into account a shape of drainage area defined via simulation. It can be calculated. The form of the area is defined as an elliptic cylinder in the model. Also it is assumed that a binomial law acts in all the volume.

It was observed during a numerical experiment that weighted-average reservoir pressure was corresponded to the length of a major semi axis which is less than a half of length of the simulated field fragment. It proves existence of an effective drainage area, sizes of which should be taken into account in gas flow equations.

References

[1] Economides M J, Hill A D and Ehlig-Economides C 1994 *Petroleum Production Systems* (NJ: Prentice Hall) p 603

[2] Goode P A and Kuchuk F J 1991 Inflow Performance of Horizontal Wells (SPE Reservoir Engineering), 319-323

[3] Ahmed T H 2010 *Reservoir engineering handbook* (Burlington: Gulf Professional Publishing) p 1454

[4] Gritsenko A I and Aliev Z S 1995 *Guidance on well tests* (Moscow: Nauka) p 523

[5] Gazprom R 086-2010 2011 *Instruction of complex tests of gas and gas-condensate wells. Part I* (Moscow: Ltd "Gazprom Expo") p 234

[6] Houz O, Viturat D and Fjaere O S 2017 *Dynamic Flow Analysis* V 5.12. Kappa Engineering Publ., 743 p

[7] Lu J, Cunha L B and Lu T 2002 *New Solutions for Well Test Analysis of Horizontal Wells* (SPE/PS-CIM/CHOA International Thermal Operations and Heavy Oil Symposium and International Horizontal Well Technology Conference) p 7

[8] Schelkachev V N and Lapuk B B 2001 Subsurface hydraulics (Ijevsk : NITs "Regular and chaotic dynamics") p 525

Integration of expert and data-driven workflows to manage reservoir and well life cycle in Arctic conditions using innovative SICLO methodology

M Antonic[1], M Solesa[1], A B Zolotukhin[2], D Rakic[3] and M Aleksic[4]

[1] Chair of Drilling and Completion Engineering, Department of Petroleum Engineering, Montanuniversität Leoben, Parkstraße 27, 8700 Leoben, Austria
[2] Gubkin Russian State University (NRU) of Oil and Gas, GSP-1, Leninsky Prospekt, 65, 119991, Moscow – Russia
[3] CMS Prodex. Novi Sad, Serbia
[4] NIS Gazprom Neft, Narodnog Fronta 12, Novi Sad, Serbia

* Corresponding author: m.solesa@cmsprodex.com

Abstract. SICLO (Source of data and information; Input data; Calculation/Analytic; Logic Analysis; Output/Value Delivery) methodology is an innovative concept for smart diagnostic, reservoir/well performance optimization and estimation of remaining reserves based on the integration of Petroleum Data Management System (PDMS) and expert rules. Implementation of SICLO methodology provides the best strategy on how to produce remaining reserves most profitably. PDMS is the foundation of SICLO methodology and provides structured and verified information that follows the Well Life Cycle. Within PDMS, data are organized and structured according to clearly defined principles and rules and filtered by different levels of quality control. Structured data allows integration of production and reservoir information with real-time data to achieve the maximum level of diagnosis of system operation performance according to reservoir and well potentials and system constraints. The built-in workflows and architecture of the whole process are automated and make the task accomplishment faster. SICLO methodology integrates expert-driven knowledge and pattern recognition tools improved by data-driven, artificial intelligence, neural network, and fuzzy logic technologies to deliver adaptive solutions for identifying locations of remaining reserves, optimizing oil and gas production, and minimizing associated operational costs.

1. Introduction

In recent years, most of the technical developments in the oil and gas industry have come from the achievements made in areas of unconventional resources. The demand for energy increases day-by-day, and thereby, the unconventional resources are of paramount importance to ensure the future energy supply. According to the estimates, the largest unconventional oil and gas reserves could be placed in one of the most unexplored and delicate world's regions, the Arctic. The Arctic presents massive potential for exploration companies, but at the same time, complex environmental and climatic conditions make the environment risky due to any unsafe operations and challenges with the implementation of existing technologies.

According to the information presented by Gubkin Russian State University of Oil and Gas [1], apart from the significant gas reserves, the noteworthy reserves in oil-bearing formations are existing. Generally, the unconventional resources have a sharp decrease in reservoir natural energy, which results in the necessity of the application of artificial lift methods.

In this paper, the SICLO methodology describes the approach of real-time Electrical Submersible Pumps (ESP) monitoring that recognizes threats in the well operation. The approach significantly enhances the pump run life and reduces workover costs. Moreover, the utilization of real-time data generated from ESP systems reduces human impact and workforce requirements in challenging Arctic conditions.

Over the past year, significant improvements in the ESP system design have enhanced the reliability of the systems in offshore locations. By refining the electrical and mechanical subsystem designs, the ESP system has reached the necessary reliability in the North Sea region, which represents the most similar conditions to those on the Arctic [2].

Although many methods are reliable for exploitation in harsh conditions and remote locations, the ESP systems hold features that could be beneficial for newly discovered reservoir formations, and they are summarized as follows:

- Downhole ESP sensors and gauges provide significant reservoir real-time information, which is necessary to reduce uncertainty at the beginning of the exploration stage (transient pressure analyses).
- ESP systems can be sized to pump all produced fluids to the processing facility miles away.
- The system is applicable in large deviations angles and diverse wellbore trajectory designs.
- The Electrical Submersible Pumps are tools that can be remotely controlled (frequency, choke size, and position), which could reduce OPEX costs, and a permanent crew would not be necessary on-site.

The presented methodology aims to interpret the pattern trends generated by downhole and surface measuring devices in order to provide a reliable model for embedding into the controller of the ESP surface control unit, which could operate in arctic conditions. Interpretation of the trends is not possible without knowing other multidisciplinary system information.

2. Levels of downhole and surface data availability

Since the ESP are becoming one of the fastest-growing artificial lift systems in the world, their application varies from low to extremely high production wells. A variety of fluid production ranges directly implicate the data acquisition systems, and data availability differs from field to field. The ESP diagnosis and monitoring solution must be flexible enough to cover different scenarios of real-time data availability. In this section, we clarify the subsurface and surface data channels, of which recordings are used as symptoms in ESP monitoring event detection.

In terms of downhole equipment, the ESP system includes downhole sensors block, motor, motor seal, gas separator, pump intake, pump, motor lead extension, cable, and check valves. Depending on the type of ESP downhole sensors, different downhole parameters are monitored (Pump Intake Pressure, Pump Discharge Pressure, Pump Intake Temperature, Motor Winding Temperature, Motor Oil Temperature, and Two or Three Axial Vibrations).

Despite general data availability from downhole gauges, the surface readings can significantly vary based on the used controllers, Variable Speed Drives (VSD), or Switchboards (SWBT). The high-level automated ESP system represents the monitoring of following surface parameters (Surface Flow Rate, Surface Tubing Pressure, Surface Casing Pressure, Current Leakage, Frequency, Cable Resistivity, and Phase Voltage) through Supervisory Control and Data Acquisition System (SCADA).

In the global picture, the ESP system diagnostic depends on a particular group of monitoring parameters, and not all of them are indispensable. Expert rules define logic relationships and how these parameters relate.

3. Methodology

The significant improvements in the data-gathering technology and the application of smart well technologies have led to the acquisition of countless amounts of data (Big Data) that require interpretation of raw data, to obtain valuable information. In recent years, the oil and gas industry has been moving forward, developing advanced diagnostics and optimization tools based on artificial intelligence techniques that are data-driven. Most autonomous workflows are built on a foundation of statistical analysis and the implementation of artificial intelligence techniques on real-time data with or without the integration of reservoir/well models. However, this raises the most crucial question: "How to predict critical dynamical conditions when the well requires long term actions (workover, chemical application, etc.) and how to maintain optimal well conditions without active downhole monitoring (without sensors)?". The started research work on the development of Automatic Adaptive Algorithm (A^3) is an innovative concept of smart diagnostic and well performance optimization based on the integration between (PDMS) and expert rules to achieve the maximum level of intelligent control of system operation according to reservoir and well performance (Figure 1).

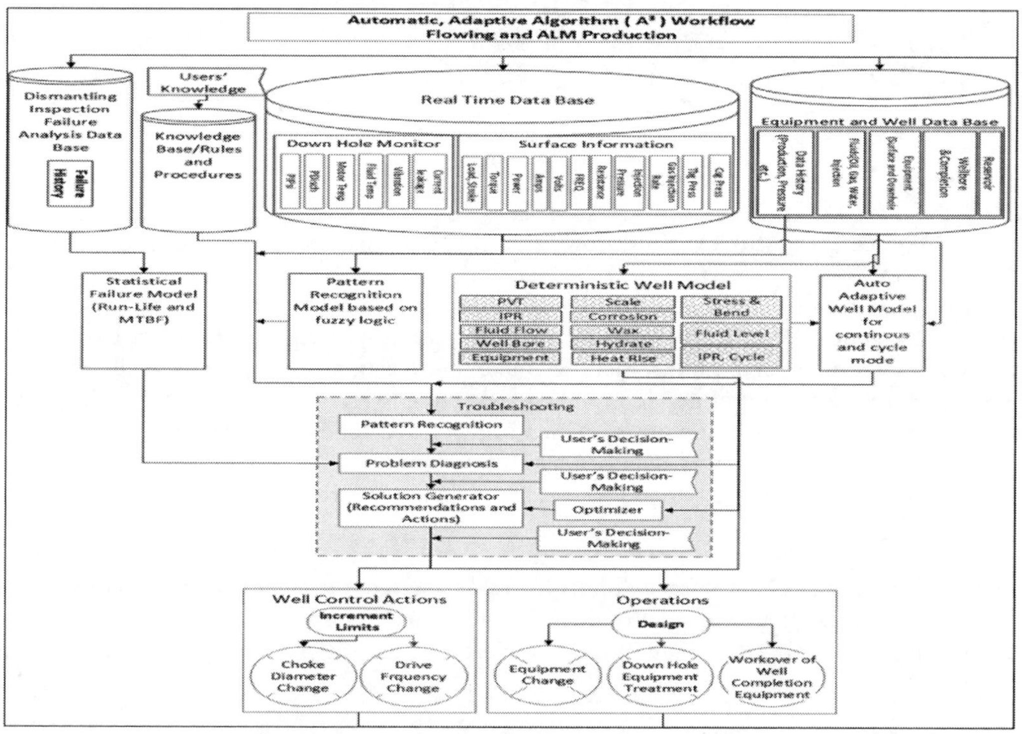

Figure 1. Structure of A^3 workflow and principal components of a diagnosis and control modules.

In this paper, the downhole sensor data recorded during operation of ESP and technical details of subsurface and surface equipment data stored in the i-ESP database are used to diagnose the problems of ESP operation (abrasive/gassy/viscous fluids, high-temperature wells, scale, intermittent/continuous operation, etc.).

In general, the concept of SICLO methodology is adapted to manage the operation of ESP, integrate and automate data collection from multiple sources, process them, and provide needed parameters for monitoring, automatic control, problem diagnosis, and recommendations for optimizing ESP operation as shown in Figure 2.

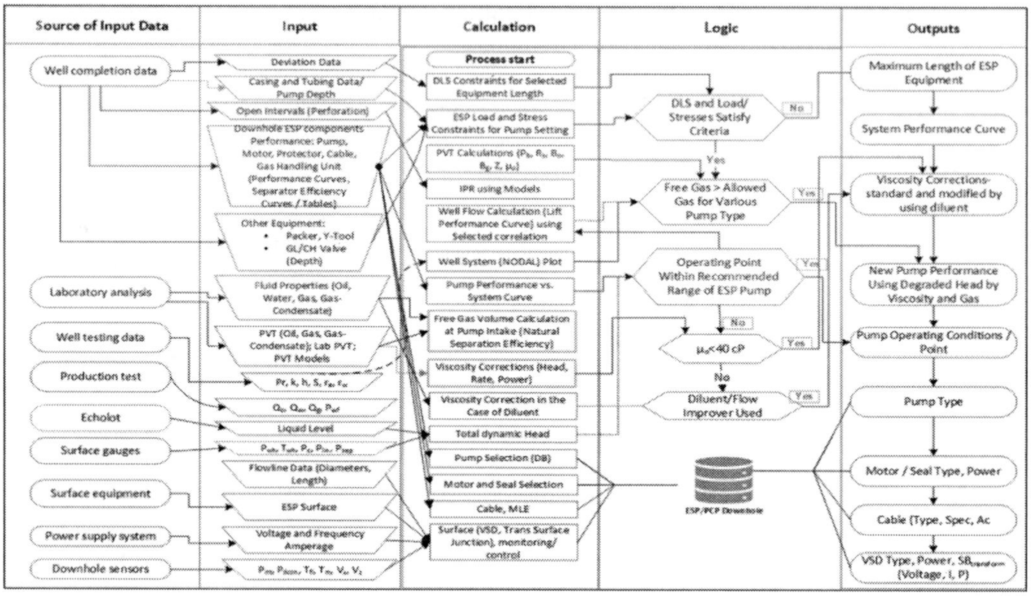

Figure 2. ESP SICLO Workflow of Design Optimization and Problem Diagnosis.

4. Data pre-processing

Sometimes downhole and surface sensors transfer the readings through the SCADA system directly to the real-time monitoring center, but sometimes the data ingestion requires additional software to read data stored at the surface controlling units and convert them to readable files formats. The A^3 algorithm requires data available in real-time to recommend preventive actions. However, data captured in real-time can contain a vast of missing data or data out of the range that could lead to misinterpretation. Therefore, this section describes two pre-processing data procedures for different applications.

4.1. Data cleansing and filtering

Firstly, the objective of data pre-processing is to create an algorithm that can eliminate erroneous data, replace missing values, and format entire data set into valuable information that will be interpreted. In the end, the algorithm works autonomously as a cleansing and filtering tool to eliminate meaningless data.

4.2. Well test data preparation

Secondly, the interpretation of data captured during the well test requires a different preprocessing procedure. Real-time monitoring of the pressure at the ESP pump intake allows acquiring a significant amount of data. These large amounts of intake pressure data are converted to the pressure at the referent depth (mid-perforation or top/bottom of the pay zone) and are used for various types of interpretation (buildup/drawdown test, inflow performance of well, virtual pressure and flow metering, etc.). The essential problems of interpretation an acquired data set from the ESP downhole sensors are conversion of the intake pressure to the static/bottom hole flowing pressure using appropriate models (two-phase correlations) and elimination of noise data as in the most situations the derivative of the recorded and converted pressure data cannot be interpreted, or interpretation is wrong. The modified workflow (Figure 3) has been developed to decrease of unavoidable noise caused by the applied derivative procedure of the measured pressure data.

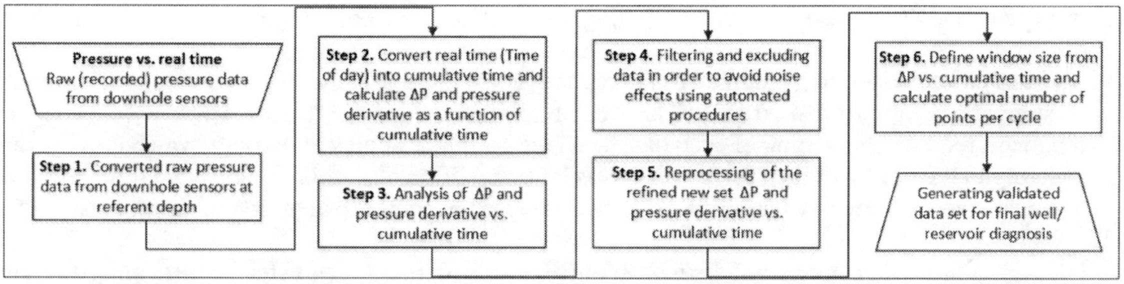

Figure 3. Application of automatic workflow for the data acquisition and processing

5. Trend (Pattern) analysis

The ESP operation parameters are dynamic, and they are changing over time. By tracking down-hole real-time operation parameters, deviation from established trends can be recognized, and actions to improve production and extend the pump life can be taken [3]. The trend analysis tool is designed to extract trend data from the time series for each of the operating parameters. As a result, all individual trends are taken simultaneously into consideration and used to understand specific ESP performance problems.

In the previous section, it was described how the pre-processed data set is created. Furthermore, the ability of the autonomous algorithm to calculate the slope highly depends on the considered period and values included in the dataset. Figure 4 depicts the process steps of the general workflow that is used to generate trends.

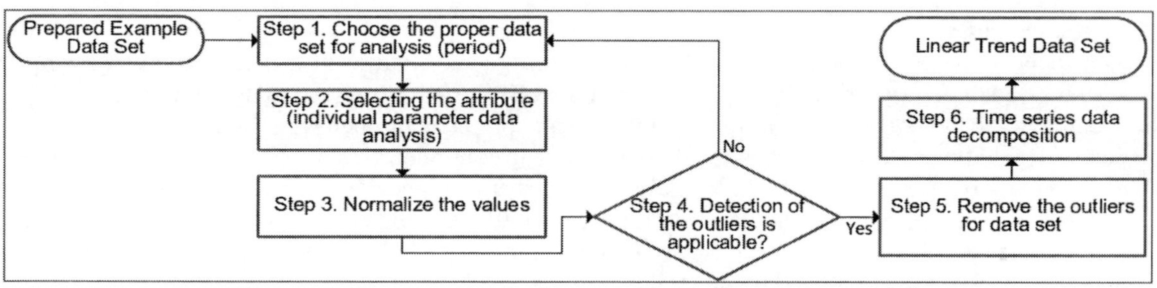

Figure 4. Process steps in the general trend extraction workflow.

Step 1. Recognizing the proper period for trend analysis. Before running the trend analysis, the period must be chosen based on the rules and valuable information from the other ESP databases described in the Methodology section. For instance, removing the period of initial production is obligatory since, at the beginning of the ESP operation, before reaching the nominal operating parameters (commissioning, the well clean up). The mechanism to remove initial well production out of the trend analysis is relying on the fluid production measurements from Production History Database. The turning point is the stable fluid flow rate. Moreover, the intermittent ESP production operation leads to significant fluctuations in operation parameters and requires a specific workflow for trend interpretation, which would be the subject of further research. Also, Run Life & Failure Database provides information on pump operation mode (continuous/intermittent).

Step 2. Data channels prioritization. Different downhole sensor units and different surface units (switchboards/VSDs) can generate a large number of data channels. Some of them are directly measured, while others represent calculated values with different levels of importance. In a variety of data channels, the expert rules are set to choose to most specific parameters that can be indicators of specific ESP problems and prioritize them based on the importance levels.

Nevertheless, the downhole sensors sometimes could have a lack of essential ESP operation data channels, such as missing the pump discharge pressure measurement. In that case, the expert rules are prioritizing data channels based on data channels availability.

Step 3. Data set normalization. Normalization methods represent compromises in achieving particular ends. Normalization involves taking values that span one range and representing them in another range [4].

Although each operating parameter is taken into account separately, the final output of the analysis is a comparison of different trends. The ESP real-time data considerably vary in range, and the comparison requires the scaling of each data set. This scaling is achieved by range value normalization. For example, the values of the voltage are between 0 and 400 volts while at the same time, the current is between 0 and 25 amps. After normalization, both data channels are in the same range, between 0 and 1.

Step 4. Outlier detection. In statistics, an outlier is a data point that differs significantly from other observations. ESP real-time data set can contain a significant number of outliers. The outlier may be due to the variability of measurement or maybe the real measured record, which is not valid for general consideration of data set and decomposition of the trend line. The trips during the pump run life are frequent in ESP operation, but the values measured do not determine the dynamic physical conditions in the well and shall be excluded from further interpretations. For example, electricity power shut-down can create a rapid increase in pump intake pressure or a sharp decrease in values of current. That kind of measurement does not represent the valid historical data for trend analysis.

Step 5. Trend decomposition - Once the time series is thoroughly prepared, and outliers, as well as missing values, do not have any influence, the time-series signal decomposes into a new trend data set. The decomposition algorithm linearly smooths data, and the newly created data set represents values of the linear trend line for each attribute. When the slope data set is generated for each attribute (current, pump intake pressure, etc.), the trend is calculated. Furthermore, the slope value is of essential importance because it provides the final input parameter for pattern recognition.

6. Expert rules

The expert Rules are created based on domain knowledge and significant experience in work with ESP systems. Each event/problem in the operation of the ESP system is defined in the Matrix of Descriptive Trend Explanations (MoDTE) by a qualitative description of the behavior of the essential working parameters in time. The SICLO methodology is based on continuous improvement, and since the failures are typically repetitive, the root cause analysis is always conducted, and MoDTE is updated. MoDTE consists of ESP operation parameters and their descriptive explanations. Each characterization has its slope range, which is characteristic for operation region or field. Figure 5 represents the example of MoDTE for the case of plugged pump intake.

Matrix of Descriptive Trend Explanations: Plugged Pump Intake	Increases Slowly	Increases Gradually	Increases rapidly	Increases Abnormally	Constantly	Decreases Slowly	Decreases Gradually	Decreases Rapidly	Decreases Abnormally
Surface Flow Rate						✓	✓	✓	✓
Current						✓	✓		
Pump Intake Pressure		✓	✓						
Motor Temperature	✓	✓							
Frequency	✓				✓	✓			
Load						✓	✓		
Pump Discharge Pressure							✓	✓	

Figure 5. Matrix of Descriptive Trend Explanations (MoDTE) - Case of the plugged pump intake.

7. Event Prediction

After converting large scale real-time data sets into trends, the value of tangent defines the trend slope, and basically, it is an indicator of the existence of potential problems. Furthermore, to better understand the trends, the algorithm automatically converts the value of tangent into a descriptive qualitative explanation (linguistic terms). The proactive troubleshooting and recommendation of corrective actions during ESP run life are achieved by matching the descriptive quality explanations with MoDTE. The trend descriptions, along with the expert rules and built-in SICLO methodology, can significantly increase the pump run life and reduce human efforts in the optimization process. Figure 6 depicts the workflow for event prediction.

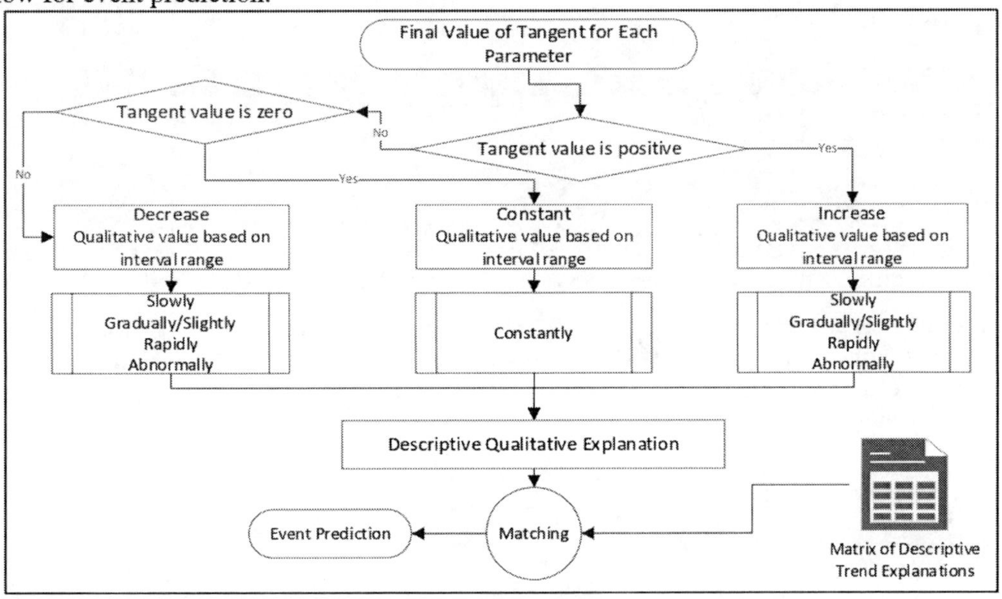

Figure 6. SICLO Methodology workflow for event prediction.

8. Field Case 1

During the well history, a significant number of the well service operations were recorded, mainly due to the well integrity issues (damaged production casing). The installation of 5" production liner resolved the casing integrity problem.

In the last ten years, on the well was performed 14 workover operations. Moreover, after the hydraulic fracturing and production intensification, the rapid increase in water cut has happened along with unexpected problems with Gravel Pack completions due to low inflow, complicated fishing of GP equipment, and low efficiency of gravel placement.

An additional factor that has affected the number of workover operations was a significant reduction in reservoir pressure. The drop-in reservoir pressure led to low inflow performance and high WOC. After the last workover operation, the sample was taken from the pump as well as from the reservoir water. The laboratory analyses proved the presence of sulfur-reducing bacteria in formation water and Ferrous carbonate ($FeCO3$) appearance on the pump.

Therefore, the SICLO approach has been applied to analyze the real-time dataset from the well. The following static information has been gathered in databases and interpreted:

a) Run Life & Failure Database, which allows the user to see the whole workover history and briefly insert new workover descriptions when necessary. All workover data are structured based on the Well Service Operation (WSO) classification. From workover and completion history, as shown in Figure 7, it is evident that the well had frequent and repetitive problems with artificial lift methods in the past.

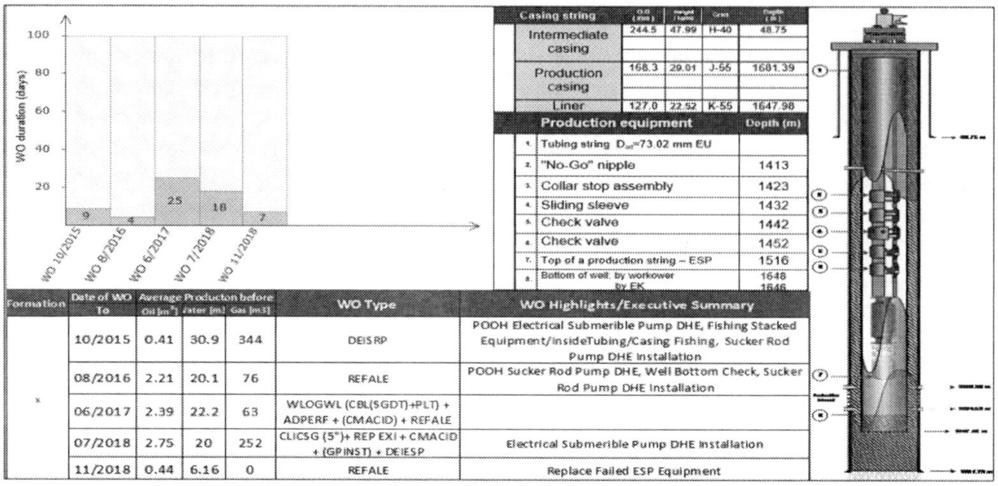

Figure 7. Workover history and well schematic data - Run Life & Failure Database.

b) Figure 8 (a) provides a complete production history of the example well. However, the period of interest is the period from the last workover job, and Figure 8 (b) shows daily fluid production since the last workover job. The well had gradually decreased in daily production.

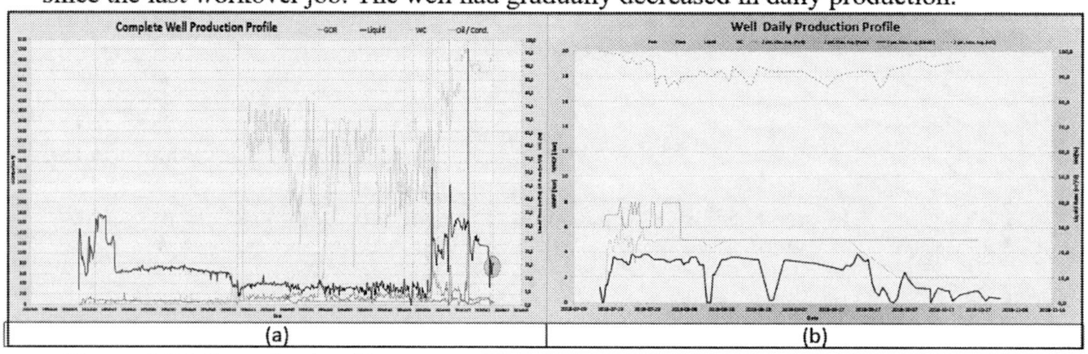

Figure 8 Well production history, a – complete; b – daily production history before failure

c) Run Life & Failure Database with Teardown Results – this database provides information about the failure root cause analysis. The database is created based on the industry standards for dismantling inspection and failure analysis: API RP 11S1 and ISO 14224. The teardown test showed that the pump failure occurred due to the precipitation of inorganic scale on stages and probably on the pump intake (Figure 9).

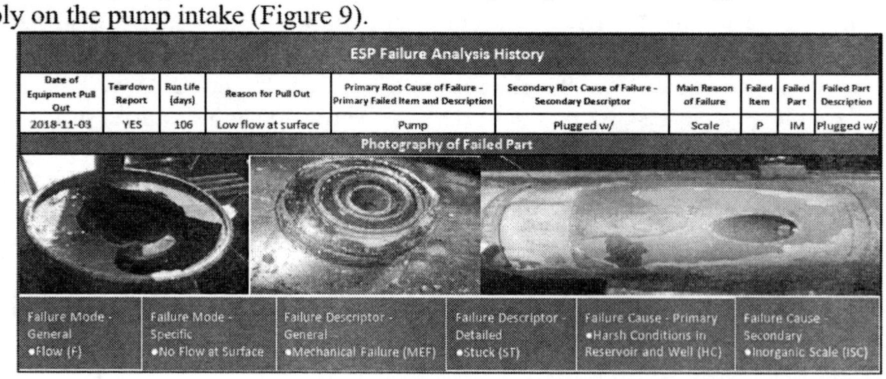

Figure 9. ESP failure analysis.

The recorded ESP parameters at the surface and downhole were used to analyze the real-time ESP data set and to compare the results with failure analysis. Figure 10 (a) gives the insight into raw ESP data set, where it is difficult to spot any pattern and interpret data, respectively. Initially, raw data were cleansed and filtered. As a result, Figure 10 (b) shows the continuous dataset without discrete values, and it is more apparent that after a period of relatively stable working conditions, the real-time parameters had a sudden change in performance before the ESP system went into intermittent operation.

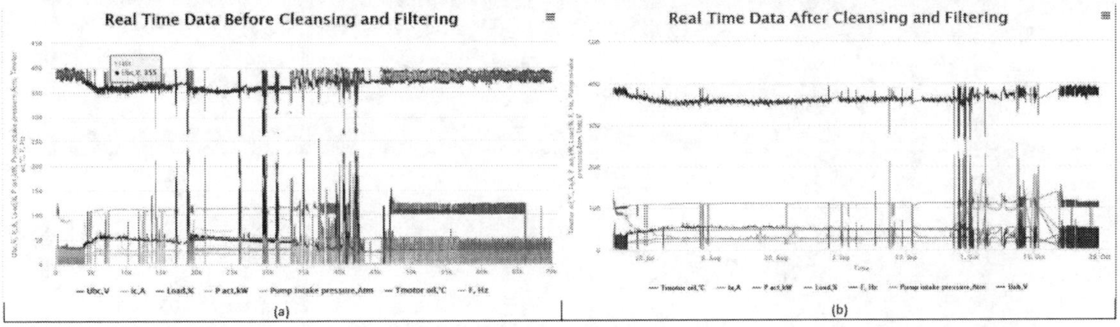

Figure 10. ESP real-time data set; a – raw data; b – cleansed and filtered data.

The trend analysis tool was applied to extract the trend line for each attribute. Figure 11 (a) depicts the first step of the general trend extraction workflow for a single attribute (motor temperature). The application of workflow for recognizing the proper period for trend analysis excluded two periods from EPS operation history, initial production, and intermittent operation period.

The general trend analysis workflow considers data scaling to obtain all data sets in the same ranges. Figure 11 (b) shows the results of range normalization for the value of the current.

The fourth and fifth steps are examining outliers. Based on input parameters, the algorithm automatically detects the outliers, and with the filtering process, it removes them. In this way, the trips and any other well interventions do not influence the trend line. Figure 11 (c) refers to outlier detection in this example. For the values of current, light blue color points out outliers that are removed from further analysis.

As a result of trend analysis, the linear trends are calculated for each attribute, and they are visualized in Figure 11 (d).

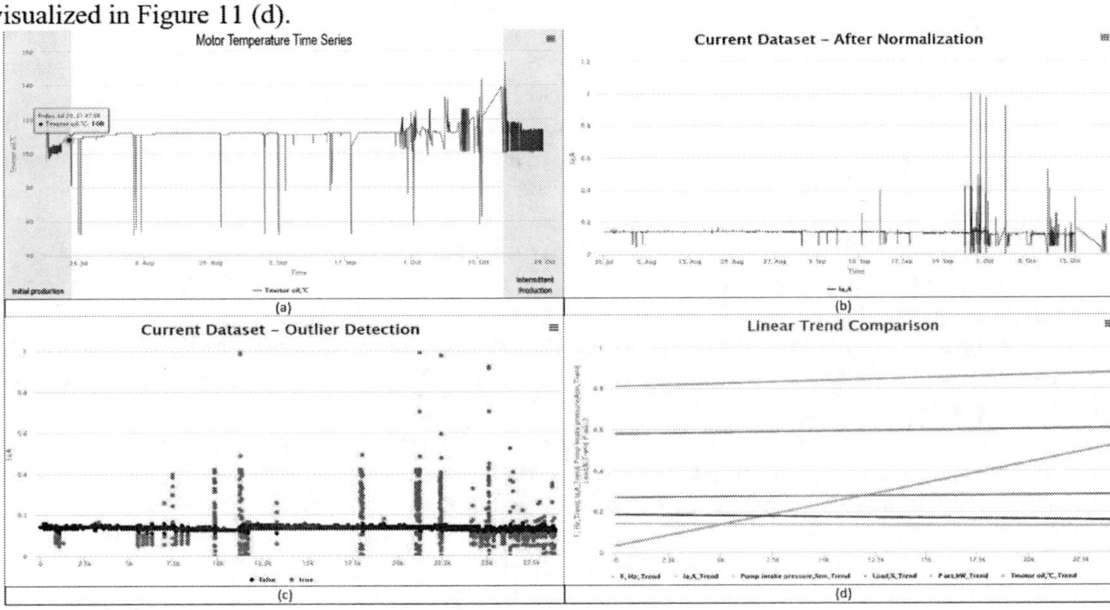

Figure 11. Trend extraction procedure; a – raw data; b – cleansed and filtered data; c – outlier detection; d – linear trends

Figure 12 contains the normalized trend slopes and qualitative descriptions of each of the ESP operating parameters. The corresponding slope descriptions are matched with the MoDTE database. The plugged pump intake is the best match for the real-time ESP operation parameters.

Time Series / Trend Description	Increases Slowly	Increases Gradually/ Slightly	Increases Rapidly	Increases Abnormally	Constantly	Decreases Slowly	Decreases Gradually/ Slightly	Decreases Rapidly	Decreases Abnormally	Slope Shape/Rate
Surface Flow Rate							✓			-0.000001022000
Current						✓				-0.000000456127
Frequency	✓									0.000002769031
Pump Intake Pressure			✓							0.000020374735
Load						✓				-0.000001163219
Motor Oil Temperature	✓									0.000001138371

Figure 12. Qualitative and Quantitative Pattern Description.

The decrease in surface flow rate was the clear indicator that something is happening with the pump performance, and if the pump performance curve is checked, it is clear that at the given frequency pump is underperforming.

The real-time performance analysis indicates that the pump had a problem with plugged pump intake before the pump went into the intermittent production regime and before the pump failed. Also, it was known from well production history that well had a problem with scale and paraffin.

The recommendation, in this case, before the failure, would be to take appropriate corrective actions such as chemical treatment since the pump had a restricted flow rate.

9. Field Case 2

The objective of field case two is to show well test data preparation and interpretation. The pressure at the ESP pump intake in the well was recorded in real-time over 460 hours. After a production period of 110 hours, well was shut-in for build-up for 350 hours. Preparation of the recorded pressure data, their filtering and validating for further analyzing in pressure transient analysis or production analysis are conducted using the A^3 algorithm throughout the following steps:

Step 1. Raw sensors pressure data in real-time at a depth of ESP installation are converted into pressure at referent depth (in this case at the middle of the perforated interval) using modified Beggs & Brill model for multiphase flow through the wellbore of any geometry (vertical, slant and horizontal), illustrated in Figure 13-a.

Step 2. In order to get a build-up type curve (basics for well test interpretation), a real-time date-time format (time of day) is converted into cumulative time upon which, *dp* and pressure derivative is calculated (Figure 13-b).

Step 3. Analysis of the shape of the build-up pressure derivative using the total number of recorded data points, it is evident, that enormous noise data hides the valid reservoir signal.

Step 4. The automated procedure for filtering and excluding redundant data is applied to avoid noise effects noticed in the previous step. The use of this procedure significantly reduces the data amount.

Step 5. Reprocessing of the refined data set gives the shape of the build-up pressure derivative (Figure 13-c) from which the reservoir model could be recognized, but it is still necessary to further improve upon the data set.

Step 6. Defining window size and calculating the optimal number of points per cycle will provide a clear and unambiguous reservoir signal with the minor influence of noise effects (Figure 13-d).

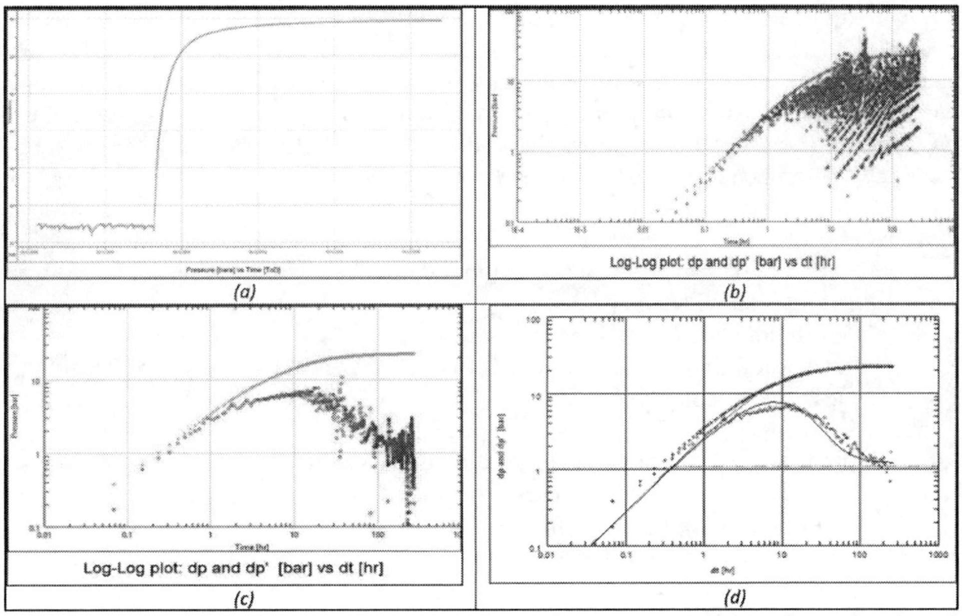

Figure 13. Well test data preparation; a - Converted raw pressure data from the downhole ESP sensor at referent depth in real-time; b - Calculated dp and pressure derivative as a function of cumulative time in log-log scale; c - Refined new set of dp and pressure derivative as a function of cumulative time in log-log scale; d - Generated valid data set and well test interpretation

Based on generated valid data using the A^3 algorithm, it is significantly easier to interpret well test data to get parameters necessary in further well/reservoir analysis (reservoir pressure, permeability, skin, distance to barriers if they exist), as shown in Figure 14.

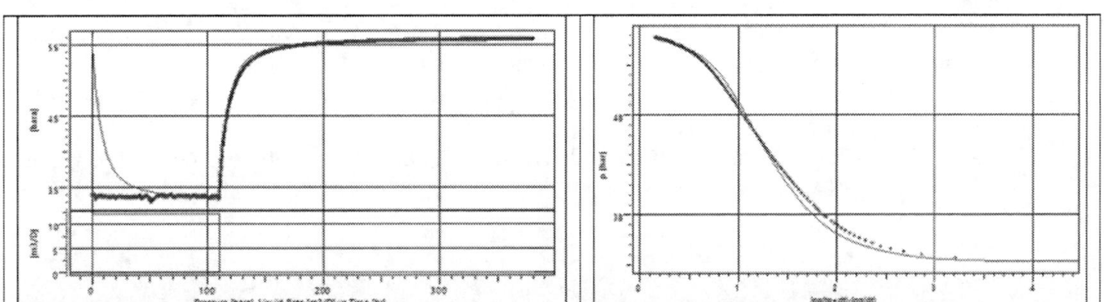

Model	Model parameters	Reservoir & boundary parameters
Constant wellbore storage	C=1.9E-6 m^3/Pa	Pi=56.11 bar
Vertical well	S=2.95	K*h=5600 md*m
Homogeneous reservoir model		K=659 md
Infinite acting		

Figure 14. Results of well test interpretation.

10. Conclusion

In challenging Arctic conditions, the need for autonomous operations is higher than ever. In this paper, the SICLO methodology suggests the approach that significantly relies on ESP real-time operation parameters, and by the interpretation and comparison of those parameters, the maximum level of intelligent control can be achieved along with the reduction in the workforce requirements in challenging Arctic conditions.

Moreover, the SICLO methodology and Automatic Adaptive Algorithm (A^3) can be embedded in the SCADA system as an advisory tool that can predict the failures, enhance the pump run life, and reduces workover costs. Nevertheless, A^3 controls the ESP system under reservoir and field performance. Field cases demonstrated the benefits of A^3. In the first case, the pattern recognition system would be beneficial to detect an event before the failure happens and to suggest the appropriate corrective actions. The second case shows the ability of the A^3 algorithm to interpret well test data to get the necessary parameters in further well/reservoir analysis.

Abbreviations

SICLO	Source of data and information; Input data; Calculation/Analytic; Logic Analysis; Output/Value Delivery
PDMS	Petroleum Data Management System
ESP	Electrical Submersible Pump
VSD	Variable Speed Drives
SWBT	Switchboards
SCADA	Supervisory Control and Data Acquisition System
i-ESP	Intelligent - Electrical Submersible Pump database
A^3	Automatic Adaptive Algorithm
MoDTE	Matrix of Descriptive Trend Explanations
WSO	Well Service Operation

References

[1] Zolotukhin A and Gavrilov V 2011 *Russian Arctic Petroleum Resources*, Gubkin Russian State University of Oil and Gas, GSP-1, Leninsky Prospekt, 65, 119991, Moscow – Russia, 2011 https://doi.org/10.4043/22062-MS

[2] Pastre L and Fastovets A 2017 *The Evolution of ESP Technology in the North Sea: A Reliability Study Based on Historical Data and Survival Analysis*, Paper SPE-187735-MS, SPE Russian Petroleum Technology Conference held in Moscow, Russia, October 16-18, 2017 https://doi.org/10.2118/187735-RU

[3] Al-Jasmi A, Nasr H et al. 2013 *ESP 'Smart Flow' Integrates Quality and Control Data for Diagnostics and Optimization in Real-Time*, Paper SPE 163809, SPE Digital Energy Conference and Exhibition held in The Woodlands, Texas, the USA, March 5-7, 2013 https://doi.org/10.2118/163809-MS

[4] Carvahal G, Maucec M and Cullick S 2018. *Intelligent Digital Oil and Gas Fields*, Elsevier, Cambridge, USA, 2018, 105

Challenges with gas hydrate formation

M A Kelland

University of Stavanger, Norway

Contact address (E-mail): malcolm.kelland@uis.no

Abstract. Many multiphase flow lines are prone to hydrate formation unless prevention methods are put in place. Chemical management of hydrate formation is traditionally done with thermodynamic hydrate inhibitors but in the last 25 years low dosage hydrate inhibitors (LDHIs) have been developed which can offer economic, environmental and other benefits. LDHIs are divided into two main categories, kinetic inhibitors (KHIs) and anti-agglomerants (AAs), both of which are successfully being used in field applications. This paper briefly reviews the hydrate management tools available to the operator. Then the review focuses on LDHIs, their structure-performance relationships and the various classes that have been designed and tested. The environmental challenges of both AAs and KHIs are also discussed.

1. Introduction

Natural gas hydrates are ice-like solids made of natural gas molecules inside the cages of a clathrate water lattice [1, 2]. Gas hydrate structures I and II are known to be formed in pipelines, with structure I being formed from very methane-rich natural gas and structure II from natural gas mixtures (Figure 1). Pipeline blockage due to gas hydrate formation is the number one problem to overcome in transporting oil, gas and water mixtures in cold and/or deep water subsea flow lines (Figure 2).

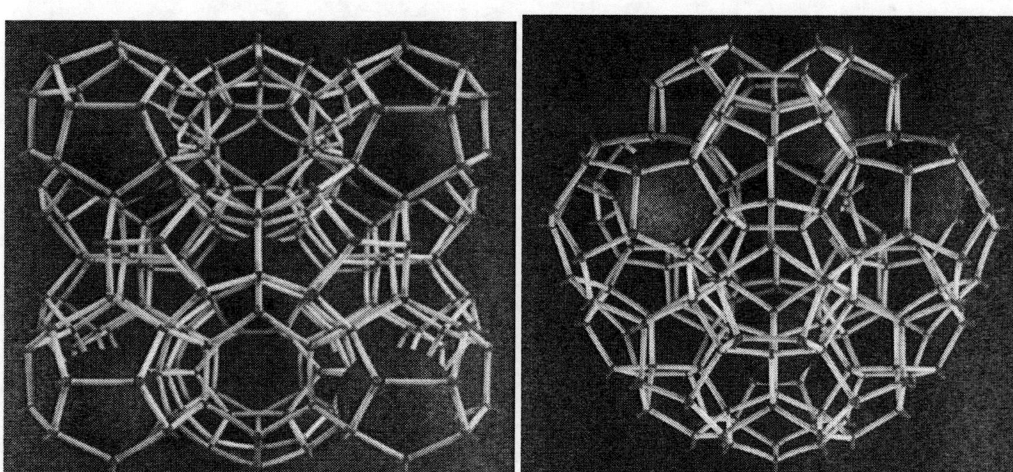

Figure 1. Gas hydrate Structure I (left) and Structure II (right).

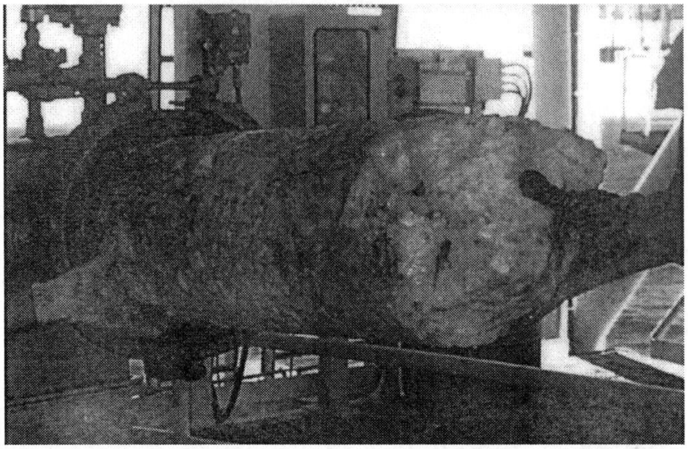

Figure 2. Gas hydrate plug being removed from a flow line.

Once the transported multiphase fluids are inside the equilibrium boundary for gas hydrate formation, hydrate crystals can occur and build up as deposits in the flow line (Figure 3). Formation of a hydrate plug, especially in a subsea flow line is a serious problem and removal of the plug requires specialist skills. Therefore, many methods to avoid hydrate plug formation have been developed (Figure 4). Due to the real fear of having to deal with a hydrate plug, the operator usually wants a method that gives total prevention of hydrate formation under all field conditions, over the field lifetime, and both during production and shut-in. These methods usually rely on keeping outside the thermodynamic region for hydrate formation, either by pressure or temperature control or by shifting the equilibrium for hydrate formation.

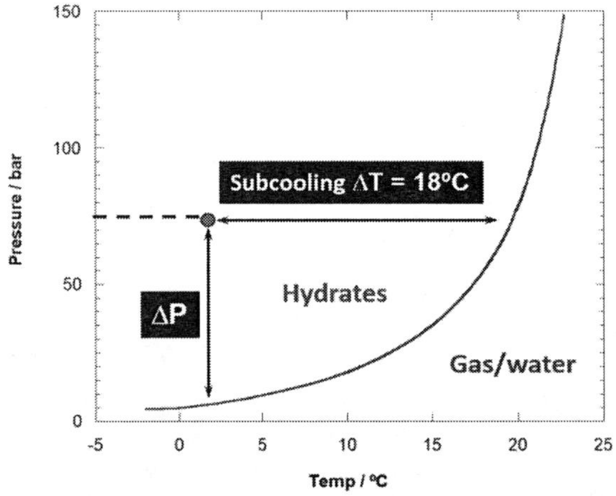

Figure 3. Pressure-temperature diagram for gas hydrate formation, illustrating determination of subcooling (ΔT).

If the field is large and water production high, chemical methods tend to be too expensive even if chemical recovery and regeneration is used. In such cases engineering solutions are usually applied based on an understanding of the flow conditions expected and the pressure and temperature changes across the whole flow line. Multiphase flow simulators are used for these calculations. If the volume of fluids is high (usually for oilfields), hydrate problems can often be avoided under normal production flow rates because the temperature does not drop sufficiently to enter the hydrate formation region. For many fields, only in planned or unplanned shut-ins will the temperature be low enough for hydrates to form. Some form of fluid heating during the shut-in, such as electrical heating, can be used. This can be fairly expensive in terms of both CAPEX and OPEX, especially for very long subsea flow lines.

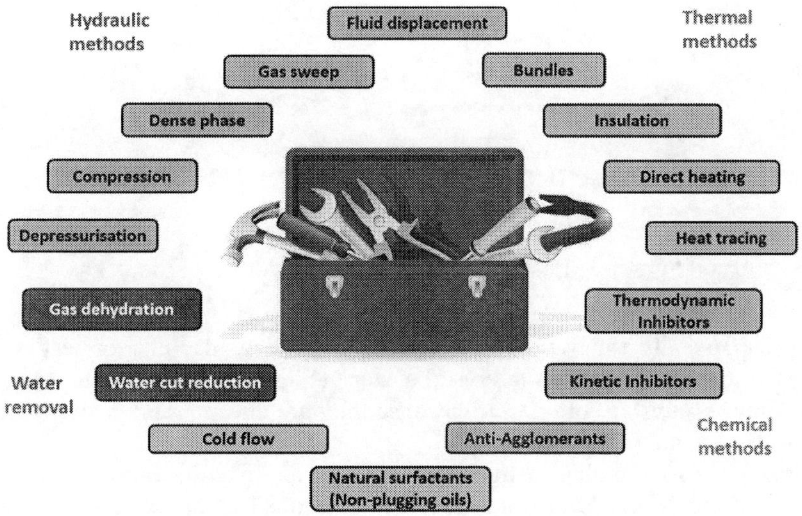

Figure 4. Gas hydrate management tools available to the operator.

2. Chemical Hydrate Inhibition

2. 1 Thermodynamic Inhibitors (THIs)

If the volume of produced water is relatively low (usually for gas or gas condensate fields), chemical methods to protect from hydrate formation can be used. The standard chemical method is to shift the equilibrium conditions for hydrate formation by adding a thermodynamic hydrate inhibitor (THI). Common examples are methanol (MeOH), ethanol or monoethyleneglycol (MEG) but even salts such as sodium chloride or potassium formate will work although they are mostly used to prevent hydrate formation in drilling or completion fluids. High concentrations of THIs are needed to shift the equilibrium, maybe 20-80 wt.% of the water phase needs to be added in some cases. Methanol is sufficiently cheap that in some parts of the world it is discharged overboard with the produced water. MEG is relatively more expensive and a less powerful THI than methanol and is often regenerated for reuse.

However, there are other tools available to the operator. For example, some oils are so-called non-plugging oils at certain water cuts [3]. These oils contain natural surfactants which prevent any formed hydrates from agglomerating and depositing in the line. The exact structure of the active components in these oils is still unknown but some characterisation of the functional groups and molecular weights of these resin fractions has been reported. Model natural surfactants have been investigated but it still looks a long way to go as to when chemists can develop AAs based on our understanding of non-plugging oils [4, 5].

2.2 Anti-Agglomerants (AAs)

The artificial equivalent of these natural surfactants are known as anti-agglomerants (AAs) and are a class of low dosage hydrate inhibitor (LDHI).[6] AAs are also surfactants but were not originally developed based on the knowledge of natural surfactants in oil; instead they came out of projects to find gas hydrate crystal growth inhibitors. These crystal-modifying AAs were first invented by Shell energy company [7, 8]. Various classes of AA surfactants are now used commercially on fields worldwide, particularly for the start-up of new subsea wells. The key feature of these AA surfactants appears to be a hydrophilic head group, which can interact with hydrate crystal surfaces, slowing down the crystal growth rate at the same time as making the hydrate particles oleophilic by coating them with the hydrophobic AA tails (Figure 5). The AAs will also accumulate at the oil-water interface where hydrates first form as well as coating the steel walls of the flow line making it less likely for hydrates to deposit. AA surfactants are therefore “hydrate-philic” [9-13].

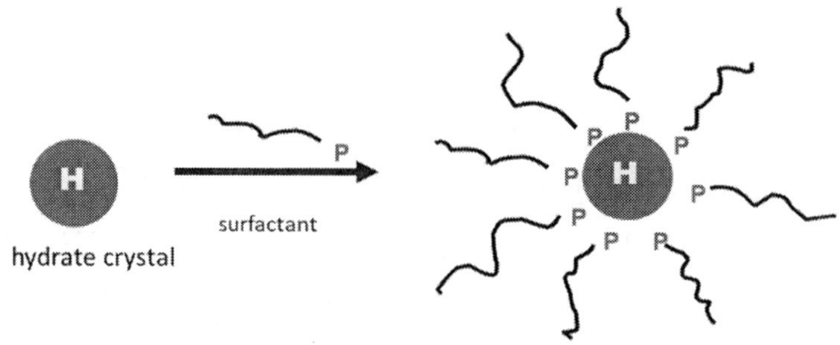

Figure 5. Possible mechanism for hydrate-philic AAs.

The classic head group used in many commercial AA surfactants is a quaternary ammonium or phosphonium group, usually with two or 3 alkyl groups of a particular size for best interaction with open hydrate cavities on the growing hydrate crystal surface (Figure 6). Typical alkyl groups are n-butyl or pentyl groups. Butyl groups are often used because they are cheaper and easier to attach to a nitrogen atom on an industrial scale.

The drawback with AAs is that to transport the crystals of hydrates you need a carrier fluid, and enough fluid so that the "slurry" of crystals is not too viscous but moves smoothly down the flow line. Strongly hydrate-philic AAs keep the crystals small, and often with no sharp edges, which helps keep viscosity low. High water cuts could be a problem if all the water converted to hydrates. Misleading laboratory results in closed systems can be obtained if all the water that could be converted to hydrates is not fully converted. However, as hydrates form the remaining free water becomes more concentrated in any produced salts and at some point can be outside the thermodynamic hydrate formation boundary. Both liquid hydrocarbon and unconverted concentrated produced water can thus be the lubricant for the hydrate crystals.

The best AAs are claimed to work at very high subcoolings, if dosed sufficiently. This makes them useful for deep water (high pressure and cold water) applications where the subcooling can be 20+ °C. However, some AAs do not perform well in fresh water. Another drawback of using AAs is the quality of the produced water, i.e. the strong surface-active property of the AA may challenge the oil-water separation processes and lead to too much oil-in-water for discharge approval. This, several AAs from different vendors may pass the hydrate tests but could fail on the water quality. However, more recent AAs have overcome this drawback [14].

As AAs are surfactants they are generally quite toxic, especially the quaternary ammonium and phosphonium surfactants. To avoid long term ecotoxicological problems it is desirable if the AA degrades quickly. Recently, an AA was developed with over 60% biodegradation in the OECD306 seawater test [15].

2.3 Hydrate slurry transportation without chemicals

The ideal method to produce a transportable slurry of hydrate particles would be to avoid the use of chemicals altogether. As mentioned earlier some crudes won't give hydrate plugs under certain conditions, thus avoiding any extra chemical need, but the majority of crudes will cause a plug. In such cases a new method has been developed which requires no additives. The idea is to get the hydrates to nucleate fast under special conditions so that the "sticky" phase when hydrates would normally aggregate and agglomerate is prevented from occurring [16-18].

The method can involve seeding to accelerate the hydrate process to such an extent that the particles have no time to agglomerate before all the water is converted. Some researchers call the method "cold flow". Field trials have been performed with partial success. Naturally, operators will need to see fully successful and reliable field data before they will be willing to risk using this method in the field. Having back-up storage of AA chemical may help reduce the risk.

IOP Conf. Series: Materials Science and Engineering **700** (2019) 012057 doi:10.1088/1757-899X/700/1/012057

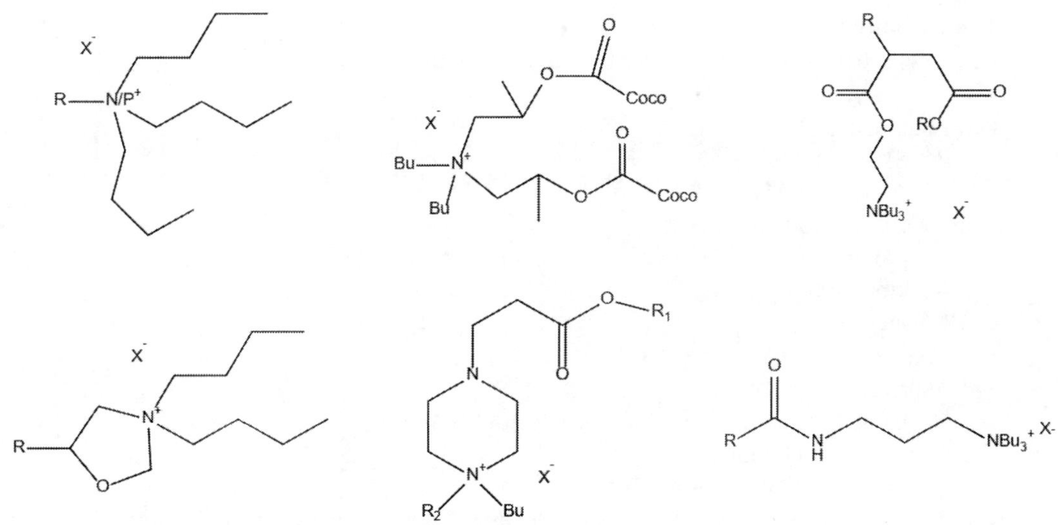

Figure 6. Typical structures of hydrate-philic AA surfactants (R = C12-18).

2.4 Kinetic Hydrate Inhibitors (KHIs)

For fields where there is no or very little liquid hydrocarbons and low produced water volumes (typically gas or gas condensate fields), an alternative LDHI known as a kinetic hydrate inhibitor (KHI) can be used.[1,6,14,19-21] KHIs typically consist of water-soluble polymers as the main component, often containing amide or imide functional groups, as formulations in solvents, and are applied at low concentrations, typically 0.1−2 wt. % based on the amount of produced water. The commonest KHI polymer classes are poly(N-vinyl lactam)s, poly(N-alkyl (meth)acrylamide)s, and copolymers thereof, and hyperbranched polyesteramides (Figure 7).

Synergists, such as other polymers or smaller nonpolymeric compounds, are often added to the formulation to enhance the KHI activity of the water-soluble polymer. Sometimes the solvent itself functions as a synergist. The mechanism by which KHIs and synergists work is still somewhat unclear, but is usually assumed to include a combination of nucleation and crystal growth inhibition.[22-23]
KHIs are fairly expensive products, the base polymers can often cost 10-20 Euro/kg. However, they can still be a cheaper method of hydrate control than more traditional methods where the water production is low. Recently, methods to capture and recycle the KHI have been reported.[24-25] This could open up the use of KHIs at much higher water production rates.

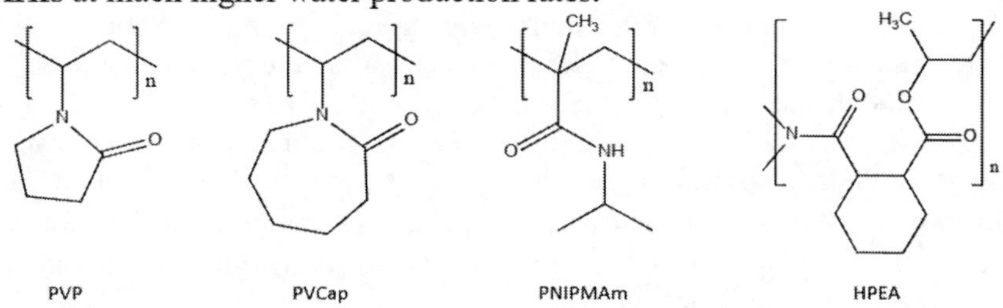

Figure 7. Common monomer units in commercial KHI polymers.

The performance usually increases with increasing concentration in the range 0.1-2.0 wt.% but the improvement generally becomes less and less. The performance limit is often quoted as a combination of time and subcooling. The best SII hydrate KHIs can prevent hydrate formation for 48 h at 12-14 °C subcooling. However, there are many other factors that will affect the performance that must be considered, including:

- Temperature profile during the residence time
- Shut-in/start-up situations

- Pressure
- Total salinity (TDS)
- Brine composition e.g. monovalent vs. divalent cations
- pH
- Gas composition
  - SI vs. SII
  - Acid gases (H_2S, CO_2)
- Condensate or crude composition
  - aromatics vs. aliphatic content, resins and natural surfactants
  - emulsion properties
- GOR, water cut
- Flow regime (laminar, slug etc.)
- Solid surfaces
  - Clay, formation fines, scale, corrosion products
- Other production chemicals
  - Particularly corrosion inhibitors

In the last 15 years, our group has studied a range of other water-soluble polymers in order to understand the key features for best KHI performance. These studies have shown that many structural features can affect KHI polymer performance. Here we only list the main factors as this is not the place for a full discussion.

- Amphiphilic functional groups
  - They should contain hydrophilic groups, probably with strong hydrogen-bonding e.g. amide or amine oxide
  - The hydrophobic groups should be bonded to or neighbouring to the hydrophilic groups
  - The size and shape of the hydrophobic groups must be optimised. 3-6 carbon atoms acyclic or cyclic aliphatic groups is most common.
  - There is recent evidence that low cloud point is beneficial if the criteria above are satisfied [26].
  - The functional groups should be part of or attached to the backbone of the polymer. This reduces the degrees of freedom and therefore loss of entropy, making the local change in Gibbs free energy less negative
- Polymer molecular weight
  - Low molecular weight is best for polyamides. PVCap with Mw as low as 500 g/mole (i.e. only 4 repeat units) appears to work well as a KHI [27]. However, the method of molecular weight determination is critical. Gel permeation (size exclusion) chromatography is often done with a calibrating polymer with very different characteristics from the KHI polymer. Absolute molecular weight determination without the need for a calibrating polymer can give very different results [28].
  - A bimodal molecular weight distribution has been shown to work better than a monomodal distribution [29].
- Maximising the hydrodynamic surface area:volume ratio appears to be important [26].
- Tacticity (of vinyl polymers) – syndiotactic is best due to maximising the surface area:hydrodynamic volume ratio [30, 31].
- Polymer end groups – type of initiator and chain transfer agent can make a difference [32].
- Copolymer – monomer distribution, block vs. statistical (random) [33, 34].
- Unknown effects from the polymerisation procedure – this is well known from the chemical companies and can cause poor quality control if the exact procedure is not adhered to each time. Even two polymer samples with similar molecular weight distributions can have different performances due to the method of manufacture, i.e. temperature, solvent, polymer concentration, initiator, method of mixing, reaction time etc.

3. Conclusion

Many methods have been developed to combat the potential of hydrate blockages in flow lines. No one method is a panacea for all fields and therefore different methods are used for different field scenarios for the best economics for the owner companies. It is akin to picking the appropriate tool from the toolbox for the job in question. For large oilfields with high produced water volumes, operators tend to use engineering methods to avoid hydrate formation. For low amounts of water production, chemicals become more of an option. Low dosage hydrate inhibitors (both AAs and KHIs) are useful tools for certain scenarios but have limitations which were discussed. KHIs that perform at higher subcoolings, are a challenge that would be very welcome to overcome, for either SI or SII hydrates. Finding low toxic AAs may be an unsurmountable challenge as they are so surface-active, but biodegradable AAs are now available. KHI polymers tend to be low in toxicity but not very biodegradable, although some are now available although their performance is not as high as the best KHIs [35-36].

4. References

[1] Sloan E D and Koh C A 2008, *Clathrate Hydrates of Natural Gases*, 3rd ed.; CRC Press: Boca Raton, FL.
[2] Carroll J J 2009 *Natural Gas Hydrates: A Guide for Engineers*, 2nd ed.; Elsevier Gulf Professional: Amsterdam.
[3] Sjöblom J, Øvrevoll B et al. 2010, Investigation of the hydrate plugging and non-plugging properties of oils. *J. Disp. Sci. Techn.* **31**(8), 1100-1119.
[4] Aman Z M, Sloan E D, Sum A K, and Koh C A 2012 Lowering of clathrate hydrate cohesive forces by surface active carboxylic acids. *Energy Fuels* **26**(8), 5102-5108.
[5] Syddall W G T, Haber A, Qin Y, Graham B, Aman Z M, May E F, Johns M J and Pickering P F, 2017, Characterization of crude oils that naturally resist hydrate plug formation. *Energy Fuels* **31**(6), 5806-5816.
[6] Kelland M A 2006, History of the development of low dosage hydrate inhibitors. *Energy Fuels* **20**, 825−847.
[7] Klomp U C, Kruka V C and Reijnhart R 1995 WO Patent Application 95/17579.
[8] Klomp U C and Rejnhart R 1996 International Patent Application WO 96/34177.
[9] Delroisse H, Torré J-P and Dicharry C 2019 Effects of a quaternary ammonium salt on the growth, wettability, and agglomeration of structure II hydrate crystals. *Energy Fuels* **32**, 12277-88.
[10] Bellucci M A, Walsh M R and Trout B L 2018 Molecular dynamics analysis of anti-agglomerant surface adsorption in natural gas hydrates. *J. Phys. Chem. C*, **122**(5), 2673-83.
[11] Chua P C and Kelland M A 2013 Study of the gas hydrate anti-agglomerant performance of a series of n-Alkyl-tri(n-butyl)ammonium bromides. *Energy Fuels* **27**, 1285−92.
[12] Sicard F, Bui T, Monteiro D, Lan Q, Ceglio M, Burress C and Striolo A 2018, Emergent properties of antiagglomerant films control methane transport: Implications for hydrate management. *Langmuir* **34** (33), 9701-10.
[13] Mehrabian H, Bellucci M A, Walsh M R, Trout B L, 2018, Effect of salt on antiagglomerant surface adsorption in natural gas hydrates. *J. Phys. Chem. C*, **122**(24), 12839- 49.
[14] Kelland M A 2014 *Production Chemicals for the Oil and Gas Industry,* 2nd ed.; CRC Press: Boca Raton, FL.
[15] Li X, Amundsen Ø, Kinnari K, Cely A, Askvik K, Kelland M A and Abrahamsen E 2017 *Discovery of an environmentally friendly low dosage hydrate inhibitor anti-agglomerant for NCS,* Proceedings of the Oil Field Chemistry Symposium 2017 Geilo, Norway, 26-29th March.
[16] Straume E O, Morales R E M and Sum A K 2019 Perspectives on gas hydrates cold flow technology, *Energy Fuels.* **33**(1), 1–15.
[17] Talley L D, Turner D J and Priedeman D K 2007 *Method of Generating a non-plugging hydrate*

slurry. International Patent No. WO 2007/095399.
[18] Larsen R, Lund A, Hjarbo K W and Wolden M 2009 *Robustness Testing of Cold Flow*, 20th International Oilfield Chemistry Symposium, Geilo, Norway, March 22–25.
[19] Kelland M A 2011 *A review of kinetic hydrate inhibitors: Tailormade water-soluble polymers for oil and gas industry applications. In Advances in Materials Science Research*; Wytherst, M. C., Ed.; Nova Science Publishers, Inc.: New York, 2011; Vol. 8.
[20] Perrin A, Musa O M and Steed J W 2013 The chemistry of low dosage clathrate hydrate inhibitors. *Chem. Soc. Rev.*, **42**(5), 1996–2015.
[21] Shahnazar S, Bagheri S, TermehYousefi A, Mehrmashhadi J, Abd Karim M S and Kadri N A 2018 Structure, mechanism, and performance evaluation of natural gas hydrate kinetic inhibitors. *Rev. Inorg. Chem.* **38**(1), 1-19.
[22] Yagasaki T, Matsumoto M and Tanaka H 2019 Molecular dynamics study of kinetic hydrate inhibitors: The optimal inhibitor size and effect of guest species. *J. Phys. Chem. C* **123**, 1806-16.
[23] Xu J F, Li L W, Liu J X, Wang X P, Yan Y G, Zhang J 2018 The molecular mechanism of the inhibition effects of PVCaps on the growth of sI hydrate: an unstable adsorption mechanism, *J. Phys. Chem. Chem. Phys.* **20**, 8326-32.
[24] Tohidi F, Anderson R and Tohidi B 2018 Evaluation of a novel water-immiscible kinetic hydrate inhibitor formulation. *Energy Fuels* **32**(6), 6518–23.
[25] Rithauddeen M A, Al-Adel S I and Mohammad A F 2019 *Removal of Kinetic Hydrate Inhibitors*, US Patent Application US20190118114.
[26] Dirdal E G and Kelland MA 2019 Does the cloud point temperature of a polymer correlate with its kinetic hydrate inhibitor performance? *Energy Fuels* **33**(8), 7127-37.
[27] Clements J, Pakulski M K, Riethmayer J and Lewis D C 2017 International Patent Application WO2017/048424.
[28] Ree LHS, Opsahl E & Kelland M A 2019 N-Alkyl methacrylamide polymers as high performing kinetic hydrate inhibitors. *Energy Fuels* **33**, 4190–201.
[29] Colle K, Talley L D and Longo J M 2005 World Patent Application WO 2005/005567.
[30] Chua P C, Kelland M A, Hirano T and Yamamoto H 2012 Kinetic hydrate inhibition of poly(N-isopropylacrylamide)s with different tacticities. *Energy Fuels* **26**, 4961-67.
[31] Chua P C, Kelland M A, Ishitake K, Satoh K, Kamigaito M and Okamoto Y 2012 Kinetic hydrate inhibition of poly(N-isopropylmethacrylamide)s with different tacticities. *Energy Fuels* **26**, 3577–85.
[32] Zhang Q and Kelland M A 2018 Study of the kinetic hydrate inhibitor performance of poly(N-vinylcaprolactam) and poly(N-isopropylmethacrylamide) with varying end caps. *Energy Fuels* **32**, 9211–19.
[33] Ree L, Kelland M A, Haddleton D and Alsubaie F 2017 Comparison of the kinetic hydrate inhibition performance of block and statistical N-Alkylacrylamide copolymers. *Energy Fuels*, **31**, 1355–61.
[34] Reyes F T, Kelland M A, Kumar N & Jia L 2015 First investigation of the kinetic hydrate inhibition of a series of poly (β-peptoid)s on structure II gas hydrate, including the comparison of block and random copolymers. *Energy Fuels* **29**(2), 695–701.
[35] Kelland M A 2018 A review of kinetic hydrate inhibitors from an environmental perspective. *Energy Fuels* **32**, 12001–12.
[36] Magnusson C, Abrahamsen E, Kelland M A, Cely A, Kinnari K, Li, X & Askvik K M, 2018, As green as it gets: An abundant kinetic hydrate inhibitor from nature. *Energy Fuels* **32**, 5772–8.

Work on structural integrity for semi submersibles exposed to bergy bits – integrated analysis of ice structure impacts

W Lu[1*], Z Yu[2], M Van D Berg[1], R Lubbad[1], E Kim[2], J Amdahl[2], L G Bjørheim[3], M A Langøy[3] and S Løset[1]

[1] Arctic Integrated Solutions (ArcISo AS) & Research Centre of Sustainable Arctic Marine and Coastal Technology (SAMCoT), Norway
[2] Department of Marine Technology, NTNU, Norway
[3] Petroleum Safety Authority, Norway

[*] Corresponding author: wenjun.lu@ntnu.no

Abstract. Bergy bits and growlers (i.e., glacial ice features with a waterline diameter < 15 m) travelling with waves are identified as risk to the integrity of offshore structure and ships operating in the high North. It is generally more difficult to detect and monitor these small ice features and apply concurrent ice management operations. Therefore, it is important to study the corresponding damage assessment of a potential impact with such glacial ice features. Traditionally, while carrying out a structural damage assessment, the ice feature is often treated as a rigid body, whereas structural deformation and fractures are simulated with, e.g., NonLinear Finite Element Method (NLFEM). However, this method is too conservative and neglects that the ice feature crushes and dissipates part of the impact energy (often obtained with the so-called external mechanics). On the other hand, it is possible to treat both the ice and the structure deformable, and a fully coupled simulation can be carried out with NLFEM. This method can capture the physical essence that the impact energy dissipation is shared by both ice crushing and structural deformation, thus leading to a less conservative structural damage assessment. However, this method is often too computationally expensive and can only take care of the so-called 'internal mechanics' part. In this paper, we propose an integrated approach. This approach is mainly based on the innovative Simulator for Arctic Marine Structures (SAMS) to calculate the impact energy following the principle of 'external mechanics'. Afterwards, the simulated results are integrated with NLFEM simulations to re-construct the shared impact energy dissipation between both the ice feature and structure. A case study regarding a glacial ice feature (diameter = 15 m) impacting a semi-submersible is carried out in this paper. This integrated simulation approach (i.e., SAMS + NLFEM) is proved to be significantly more effective than the fully coupled analysis; more versatile and less conservative than the limiting scenario analysis when dealing with the impact between ice features and man-made structures.

1. Introduction

The objective of this paper is to present background for the work initiated by the Petroleum Safety Authority (PSA) as a part of the NORD-initiative to assess challenges and possible improve the safety level of operation of oil and gas installation in the high north. Two of the main activities have been to: Achieve a better understanding of the extent of and uncertainty about icebergs and growlers in the Barents Sea, including methods for mapping and forecasting snow and ice. And to assess the risk level of loss of structural integrity for semi submersibles exposed to bergy bits, and identification of possible gaps in knowledge

In this paper the consequences of ice to the structural integrity is presented. Ice actions should be taken into the design considerations for structures to be deployed/installed in ice-prone areas. The calculation of ice action depends on a wide spectrum of parameters [1], among which, the design ice condition is of paramount importance. The Petroleum Safety Authority Norway (Petroleumstilsynet) recently initiated a series of projects to study the structural safety in the Northern areas (of Norway). One of the most recent projects was carried out by ArcISo AS and the study site was chosen at the northernmost Block A in the 23rd licensing round (located at N74°, E35.67°). For the ice condition in this selected study site, it is bergy bits and growlers (i.e., glacial ice features with a waterline diameter less than 15 m) that are of major concern. This is because these relatively small glacial ice features are more difficult to detect and monitor by concurrent ice surveillance system; and more difficult to perform existing ice management operations due to their limited sizes. Given this selected ice features' potential threat, their impacts with offshore structures and consequent damage assessment should be carried out. This paper describes an effective yet relatively accurate approach to perform such damage assessment with respect the impact between an ice feature and an offshore structure. This method is based on integrated analyses using the Simulator for Arctic Marine Structures (SAMS) and NonLinear Finite Element Method (NLFEM); and will be described in the following sections.

2. Background

From a regulators perspective the attention to bolted connections is associated with the risk it imposes. The PSA definition of risk is defined in guideline to Framework regulation, Section 11: "*Risk means the consequences of the activities, with associated uncertainty*".

Two of the main activities have been to: Achieve a better understanding of the extent of and uncertainty about icebergs and growlers in the Barents Sea, including methods for mapping and forecasting snow and ice. This has been reported in the project "Challenges linked to ice and snow – Icebergs in the Barents Sea" (24). And second to assess the risk level of loss of structural integrity for semi submersibles exposed to bergy bits. This has been reported in the projects "Aspects of structural safety in the Barents Sea" (25) and Loads, design and operation of floaters in the Arctic (26-27). The conclusion of this effort is that the risk of ice hazards to structural integrity has to be further studied.

3. Method

For the considered ice feature – structure impact study, ice crushing plays an important role. There are many ways to characterize the ice crushing properties, e.g., the design Pressure-Area (P-A) curves and process P-A curves. The design P-A curves are often viewed in the context of local or global ice loads and in the context of probabilistic or deterministic approaches to design. The process P-A curves (termed by Frederking [2]) describe the process of a structure penetrating into an ice feature or of an ice feature hitting a structure. It is a continuous plot of pressure versus nominal contact area variation during an ice-structure interaction process.

The choice on how we characterize the ice crushing process is again influenced by the chosen design limiting state. Generally, we have the Ultimate Limit State (ULS), in which, the designed structure is supposed to undergo only elastic or small plastic deformations for the given impact. For the internal mechanics (i.e., local damage assessment), when the structure is not allowed to deform, localized high pressures can develop anywhere in the contact area and they govern the local design for the ULS state. This contact pressure can be characterized by the design P-A relationships.

On the other hand, for the Accidental Limit State (ALS), the structure is permitted to undergo substantial plastic deformations given an ice impact. In this case, it is the crushing process, not the peaks of the crushing force that are of interest. Because the structure in direct contact are most likely to 'give in', the contact force peaks due to localized high pressures, as in the ULS scenario, are not applicable herein. Instead, it is the overall process (without the formation of localized high pressure zones) that governs the contact force. For this paper, it is the ALS condition that is focused. Therefore, the impact studies carried out in this project are following the ALS design principles when it comes to characterizing how the ice crushes and its associated contact force.

According to NORSOK code [3] for ALS designs, several design formats can be found for ice actions. For example, the ductile design, in which the ice feature can be assumed as a rigid body and the collision energy is mainly absorbed by the structure. For its simplicity and convenience, this approach is adopted in a previously related study project [4].

4. SAMS simulation

For the first limiting scenario (i.e., impacting between a crushable ice feature and rigid structure), SAMS is adopted to offer a unique and effective solution. SAMS can construct *contact force – penetration depth* curves for each impact cases with varying impact direction and location. In addition, for each impact case, SAMS also yields the solutions to external mechanics to determine the demand for energy dissipation. It is how much of the kinetic impact energy must be dissipated by deformation of the structure and/or the ice feature and how much that will remain as kinetic energy, possibly transferred to other rigid body motions and surrounding fluid field. The demand for energy dissipation only specifies the total amount of deformation energy, not how this is distributed/shared between the structure and the ice feature. Further quantification of energy sharing is determined by the proposed integrated analysis when incorporating both limiting scenario analyses.

4.1. A Brief Introduction to SAMS

SAMS is a time-domain simulation product of Arctic Integrated Solutions (ArcISo AS) and it is largely based on the non-smooth Discrete Element Method (DEM) whose contact algorithm is tailor-developed for ice materials. Until recently, time domain models of sufficient quality to perform numerical simulations of ice and marine structures interactions have not existed. Today, this has changed, partly through the efforts at the Norwegian University of Science and Technology (NTNU) hosting SAMCoT (Centre for Research-based Innovation - Sustainable Arctic Marine and Coastal Technology), laying the foundation of a versatile and accurate numerical simulator for fixed and floating structures in various ice conditions. As continuum-mechanics-based methods are not valid for representing realistic ice conditions in major parts of the Arctic, SAMCoT has placed a considerable emphasis on developing DEMs that enable modelling the interactions between individual ice blocks (e.g., floe ice, level ice, ridges and icebergs) and the structure of interest [5].

The DEM can broadly be divided into two main categories: smooth discrete element modelling (SDEM) and non-smooth discrete element modelling (NDEM). The difference between the two can be seen as the difference between explicit and implicit time integration, allowing much larger simulation time steps, while maintaining stable simulations, when using NDEM. SAMS falls under the NDEM category, but it applies a novel implicit time stepping scheme and an improved contact model, enabling general visco-elastic contacts. SAMS distinguishes two types of contacts: the rigid contacts and the compliant contacts. The earlier does not adopt any upper-limit to the contact force resulting in computationally inexpensive contact model that can properly estimate the average contact force, but not the exact contact behavior. This limitation makes rigid contacts inadequate if ice fracturing is to be encountered and thus they should only be used to model contacts between small ice fragments and the structure. Compliant contacts, on the other hand, are able to predict the exact contact behavior. They consider the contact crushing force as well as the force-penetration gradient, leading to highly accurate contact force predictions [6]. For this study, i.e., one ice feature impacting an offshore structure, it is the compliant contact model that is utilized to simulate the contact between the ice feature and the structure; and the crushing process of the ice feature at the contact zone is characterized by such compliant model.

Aside from the novel contact model SAMS built upon [6], it is also worth mentioning other modules SAMS contains, such as the ice fracture module characterizing the fracture of (mainly) floe ice [7-10] and the hydrodynamic module [11, 12] calculating multiple floating objects' motion in waves, wind and currents.

4.2. Derivation of CSE for Ice Crushing Simulations

As introduced previously, the simulations of an ice feature impacting a structure by SAMS yield results including the contact energy dissipation as a function of time, as well as the occurring contact forces. The contact energy dissipation can be used in combination with the contact forces to derive a force-penetration curve from the SAMS results for the ice feature. Both the contact energy dissipation as well as the contact force depend on the crushing specific energy (CSE) of the ice feature, which is a necessary input parameter of SAMS. The CSE is the amount of energy needed to crush a unit volume of ice (unit: kJ/m3). It is obtained by analyzing full-scale data from in-situ indentation experiments. Through personal communication, we were granted access to the Pond Inlet test data (Geotech Arctic Services, 1985), and we re-analyzed the data set. Force-time histories for the spherically-ended indenters with radius R = 900 mm, 1280 mm and 2300 mm are gathered (ten sets in total). Negative force values are excluded from the analysis. The data sets have been converted from force-time histories to force-displacement history. The indenter displacement, u, during each test was controlled according to the following (1), where t is time and ω is $\frac{100\ mm/s}{0.1R\ mm}$ [13].

$$u = 0.1R\sin(\omega t) \tag{1}$$

To avoid scale effects caused by small strained volumes in relation to the grain size, data points from outmost 5.0 cm of ice have been excluded from the analysis. Penetration depth at the conclusion of the test is 0.1R. For each data point, the crushing specific energy is calculated as in (2),

$$\psi(u_j) = \frac{\int_0^{u_j} F(u)du}{\frac{1}{3}\pi u_j{}^2(3R - u_j)\rho} \tag{2}$$

where ψ is the crushing specific energy (J/kg), F is the measured force (N), u is displacement of the indenter (m), R is the indenter radius (m) and ρ is the ice density (ρ = 900 kg/m^3). For NN scalar data points, the average is defined as in (3)

$$\psi_\mu = \frac{1}{NN}\sum_{k=1}^{NN} \psi_k \tag{3}$$

The resulting average value $\psi_\mu = 2.96$ kJ/kg , range is from 1.30 kJ/kg to 6.09 kJ/kg.

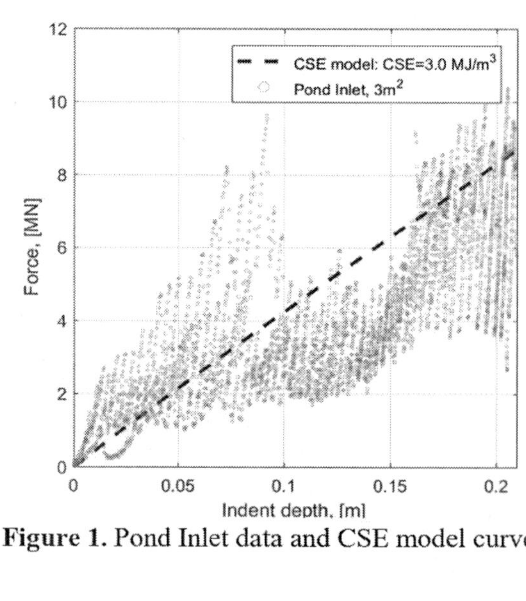

Figure 1. Pond Inlet data and CSE model curve.

Based on the above results, CSE in SAMS is chosen as 3.0 MJ/m^3 (Figure 1). Note that with this approximation method, the high-frequency variation in the contact force, caused by local failure events such as spalling, recrystallization and cracking will not be visible in the SAMS results.

5. NLFEM simulations

After solving the external mechanics part in the previous section, for the other limiting scenario, i.e., a rigid ice feature impacting a deformable structure, numerical simulations were carried out by using explicit NLFEM software LS-DYNA 971. This is also the process to solve for the internal mechanics and conduct local damage assessment.

5.1. Selection of Finite Elements

Since only the local damage is concerned, only part of the structure needs to be modelled into detail. In our case, only the column of the chosen offshore structure (in Figure 6 and to be described below) is modelled with shell elements. More specifically, the four-node Belytschko-Lin-Tsay shell element with reduced integration was used with 5 integration points through the thickness. Hourglass stiffness is added using the stiffness based form (option 4 in LS-DYNA). This is very efficient and gives low dissipation of spurious hourglass energy of less than 2-3%.

5.2. Material and Fracture Modelling

Material and fracture modelling are crucial to determine the structural strength in the ice impact analysis. Material fracture will degrade the structural strength to and beyond the point of collapse. Another effect is leakage and flooding of compartments, leading to stability problems. It is very challenging to accurately simulate fracture initiation and propagation with large scale shell elements. The complexity lies in that fracture is a localized phenomenon in the length scale of plate thickness, and is difficult to capture with large shell elements, the sizes of which are several times larger than the plate thickness. In addition, fracture depends highly on the stress state, material deformation history and is sensitive to the mesh size adopted. It is essential to correctly calibrate the material properties in order to capture accurately strain localizations and the subsequent fracture. The probabilistic nature of material properties makes the fracture modelling even more complicated [14]. Due to the significance of material and fracture modelling and its physical complexity, a proper model is necessary for achieving realistic results.

For material modelling, the power law hardening with a yield plateau is used to model the material. The hardening is described by the yield criterion in (4).

$$f = \sigma_{eq} - \sigma_f(\varepsilon_{eq}) = 0 \tag{4}$$

where σ_{eq} is the von-Mises equivalent stress. The current flow stress σ_f is a function of the equivalent plastic strain ε_{eq} via the Hollomon-type power law hardening rule in (5).

$$\begin{aligned} \sigma_f(\varepsilon_{eq}) &= \sigma_0 \qquad \text{if } \varepsilon_{eq} \le \varepsilon_{plateau} \\ \sigma_f(\varepsilon_{eq}) &= K(\varepsilon_{eq} + \varepsilon_{0,eff})^n \text{ if } \varepsilon_{eq} > \varepsilon_{plateau} \end{aligned} \tag{5}$$

in which, K and n are the hardening parameters and σ_0 is the initial yield stress. To account for the existence of a strain plateau, the hardening is delayed until the plastic strain reaches the plateau strain $\varepsilon_{plateau}$. Thus, $\varepsilon_{0,eff}$ is defined by the relation in (6),

$$\varepsilon_{0,eff} = \varepsilon_0 - \varepsilon_{plateau} = \left(\frac{\sigma_y}{K}\right)^{\frac{1}{n}} - \varepsilon_{plateau} \tag{6}$$

where ε_0 is the strain at initial yield. The BWH (Bressan-Williams-Hill) instability criterion is used to model fracture in the ice impact simulation. The BWH instability criterion was proposed by Alsos, Hopperstad, Törnqvist and Amdahl [15], which combines Hill's local necking model [16] and the Bressan-Williams shear stress criterion [17]. The BWH criterion considers that fracture occurs at the onset of local necking instability neglecting the post-necking regime, and this is conservative. The BWH criterion has been validated of good accuracy by comparison with various impact experiments [18, 19].

Fracture is simulated by eroding the failed elements when the fracture criterion is fulfilled. A through-thickness integration point is failed by setting the stresses to zero once a failure criterion is satisfied. Final element erosion occurs once the middle integration point fails. This approach is preferred over requiring all integration points to fail prior to erosion because nodal fiber rotations in elements undergoing large strains may limit the strains in the remaining integration points, thus resulting in no erosion of the element.

Two kinds of steel material grades are used for the structure materials, and the material properties are shown in Table 1. The outer shell material is with a yield stress of 420 MPa, while the stiffeners are modelled with a 355 MPa yield stress steel.

Table 1. Properties of steel material.

Steel Grade	Plate	HP
Young's Modulus [MPa]	2.07×105	2.07×105
Yield Strength [MPa]	420	355
Poisson Ratio [-]	0.3	0.3
Power law K [MPa]	860	780
Power law n	0.16	0.22
$\varepsilon_{plateau}$	0.0	0.0

6. Integrated analysis

In order to calculate the distribution of energy dissipation between the ice and the structure we will adopt the same principles as those used for ship impact according to NORSOK N-004 Appendix A [20]. The principle is sketched in Figure 2. The ship may represent the ice feature in the present context. The force-deformation curve for the installation is established assuming the ship to be rigid. Likewise, the force-deformation curve for the ship is established assuming the installation to be rigid. The resulting damage is determined when the energy dissipation (equal to the area under force deformation curves) reaches the *demand* for energy dissipation, as determined by the external mechanics analysis in Section 4.

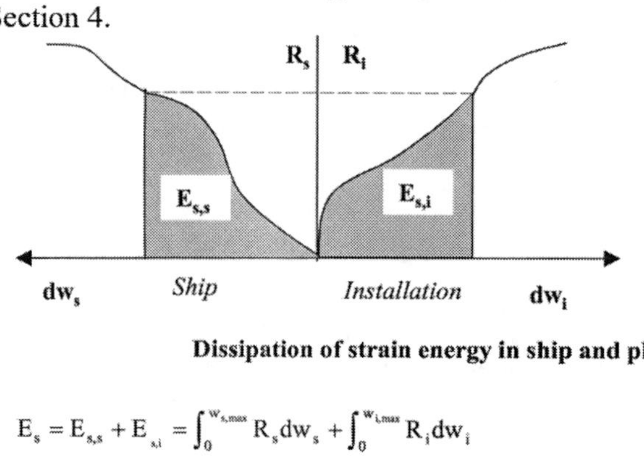

$$E_s = E_{s,s} + E_{s,i} = \int_0^{w_{s,max}} R_s dw_s + \int_0^{w_{i,max}} R_i dw_i$$

Figure 2. Estimation of force and damage (deformation) for ship impact against an installation according to NORSOK N-004.

IOP Conf. Series: Materials Science and Engineering **700** (2019) 012058 doi:10.1088/1757-899X/700/1/012058

It is noted that this approach does not take the coupling of interaction effects into account; deformation of the ship bow increases the contact area and may hence increase the resistance of the installation. Simplified methods to account for this interaction are proposed in a proposed revision of DNV-GL RP-C204, (which is virtually identical to NORSOK N-004).

Similarly, interaction effects exist for ice-structure impacts, but these are associated with significant uncertainties at present. Hence, coupled ice-structure interaction will be neglected in the current investigation.

In summary, the following procedure will be used to determine the damage and energy dissipation:

- Analysis principles for ship-installation impacts in NORSOK N-004/DNV-GL RPC204 will be adopted.
- Extreme local pressures over limited areas will not be considered; it is the resistance to the total force that is essential in the ALS.
- Coupled analysis of the interaction between structural deformation and ice crushing is at present stage not viable and will be disregarded.
- Only the two limiting scenarios described in Section 3 will be analyzed and integrated in this paper for assessment of structural damage. The analysis procedure illustrated in Figure 3 and are summarized below:
  - *Contact force – penetration depth* relationships for the ice will be established using SAMS, i.e., by analyzing ice impacts against a rigid structure.
  - Force-deformation relationship for the structure will be established using NLFEM analysis in LS-DYNA, i.e., by analyzing rigid ice impacts against the structure.
  - The crushing of the ice and the structure damage is determined such that the total energy dissipation is equal to the demand for energy dissipation, as determined by the external mechanics analysis in Section 2.
- The above procedure will be sufficiently accurate as long as the impact period is small compared to the relevant eigenperiods, i.e., the impact force and inertia forces predominate the response.

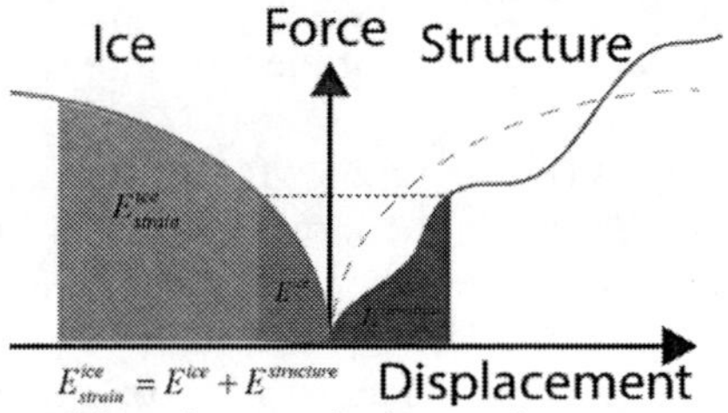

Figure 3. Determination of resulting damage in ice and structure. Pond Inlet data and CSE model curve.

7. Ice impacts case studies

In this section, we will demonstrate the above introduced integrated approach by a case study. In this case study, we pre-define an ice feature and an offshore structure. By virtue of SAMS' efficient simulations, we are able to run a large number (≈1800) impact simulations with varying impact direction and location. Among all these simulations, solutions to the external mechanics are achieved. This allows us to create an energy map showing sensitive locations around the structure, where collision energy is critical (see Section 4). In addition, *contact force – penetration depth* curves for each impact case are also generated by SAMS. Detailed damage assessment (i.e., internal mechanics) are thereby carried out using NLFEM at selected critical locations.

7.1. Ice Feature and Structure Modelling

In this case study, we choose a spheroidal ice feature in Figure 4 for both the external and internal mechanics analyses. It is cautioned here that a different geometry can have a significant influence on the local damage assessment. This is because the local sharpness of an ice feature at the contact location is decisive regarding if the collision energy is absorbed mainly by the ice feature or by local deformation of the structure. For continuation of the study (i.e., a previous study has chosen the spheroidal geometry for local damage assessment [4]) and without deviating from the purpose of demonstrating this method, we chose the same ice feature geometry (in Figure 4) in this case study.

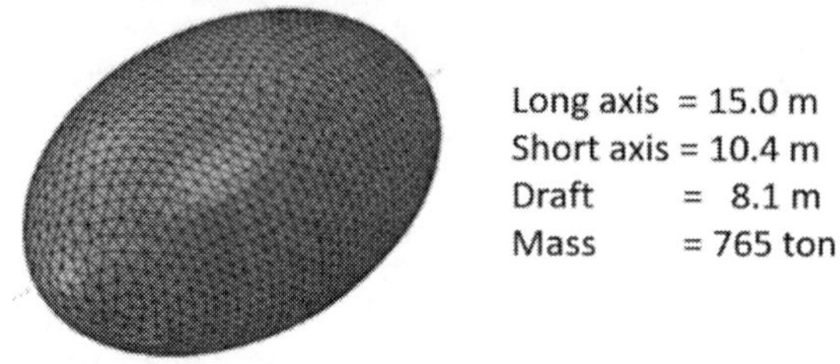

Figure 4. Geometry and size of the spheroidal ice feature.

For the external mechanics analysis (i.e., a crushable ice feature + a rigid structure), we need to explicitly model the global geometry of the structure. In this regard, we choose the global geometry of the Exwave semi-submersible as described in a previous literature [21] (see Figure 5). Its global geometry is digitalized into .obj format and imported in SAMS. In addition, available mass and hydrodynamic properties are also fed into SAMS as inputs for external mechanics analysis so as to construct the energy map and associated contact force – penetration depth curves.

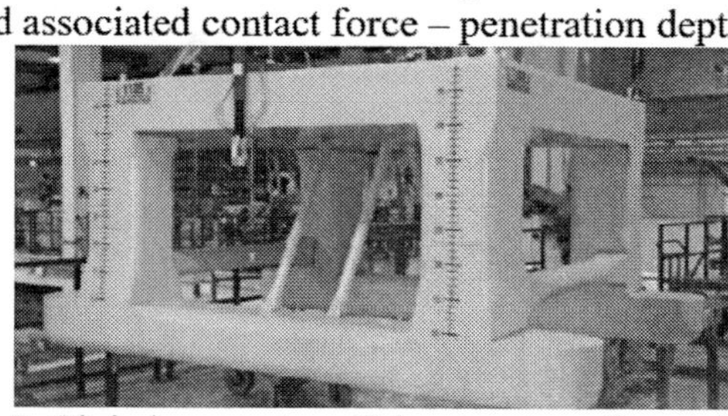

Figure 5. Global geometry of the structure (in model scale).

For the internal mechanics analysis (i.e., a rigid ice feature + a deformable structure), we need not to model the entire structure. Only the part of the structure that is involved in the impact needs to be modelled into detail. In this regard, we choose the 'Midgard' semi-submersible's column structure for the local damage analysis (see Figure 6). The column leg (C10, S10) of the Midgard structure was modelled by Tavakoli and Amdahl [22] for the assessment of structure strength against supply vessel impacts. Detailed FEM modelling of the column is performed in LS-DYNA with shell elements (see Section 5 for details). The rear side, the top and the bottom of the column are constrained in all degrees of freedom (translation in direction of x-, y- and z-axis and rotation around x-, y- and z-axis).

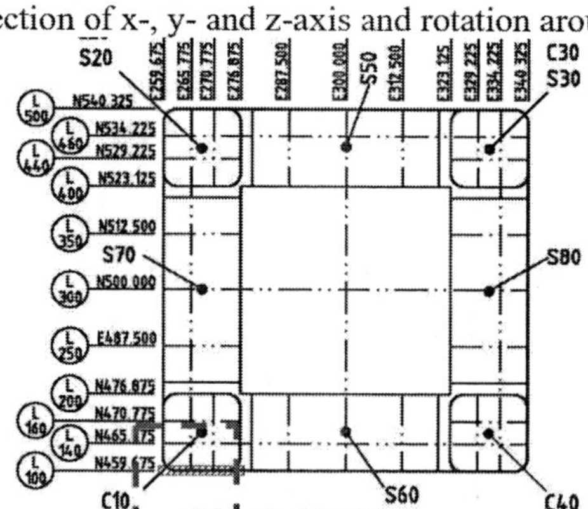

Figure 6. The column (in red dashed square) of the Midgard semi-submersible chosen for local damage assessment.

IOP Conf. Series: Materials Science and Engineering **700** (2019) 012058 doi:10.1088/1757-899X/700/1/012058

Figure 7 illustrates the finite element model of the structure. The column outer shell is in the range of 16-18 mm. The vertical stiffeners used in the column are HP320x12, HP300x11 and HP240x10. These stiffeners were modelled as L-bars with dimensions 320x50x40x12 (mm), 300x50x50x11 (mm) and 240x40x30x10 (mm). This gives nearly the same height, width and the cross sectional area as the HPs. The column model was meshed using approximately 245,000 4-noded shell elements. The general element size is 120 mm.

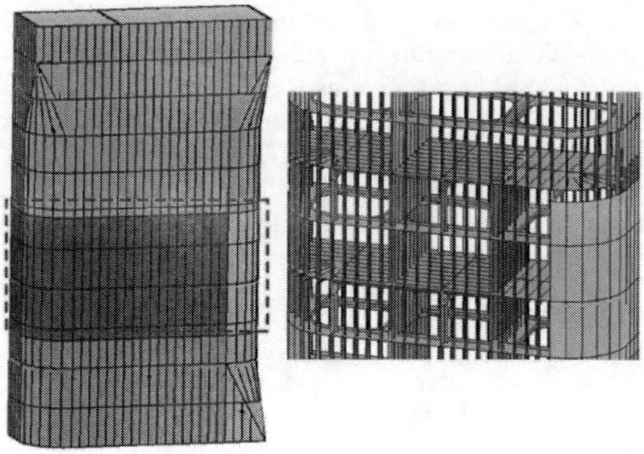

Figure 7. The finite element model of the column structure of the Midgard semi-submersible.

7.2. Impact Simulations by SAMS and NLFEM

With the available modelling of the ice feature and both global and local geometry of the structure, SAMS and LS-DYNA are employed to analyze the two limiting impact scenarios separately.

For the external mechanics analysis with SAMS, ice crushing against a rigid structure is simulated extensively with varying impact locations and directions. In total, around 1800 simulations were conducted. In all simulations, as the structure is symmetric, we only simulate scenarios when the ice feature is approaching the structure from the North-West direction (i.e., only a quarter of the structure is directly exposed to the ice feature's impacts). Detailed simulation matrix is presented in an accompanying report [23]. Figure 8 illustrates only two critical impacting scenarios. We see from Figure 8 that both the ice feature and the entire structure is modelled. The left snapshot of Figure 8 also shows that although the ice feature is travelling from the North-West direction, it is possible that the ice feature manages to travel through the semi-submersible and only impacts the downstream column at the South-East direction. For each of the 1800 impact scenarios, SAMS yields the collision energy and its associated contact force – penetration depth curves.

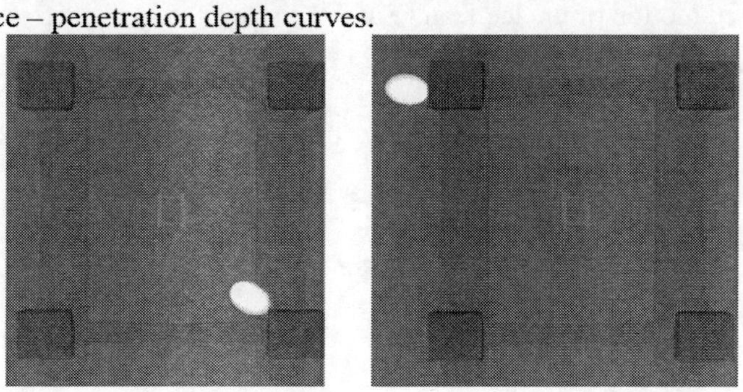

Figure 8. The SAMS simulation snapshots of the two selected impact scenarios of a crushable ice feature impacting a rigid structure (external mechanic analysis).

For the internal mechanics analysis with NLFEM, the simulation is more focused on the deformable structure under the impact of a rigid ice feature. This is achieved with LSDYNA simulations. Aside from detailed structure modelling presented in Section 7.1, the rigid ice feature is given a prescribed motion velocity of 3 m/s (attained through motion analysis in the accompany report [23]), and any strain rate effect is not taken into account. Two kinds of contacts are defined in this analysis, which are the self-contact and master-slave contact. For the rigid ice-column impact, the master-slave contact is used with the column being the slave part. Self-contacts are defined for the column model to detect possible contacts due to deformation. A static friction coefficient of 0.3 was used for all the contacts.

Depending on the angle of impact and location of impact, different simulations can be defined in LS-DYNA. Figure 9 illustrates three possible impact simulations between a rigid ice feature and a deformable local structure column. In practice, as SAMS simulation is much more efficient than the NLFEM analysis, we first carried out a large amount (≈1800) of external mechanics analysis by SAMS and then identify critical locations (≈7) for further NLFEM analysis.

With both limiting scenarios analyzed, for each impact case, we can create the contact force – penetration depth /deformation curve. This further allows us to quantify collision energy dissipations following the shared-energy principle.

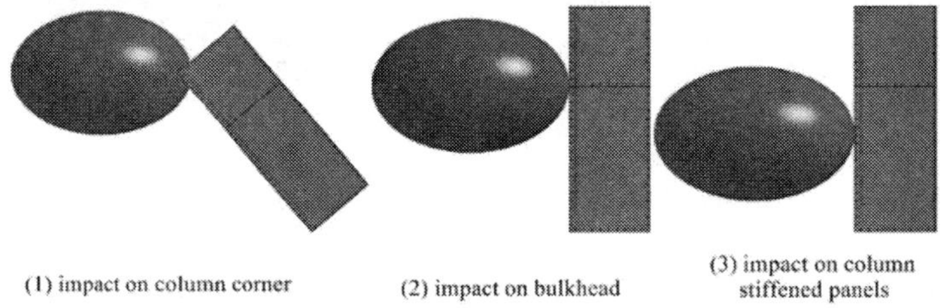

Figure 9. LS-DYNA simulation snapshots of the three selected impact scenarios of a rigid ice feature impacting a deformable column structure (internal mechanic analysis).

8. Results and discussion

This section presents the results of the case studies. The results include the external mechanics analysis by SAMS and internal mechanics analysis by LS-DYNA following the NLFEM.

8.1. Simulation Results by SAMS

The external mechanics (i.e., mainly solving for the collision energy) is obtained by SAMS simulations. The large amount of simulations under various impact scenarios (i.e., impact direction and location) enable us to construct an energy map showing sensitive location of impact. Figure 10a) first illustrate the ice feature's trajectory both before (in red) and after (in green) impacting with structure. Note here that impact scenarios with only a quarter of the structure are illustrated in the figure given various initial location and direction of the advancing ice feature in waves. The impact velocity and probability depends on the coupled motion of the ice feature and the structure in chosen wave conditions. This part of the study is conducted in the accompany project [23].

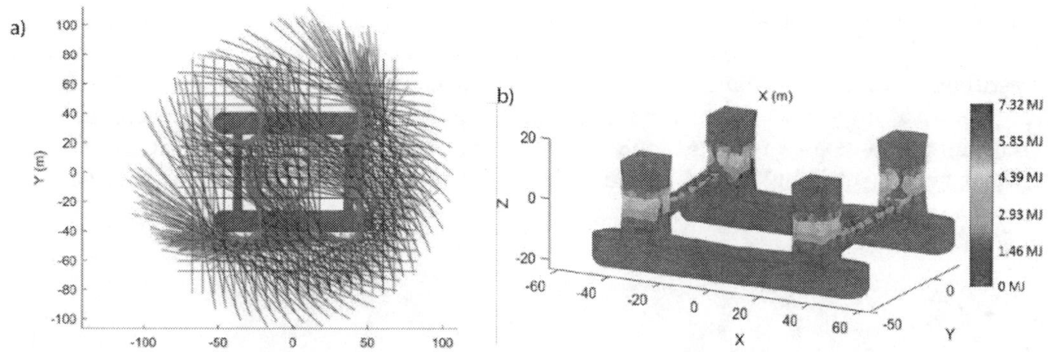

Figure 10. SAMS simulation showing a) The ice feature's impact trajectories (red before impact, green after impact); and b) collision energy distribution at different locations of the structure.

Combining the SAMS simulation results (mainly the collision energy) with the probability of impact and impact velocity, we are able to construct an energy map in Figure 10b) showing sensitive locations that is most exposed to high collision energy. Out of the 1800 simulations, it is found that critical impacts mainly occur on the structure columns. Impacts on the pontoons have a very low probability – only 0.6% of impacts will occur on the pontoons under the considered wave and current conditions. We see from Figure 10b) that most of the impact take place at around the Still Water Level (SWL at Z = 0 m in the figure). Moreover, higher collision energy (to maximum at around 7.32 MJ) are expected as the impact location gets higher (at around +5 m above the SWL) on the structure column. However, high energy impacts have a low probability. 90% of the simulated open water impacts result in an impact energy lower than 4.3 MJ.

With the energy map presented in Figure 10b), we can select several critical locations for further NLFEM analysis. We take a combined reasoning in selecting a critical location on the structure, i.e., a location is qualified as critical only if relatively large collision energy is encountered at a relatively less re-enforced structural locations.

With this combined reasoning, we chose 7 critical locations around the column structure shown in Figure 11.

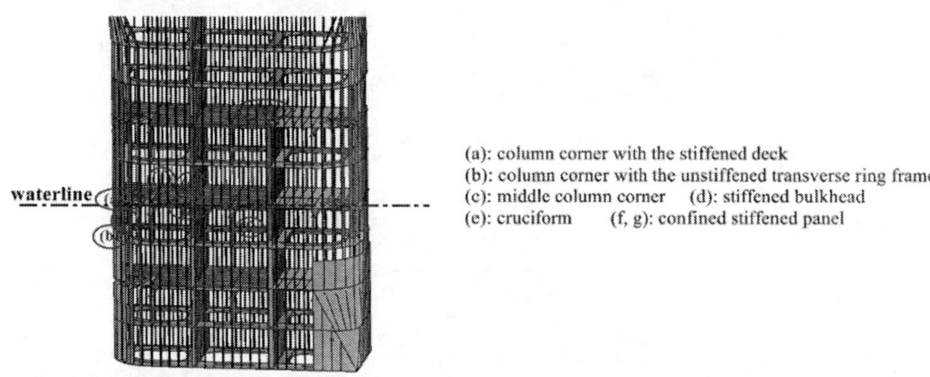

Figure 11. Selected critical locations for further damage assessments by NLFM (internal mechanic).

8.2. Simulation Results by NLFEM

Based on the SAMS simulation results presented in Section 8.1, seven critical locations (together with its associated impact scenarios) were chosen for further local damage assessment by the NLFEM using LS-DYNA. In this section, for exemplifying purpose, we only present the simulation results of impact scenarios (a), (b) and (c) in Figure 11. The entire simulation results can be found in the accompanying report [23]. In addition, it should also be emphasized that the results presented here is only one of the limiting scenarios (i.e., a rigid ice feature impacting a deformable structure). It is incomplete by itself.

IOP Conf. Series: Materials Science and Engineering **700** (2019) 012058 doi:10.1088/1757-899X/700/1/012058

Only after integrating the results from both Sections 8.1 and 8.2 can we establish the complete picture with respect to how the collision energy is shared between the ice feature and the structure.

A rigid ice feature impacts on the structure's column corner are presented here. This includes impacts on (a) stiffened deck, (b) the transverse frame and (c) the column corner in between (a) and (b) (refer to Figure 11). Take impact scenario (a) as an example, and the sectional cut of column deformation at different time instants are given in Figure 12. Note that the impact velocity is 3 m/s. As the crushing distance increases, buckling of the stiffened deck and HP stiffeners are clearly observed until final fracture. Deformation and fracture of the column corner structure at an indentation of 1.5 m is plotted in Figure 13.

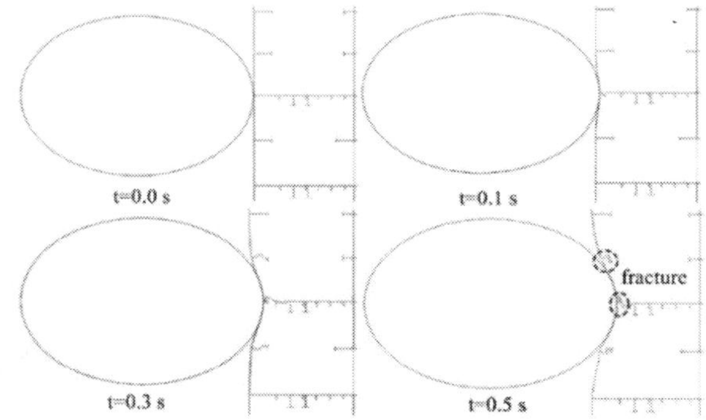

Figure 12. Snapshots of the impact scenario (a) on the stiffened deck (side view).

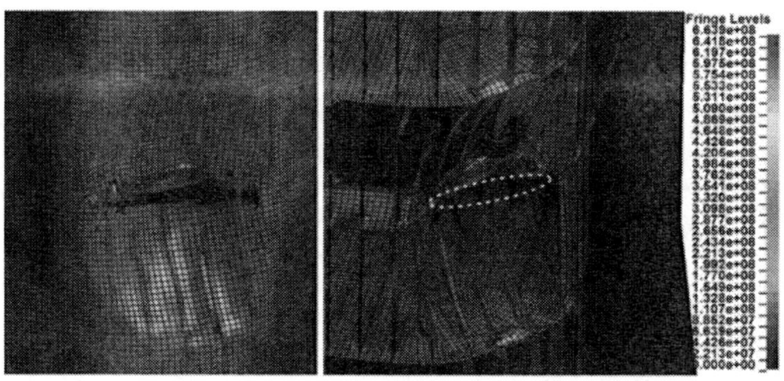

Figure 13. Deformation and fracture of the column corner in scenario (a) at an indentation of 1.5m.

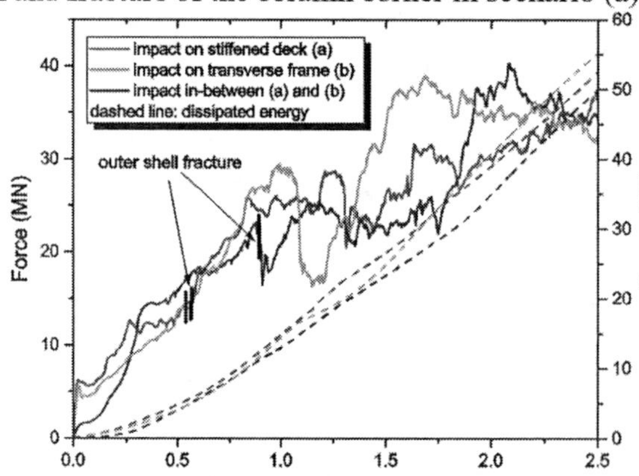

Figure 14. Internal mechanics analysis results: Rigid ice crushing resistance and energy versus crushing distance for scenarios a) – c).

The rigid ice crushing resistance and energy with increasing crushing length is plotted in Figure 14, and the outer shell fracture initiation is marked with short bold lines. Results show that the initial bending capacity of the column corner on the stiffened deck and the transverse frame is about 5 MN. As crushing continues, the indentation resistance increases. Initial outer shell rupture occurs at a crushing length of 0.6 m-0.9 m.

8.3. SAMS and NLFEM Integrated Analysis

We have analyzed two limiting impacting scenarios (i.e., #1 a crushable ice feature impacting a rigid structure and #2 a rigid ice feature impacting a deformable structure) at seven critical locations hitherto. For each impacting scenario, we constructed two limiting curves characterizing the *contact force – penetration depth/deformation* during the impact. Next, we need to integrated these results and quantify how the collision energy is dissipated between the ice feature and the structure. This is achieved by the plot in Figure 15.

Figure 15 plots on the right-hand side, the structural impact resistance on seven different locations from NLFEM simulations assuming rigid ice, while on the left-hand side of Figure 15, the ice crushing resistance calculated from SAMS assuming rigid structures is plotted. The ice load of$F_{ice} = 3.2A^{0.9}$ calculated using the process P-A model (as described in Section 4) is also plotted. It is assumed that under the same force level, both the ice and the structure should deform and absorb energy in accordance with the resistance curves, and the area below the curves are the corresponding energy that is dissipated in ice and the structure, respectively.

From Figure 15, the column bulkhead and cruciform represent hard points of the structure, and are capable of crushing ice significantly. Below a certain force level (around 6 MN for bulkhead and 12 MN for the cruciform), all the deformation goes to the ice. The column stiffened panels have similar force level with that of the ice, and therefore both the ice and the stiffened panels deform and absorb around 50% of the total energy. The resist force of the column corner is much lower than the column side, but it is interesting to observe that the ice crushing force on the corner defined by SAMS is also lower than the ice crushing force from the column side. This makes the strength of ice and structure on the column corner very close, and both objects will deform to absorb the energy.

Figure 16 summarizes the energy absorption with increasing crushing distances. In Section 5, we have calculated a maximum energy dissipation of 7.32 MJ. For simplicity, we take 7.5 MJ here as an example to explain Figure 16. If all the energy 7.5 MJ goes to the structure, this will result in a crushing distance of 0.4-0.7 m. In reality, the total energy should be shared by both ice crushing and structural deformations, and the resulting deformation of the structures will be less. Considering shared energy in ice and the structure, the permanent deformation of the structures with a total energy dissipation of 7.5 MJ is summarized in Table 2.

It is found that given a total energy of 7.5 MJ, both the ice and the column structure should deform and dissipate part of the energy. The structural deformation can vary from 0.25 to 0.55 m depending on the impact locations. These numbers are much smaller than their corresponding values in Figure 16 (between 0.4 to 0.7 m), signifying the shared energy approach analysis as carried out in this paper is less conservative when it comes to structural damage assessment. More specifically, as initial outer shell rupture generally occurs at 0.6 to 0.9 m, the structure is considered safe from compartment flooding for the given impact energy (7.5 MJ). It is noted that the markers on Figure 15 represent initial fracture of the outer shell when one element is eroded. There is still considerable capacity from one element erosion to large outer shell opening.

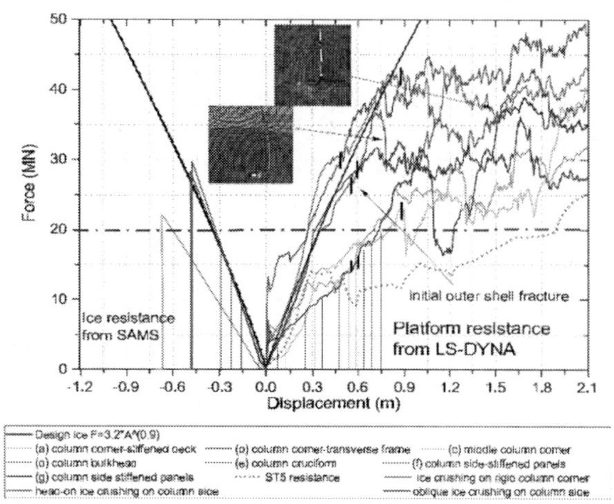

Figure 15. Ice-structure impact resistance.

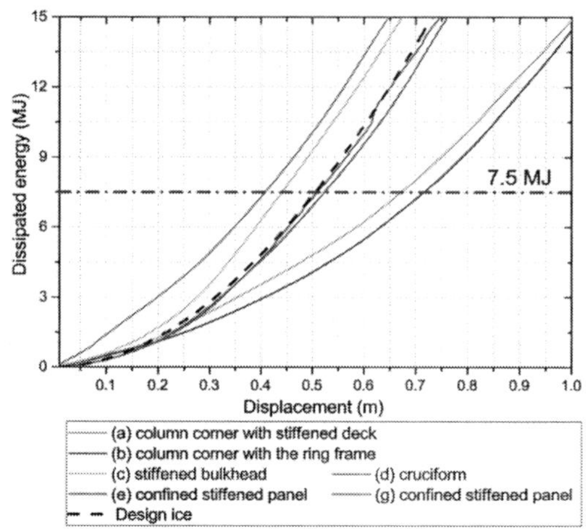

Figure 16. Energy absorption in the ice-structure impacts.

Table 2. Ice crushing depth and structural deformation with a total energy of 7.5 MJ.

Location	Ice crushing [m]	Structure deformation [m]	Energy in Ice [MJ]	Energy in Structure [MJ]
(a)	0.41	0.51	2.5	5.0
(b)	0.42	0.55	2.7	4.8
(c)	0.48	0.50	3.55	3.95
(d)	0.37	0.29	4.1	3.4
(e)	0.33	0.27	3.2	4.3
(f)	0.33	0.38	3.2	4.3
(g)	0.36	0.36	3.75	3.75

9. Conclusions

Following a same principle to analyze ship impacts as in NORSOK N-004 Appendix A, we propose an efficient method to carry out ice – structure impact analysis. This method allows us to quantify collision energy's dissipation in both the ice feature and the structure, thus leading to a less conservative structural damage assessment comparing to the ductile design criteria, in which, the ice feature is normally considered as rigid.

In this proposed method, we integrate the simulation results of two limiting scenarios. These are: 1) external mechanics, in which impact analyses between a crushable ice feature and a rigid structure using the Simulator for Arctic Marine Structures (SAMS); and 2) internal mechanics, in which, impact analyses between a rigid ice feature and a deformable structure using NonLinear Finite Element Method (NLFEM). Important modelling inputs for both SAMS (i.e., the Crushing Specific Energy) and NLFEM (i.e., element types, material modelling and fracture criteria) are described into detail. To demonstrate this integrated method, a case study (i.e., a spheroidal ice feature impacting the Exwave semi-submersible) is carried out.

By virtue of SAMS' efficient simulation to solve for the external mechanics, a large amount of impact scenarios (≈1800) were simulated. This leads to the creation of an energy map illustrating the distribution of collision energy around the structure. Major conclusion in the results to SAMS simulations are:

1. Out of the 1800 simulations, it is found that critical impacts mainly occur on the structure columns. Impacts on the pontoons have a very low probability, i.e., only 0.6% of impacts will occur on the pontoons under the considered wave and current conditions.
2. The maximum impact energy resulting from the simulations is 7.32 MJ. However, high energy impacts have a low probability. 90% of the simulated open water impacts result in an impact energy lower than 4.3 MJ.

Damage assessments were carried out at seven selected critical locations on the structure's column. The integrated SAMS and LS-DYNA analysis results yield rather rich information to quantify impact energy's dissipation in a shared manner. As an example, for a conservatively chosen total impact energy of 7.5 MJ, it is found that:

a. Both the ice and the column structure should deform and dissipate part of the impact energy.
b. The indentation depth into the structure can vary from 0.25 m to 0.55 m depending on the impact locations. As the outer shell rupture generally is predicted to occur at 0.6 m-0.9 m, the structure is considered safe from compartment flooding risks with the given impact energy.
c. For any other impact energy levels, one can refer to Figure 15 to obtain information about the shared energy dissipation, structural deformation and ice crushing depth respectively.

Through the above case study, the simulation efficiency of SAMS is demonstrated. Given the time constraint of the project, SAMS is capable of simulating a large amount of impacting scenarios (i.e., around 1800 cases), whereas NLFEM analysis can only be conducted in selected scenarios (i.e., 7 cases). Moreover, integrating the simulation results from both SAMS and NLFEM, we are able to quantify the collision energy's dissipation between both the ice feature and the structure. As comparisons between Figure 15 and 16 shows, the shared energy approach yields much less conservative damage assessment comparing to if we assume the structure absorb all the collision energy. PSA will promote continued work on "Snow and ice" and "Structural integrity" for the design and operations of floaters for oil and gas activity in areas open for exploration in the north.

Acknowledgements
The authors would like to thank the financial sponsorship from the Petroleum Safety Authority Norway (Petroleumstilsynet) to carry out this study. The first author would also like to thank VISTA – a basic research program in collaboration between The Norwegian Academy of Science and Letters, and Equinor (former Statoil), for financial support in writing this paper.

References

[1] Løset S, Shkhinek K, Gudmestad O T and Høyland K 2006 *Actions from Ice on Arctic Offshore and Coastal Structures*, Krasnodar, St Petersburg, Russia, 2006.
[2] Frederking R 1998 The pressure area relation in the definition of ice forces, *The Eighth Int. Offshore and Polar Eng. Conf., Int. Society of Offshore and Polar Engineers*, 24 - 29 May, Montreal, Canada.
[3] NORSOK, NORSOK Standard N004. Design of steel structures, Appendix A, design against accidental actions. Det Norske Veritas 2004, 2004.
[4] Ekeberg O-C, Shipilova O, Birknes-Berg J and Johansen A 2018, *Glacial Ice Impact*, DNV GL Report No. 2017-0425, Rev. 2.
[5] Lubbad R, Løset S, Lu W, Tsarau A, Van den Berg M 2015 An overview of the Oden Arctic Technology Research Cruise 2015 (OATRC2015) and numerical simulations performed with SAMS driven by data collected during the cruise, Cold Regions Science and Technology.
[6] van den Berg M, Lubbad R and Løset S 2018 An implicit time-stepping scheme and an improved contact model for ice-structure interaction simulations, *Cold Reg. Sci. Technol.* **155** 193-213.
[7] Lu W, Lubbad R and Løset S 2015 In-plane fracture of an ice floe: A theoretical study on the splitting failure mode, *Cold Reg. Sci. Technol.* **110**(0) 77-101.
[8] Lu W, Lubbad R and Løset S 2015 Out-of-plane failure of an ice floe: Radial-crack-initiation-

controlled fracture, *Cold Reg. Sci. Technol.* **119** 183-203.
[9] Lu W, Lubbad R, Løset S and Kashafutdinov M 2016 Fracture of an ice floe: Local out-of-plane flexural failures versus global in-plane splitting failure, *Cold Reg. Sci. Technol.* **123** 1-13.
[10] Lubbad R and Løset S 2011 A numerical model for real-time simulation of ship-ice interaction, *Cold Reg. Sci. Technol.* **65**(2) 111-27.
[11] Tsarau A, Løset S and Grindstad T 2014, Propeller wash by an icebreaker, *22nd IAHR International Symposium on Ice Singapore*, 2014.
[12] Tsarau A 2015 Numerical Modelling of the Hydrodynamic Effects of Marine Operations in Broken Ice, Civil and Environmental Engineering, Norwegian University of Science and Technology, Trondheim, 2015, p. 159.
[13] GEOTECH, Medium Scale Iceberg Impact Simulation Test Program, 1985.
[14] Yu Z and Amdahl J 2018 A review of structural responses and design of offshore tubular structures subjected to ship impacts, *Ocean Eng.* **154** 177-203.
[15] Alsos H S, Hopperstad O S, Törnqvist R and Amdahl J 2008 Analytical and numerical analysis of sheet metal instability using a stress based criterion, *Int. J. Solids Struct.* **45**(7) 2042-55.
[16] Hill R 1952 On discontinuous plastic states, with special reference to localized necking in thin sheets, *J. Mech. Phys. Solid* **1**(1) 19-30.
[17] Bressan J and Williams J 1983 The use of a shear instability criterion to predict local necking in sheet metal deformation, *Int. J. Mech. Sci.* **25**(3) 155-68.
[18] Storheim M, Amdahl J and Martens I 2015 On the accuracy of fracture estimation in collision analysis of ship and offshore structures, *Mar. Struct.* **44** 254-87.
[19] J.N. Marinatos, M.S. Samuelides 2013, *Material characterization and implementation of the RTCL, BWH and SHEAR failure criteria to finite element codes for the simulation of impacts on ship structures*, 6th Int. Conf. on Collision and Grounding of Ships and Offshore Structures, ICCGS 2013, Trondheim, 2013, p 57-67.
[20] NORSOK, Actions and action effects, N-003, Oslo: Norwegian Technology Standards Institution (2017).
[21] Fonseca N and Stansberg C T 2017 *Wave Drift Forces and Low Frequency Damping on the Exwave Semi-Submersible*, Proc. of the ASME 2017 36th Int. Conf. on Ocean, Offshore and Arctic Engineering, Trondheim, Norway.
[22] Tavakoli M T and Amdahl J 2010, Analysis of collision between Midgard platform and 8000 tonnes displacement ship (2010).
[23] Lu W, Yu Z, Van den Berg M, Lubbad R, Amdahl J, Løset S and Kim E 2018 *Assessment of Structural Damage due to Glacial Ice Impact*, in: P.P.S.A. Norway) (Ed.) PTIL-Konstruksjonssikkerhet i Nordområdene, Petroleumstilsynet, Stavanger, 2018, p. 96.
[24] www.psa.no/Challenges linked to ice and snow – Icebergs in the Barents Sea (ST3)
[25] www.psa.no/Aspects of structural safety in the Barents Sea (ST5)
[26] www.psa.no/Loads, design and operation of floaters in the Arctic (ST20-2018)
[27] Ommani B, Berthelsen P A, Lie H, Aksnes V and Løland G 2019, OMAE2019- 95798.

Comparative study of wax inhibitor performance for pour-point reduction of oil from Sirikit Oilfield in Thailand

K Maneeintr*, K Jongkittinarukorn and T Boonpramote

Carbon Capture, Storage and Utilization Research Group, Department of Mining and Petroleum Engineering, Faculty of Engineering, Chulalongkorn University, Bangkok 10330, Thailand

* Corresponding author: Krengkrai.M@chula.ac.th

Abstract. In petroleum industries, waxy crude oil is normally found in petroleum production. Wax formation and precipitation during pipeline transport of waxy crude oils can cause several challenges, including wax deposition and flow reduction which adversely impacts pipeline performance in oil and gas production. To overcome the wax deposition problem, the small amount of chemicals like wax inhibitors can provide an effective preventative measure. One of this mechanism is to reduce the pour point temperature. In this work, oil from Sirikit Oilfield in Thailand is selected for this study. The wax inhibitors are selected and studied the performance for pour-point reduction. These chemicals are Polyalkyl methacrylate (PAMA), Poly(maleic anhydride-alt-1-octadecane) (PMAO), Polyalkyl methacrylate-co-ethylene glycol (PAMAEG) and Copolymers of maleic anhydride (CPMA) with the concentration ranging from 100 to 1000 ppm. Also, n-pentane with the concentration of 5-20 % by weight is used as a solvent for wax inhibitors. The results from this study show that n-pentane can reduce the pour point to 39.41 % compared to the original oil. Furthermore, among the polymer group, PMAO can reduce the lowest pour point to 52.82 %. In addition, the mixed solvents of polymer with n-pentane can provide relatively less effect on pour-point reduction. The mixture of n-pentane with PMAO can lower the pour point to a certain level of 54.96 % compared to the original oil. This study can be applied to use in the real field to prevent wax deposition in the pipeline transportation for oil production.

1. Introduction

Waxy crude oils contain some amounts of wax and paraffin. When the temperature decreases, the wax starts to precipitate and forms the crystals [1]. In the presence of other forms of hydrocarbon such as the cyclic hydrocarbons, the structure of wax molecules can change from linear and/or branching to cyclic form including aromatic ones. This can result in the serious effects on the melting points and on the wax solubility in crude oil [2]. An evaluation of the effect of some parameters in wax precipitation can mitigate this problem [3]. The crystal formation of these molecules could be the reason for the solid wax formation in crude oil; thus, resulting in the reduction of its fluidity [4]. Furthermore, the severity of wax deposition problem is also relied on the type of crude oil and on the composition of the wax molecules in the crude oil. The pour point of the oil portion is also the important characteristic of crude oil that can lead to flow assurance in oil production and transportation.

The wax formation in crude oil with high pour-point temperature can cause major problems in production, transportation and storage of oil including an increase in of oil viscosity and pressure drop in pipeline, a high yield stress for flow return, wax deposition on the pipeline surface and a decrease in equipment performance for oil production; thus increasing the costs of oil production [5].

Recently, some studies have discussed on pour point reduction, but generally it is on the basis on wax deposition and pour-point reduction that researchers have divided the commonly used methods to solve the problem of wax deposition into three groups: 1) mechanical methods, 2) thermal method by using heat and 3) chemical method by using solvents to removal wax [6]. In these current years, most research works have focused mainly on chemical methods to prevent wax precipitation even though they cannot completely prevent wax precipitation in pipelines. Therefore, chemical additives will be the one of the main factors for the performance of chemical method especially with the mechanism of wax crystal modifiers (inhibitors) and pour point depressants [7].

In recent studies, many researchers have investigated the effect of additives on crude oils or a mixture of them as discussed above. For this study, many additives such as Poly(maleic anhydride-alt-1-octadecane) (PMAO), Polyalkyl methacrylate (PAMA), Polyalkyl methacrylate-co-ethylene glycol (PAMAEG) and Copolymers of maleic anhydride (CPMA) with the concentration ranging from 100 to 1000 ppm will be studied as a wax inhibitors for pour-point reduction. Also, n-pentane with the concentration of 5-20 % by weight is used as a solvent in crude oil.

Therefore, the objective of this research is to investigate and compare the effects of different additives on the pour-point temperature of oil sample from Sirikit Oilfield. Moreover, the effect of concentration of each additives on pour-point reduction is also studied. The result of this study can be used as a fundamental data for additives screening for pour-point reduction in the real field in the future.

2. Materials and methods

2.1 Materials

The oil for this study is obtained from the PTT Exploration and Production Public Company Limited, Sirikit Oilfield in Thailand with the density 0.85 g/cm^3 at 70 °C. The oil composition from Sirikit Oilfield is presented in Table 1. The main oil composition is C16-C20 ranging for 21.63 %. N-pentane is purchased from Ajax Company with AR grade. Polymers like poly(maleic anhydride-alt-1-octadecane) (PMAO), polyalkyl methacrylate (PAMA), polyalkyl methacrylate-co-ethylene glycol (PAMAEG) and copolymers of maleic anhydride (CPMA) for this work are ordered from Sigma Aldrich Company. The chemical structure of these additives are illustrated in Figure 1.

Table 1. Oil composition from Sirikit Oilfield.

Percent by weight (%)	Percent by weight (%)
C6-C10	17.6
C11-C15	16.8
C16-C20	21.6
C21-C25	18.1
C26-C30	16.2
C31-C35$^+$	9.7

IOP Conf. Series: Materials Science and Engineering **700** (2019) 012059 doi:10.1088/1757-899X/700/1/012059

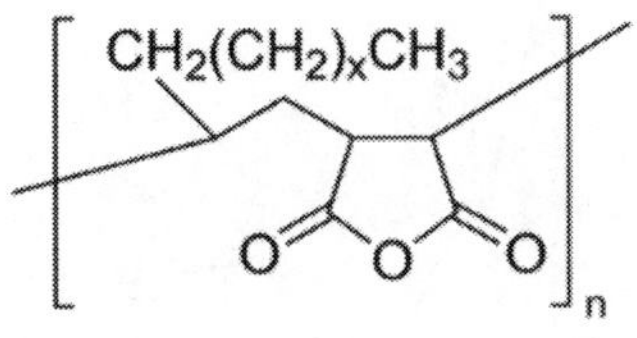

a) PMAO

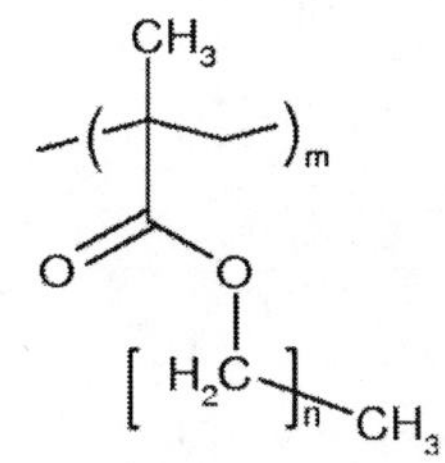

b) PAMA

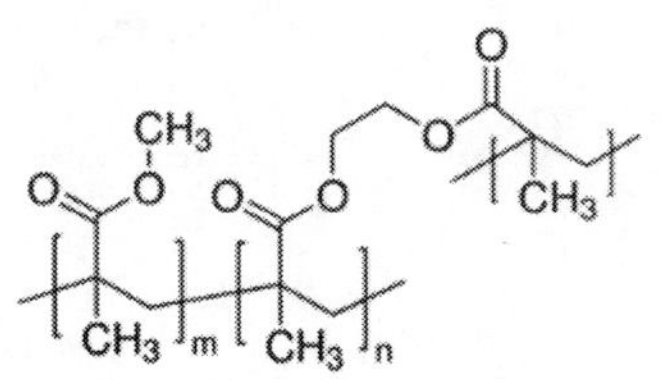

c) PAMAEG

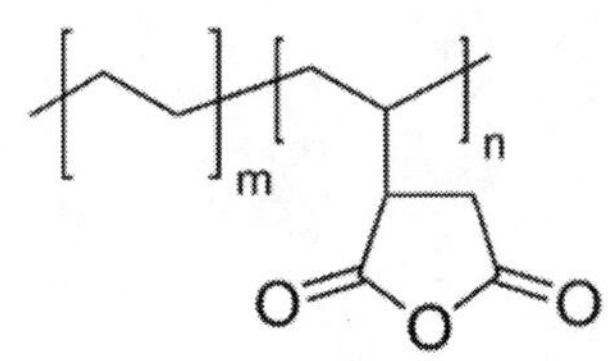

d) CPMA

Figure 1. Chemical structure of additives a) PMAO b) PAMA c) PAMAEG d) CPMA

2.2 Experimental procedures
Initially, to do the pour-point measurement, the procedure is arranged to follow the Standard Test Method for Pour Point of Crude Oils or ASTM D97, as presented below:

The crude oil sample is set to pretreat the oil for testing and it is cooled inside a cooling bath in order to form the paraffin wax crystals. It is kept at about 9 °C above the expected pour point. For every 3 °C, the test container is removed and tilted to check for the surface movement. When the sample does not flow when tilted, the container is held horizontally for 5 sec. If it does not flow, 3 °C is added to the corresponding temperature and the result is the pour point temperature. However, it should be warned that the flow of the sample can be affected by its viscosity. Therefore, the pour point may give a misleading result of the properties of the oil.

3. Results and discussion

3.1 Effect of N-pentane on pour-point reduction
The concentration of n-pentane for this study is ranged from 5-20 % weight. The results is presented in Figure 2. From the results, it is clearly seen that the pour-point temperature is reduced from 37.3 °C to 22.6 °C or 39.4% as the n-pentane concentration increases because of the wax and heavier fractions in the crude oil can dissolve more in n-pentane; thus making the oil more flowable.

3.2 Effect of additives on pour-point reduction
The concentration of polymeric additives for this study are varied from 100-1000 ppm. The effect of additives on pour-point reduction are shown in Figure 3-6 for single additives and a mixture with n-pentane. It can be seen that among these polymers, PMAO can reduce the pour-point temperature to the lower level for 52.8% at 1,000 ppm concentration followed by PAMA, CPMA, PAMAEG and EVAGMA for 40.8%, 37.5%, 30.8% and 29.5%, respectively.

Furthermore, the effects of polymer combined with solvent, 20% n-pentane, are also investigated and presented in the figure. From the results, the pour-point temperature is reduced with an increase in polymer concentration. Moreover, like that of single additive, the pour-point temperature can be lowered with the mixed additives at the concentration of 1000 ppm of PMAO for 55.0% followed by PAMA, CPMA, PAMAEG and EVAGMA for 48.5%, 46.4%, 46.1% and 43.7%, respectively.

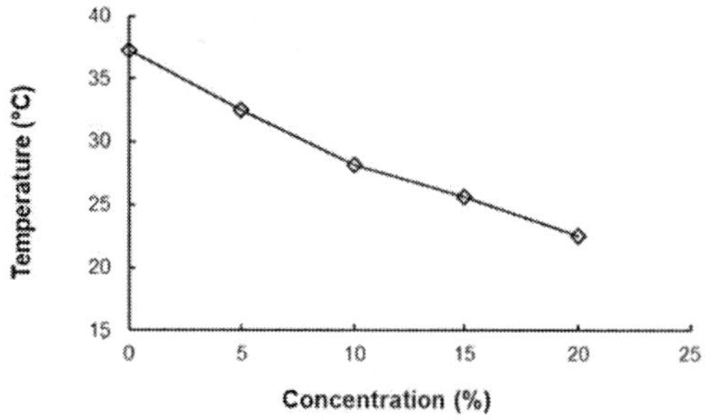

Figure 2. Effect of N-pentane on pour-point reduction.

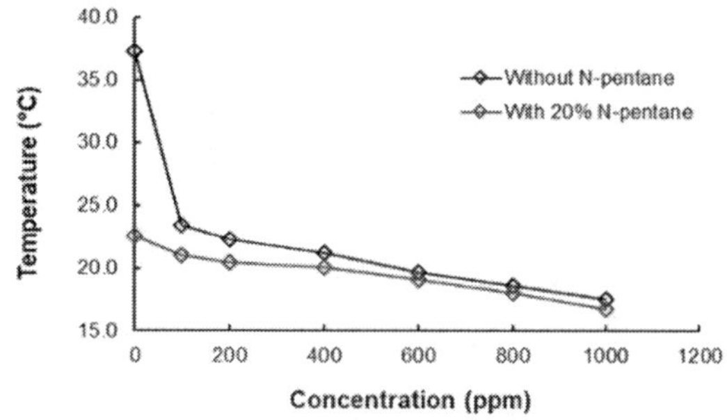

Figure 3. Effect of PMAO on pour-point reduction.

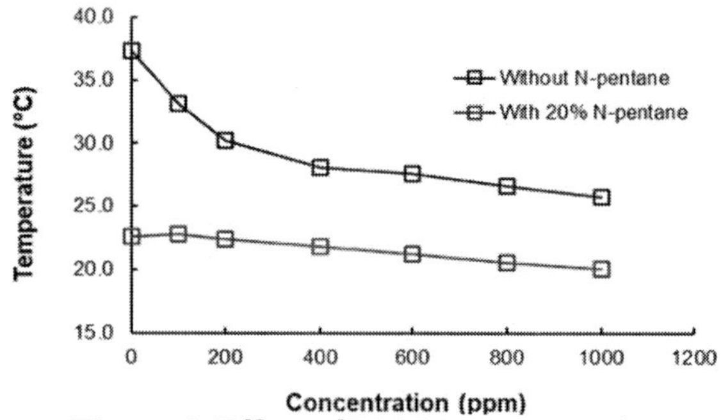

Figure 4. Effect of PAMA on pour-point reduction.

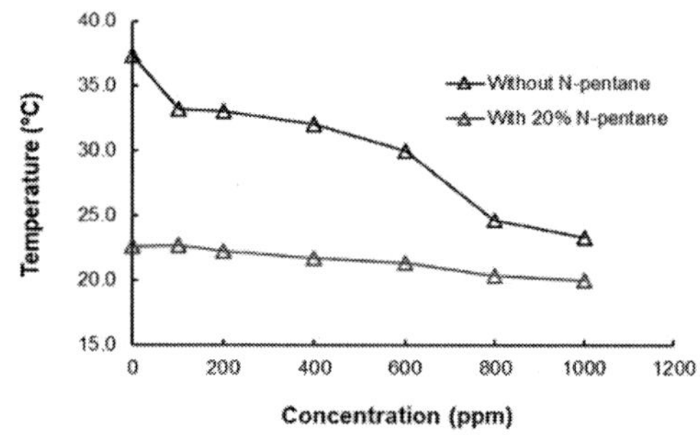

Figure 5. Effect of PAMAEG on pour-point reduction.

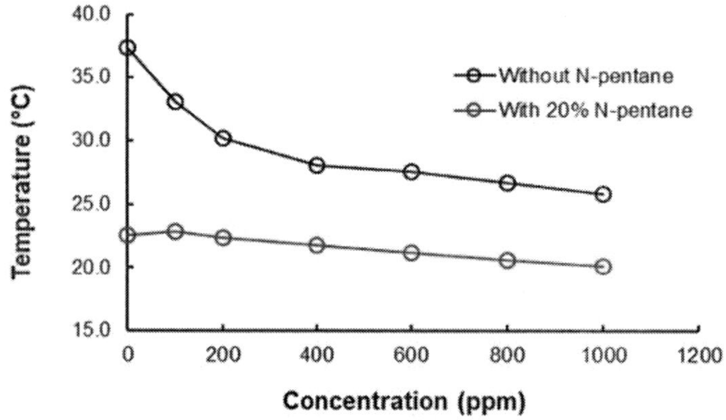

Figure 6. Effect of CPMA on pour-point reduction.

It can be concluded that adding n-pentane can increase the performance of polymer additive for pour-point reduction. The reason is that polymer can prohibit the formation of wax. At the same time, the solvent can dissolve the heavier fraction into liquid. Therefore, the oil can flow in the pipeline.

4. Conclusions

In the petroleum production, wax formation and precipitation can be found in production and transportation from waxy crude oil causing oil flow reduction. To solve this problem, the small amount of wax inhibitors or solvent can prevent wax deposition, reduce the pour point temperature and increase flow assurance. In this study, oil from Sirikit Oilfield in Thailand is selected to evaluate and compare the performance of four polymers for reducing pour-point temperature. These chemicals are Polyalkyl methacrylate (PAMA), Poly(maleic anhydride-alt-1-octadecane) (PMAO), Polyalkyl methacrylate-co-ethylene glycol (PAMAEG) and Copolymers of maleic anhydride (CPMA) with the concentration ranging from 100 to 1000 ppm. In addition, n-pentane with 5-20 % by weight is provided as a solvent for wax inhibitors. The results present that n-pentane can reduce the pour point to 39.41 % compared to the original oil. Moreover, among the polymer additives, PMAO can provide the lowest pour point reduction to 52.82 % for a single chemical. Also, the mixed solvents of polymer with n-pentane can provide relatively less effect on pour-point reduction. The mixture of n-pentane with PMAO can lower the pour point to a certain level of 54.96 % but can increase more performance with other polymers. This work can be used as a fundamental data to apply in the real field to prevent wax deposition in the pipeline transportation.

5. Acknowledgement

The authors would like to acknowledge the Chevron Thailand Exploration and Production, Co., Ltd. for financial support. Also, the authors would like to thank PTT Exploration and Production Public Company Limited for oil sample.

6. References

[1] Gudmestad O Kellan M A 2014 *Production Chemicals for the Oil and Gas Industry* (Boca Raton: CRC Press).

[2] Sanjay M, Simanta B and Kulwant S 1995 Paraffin problems in crude oil production and transportation, a review, *SPE Production Operation*, **10**, 50–54.

[3] Machado A L and Elizabete F L 2001 Poly (ethylene-co-vinyl acetate) (EVA) as wax inhibitor of a Brazilian crude oil, *J. Pet. Sci. Eng.* **32**, 159–165.

[4] Hammami A and Raines A 1999 Paraffin deposition from crude oils. Comparison of laboratory results with field data, *SPE J.*, **4**, 9–18).

[5] Al-Yaari M 2011 Paraffin wax deposition: mitigation and removal techniques *SPE Saudi Arabia section Young Professionals Technical Symposium* (Dhahran, Saudi Arabia: March 14-16, 2011).

[6] Thant M M and Sallehud-Din M T 2011 Mitigating flow assurance challenges in deep water fields using active heating methods, *SPE Middle East Oil and Gas Show and Conference* (Manama, Bahrain: September 25-28, 2011).

[7] Pedersen K S and Rønningsen H P 2003 Influence of wax inhibitors on wax appearance temperature, pour point, and viscosity of waxy crude oils. *Energy & Fuels*, **17**, 321–328.

IOP Conf. Series: Materials Science and Engineering **700** (2019) 012060 doi:10.1088/1757-899X/700/1/012060

About a method of acoustic impact on high viscosity oil fields

Z N Alisheva[1], N A Eremin[2, 3*] and G P Metaksa[1]

[1]JSC "Kazakh National Research Technical University named after K.I. Satpaev", Almaty, Kazakhastan
[2]Oil and Gas Research Institute of Russian Academy of Sciences, Moscow, Russia
[3]Gubkin Russian State University of Oil and Gas, Moscow, Russia

* Corresponding author: ermn@mail.ru

Abstract. The report describes the method of the acoustic impact on the oil reservoir. Chemical decomposition of formation water molecules occurs with the help of energy-accumulating substances. Water becomes a supplier of hydrogen and oxygen. This allows the processes of hydrogenation and synthesis of new compounds. The decomposition of water is carried out using activated aluminum. As a result of the reaction, hydrogen is released, which provides the reaction of hydrogenation of all fractions of heavy oils. Industrial tests of the new method of acoustic impact were carried out at one of the fields with high viscosity of the Embamunaigas company in 2016. Pilot tests revealed an increase in the flowability of heavy oil during the hydrogenation of its components. The increase in well productivity was recorded during the first days of testing a new method. The negative side of the tests was the colmatation of the well with the released aluminum oxide. Physical and chemical processes on the surface of the reacting components were studied. In the course of the research, the state of the phase interface was studied. The spectral composition of the responses at the phase interface was measured. The report presents the results of measurements of the spectral response of water to external influences. The test results of the new acoustic method showed the need for additional research.

1. Introduction

The oil and gas field development system is formed at the initial stage of the field life cycle. One of the key elements of creating an oil field development system is the selection of primary, secondary, and tertiary methods. Primary methods of development of an oil and gas reservoir are the set of natural forces (types of energy) that ensure the movement of oil or gas in the porous system of the reservoir to the bottom of producing wells. Secondary development methods to maintain reservoir pressure include methods of injecting working agents in the form of water (waterflooding) and natural gas. Tertiary development methods in terrigenous and carbonate reservoirs are conventionally divided into two types - the method of enhanced oil recovery (EOR) and the method of improved oil recovery (IOR), depending on the prevailing influence on the component multipliers in the equation for the oil recovery coefficient: coverage ratio and displacement factor, respectively [1-5].

A new direction in the methods of the acoustic impact on the reservoir is the chemical decomposition of the molecules of produced water using energy-accumulating substances [6-10]. In this case, water is a supplier of hydrogen and oxygen, which allows the processes of hydrogenation and synthesis of new compounds. The acoustic impact was carried out using a pulse generator that operated at the frequency of decomposition of water, i.e. 42.8 kHz. These studies should be aimed at preventing the negative effects of the processes of oil hydrogenation under natural conditions with the help of energy-accumulating substances.

The pilot tests of a new acoustic method of enhanced oil recovery in fields with high viscous oil showed its efficiency. The set of experimental results shows that the layered structure of the "water-solid" interface is the basis for controlling the chemical composition of heavy oil under acoustic impact.

The change in the chemical composition of heavy oil by the hydrogenation of oil under conditions of natural occurrence is an acoustic effect, carried out using a chemical reaction of water decomposition with activated aluminum [11-13]. The released hydrogen provides the reaction of hydrogenation of all fractions of heavy oils [4, 7, 14 and 15].

2. Pilot tests of the method of the acoustic impact on the field with high oil viscosity

Pilot tests of the acoustic method to change the viscosity of oil by hydrogenating its components were carried out in Embamunaigas (Kazakhstan) in 2016 year [7]. As a result of industrial tests, a positive result was obtained for reducing the viscosity of oil and increasing the productivity of wells during several days of testing. At the same time, the colmatation of wellbore zones with the released aluminum oxide was noted. It took from two to four weeks to clean the wellbore zones from the dropped alumina compounds. The study of the state of the phase boundary was performed using measurements of the spectral composition of the responses at the phase interface. The work used the frequency of decomposition of water, which was experimentally detected by A. Pucharich [16]. The report presents the results of measurements of the spectral responses of water to an external acoustic effect, which was carried out from a pulse generator operating at the frequency of water decomposition, i.e. 42.8 kHz. To study the processes of oil decomposition at the interface, an experimental test bench was made, the functional diagram of which is shown in Figure 1.

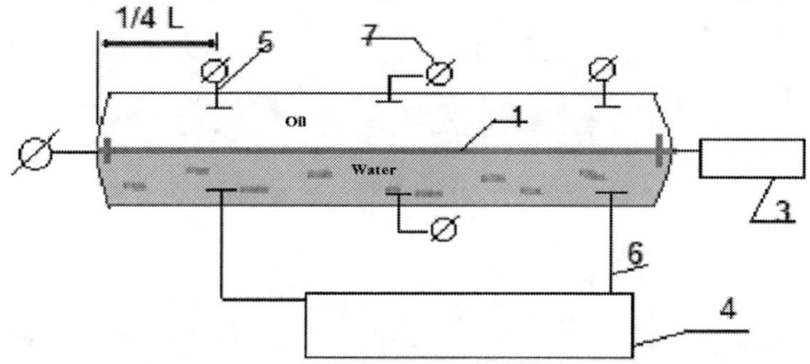

Figure 1. Functional diagram of the experimental setup for the evaluation of the acoustic response at the interface, where 1 - oil-water phase boundary; 2 - quartz cuvettes; 3 - pulse generator; 4 - USB spectrograph; 5 - 1 channel of the spectrograph; 6 - 2 channel spectrograph; 7 - a device for measuring the electrical resistance of the fluid at the interface.

Oil samples were investigated by infrared spectroscopy at the Institute of chemistry of the Academy of Sciences of the Republic of Kazakhstan on the spectrograph. The characteristic features of the oil spectrum from Zhetybay field are shown in Figure 2.

Figure 3 shows the spectrogram of the source water, from which it is clear that the response spectrum includes the fundamental frequency (42.8 kHz), which has a maximum amplitude and the first two harmonics, the amplitude values of which are much less (2-3 times) basic.

Analysis of the spectrogram indicates that the oil sample contains a small amount of amines, many thiophosphochloride halides. Along with saturated hydrocarbons, there are complex compounds of the CH and CH_2 groups. Carbometaethers and aliphatic hydrocarbons with linear chains were identified with the CH_3 group. In addition, ring structures in the form of benzene derivatives with various forms of substitution are present in sufficient quantity. Heavy oils contain a large amount of sulfur and its compounds, therefore for the model experiment there were pressed sulfur tablets (Figure 4).

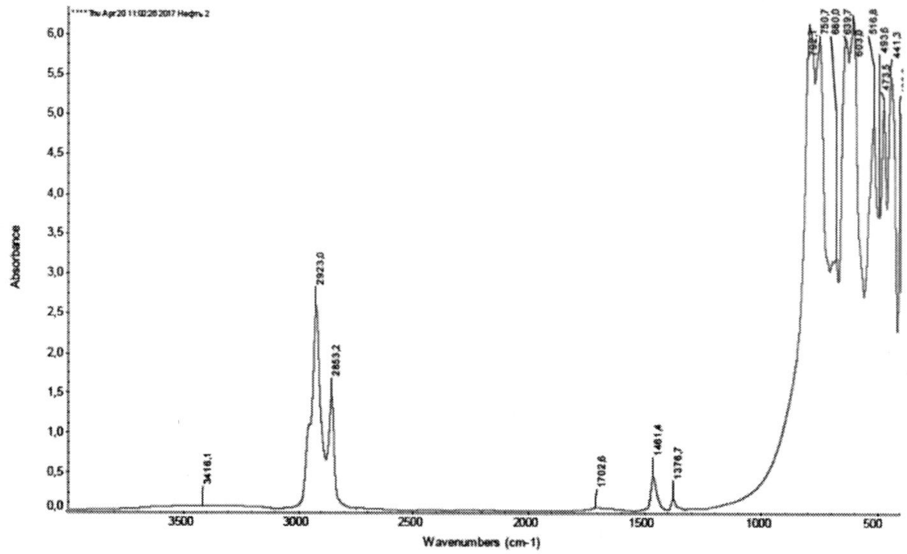

Figure 2. An infrared spectrum of oil from the Zhetybai field.

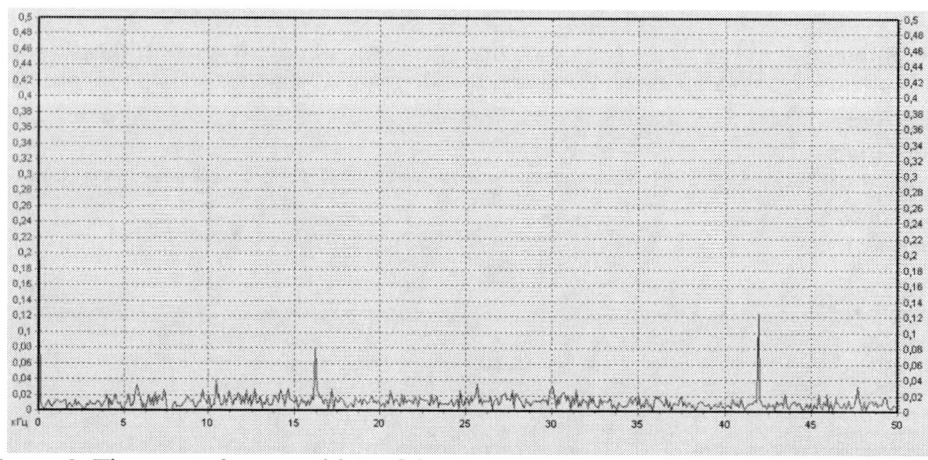

Figure 3. The spectral composition of the response of water at a frequency of 42.8 kHz.

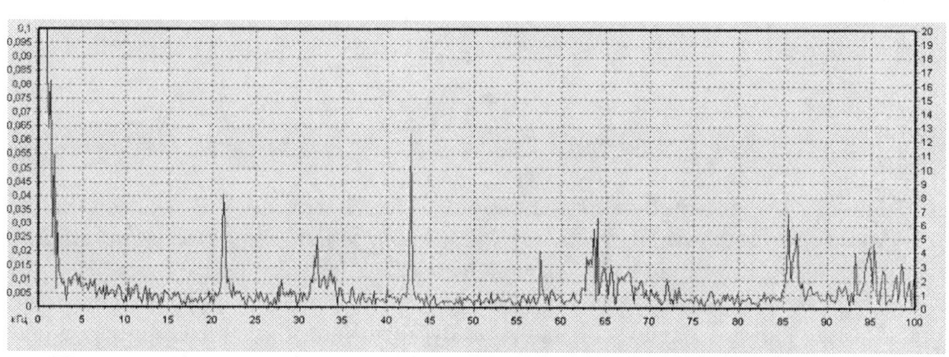

Figure 4. The spectral composition of the response of pressed sulfur, covered with water at a frequency of 42.8 kHz.

Sulfur pressed, covered with a film of water gives the maximum response to the frequency of exposure, and the harmonic components of this response have a 3-braided structure. Resonance conditions correspond to a different dimension of the structural elements of the response. For a triplet response, the model of the layered structure of the phase boundary can be represented as a diagram shown in Figure 5.

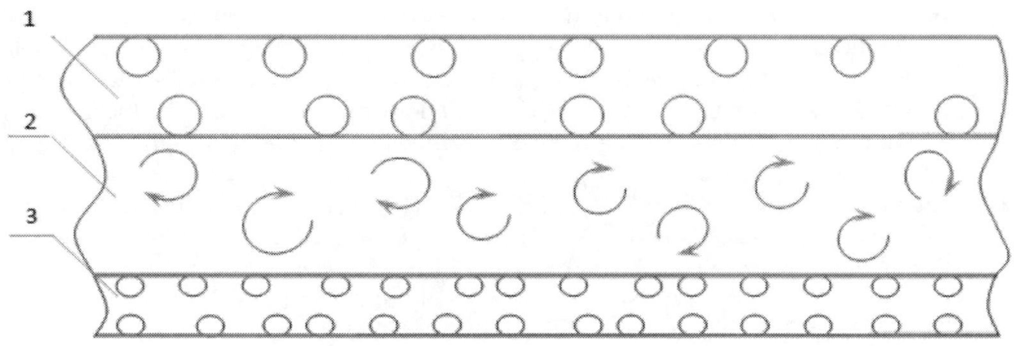

Figure 5. The hypothetical structure of the phase boundary, where 1 is the low-frequency response; 2 is the main signal and 3 is the high-frequency response.

With a similar structure of the phase boundary, the conductivity of the main signal depends on the state of the upper and lower boundaries and, accordingly, in the case of hydrogenation, at least three reaction products may be present at the output, which is confirmed by the results of [7, 8]. The report shows that this effect can be used to obtain desired properties in petroleum compositions by an external impact on the interface by an electromagnetic pulse effect on the frequency of water decomposition. Thus, additional opportunities have been revealed for controlling the structure of the interface by using powders of different dispersion. This is a new effect based on the achievements of such branches of science as optics, metallurgy and tribotechnology.

The set of experimental results show that the layered structure of the "water-solid" interface is the basis for controlling the chemical composition of heavy oil under acoustic impact, and indicate that the state of the oil-water phase interface determines the conditions for the splitting of heavy oil into new compounds, providing for the processes of hydrogenation and structuring of the resulting hydrocarbons.

3. Conclusions

Experimental evidence of the existence of the processes of synthesis and decomposition of hydrocarbons during an acoustic impact on the oil-water phase boundary has been obtained. To identify the mechanism of controlling the phase boundary in the oil-water compositions, an experimental setup was created to measure the spectral composition of the responses when acting on the phase boundary at the frequency of water decomposition.

The results of physical modeling to study the structure of the interface between the phases "water - solid" showed that any inhomogeneity that prevents the passage of the acoustic signal causes a response at the fundamental frequency of decomposition of water and at the frequencies of the harmonic series. The components of more viscous oil at the phase boundary play an important role.

The effect of the appearance of a layered structure of the "water – solid" interface appears, allowing one to control the mechanism of interaction between the separated phases during an external acoustic effect at the frequency of water decomposition, hydrogenation processes occur and, accordingly, the synthesis of hydrocarbons with benzene group derivatives.

Acknowledgements

The authors would like to thank management of the Kazakh National Research Technical University named after K.I. Satpaev and the Oil and Gas Research Institute of Russian Academy of Sciences for the permission to present and publish this paper based on the results of the project "The Fundamental Basis of Innovative Technologies in the Oil and Gas Industry", No. AAAA A16-116031750016-3.

References

[1] Basnieva I, Zolotukhin A, Eremin N and Udovina E 1994 Comparative Analysis of Successful Application of EOR in Russia and CIS *Proc. of University of Tulsa Centennial Petroleum Engineering Symposium*. https://search.crossref.org/?q=10.2523%2F28002-ms

[2] Zolotukhin A, Eremin N, Nazarova L and Ponomarenko Ev 1991 The theory of fuzzy sets for the problem of the selecting recovery methods on the oil field *Neftyanoe Khozyaystvo - Oil Industry,* **3** 21-23.

[3] Eremin N, Zheltov Yu and Makarova E 1993 The well-spacing ratio and the application of the oil recovery methods EOR/IOR. *Neftyanoe Khozyaystvo - Oil Industry* **11** 28-31.

[4] Zolotukhin A and Eremin N 1986 *Design of development of oil fields with the use of in-situ combustion* (Gubkin Russian State University of Oil and Gas, Moscow, Russia) 73. https://www.researchgate.net/publication/330669021_Design_of_development_of_oil_fields_with_the_use_of_in-situ_combustion

[5] Sitnikov A, Eremin N and Ibatullin R 1994 A Mathematical Model of Microbial Enhanced Oil Recovery MEOR Method for Mixed Type Rock *Society of Petroleum Engineers*. https://search.crossref.org/?q=10.2118%2F28903-MS

[6] Ahmed Y et al 2019 Acoustic Characterisation of Phase Inversion in a Water-Oil System. *Society of Petroleum Engineers*. https://search.crossref.org/?q=10.2118%2F194885-MS&page=

[7] Alisheva Z and Eremin N 2018 Obtaining the preset properties during hydrogenation process by controlling the state of the phases division boundary *Geology, Geophysics and Development of Oil and Gas Fields* **8** 66–70.

[8] Alisheva Z and Metaksa G 2017 Experimental evidence of the existence of natural processes of resource reproduction *Commonwealth,* 16(2) 4-9.

[9] Wang Z *et al* 1988 Acoustic Velocities in Petroleum Oils *SPE Annual Technical Conference and Exhibition*. https://search.crossref.org/?q=10.2118%2F18163-MS

[10] Yeo L, Matar O K, Perez de Ortiz E S and Hewitt G F 2002 Phase inversion and associated phenomena *Multiphase. Sci Technol,* **1**(1) 12–51.

[11] Johnson J, Olsson B et al 2019 High Temperature Hydrogen Attack Life Assessment Modeling and Inspection. *NACE Int. Corrosion* 2019, 24-28 March, Nashville, Tennessee, USA

[12] Metaxa G *et al* 1977 *The method of processing metals and alloys*. Certificate of authorship. The USSR. 12.09.1977.

[13] Sokolsky D 1979 *Hydrogenation in solutions* (Nauka, Alma-Ata) 364.

[14] Mancilla-Polanco A et al 2019 Phase Behavior of Heavy-Oil/Propane Mixtures *Society of Petroleum Engineers*. https://doi.org/10.2118/184988-pa

[15] Marcano G, Basnieva I, Eremin N, Sardanashvili O and Kraus Z 2019 About thermal oil recovery during the extraction of heavy oil and bitumen for the fields of Venezuela *Actual Problems of Oil and Gas* **24**(1).

[16] Pucharich A 1983 Patent USD №9.394.230 issued 19.07.1983.

Study on application of colloidal particles of metal oxides to increase the oil recovery factor

V I Lesin[1,*] and N A Eremin[1,2]

[1] Oil and Gas Research Institute of Russian Academy of Sciences, Moscow, Russia
[2] Gubkin Russian State University of Oil and Gas, Moscow, Russia

* Corresponding author: vilesin@inbox.ru

Abstract. The extraction of heavy oil, including bitumen, is complicated by the extremely high viscosity of the fluid in the reservoir. The adsorption of heavy oil fractions on the surface of minerals leads to the hydrophobization of the pore space. The magnetic colloidal particles of iron oxides present in the composition of the water remove adsorbed oil from the surface of pores, which manifests itself as an increase in the oil recovery factor and the injectivity of injection wells. Iron particles of submicron size, located on the surface of an electrically charged gas bubble, are concentrated at the water-oil interface. Due to the high adsorption energy on the surface of the iron particle, oil is deposited on the iron oxide particles. The drop-in bottom pressure of injection and production wells contributes to the movement of the gas bubble with oil and iron oxides to the bottom of production wells. The study of the mechanism of exposure to electromagnetic radiation showed that the electromagnetic field selectively heats the particles of iron oxide, causing catalytic cracking of oil, and contributes to an increase in the oil recovery factor and well productivity index.

1. Introduction

Interest in the use of nanotechnology in the development of oil and gas fields is growing. Traditional methods, such as nuclear magnetic logging and indicator injection, do not provide reliable information on the distribution of residual oil in the reservoir. Development of heavy oil fields by using aqueous solutions is complicated by the high viscosity of the oil and gas fluid in the oil reservoir [1-8]. In addition, adsorption of asphaltenes and resins on the surface of minerals leads to hydrophobization of the pore space, which also prevents the process of oil displacement.

Ferrofluid (or magnetic fluid) is a stable dispersion of paramagnetic nanoparticles, coated with a dispersant, in a liquid carrier. These magnetic particles can be magnetized and have sizes from 3 to 15 nanometers. The behavior of the magnetic fluid depends on the presence or absence of a magnetic field. The suspension is stable as the dispersant prevents agglomeration of the particles and the Brownian motion keeps the particles in the carrier fluid. Ferrofluid can be injected into oil fields without the significant decreasing of permeability. Surface-coated nanoparticles are able to penetrate through micron-sized pores over long distances in the reservoir with small losses of nanoparticles in the rock. Nanoparticles change the magnetic permeability of the flooded area of the field. Paramagnetic particles are particles that are oriented parallel to the applied magnetic field and do not retain the magnetic moment after the field is removed, i.e. soft magnetic particles. The mass force acting on the liquid is due only to the magnetic polarization of the particles. Paramagnetic nanoparticles retain their mobility in intense magnetic fields. While a liquid with paramagnetic particles of micron size or more solidifies

COTech IOP Publishing
IOP Conf. Series: Materials Science and Engineering **700** (2019) 012061 doi:10.1088/1757-899X/700/1/012061

in the presence of an applied magnetic field. Paramagnetic nanoparticles have unique properties. Ferrofluids can be easily filtered in the pore space of the formation. The properties of paramagnetic nanoparticles depend on their size due to the large ratio of surface area to volume of nanoparticles. Ferrofluids can be designed in such a way that they have certain magnetic and interfacial properties (long-term dispersion stability in an aqueous solution with minimal losses in the reservoir and adsorption at the interface of the oil-water phases). Paramagnetic nanoparticles are either adsorbed at the oil-water interface or dispersed in one of the two liquid phases coexisting in pores of the reservoir rock [9-13].

The magnetization of the ferrofluid is easily controlled. Stable water dispersions of paramagnetic nanoparticles moving under the influence of the applied magnetic field can be potentially useful for monitoring the process of field development. The interfacial tension acts as a restoring force, which increases the interphase oscillations and phase pressures, which helps to detect the location of the phase boundary, determine the distribution of multiphase fluids in the reservoir rock and the location of the lenses with residual oil. There is a potential application of the method of injection of aqueous solutions of ferrofluid as Improved Oil Recovery (IOR) and Enhanced Oil Recovery (EOR) methods for mature, heavy oil, High Pressure High Temperature and highly water saturated fields [9-11, 14-19].

Magnetic pressure depends on the pore geometry, the strength and direction of the magnetic field. The magnetic pressure created by the magnetization of the wetting phase helps to displace the fractal aggregates (or ganglia) of the non-wetting oil and thereby reduce the residual oil saturation. Magnetic stresses are amplified near the contacts between the fractal aggregates of oil and the wall of the rock matrix twice. Fractal aggregates of oil have a tendency to adapt to the configuration of the voltage by disconnecting it from the walls of the matrix of rock. Stresses caused by magnetization of ferrofluid can help overcome strong capillary pressures and displace fractal aggregates of residual oil [13].

Monitoring the movement of reservoir fluids in a porous medium can be based on inter-well electromagnetic (EM) conductivity monitoring technology. Reservoir fluids and injected water have different magnetic permeability. Magnetic response or sensitivity of inductive measurements from heterogeneous fluids is best detected at high frequencies. As the injected slug of ferrofluid moves through the reservoir, the dispersion leads to a decrease in the concentration of nanoparticles, and the spatial contributions of magnetic measurements become more scattered [11].

The effect of the magnetic field on paramagnetic nanoparticles leads to an increase in the velocity of their motion in liquid phases and to a shift in the phase boundary. Measurements of the interfacial motion between an aqueous solution of paramagnetic nanofluid and a non-magnetic fluid using phase-sensitive optical coherence tomography showed a doubling of frequency and magnetic susceptibility. The application of an innovative method of magnetic resonance imaging (MRI) using paramagnetic nanoparticles makes it possible to improve the accuracy of determining the location of residual oil in the reservoir. The new method is based on the concept of detecting the acoustic response (and resonance) of the interface between oil and ferrofluid in the porous space of rock exposed to an oscillating magnetic field [12, 19].

The objective of this article is to increase the injectivity of wells due to the magnetic field treatment of working agents represented by aqueous solutions of colloidal particles of iron oxides.

2. Study of the influence of colloidal magnetic particles of iron oxides on the process of oil displacement

The study of water displacing properties showed that in addition to temperature, impurities of colloidal magnetic particles of iron oxides in water play a huge role in the process of oil displacement. Superparamagnetic iron oxides magnetic particles image were really registered in oil by using Transmission Electronic Microscopy (TEM) technique [20, 21]. Electronic Magnetic Resonance (EMR) studies of water and oil have also shown the presence of nano-sized superparamagnetic iron oxide particles in water and oil [20, 22 and 23].

In articles [20, 21] the characteristic sizes of such particles in oil were determined by Transmission Electronic Microscopy (TEM): length 50-100 nm, diameter 10-20 nm. Superparamagnetic particles form aggregates of fractal structure (FA) of sufficiently large size: 10^3 -10^5 nm. Studies of tap water by

IOP Conf. Series: Materials Science and Engineering **700** (2019) 012061 doi:10.1088/1757-899X/700/1/012061

Electronic Magnetic Resonance (EMR) method showed the presence of such particles: in the EMR spectrum, the presence of nanoparticles with a characteristic size of about 10 nm is recorded as a single line with g factor 2, 186 (+/- 0.003) and a width ΔH=70 Oe.

In Figure 1, an image of a superparamagnetic magnetite/maghemite particle (a rod-like particle) as part of an aggregate of colloidal oil particles (rounded gray particles) is presented. The image was obtained by high resolution of Transmission Electronic Microscopy (TEM) method.

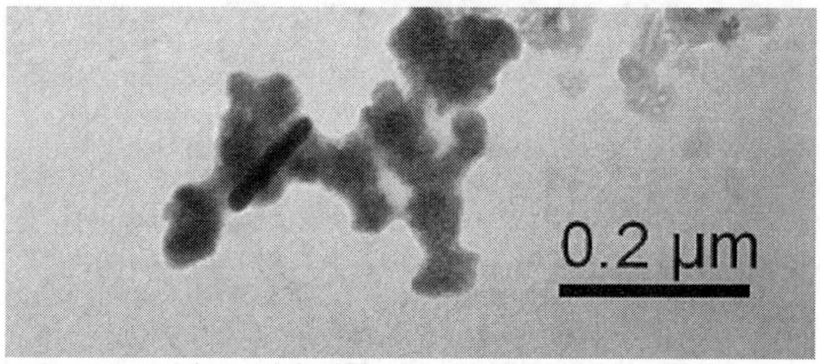

Figure 1. Superparamagnetic particle of iron oxide (rod like particle) as part of the unit of colloidal oil particles aggregate.

In articles [21, 24], it was shown that the treatment of water by flowing through a pipeline, inside which a magnetic field is specially formed, leads to the destruction (crushing) of magnetic particle aggregates. To establish this fact, water filtration through a filter with a pore size of 2-3 microns was used. It was found that after magnetic treatment of water, the amount of iron particles retained by the filter decreased more than twice. Magnetically untreated retained particles Electronic Magnetic Resonance (EMR) spectrum contained line with g = 2,186, however, in the treated retained EMR spectrum this line were absent. Thus, in the spectrum, EMR has disappeared line of particles of small size, i.e. FA fragmentation occurred, as a result of which fragments of the initial aggregates of fractal structure (FA) and small particles were able to pass through pores with a diameter of 2-3 microns.

In the EMR spectra of both treated and untreated iron particles, narrow (less than 5 Oe wide) single lines were found that correspond to organic radicals adsorbed on the particles. Further studies have shown that upon contact of water with air, gas bubbles of a wide range of diameters from 10 to 100 nm are formed on iron oxide particles [21].

In article [25], the presence of electric charges on the surface of gas bubbles - bubstone, the size of about 100 – 200 nm in electrolyte aqueous solutions of was established. Bubstones form aggregates of fractal structure, consisting of tens - hundreds of microbubbles. The presence of an electrically charged surface prevents these bubbles from collapsing even at pressures of several tens of atmospheres, which allows them to migrate in the porous space of the formation, interacting with electrically active heavy components of oil.

It is known that the adsorption energy of asphaltenes and resins on the surface of magnetic particles of iron oxides exceeds the energy of their intermolecular attraction, which allows, for example, to purify water emulsions from oil impurities [26]. In article [24], it is shown that the magnetic colloidal iron oxide particles present in the composition of water remove adsorbed oil from the pore surface, which is manifested as an increase in the extraction coefficient and an increase in the pickup of injection wells. This effect occurs due to the fact that iron particles of submicron size located on the surface of an electrically charged gas bubble are concentrated at the water-oil interface. Due to the high adsorption energy on the surface of the iron particle, oil is deposited on the iron oxide particles, and due to the gas bubble is involved in the process of movement.

Summing up the above, it can be argued that the role of impurities of nanoscale magnetic iron oxides in water is to form flotation (washing) components representing FA and colloidal particles that are part of gas bubbles (or the surface of gas bubbles). Such a gas bubble, located at the water-oil interface,

COTech IOP Publishing
IOP Conf. Series: Materials Science and Engineering **700** (2019) 012061 doi:10.1088/1757-899X/700/1/012061

collides with a colloidal particle of oil, adsorbs it in its volume and is filtered together with the water flow.

As noted above, colloidal particles and their aggregates of FA are usually present in water in concentrations of the order of units – tens of grams per ton. For example, in purified tap water, which was studied in [22, 24], the total concentration of iron atoms was approx. 0.3 g/ tone. With this concentration, the sizes of the significant part of FA was greater than 3 microns. The large size of aggregates of FA particles of iron oxides present as micro-impurities in water, prevent the penetration of FA into the pores of small size, reducing the coverage of the formation by the action of iron oxides. Treatment of water by a magnetic field makes it possible to split such aggregates of FA into smaller particles capable of penetrating into pores of small size, providing an increase the rate of the water flooding in injection wells. It should be noted that the addition of magnetic iron nanoparticles in oil is accompanied by a decrease in its viscosity by tens of percent [14]. The penetration of iron nanoparticles in oil helps to increase the efficiency of water flooding.

The presence of iron impurities in water sources allows to increase the concentration of nanoscale magnetic particles of iron oxides by magnetic treatment of water, without the introduction of particles. Industrial tests of magnetic devices were carried out at the field of the Republic of Tatarstan. Installation of magnetic devices on two water injection wells with a injection index of about 200-300 cubic meters per hour allowed to increase the average annual pick-up rate of wells for fresh water by about 110%, and for salt water – by 40%. The growth effect developed within 2-3 months. Industrial tests of magnetic devices were carried out at the field of the Republic of Tatarstan.

Experiments on the displacement of oil by water with metal oxide particles are interpreted as a manifestation of a decrease in surface tension energy at the oil/water boundary [17]. This interpretation does not contradict the proposed mechanism of action of the complex particle – iron oxide + gas bubble.

Indeed, thanks to the electric charge, such a particle will experience the forces of attraction to water, which has a high dielectric permeability and electrical conductivity. Due to adsorption of oil particles, the oil/water boundary changes to the water/rock boundary. Because the rock pore surface is covered by an adsorbed layer of water, the apparent surface tension energy of water/oil is reduced.

The properties of natural impurities of metal oxide nanoparticles to reduce the energy of surface tension and reduce viscosity [14, 17] led to intensive studies of the effect of artificial introduction of impurities of various metal oxides, including magnetic particles of iron oxides on the processes of the oil displacement when the formation is heated by electromagnetic radiation. In articles [18, 23 and 27], a significant increase in the output is reported when using electromagnetic radiation of the 10^6 Hz frequency range. Moreover, the injection of magnetic iron particles allows cracking and oil in the reservoir in the presence of water [23, 27], achieving a decrease in viscosity and an increase in the proportion of light oil fractions in the reservoir. The study of the mechanism of exposure to electromagnetic radiation [23, 27] showed that the electromagnetic field selectively heats iron oxide particles, causing catalytic cracking of oil.

3. Conclusions

The injection of magnetic particles of iron allows cracking and oil in the reservoir in the presence of water, achieving a reduction in viscosity and increasing the proportion of light oil fractions in the reservoir. The destruction of aggregates of magnetic colloidal particles of iron oxides and a multiple increase in the number of the above single colloidal particles as a result of electromagnetic action is accompanied by: an increase in the rate of water injection into the reservoir and the oil recovery factor; a decrease in the rate of growth of deposits of paraffins and salts; an acceleration of oil and water separation; an improvement in the quality of cement stone; an increase in the pH of the medium (alkalization) and, as a consequence, a decrease in the rate of corrosion.

Intensive studies of the use of metal oxide nanoparticles for the intensification of heavy oil recovery processes, which are conducted by research groups in Canada and the United States and other countries, show the prospects for the use of nanoparticles and electromagnetic fields in the oil field development.

IOP Conf. Series: Materials Science and Engineering **700** (2019) 012061 doi:10.1088/1757-899X/700/1/012061

Acknowledgments
The authors would like to thank management of the Oil and Gas Research Institute of Russian Academy of Sciences for the permission to present and publish this paper based on the results of the project "The Fundamental Basis of Innovative Technologies in the Oil and Gas Industry", No. AAAA A16-116031750016-3.

References
[1] Askarian M, Vatani A and Edalat 2016 Heavy oil upgrading in a hydrodynamic cavitation system: CFD modelling, effect of the presence of hydrogen donor and metal nanoparticles, *Can. J. Chem. Eng.* **954** 670–79.
[2] Basnieva I, Zolotukhin A, Eremin N and Udovina E 1994 Comparative Analysis of Successful Application of EOR in Russia and CIS *Proc. of University of Tulsa Centennial Petroleum Engineering Symposium*. https://search.crossref.org/?q=10.2523%2F28002-ms
[3] Eremin N, Surina V, Basnieva I and Eliseenko E 2000 Integrated approach on hydrocarbon fields' development planning *Neftyanoe Khozyaystvo - Oil Industry* **3** 15-18.
[4] Hart A 2014 A review of technologies for transporting heavy crude oil and bitumen via pipelines *J. Pet. Explor. Prod. Technol.* **43** 327-36.
[5] Ponomorenko E and Eremin N 1996 The method of the determination of reservoir's similarity *Neftyanoe Khozyaystvo - Oil Industry* **7** 42-44.
[6] Pinzón D 2018 Rheological Demonstration of Heavy Oil Viscosity Reduction by NiO/SiO2 Nanoparticles-Assisted Ultrasound Cavitation *Society of Petroleum Engineers*. https://search.crossref.org/?q=10.2118%2F194030-STU
[7] Taborda E, Franco C, Ruiz M, Alvarado V and Cortes F 2017 Anomalous heavy-oil rheological thinning behavior upon addition of nanoparticles: Departiue from Einstein's theory, *Chem. Eng. Commun.* **2046** 648-657.
[8] Taborda E, Alvarado V and Cortes F 2017 Effect of SiO 2-based nanofluids in the reduction of naphtha consumption for heavy and extra-heavy oils transport: Economic impacts on the Colombian market *Energy Convers. Manage.* **148** 30-42.
[9] Jauhari S, Parekh K, Pai K and Upadhyay R 2009 Corrosion Inhibition of Mild Steel in an Acidic Media Using Ferrofluid as a Novel Approach to Superior Corrosion Resistance *NACE International, Corrosion 2009 Conf. and Expo*.
[10] Lau H C, Yu M and Nguyen Q 2017 Nanotechnology for oilfield applications: Challenges and impact, *J. Petrol. Sci. Eng.* **157** 1160 - 69
[11] Rahmani A, Bryant S, Huh C, Athey A, Ahmadian M, Chen J and Wilt M 2015 Crosswell magnetic sensing of superparamagnetic nanoparticles for subsurface applications, *SPE J.* **20**(5) SPE-166140-PA
[12] Ryoo S, Rahmani A, Yoon K, Prodanovic M, Kotsmar C, Milner T *et al* 2010 Theoretical and Experimental Investigation of the Motion of Multiphase Fluids Containing Paramagnetic Nanoparticles in Porous Media, *J. Petrol. Sci. Eng.* **81** 129-44
[13] Soares F, Prodanovic M and Huh C 2014 Excitable Nanoparticles for Trapped Oil Mobilization *SPE Improved Oil Recovery Symposium*, 12-16 April, Tulsa, Oklahoma, USA
[14] Aristizábal-Fontal J, Cortés F and Franco C 2017 Viscosity reduction of extra heavy crude oil by magnetite nanoparticle-based ferrofluids, *Adsorpt.Sci. Technol.* **361**(2) 23–45.
[15] Eremin N and Ponomarenko E 1994 Signed structures in the allocation of operational objects *Neftyanoe Khozyaystvo - Oil Industry* **8** 35-37.
[16] Sitnikov A, Eremin N and Ibatullin R 1994 A Mathematical Model of Microbial Enhanced Oil Recovery MEOR Method for Mixed Type Rock, *SPE European Petroleum Conference*, 25-27 October, London, UK
[17] Bera A & Babadagli T 2015 Status of electromagnetic heating for enhanced heavy oil/bitumen recovery and future prospects: A review *Appl. Energy* **151** 206–26.

IOP Conf. Series: Materials Science and Engineering **700** (2019) 012061 doi:10.1088/1757-899X/700/1/012061

[18] Hashemi R, Nassar N and Pereira A 2014 Nanoparticle technology for heavy oil in-situ upgrading and recovery enhancement: Opportunities and challenges *Appl. Energy* **133** 374–87.
[19] Shekhawat D, Aggarwal A, Agarwal S and Imtiaz M 2016 Magnetic Recovery-Injecting Newly Designed Magnetic Fracturing Fluid with Applied Magnetic Field for EOR, *SPE Asia Pacific Hydraulic Fracturing Conf.*, 24-26 August, Beijing, China.
[20] Lesin V, Koksharov Yu and Khomutov G 2010 Magnetic nanoparticles in petroleum, *Petrol. Chem.* **50**(2) 102-5.
[21] Lesin V, Koksharov Yu and Khomutov G 2011 Viscosity of liquid suspensions with fractal aggregates: magnetic nanoparticles in petroleum colloidal structures, *Colloids Surf. A* **392** 88-94.
[22] Lesin V 2001 Physico-chemical model of change of oil-displacing properties of water after its magnetic treatment, *Neftepromislovoye delo (Oilfield Eng.)* **3** 15-17.
[23] Lesin V and Eremin N 2018 The natural and synthesized nanoscale iron oxides - nanobots in the control processes of the production, the transportation, the preparation and the refining of oil by using the magnetic field, *Oil Gas Innovation.* **1** 18-22.
[24] Lesin V, Dunin A and Khavkin A 1993 Changes in Physical and Chemical Properties of Aqueus Solutions Induced by an Electromagnetic Field, *Russ. J. Phys. Chem.* **67**(7) 1403 -8.
[25] Bunkin N and Bunkin F 2016 Bubston structure of water and electrolyte aqueous solutions, *Physics-Uspekhi* **599** 846–65.
[26] Mirshahghassemi S and Lead J 2015 Oil Recovery from Water under Environmentally Relevant Conditions Using Magnetic Nanoparticles Environmental, *Sci. Technol.* **4919** 11729–36.
[27] Lesin S and Lesin V 2018 Mechanism of the Effect of Alternating Electromagnetic Field on Petroleum Dispersed Systems, *Petrol. Chem.* **58**(7) 553–6.

Applying of the associative polymer solutions to enhance oil recovery

R Zh Abirov[1] and N A Eremin[2, 3, *]

[1]JSC "Kazakh-British Technical University", Almaty, Kazakhstan
[2]Oil and Gas Research Institute of Russian Academy of Sciences, Moscow, Russia
[3]Gubkin Russian State University of Oil and Gas, Moscow, Russia

* Corresponding author: ermn@mail.ru

Abstract. Purpose of this report to inform the petroleum readership of recent advances in areas of the associative polymer flooding. Polymer flooding is the industrial enhanced oil recovery method, which successfully has applied in various types of reservoirs. The polymer flooding for high-temperature and high-salinity reservoir faces enormous challenges. Hydrolyzed polymer has been used as the main polymer type for the polymer flooding. Now researchers have turned their attention to associative polymer solutions. This report presents an overview of the associative polymer flooding. Some recent projects have shown an expanding technology. This report reviews published results of the use of associative polymers to enhance oil recovery in recent years. The associative polymer flooding is one enhanced oil recovery (EOR) method being developed recently to increase the oil supply. Polymer flooding application in high-temperature and high-salinity reservoirs is currently limited by polymers chemically and thermally degradation at high temperature and salinity. The associative polymer solutions improve the over-all reservoir conformance by providing mobility control. Water is injected after the polymer solution. Based on the number of new laboratory and pilot tests, field activity is continuing at a moderate level.

1. Introduction

In this report, the evaluation of several associative polymers is presented. In associative polymer flooding, a water-soluble associative polymer is added to the flood water. The applicability of the associative polymer flooding is carefully controlled by lab and pilot experiments. The associative polymer flooding can make the oil production process more efficient due to decreasing the water/oil mobility ratio. The processes associative polymer flooding have been described theoretically [1-3] and examined experimentally [4, 5]. There are multiple field applications of associative polymer flooding in China, Saudi Arabia and other countries [6].

The novelty of this report is that for the first time a review of laboratory work, pilot tests and industrial implementation of an innovative method of injection of associated polymers into oil reservoirs has been prepared. Injection of chemical reagents causes a range of physical and chemical mechanisms of oil displacement. Injection of aqueous solutions of polymers leads to a change in the properties of the reservoir water, a decrease in the relative mobility parameter and an improvement in the oil-washing properties of water. The decrease in the relative mobility of water and oil increases the coverage of the reservoir effect and the coefficient of oil displacement [7].

The method of injection of an aqueous polymer solution is the injection of a low-concentrated solution of a high-molecular polymer. Polymers are substances with a high molecular weight of about

IOP Conf. Series: Materials Science and Engineering **700** (2019) 012062 doi:10.1088/1757-899X/700/1/012062

10^4-10^6. These are the giants of the chain structure. They have the ability to significantly increase the viscosity of water, thereby reducing its mobility, which leads to an increase in the sweep efficiency of the reservoir compared with conventional water-flooding [8, 9]. Polymer flooding is used on oil-containing reservoirs with a relatively high viscosity of oil (up to 100 MPa*s) and the ratio of the coefficients of mobility of water and oil and moderate heterogeneity. The method of polymer flooding is not used for the development of oil fields with gas caps, fractured reservoir, high permeability and active pressure of the bottom water. The increase in oil recovery in polymer flooding averages 3 - 10%. The size of the polymer slug varies from 0.1 to 0.4 porous volumes. When using the polymer, the ratio of mobility coefficients decreases and, accordingly, the coefficient of the sweep efficiency by area and net pay increases.

Basic mechanisms of enhanced oil recovery during the injection of aqueous solutions of polymers are water thickening, which leads to a decrease in the ratio of the mobilities of oil and water and reduce the possibility of water breakthrough to production wells, and the blockage of high permeability channels due to the adsorption of polymers on the surface of the rock [10]. It should be noted that the apparent viscosity of aqueous solutions of polymers μ^* increases with increasing filtration rate and with decreasing size of pore channels. The apparent viscosity can be 10 - 20 times higher than measured. The apparent viscosity of the polymer solution is defined as

$$\mu^* = \tau/\gamma \quad (1)$$

where τ shear stress, γ shear gradient.

At high shear gradients there is a mechanical destruction of polymer chains. This leads to a decrease in the viscosity and the resistance factor, and the resistance factor decreases faster than the viscosity of the polymer [11-15].

The adsorption of polymers by the surface of the porous medium increases with increasing the salinity of the reservoir water and reducing the permeability and can range from 0.007 to 0.75 kg/m^3. The amount of adsorbed polymer depends on the structure of the porous medium, lithology of rocks and component composition, specific weight, filtration rate in a porous medium, temperature and pH of the medium [13]. Adsorption on the rock surface depends on the type of polymer.

Associative polymers, or multiply adherent polymers are a variety of polymers that have a higher thickening ability than traditional water-soluble polymers, such as polyacrylamide. Associative polymers are copolymers of a hydrophilic acrylamide monomer with a small amount of a hydrophobic monomer (0.3 – 4 mol%), such as a partially hydrolyzed acrylamide [16]. Copolymers have a block structure with two or more different structural blocks of a hydrophobic monomer located at a considerable distance from each other on a hydrophilic polymer chain. Copolymers are divided into regular and irregular copolymers. In the aqueous solution that has chemical polarity, the hydrophobic areas of polymer chains interact with each other and form associative links. Copolymers used as mobility control agents in EOR/IOR methods [2].

Vinylpyrrolidone (VP) is one of the important hydrophobic monomers used for copolymerization with acrylamide as working agents for the polymer flooding. Polyvinylpyrrolidone is a water-soluble polymer composed of monomeric units of N-vinylpyrrolidone (NVP). Polyvinylpyrrolidone was first synthesized by Walter Reppe. The patent for polyvinylpyrrolidone was registered in 1939. Polyvinylpyrrolidone is soluble in water and other polar solvents. In its dry form, polyvinylpyrrolidone has the appearance of a white or light-yellow puff of hygroscopic powder, which easily absorbs up to 40% of its weight in atmospheric water (figure 1). The molecular formula of polyvinylpyrrolidone is C_6H_9NO, the molar mass is 2.500 - 2.500.000 g / mol, the density is 1.2 g / cm^3, the melting point is 150-180 °C. Polyvinylpyrrolidone is readily soluble in water, forming water-soluble complexes, and is used to thicken water during polymer flooding. The polymer flooding is one of the key methods of EOR.

COTech IOP Publishing
IOP Conf. Series: Materials Science and Engineering **700** (2019) 012062 doi:10.1088/1757-899X/700/1/012062

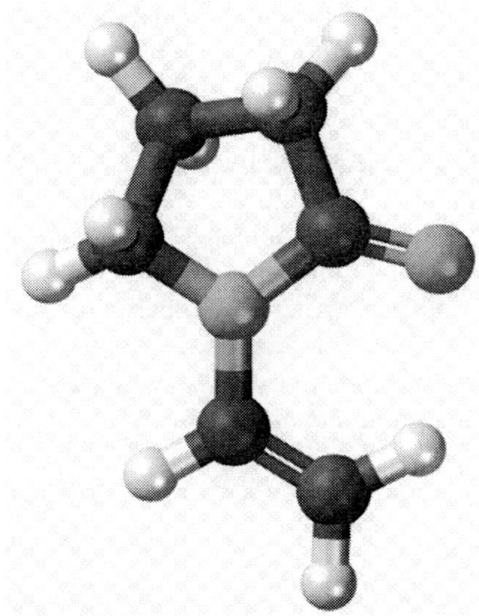

(a)

(b)

Figure 1. Ball-and-stick model (a) and appearance in powder form of the N-vinylpyrrolidone molecule (b), a compound used in organic synthesis. Colour code: Carbon, C: black; Hydrogen, H: white; Oxygen, O: dark grey; Nitrogen, N: light grey. Source: *Discovery Studio Visualizer*.

Homogeneous polymer vinylpyrrolidone has good thermal stability in aqueous and saline solutions. Vinylpyrrolidone polymer is not a good agent for increasing the viscosity of an aqueous polymer solution. To achieve a viscosity equivalent to a viscosity similar to the viscosity of hydrolyzed polyacrylamide, working compound vinylpyrrolidone requires almost as 10 times concentration as hydrolyzed polyacrylamide [1, 3 and 6].

Synthetic polymers such as partially HPAM are not thermally stable. Thermal stability of hydrolyzed polyacrylamide can be improved by incorporating monomer such as NVP. The development of terpolymers with the incorporation of NVP brought stability of compound up to 120°C. The use of NVP increased the cost of the polymer and limited its molecular weight [15].

The cost of the polymer vinylpyrrolidone is three times higher than that of the polymer PAM101. The adsorption of the vinylpyrrolidone polymer is very high on the surface of the rocks. These two factors - cost and adsorption make the polymer vinylpyrrolidone economically unreasonable for application as a working agent for polymer flooding.

A copolymer of acrylamide (AM) / N-Vinylpyrrolidone was synthesized in an aqueous solution. As shown in [14], the addition of VP to AM can reduce the thermal hydrolysis of AM in the presence of divalent cations. The decrease in hydrolysis is directly proportional to the content of VP in the copolymer. A copolymer of acrylamide/ N-Vinylpyrrolidone with an equal weight ratio of monomers was stable at 120 °C in the presence of seawater for several months. The copolymerization provides a hydrogen bond between AM and VP [14].

The adsorption AM/ VP in the core, represented by sandstone, was lower compared to hydrolyzed polyacrylamide. Poly 1-vinyl-2-pyrrolidone (partially hydrolyzed acrylamide) copolymer that was obtained later had a higher decomposition temperature up to 23 °C than the homopolymer of acrylamide. The inclusion of VP in the AM chain increases shear resistance compared to the AM due to the large size of VP [11, 17 and 18]. VP helps better protect acrylamide from degradation, but a polymer with high molecular weight is difficult to obtain due to low polymerization activity [9, 10, and 19].

Currently, laboratory studies of associative polymers and HPAM are being actively conducted with the aim of preserving properties at high temperatures and salinity over a long period of time; corefloodings for different field conditions and mathematical modeling of the process of oil

IOP Conf. Series: Materials Science and Engineering **700** (2019) 012062 doi:10.1088/1757-899X/700/1/012062

displacement with copolymer aqueous solutions are carried out. During corefloodings, the influence of various factors on the apparent viscosity, such as polymer concentration, copolymer composition, electrolyte concentration, temperature, shear rate, and surfactant concentration, was investigated.

2. Coreflooding by the associative polymer solutions

Below is a brief overview of polymer flooding in high-saline and high-temperature reservoirs. During polymer flooding in the pilot section of Guan-109 (7 injection and 12 producing wells) from December 2016 to December 2017, 0.102 PV of an aqueous polymer solution was injected. The viscosity of the associative polymer AP-P7 changed from 31.7 to 88.5 mPa * s at a concentration of 1500-2500 mg / l. The formation water mineralization reached 26983 mg / l, and the formation temperature was 78 ° C. Oil production increased from 27.7 to 51.7 m^3 / day, and the water cut decreased from 92.3 to 90.1%. The binary system made it possible to improve the oil recovery coefficient by 26.3% after water flooding [5]. Table 1 presents studies of the thermal stability of various polymers and copolymers.

Table 1. The thermal stability of various polymers and copolymers [9].

Polymers and copolymers	Salinity (ppm)	T (^{0}C)	Time (days)	Comments	References
HPAM	30 000	85	100	-	[7]
HPAM	20 000	90	619	-	[20]
AMPS-AM	30 000	100	200	Higher cost	[7]
AMPS-AM	33 560	93	100	Higher cost	[14]
NaAMPS	33 560	121	100	Higher cost	[14]
VP-AM	33 756	121	100	Lower molecular weight, lower viscosity	[21]

Interesting results in core flooding were obtained by testing the polymer Zetag 8187G, which is a co-polymer of acrylamide and a quaternized cationic monomer [12]. In laboratory studies on polymer flooding of the core of low-permeable sandstone at high temperatures (85 °C) and in high-saline brine (150 000 ppm, Ca^{2+}, Mg^{2+}), an oil displacement coefficient of 60.95% was achieved when pumping an aqueous polymer solution of 33.68% of the total pore volume.

A ter-polymer containing N-vinylpyrrolidone, which is intended for polymer flooding in high-salt and high-temperature collectors, was investigated during core flooding (table 2).

Table 2. Experimental layout of core displacement study at 82 ^{0}C.

Test	Core ID	1st flood	Recovery (% OOIP)	Polymer flood	Oil displacement ratio (% OOIP)	Crude type
Test 3	A3	Fresh Water	16.7	Superpusher, 20 cc	5.8	A

Crude oil type A had viscosity ranged from 1 to 1.68 mPa*s (1.6 mPa*s at 60 ^{0}C) and density of oil was 0.78 g/cm^3. During the experiments, it was found that at a concentration of 400 ppm of N-vinylpyrrolidone ter-polymer at a concentration of 92,000 ppm and a temperature of 82 °C modeled in the sample, the maximum displacement coefficient reaches 22.5% of the total oil volume (table 3). In previous studies of a polymer such as VP, with a polymer concentration of 2000 ppm and a temperature of 85 °C in a core sample of sandstone, the displacement coefficient was 13.5% [22].

Laboratory studies of the terpolymer N-vinylpyrrolidone showed that oil recovery may increase up to 22.5% (Figure 2).

Table 3. Core polymer flooding in high-saline and high-temperature core samples.

Polymer	Polymer concentration (ppm)	Temperature (°C)	Core type	Oil displacement ratio (%)	References
AP-P7	1500-2500	78	Sandstone	26.4	[5]
Copolymer of acrylamide and quaternized cationic monomer	2000	85	Sandstone	60.9	[12]
Terpolymer containing N-vinylpyrrolidone	400	82	Sandpack	22.5	[23]

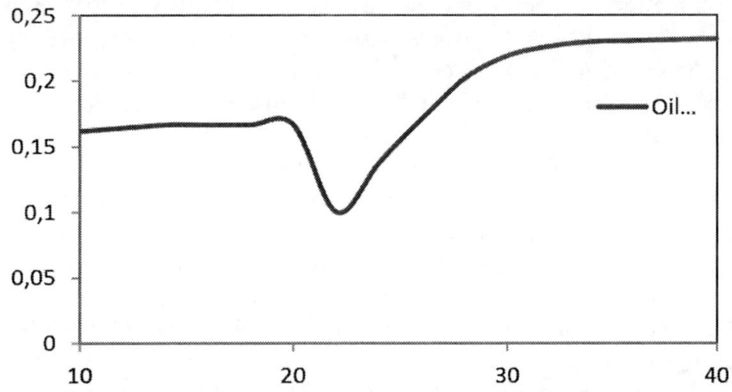

Figure 2. Oil recovery in case of flooding with a terpolymer vinylpyrrolidone.

3. Conclusions

Injection of aqueous solutions of traditional polyacrylamide polymers under conditions of increased salinity of formation water and the content of calcium and magnesium salts becomes unstable, disruption of the structure of polymers occurs and the effect of water layer thickening disappears. In contrast to traditional polymer flooding, in such conditions, polymer flooding with associative polymers, that is, polymers that are polymerized with several monomers, can find effective application. The viscoelastic system of associative polymers is created by the interaction of hydrophobic parts of the chain, and not due to the introduction of an additional crosslinker. Polymer flooding by associative polymers is applicable in a wider range of reservoir temperatures and salinity compared to traditional polymer flooding. In this regard, in recent years, interest in the use of polymer flooding associative polymers has been growing. In this article, we considered an associative polymer and previous experiments with polymers in reservoir conditions with high salinity and high temperature. Polymers with polymerization of vinylpyrrolidone contribute to an additional increase in oil recovery when the core is flooded at a temperature of 86 °C and a salinity of 92,000 ppm. Polymer flooding by associative polymers of a new type has not been tested at the proper level in oil fields. Laboratory studies and comparing them with the latest research in this area are very necessary. With proper planning, logistics and in conditions of low crude oil prices, the injection of aqueous solutions of associative polymers can be cost-effective.

Acknowledgments

The authors would like to thank management of the Kazakh-British Technical University and the Oil and Gas Research Institute of Russian Academy of Sciences for the permission to present and publish this paper based on the results of the project "The Fundamental Basis of Innovative Technologies in the Oil and Gas Industry", No. AAAA A16-116031750016-3.

COTech IOP Publishing
IOP Conf. Series: Materials Science and Engineering **700** (2019) 012062 doi:10.1088/1757-899X/700/1/012062

List of Abbreviations

EOR – Enchanced Oil Recovery

IOR - Improved Oil recovery

VP - Vinylpyrrolidone

NVP - N-vinylpyrrolidone

HPAM – Hydrolyzed Poly-Acrylamide

AM - Acrylamide

OOIP – Original Oil in Place

References

[1] Akbari S, Mahmood S, Tan I, Bharadwaj A and Hematpour H 2016 Experimental investigation of the effect of different process variables on the viscosity of sulfonated polyacrylamide copolymers, *J. Pet. Explor. Prod. Technol.*7(1) 87–101.

[2] Gogoi S and Gogoi S B 2019 Review on microfluidic studies for EOR application, *J. Petrol. Explor. Prod. Technol.* **9**(3), 2263 - 77.

[3] Han X, Zhang G, Yu J, Chen Z and Kurnia I 2018 An Investigation of Retention and Unusually High Apparent Viscosity of Hydrophobically Associative Polymer in Porous Media, *Society of Petroleum Engineers.* https://search.crossref.org/?q=10.2118%2F190330-MS.

[4] Cheraghian G and Khalili Nezhad S 2015 Experimental Investigation of Polymer Solutions Used in Enhanced Oil Recovery - Thermal properties Improved by Nanoclay *77th EAGE Conference and Exhibition.* https://doi.org/10.3997/2214-4609.201412504

[5] Guo Y, Zhang J, Zhang X, Hu J, Wang W and Liang Y 2018 Investigation and Application of an Associative Polymer-Surfactant Binary System for a Successful Flooding Pilot in a High-Temperature, High-Salinity, Ordinary Heavy Oil Reservoir *Society of Petroleum Engineers.* https://doi.org/10.2118/190411-ms

[6] Aktas F, Clemens T, Castanier L and Kovscek A 2008 Viscous Oil Displacement via Aqueous Associative Polymers *SPE Symposium on Improved Oil Recovery.* https://search.crossref.org/?q=10.2118%2F113264-MS

[7] Levitt D and Pope G 2008 Selection and Screening of Polymers for Enhanced-Oil Recovery *SPE Symposium on Improved Oil Recovery. https://doi.org/10.2118/113845-ms*

[8] *Eremin* N, Surina V and Prikazchikova M 1994 Estimation of polymer flooding using a fuzzy set theory *Neftyanoe Khozyaystvo - Oil Industry* **4** 54-57.

[9] Kamal M, Sultan A, Al-Mubaiyedh U and Hussein I 2015 Review on Polymer Flooding: Rheology, Adsorption, Stability, and Field Applications of Various Polymer Systems *Polymer Reviews* **55**(3) 491–530.

[10] Zhang P, Bai S, Chen S, Li D, Jia Z and Zhou C 2018 Preparation, solution characteristics and displacement performances of a novel acrylamide copolymer for enhanced oil recovery (EOR) *Polymer Bulletin* **75**(3) 1001–11.

[11] *Zaitoun* A, Makakou P, Blin N, Al-Maamari R, Al-Hashmi A and Abdel-Goad M 2012 Shear Stability of EOR Polymers *SPE Journal* **17**(02) 335–9.

[12] Mohsenatabar Firozjaii A, Zargar G and Kazemzadeh E 2019 An investigation into polymer flooding in high temperature and high salinity oil reservoir using acrylamide based cationic co-polymer: experimental and numerical simulation *J. Petrol. Explor. Prod. Technol.* **9** 1485.

[13] Eremin N and Nazarova L 2001 *Enhanced Oil Recovery Methods* (Gubkin Russian State University of Oil and Gas, Moscow, Russia) 153. https://www.researchgate.net/publication/297392172_Enhanced_Oil_Recovery_Methods

[14] Moradi-Araghi A, Cleveland D and Westerman I 1987 Development and Evaluation of EOR

COTech IOP Publishing
IOP Conf. Series: Materials Science and Engineering **700** (2019) 012062 doi:10.1088/1757-899X/700/1/012062

Polymers Suitable for Hostile Environments: II-Copolymers of Acrylamide and Sodium AMPS *SPE International Symp. on Oilfield Chemistry*. https://doi.org/10.2118/16273-ms

[15] Rodriguez L, Antignard S, Giovannetti B, Dupuis G, Gaillard N, Jouenne S, Bourdarot G, Morel D, Zaitoun A and Grassl B 2018 A New Thermally Stable Synthetic Polymer for Harsh Conditions of Middle East Reservoirs *Society of Exploration Geophysicists. https://doi.org/10.1190/rdp2018-41754597.1*

[16] Abirov R, Mukhamedova A, Panabekkyzy B, Eremin N, Abirov Z and Nestyerkin A 2016 Pilot introduction of the polymer flooding in the oilfield of South Turgay basin, *Oilfield Eng.* **5** 15-19.

[17] Kiatkamjornwong S and Suwanmala P 1998 Partially hydrolyzed polyacrylamide–poly (Nvinylpyrrolidone) copolymers as superabsorbents synthesized by gamma irradiation, *J. Appl. Polym. Sci.* **68**(2) 191–203.

[18] Liu R, Pu W, Peng Q and Yao F 2012 Synthesis of AM-co-NVP and Thermal Stability in Hostile Saline Solution, *Adv. Mater. Res.* **602-604** 1349–54.

[19] Xu L, Che L, Zheng J, Huang G, Wu X, Chen P, Zhang L and Hu Q 2014 Synthesis and thermal degradation property study of *n*-vinylpyrrolidone and acrylamide copolymer *RSC Adv* **4**(63) 33269–78.

[20] Ryles R 1988 Chemical stability limits of water-soluble polymers used in oil recovery processes *SPE Reservoir Engineering* **3**(01) 23–34.

[21] Doe P, Moradi-Araghi A, Shaw J and Stahl G 1987 Development and evaluation of EOR polymers suitable for hostile environments Part 1: Copolymers of vinylpyrrolidone and acrylamide. *SPE Reservoir Engineering* **2**(04) 461–467.

[22] Chen Q, Wang Y, Lu Z and Feng Y 2012 Thermoviscosifying polymer used for enhanced oil recovery: rheological behaviors and core flooding test *Polymer Bulletin* **70**(2) 391–401.

[23] Abirov R, Ivakhnenko A, Abirov Z and Eremin N 2019 The associative polymer flooding: an experimental study, *J. Petrol. Explor. Prod. Technol.* 1 – 8.

A summary of adapting Industry 4.0 vision into engineering education in Azerbaijan

A Ahadov[1], E S Asgarov[2] and I El-Thalji[1]

[1] Department of Mechanical and Structural Engineering and Material Science, University of Stavanger, 4036 Stavanger, Norway.
[2] The State Oil Company of the Azerbaijan Republic (SOCAR), AZ 1075, Baku, Azerbaijan.

Corresponding author: ahadovanar@gmail.com

Abstract. Industry 4.0 vision and associated technologies are rapidly adopted in several industrial sectors to gain the benefits of creating smart cyber-physical systems and operations. Some sectors, e.g. manufacturing, oil and gas, offshore wind energy, have progressed in developing digitization strategies, executing pilot projects and progressing toward mature implementation of industry 4.0 vision. Offshore Oil and Gas industry highly believes in the potential industrial and societal impacts of digital transformation, due to the need for stochastic and remote operations. Azerbaijan as one of the countries that heavily depend on the Oil and Gas industry is developing more projects in the Caspian Sea. There are several worldwide challenges, mainly, lack of standards, business models, ready products/services and competent and skilled employees. Fortunately, specific developed countries are working hard to standardize industry 4.0 architecture. Moreover, large-scale companies are creating alliances to create a trustful and long-term business model. Furthermore, large-scale companies of information and operational technology are creating robust products and services to be commercially available off the shelf. In terms of education and training, many worldwide universities are upgrading their programs, curriculums, teaching approaches with the goal to support the industry with competent future employees and entrepreneurs. Therefore, the purpose of this paper is, to present the status of engineering education programs in adapting the industry 4.0 vision in Azerbaijan and address the skills that are required for future employment. In order to present the targeted status, the curriculums of all engineering education programs at the master level were collected and analyzed. However, five of them were directly adapting industry 4.0 vision and relevant for industry 4.0. Moreover, a semi-structured interview with industrial managers was applied to extract the future required skills. This study can be considered as a first step in developing a roadmap for engineering education, particularly industrial engineering, to adopt industry 4.0 vision at the national level.

1. Introduction

The industrial value chains are complex. The development of technology creates several advantages for the business world. The new concept such as Internet of Things (IoT), digitalization and Cyber-Physical Systems (CPS) gain important roles across industries. These terms are used in defining the Fourth Industrial

Revolution (Industry 4.0). Industry 4.0 is originally a future vision described in the high-tech strategy of the German government that is conceived upon the information and communication technologies including initiatives such as the Industrial Internet, Factories of The Future, Internet of Things, Physical Internet, Internet of Services, and Cyber-Physical Systems, to achieve a high degree of flexibility in production, higher productivity rates through real-time monitoring and diagnosis, and a lower wastage rate of material in production. Cyber-connected manufacturing systems improve efficiency and optimize operations, but also have the potential to change the way manufacturers and industrial companies run their business. In other words, Industry 4.0 will play a significant role in transforming traditional companies into Smart Factories with the help of Internet of Things (IoT) and Cyber-Physical Systems (CPS)[1]. Industry 4.0 vision and associated technologies are rapidly adopted in several industrial sectors to gain the benefits of creating smart cyber-physical systems and operations. Some sectors, e.g. manufacturing, oil and gas, offshore wind energy, have progressed in developing digitization strategies, executing pilot projects and progressing toward mature implementation of industry 4.0 vision. Offshore Oil and Gas industry highly believes in the potential industrial and societal impacts of digital transformation, due to the need for stochastic and remote operations.

Azerbaijan is developing more projects in the Caspian Sea, as one of the countries that heavily depend on the oil and gas industry. The oil and gas industry remains by far the most crucial contributor to Azerbaijan's economy and investment. According to the consultancy group Wood Mackenzie, almost $20 billion has been invested in the oil and gas industry from 2018 to 2025. For example, offshore Azeri Chirag Gunashli (ACG) and the deepwater oil field by the Azerbaijan International Company (AIOC) were developed [2]. Moreover, the exploration of the Shah Deniz 2 gas field, including the construction and development of gas export pipelines to Europe, is the second major project in Azerbaijan's hydrocarbon sector. The State Oil Company of Azerbaijan Republic (SOCAR) and British Petroleum (BP) are jointly going to operate the Shah Deniz gas field, as the agreement on the second stage development of the gas field was signed at the end of December 2013. It is expected to supply that 6 billion cubic meters of gas to Turkey and 10 Billion cubic meters of gas to Europe by 2020 [2]. Furthermore, there are several exploration and development projects that are planned, for example, Absheron (foreign partner: Total), MuradkhanlyJafarly-Zardab (Zenith Energy), Govsany-Zykh (Global Energy) and Block D230 / North Absheron (BP), as well as the Ashrafi-Dan Ulduzu-Aypara and Karabakh (Equinor) oil and gas fields [2].

The growth of the oil and gas industry has also developed the non-oil sector and economic development in the country's regions [5]. Currently, Azerbaijan demonstrates positive dynamics of development in ICT and is the leader among the South Caucasus countries. According to the World Bank's Doing Business 2019 report, the country is in the top 10 most reformist countries in Europe and Central Asia and is ranked 25th among 190 countries worldwide. A record number of reforms now allow Azerbaijan to develop technologically [2]. Developing new offshore oil and gas projects requires adopting new facility designs and operating technologies. Therefore, the Azerbaijani operating companies have also defined digitalization and working toward the industry 4.0 vision, not only for the newly developed green fields but also to upgrade/modernize the brown fields. There are several worldwide challenges regarding the technological and digitalization transformations, mainly, lack of standards, business models, ready products/services and competent and skilled employees. Fortunately, specific developed countries are working hard to standardize industry 4.0 architecture. Moreover, large-scale companies are creating alliances to create a trustful and long-term business model. Furthermore, large-scale companies of information and operational technology are creating robust products and services to be commercially available off the shelf. In terms of education and training, many worldwide universities are upgrading their programs, curriculums, teaching approaches with the goal to support the industry with competent future employees and entrepreneurs. Sackey and Bester [4] described the importance of reform the curriculum in order to prepare engineers for industries and avoid greater shock.

The main strategic challenge for adopting industry 4.0 in Azerbaijanis is to educate and recruit skilled workers. The universities are cores actor to educate graduate students to be competent for future work profiles and that requires upgrading curriculum, teaching methods, and public/industrial collaboration. There several frameworks of how to adopt industry 4.0 vision into educational programs, for example, Coskun et al [1], highlighted that the upgrading process shall involve changes in curriculum, laboratories, and student clubs and get the feed from research activities, as illustrated in Figure 1. The curriculum and its supportive learning activities shall cover as wide and deep as possible several technologies and methods related to industry 4.0 framework. For example, technologies such as Additive Manufacturing, Internet of Things, Artificial Intelligence, Cloud Computing, Predictive Maintenance, Robotics, and other advancements have already influenced today industry and students shall understand such technologies and their applications.

Therefore, the purpose of this paper is to present the status of engineering education programs in adapting the industry 4.0 vision in Azerbaijan and address the skills that are required for future employment.

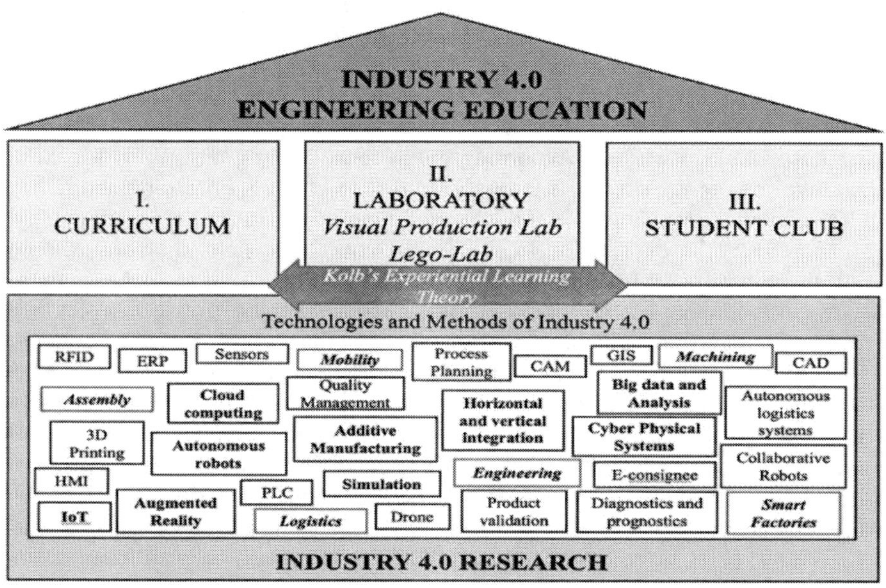

Figure 1. The generic framework of Industry engineering education [1].

2. Methods

This section discusses how a summary of adapting industry 4.0 vision into engineering education in Azerbaijan was conceptualized. The methodology included three iterative steps:(1) identification of the curriculums of all engineering programs at the master level;(2) The collect required data and analyze universities' adaption into Industry 4.0 at the master level;(3) Applying a semi-structured interview with two industrial managers. The curriculums for all engineering education programs at the maser level were collected and analyzed, and all the required data have been taken from the universities website. Semi-structured interviewes with two industrial managers from the National operating company were conducted via skype, the recommended qualifications and skills are extracted and summarized. Finally, some conclusions and future recommendations are drawn up.

IOP Conf. Series: Materials Science and Engineering **700** (2019) 012063 doi:10.1088/1757-899X/700/1/012063

3. Engineering education in Azerbaijan in Industrial 4.0 context

The framework, shown in Figure 1, depicts relevant engineering education that embraces Industry 4.0 vision in Azerbaijani Universities. This is a generic framework that comprises three main parts: concretely curriculums, laboratories, and student clubs. These three main parts are interrelated and even dependent on each other. It is worth to highlight, that this study is based on the collected curriculums of engineering education programs at the master level. Therefore, it was hard to get information regarding the availability of the laboratories and student clubs without direct contacts with these specific program leaders. This is something shall be done in the future to map all aspects of industry 4.0 education. Thus, the focus is on curriculum contents in this study.

Five of the collected curriculums of engineering education programs were directly adapting industry 4.0 vision and relevant for industry 4.0 [6 - 10] as summarized in Table 1. In Table 1, the courses that are taught in Engineering programs, which are related to industry 4.0 are listed. The sign '+' in the table indicates the current state of the courses at the universities. The blanks depict the need to enhance curriculums at engineering departments.

Table 1. Current state of some Azerbaijani engineering education programs in adopting Industry 4.0 vision

Key areas	Master Programme in Information and Control in Technical systems (Baku high oil School)	Master Programme in Computer science and Electronics, Mechanical (Khazar Univ.)	Master of a program in Computer engineering (Univ. of ADA)	Master program in Robotic engineering and Computer engineering (Azerbaijan Technical univ.)	Master Program in Mechatronics and Robotics,Process automazation,Oil and gas engineering, Computer science(Azerbaijan Stae Oil and Industry Univ.)
Modern control systems	+		+	+	+
Machine Learning	+	+	+	+	+
Internet of Things	+	+			
Robotics and Autonomous systems	+	+	+	+	+
Automatized control system of Industry	+			+	+
Digital signal and data processing	+	+	+	+	+
Design for maintenance		+			
Condition monitoring					
Predictive maintenace					
Artificial Intelligence	+	+	+	+	
Additive Manufacturing					
Cyber-physical manufacturing					
Cloud computing	+				
Operation Optimization					+

IOP Conf. Series: Materials Science and Engineering **700** (2019) 012063 doi:10.1088/1757-899X/700/1/012063

Some of the courses are aligned with the current concept as well as the future concept of Industry 4.0. Moreover, new study models are designed in order to adapt the vision into engineering education. Furthermore, the courses that explicitly intersect with industry 4.0 vision are determined and the courses are afterward with practical examples in the laboratories. Finally, the curriculums with regard to industry 4.0 are prepared in order to educate students. Bringing together theoretical and practical units, the curriculum enables students to obtain the basic knowledge of industry 4.0 and its technologies (AM, IoT, AI, PdM, Robots, Cobots, etc.), and to provide hands-on exercises at the laboratory stage.

It is clear that the studied programs are moving toward more analytics and data science. Several programs introduce more advanced artificial intelligence and data science packages, e.g. Python and R. This can be considered as the main adaption practice for several curriculums to adapt them to industry 4.0. Other relatively common courses are statistical modeling and data analysis for engineering students. These courses lay a foundation for understanding machine learning and preparing students for the manufacturing environment in the era of industry 4.0, where a tremendous amount of data flow occurs between production resources and cloud systems. There is a clear lack in the studied programs in adapting curriculums related to predictive maintenance, condition monitoring, additive manufacturing, etc.

4. Industrial qualification and skills requirements

The Azerbaijani industrial sector believes that the fourth industrial revolution will bring new approaches and new technologies to foster the Azerbaijan oil and gas industry. Consequently, the technologies needs educated and skilled employees to operate and gain potential benefits. To explore what are the main qualifications and skills required for developing and operating future oil and gas that complies with industry 4.0 vision and requirements, a semi-structured interview with industrial managers was applied The required qualifications and skills for industry 4.0 job profiles in Azerbaijan oil and gas industry are summarized in Table 2. In summary, the industrial managers indicate the need for computing competences and skills for future job profiles. The computing competence covers analytics and programming skills.

Table 2. Qualifications and skills for future employability recommended by Azerbaijan oil and gas company

	Qualification	Skills
PLC Programmer	Proven experience in programming of machinery and knowledge of PLC	Language skills: English, Russian Responsibility, Communications
Robot Programmer	Knowledge of offline and on-line robot programming	Language skills: English, Russian Responsibility,flexibility,Reliability
Robotic engineer	Knowledge in numerically controlled systems, CADD/CAM systems.	Knowledge in numerically controlled systems, CADD
Data Analyst	Knowledge of Statistical modeling and Data analyzing. High knowledge Python and R programming	Language skills - English, Russian, Autonomy; Creativity; Flexibility; Knowledge of working with a spreadsheet (Excel) Basic knowledge statistically; Problem solving.
Predictive maintainer	Experience in machine learning, AI. Knowledge of Python programming and Statistical modeling.	Language skills: English, Russian Flexibility, Analytical/Logic thinking
Condition monitoring engineer	Knowledge and skills in vibration measurement & analysis	Language skills: English, Russian With good planning & scheduling, time management, coordination.
Digitalization engineer	Degree in Electric Electronic or Electronic Technology	Language skills: English, Russian Responsibility, Communications

Machine learning engineer	Knowledge in Probability and Statistics, Data modelling and evaluation .	Language skills: English, Russian Autonomy; Creativity; Flexibility
Cloud computing engineer	Degree in Computer engineering, Computer science, Information Technology or Engineering.	Language skills: English, Russian Strong Analytical skills Highly detail oriented

5. Conclusions

Even though this study can be considered as an overview of an emerging situation, it was helpful to explore how the Azerbaijani higher education programs react to industry 4.0 vision. It is worth highlighting that Industry 4.0 is about cyber-physical systems and the current status of the Azerbaijani education reactions are dominated toward the cyber part, i.e. machine learning, big data, cloud computing, and so on. Therefore, computer science programs are dominating the adoption of industry 4.0. However, it is important to indicate the lack of adoption-related to the physical part, which should be taken by mechanical and industrial engineering programs, as they will educate future employees to operate these systems. Thus, the mechanical, industrial and petroleum engineering programs should align their curriculums with the core concepts of industry 4.0.

The educational programs lack some core bodies of knowledge for operating future cyber-physical systems (oil and gas platforms) such as predictive maintenance, operation optimization, machine learning, additive manufacturing. The qualifications and skills recommended by industrial managers clearly indicate that managers within the Azerbaijani context are aware about the industry 4.0 vision and working toward achieving it. It also indicates the need for operating technologies, e.g.actors predictive maintenance and operation optimization. The study shows that universities should collaborate with industry to accelerate the education upgrading and adoption process. Universities should put more efforts into educating the new generations with the latest knowledge, skills, and technology.

References

[1] Coskun, S , Y Kayikci and E Gençay 2019, Adapting Engineering Education to Industry 4.0 Vision. *Technol.*, **7**(1) 10.3390/technologies7010010.
[2] Nigar, B and B Tabous, *Market Analyis Azerbaijan 2019. GTAI*, 2019.
[3] Group, W B , *Doing Business 2019, Training for reform*. 2019: Washington ,USA.
[4] Sackey, S and A Bester 2016, Industrial engineering curriculum in Industry 4.0 in a South African context. *S. Afr. J. Ind. Eng.*, **27**(4) 101-14.
[5] Mehdiyev, A, [Online]. Available: https://bakuresearchinstitute.org [Accessed: 12.02.2019]
[6] Baku Higher oil school [Online]. Available: http://www.bhos.edu.az/en/index [Accessed: 30.09. 2019]
[7] Khazar University [Online]. Available : http://khazar.org/ [Accessed: 30.09. 2019]
[8] ADA University [Online]. Available: https://www.ada.edu.az/en [Accessed: 30.09.2019]
[9] Azerbaijan Technical University [Online]. Available: http://www.aztu.edu.az/azp/ [Accessed: 30.09.2019]
[10] Azerbaijan State Oil and Industry University [Online]. Available: http://asoiu.edu.az/en [Accessed: 30.09.2019]

IOP Conf. Series: Materials Science and Engineering **700** (2019) 012064 doi:10.1088/1757-899X/700/1/012064

Scalability and compatibility analyses of airborne wind technology for maritime transport

D A Domínguez Santana* **and I El-Thalji**

Department of Mechanical and Structural engineering and material science, University of Stavanger, Stavanger, Norway.

* Corresponding author: da.dominguezsantana@stud.uis.no

Abstract. Wind energy is a source freely available in the oceans. New developments in the wind industry are towards high altitude wind turbines, known as Airborne Wind Energy Systems. The airborne wind technology is rapidly developing with several different design concepts, e.g. Skysail, Makani, KiteMill. There are mainly two types of airborne wind technology: propulsion-supporting and electricity-generating. Electricity generating airborne technology is mainly used for land-based buildings and fixed offshore structures and with small-scale airborne turbines. Thus, the applications of this technology in the maritime industry are limited to ship propulsion-support only and no scalable airborne wind turbine is used on board of a vessel. Therefore, the objective of this study is to explore the application of electricity-generating airborne turbines on the board of a ship. The exploration is mainly to determine how scalable these turbines shall be to satisfy a ship's electricity needs and how compatible the design of these turbines can be with the ships design. . Therefore, a scalability analysis based on crosswind kite power model is applied to determine the technical requirements to scale up the existing airborne turbine to meet the electricity demand of a ship. Moreover, a compatibility analysis is applied to determine the technical interfaces and constrains between the airborne turbine configuration and the ship architecture. In this case, a commercial electricity-generating airborne turbine and a commercial floating storage unit are considered as part of the case study. The developed scalability model indicates that the traction force is the most critical design parameter to scale up a wind turbine. On the other hand, the developed compatibility model shows that there is a notorious complexity in merging airborne and ship technology due to their contexts. Consequently, this research appears to be relevant for both, the airborne technology industry and the maritime industry toward more innovative and cost- effective applications.

1. Introduction

Wind energy is a source freely available in the oceans. New developments in the wind industry are towards high altitude wind turbines, known as Airborne Wind Energy Systems (AWES). The Maritime industry is facing a challenge in reducing its dependence on fossil fuels. New established regulations to reduce the greenhouse gas (GHG) emissions by maritime transport force the sector stakeholders to adopt rigorous performance indicators and explore new technologies for propulsion and electricity generation on board of sea-going vessels.

Nowadays, there is a wide range of modern marine green technologies available on the market, used to improve the performance and sustainability of the ocean-going vessels. The Airborne Wind

IOP Conf. Series: Materials Science and Engineering **700** (2019) 012064 doi:10.1088/1757-899X/700/1/012064

Technology (AWT) as a mean to harvest wind in high altitudes have been explored since the seventies. It is nowadays when clear advantages have been shown and prototypes have been developed to prove that AWES are a feasible and competent solution to bare the worldwide emissions challenge [1]. There are two types of AWES, as shown in Figure 1. First, Ground- Gen AWES where the electrical energy is produced on the ground caused by a mechanical traction force [1]. Second, the Fly- Gen AWES where the electrical energy is generated on the aircraft and it is connected by an electrical cable to the ground station [1].

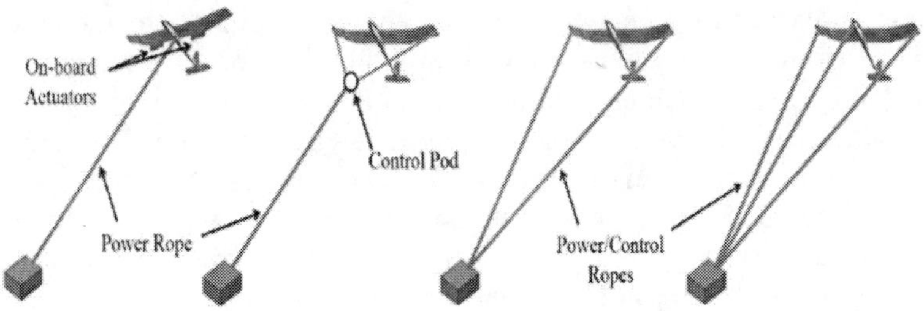

Figure 1. Fly-Gen AWES (left)and Ground- Gen AWES [1].

Currently, there are mainly two types of airborne wind technology: propulsion-supporting and electricity-generating. Few studies and patents related to AWES propulsion-supporting AWES, such as Sky Sails [2], Makani [3] and hybrid of kite and flettner-rotor systems [4]. Implementations on vessels with a specific operation location shall be performed to conclude in the contribution that this technology may provide.

The research and development that focuses on propulsion-support AWES is toward improving the design aspects such as take-off and landing techniques [5], [6], airfoil design like a rigid wing or soft kites, and Computational Fluid dynamics (CFD) and numerical modeling for power production [7]. However, there are no attempts yet to explore the electricity-generating AWES for maritime industry and specially for floating storage units, as shown in Table 1. Electricity generating AWES might have advantages over the propulsion support AWT as the produced energy can be shared for other purposes and not just for propulsion system. In fact, the floating storage unit has limited benefit from propulsion supporting AWT as it remains allocated at specific geographical site. Today, the electricity generating AWES is existing and is mainly explored and tested for land-based applications, e.g. Makani and Kitemill. The electricity- generating AWES can be still considered as small-scale systems that requires scaling up efforts to satisfy the electricity demand of floating storage units. Moreover, this technology interferes with the operational and physical (space, layout) architectures of the ship, where a compatibility and impact studies shall be performed.

Table 1. Airborne wind technologies and applications.

	Airborne Wind Technology	
	Propulsion supporting	Electricity-generating
Land Base	Not Valid	Kitemill Makani
Seagoing vessels	Sky Sails Makani	Not existing yet

Therefore, this study aims to develop the scalability and compatibility models of a rigid wing airborne technology energy to be used as a power plant in maritime transport i.e. floating storage unit. Therefore, a scalability analysis based on crosswind kite power model is applied to determine the technical requirements to scale up the existing airborne turbine to meet the electricity demand of a ship. Moreover, a compatibility analysis is applied to determine the technical interfaces and constrains between the

IOP Conf. Series: Materials Science and Engineering **700** (2019) 012064 doi:10.1088/1757-899X/700/1/012064

airborne turbine configuration and the ship architecture. In this case, a commercial electricity-generated airborne turbine and a commercial floating storage unit are considered as part of the case study.

The remainder of this article is organized into four sections. Section two corresponds to the theoretical background; section three presents the case study, scalability and compatibility analyses and results; section four discusses the critical issues before the main conclusions are drawn up.

2. Simplified crosswind power model

A kite's aerodynamic surface (airfoil) converts wind energy into motion of the kite. This motion may be converted into useful power by driving turbines on the kite or by pulling a load on the ground. Some developments have converted the kite motion into useful work pulling a load on the ground with a tether. To support the dimensioning and scaling up the prototype used for this research the following model is used. A kite is an aerodynamic vehicle restrained by a tether. Like an airplane, a kite produces lift ($\overline{L}$) and drag ($\overline{D_k}$) as it moves relative to the air. The kite is characterized by the reference area *(A)* of its wing, by its coefficient of lift (C_L) and by its lift- to- drag ratio [10].

In addition, the strength (S) of the kite must be sufficient to transfer the aerodynamic forces to the tether. This strength and the ratio of strength to weight (S/W) determine the necessary weight of the kite. The tether is characterized by length (R), tether cross- sectional area (AT), working stress (σ), mass density (ρ_T), and coefficient of drag (CDT). The resulting drag of the tether is ($\overline{D_T}$). As the kite moves through the air, power may be generated by the tether traction force ($\overline{T}$) pulling a load at a velocity ($\overline{V_L}$) [9]. Calculation of the power generated by a cross- wind flight mode kite is simplified as follows. In this simple model the weight of the kite and the characteristics of the tether, including drag are neglected. The power generated is expressed in terms of A, CL, the wind power density Pw, and the function F, representing the specific model. The final result is of the form:

$$P = P_w A C_L F \tag{1}$$

The magnitude of the relative wind velocity is VW and the air density ρ. Kites are commonly maneuvered by roll control. When one is flown to a position where the tether is parallel to the wind, the motion is directly crosswind. The speed through the air is increased above the wind speed, and the resulting power that may be generated is increased. The forces and velocities are shown in Figure 2.

The total drag $\overline{D}$ is $\overline{D_k}$, $\overline{V_w}$ and $\overline{V_c}$ are the kite velocity, which is normal to the wind. Power is generated by pulling a load downwind at $\overline{V_L}$, so the effective wind speed at the kite is reduced to $\overline{V_w} - \overline{V_L}$. Since, $\overline{T}$ is parallel to $\overline{V_w}$, and $\overline{D_k}$ is parallel to $\overline{V_A}$, and since $\overline{L}$ and $\overline{D_k}$ are perpendicular and $\overline{V_w}$ and $\overline{V_c}$ are perpendicular, the velocities and forces form similar right triangles [9]. Thus,

$$V_C = (V_W - V_L)L/D_k \tag{2}$$

If L/D_k is large, $\overline{V_c}$ and $\overline{V_A}$ are approximately equal in magnitude so that

$$V_A = (V_W - V_L)L/D_k \tag{3}$$

The lift of the kite is given by

$$L = \frac{1}{2}\rho C_L A V_A^2 \tag{4}$$

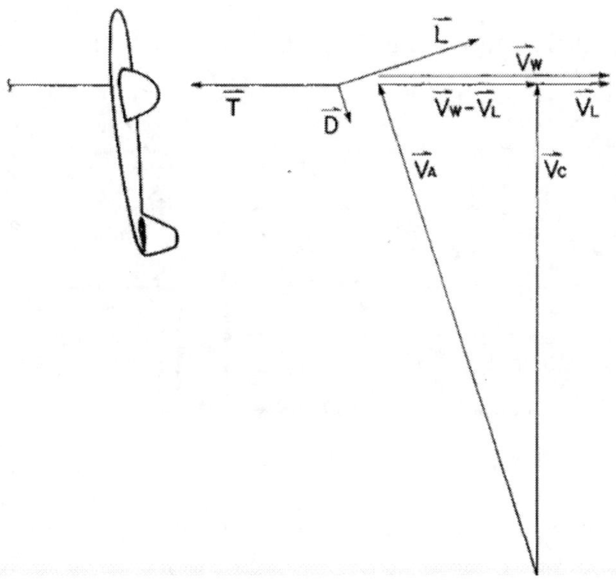

Figure 2. Aerodynamic forces on the AWE [9].

Which becomes

$$L = \frac{1}{2}\rho C_L A[(V_W - V_L)L/D_k]^2 \tag{5}$$

Since $\overline{T}$ is colinear with $\overline{V_L}$, and the magnitude the power produced is

$$P = TV_L \tag{6}$$

However, the Lift and the Traction force are approximately of the same magnitude. The function F becomes, F_C,

$$F_C = (L/D_k)^2(V_L/V_W)(1 - V_L/V_W)^2 \tag{7}$$

The maximum value of Fc is

$$F_{Cmax} = 4/27(L/D_k)^2 \tag{8}$$

Which occurs at

$$\frac{V_L}{V_W} = \frac{1}{3} \tag{9}$$

3. Scalability and compatibility analyses

The scalability concept refers to the ability of a system to increase the capacity or dimensions of its design parameters in order to supply or withstand a larger demand. A system is scalable when its design parameters increase the size of the system by keeping its performance and function and retaining all its desired properties without a corresponding increase in its internal complexity [8]. The compatibility concept refers to the ability of two system with different or same purpose to work together without interfering on their functions or performance. A system is compatible with other when its main characteristics or functions do not need to be altered [8]. If the system is scalable to a larger power demand, then the scaled-up AWES will be studied in order to be fit in an existing ship, therefore the compatibility of ship technology and AWET will be modelled.

The case to be studied is based on an AWES prototype of 30 Kw developed by one company in Norway and the Floating Storage Unit Njord Bravo, which operates in the North Sea in the Norwegian Njord field. The study consists of developing a model to scale-up the capacity of the AWES prototype

IOP Conf. Series: Materials Science and Engineering **700** (2019) 012064 doi:10.1088/1757-899X/700/1/012064

following the crosswind kite power model written by M. Lloyd. Once the design parameters are scaled-up, the compatibility of the new AWES with the FSU Njord Bravo will be studied. To do so two models are developed. The technical specifications of the AWES and the vessel have been provided by the owners of both systems.

3.1 Scalability analysis

First of all, based on the information provided by the crosswind kite power a scalability model is proposed in this section in pursuance of estimating the scalable design parameters of the AWES prototype so that it provides the power demand of the vessel. The model is proposed in Figure 3:

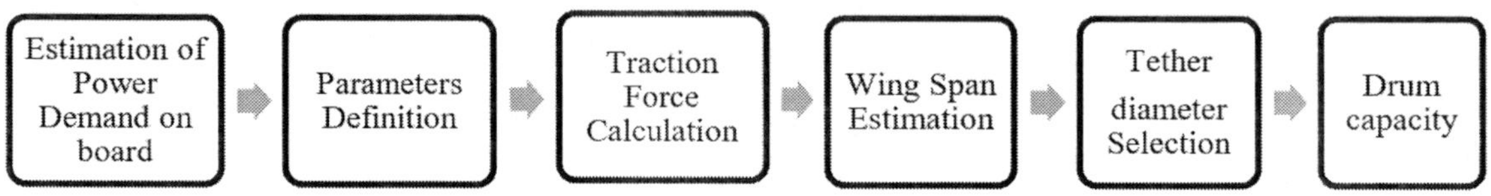

Figure 3. Steps of scalability analysis for AWT.

The model is divided into six steps, each of them is developed in order to achieve the desired AWES that should satisfy the power demand of the FSU Njord Bravo. In step 1, for this particular case, it is a good approach to assume that the power consumed onboard is around 60-70% of the total capacity of the main generator. Since a generator must not run more than 30 minutes at its maximum capacity. With this assumption, a safety margin of 30-40% is considered. The power demand onboard of the FSU Njord Bravo is 1222 kW. Due to the safety measure mentioned above, it is a safe approximation to consider the power output of the AWES as 1500 kW. In step 2, it is worth listing the assumptions considered to justify the value of some input parameters shown in table 2.

- Assumption 1: lift- to- drag ratio values for kite airfoils vary from 7 to 10, the most common values used for the AWES developers airfoil design are from 7 to 9, but in several design models value is 10 [15] & [26].
- Assumption 2: the most common value used for the lift coefficient is 1[10].
- Assumption 3: the optimal wind speed for kite power production from 4 to 12 m/s [11].
- Assumption 4: according to the crosswind kite power simplified model the load velocity is assumed to be one-third of the wind speed, so that $V_L = \frac{1}{3} V_W$[10].
- Assumption 5: The motion of the kite and the vessel are neglected.

Table 2. Input design parameters.

Definition	Parameter	Value	Units
Lift- to- drag ratio	L/D_k	[7,10]	[-]
Lift coefficient of the kite	C_L	1	[-]
Atmospheric Air density	ρ	1.2	[kg/m^3]
Wind velocity	V_w	[4,12]	[m/s]
Load velocity of the tether	V_L	1/3 of V_w	[m/s]
Maximum theoretical power coefficient [7]	C_{pmax}	4/27	[-]
Tether length	R	[250-1200]	[m]
Power Output	P	1.5	[MW]

In step 3, the traction force is calculated by applying the equation (6), considering the power output of 1500 kW and the wind speed of 12 m/s. This wind speed is the most repeated wind velocity in the operation zone of the FSU Njord Bravo at 500 m. These inputs result in a traction force of 375000 N. So that for a power output 50 times bigger than the prototype, a traction force 50 times higher is required. Step 4, consists in the wingspan estimation. The wingspan is the distance from tip to tip of the wings. For this calculation, equation (3), equation (4) and equation (5) are implemented respectively. Moreover, the fly mode of the AWES is considered to have a circular path trajectory as a traditional HAWT. Thus, the following equation is taken:

$$A = \frac{\pi D^2}{4} \tag{10}$$

Consequently, different wingspan (D) have been calculated for several lift- to- drag ratios (L/D_k) used by the aviation industry. For a wind speed of 12 m/s, and a L/D_k of 8, the resultant wingspan is 14 m. It is twice bigger than the prototype of 30 Kw. The next step consists of the selection of the tether. It has been based on manufacturers of synthetic fiber tethers for offshore applications. The preferred manufacturer is *Lankhorst Offshore.* The selected diameter of the tether is 30 mm, which is able to withstand 380000 N for a length of 1200 m. The last step comprises the calculation of the drum capacity or how much tether the winch can allocate for the selected diameter. This calculation is based on DNV-GL Rules for classification and construction for loading gear on seagoing ships and offshore installation, section 8, C.4.4 and C.4.5. To add up, the AWES developer has the practice to assume that the winch drum diameter should be 40 times bigger than the tether diameter. Therefore, a winch of 1.2 meters of diameter can allocate 1523.47 m of tether, which is sufficient to reach wind fields between 500-1000 m of altitude.

3.2 Compatibility model

Firstly, in pursuance of assessing the compatibility of the AWES on board of the FSU Njord Bravo for electricity generation, it is worth to mention the context of the application, figures 4a and 4b show the framework of the airborne wind technology and the ship technology respectively.

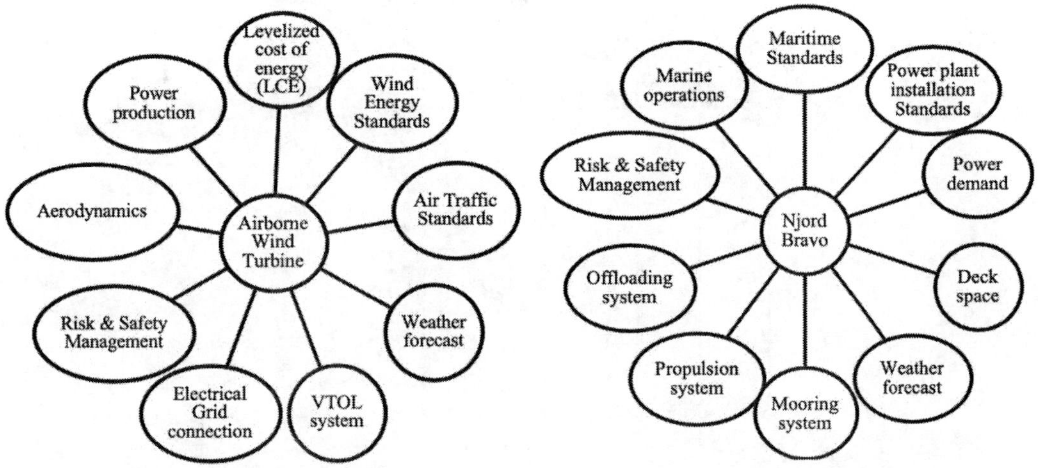

Figure 4 (a). AWES context (b). FSU Njord Bravo context

The compatibility model is proposed (as illustrated in Figure 5) to check if the up-scaled AWES is compatible with the vessel where it will be installed on. It consists of three steps that have to comply with the standards and stability requirements of the vessel.

IOP Conf. Series: Materials Science and Engineering **700** (2019) 012064 doi:10.1088/1757-899X/700/1/012064

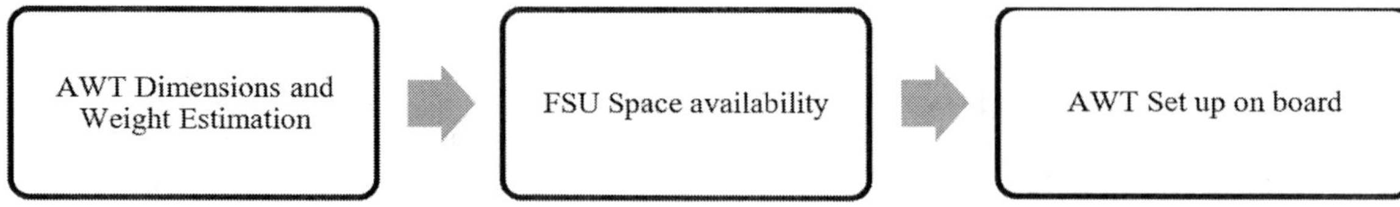

Figure 5. Steps of compatibility analysis for AWT.

On step 1, the dimensions and weight of every component of the system have to be estimated, so that the stability and operation of the vessel are not compromised at any time.

Table 3. Weight and Dimensions estimation.

Component	Weight [Kg]	Dimensions [m]
Generator	5205	1.79x1.33x0.97
Winch	8878.14	1x1.2
Rope/tether	947.78	30x1523.67
Aircraft	70-90	14x8
Take- off and landing platform	-	16x10

On step 2, the General Arrangement of the FSU Njord Bravo has been provided. It is appropriate to accommodate the winch and generator on the engine room of the vessel since it is already prepared according to standards to allocate this type of machinery. In terms of the tether and the aircraft, the Poop-deck is proposed for the installation of the equipment. On the last step, a new set up for the AWES onboard of the FSU Njord Bravo is proposed, as shown in Figure 6.

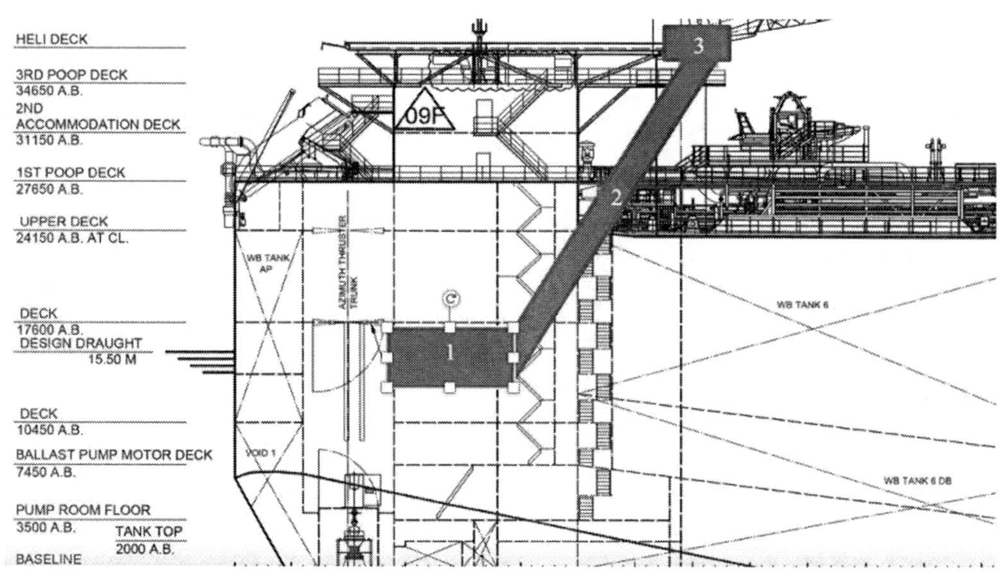

Figure 6. AWES set-up on FSU Njord Bravo, section 1 is the winch and generator; section 2 is the tether and the pipeline that contains it and section 3 is the aircraft and take-off and landing platform.

IOP Conf. Series: Materials Science and Engineering **700** (2019) 012064 doi:10.1088/1757-899X/700/1/012064

4. Discussion and conclusion

It is important to note that both models are established for the Technology Readiness Level 2 (Invention and research). Table 4 shows the comparison between the currently developed prototype of 30 kW and the up-scaled one.

Table 4. Comparison between 30 Kw and 1.5 Mw AWES.

Parameter	Current prototype	Upscaled Design	Unit
Wind speed	12	12	[m/s]
Reel speed	4	4	[m/s]
Power output	30	1500	[KW]
Traction force	7.5	375	[KN]
Wingspan	7.5	14	[m]
Tether Diameter	2	30	[mm]
Drum diameter	0.25	1.2	[m]

Regarding the compatibility model, the results conclude that there is the availability of space and weight in order to install the new system for electricity generation without compromising the stability, operation, and integrity of the vessel. In order to check if the wingspan obtained is sufficient in terms of aerodynamic efficiency, CFD analysis shall be performed. Lastly, the tether diameter and drum capacity were calculated based on manufacturers. Both parameters are related to each other by the traction force so that the higher the traction force the bigger the tether diameter should be. Getting a higher tether diameter implies a bigger drum capacity, which may affect the compatibility of the system with the ship. To add up, the bigger is the tether diameter the higher is the drag force. The drag force will reduce the crosswind power output and will increase the loading on the system. For future cases, a drag model for the tether shall be proposed.

On the other hand, the compatibility model is also reliable, it shows that structural changes shall be performed in order to install the AWES on board. These changes may affect the function of the vessel and the system. For example, the set-up onboard shall ensure good stability of the vessel, not overlap between operations on board, not excessive ship motions and the safety shall never be compromised. In the model, it is assumed that the flight mode of the kite is static and does not contribute to the propulsion or motions of the vessel. Although the mooring turret together with the DP system are in operation such that the vessel remains in position under strong wind and sea states, the AWES will tend to propel and move the vessel if the traction force is high enough to do so. Nevertheless, the FSU Njord Bravo presents a high potential to allocate the AWES as an electricity generator onboard. Overall, it can be concluded that AWT is scalable and compatible in size for floating storage units. It is vital to mention that this technology is in an early stage of development and the power supply for floating storage units is not as demanding as in other ships in the industry.

Essentially, the models might be applied to any crosswind airborne wind system and any commercial vessel. The main lessons learned of this study are:

- The traction force is the most critical design parameter for the scalability of the AWES. More complex crosswind kite power model accounting the tether drag shall be developed and studied for this case.
- It exists a big complexity merging the airborne wind technology, especially when it comes to the AWES motion and the vessel motion in terms of relative wind speed. Both dynamic and risk analysis shall be performed.

IOP Conf. Series: Materials Science and Engineering **700** (2019) 012064 doi:10.1088/1757-899X/700/1/012064

References

[1] Antonello C, Andrea P, Rocco V and Marco F 2015 Airborne Wind Energy Systems: A review of technologies, *Renewable Sustainable Energy Rev.* **51** 1461-76.

[2] SkySails, [Online]. Available: https://www.skysails.info/en/. [Accessed: 12.02.2019].

[3] Makani [Online]. Available: https://makanipower.com. [Accessed: 12.02.2019].

[4] Michael T, Paul G, Conor W, Alice B, Antonio F, Peter S and Ruth W 2013 Propulsive power contribution of a kite and a Flettner rotor on selected shipping routes, *Appl. Energy.* **113** 362 -72.

[5] Fagiano L and Schnez S 2017 On the take-off of airborne wind energy systems based on rigid wings, *Renewable Energy*. **107** 473-88.

[6] Jonas K Paul W and Soeren S 2017 Modeling of an airborne wind energy system with a flexible model for the optimization of landing trajectories, *IFAC-PapersOnLine*, **50**(1) 11944-50.

[7] Marcello D L, Romeu R, Ramiro S and Alexandre T 2018 The Betz Limit applied to Airborne Wind Energy, *Renewable Energy,* **127** 32-40.

[8] Anthony S and Manuel FDO 2010 *Networked Graphics: building networked games and virtual environments*, Morgan Kaufmann Publishers, MA, USA

[9] Loyd M L 1980 Crosswind Kite Power (for large-scale wind power production), *J. Energy*. **4**(3) 106.

[10] Kitemill, [Online]. Available: http://kitemill.com. [Accessed: 01.02 2019].

IOP Conf. Series: Materials Science and Engineering **700** (2019) 012065 doi:10.1088/1757-899X/700/1/012065

Context analysis of Offshore Fish Farming

I El-Thalji

Associate Professor, Department of Mechanical and Structural Engineering and Materials Science, University of Stavanger, Norway

Contact E-mail: idriss.el-thalji@uis.no

Abstract. The in-land and nearshore fish farming is facing capacity limitation and onshore push-out regulations. Huge technological innovations are rapidly evolving toward developing competitive Offshore fish farming. These technological innovations are mainly targeting to innovate new farming concepts that dynamically stable, reliable and compatible with offshore environmental loads and conditions. The dynamic operational behaviour of each farming concept is quite complicated. It is a combination of reinforcing behaviours (Loads, cage deformations, welfare issues, e.g. escaping, stress-related disease) and leveraging behaviours (Biofouling-cleaning, Deterioration-maintenance) and all influenced by fluctuating and harsh environmental loads and conditions. Therefore, the purpose of this paper is to analyse the context of offshore fish farming and explore quantitative descriptions of its reinforcement and leveraging behaviours. The context analysis is a well-known method within systems engineering methodology to illustrate and extract critical interfaces. The context analysis is considered as the first step in building simulation model to quantify the impact of systems interfaces on the entire farming economics, i.e. income and cost.

1. Introduction

The profitability of aquaculture production is highly dependent on the accumulated healthy bio-bass at the end of the production season, and on the other side, operating cost (Crew, feeding, energy, net-related services), maintenance cost (failures, inspections), and the associated economic and environmental impacts [1]. At the moment, the near-shore, coastal and fjords-located fish farms are more profitable than land-based farms due to investment, water treatment, etc [2]. However, the chances to install more fish farms at near-shores are getting lower over time due to the capacity limitations and high social and environmental issues. Therefore, the trend is to move offshore (at sea, open sea and ocean) which is a new era for the cultured fish industry. From a technical and economic perspective, offshore farms shall withstand the harsh environmental loading at exposed sites, which requires high investment cost. However, the scale of offshore fish farms might ensure the profitability of such investment.

Innovative concepts for offshore fish farming are rapidly increasing [3], [4]. The platform-like farm concept (Ocean farm 1, SalMar) is in operation. The Vessel-like farm concept (Havfarm) is planned to be installed in 2019 and has functions to provide optimal water exchange and farm mobility. The Submarine-like farm concept (Preline) is planned to be constructed in the future and aims to avoid high loads at sea surface. Closed deep-water sphere farm concept (Aquapods) is also planned to be constructed and aims to avoid the spread of pollutants and reduce the installation impacts of the surroundings, collect organic faeces and prevent escapes [5]. The profitability model (revenue and cost) of offshore fish farming is it is needed to evaluate such concepts. However, it is complex and relying heavily on the environmental variables and farming concept (on-surface, submersed). It is needed, not only for feasibility and cost-volume analysis studies, but for design concept review and modification (functions, design parameters) and mapping the impact of new technologies [4], [6], at the design phase. It also helps to determine the optimal operating policy at utilisation phase. The design concepts shall ultimately aim to increase the production rate (at least reduce losses) and cut-down the operating cost (energy, cleaning and maintenance), besides ensuring the environmental compatibility to withstand the harsh loading and operating conditions. The complexity of developing profitable model that represent the offshore fish farming concepts and operations is related to the dynamic relationships among five main processes [7]: fish welfare [8] and growth process (eating, swimming [9], growing [10], [11], [12], infection [13], [14], [15], sea-lice growth [16], [17], escaping [18]), the production process (supply, storage, feeding [19], control the artificial conditions, energy consumption [20]), farm assets (stability, deterioration [21], bio-fouling [22]), farm services (net cleaning, infection treatment, maintenance), and the farm site and its environmental loads and conditions [23](waves, current velocities, water temperature, salinity rate, diseases, access-ability).

The impact of all previously mentioned dynamic processes shall be considered to develop a profitability model of offshore fish farming. There are several models developed to estimate the impact of one or more of these dynamic processes. Anyadike et al.,[24] have reviewed the aquaculture production models. It was highlighted that there is a lack of utilising the systems dynamic simulation to estimate the impact of fish farms within their environmental context and especially sea-based ones [25]. Hamze et la.,[17] used system dynamics simulation, and Alaliyat and Yndestad [15] agent-based simulation to predict the impact of diseases within fish farms. Deterioration of assets is vital to predict the potential failures, unintended maintenance visits and associated costs. Deterioration estimation depends on physical modelling and simulation where the applied stresses due to environmental loading are accumulated and used for failure predictions, as done in [26] and [27]. In this regard, there are several structural assessment studies that can be used for deterioration modelling, for example, Mooring system [28], Cage system [29] Feeding pipes [30]. Moreover, there are recent studies for structural assessment for offshore fish farm concepts for example Havfarm concept [31], [32] and Ocean farm 1 [33].

It is quite clear that each previously mentioned dynamic processes is related to a specific domain of science and engineering and requires specific modelling and simulation method. Fish growth and welfare is a biological process and requires specific biological modelling and simulation tools. Production process is more related to be an industrial engineering and management issue. Fish asset is related to mechanical and structural engineering where several engineering modelling and simulation tools are used FEM, CFD, Systems dynamic. Farm services is a combination of maintenance management, treatment management and bio-fouling management. Therefore, multi-method modelling and system dynamic modelling have been used to integrate the estimations of the impact for one or more of these dynamic processes. For example, system dynamics modelling is used in aquaculture (Shrimp [34], [35], Tuna ranching [36]) and fishery [37] industries to estimate the the impact of environmental changes, consumer behaviours and over-fishing management. For marine cage aquaculture, Chateau and Chang [25] developed a system dynamics model to estimate the profitability. Even though this model ignores cost factors which are required to estimate profitability as a surplus remaining after total costs are deducted from total revenue, it illustrates the impact of availability of dissolved oxygen in the water, feed floatability, delays between production seasons on overall profitability.

Offshore fish farm concepts and operations affect and affected by all these dynamic processes. An integrated model is required to enable the estimation of the cost and revenue over the lifetime of the offshore fish farm. Therefore, the objective of this paper is to conceptualise profitability model that are affected by site conditions, environmental loading, production process, fish growth and welfare process, asset deterioration process and farm services. The model conceptualisation is the first step to develop a robust simulation model in the future using a multi-method simulation package, e.g. Anylogic, Numerus. The simulation architecting process is used to conceptualise the simulation model that considers context and interfaces.

2. Purpose analysis

As mentioned in the introduction section, the objective is to conceptualise a profitability simulation model for offshore fish farming. The purpose of this model is to predict the potential income and costs over the entire farm lifetime "utilisation period". The income is mainly related to the total biomass (in kg) that are accepted to enter the production seafood process, which means the dead and unhealthy fishes will be deducted from the harvested biomass. The harvested biomass is affected by the number of fishes and their sizes. The number of fishes is affected by survival rate where there is a losses percentage due to death (injuries), infection, and escaping. While, the size of the fish is affected by the physical and metabolic performance. The physical and metabolic performance are both affected by environmental conditions, cage conditions, and feeding regimes and rates. Therefore, income has a clear relationship with fish growth which is also related to the feeding pattern. However, there are influencing factors that might reinforce or dilute the fish growth.

The operating cost for any offshore fish farm is a summation of feeding cost, energy consumption cost, net exchange and cleaning cost, fish treatment cost, maintenance cost, and salaries. Among these cost categories, the treatment and maintenance are considered as the most variable ones. Feeding, energy consumption, net services are predictable activities and almost scheduled with some variation due to weather and environmental conditions (eating rate, swimming and physical activity rate, bio-fouling rate). Treatment events are stochastic events that depend on the probability of infection and spreading rate. Maintenance visits are also stochastic events that rely on the probability of failure (structures, cage, mooring, pipes, equipment). Failures depend on accumulated stresses and material strength. Therefore, cost categories have a clear relationship with the environmental loads and conditions that might increase or decrease their levels (frequency and consequences).

3. Context and content modelling of Offshore Fish farming

The context of offshore fish farming is vital to understand the external influencing factors on an entire offshore fish farm, and how internal systems of the offshore fish farming interact with each other (interfaces and feedbacks). Offshore fish farming context can be decomposed into five systems related to the main five processes mentioned in [7]. Context analysis using N^2 interface diagram is an effective method to illustrate interfaces and feedbacks between these five systems, as shown in the diagonal blocks in Figure 1. Moreover, the interface diagram shows the external inputs (Zone 1) for each system, outcomes (Zone 2), interfaces (Zone 3) and feed-backs (Zone 4) with other systems. To fill the interface diagram, the behaviour of these systems and their interactions with each other shall be explored with the help of literature, as in the following subsections.

3.1. Site ecosystem and environmental loading

First, the site ecosystem and loads mainly affect the fish. Oppedal et al., [23] defined several environmental variables that primarily affect the fish as follows: temperature, salinity, dissolved oxygen, water currents, light, and chemical treatments. Spatial conditions such as diseases organisms, parasites, water exchange (e.g. effect of freshwater runoff), pollution are significantly crucial for fish welfare [44]. These environmental variables are highly depending on depth, space and time (seasons). Each fish type has preferable, related physiological structures and metabolic needs, ranges and combinations of these environmental variables. The environmental variables, together with feed and threats sources, influence the fish responses to determine the location

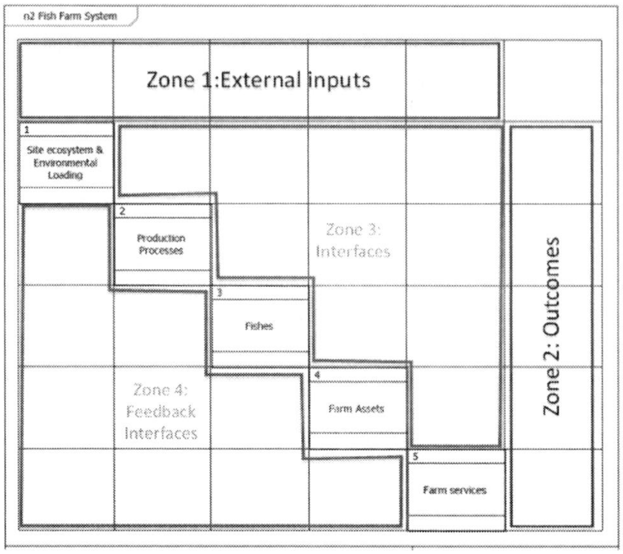

Figure 1. N^2 interface diagram of Offshore Fish farming

(depth), density and swimming pattern to survive. Temperature affects the fish metabolic processes such as circulation, food intake, digestion, growth, bio-energetical re-acclimation and scope of activities [23]. Johansson et al., [9] observed that the preferred temperature range of Salmon was $16 - 18°C$ within a range of $11 - 20°C$. Oppedal et al., [23] highlighted that there is a potential influence of current velocities of strong currents on schooling structure, swimming speeds, directions and depth, even though we have a limited understanding about that.

Second, the site ecosystem and loads mainly affect the farm asset. Toffoli and Bitner-Gregersen [42] have summarised types of ocean waves based on frequency and period of the wave: Capillary, Ultra-gravity, gravity (wind sea, swell), infra-gravity, long-period, ordinary tidal and trans-tidal waves. Depending on the offshore fish farm location, different types of ocean waves might influence its structure, stability, and the cultured organisms inside. Salinity and water currents, shelter (waves), fouling, and sea bed conditions have also effect on the farm. The environmental loading of specific fish farm site influences the fish farm structures (stability, cage deformation, storms, deterioration). Moreover, biofouling, species richness initially increased with depth, but decreased after 1520 m. Species richness decreased significantly with increasing distance from shore [38]. Third, the site conditions over seasons also affect production and service operations in terms of delays and shifting schedules. The harsh weather conditions of offshore sites might cause lack of access-ability for some service operations or more costly operations to due to the longer time for such operations to be accomplished (e.g. Mean time to repair) or due to the need for special equipment and tools. Fourth, the fish faeces, uneaten food, escaped fishes and chemicals/antibiotics have accumulated effect on the site environment [41]. It hard yet to quantify the impact of fish faeces and uneaten food on the water quality of the open sea, but fish escape is a very critical measure that shall be considered and assesses for each fish farming concept.

3.2. Production Process
The production process is a combination of several functions to produce the market-size fishes. It includes smolts-migration, feeding, maintaining (e.g.dead-fish removal, not repair and replace), and harvesting. Feeding process can be decomposed into sub-functions: supply, store, power

supply, feed and control. We have considered the daily services as part of the production process. Whereas other medium or low-frequent services are regarded as farm services, as it is outsourced for many fish farms. The production process is affecting fish growth and welfare. However, the fish behaviour might also affect the production process, e.g. feeding rate/regimes and time-to-harvest.

The feeding process has a reinforcement relationship with the fish growth process. The daily feed is affected by the fish size/weight, water temperature, available oxygen in the water, and feeding policy [17]. The feeding regimes and rates are well defined and controlled either by automatic feeder or manual feeding. In general, each kilogram of salmon requires approximately 1.2 kilograms of fish feed. In its total lifespan, a salmon will, therefore, consume 6-7 kilograms of fish feed [39]. Jørgensen and Jobling [19] have studied the feeding rates (food intake per hour) during night and day, feeding modes under different light conditions and the effects of the length of the daily feeding period on growth for Atlantic salmon. Feeding Time has also effect on the growth, survival, and food conversion ratio. The production is also affected by the fish death rate where dead-fishes are pumped and removed in a daily bases. Moreover, the output is affected by infection and escaping events once they occur.

3.3. Fishes: Growth and welfare

Hamza et al., [17] described the life stages of farmed salmon as eggs, fry, and smolts, adults and harvesting. First three stages are at the freshwater phase, and the remaining three are an at-sea phase. Fishes eat food to gain energy to swim and to grow. Fishes dissipate energy and produce waste. Fishes need oxygen and can live in a specific internal of water temperature and current velocity. Fishes need space to swim in individual and group patterns. Fishes as a living body might get diseases or injuries (hitting the cage wall, net, or each other).

First, *the growth rate is directly affected by the feeding rate, farming conditions (dissolved oxygen, Temperature).* AKVA [12] illustrated the relations between daily growth (in gram)and available oxygen in water and temperature. The curve shows that salmon can gain 8-9 grams per day when the required amount of oxygen is available, and the temperature range is between $10 - 16°C$. The von Bertalanffy growth curve is a more generic model used to estimate growth over time. The growth curve is modelled as a C-shaped curve and depends on growth coefficient, which should be determined for a specific farming condition. There are other fish growth models as Gompertz, inverse logistic, and Schnute models. Føre et al., [11] modelled the growth performance and feeding behaviour of Atlantic salmon (Salmo salar L.) in sea cages.

Second, the growth of Atlantic salmon may be affected by *cage space, current speed and fish density* cage space, current speed and fish density even while the temperature, light and feeding regimes are stable. These factors affect the physical behaviour and food conversion ratio. Johansson et al., [9] studied the swimming patterns of farmed salmon under low and high water current velocities. It is expected that fishes will swim faster (speeds of 0.3-0.9 Body Length s^{-1}) and spend more energy to speed up (double, 2 X 0.7 Body Length s^{-1}) and maintain the group structure. However, it was observed that fish experience new swimming pattern during high water current velocities to save energy. Thus, this indicates a shift from circular schooling pattern to stationary swimming against the current in a group. Moreover, the swimming depth for sea trout Salmo Trutta in fjord areas is observed to be influenced by habitat, time (day and season), water temperature and body length at the time of tagging [41]. High current velocities might injure the fishes as the probability of collisions with each other and the cage wall increases. This also increases their physical activities and food conversion ratio (probably more feed and lower growth rate).

Third, *disease and treatment process* is a vital issue that shall be considered as it affects the overall production and costs. Fish suffer from diseases and parasites. Hamze et la.,[17] used system dynamics simulation, and Alaliyat and Yndestad [15] used agent-based simulation to predict the impact of diseases within fish farms. It is expected that disease problems to be much lower for farming offshore, compared to in-land or near-shore aquaculture. However, there is limited knowledge of the probability of diseases and parasites for offshore farmed fish.

3.4. Farm asset
As mentioned early in the introduction, there are many contributions and studies of a structural assessment of near-shore fish farms and mostly for high-density polyethylene (HDPE) cages. The HDPE cages are preferable as Cardia and Lovatelli [44] highlighted "due to its limited investment capital, the relative simplicity in the performance of the various farming operations, versatility of the materials used." However, the existing offshore fish farming concepts have a steel structure, e.g. Ocean farm 1 and HavFarm. Even though the offshore oil and gas industry and maritime industry have long and robust experience in designing and assessing offshore structures, the failures might occur, and maintenance is needed, which in total affect the operation and maintenance cost and income losses. Human errors are also important to be considered, for example, the Ocean Farm 1 structure was bent due to water coming in through a hatch in a tank system, and human error was reported in [45] "The closure was opened for inspection the day before and by mistake had been left open". Therefore, it is vital to consider the potential failures, human errors and maintenance events over the farm lifetime in any cost-benefit studies. The failures might occur for any physical asset within the fish farm which generally consists of several sub-systems:(1) Steel structures, (2) Cage/net system, (3) Mooring and anchoring system, (4) Environmental control system (Fish escape control, chemical and waste dispersion mechanisms), (5) Feeding system (Automatic feeder, Barge), (6) Support system (treatment, dead fish collection, vessel support), and (7) Monitoring system (Video, Audio, structural health monitoring).

The first issue related to farm structure is *stability of Offshore fish farm.* Therefore, there are several studies to understand how stable and environmental compatible are the offshore farm structures. Most of the Offshore installations are well designed to withstand and be stable during extreme conditions. However, the instability situations might happen as a combination of technical and human interactions as happened for Ocean Farm 1 case, reported in [45]. Such a situation lead to economic losses (Maintenance cost, capture fishes, etc.). As the purpose of our analysis to conceptualise a simulation model, it is important to consider the impact of potential instability situations. The role of Mooring is to hold cages against the forces generated by wind current and waves and to avoid the transfer of excessive forces (static, i.e. gravity, buoyancy and dynamic, i.e. currents, winds and wave reaction) to the cages [46]. Thus, the incident forces that are act on the mooring lines and anchors shall be quantified.

The second issue related to farm asset is the *structures deterioration process.* Environmental loading exposed to Offshore installations causes cyclic stress variations in its structural members. These stresses accumulate over a lifetime and cause fatigue damages in these members. For example, a wave loading over a 20 years service life will result in about 10^8 cycles of stress variation [47]. Etube [48] illustrates the sea-state probability model to predict crack growth in high strength steel (SE702). This model is based on Gumbel distribution, which is accurately describing the distribution of significant wave height at North sea. Thus, the sea-state equivalent stresses can be accumulated to predict the crack growth and then the time between failures can be determined and used for maintenance cost estimation.

The third issue related to farm asset is the *cage deformation.* Kumar and Karnatak [49] highlighted that "the water exchange inside the cage is inversely proportional to the volume of the net and depends on the speed of the current and the distance between the opposite walls". Therefore, the cage deformation due to currents affects the water exchange and the level of dissolved oxygen. Bi et al., [50] estimated the flow velocity around the fish farm and the effects of cage height, current type (constant and tidal), incidence angle current and level of bio-fouling on the net.

The fourth issue related to farm asset is the *bio-fouling and cleaning process.* When artificial assets are submerged in the sea, the benthic organisms might accumulate on the asset surfaces, which is named bio-fouling. Bio-fouling is a process that is influenced temperature, salinity, water quality, depth, current and distance to shore. Bi et al., [50] listed four impacts of bio-fouling: (1) increase the disease risk, (2) increase structural failure risk due to hydrodynamic forces acting on the net, drastically affects the coefficient of drag (3) increase the net damage risk and escaped fish, with large cages the losses are greater if the net is torn accidentally [51] and (4) increase the decease the water exchange and oxygen supply inside the cage. Bi et al.,[50] studied and shown the effect of level of bio-fouling on the net. The offshore industry has long experience of bio-fouling and how to estimate the formation and accumulation of biofouling communities on steel structures [38].

3.5. Farm Services
The farm services are affected by the production and maintenance operations, e.g. repair, bio-fouling removal, diseases treatment, net exchange. Besides, routine maintenance and certain management activities such as net exchange or fish sorting can become more complicated for large and offshore cages [51].

4. Conclusions
In the previous section, we tried to explore the interfaces between the interacted systems that are within any offshore fish farming context. These interfaces are compiled and illustrated in Figure 2. Moreover, we tried as much as possible to explain the quantify relationships of these interfaces. The impact of some interfaces are quantified be well/known formula in literature, and some need further studies. Furthermore, the context analysis and the interfaces clearly show the influence of the farming concept [40], [52] and design parameters [53], [49] on total farm income and costs. It also helps to determine the critical acceptance criteria to evaluate offshore fish farming concepts. Capacity and capital cost is vital criteria, e.g. ocean Farm 1 has the capacity of 1.5 million salmon, and it is a 300 million US$ project and includes 61 meter high (approx. 40 meters in seawater) by 91-meter diameter [54]. The mesh-wire frames and nets are designed to disperse wastes and support higher fish packing density [54]. It becomes clear based on the context analysis that farming concept is also responsible for providing the right welfare environment for fishes to grow, even though the growth process is related to feeding and metabolic processes.

The farming concepts must enable the fish to get the required water exchange and prevent high current velocities from being exposed directly to the fishes. The farming concept shall be proactive by design-out potential failures, predictive to detect unexpected failures, easy to be maintained and preventive for bio-fouling communities. These design requirements aim to cut-off the maintenance cost. One of the most critical criteria is the escaped fish rate. There is a Norwegian Standard 9415 for cage farming equipment to prevent fish escape. However, farming concepts might differ in their mechanism to avoid or mitigate any escaping events.

It is worth to highlight that at the moment and due to limited experience with offshore fish farming, there are several relationships between these dynamic process not quantified, i.e. not an explicit equation, and several behaviours and influencing parameters are still unknown for offshore farming applications.

Therefore, it is the time to conceptualise a valid profitability model for offshore fish farming and determine which influencing factors and impact relationships shall be studied and measured.Then, it might be possible to develop an integrated simulation model with valid and reliable outcomes.

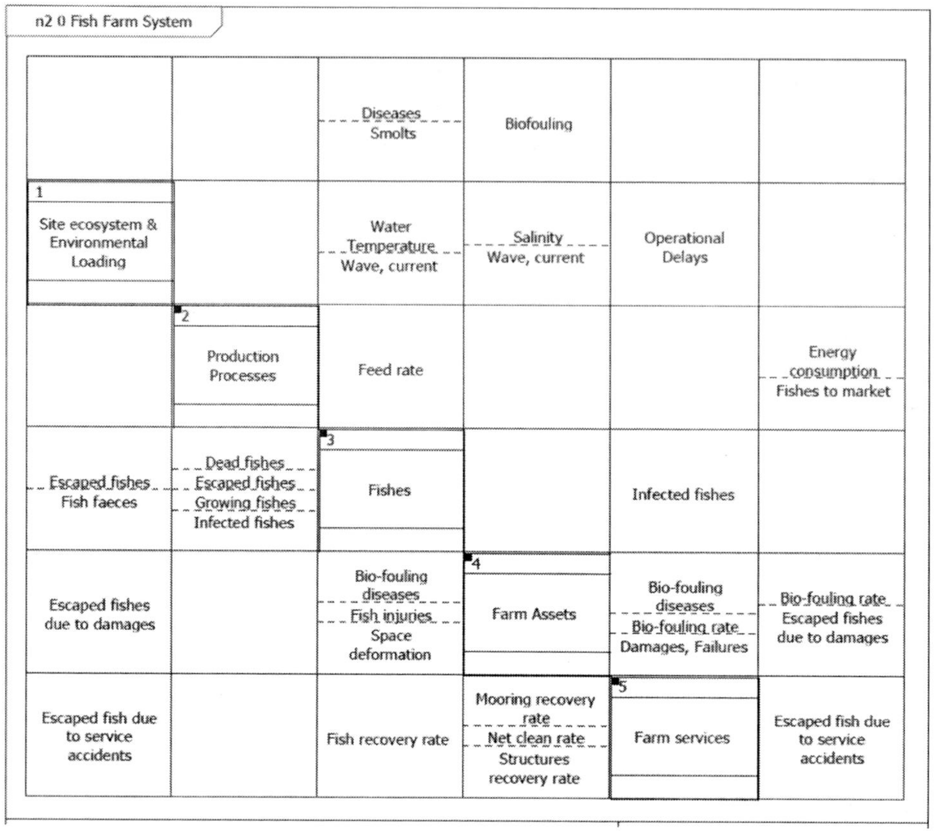

Figure 2. N^2 interface diagram of Offshore Fish farming

References

[1] Slater, M J 2017. "Societal and Economic Impacts of Aquaculture." Journal of the World Aquaculture Society **48**(4): 539-541.

[2] Bjørndal, T and A Tusvik (2017). Land based farming of salmon: economic analysis. Department of International Business NTNU. lesund, Norway.

[3] Joffre, O M, (2017). How is innovation in aquaculture conceptualized and managed? A systematic literature review and reflection framework to inform analysis and action. Aquaculture 470: 129-148.

[4] Føre, M, K Frank T Norton E Svendsen J A Alfredsen T Dempster H Eguiraun W Watson A Stahl L M Sunde C Schellewald K R Skien M O Alver D Berckmans (2017). "Precision fish farming: A new framework to improve production in aquaculture." Biosystems Engineering.

[5] Pedersen, C U (2016). The economic potential for closed cage fish farming and the total economic value of the wild salmon stock: A literature review. Master Thesis, The Arctic University of Norway.

[6] Kongsberg Group (2016). World's first Offshore aquaculture development project receives green light from Norwegian Goverment. https://www.kongsberg.com/maritime/about-us/news-and-media/news-archive/2016/worlds-first-offshore-aquaculture-development-project-receives-green-light/.

[7] Thomassen, M, Aquaculture research: from cage to consumption, in Aquaculture: Production of aquatic organisms. 2007, The Research Council of Norway. p. 6-25.

[8] Stien, L H, M B M Bracke, O Folkedal, J Nilsson, F Oppedal, T Torgersen, S Kittilsen, P J Midtlyng, M A Vindas, verli and T S Kristiansen (2013). Salmon Welfare Index Model (SWIM 1.0): a semantic model for overall welfare assessment of caged Atlantic salmon: review of the selected welfare indicators and model presentation. Reviews in Aquaculture **5**(1): 33-57.

[9] Johansson, D, F Laursen, A Ferno, J E Fosseidengen, P Klebert, L H Stien, T Vagseth and F Oppedal (2014). The interaction between water currents and salmon swimming behaviour in sea cages. PLoS One **9**(5): e97635.

[10] From, J and G Rasmussen (1987). A Model of Fish Growth in Aquaculture. IFAC Proceedings Volumes

20(7): 87-96.
[11] Føre, M, M Alver J A Alfredsen G Marafioti G Senneset J Birkevold F V Willumsen G Lange Espmark B F Terjesen (2016). Modelling growth performance and feeding behaviour of Atlantic salmon (Salmo salar L.) in commercial-size aquaculture net pens: Model details and validation through full-scale experiments. Aquaculture 464: 268-278.
[12] AKVA Group (2017). Cage Farming Aquaculture. AKVA Group, https://www.akvagroup.com/Downloads/.
[13] Yndestad, H, System dynamics of the Barents Sea capelin. ICES Journal of Marine Science, 2002. 59(6): p. 1155-1166.
[14] Thorarinsson, R and D B Powell (2006). Effects of disease risk, vaccine efficacy, and market price on the economics of fish vaccination. Aquaculture **256**(1-4): 42-49.
[15] Alaliyat, S and H Yndestad (2015). An Agent-Based Model to Simulate Infectious Disease Dynamics in an Aquaculture Facility. 2015 17th UKSim-AMSS International Conference on Modelling and Simulation (UKSim): 131-136.
[16] Revie, C W, C Robbins, G Gettinby, L Kelly and J W Treasurer (2005). A mathematical model of the growth of sea lice, Lepeophtheirus salmonis, populations on farmed Atlantic salmon, Salmo salar L., in Scotland and its use in the assessment of treatment strategies. *J. Fish Diseases* **28**(10): 603-613.
[17] Hamza, K, K M Rich and I D Wheat (2014). A System Dynamics Approach to Sea Lice Control in Norway. *Aquaculture Economics & Management* **18**(4) 344-368.
[18] Høyli, R (2016). Assessing the Risk of Escape from Marine Fish Farms: Improving Data Collection Strategies and Development of Risk Indicators. Master Thesis, The Arctic University of Norway.
[19] Jørgensen, E H and M Jobling (1992). Feeding behaviour and effect of feeding regime on growth of Atlantic salmon, Salmo salar. *Aquaculture* **101**(1) 135-146.
[20] Buck , B H and R Langan, (2017). Aquaculture Perspective of Multi-Use Sites in the Open Ocean: The Untapped Potential for Marine Resources in the Anthropocene, Springer Open.
[21] Lee, C W , J Lee, and S Park 2015 Dynamic behavior and deformation analysis of the fish cage system using mass-spring model. *China Ocean Eng.* **29**(3) 311-324.
[22] Fitridge, I T Dempster , J Guenther , R de Nys (2012). The impact and control of biofouling in marine aquaculture: a review. *Biofouling* **28**(7) 649-669.
[23] Oppedal, F, T Dempster and L H Stien (2011). Environmental drivers of Atlantic salmon behaviour in sea-cages: A review. *Aquaculture* **311**(1) 1-18.
[24] Anyadike, C C, C C Mbajiorgu, and G N Ajah 2016 Review of Aquacultural Production System Models. Nigerian *J. Technol.* **35**(2).
[25] Chateau, P A and Y C Chang, A system dynamics model for marine cage aquaculture, in The 28th International Conference of the System Dynamics Society. 2010: Seoul, South Korea.
[26] Bai, Xi, Y Zhao, G Dong and C Bi (2016). Fatigue Analysis for the Floating Collar of Fish Cage in Waves. The 26th International Ocean and Polar Engineering Conference: 7.
[27] Hou, H M, G H Dong, T J Xu, Y P Zhao, C W Bi and F K Gui (2018). Fatigue reliability analysis of mooring system for fish cage. *Appl. Ocean Res.* **71** 77-89.
[28] Huang, X, H Liu, Q Tao, Y Hu, S Wang and T Yuan (2019). Numerical Analysis of the Dynamic Response of a Single-Point Mooring Fish Cage in Waves and Currents. *Aquaculture Studies* **19**(1) 25-35.
[29] Bai, X, Y Zhao, G Dong and C Bi (2018). Probabilistic analysis and fatigue life assessment of floating collar of fish cage due to random wave loads. *Appl. Ocean Res.* **81** 93-105.
[30] Gjuka, A (2017). Dynamic Analysis of Feed Pipes for Fish Farming in Open Sea. Master Thesis, University of Stavanger.
[31] Ottersen, P A (2017). Structural assessment of an off shore ship-shaped fish farm. Master Thesis, Technical University of Denmark.
[32] Li, L, Z Jiang, A V Hiland and M C Ong (2018). Numerical Analysis of a Vessel-Shaped Offshore Fish Farm. *J. Offshore Mech. Arctic Eng.* **140**(4): 041201-041201-041211.
[33] Zhao, Y, C Guan, C Bi, H Liu and Y Cui (2019). Experimental Investigations on Hydrodynamic Responses of a Semi-Submersible Offshore Fish Farm in Waves. *J. Mar. Sci. Eng.* **7**(7).
[34] Arquitt, S P and T B Cornwell (2007). Micro-Macro Linking Using System Dynamics Modeling: An Examination of Eco-Labeling Effects for Farmed Shrimp. *J. Macromarketing* **27**(3) 243-255.
[35] Safiih, L M, M N Afiq, A Naeim, A M Ikhwanuddin, H Madzli, Z. Syerrina, I Marzuki , A system dynamics model for analyzing the eco-aquaculture system with policy recommendations: Case study on Integrated Aquaculture Park (i-Sharp), Setiu Terengganu, Malaysia. 2016.
[36] Robadue, D D and R d M Simanek (2007). A system dynamics perspective on a global fishing enterprise: The case of tuna ranching in Mexico. The 25th International Conference of the System Dynamics Society Boston, Massachusetts, July 29 August 2, 2007.
[37] Garrity, E J (2011). System Dynamics Modeling of Individual Transferable Quota Fisheries and Suggestions

for Rebuilding Stocks. *Sustain.* **3**(1): 184-215.
[38] van der Stap, T, J W Coolen and H J Lindeboom (2016). Marine Fouling Assemblages on Offshore Gas Platforms in the Southern North Sea: Effects of Depth and Distance from Shore on Biodiversity. PLoS One 11(1): e0146324.
[39] Norwegian Seafood Council. "The Truth About Norwegian Farm-Raised Salmon." from https://seafoodfromnorway.us/origin/Norway-the-worlds-leader-in-aquaculture/the-truth-about-norwegian-farm-raised-salmon/.
[40] Shainee, M, C Haskins, H Ellingsen and B J Leira (2012). Designing offshore fish cages using systems engineering principles. *Syst. Eng.* **15**(4) 396-406.
[41] Gentry, R R, S E Lester, C V Kappel, C White, T W Bell, J Stevens and S D Gaines (2017). Offshore aquaculture: Spatial planning principles for sustainable development. *Ecol. Evol.* 7(2) 733-743.
[42] Toffoli, A and E M Bitner-Gregersen (2017). Types of Ocean Surface Waves, Wave Classification. Encyclopedia Mar. Offshore Eng. 1-8.
[43] Eldøy, S H, J G Davidsen, E B Thorstad, F G Whoriskey, K Aarestrup, T F Naesje, L Rønning, A D Sjursen, A H Rikardsen and J V Arnekleiv (2017). Marine depth use of sea trout Salmo trutta in fjord areas of central Norway. *J. Fish. Biol.* **91**(5) 1268-1283.
[44] Cardia, F and A Lovatelli (2015). Aquaculture operations in floating HDPE cages A field handbook. Rome, Italy, FAO FISHERIES AND AQUACULTURE TECHNICAL PAPER 593.
[45] Berge, A (2018). Ocean Farm 1 hit by salmon leak after structure accidentally tilts. from https://salmonbusiness.com/ocean-farm-1-hit-by-salmon-leak-after-structure-accidentally-leans/.
[46] Rao, S G (2013). Cage aquaculture in India, Cochin (India) CMFRI.
[47] Lieurade, H P (1989). Fatigue Analysis in Offshore Structures. Advances in Fatigue Science and Technology. C. M. Branco and L. G. Rosa. Dordrecht, Springer Netherlands: 585-625.
[48] Etube, L S (2001). Fatigue and fracture mechanics of offshore structures, Professional Engineering.
[49] Kumar, V and G Karnatak (2014). Engineering consideration for cage aquaculture. *IOSR J. Eng.* (IOSRJEN) **4**(6): 11-18.
[50] Bi, C W and T J Xu (2018). Numerical Study on the Flow Field Around a Fish Farm in Tidal Current. *Turkish J. Fisheries Aquatic Sci.* **18** 705-716.
[51] Piccolotti, F and A Lovatelli (2013). Construction and installation of hexagonal wooden cages for fish farming: A technical manual Rome, FAO Fisheries and Aquaculture No. 576: 76
[52] DNVGL-RU-OU-0503 (2017). Rules for Classification Offshore units: Offshore fish farming units and installations. DNV GL AS.
[53] ISO16488 (2015). Marine finfish farms Open net cage Design and operation. ISO International Organization for Standardization, Geneva, Switzerland.
[54] Wang, B. (2018). First Ocean fish farm raising 1.5 million salmon three miles off Norway. Next big future, from https://www.nextbigfuture.com/2018/08/first-ocean-fish-farm-raising-1-5-million-salmon-three-miles-off-norway.html.

Feasibility study for utilization of solar energy in the arctic areas

S Tamrakar*, M Mustafa and R Riise

UiT – The Arctic University of Norway, Faculty of Engineering Science and Technology, Institute of Building, Energy and Material Technology, Narvik, Norway.

* Corresponding author: stutee.tamrakar@uit.no

Abstract. This research investigates a solar thermal system and a solar photovoltaic system which produces local energy, by incoming solar radiation, to meet the energy consumption demand of a residential building in the arctic region. The study on the use of the solar systems in the arctic areas is rare and often doubted. So, to analyze the potential of solar energy, a study case of an existing building block in Narvik, Norway was selected. The performance and function of both the systems were studied and achieved by calculation and simulation models of the thermal system and the PV system separately. The solar systems met the energy demand of *DHW and space heating* during summer due to the availability of the sun for long hours. However, in winter, especially in December and January, the energy production was zero due to snow accumulation and minimum sunlight. The results showed that the solar thermal collector produced about 14315 kWh throughout a year, whereas, the PV system generated an annual energy output of 18640 kWh. Thus, the results suggest considerable potential for solar utilization, however in the current context cost can be a limitation.

1. Introduction

Ever-increasing energy consumption and greenhouse gas emissions from energy production have led to the need for measures that can reduce emissions. The Renewable Energy Directive is a measure of the European Union (EU), with goals for reducing greenhouse gas emissions, increasing renewable energy and improving energy efficiency within 2020 [1]. Norway has a total energy consumption of about 80 TWh in a year with about 40% of the total energy consumption in households and buildings [2]. The high energy consumption is an increasingly discussed topic, especially in old residential buildings, which need to be limited to some extent with focus on energy efficiency, building standards and increased integration of local renewable energy production. Cumulative focus on the integration of renewable production and energy efficiency of buildings implies that an increasing number of buildings will install the local power supply. Solar energy is a renewable energy source that can be utilized for such a local power supply.

The study on utilization of solar energy in the cold far north is rare and often doubted due to the climatic challenges and varying solar availability. However, the solar radiation in the arctic areas is abundant when the radiation of the whole year is measured. So, this study strives to prove the feasibility of solar energy in the arctic region.

This paper focuses on the possibility of utilization of solar energy for an existing residential building based on calculation and simulation results for the solar thermal system and the solar photovoltaic (PV)

IOP Conf. Series: Materials Science and Engineering **700** (2019) 012066 doi:10.1088/1757-899X/700/1/012066

system respectively. Various parameters that influence the design and operation of the systems such as orientation, inclination angle, solar irradiation, solar hours and collector area for both the systems are considered in the study.

2. Literature review

The Arctic region is located at the northernmost part of the Earth with several definitions according to AMAP [3]. The first definition states that the arctic circle runs above the latitude of 66°33′ N, the second is defined as the area north of the region which has an average temperature of 10°C in July. The third definition is based on vegetation boundaries where arctic is further categorized into high arctic, subarctic and low arctic. With reference to the definitions, Narvik is located in Northern Norway, just above the arctic circle and within the subarctic region.

Weather conditions and varying solar availability throughout the year in the subarctic region is low compared to other regions. However, low temperatures and snow are considered beneficial for photovoltaic solar systems, as the solar cells operate efficiently at a lower temperature than in higher temperatures, which as a result minimizes heat loss and the wearing of the system. The standard test condition for determining the efficiency of solar cells is 25°C. Hence, at lower temperatures, the efficiency of the solar cell increases by 0.2% - 0.5% per degree [4]. This shows that conditions in the arctic region are favorable for solar energy.

The solar radiation passes through a thick atmosphere in northern latitudes and hits the surface of the earth at a low angle. For maximum utilization, the receiving surface must be installed at an optimum angle facing southwards. In Northern Norway, the solar radiation on a horizontal surface spans from 700 kWh/ m^2 per year to 900 kWh/ m^2 per year [5]. Whereas in southern Norway, the maximum solar radiation on a horizontal surface spans upto 1100 kWh/ m^2 per year [5].

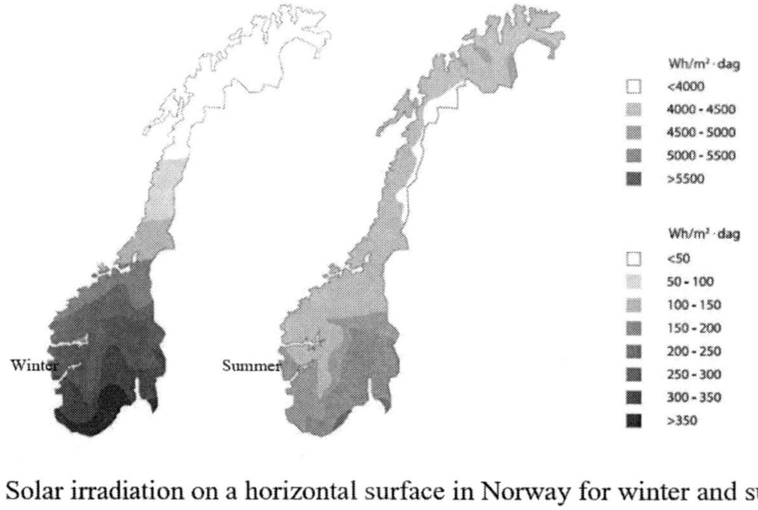

Figure 1. Solar irradiation on a horizontal surface in Norway for winter and summer *[5]*.

The possible utilization of solar energy in buildings is harnessed by three main types of technology: passive solar energy, solar thermal collector and photovoltaics (PV) system.

- Passive solar energy: utilization of solar energy for heating purposes of building via solar heat gains through large windows and thermal walls.
- Solar thermal collector: directly converts radiation from the sun to thermal energy or convert that thermal energy to electricity through a device.
- Photovoltaic system: also known as PV system, directly converts photons from the sunlight into electricity using a semiconductor device.

IOP Conf. Series: Materials Science and Engineering **700** (2019) 012066 doi:10.1088/1757-899X/700/1/012066

2.1. Solar thermal system

The solar thermal heating system transforms the energy from the sun to usable heat, which can be used for space heating and domestic hot water (DHW). This system consists of thermal solar collectors, a distribution system, an accumulator tank for heat storage and a control system as shown in figure 2. The solar collector can be designed as stationary or mobile to track solar radiation, there are two types of tracking systems: single-axis tracking and two-axis tracking. Only the stationary collector is considered for this study as the addition of a tracking system will incur additional cost and complexity to the planned installation. The most common types of stationary thermal solar collectors are flat solar collectors and vacuum tube collectors, among which flat solar collector will be discussed in the following paragraphs.

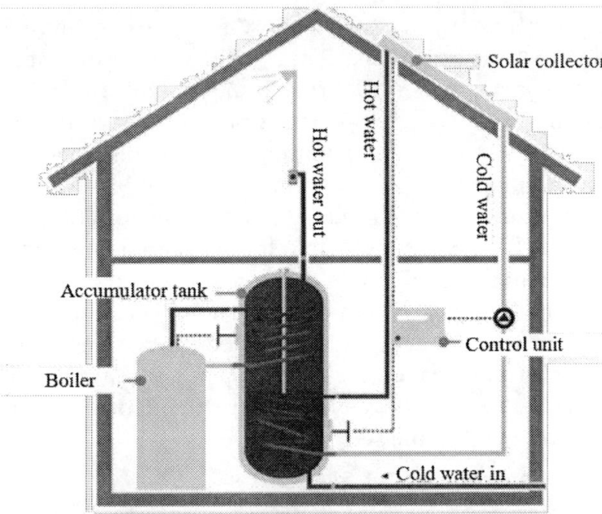

Figure 2. Principle diagram of a solar thermal collector in a building *[6]*.

2.1.1. Flat plate collector. The flat plate collector consists of a transparent front cover, channels and absorber. The collector can be either glazed or unglazed. Glazed collectors are sealed in a tight insulated container with a glazed front in order to prevent thermal losses by convection, while the unglazed collectors are exposed to the surrounding environment and are prone to lose thermal energy due to convection. Whereas unglazed collectors are preferable in warmer climates due to their reduced cost. According to Newton's law of cooling, heat transfer depends on the temperature gradient, so this gradient is reduced due to convection when the temperature of the absorbing medium is increased. Then, the heat losses to the surrounding increases, similarly, the heating medium circulates through the channels in a flat absorber under the absorber surface. In Norway, a proper system for residential buildings can produce up to 300 – 700 kWh/ m^2 *[6]*.

2.1.2. Collector's performance and efficiency. According to Zijdemans *[7]*, the efficiency of a solar collector, η_{Sc}, can be calculated using the following equation:

$$\eta_{sc} = \eta_0 - a_1 \cdot \frac{(T_L - T_A)}{G} - a_2 \cdot \frac{(T_L - T_A)^2}{G} \quad (1)$$

Where, η_0 is the efficiency of the collector without convection and radiation losses known as the optical efficiency, a_1 is the heat loss coefficient as a result of convection and conduction [W/m^{2}K], a_2 is the heat loss coefficient as a result of radiation [W/m^{2}K], G is the solar irradiance of the location [W/m^2], T_L is the average liquid temperature within the solar collector [K] and T_A is the ambient air temperature [K].

IOP Conf. Series: Materials Science and Engineering **700** (2019) 012066 doi:10.1088/1757-899X/700/1/012066

2.1.3. Solar collector area. Table 1 illustrates the simplified estimation of the required solar collector area.

Table 1. Estimation of solar collector area.

	DHW heating	DHW and space heating
Per person in a multi-dwelling building	1 – 1.5 m^2	1.5 – 2 m^2
Per 100 m^2 dwelling in a multi-dwelling building	3 - 4 m^2	4- 5 m^2

2.2. Solar photovoltaic (PV) solar system

The solar photovoltaic system converts solar energy to electricity by means of photovoltaics, where photo means light and voltaic means voltage. The system consists of several componensts including PV array and balance of system components. PV array is an ensemble of PV modules that operate as a single electricity generating component. Whereas balance of system signifies all the components except for the PV modules, such as solar inverter, mounting, wiring, instrumentation and control systems to assemble a functioning system [8].

PV systems range from small building-integrated systems to large-scale power stations. There are generally two types of solar PV system preferable for residential buildings: Stand-alone PV system and grid-connected PV system. In this study, only the grid-connected PV system is considered as these are commonly used in residential units. Though only about 1.5% of the solar panels are connected to the grid in Norway [9]. There are two types of grid-connected PV systems available, one with battery storage and the other without battery storage facility. In this paper, a PV system with batteries as energy storage will be focused since the battery stabilizes the electrical fluctuations that occur in a household and improves the overall performance of the solar system.

2.2.1. Grid-connected photovoltaic (PV) solar system

The grid-connected PV systems are composed of various components with specific purposes such as utility grid, solar modules/ panels, inverter, battery bank and loads [9]. The principle mechanism of the PV system connected to the grid is such that PV cells produce DC when they react to solar radiation. The DC-AC inverter changes the received electric current from DC to AC, where AC current can be utilized by the building load/ appliances or fed into the utility grid as shown in figure 3. This system is regulated under what is known as a feed-in tariff.

During a sunny day, PV modules generate a higher amount of electricity which is utilized by the building and the excess energy is sold to the grid. The customers who sell surplus energy back to the grid (below 100 kW at any time) are known as plus-customers. And when there is no sun, the electricity is taken either from the utility grid or from the battery. The battery system stores electrical power which is later used when sunlight is not available to meet the energy demand of the building.

2.2.2. Solar cell efficiency. The efficiency of the solar cell (η) is the ratio of power emitted (P_{max}) from the solar cell and the effect of the incident light (P_{in}) measured under standard conditions *[4]*. In terms of MPP voltage (V) and current (I), the efficiency can be expressed as:

$$\eta = \frac{P_{max}}{P_{in}} = \frac{V_{mpp} \cdot I_{mpp}}{P_{in}} \quad (2)$$

The performance of a solar system can be measured by electrical energy and expressed as follows:

$$Total\ Energy\ [kWh] = Total\ power\ [kW] * time\ [h] \quad (3)$$

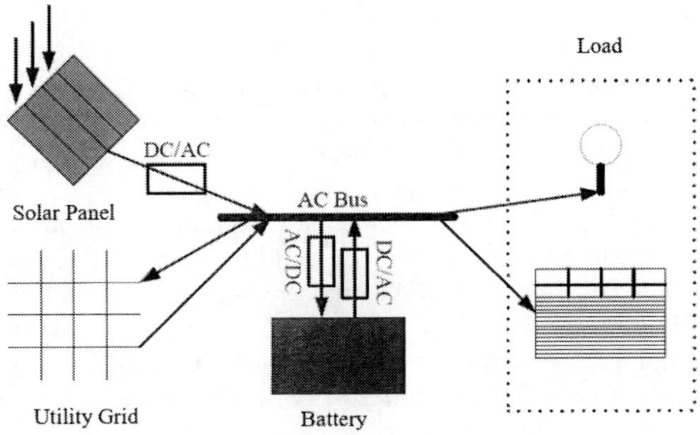

Figure 3. Standard grid-connected solar system *[10]*.

2.2.3. Battery for energy storage. The most relevant energy storage alternative in a grid-connected PV system that saves surplus electricity in a household is a battery and can be used as a backup when there is power interruption in the grid. When the price of the grid electricity is too high, the battery is used since batteries can store power output in a low demand period and deliver power in high demand to the household. In a grid-connected system, charging and discharging of the battery occurs frequently and lithium-ion (Li-ion) batteries are the best option among other battery types for the grid-connected system [11]. Li-ion batteries are light and compact battery with long cycle life, high energy density, deep recycling characteristics and safe use.

2.2.4. Solar cell technologies today. The solar cell technologies that lead the market today are monocrystalline silicon (mono-Si), multi-crystalline silicon (multi-Si) and several types of thin-film cells. The most predominant of solar cell technologies are wafer-based silicon solar cells. The production of crystalline silicon cells is either single crystal (mono-Si) or polycrystalline (multi-Si) cells. The multi-Si cell comprises of numerous crystal gains which require less energy to produce, resulting in less efficient cells. In mono-Si cells, the silicon has only one continuous crystal lattice with the least defects and impurities, thus providing comparatively high efficiencies. Due to the advanced production process, the price of mono-Si is relatively higher than other solar cells in the market.

2.3. Obtaining data

The amount of solar radiation reaching the surface of the earth is dependent upon solar hours, sun peak hours, solar path, local solar irradiation and orientation of the solar collector. The accuracy of these parameters plays a vital role in both calculation and simulation. Temperature and climate data were collected from the Norwegian Meteorological Institute, solar hours and peak sun hours were acquired from Suncurves AS and solar path and solar radiation per month were simulated using PVsyst V6.77 software. The optimal inclination angle and optimal azimuth angle for solar collectors were simulated from PVGIS software. The technical software SIMIEN was used for simulation of PV system which simulates energy use, ventilation and indoor climate in buildings.

3. Research method

The main motive of this study is to analyze whether the installation of a solar system in an existing building will be economically viable in the arctic region. The selected building is considered for the calculation of local energy production through convenient solar installations, in this case, solar thermal collectors and PV modules. The building is a three-story apartment block to be renovated and upgraded that follows the latest Norwegian building engineering regulation standard – TEK17 [12].

IOP Conf. Series: Materials Science and Engineering **700** (2019) 012066 doi:10.1088/1757-899X/700/1/012066

Simulation of energy output by solar thermal collectors was executed in the solar calculator which is based upon Bird and Hulstrom's model [13], and Ryan and Stolzenbach's model [14]. The flat collector was selected for the building because it has good insulation property, a larger area of absorption and an aesthetic appearance. However, the drawbacks could be its relatively high heat losses and a higher degree of reflection. The area of the solar collector was calculated by simple calculation techniques using table 1 in section 2.1.3 provided by Zijdemans [7].

In the case of the PV system, the area of the PV modules was determined from a simplified method developed by IEA-PVPS Task 7 [15] known as the estimation of architecturally suitable areas for the solar installation with respect to the floor area of the building. The SIMIEN model of the building that follows TEK17 was modelled by Enerconsult AS, and the model was used for further simulation of the energy output by the PV system. Finally, for the economic analysis, costs were retrieved from Catch Solar and STS solar technologies Scandinavia for the solar thermal collector system and the PV system respectively. The payback period and profitability of both solar systems were calculated using the net present value formula.

4. The case study building

The characteristics, energy demand and placement of the solar collectors/ modules in the building are studied and analyzed as it affects the overall energy production from both the solar systems.

4.1. Characteristics of the building

The building is a three-story apartment block with a built-up area of 312 m^2. The block comprises of 10 apartments with two staircases and four apartments on each upper floor and two on the basement. The basement also includes storage units for each apartment and laundry. The basement is casted in concrete whereas the upper floors has wooden structure. The walls are cladded with steel plates and the windows are two-layered glass with PVC frame [16].

4.2. Placement of the solar collectors/ modules

The narrow side of the building faces north and south direction, with only possibility to place the system to the east and west sides of the roof which is 27° inclined. Regarding façade of the building, south and west facades are considered for PV installations. West façade is selected since there is an open space in the west direction, and maximum solar radiation is received at west facades due to low angle of the sun [17]. The building is not shaded by any nearby building or hill but the shadow from nearby collectors and chimney shafts can make a difference in the output from the system.

4.3. Energy demand of the case study building

The total energy demand and specific energy demand for the building is simulated in SIMIEN with the value of 70534 kWh and 98.8 kWh/m^2 respectively per year. The required DHW of the building is 21264 kWh and space heating demand is 19239 kWh.

Figure 4. The case study building.

IOP Conf. Series: Materials Science and Engineering **700** (2019) 012066 doi:10.1088/1757-899X/700/1/012066

5. Result

5.1. Solar thermal collector system

To determine the feasibility of the system, the area of the solar collector was calculated, which provides for DHW and space heating in the building. The required daily DHW is taken 50 liters per person. The water inlet temperature and outlet temperature in the system are considered as 5°C and 50°C respectively. Depending upon the number and size of apartments, 32 people are assumed to reside within the building. According to Zijdemans [7] estimation of solar area collector as shown in table 1, the utilization area of DHW and space heating for each person is 2 m^2, resulting in the total solar thermal collector area to be 64 m^2. The amount of energy demand that solar heat can cover depends upon the temperature level on the transport medium. Waterborne floor heating and low-temperature radiators provide better utilization of solar energy than traditional radiator systems.

Table 2. Area of the solar thermal collector.

No. of residents	DHW and space heating in a multi-dwelling	Area of solar thermal system
32	2 m^2	64 m^2

Figure 5 illustrates the total energy production from the proposed solar collector at the roof inclination of 27° facing towards west direction. Also, the total energy production of hypothetical solar collectors at façade inclination of 90° facing west direction is simulated to better understand production increase with respect to placement and inclination of the building-integrated solar collector. The total output produced from the solar collector is 14315 kWh on the roof. Though, according to the calculation, façade placement provides a marginally better result with total output production of 15620 kWh.

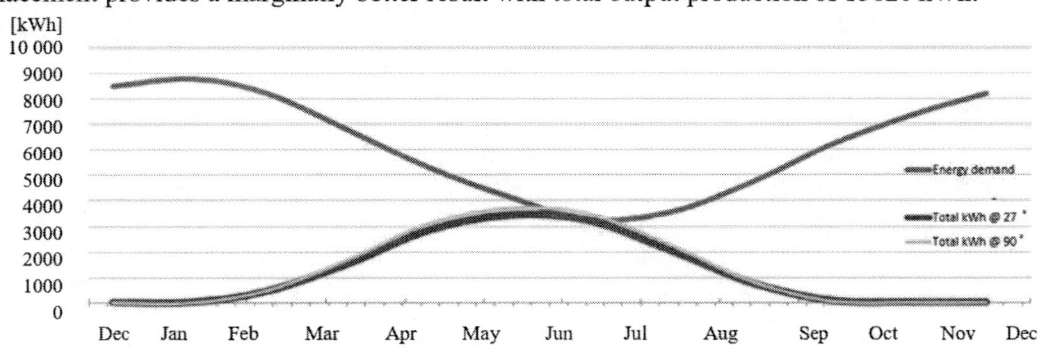

Figure 5. Solar thermal energy production with respect to the energy demand of the building.

5.2. PV system

The viability of the system was first analyzed by the area of the PV modules with respect to a simplified method developed by IEA-PVPS Task 7 [15], where generalized utilization factors for roofs are 0.4 and facades are 0.15 per m^2 built-up area. The built-up area of the building is 312 m^2. With respect to the architectural suitability, the total area of PV panels is 172 m^2. Monocrystalline panels are recommended for the system as panels have higher efficiency for diffused radiation, which will be an advantage in the arctic region where the period of direct radiation is shorter [18]. Lithium-ion phosphate Powerwall battery bank of 13.5 kWh by Tesla is proposed for the building with a capacity of peak power up to 5 kW continuous power. Table 3 shows the number of PV modules, nominal power (Wp = watt peak) and total annual energy produced by the system. The nominal power is calculated by multiplying the PV area by the efficiency of the Mono-Si cell. Here, the efficiency of Mono-Si is considered to be 20% [19]. The module size of 1.7 m^2 is considered and attached to respective building elements.

Table 3. PV system area and nominal power.

Placement	PV Area [m²]	No. of modules	Nominal power [kWp]
Roof – west orientation	125	74	20
South facade	27	16	5.4
West facade	20	12	4
Total Sum	172	102	34.4

The total energy production from the PV modules at roof inclination of 27° and vertical façades of 90° was simulated from SIMIEN. The panels are oriented to west direction since there is no shading and open space is available at the western side of the building. The PV modules placed at the south façade harness more solar radiation than in the west facade. Proposed 102 numbers of PV modules generate the total output production of 18639 kWh, where around 65% of the production satisfies the demand of the building and rest is fed back to the grid.

Figure 6 illustrates a comparison of total energy production from the PV system and the total electricity demand of the building. The variation of solar radiation leads to the over-production of electricity in summertime, and a lack of production to satisfy the energy demand in wintertime. Due to minimum hours of sunlight and accumulation of snow, in winter, the solar panels produce no energy in December and January but in November little energy is produced through the vertical PV panels. The maximum energy production is during May, June and July. As a result, there is a transmission of energy back to the grid when the demand is met. It can be seen in April and August that the electricity is sent back to the grid even though the demand is not achieved. Here, energy storage technology would play a significant role in increasing the reliability of the solar PV system and maximizing solar PV energy usage. In SIMIEN, energy storage batteries are not taken into consideration, such that electricity is fed to the grid rather than be utilized later in need.

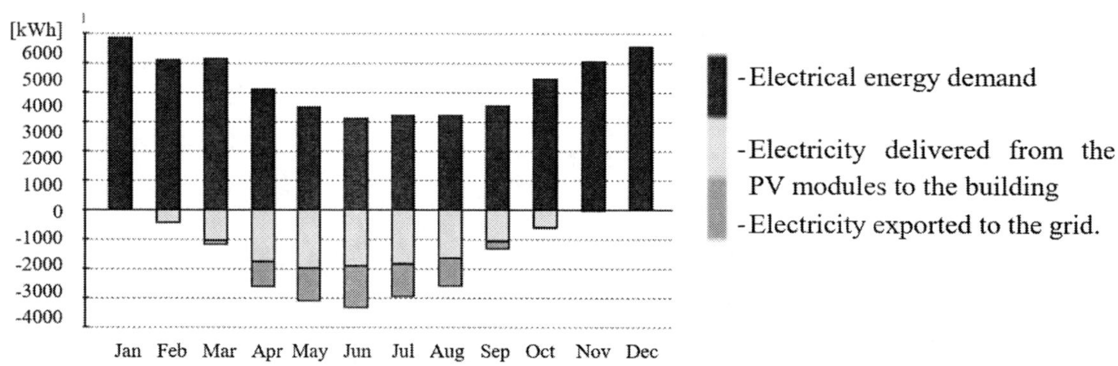

Figure 6. Energy production from PV panels drawn by SIMIEN.

5.3. Economic analysis of the solar systems

The price for the solar collectors is estimated to be 1674 NOK/m² [20], and the current subsidy given by Enova SF is 15,000 NOK for the solar thermal collector. The heat production from the solar collector is the amount of the cost saved and utilized by the building. For 64 m² area coverage of the solar thermal collectors, the total sum investment cost is about 225,000 NOK. The loan is paid at an interest rate of 2% per year. The expected lifetime of the solar thermal systems is 20 years, while supplementary equipment, such as pumps, steering and tanks can have a lifetime of around 10-20 years. This system is profitable with a payback period of 15 years and can double the initial investment in about 24 years.

The estimated investment price for the PV panels, with the cost of a fully assembled solar PV system presumed to be 16 NOK/Wp (p=peak), is 440,000 NOK. The installed PV systems produce electricity of 18640 kWh which reduces the electricity bill of the household by consuming the energy of 12600 kWh generated from the PV system and gets earnings of 6040 kWh monthly by exporting generated electricity to the grid. According to Statistics Norway [21], the total electricity cost is 1.23 NOK/ kWh. All the energy that is delivered to the building saves the extra cost. And, the energy cost for purchase of each kWh from the household by the grid company is assumed to be 1 NOK/kWh [22], which is the maximum purchase rate by power company till date. At an interest rate of 2%, the system will have a payback period of within 20 years.

6. Discussion

During the process of calculation, simulation and economic analysis some aspects affecting the discrepancies were identified and are as follows:

- Simple calculations and simulation models have been used for determining the size and energy output production of both the systems. The data of solar hours and peak sun hours were entered for the calculation of both the systems. However, the data are not guaranteed. Data from meteorological institute would have been more accurate, but the meteorological institutes have limited stations installed in Norway.
- In the case of the PV system, the SIMIEN model of the case study building was used, which was provided by Enerconsult that followed TEK17. The solar PV system inputs were done based on area calculations, PV cell efficiency, orientation and inclination of the building. But whether the building satisfied the standard TEK17 was not validated.
- Regarding the cost of the systems, they varied depending upon the companies. The costs ranged from 10 NOK/ Wp to 30 NOK/Wp for a complete setup of the PV system. Finally, the price provided by STS solar technologies Scandinavia was chosen.
- The prices provided by different companies had different units, for example, NVE stated that the cost for solar cell installation is 1.7 NOK/ kWh for 15 kWp system [23]. Nevertheless, the cost of the PV modules continues to decline year after year.
- The payback period used in this study is a simple calculation with a defined interest rate that can vary with future years. The payback period also fails to feature several important factors in estimation such as the lifetime of the system, operation and maintenance costs. These parameters are considered in lifecycle costing (LCC), which is the cost of using a PV system during its lifetime.

7. Conclusion

This study aimed to suggest the possible advantage of solar energy in existing residential buildings in the arctic region. To reach this aim, calculation and simulation models that take solar energy systems into account have been investigated. In both the systems, solar thermal collector and PV modules, the energy demand of the building was met during summer. The solar collector was designed such that the area of the collector was properly sized with respect to energy demand in summer to avoid wastage of generated energy. For building applied PV modules, the architectural suitability area was taken into consideration since it is unfit to install the whole building with PV panels. The structure and orientation of the case study building affect the maximum amount of possible solar utilization since the minimum area is exposed to the south direction, resulting in lower energy generation than expected for the number of PV panels installed.

The solar thermal collector produces 14315 kWh which satisfies around 68% of DHW demand throughout a year whereas the proposed installation of PV system size generates 18640 kWh where around 67% of the production is self-consumed. When battery storage is considered for the PV system almost 90% of the production can be utilized by the building. Storage of energy during low demand and utilization during high demand can reduce electricity bill, besides that, batteries aid in controlling energy fluctuations. Both the systems contribute to the building in self-sufficiency and less dependency on external power consumption.

With respect to both the solar systems, the solar thermal system is relatively cheaper and produces more output energy production of 335 kWh/m^2 than PV system with 109 kWh/m^2, even though the size of PV panels is more than double the size of the solar thermal system. The cost of PV panels and even batteries decline year after year providing increasing motivation for the installation of solar systems. Enova supports both the solar systems for individual customers who invest in renewable energy, though when a building is owned by real-estate companies or other companies PV system is not subsidized. Nevertheless, solar technologies create a positive impact on residential buildings by meeting the energy demand of the building and is profitable in the long run.

Acknowledgement

The authors would like to thank Sigurd Leiros, the project leader from Ofoten Midt-Troms Boligbyggelag (OMTBBL) and Lars Kimo Jørgensen, the energy advisor from Enerconsult for providing all the necessary and helpful information about the project. Furthermore, we would like to also thank the team involved in the Smart Arctic Building workshop organized by OMTBBL which resulted in additional development of the work. Also, thanks to Eirik Lockertsen and Trond Øines for taking time to answer queries and assisting with valuable information regarding solar systems.

References

[1] C. EU, "Climate Action 2020," EU, 2019.
[2] Dokka, H et al. 2009 Energieffektivisering i bygninger, Oslo: SINTEF Report [in Norwegian].
[3] Murray J et al. 2009 Geographical characteristics of the Arctic region, AMAP, Tromsø, 2009.
[4] Jha A 2009 Solar cell technology and applications, New York: Auerbach Publications, 2009.
[5] Hagos D, Gebremedhin A and Zethraeus B 2014 Solar water heating as a potential source for inland Norway energy mix, Renewable Energy. (3) 11.
[6] Halvorsen U M et al. 2011 Mulighetsstudie solenergi i Norge, SINTEF and KanEnergi [in Norwegian], Research Report, Oslo, Norway.
[7] Zijdemans D 2012, Vannbaserte oppvarmings og kjølesystemer [in Norwegian], Oslo: Skarland Press AS.
[8] Andrews J et al. 2013 Energy science – principles, technologies and impacts, Oxford: Oxford University Press.
[9] Yang B, Li W, et al. 2010 Design and analysis of a grid-connected photovoltaic power system," *IEEE Trans Power Electron.* **25**(4), 992-1000.
[10] Ru Y, Kleiss J, et al. 2013 Storage size determination for grid-connected photovoltaic systems, *Trans Sustain. Energy*, **4**(1) 68-81.
[11] Dogger J D, Rossien B, et al. 2011 Characterization of Li-Ion batteries for intelligent management of distributed grid-connected storage, *IEEE Trans Energy Convers.* **26**(1), 256-63.
[12] Ministry of Municipalities and Modernization, "*Building Technical Regulations (TEK17)*," Municipality, Oslo, 2017.
[13] Bird R and Hulstrom R 1981 A simplified clear sky model for direct and diffuse insolation on horizontal surfaces, *Solar energy reasearch institute,* Virginia, 1981.
[14] Ryan P and Stolzenbach K 1972 Environmental heat transfer, in Engineering aspects of heat disposal from power generation, vol. 43, Massachusetts, MIT.
[15] Gutschner M et al. 2002, Potential for Building Integrated Photovoltaics, *The Int. Energy Agency*, Saint Ursen, 2002.
[16] Leiros S 2018 *Condition assessment report of Beisfjordveien 88-90*, OMTBBL, Narvik, Norway.
[17] Page J 2012 Chapter IIA-1 - *The Role of Solar-Radiation Climatology in the Design of Photovoltaic Systems*, in Practical Handbook of Photovoltaics , Manchester, Academic Press, p 573-643.
[18] Fraunhofer ISE 2019 *Photovoltaics report*, Fraunhofer Inst. for Solar Energy Systems, Germany.
[19] Green M and Emery K 2018 Solar cell efficiency tables, *Prog. photovoltaics*, **26**(7) 427-36.

[20] Solar C 2019 *Solenergikalkulator* [in Norwegian] Available: http://www.catchsolar.com/.
[21] SSB, "Statistics Norway," 15 03 2019. [Online]. Available: https://www.ssb.no/statbank/table/06266/tableViewLayout1. [Accessed 15 03 2019].
[22] Barstad H 2017, *Priskrig om solstrøm - betaler 1 krone per kWh,*" April, Oslo [in Norwegian].
[23] Sidelnikova, M et al. 2015 *Kostnader i energisektoren,* Norges vassdrags- og energidirektorat, Oslo [in Norwegian].

AUTHOR INDEX

Institute of Physics
Dirac House, Temple Back
Bristol BS1 6BE UK

ISSN: 1757-8981
ISBN 978-1-7138-1030-8

9 781713 810308

CURRAN ASSOCIATES INC.
proceedings.com